NEUROTHERAPEUTICS IN THE ERA OF TRANSLATIONAL MEDICINE

NEUROTHERAPEUTICS IN THE ERA OF TRANSLATIONAL MEDICINE

Edited by

RICHARD A. SMITH
Director, Center for Neurologic Study, La Jolla, CA, United States

BRIAN K. KASPAR
Scientific Founder, Milo Biotechnology, San Diego, CA, United States

CLIVE N. SVENDSEN
Executive Director, Board of Governors Regenerative Medicine Institute, Cedars-Sinai, Los Angeles, CA, United States

ELSEVIER

ACADEMIC PRESS
An imprint of Elsevier

Academic Press is an imprint of Elsevier
125 London Wall, London EC2Y 5AS, United Kingdom
525 B Street, Suite 1650, San Diego, CA 92101, United States
50 Hampshire Street, 5th Floor, Cambridge, MA 02139, United States
The Boulevard, Langford Lane, Kidlington, Oxford OX5 1GB, United Kingdom

Notices
Knowledge and best practice in this field are constantly changing. As new research and experience broaden our understanding, changes in research methods, professional practices, or medical treatment may become necessary.

Practitioners and researchers must always rely on their own experience and knowledge in evaluating and using any information, methods, compounds, or experiments described herein. In using such information or methods they should be mindful of their own safety and the safety of others, including parties for whom they have a professional responsibility.

To the fullest extent of the law, neither the Publisher nor the authors, contributors, or editors, assume any liability for any injury and/or damage to persons or property as a matter of products liability, negligence or otherwise, or from any use or operation of any methods, products, instructions, or ideas contained in the material herein.

Library of Congress Cataloging-in-Publication Data
A catalog record for this book is available from the Library of Congress

British Library Cataloguing-in-Publication Data
A catalogue record for this book is available from the British Library

ISBN 978-0-12-816475-4

For information on all Academic Press publications
visit our website at https://www.elsevier.com/books-and-journals

Publisher: Nikki Levy
Acquisitions Editor: Melanie Tucker
Editorial Project Manager: Kristi Anderson
Production Project Manager: Punithavathy Govindaradjane
Cover Designer: Matthew Limbert

Typeset by SPi Global, India

Contents

Contributors

Karen Aboody
Department of Developmental and Stem Cell Biology, Beckman Research Institute of City of Hope, Duarte, CA, United States

Olubankole Aladesuyi Arogundade
Department of Neurosciences, University of California at San Diego, La Jolla, CA, United States

Giovanni Appendino
Department of Pharmaceutical Sciences, University of Eastern Piedmont, Novara, Italy

Weiqi Bao
PET Center, Huashan Hospital, Fudan University, Shanghai, China

Marco A.S. Baptista
The Michael J. Fox Foundation for Parkinson's Research, New York, NY, United States

Danielle Beaulieu
Origent Data Sciences, Inc., Vienna, VA, United States

Frank Bennett
Ionis Pharmaceuticals, Carlsbad, CA, United States

Emma Brand
Genetics and Aging Research Unit, MassGeneral Institute for Neurodegenerative Disease, McCance Center for Brain Health, Massachusetts General Hospital, Harvard Medical School, Charlestown, MA, United States

Don Cleveland
Ludwig Institute at the University of California, San Diego, La Jolla, CA, United States

Stanley L. Cohan
Providence Multiple Sclerosis Center, Providence Brain and Spine Institute, Portland, OR, United States

Jonavelle Cuerdo
Origent Data Sciences, Inc., Vienna, VA, United States

Jim DeMesa
Emerald Health Pharmaceuticals, San Diego, CA, United States

David L. Ennist
Origent Data Sciences, Inc., Vienna, VA, United States

Meryem Guler
Cogent Family Healthcare, Mountain View, CA, United States

Nada Hindiyeh
Department of Neurology, Stanford University, Stanford, CA, United States

Hanna Hovren
Center for Neurologic Study, La Jolla, CA, United States

Yiyun Henry Huang
PET Center, Department of Radiology and Biomedical Imaging, Yale University School of Medicine, New Haven, CT, United States

Hongmei Jia
Ministry of Education Key Laboratory of Radiopharmaceuticals, College of Chemistry, Beijing Normal University, Beijing, China

Jie Jiang
Department of Cell Biology, Emory University, Atlanta, GA, United States

Brian K. Kaspar
Milo Biotechnology, San Diego, CA, United States

Petra Kaufmann
VC R&D Translational Medicine R&D, AveXis, A Novartis Company, San Diego, CA, United States

Randall Kaye
Center for Neurologic Study, La Jolla, CA, United States

Jeffery W. Kelly
Department of Chemistry; The Skaggs Institute for Chemical Biology, The Scripps Research Institute, La Jolla, CA, United States

Mike Keymer
Origent Data Sciences, Inc., Vienna, VA, United States

Doo Yeon Kim
Genetics and Aging Research Unit, MassGeneral Institute for Neurodegenerative Disease, McCance Center for Brain Health, Massachusetts General Hospital, Harvard Medical School, Charlestown, MA, United States

Kiren Kresa-Reahl
Atara Biotherapeutics Inc., Thousand Oaks, CA, United States

Ikjae Lee
Department of Neurology, University of Alabama at Birmingham, Birmingham, AL, United States

Melanie Leitner
Accelerating NeuroVentures, LLC, Boston, MA, United States

Stefanie N. Linch
Cancer Early Detection Advanced Research Center (CEDAR), Knight Cancer Institute, Oregon Health and Science University, Portland, OR, United States

Elisabeth B. Lucassen
Providence Multiple Sclerosis Center, Providence Brain and Spine Institute, Portland, OR, United States

Timothy Miller
Department of Neurology, Washington University in St. Louis-School of Medicine, St. Louis, MO, United States

Rachael Mooney
Department of Developmental and Stem Cell Biology, Beckman Research Institute of City of Hope, Duarte, CA, United States

Gaetano Morello
Emerald Health Naturals, Vancouver, BC, Canada

Eduardo Muñoz
Emerald Health Pharmaceuticals, San Diego, CA, United States; Maimonides Biomedical Research Institute of Córdoba (IMIBIC); Department of Cellular Biology, Physiology and Immunology, University of Córdoba; University Hospital Reina Sofía, Córdoba, Spain

Michael Murray
Emerald Health Naturals, Vancouver, BC, Canada

Hoi Wa Ngai
Department of Developmental and Stem Cell Biology, Beckman Research Institute of City of Hope, Duarte, CA, United States

Joseph Park
Genetics and Aging Research Unit, MassGeneral Institute for Neurodegenerative Disease, McCance Center for Brain Health, Massachusetts General Hospital, Harvard Medical School, Charlestown, MA, United States

Dustin Pierce
Origent Data Sciences, Inc., Vienna, VA, United States

Jana Portnow
Department of PS Medical Oncology, Beckman Research Institute of City of Hope, Duarte, CA, United States

Evan T. Powers
Department of Chemistry, The Scripps Research Institute, La Jolla, CA, United States

Luisa Quinti
Genetics and Aging Research Unit, MassGeneral Institute for Neurodegenerative Disease, McCance Center for Brain Health, Massachusetts General Hospital, Harvard Medical School, Charlestown, MA, United States

John Ravits
Department of Neurosciences, University of California at San Diego, La Jolla, CA, United States

Meghan Romba
Department of Neurology, School of Medicine, University of Washington, Seattle, WA, United States

Rajesh Sampath
Heller School for Social Policy and Management, Brandeis University, Waltham, MA, United States

S. Pablo Sardi
Rare and Neurological Diseases Therapeutic Area, Sanofi, Framingham, MA, United States

Mark Schactman
Origent Data Sciences, Inc., Vienna, VA, United States

Mohammad Shahnawaz
Mitchell Center for Alzheimer's disease and Related Brain Disorders, Department of Neurology, University of Texas McGovern Medical School, Houston, TX, United States

Richard A. Smith
Center for Neurologic Study, La Jolla, CA, United States

Kyle E. Smoot
Providence Multiple Sclerosis Center, Providence Brain and Spine Institute, Portland, OR, United States

Claudio Soto
Mitchell Center for Alzheimer's disease and Related Brain Disorders, Department of Neurology, University of Texas McGovern Medical School, Houston, TX, United States

Clive N. Svendsen
Board of Governors Regenerative Medicine Institute, Cedars-Sinai, Los Angeles, CA, United States

Soshana P. Svendsen
Board of Governors Regenerative Medicine Institute, Cedars-Sinai, Los Angeles, CA, United States

Rudolph E. Tanzi
Genetics and Aging Research Unit, MassGeneral Institute for Neurodegenerative Disease, McCance Center for Brain Health, Massachusetts General Hospital, Harvard Medical School, Charlestown, MA, United States

Albert A. Taylor
Origent Data Sciences, Inc., Vienna, VA, United States

Gil I. Wolfe
Department of Neurology, Jacobs School of Medicine and Biomedical Sciences, University in Buffalo, Buffalo, NY, United States

Clinton B. Wright
Division of Clinical Research, National Institute of Neurological Disorders and Stroke, National Institutes of Health, Bethesda, MD, United States

Gerald Yakatan
Center for Neurologic Study, La Jolla, CA, United States

Neta Zach
Prize4Life, Haifa, Israel

Preface

As this book goes to press, the world is in the midst of a global pandemic. This is of biblical proportion and draws comparisons to the 14th-century bubonic plague and the early 20th-century "Spanish flu." Efforts to control the spread of disease now are remarkably similar to strategies then: the imposition of social policies such as the use of quarantine and "social distancing." In the past, faced with a contagion, people would have looked to religion for miracles. In modern times, science has mostly displaced religion in this regard. In a way, the research ecosystem that has rallied in an effort to curtail the current contagion is a microcosm of the endeavor to develop "cures" for a range of disorders, including neurological ones which have proven incredibly intractable to treatment.

Unlike COVID-19, which is caused by a highly infectious and often lethal virus, for most chronic neurologic diseases, the cause is unknown. With no specific target, researchers and clinicians have had to be creative in coming up with potential remedies. The imagination and creativity of the contributors to this volume are a testament to this. When the idea for this book was proposed to us we agreed, with one condition: that contributors be permitted to summarize their major achievements in a narrative style, in other words, to tell their story. We thought readers would welcome the chance to learn firsthand about the experiments that worked and the ones that didn't, the years of effort, and the barriers to success. These aspects of research are not often communicated in journal formats.

Much of the basic research underlying neurotherapeutics has been sponsored by governments, foundations, academia, pharmaceutical companies, and novel funding sources (Chapters 18 and 19). Due to continued failure, "Big Pharma" in many instances has shuttered their neurological research programs, preferring to focus on more productive and lucrative opportunities such as the treatment of cancer, pain, and now, of course, infectious diseases, i.e., SARS, MERS, and COVID-19. Central nervous system (CNS) drug discovery is now, for the most part, represented by a profusion of small companies which have filled the space previously occupied by traditional drug companies who prefer to in-license drug candidates rather than build-out brick and mortar facilities. It turns out, with the advent of this "therapeutic ecosystem," there has never been a time in history when so many resources have been focused on the discovery of novel neurotherapeutics. For sure, this book would not have been possible in the recent past. But paradoxically, it is currently impossible, in one volume, to fully represent the vastness and diversity of the drug discovery effort centered on diseases of the nervous system.

If one were to randomly read almost any chapter in this book, it would be immediately apparent that drug discovery and development is a lengthy process, often extending over decades, and one that is fraught with uncertainty, risk of failure, and exorbitant costs. Further, it is well known that getting a CNS therapeutic to market takes much longer than it does in the instance of other disorders such as cancer or cardiovascular disease. And the cost of getting a drug to market approaches or exceeds a billion dollars by the time it is approved for clinical use. In part, this may account for the extraordinary expense, now or in the future, associated with the treatment of chronic disorders, such as amyotrophic lateral sclerosis (ALS) or spinal muscular atrophy with either antisense oligonucleotides or gene therapy. Considering this, society inevitably faces a number of choices, for example, treat a privileged few or treat the many (Chapter 17). A strategy that offers the possibility of lessening both the development time and associated cost is the repurposing of currently approved medications as discussed by Smith et al. (Chapter 16).

Increasingly, diseases such as Alzheimer's, amyotrophic lateral sclerosis, Parkinson's, etc., have been found to have a genetic component. In the instance of ALS, the familial form of the disease represents upward of 10% of cases (Chapter 1). Surprisingly, a minority of sporadic cases of ALS (7%) are due to a unique mutation, a C9ORF72 variant which may be the most biologically complex neurological disorder yet described (Chapter 2). For Parkinson's disease, a similar story exists in that both familial and sporadic forms of the disease result from an intricate interaction of susceptibility genes and the environment. In fact, genome-wide association studies (GWAS) have generated a new set of polygenic risk scores for Parkinson's Disease and other major neurological diseases (Chapter 6). However, in spite of having identified a genetic basis for numerous neurodegenerative diseases, it is not at all certain how these mutations ultimately lead to disease. Furthermore, even in the instance of Huntington's disease, there is still no effective drug therapy two decades after the cause of the disease was identified. In spite of this lack of overall progress, therapies that downregulate or alter mutant genes are under development or in one instance has already been approved, i.e., spinal muscular atrophy (Chapter 3). In short, it may be possible to effectively alter the course of genetically driven neurodegenerative disorders without knowing as much as one would like about the pathophysiology of these conditions. This may be an interim solution if, in the future, CRISPR technology allows one to edit deleterious human mutations. A more traditional approach to therapy is to target a gene product (downstream proteins) that are the malevolent actors in disorders that result from protein aggregation, a feature of amyloidosis, Parkinson's disease, and Alzheimer's disease. The development of Tafamidis for the treatment of amyloid polyneuropathy is the most striking example of this strategy (Chapter 4).

While inflammation is characteristic of most neurodegenerative disease, including Alzheimer's disease, Parkinson's disease, and ALS, its exact role remains elusive. In some cases it may be protecting cells against degeneration, in others it may be the villain—speeding up the demise of innocent cellular bystanders. In multiple sclerosis (MS) and myasthenia gravis (MG), it is almost certain to be the principle driver of disease. While MS has long been considered to be an immunologic disorder, initially because of its similarity to experimental autoimmune encephalitis, the realization that MG is also an autoimmune disorder was appreciated later. In both instances, nonspecific immunotherapies have been utilized, primarily the use of corticosteroids, cyclophosphamide, and methotrexate. Next-generation therapies for MS and MG now target specific cell types in the inflammatory cascade and other molecular pathways (Chapters 9 and 10). These advances have enhanced the outcome for these patients.

Other sporadic neurological disorders have not been so fortunate with regard to small molecule drug development. For ALS the history of failed drugs is long and depressing and only two have attained FDA approval (Riluzole and Edaravone) and their treatment effect is modest, at best. It has been a challenge to decide what to do next. One creative idea is to transplant neural progenitor cells modified to release GDNF, a potent growth factor, in an effort to provide support to dying motor neurons in ALS or dopaminergic neurons in the instance of Parkinson's disease (Chapter 5). A similar approach has been utilized in the instance of glioblastoma where cell therapy has been used to deliver pharmaceuticals encapsulated in exosomes, secreted by engineered cells, for the treatment of malignant brain tumors (Chapter 7).

Two chapters, one dealing with cannabinoids (Chapter 11) and the other with migraine therapies (Chapter 8), may seem unrelated, but from a historical perspective share similarities. Almost 100 years ago, ergotamine, a derivative of ergot—responsible for the bizarre mental perturbations that led to the afflicted being considered "witches"—was discovered to be an effective treatment for migraine. Interestingly, the major advances since then, the development of triptans and drugs that target CGRP, can all be seen as having their origin in what was, at one time, essentially a plant-based therapy (Chapter 8). And to a similar degree, cannabis, which has been consumed in one form or the other over eons, has proven to be a potential therapeutic target. This is primarily based on the discovery of cannabinoid receptors in the brain. As described in Chapter 11, these have been detected in a variety of cell types, including neurons, microglia, and astrocytes, and are widely distributed throughout areas of the brain such as the hippocampus, basal ganglia, etc. Considering that hundreds of cannabinoids have been discovered and that their receptors are promiscuous, the opportunities for drug discovery seem almost limitless. A forerunner, the first medically approved application of a cannabinoid (Epidiolex) was approved in 2018 for the treatment of Dravet syndrome, a form of epilepsy.

The lack of animal models that faithfully reflect the complexity of neurologic disease has been a limiting factor in drug discovery and development. Considerable effort has been made toward remedying this. Depending on one's point of view, this has not been entirely satisfactory, but the development of CNS organoids is an exciting and recent contribution to the modeling effort (Chapter 13). They provide a window into the complex biology of disease since they incorporate a panoply of tissue types that are miniature representatives of the human nervous system. The development and utility of these in the instance of Alzheimer's disease will be of interest to anyone who is trying to understand the pathogenesis and treatment of this malady. Another key advance in understanding neuroproteinopathies is the ability to detect minute amounts of offending proteins in the spinal fluid and, potentially, plasma. Shahnawaz et al. have illustrated the power of this technique to discriminate diseases which clinically can be difficult to differentiate, i.e., multisystem degeneration versus Parkinson's disease (Chapter 12). Not only should this technology be useful for diagnosis, but it is certain to be useful for monitoring traditional and experimental treatments for these and other conditions.

Well-needed advances in clinical trial methodology have been the focus of recent research. The emphasis on this aspect of drug development results from the fact that the majority of clinical trials in the instance of Alzheimer's disease and ALS, to name a few examples, have been dismal failures. This of course could be due to the lack of meaningful therapies, but it is clear that outdated clinical methodology may, in part, be responsible. One way out of this dilemma is to improve clinical trial design by interrogating clinical data sets utilizing machine learning. Utilizing this strategy, it is now possible to predict the course of a neurodegenerative disease with greater precision. The use of these algorithms allows researchers to design better research protocols and refine the enrollment criteria for inclusion of patients in a clinical trial. Additionally, this has the effect of reducing the costs while increasing the efficiency of the trial paradigm (Chapter 14). An added adjunct to clinical trial design is the development of imaging techniques that target specific aspects of a disease process. This too allows for better selection of research subjects and provides a radiographic window that can serve as a meaningful clinical endpoint and possibly to the holy grail of endpoints, a validated biomarker (Chapter 15).

The current pandemic due to COVID-19 reminds us how vulnerable we are as a species and that our current and future health is dependent on medicine, science, and social conditions. If we have succeeded, the reader will agree that the future of neurotherapeutics, having lagged in the 20th century, is destined to be considered among the 21st century's greatest achievements.

Acknowledgments

One way of emphasizing our appreciation is to acknowledge the effort made by authors to submit chapters to us, considering the demands upon them and the current conditions under which they were forced to work. In their case and in ours, the support of colleagues and families kept us going. To them, and to Roya, and to Soshana, and to Rita, we dedicate this book.

Richard A. Smith
Clive N. Svendsen
Brian K. Kaspar

Prologue

The near-term future clearly holds real promise of significantly beneficial therapies for an increasing number of neurological disorders. Powerful new tools and capabilities such as gene sequencing, access to expanded drug libraries, protein and gene replacement, and the possibility of gene editing are driving breakthrough advances in the treatment of neurological diseases previously perceived to be largely intractable. Many of these tools, capabilities, and innovations result from multistakeholder collaborations involving governments, academia, and the pharmaceutical industry. The most recent contribution to this stakeholder mix is the advent of patient advocacy organizations. This is briefly illustrated below with the story of one such foundation, which has provided the Friedreich ataxia patient community genuine hope and, now, confidence that treatment and cures are coming within their reach.

In March 1997, my wife Raychel and I received the diagnosis of our son Keith, then 11 years old, with Friedreich ataxia (FA). We asked the diagnosing physician what we should do to treat the disorder and were told there was no treatment—there was nothing we could do. Having never heard of the disorder, we searched the web and learned that our son was going to get very sick quite quickly. We saw that FA was rare (about 5000 American patients and probably about 15,000 worldwide) and that it involved relentlessly degenerative symptoms such that Keith would lose strength and coordination in his arms and legs (probably requiring use of a wheelchair by his mid-teens); would lose some of his ability to see, hear, and speak; would develop severe scoliosis requiring surgical implantation of titanium rods to straighten his spine; would have a much-increased risk of diabetes; and would likely develop severe cardiomyopathy that would result in congestive heart failure that would take his life by early adulthood.

That initial web search carried additional distressing news. We could find very few publications on FA, could identify very few active FA research projects, no active clinical trials, and no organization devoted entirely to supporting research on the disorder. We did find one piece of encouraging news—the gene carrying the recessive mutation that causes FA had been identified a year earlier. We seized upon that encouragement. Raychel and I had met on Capitol Hill and were still working on congressional staff. I had earlier served for many years in the federal executive branch and we lived within 30 min of the National Institutes of Health (NIH) and 45 min of the Food and Drug Administration (FDA). We considered all the above and concluded that the world needed an organization devoted entirely to supporting FA research and that if that organization could create and nurture a global FA community, build sufficient awareness of the disorder, and assemble sufficient resources, it might be able to make a significant difference for FA patients. We also concluded that we were fairly well positioned to form such an organization and committed ourselves to doing so.

We soon had our first engagement with the NIH. We fortuitously met the program director for pediatric neurodegenerative disorders at the National Institute of Neurological Disorders and Stroke (NINDS). She said NIH would benefit from a nonprofit organization able to help with two-way communications between the NIH and the FA scientific and patient communities. She encouraged us to consider establishing such an organization and took us upstairs to sit with her through the public meeting of the NINDS National Advisory Neurological Disorders and Stroke (NANDS) Council. She also made suggestions about FA scientists and a few other FA parents we might want to meet. We had already decided we would work to establish an FA foundation and we were greatly heartened by the program director's suggestions. We were later able to identify and meet with five other FA parents, one adult FA patient, and three leading FA scientists, all of whom agreed to serve on the board of directors. So, we prepared bylaws and articles of incorporation and established the Friedreich's Ataxia Research Alliance (FARA) in September 1998. What did this all-volunteer organization set out to do?

Collaborate, collaborate, collaborate—We knew from the beginning that our small, new organization could not get it done alone, so we adopted an additional expansion of the FARA acronym, Forging A Research Alliance, and crafted the slogan that FARA continues to live by, "Acting alone, there is very little we can accomplish, while acting together, there is very little we will NOT accomplish." FARA began immediately applying that approach to all the relationships it needed to build—with patient families, with academic investigators, with government partners such as the NIH, FDA, and Congress, with prospective industry partners and with other advocacy organizations that shared an interest in our disease.

FARA leadership listened well to those early collaborators and learned quickly about the numerous ways in which we could lay important pieces of the foundation from which our partners could make more rapid progress toward effective therapeutics for FA. For example, FARA's scientific directors urged us to prioritize an effort to grow the FA scientific field and increase collaborations among FA scientists. They suggested that, to begin doing so, FARA should "get the scientists together—it's never been done before." We prepared an application for an NIH workshop grant and submitted it the day after FARA was incorporated. That grant enabled FARA to organize and conduct the world's First International FA Scientific Research Conference, which we cohosted with NINDS in the spring of 1999.

FARA has since assembled and grown the FA scientific field by conducting seven such international conferences in which participation has grown from the 1999 number of 80 academic and government scientists to the 2019 number of 450, including academic investigators, NIH and FDA leadership, more than three dozen pharmaceutical industry partners and advocates from 16 other patient advocacy organizations. This consistent and intensive collaborative approach has been successful in building a remarkable "culture of collaboration" in which the growing number of academic investigators collaborate early and often in their research projects, an increasing number of industry partners collaborate effectively in the precompetitive space, and even co-fund with FARA the development of important preclinical tools such as cell and animal models as well as biomarkers and neuroimaging capabilities, and patient advocacy partners frequently co-fund FA research projects with FARA. What other pieces of the foundation for therapeutic progress has FARA identified and worked to set in place?

The reason for FARA's existence, the foundation for and *sine qua non* of everything FARA has done and is doing, of course, is the FA patient families. FARA was founded by FA families and began immediately to grow and nurture an FA community that is well educated regarding the disorder and highly motivated to support and participate in the research aimed at treatments and a cure. Recognizing the importance of communications in this effort, Raychel and I were among the pioneers of the Friedreich's Ataxia Parents' Group—a Listserv on which hundreds of FA families worldwide continue to converse daily—and initiated the FARA website. Grassroots fundraising within that expanding FA community provided FARA with the vast majority of its resources over the organization's first 6 years and continues to provide a major part of its annual budget.

With that grassroots support, FARA began a research grant program that adhered to the NIH gold-standard, two-tiered review process—each application is peer-reviewed and, if found meritorious by peers and FARA's Scientific Review Committee, is submitted to the FARA Board of Directors for approval. Initially, all FARA grants were devoted to basic science aimed at characterizing the disease and understanding the gene, function of the protein, mechanisms of damage and action, etc. Over time, FARA began to devote some grant funds to translational efforts such as assays, cell and animal models and tissue repositories, for example, and then to funding some clinical research. From the beginning, FARA has been committed to a diverse pipeline of therapeutic approaches to which new candidates will continue to be added until fully effective treatments are available.

FARA leadership also realized early that a contact patient registry would be an important communication link. The organization began simply with online entries of contact information onto a spreadsheet that has evolved into a global digital platform that fully conforms with international confidentiality requirements and contains sufficient medical and staging information on each enrolled patient to help identify the patients best suited for each clinical research program and to communicate with them. As of July 2020, this registry contains about 3600 FA patients and has enabled FARA to assist in fully recruiting large FA clinical trials in a matter of a few hours versus the more normal weeks, months, or years.

The cornerstone of the tremendous progress FARA has made with its industry partners in FA clinical research is the natural history database that results from its continuing, prospective natural history study. That extremely important study had its roots in a decision made 20 years ago. FARA had learned in that First International FA Scientific Research Conference about a small molecule that a French clinician had administered with some apparent benefit to a small number of FA patients in a very short, open-label trial. FARA wanted to put that molecule through a larger, controlled trial but learned that would not be possible without adequate clinical outcome measures such as an ataxia rating scale. In all, 14 FA patients were assembled at the NIH. These were evaluated by research clinicians who conducted a standard neurological examination along with several additional tests borrowed from other conditions that might prove useful in measuring neurological function in FA. Eighteen months later, with no medical intervention in the meantime, the same clinicians conducted the same exams on the same patients. Analysis of the data from those two events identified the tests most sensitive to change over 18 months. The results of these studies led to the development of the Friedreich's Ataxia Rating Scale (FARS).

FARA then identified five clinics to which it awarded grants co-funded by our colleagues at the Muscular Dystrophy Association, to administer the FARS to as many FA patients as they could enroll and to enter the resulting data into the computer servers FARA installed at one of those clinics. Before long, the clinicians recognized that the data being collected to design effective outcome measures, if continued over time, could provide the basis for a very helpful natural history study. The result has been multiple benefits for FA patients and FA research. The five original clinics have grown to more than a dozen sites in the United States, Canada, Australia, New Zealand, Brazil, and soon in India that collaborate fully with a number of FA clinics in Europe. FA patients worldwide can consequently go to physicians who know them and their disease. The data collected at these sites now comprise a robust natural history database with longitudinal data on about 1200 patients over as many as 16 years and, now, on another cohort of 100 pediatric FA patients being followed under a recent grant from the FDA. Also, most of the FA clinical trials, well informed by this natural history database, are being conducted by those same physicians at those same clinical sites (the Collaborative Clinical Research Network in FA) and the primary end point being used in many of them is the modified Friedreich's Ataxia Rating Scale (mFARS) that resulted from FARA's adoption of FARS revisions recommended by the FDA.

We had begun engaging the NIH even before FARA was established and FARA has continued to nurture and grow its partnerships with government agencies ever since. I had the good fortune of serving on the NINDS NANDS Council for 4 years, continue to serve on the planning committee for the annual NINDS Nonprofit Forum, serve currently on the NIH National Center for Advancing Translational Sciences (NCATS) National Advisory Council, and as vice chair of the affiliated Cures Acceleration

Network Review Board. These activities have helped enable FARA to keep its academic investigators and industry partners informed of pertinent funding and collaboration opportunities, to support their applications and to help advance innovative programs that hold significant promise, not just in FA, but for all disease groups (e.g., a workshop series aimed at resolving issues that currently obstruct progress in all gene therapy programs and a multistakeholder effort to facilitate effective repurposing of off-patent therapeutics).

As FARA progressed from its initial concentration on basic and translational science and began to prepare the FA community for clinical trials, FARA leadership began to approach the FDA and to develop the rewarding relationship it continues to enjoy with the agency. Because of that relationship, our industry sponsors invite FARA leadership and FA clinicians to help them prepare for and participate in their pre-IND meetings and other types of FDA engagements. The FDA has asked FARA leadership to make presentations at a number of the agency's public fora to help model how nonprofit organizations can best contribute to an efficient and effective regulatory process. FDA personnel have presented at and chaired sessions of a number of FARA's international scientific meetings. In June 2017, FARA conducted a highly successful Externally Led Patient-Focused Drug Development meeting in which a large number of FDA personnel participated and subsequently touted our meeting's Voice of the Patient Report as a model for other disease groups. FARA's relationship with the FDA, built on mutual respect, trust, and good science, has helped demonstrate that the agency is genuinely listening to the patient's voice, and FARA's industry partners are eager to have FARA accompany them in their dealings with the FDA so as to feature that voice.

The other federal government entity with which FARA continues to work closely is the US Congress. FARA leadership has testified before congressional committees and has frequently met with multiple congressional staffs in constant advocacy efforts. FARA does not advocate for the FA community only but, rather, for all rare disease patients and medical research and development generally. For example, FARA most frequently advocates for resources and helpful policies for the NIH and FDA. In that advocacy, FARA always applies the principle that, if you want to accomplish something at the NIH or FDA, it is far more effective to try to "do it WITH them rather than to try to do it TO them." So, whether advocating alone or, as we do far more frequently, with our colleagues in other advocacy organizations such as the National Organization for Rare Diseases (NORD), the Alliance for a Stronger FDA, the Alliance for Regenerative Medicine, and the EveryLife Foundation, we collaborate fully with the NIH and FDA beforehand to determine the resources and policies they most need to advance therapeutic development.

In sum, FARA has taken the "build it and they will come" approach. FARA has worked to develop the assets its development partners need to succeed. The organization has funded much of the basic research required to characterize the disease, to identify the mechanisms of damage and action along with a growing number of targets in a widely diverse treatment pipeline and to provide a compelling basis for the conviction that FA is a treatable disorder. FARA has also developed important translational tools, a well-educated and highly motivated patient community, a robust and growing natural history database, a clinical research network with highly qualified medical teams, an extensive patient registry—all guided by excellent staff, directors, and scientific advisers. Because of the "culture of collaboration" FARA has built, FA research can be viewed as a "family affair" in which all the members of this FA family (patients, academic investigators, government colleagues, industry partners, and other advocacy organizations), with whom FARA is fully engaged throughout the R&D continuum, genuinely care about FA patients and want to play a part in developing treatments for them.

FARA has clearly laid a solid foundation for truly significant breakthroughs and the FA community is on the threshold of those transformative advances. Although our son, Keith, was our initial inspiration for founding FARA and we lost him to FA 10 years ago, FARA and the FA community are now driven by the boundless inspiration of the thousands of FA patients and families around the world. FARA and those thousands of FA families are, now, not just hopeful—they are confident—that, with the progress being made in small-molecule therapeutics, the near-term prospects for gene therapy and protein replacement strategies and the kind of brilliant science and collaborative effort you will see illustrated on the following pages, we will, together, treat and cure FA and numerous other neurological conditions.

Ronald J. Bartek
Co-Founder/President, Friedreich's Ataxia Research Alliance (FARA),
Springfield, VA, United States

CHAPTER 1

Development of a molecular therapy for the SOD1 familial variant of ALS

Richard A. Smith[a], Timothy Miller[b], Don Cleveland[c], and Frank Bennett[d]

[a]Center for Neurologic Study, La Jolla, CA, United States
[b]Department of Neurology, Washington University in St. Louis-School of Medicine, St. Louis, MO, United States
[c]Ludwig Institute at the University of California, San Diego, La Jolla, CA, United States
[d]Ionis Pharmaceuticals, Carlsbad, CA, United States

Introduction

The therapeutic potential of antisense therapy was first recognized by Zamecnik et al. in a seminal publication in 1978.[1] Having demonstrated that a short DNA strand (13-mer) complementary to RNA of the Rous sarcoma virus could be used to inhibit viral replication, several mechanisms were postulated to explain the result. This did not include recognition that the effect was likely mediated by an intrinsic endonuclease (RNase H) (reviewed in Ref. 2). At the time, he suggested that his discovery might be useful for the treatment of cancer. The potential of antisense-based therapies to treat neurological disorders was considered by several of us in the early 1990s, but the immediate concern of anyone who thought about this was the inherent limitations of administering drugs to the CNS due to the blood–brain barrier. Further, there was concern about the uptake of a charged molecule like DNA after having been administered directly into the spinal fluid. In this regard, it was suggested that this might be overcome by delivering antisense oligonucleotides (ASOs) under pressure so as to enhance uptake into the brain parenchyma.[3, 4] But the situation looked brighter when it was reported by Chauhan et al. (2002)[5] that ASOs could be avidly taken up in both the cytoplasm and nucleus of neurons following intraventricular administration, albeit in these studies they were rapidly cleared (within 8 h). No mention was made, however, as to whether the injected ASOs exerted a biologic effect, i.e., reductions in targeted mRNA and/or protein.

In spite of Chauhan's report, there were reasons to be concerned about the use of antisense for treating neurological disorders. Firstly, it was not certain that any therapy could arrest a neurodegenerative disease such as ALS or Parkinson's since about half of the vulnerable neurons associated with these diseases may have already been lost by the time of diagnosis. Moreover, little was known as to the relative uptake of ASOs into the various anatomic structures within the CNS (e.g., hippocampus, basal ganglia, etc.) or the cellular constituents of the brain (i.e., neurons, microglia, oligodendrocytes, etc.). This information is critical if one is to target a neurodegenerative disease that preferentially impacts specific

Neurotherapeutics in the Era of Translational Medicine
https://doi.org/10.1016/B978-0-12-816475-4.00006-9

areas of the brain. Secondly, there was still much to be done to develop an ideal oligonucleotide chemistry that would enhance the half-life, safety, and tissue distribution of these molecules.

Given the limitations of administering drugs to the CNS and the reported short half-life in CNS tissues, we envisioned that an ASO treatment strategy would obligatorily require the use of a pump to deliver drug intraventricularly or intrathecally. This was not an insurmountable problem since the technology for long-term intrathecal administration had been established. By the time we began to work toward developing an intrathecal administered ASO to treat ALS, the FDA had approved (in June of 1996) the use of a Medtronic pump for long-term administration of baclofen for the treatment of spasticity in patients with multiple sclerosis and other neurological conditions.[6] So while our initial treatment approach might have been cumbersome, it did not discourage us from going forward since the outlook for patients with neurodegenerative diseases was uniformly grim.

While there were a number of CNS diseases that we might have focused on, the most compelling for us was ALS, not only because of its uniformly dire prognosis but also because it was the principle clinical and basic research interest of several of us. When we initiated our studies one of the best validated targets for ALS was superoxide dismutase-1 (SOD1), the first known hereditable cause of ALS.[7] Since the original description of the disease in the 1860s, there never had been a better opportunity to confront a disorder which heretofore had failed to benefit from the medical advances that have characterized modern times. Our effort to treat ALS with an antisense oligonucleotide, chronicled herein after 20 years of research, may well presage a revolution in the treatment of presently incurable disorders of the central nervous system.

Amyotrophic lateral sclerosis

To understand why a targeted therapy for ALS is precedent setting, it is helpful to recall what was generally known about the disease at the time we initiated our research. By the mid-20th century, epidemiologists had established that the disease was widely distributed and occurred throughout the world. In Europe and North America, where most of the epidemiologic studies have been conducted, the median incidence was found to be approximately 2 per 100,000 people.[8] Although it is considered an orphan disease, it is the most common adult motor neuron disorder. The typical lifespan is 3–5 years. Rather early, it was discerned that a small proportion of cases had a familial history whereas the great majority of patients appeared to develop the disease spontaneously.[9]

For the most part, ALS is readily diagnosed. Onset is invariably asymmetric, involving the limbs and commencing either as a foot drop or in the instance of the upper extremities, as weakness in a hand. The presence of either of these manifestations can suggest alternative diagnoses such as an entrapment neuropathy involving the hand or a lumbar disk herniation involving the foot. A minority of patients present with impairment of speech and/or swallowing associated with difficulty handling oral secretions, so-called bulbar onset ALS. But clinically, the occurrence of other symptoms or physical signs leads an experienced clinician to suspect that the diagnosis will ultimately be ALS. These clues include the presence of exaggerated reflexes, the occurrence of emotional lability, widespread fasciculations, and neurophysiological evidence for denervation of muscles manifested by fibrillations and positive waves. More recently, it has been appreciated that cognitive impairment is a relatively common feature of ALS.[10] Historically, this was likely overlooked because motor manifestations of the disease overshadowed and possibly masked cognitive decline. Histologically, it has long been observed that the brunt of the disease process is evident in the loss of motor neurons. But the discovery that a DNA-binding protein (TDP-43) accumulates in the cytoplasm of almost all patients is a striking recent histopathologic observation.[11] For unknown reasons, aggregates of TDP-43 are not present in the instance of familial ALS secondary to SOD1 mutations.[12]

Following World War II, endemic foci of the disease received prominent attention, the most dramatic being the recognition that the incidence of ALS in Guam was approximately 50-fold above that of the worldwide burden.[13] More than likely this was due to the consumption of cycad, a common food source in the Marianas. On its own, likely because islanders changed their diet, the incidence of ALS in Guam fell precipitously, approaching the worldwide incidence.[13] This suggested that avoidance of a neurotoxin, currently unknown, is responsible for the Guamanian form of ALS. We know of no other population where the incidence of ALS has subsided. While an environmental etiology for ALS is still a consideration, the idea that Guam would prove to be the Rosetta stone for the answer to ALS has not panned out.[14]

An alternative to an environmental etiology for ALS and kindred diseases is rooted in the work of John Olney and others, who postulated that "excitotoxicity" is the causative factor.[15–17] This was the basis for the development of riluzole, which was found to modestly increase survival in patients in two multinational placebo-controlled trials.[18, 19]

Additional treatment options that researchers have explored were based on concepts as diverse as the idea that loss of neurotrophic support, inflammation, and/or autoimmunity were the principle factors responsible for ALS. An example of the latter therapeutic approach was undertaken by Drachman et al.,[20] who irradiated the lymphoid system in patients. In a second example, an extensive study with a COX-2 inhibitor (celecoxib) was undertaken in an effort to dampen the inflammatory component the disease.[21] Neither of these studies worked, nor did a succession of trials utilizing the neurotrophic factors BDNF, CNTF, and human growth hormone (Protropin).[22, 23] These and other therapeutic efforts, more or less, exemplified the generic approach to treating ALS toward the end of the 20th century. The bridge to the modern epoch of ALS research and discovery was born with the advent of gene sequencing, the elucidation of the human genome, and ultimately with the identification of a mutation in SOD1 as a cause of familial ALS.

The familial form of ALS constitutes 10% of all cases.[24] In general, the C9ORF72 mutation represents the largest segment of familial cases (see Chapter 2), with SOD1 patients comprising approximately 15% of inherited forms of the disease. Since first identifying the SOD1 mutation in 1993[7] more than 180 mutations in the SOD1 gene have been described. This is remarkable given that the protein only has 153 amino acids. To understand the natural history of SOD1 ALS, 15 North American research centers conducted a retrospective review of 175 individuals, all of whom had confirmed SOD1 mutations. A total of 36 missense mutations were identified, the most common being A4V (36.4%), I113T, G41D, and E100K. Average age of disease onset was 49.7 ± 12.3 years. Surprisingly, disease onset in the A4V subgroup was similar to other mutations, but progression of the disease was far more rapid in this group (see Fig. 1).[25]

A limitation of the data set was that not all cases were studied prospectively. From a clinical trial perspective, these data are relevant in that they lead to the conclusion that a trial conducted with a homogeneous group of A4V participants (A5V in current nomenclature) would require fewer subjects than a heterogeneous group of SOD1 patients. These data were overall similar to previous retrospective studies.[26]

To date, SOD1 mutations have been described throughout the gene, including point mutations, insertions, deletions, and truncations. However, these mutations are poorly correlated with clinical presentation. Almost all of the features of ALS can be observed with the various SOD1 variants. Furthermore, the clinical phenotype may vary between members of the same family. Although it is relatively rare, the G93A mutation (glycine at position 93 changed to alanine) has been extensively studied because it has been recapitulated in transgenic rodent models where it was found to generate a motor neuron syndrome.[27] Worldwide, the D90A (alanine for aspartic acid substitution) mutation is the most common, and curiously, can be inherited either as a dominant or recessive trait.[28]

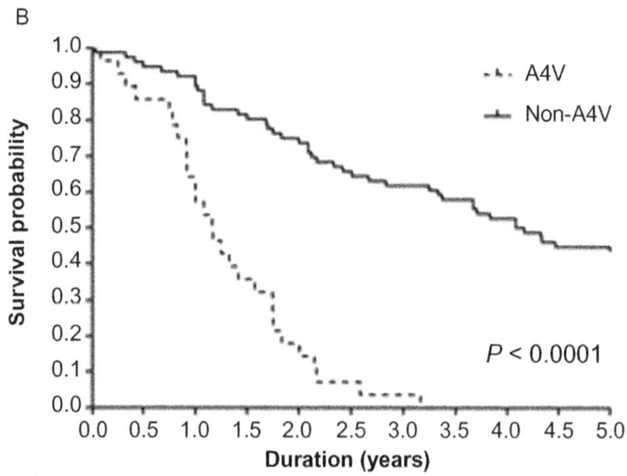

Number at risk											
A4V	35	31	20	11	5	2	1	0	0	0	0
Non-A4V	85	81	73	67	60	53	50	47	42	36	36

Fig. 1 Plot for the survival probability of patients with A4V versus non-A4V SOD1 mutations.[25]

A4V (alanine at codon 4 changed to valine) is the most common ALS-causing mutation in the United States where it is responsible for approximately 50% of the familial cases.[29] Interestingly, it has been suggested that the mutant gene is of both Amerindian and European descent.[29]

Superoxide dismutase-1 (SOD1)

Human Cu—Zn superoxide dismutase-1 (SOD1) is a 32 kDa homodimeric metalloprotein, 153 amino acids in length with its two subunits linked non-covalently. It can be found throughout the cell, including in the mitochondria, lysosomes, and nucleus, but is mainly located in the cytoplasm.[30] The enzyme is an antioxidant, alternately catalyzing the partitioning (or dismutation) of the superoxide radical (O_2-) into either molecular oxygen (O_2) or hydrogen peroxide (H_2O_2). Superoxide, a by-product of oxygen metabolism, is highly toxic. Hydrogen peroxide is degraded by other enzymes such as catalase and therefore is not as damaging as the superoxide radical.

Since SOD1 is present in every cell and is an essential component of the cell's defenses against oxidative stress, after the genetic mutation was reported in 1993, it was understandably assumed that a mutation in SOD1 would lead to loss of function which would be detrimental to the cell's integrity and health. However, this explanation did not survive scrutiny when it was demonstrated that most SOD1 mutations maintain their catalytic function. For example, the dismutase activity of the G93A mutation proved to be as robust as that of wild-type SOD1.[31] Ideally, over the course of some 25-plus years of research, it would have been satisfying had the toxicity of mutant SOD1 been revealed. An initial effort to do so was undertaken by Bruijn et al.[31–33] In their initial experiments, they demonstrated that overexpression of wild-type SOD1 in a G85R mouse did not delay disease onset or extend survival. These studies convincingly proved that loss of dismutase activity did not explain the disease process. Subsequently, in animals in which SOD1 was knocked out in different cell types, it was later demonstrated that the expression of mutant SOD1 in microglia was more deleterious than the expression of mutant SOD1 in motor neurons.[34–36]

Putting aside whether oxidative stress is an etiologic factor in ALS, it remains controversial as to whether aggregation of SOD1 is deleterious to the survival of the cell or represents a protective response. Even more controversial is whether wild-type SOD1 is involved in the pathogenesis of sporadic ALS. For example, overexpression of wild-type SOD1 in a G85R transgenic mouse accelerated disease progression.[37] Further, Bosco et al.[38] demonstrated that wild-type and mutant SOD1 share similar perturbations in their biophysical conformation, though this has not been seen by all groups.[39] If SOD1 is shown to be clearly involved in sporadic ALS, there will be an irresistible temptation to target SOD1 in the case of both inherited and sporadic ALS.

SOD1 animal models of ALS

Rodent models

Rodent models have been the principal laboratory workhorse in the effort to understand and treat ALS since 1993 when mutations of the SOD1 gene were identified as a cause of familial motor neuron disease. Research has understandably gone in two directions. It was presumed that an animal model would provide insight into the cellular perturbations that lead to motor neuron degeneration in ALS and that animal models would provide a unique opportunity to test therapeutic agents for their efficacy in the instance of the human disorder. Depending on one's point of view, neither of these expectations has been fully realized. Despite years of elegant experimentation, the pathogenesis of sporadic and familial ALS is not known, although a number of interesting hypotheses have been generated. For the most part, the reported success of treatment in animal models of ALS failed to slow disease progression after onset. When taken to human trial, none had been successful, replicating the prior outcome in mice. For brevity, we will restrict the voluminous literature regarding transgenic ALS models to the most widely used one, the G93A mouse (by Gurney et al.).[27] These mice develop a clinical phenotype at 3–4 months of age, at the outset manifested by weakness of the hind limbs.

After disease onset there is inexorable progression, leading to quadriplegia and inability to feed and groom, features which by convention are considered terminal end points. At this stage, histopathologic examination reveals severe loss of motor neurons, and aggregated human SOD1 is visualized in those remaining in the ventral horn. As seen in human ALS, reinnervation is readily demonstrated in muscle fibers. Interestingly, the A4V mutation, virulent in humans, is benign in a mouse model. Of further interest is the finding that SOD1 knockout mice do not develop ALS.[40] As they age, they do exhibit distal motor axonopathy. However, based on a single case report in an infant with a homozygous deletion of the SOD1 gene, loss of SOD1 activity was associated with a severe upper motor neuron disorder.[41]

The later development of a G93A transgenic rat model,[42] similar to the mouse model, proved to be more convenient for our studies, which involved intraventricular injections of antisense oligonucleotides (see Fig. 2).

Non-rodent models

The G93A transgenic mouse is by far the most commonly used animal model for studying ALS. Nonetheless, a wide variety of other models are available, ranging in phylogenetic complexity from large mammals to simple nematodes. Each of these models imparts certain advantages for studying the disease and testing therapeutic interventions, yet none is a perfect duplication of the human disease process. This, of course, is true for nearly every animal model of a human disease.

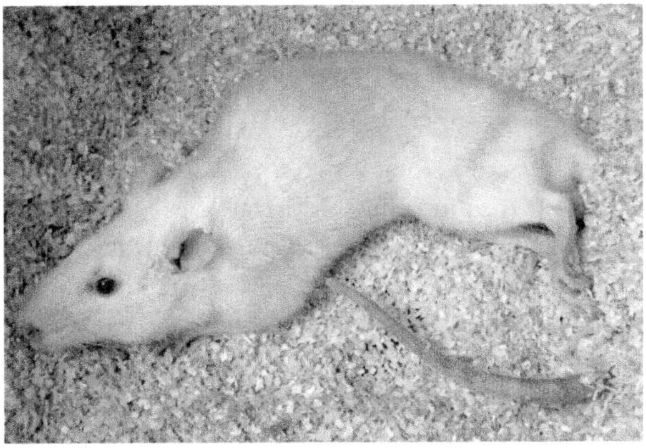

Fig. 2 ALS G93A Rat Model.

A spontaneously occurring degenerative myelopathy, an ALS-like disease, has been recognized in older dogs[43] with homozygous SOD1 mutations. Several of these dogs with SOD1 mutations share pathologic features with human ALS, including axonal degeneration, astrocytosis, and SOD1 inclusions. However, motor neuron loss has not been clearly established. Neurophysiological similarities to ALS include the presence of fibrillation and positive sharp waves, demonstrated by electromyography.

Another large mammal model of ALS involves SOD1 transgenic pigs. Pigs are useful in modeling the disease because of certain anatomical as well as genetic and biochemical similarities to humans. In particular, the CNS anatomy of a pig is much closer to human anatomy than is the rodent CNS. Minipigs have been engineered to express the human SOD1 G93A mutant under control of a CMV promoter.[44]

At the simpler end of the phylogenetic spectrum, SOD1 mutations have been introduced into flies, worms, and fish,[45–47] perhaps underscoring the widespread interest in these genetic aberrations and their convenience as laboratory models. The most salient observation in this medley of animal models is that overexpression of wild-type human SOD1 in *C. elegans* results in motor deficits leading to paralysis within 10–20 days.[46]

Development of an antisense therapeutic

By the 1990s, it was well known that systemically administered ASOs were widely distributed in the liver, kidney, and bone marrow, whereas uptake in muscle was minimal and distribution to the brain was virtually nonexistent.[48] This disparity was understandably attributed to the effect of the blood–brain barrier. In discussions with Ionis Pharmaceuticals, it was mutually agreed that it would be of interest to know more about the pharmacokinetics and pharmacodynamics of instilling ASOs into the spinal fluid, either by intracerebral or intrathecal injections. This ultimately dovetailed with the interests of the company, the Center for Neurologic Study, and Don Cleveland at the University of California at San Diego. Thus, one of the more promising therapeutic collaborations in the field of ALS was born, involving industry, academia, and practicing neurologists specializing in neuromuscular disease.

As previously noted, the blood–brain barrier is an impediment to the administration of drugs in the CNS. While the mouse has historically been utilized in most ALS research studies, it was fortunate that the G93A rat model was available by the time our research was undertaken since it was our intention to instill ASOs intraventricularly. A larger animal made the task easier, although this was not without challenges. It was customary at the time to drill a large burr hole to gain access to the brain and to subsequently place the delivery cannula stereotactically into the right or left lateral ventricles. The coordinates for this are readily available, so placement is not an obstacle for a skilled operator. The challenge is to fix the cannula to the skull. The usual practice was to accomplish this by cementing the cannula to the skull using the surgical glue methyl methacrylate. To avoid subjecting the operator to noxious fumes, the procedure was altered.

The solution was to drill a much smaller burr hole, which was placed at the correct coordinates on the skull. This accomplished two things. First, the appropriately placed burr hole served as a guide, which guaranteed accurate placement of the cannula. More importantly, the skull acted as a pincer providing most of the purchase needed to fasten the cannula securely in place. Further bonding was attained by applying a miniscule amount of surgical glue to the disk that, together with the cannula, comprises the delivery element of the Alzet mini-pump system for administering pharmaceuticals to small animals such as rodents and guinea pigs. The remaining procedures, mostly standard, also required attention. The osmotic pumps needed to be loaded well in advance of the procedure and subsequently incubated in normal saline at 37°. To ensure delivery of the correct dose, the pumps were weighed before and after filling. At the time of implantation, scissors were employed to create a subcutaneous tunnel under the skin overlying the back of the animal. To complete the procedure, pumps were burrowed into the tunnel and connected by tubing to the cannula. Depending on the experiments, animals were administered drug for 14 days in the instance of short-term experiments, or for longer periods in the instance of survival experiments. For the latter studies, larger pumps were utilized, and these were replaced monthly as needed. Animals were observed daily, and those that exhibited limited mobility, marked weight loss, inability to groom, etc. were euthanized. Subsequently, the brain and spinal cord were removed and processed as required. As a final step, methylene blue was injected through the cannula and tubing that had previously been connected to the Alzet pump. If obstructed, it was presumed that the animal did not receive the full dose of the ASO. Data from these animals were excluded.

Two animal models were employed: normal laboratory rats (*Rattus norvegicus*) and G93A SOD1 transgenic rats that were customarily studied at 2–3 months of age.

To address the distribution of ASOs administered intraventricularly, an oligonucleotide capable of being visualized by immunohistochemistry was instilled for 2 weeks by continuous infusion in the right lateral ventricles of normal rats using Alzet mini pumps. At the end of the experiments, animals were sacrificed and representative sections from the entire CNS axis were examined. Since the brunt of the pathology in ALS is born by motor neurons in the spinal cord, it was tempting to start the histologic examination at the lumbar spine, but the decision was made to start the examination at the level of the cerebral hemispheres. Going through the sections took time, but at the conclusion of the review it was reassuring to observe robust uptake in both the hippocampus (Fig. 3) and motor neurons at the level of the lumbar spinal cord (Fig. 4).

The next objective was to demonstrate that the ASO in neuronal cells was pharmacologically active by documenting a reduction in SOD1 mRNA. After screening a

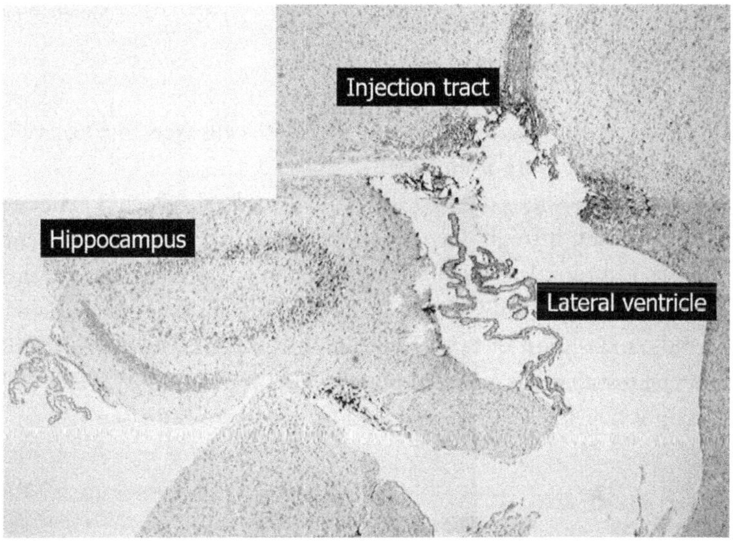

Fig. 3 Localization of a tracking oligonucleotide in the hippocampus following ICV injection into the right lateral ventricle of a normal laboratory rat.

Fig. 4 Delivery to spinal motor neurons following ICV administration in the lumbar ventral horn.

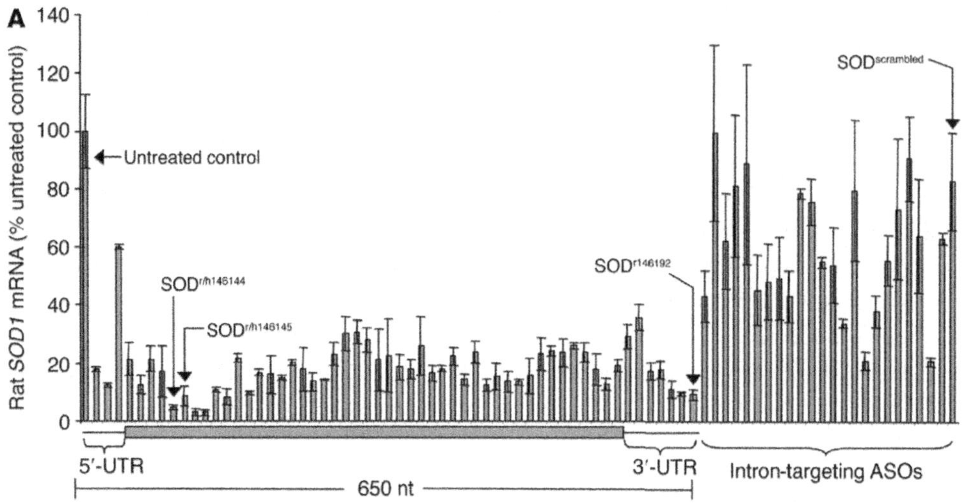

Fig. 5 Identification of Candidate Oligos.

series of SOD1 targeting oligonucleotides in A549 cells (see Fig. 5), two candidate ASOs specific for SOD1 were identified.

Frustratingly, the two oligos (144 and 145) that had been selected generated inconsistent results. At this point, it might have been expected that Ionis Pharmaceuticals would abandon the project but instead the company was convinced to give it another shot, synthesizing a second panel of 80 ASOs. This generated several lead oligos that convincingly downregulated the expression of SOD1-specific mRNA in wild-type rats following the same experimental paradigm as undertaken with the original set of ASOs.[49] At all levels of the CNS axis, SOD1 mRNA was reduced from 50% to 80% (see Fig. 6).

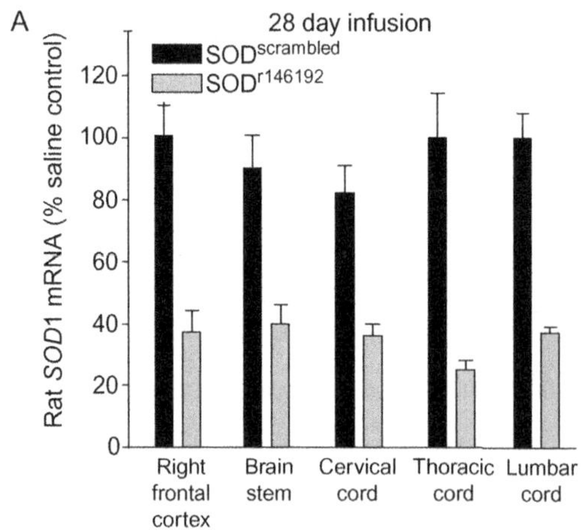

Fig. 6 Antisense oligo targets SOD1 mRNA in the brain and the spinal cord.

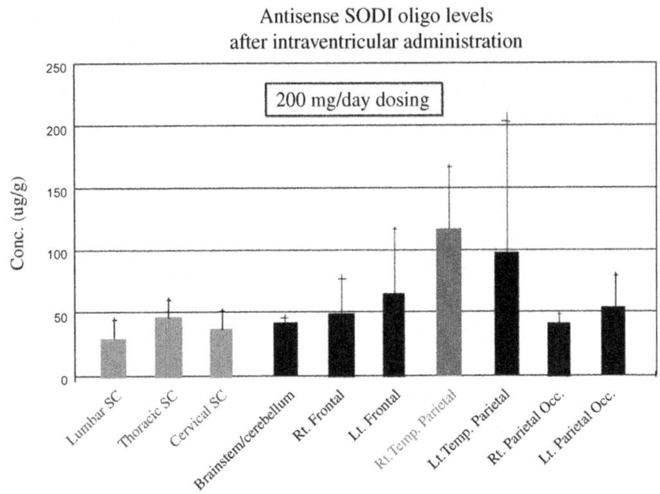

Fig. 7 Antisense SOD1 oligo levels after intraventricular administration.

In the next experiment, the concentration of the SOD1 targeting oligo was measured throughout the CNS following intraventricular administration. At the time, the result was satisfying since it was known that an ASO concentration as low as 10 μg/g tissue was sufficient to downregulate the expression of targeted proteins (Fig. 7).

At a subsequent laboratory meeting, it was questioned whether the 50%–60% reduction of SOD1 mRNA we had just demonstrated (see Fig. 6)[50] would result in a reduction of mutant SOD1 protein that would be sufficient to alter the course of ALS in an experimental or human setting. The next obvious experiment was to demonstrate that the treatment effect would favorably enhance survival in the rat model of familial ALS. However, this was thought to be a daunting challenge since the G93A transgenic rat carried at least 15 copies of the mutant human SOD1 gene. As a result, based on inference from studies with the G93A mouse in which an increased copy number is associated with increased amounts of mutant protein, we assumed that the level of mutant SOD1 protein in our rat model would be similarly increased. Until we conducted the experiment, it was not certain that our treatment strategy would be sufficient to demonstrate an effect in the animal model.

In all, 24 animals were either treated with saline or the most potent oligonucleotide identified in the second round of synthesis (ASO 333611). Treatment was initiated at 65 days because we had previously determined that it took about 30 days to see meaningful reduction in SOD1 protein following ASO treatment and disease onset occurred approximately 30 days before animals succumbed to their illness. To our surprise and delight, the oligo-treated animals lived 10 days longer than the placebo group (132 vs 122 days) (see Fig. 8).

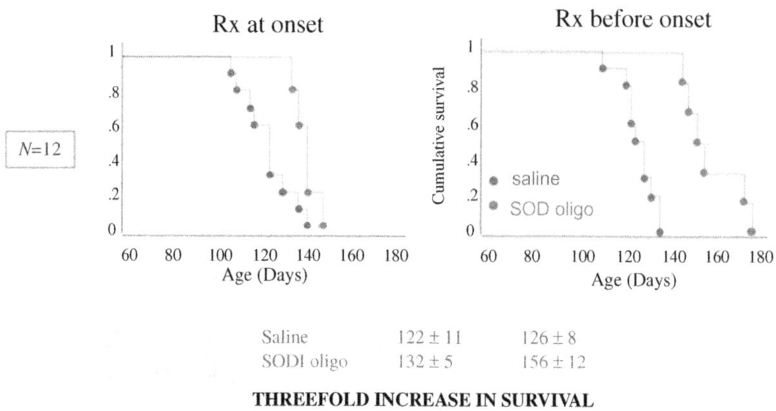

Fig. 8 Treatment with SOD1 oligo extends survival in SOD1 G93A rat.

Following up on the initial experiment, it was decided that the treatment effect might be more dramatic if the animals were dosed at a younger age. The follow-up experiment was similar in all respects to the initial survival study, except that the animals were treated at 30 days. Both the animal scoring and data analysis were conducted in a blinded fashion. The result was even more dramatic. The treated group enjoyed a 30-day survival benefit versus the controls (see Fig. 8, unpublished data). More recently, an updated version of this experiment was undertaken using an ASO with increased potency that is currently being tested in a human familial ALS treatment trial. In two rodent models of ALS (G93A mice and rats), survival was markedly increased in comparison with earlier experiments using ASO 333611. And importantly, treatment reversed the neuromuscular signatures associated with motor unit denervation.[49]

Investigational new drug (IND)-enabling studies

As required, the IND-enabling studies were undertaken in two species: rats and subhuman primates, with the understanding that the outcome in these studies would be necessary for approval of the proposed clinical studies. As is customary, the primary objective was to demonstrate the safety and the pharmacokinetic behavior of our candidate therapy, in the belief that results in animals would provide guidance for the subsequent studies in humans.

Pharmacokinetic and safety studies in rhesus monkeys addressed the concern that the ASO distribution results in rodents might not be reproducible in a larger animal. Although we intended to treat patients utilizing the intrathecal route of administration, for technical reasons the studies in rats were performed by administering the ASO by ICV infusion. However, we had to convincingly demonstrate that we could achieve adequate distribution by intrathecal delivery. To do this, spinal fluid, plasma, and tissue concentrations

Parameter	0.1 mg/day	0.6 mg/day	3 mg/day
CSF μg/mL	1.0 ± 1.2	7.8 ± 5.5	28 ± 35
Plasma μg/mL	0.003 ± 0.004	0.042 ± 0.025	0.180 ± 0.115
CSF: Plasma	333	186	156
Cervical cord levels	10 μg/gm	40 μg/gm	75 μg/gm
Half life	14–28 days	14–28 days	14–28 days

Fig. 9 Pharmacokinetic summary.

were measured following administration of ASOs at three doses ranging from 0.1 to 3 mg per day for 28 days (see Fig. 9) in nonhuman primates.

The results in primates were especially gratifying since the lead oligo (IONIS 333611) was robustly taken up by motor neurons in the lumbar spinal cord and interestingly, in the cingulate gyrus of the cortex, demonstrating that even intrathecal administration resulted in fairly widespread distribution of ASOs in the CNS (see Fig. 10). Had this experiment failed, it likely would have eliminated the rationale for proceeding to human trials.

An unexpected observation in subhuman primates that proved to be of considerable importance was the fact that blood levels of ASOs tracked the levels attainable in the spinal fluid, albeit at a much lower range (see Fig. 11). This was of potential utility, in that it suggested that one could monitor drug levels in the brain without having to sample the spinal fluid.

The correlation between plasma and CSF ASO levels should facilitate monitoring in clinical trials.

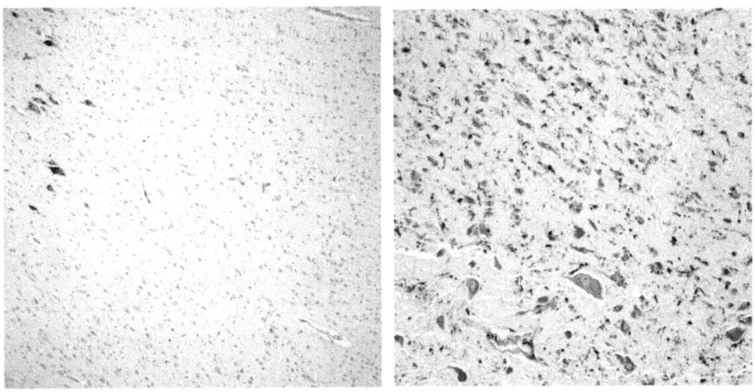

Cortex (Cingulate gyrus) Lumbar cord

Fig. 10 Uptake in the monkey brain and spinal cord following a 13-week intrathecal infusion.

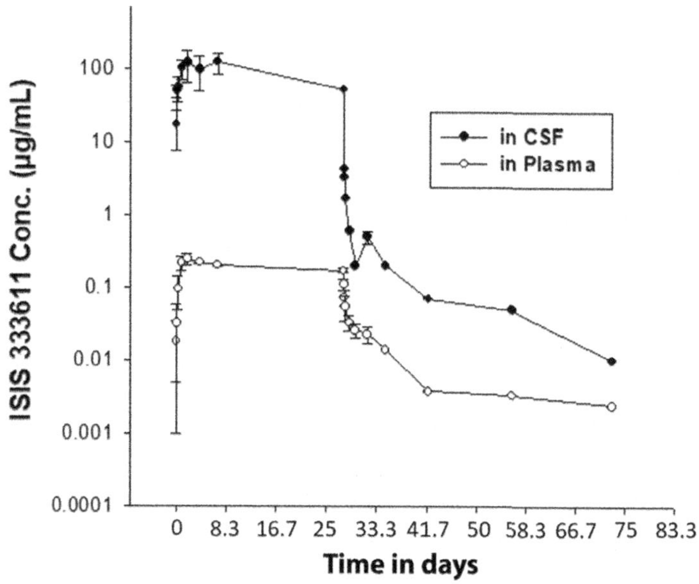

Fig. 11 Plasma and CSF concentration during and after 28-day intrathecal infusion.

Clinical studies

An IND submission to the FDA was completed in 2010. The goals of the Phase 1 study, as expected, were to gain insight into the pharmacokinetic properties of administering an antisense oligonucleotide intrathecally and to determine the safety of doing so. To accomplish this, a dose-ranging study was conducted in which patients were infused intrathecally with either saline or 333611 over 12 h to simulate a single dose of the drug. If successful, it was planned that patients would have a continuous infusion pump implanted for longer-term treatment. Based on the tolerability, the dose was subsequently increased from 0.15 mg in the first cohort to 0.5 mg for the second, 1.5 mg for the third, and 3 mg in the fourth and final cohort. Some patients who were enrolled in the low-dose group were later enrolled in a higher dose group. Spinal fluid levels of the drug and SOD1 protein were assessed following the completion of each infusion (see Fig. 12). This "First-in-Human" clinical trial was published in Lancet Neurology in 2013.[51]

At each dose escalation, CSF levels of the oligonucleotide were found to be increased. Although only a limited number of subjects were studied, it was reassuring that the side effects were limited, primarily due to the occurrence of headache which is commonly associated with lumbar puncture. In the instance of a patient who died of his disease approximately 2 months after receiving the final infusion, concentrations of the ASO in the spinal cord ranged from 218 ng/g in the lumbar region to 39 ng/g in the cervical spinal cord. These data were particularly helpful in understanding tissue half-life of the ASO and matched well with previous modeling of clearance.

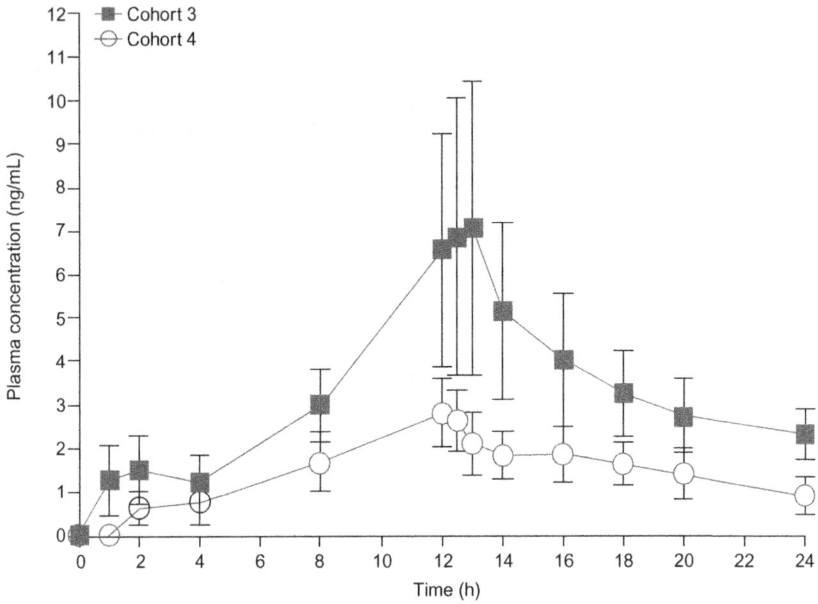

Fig. 12 Plasma concentrations of SOD1 antisense oligonucleotide following intrathecal infusion.

Ultimately, Ionis Pharmaceuticals chose not to proceed with development of the 333611 oligonucleotide because they identified a more potent ASO compared to 333611. The improved ASO targeting SOD1 (tofersen) markedly extended survival in the SOD1 G93A mouse and rat models and reversed the loss of muscle response upon stimulation of motor nerves.[49] While it was initially thought that the ASOs would need to be delivered by constant infusion as had been done in the initial Phase I ASO trial, subsequent PK studies in animal models demonstrated that repeated bolus injections provide similar distribution and potency, with single injections producing target RNA modulation for more than 6 months.[52]

The results of a Phase I/II ascending dose trial were published a few months in advance of this chapter going to press.[53] The primary objective of the study was to demonstrate the tolerability, pharmacokinetics, and exploratory efficacy of administering 20–100 mg intrathecal boluses of tofersen, the generic name assigned to the ASO. Noteworthy was the finding that the level of SOD1 protein in the CSF of patients receiving the highest dose was reduced, and, in a subset of fast-progressing patients, several measures of disease activity were ameliorated. Based on encouraging results in the Phase I/II trial, a Phase 3 trial (VALOR) has been initiated in 80 ALS patients with confirmed SOD1 mutations. Assuming that the preliminary findings noted in the earlier trial are confirmed, one presumes that the drug would be rapidly approved by regulatory authorities for the treatment of the SOD1 variant of ALS.

Conclusion

After years in development, antisense drugs have been recognized as one of the most promising advances in the history of CNS therapeutics. There are numerous advantages to this technology, the first being that, in contrast to "gene therapy," ASOs are similar to conventional drugs since the dose can be increased to enhance efficacy or can be lowered or terminated in the instance of adverse events. Second, the technology is readily adaptable in that the gene products can be upregulated or downregulated or, as in the instance of spinal muscular atrophy, the technology can be utilized to alter RNA splicing. Finally, in spite of not knowing the molecular events that lead to neurodegeneration in the instance of familial disorders such as Huntington's disease, the use of antisense has provided a treatment strategy that does not require a complete understanding of disease pathogenesis.

Acknowledgments

The research described herein was supported by the ALS Association and the Muscular Dystrophy Association. Dr. Smith was supported, in part, by a Skaggs Clinical Scholar Award (Scripps Research Institute) and by Ionis Pharmaceutical Corporation. Melissa McAlonis-Downes provided essential research assistance and husbandry services. Many of the patients enrolled in the clinical trials were, for the most part, selected because their underlying mutation was known to result in an aggressive course of ALS. The participation of all subjects required considerable personal sacrifice as well as that of their families. The authors gratefully acknowledge this and dedicate this chapter to them and their families. Finally, Kathleen Myers, PhD, Meryem Guler, and Hanna Hovren's efforts on our behalf were indispensable, as was the feedback provided to us by Brett Monia, PhD, CEO of Ionis Pharmaceuticals.

References

1. Zamecnik PC, Stephenson ML. Inhibition of Rous sarcoma virus replication and cell transformation by a specific oligodeoxynucleotide. *Proc Natl Acad Sci U S A*. 1978;75(1):280–284.
2. Lima W, Wu H, Crooke ST. The RNase H mechanism. In: Crooke ST, ed. *Antisense Drug Technology: Principles, Strategies, and Applications*. 2nd ed. Boca Raton, FL: Taylor & Francis Group, CRC Press; 2008:47–74.
3. Broaddus WC, Prabhu SS, Wu-Pong S, Gillies GT, Fillmore H. Strategies for the design and delivery of antisense oligonucleotides in central nervous system. *Methods Enzymol*. 2000;314:121–135.
4. Smith RA, Miller TM. Targeting neurological disorders with antisense oligonucleotides. In: Crooke ST, ed. *Antisense Drug Technology Principles, Strategies, and Applications*. 2nd ed. Boca Raton, Florida: CRC Press: Taylor and Francis Group; 2008:721–746.
5. Chauhan NB. Trafficking of intracerebroventricularly injected antisense oligonucleotides in the mouse brain. *Antisense Nucleic Acid Drug Dev*. 2002;12(5):353–357.
6. Sawa GM, Paty DW. The use of baclofen in treatment of spasticity in multiple sclerosis. *Can J Neurol Sci*. 1979;6(3):351–354.
7. Rosen DR, Siddique T, Patterson D, et al. Mutations in Cu/Zn superoxide dismutase gene are associated with familial amyotrophic lateral sclerosis. *Nature*. 1993;362(6415):59–62.
8. Chio A, Logroscino G, Traynor BJ, et al. Global epidemiology of amyotrophic lateral sclerosis: a systematic review of the published literature. *Neuroepidemiology*. 2013;41(2):118–130.

9. Hirano A, Kurland LT, Sayre GP. Familial amyotrophic lateral sclerosis. A subgroup characterized by posterior and spinocerebellar tract involvement and hyaline inclusions in the anterior horn cells. *Arch Neurol.* 1967;16(3):232–243.
10. Strong MJ, Grace GM, Orange JB, Leeper HA, Menon RS, Aere C. A prospective study of cognitive impairment in ALS. *Neurology.* 1999;53(8):1665–1670.
11. Neumann M, Sampathu DM, Kwong LK, et al. Ubiquitinated TDP-43 in frontotemporal lobar degeneration and amyotrophic lateral sclerosis. *Science.* 2006;314(5796):130–133.
12. Mackenzie IR, Bigio EH, Ince PG, et al. Pathological TDP-43 distinguishes sporadic amyotrophic lateral sclerosis from amyotrophic lateral sclerosis with SOD1 mutations. *Ann Neurol.* 2007;61(5):427–434.
13. Plato CC, Garruto RM, Galasko D, et al. Amyotrophic lateral sclerosis and parkinsonism-dementia complex of Guam: changing incidence rates during the past 60 years. *Am J Epidemiol.* 2003;157(2):149–157.
14. Steele JC, McGeer PL. The ALS/PDC syndrome of Guam and the cycad hypothesis. *Neurology.* 2008;70(21):1984–1990.
15. Olney JW. Glutamate, a neurotoxic transmitter. *J Child Neurol.* 1989;4(3):218–226.
16. Choi DW, Rothman SM. The role of glutamate neurotoxicity in hypoxic-ischemic neuronal death. *Annu Rev Neurosci.* 1990;13:171–182.
17. Shaw PJ, Ince PG. Glutamate, excitotoxicity and amyotrophic lateral sclerosis. *J Neurol.* 1997;244(Suppl 2):S3–14.
18. Lacomblez L, Bensimon G, Leigh PN, et al. A confirmatory dose-ranging study of riluzole in ALS. ALS/Riluzole Study Group-II. *Neurology.* 1996;47(6 Suppl 4):S242–S250.
19. Bensimon G, Lacomblez L, Meininger V. A controlled trial of riluzole in amyotrophic lateral sclerosis. ALS/Riluzole study group. *N Engl J Med.* 1994;330(9):585–591.
20. Drachman DB, Chaudhry V, Cornblath D, et al. Trial of immunosuppression in amyotrophic lateral sclerosis using total lymphoid irradiation. *Ann Neurol.* 1994;35(2):142–150.
21. Cudkowicz ME, Shefner JM, Schoenfeld DA, et al. Trial of celecoxib in amyotrophic lateral sclerosis. *Ann Neurol.* 2006;60(1):22–31.
22. A controlled trial of recombinant methionyl human BDNF in ALS: The BDNF Study Group (Phase III). Neurology. 52(7):1427–1433. https://n.neurology.org/content/52/7/1427.long.
23. Miller RG, Petajan JH, Bryan WW, et al. A placebo-controlled trial of recombinant human ciliary neurotrophic (rhCNTF) factor in amyotrophic lateral sclerosis. rhCNTF ALS Study Group. *Ann Neurol.* 1996;39(2):256–260.
24. Taylor JP, Brown Jr RH, Cleveland DW. Decoding ALS: from genes to mechanism. *Nature.* 2016;539(7628):197–206.
25. Bali T, Self W, Liu J, et al. Defining SOD1 ALS natural history to guide therapeutic clinical trial design. *J Neurol Neurosurg Psychiatry.* 2017;88(2):99–105.
26. Cudkowicz ME, McKenna-Yasek D, Sapp PE, et al. Epidemiology of mutations in superoxide dismutase in amyotrophic lateral sclerosis. *Ann Neurol.* 1997;41(2):210–221.
27. Gurney ME, Pu H, Chiu AY, et al. Motor neuron degeneration in mice that express a human Cu,Zn superoxide dismutase mutation. *Science.* 1994;264(5166):1772–1775.
28. Parton MJ, Broom W, Andersen PM, et al. D90A-SOD1 mediated amyotrophic lateral sclerosis: a single founder for all cases with evidence for a Cis-acting disease modifier in the recessive haplotype. *Hum Mutat.* 2002;20(6):473.
29. Saeed M, Yang Y, Deng HX, et al. Age and founder effect of SOD1 A4V mutation causing ALS. *Neurology.* 2009;72(19):1634–1639.
30. Crapo JD, Oury T, Rabouille C, Slot JW, Chang LY. Copper,zinc superoxide dismutase is primarily a cytosolic protein in human cells. *Proc Natl Acad Sci U S A.* 1992;89(21):10405–10409.
31. Borchelt DR, Guarnieri M, Wong PC, et al. Superoxide dismutase 1 subunits with mutations linked to familial amyotrophic lateral sclerosis do not affect wild-type subunit function. *J Biol Chem.* 1995;270(7):3234–3238.
32. Bruijn LI, Houseweart MK, Kato S, et al. Aggregation and motor neuron toxicity of an ALS-linked SOD1 mutant independent from wild-type SOD1. *Science.* 1998;281(5384):1851–1854.

33. Lobsiger CS, Boillee S, Cleveland DW. Toxicity from different SOD1 mutants dysregulates the complement system and the neuronal regenerative response in ALS motor neurons. *Proc Natl Acad Sci U S A*. 2007;104(18):7319–7326.
34. Boillee S, Vande Velde C, Cleveland DW. ALS: a disease of motor neurons and their nonneuronal neighbors. *Neuron*. 2006;52(1):39–59.
35. Boillee S, Yamanaka K, Lobsiger CS, et al. Onset and progression in inherited ALS determined by motor neurons and microglia. *Science*. 2006;312(5778):1389–1392.
36. Yamanaka K, Boillee S, Roberts EA, et al. Mutant SOD1 in cell types other than motor neurons and oligodendrocytes accelerates onset of disease in ALS mice. *Proc Natl Acad Sci USA*. 2008;105(21): 7594–7599.
37. Wang L, Deng HX, Grisotti G, Zhai H, Siddique T, Roos RP. Wild-type SOD1 overexpression accelerates disease onset of a G85R SOD1 mouse. *Hum Mol Genet*. 2009;18(9):1642–1651.
38. Bosco DA, Morfini G, Karabacak NM, et al. Wild-type and mutant SOD1 share an aberrant conformation and a common pathogenic pathway in ALS. *Nat Neurosci*. 2010;13(11):1396–1403.
39. Da Cruz S, Bui A, Saberi S, et al. Misfolded SOD1 is not a primary component of sporadic ALS. *Acta Neuropathol*. 2017;134(1):97–111.
40. Shefner JM, Reaume AG, Flood DG, et al. Mice lacking cytosolic copper/zinc superoxide dismutase display a distinctive motor axonopathy. *Neurology*. 1999;53(6):1239–1246.
41. Andersen PM, Nordstrom U, Tsiakas K, et al. Phenotype in an infant with SOD1 homozygous truncating mutation. *N Engl J Med*. 2019;381(5):486–488.
42. Matsumoto A, Okada Y, Nakamichi M, et al. Disease progression of human SOD1 (G93A) transgenic ALS model rats. *J Neurosci Res*. 2006;83(1):119–133.
43. Nardone R, Holler Y, Taylor AC, et al. Canine degenerative myelopathy: a model of human amyotrophic lateral sclerosis. *Zoology (Jena)*. 2016;119(1):64–73.
44. Crociara P, Chieppa MN, Vallino Costassa E, et al. Motor neuron degeneration, severe myopathy and TDP-43 increase in a transgenic pig model of SOD1-linked familiar ALS. *Neurobiol Dis*. 2019;124:263–275.
45. Watson MR, Lagow RD, Xu K, Zhang B, Bonini NM. A drosophila model for amyotrophic lateral sclerosis reveals motor neuron damage by human SOD1. *J Biol Chem*. 2008;283(36):24972–24981.
46. Ogawa M, Shidara H, Oka K, Kurosawa M, Nukina N, Furukawa Y. Cysteine residues in Cu, Zn-superoxide dismutase are essential to toxicity in *Caenorhabditis elegans* model of amyotrophic lateral sclerosis. *Biochem Biophys Res Commun*. 2015;463(4):1196–1202.
47. Lemmens R, Van Hoecke A, Hersmus N, et al. Overexpression of mutant superoxide dismutase 1 causes a motor axonopathy in the zebrafish. *Hum Mol Genet*. 2007;16(19):2359–2365.
48. Levin AA, Yu RZ, Geary RS. Basic principles of the pharmacokinetics of antisense oligonucleotide drugs. In: Crooke ST, ed. *Antisense Drug Technology Principles, Strategies, and Applications*. 2nd ed. Boca Raton, Florida: CRC Press/Taylor and Francis Group; 2008:183–216.
49. McCampbell A, Cole T, Wegener AJ, et al. Antisense oligonucleotides extend survival and reverse decrement in muscle response in ALS models. *J Clin Investig*. 2018;128(8):3558–3567.
50. Smith RA, Miller TM, Yamanaka K, et al. Antisense oligonucleotide therapy for neurodegenerative disease. *J Clin Investig*. 2006;116(8):2290–2296.
51. Miller TM, Pestronk A, David W, et al. An antisense oligonucleotide against SOD1 delivered intrathecally for patients with SOD1 familial amyotrophic lateral sclerosis: a phase 1, randomised, first-in-man study. *Lancet Neurol*. 2013;12(5):435–442.
52. Rigo F, Chun SJ, Norris DA, et al. Pharmacology of a central nervous system delivered 2'-O-methoxyethyl-modified survival of motor neuron splicing oligonucleotide in mice and nonhuman primates. *J Pharmacol Exp Ther*. 2014;350(1):46–55.
53. Miller T, Cudkowicz M, Shaw PJ, et al. Phase 1–2 trial of antisense oligonucleotide tofersen for *SOD1* ALS. *N Engl J Med*. 2020;383:109–119.

CHAPTER 2

Unraveling molecular biology of C9ORF72 repeat expansions in amyotrophic lateral sclerosis-frontotemporal dementia: Implications for therapy

Olubankole Aladesuyi Arogundade[a,*], Jie Jiang[b,*], and John Ravits[a]
[a]Department of Neurosciences, University of California at San Diego, La Jolla, CA, United States
[b]Department of Cell Biology, Emory University, Atlanta, GA, United States

Introduction

The literature bringing together amyotrophic lateral sclerosis (ALS) and frontotemporal dementia (FTD), two devastating, adult-onset neurodegenerative disorders with distinct clinical presentations, can be traced nearly to their original descriptions. ALS was first described by the French neurologist Jean-Martin Charcot in 1869 and is the most common motor neuron disease. It is characterized by rapid progression of muscle weakness leading to paralysis and respiratory failure caused by the loss of upper and lower motor neurons and is usually fatal within 3–5 years.[1] FTD was first described by Arnold Pick in 1892 and was originally called Picks Disease, a term now reserved for cases with pathological verification of abnormal tau aggregates called Pick bodies. FTD is characterized by behavioral and personality changes, deficits in executive functions, and/or language impairments due to the degeneration of neurons in the frontal and temporal lobes.[2] While most cases of either ALS or FTD are sporadic, about 10 or more percent of each occur in families, almost always in a dominant pattern.[3] Some of the genes that cause ALS, notably SOD1, usually do not cause dementia, and some of the kinships that have FTD, notably caused by granulin (GRN) mutations, infrequently have ALS. But a large group of families with either ALS or FTD or overlap ALS-FTD in an autosomal genetic pattern have been identified and found to be caused by a hexanucleotide repeat expansion in the C9orf72 (abbreviated from its genomic loci in chromosome 9 open reading frame 72) gene, the most common genetic cause of both ALS and FTD.[4–6] This mutation accounts

[*] Equal contributions.

Neurotherapeutics in the Era of Translational Medicine
https://doi.org/10.1016/B978-0-12-816475-4.00009-4

for 40% of familial ALS, 5%–10% of sporadic ALS, 12%–25% of familial FTD, and 6%–7% of sporadic FTD cases.[7] It is now recognized that ALS and FTD belong to a continuous disease spectrum with clinical, pathological, and genetic overlaps.

The identification of the C9orf72 locus added ALS/FTD (referred to as c9ALS/FTD hereafter) to more than 40 other repeat expansion disorders (e.g., myotonic dystrophy, Huntington's disease, and several spinocerebellar ataxias).[8] Based on lessons learned from these other repeat expansion disorders and initial pathological characterization of C9orf72 postmortem tissues, three disease mechanisms have been proposed: (1) loss of C9orf72 function, (2) toxicity from repeat-containing RNAs, which are bidirectionally transcribed, form foci, and perhaps sequester key RNA binding proteins (RBPs), and (3) toxicity from the production of dipeptide repeat (DPR) proteins through a noncanonical repeat-associated non-AUG dependent (RAN) translation which also happens bidirectionally. Although the C9orf72 mutation was only identified 8 years ago, research progress has been rapid, and significantly influenced current thinking about the pathogenesis and treatment of neurological disorders. A number of trials directed at c9ALS/FTD are now underway or in the early stages of development.

Pathogenic mechanisms of c9ALS/FTD

C9orf72 hexanucleotide repeat expansions

The hexanucleotide repeat expansion is localized in the first intron of the C9orf72 gene (Fig. 1A). The size of the repeats varies dramatically among C9orf72 patients. The threshold pathogenic repeat size is unknown, but 30 is used in most publications. Healthy individuals have typically fewer than 11 repeats, whereas most patients have several hundreds to thousands.[4,9–13] A small percentage of patients have shorter expansions, 45–80 repeats.[14] Unlike Huntington's disease and other CAG repeat expansion disorders where a phenomenon called anticipation causes increased size and a more severe phenotype in offspring, increased C9orf72 repeat size in offspring is only reported in a few families.[15,16] The correlation of the repeat size and clinical and pathological characteristics has also been explored, and the results are variable, depending on patient size, ethnic background, and tissues in which expansion size was measured.[9,12,14,17–21] Longer repeat sizes seem to be associated with a survival disadvantage[18] and an earlier onset of disease.[9,12,14] Interestingly, individuals with an intermediate repeat length in blood or fibroblast DNAs have significant increases of repeat size (repeat instability) and variation in repeat size (mosaicism) in different regions of the central nervous system.[14,18,20] The mechanisms of such instability and somatic mosaicism are not well understood but need to be considered when using repeat size from peripheral blood and fibroblast cells for genetic counseling.

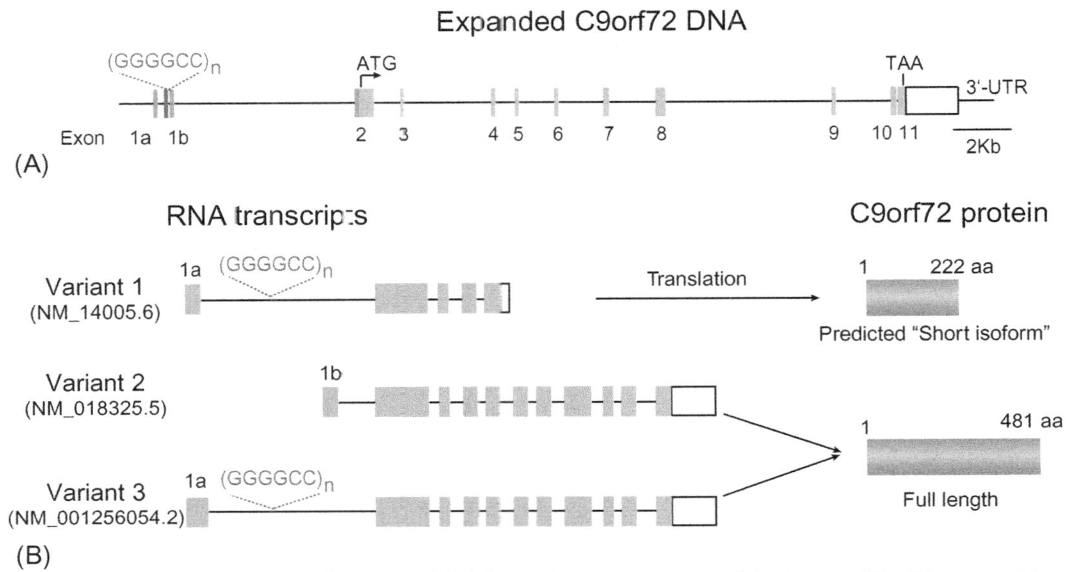

Fig. 1 C9orf72 gene structure, RNA transcripts, and proteins. (A) Schematic representation of the human C9orf72 gene with expanded GGGGCC hexanucleotide repeats. (B) Three RNA transcripts can be produced from the C9orf72 gene. Variant 1 is predicted to result in a "short isoform" C9ORF72 protein of 222 amino acids, whereas variants 2 and 3 encode a long C9ORF72 protein of 481 amino acids.

Loss of C9orf72 protein function

In the original discovery, C9ORF72 was described to have three transcript variants, variant 1 (NM_145005.6), variant 2 (NM_018325.5), and variant 3 (NM_001256054.2) although this naming scheme has been inconsistent in the literature (Fig. 1B). Transcript variants 2 and 3 are transcribed from two alternatively spliced exons, exon 1B and exon 1A, respectively, and encode a full-length C9orf72 protein with 481 amino acids. The expanded GGGGCC repeat is located in an intron between these two exons. Variant 1 comprising 5 of 11 exons produces a predicted "short isoform" protein of 222 amino acids, which harbors a unique lysine residue at the C-terminus. However, it remains controversial whether the predicted C9orf72 short isoform is expressed in nature. Xiao et al. developed antibodies that selectively detect either the full-length or the predicted short C9orf72 protein isoforms and showed that the full-length protein was expressed within the cytoplasm whereas the short isoform was preferentially expressed on the nuclear membrane.[22] However, using an antibody against the N-terminal region of C9orf72 that could recognize both the full-length and the predicted short isoform C9orf72, we demonstrated that C9orf72 is predominantly, if not exclusively, expressed as the full-length protein in human and mouse tissues.[23] This is further confirmed by another study using novel knock-out validated C9orf72 monoclonal rat and mouse antibodies.[24]

While the expression of C9orf72 protein isoforms in human cells has yet to be unequivocally established, our understanding of the physiological function of the C9orf72 protein is emerging. Bioinformatic analysis shows that C9orf72 has high homology with the DENN (differentially expressed in normal and neoplastic cells) protein family, which function as guanine nucleotide exchange factors (GEFs) activating RAB GTPases to regulate membrane trafficking.[25,26] In agreement with this, C9orf72 interacts with Smith-Magenis syndrome chromosomal region candidate gene 8 (SMCR8) and WD repeat-containing protein 41 (WDR41), and this complex possesses GEF activity for different Rab proteins and plays a role in multiple steps of autophagy/lysosomal pathways.[27–34] Several Rab proteins are substrates of C9orf72, including Rab1, Rab3, Rab5, Rab7, Rab8a, Rab11, and Rab39.[28,30,31,33,34]

Early on, it was proposed that the presence of an enormous GGGGCC repeat expansion in the intron region may lead to abortive C9orf72 transcripts, thus depriving cells of an important player in autophagy/lysosomal functions and leading to neurodegeneration (Fig. 2A). Indeed, reduced transcript expression was the case for variants 2 and 3 as they were found to be attenuated in patient lymphoblast cells and frontal cortex.[4] Alternative explanations for reduced C9orf72 transcript expression have also been proposed. For example, it has been suggested that CpG island regions 5′ of GGGGCC repeats cause cytosine hypermethylation of CG dinucleotides and silencing of C9ORF72. This is similar to epigenetic silencing by CGG expansion in the 5′UTR of FMR1 causing fragile X mental retardation syndrome.[35] In C9orf72 patients, the CpG islands are hypermethylated,[36] as is the repeat itself.[37] Furthermore, trimethylation of histones H3 and H4 at the C9orf72 locus has been detected in C9orf72 patient blood,[38] providing a third

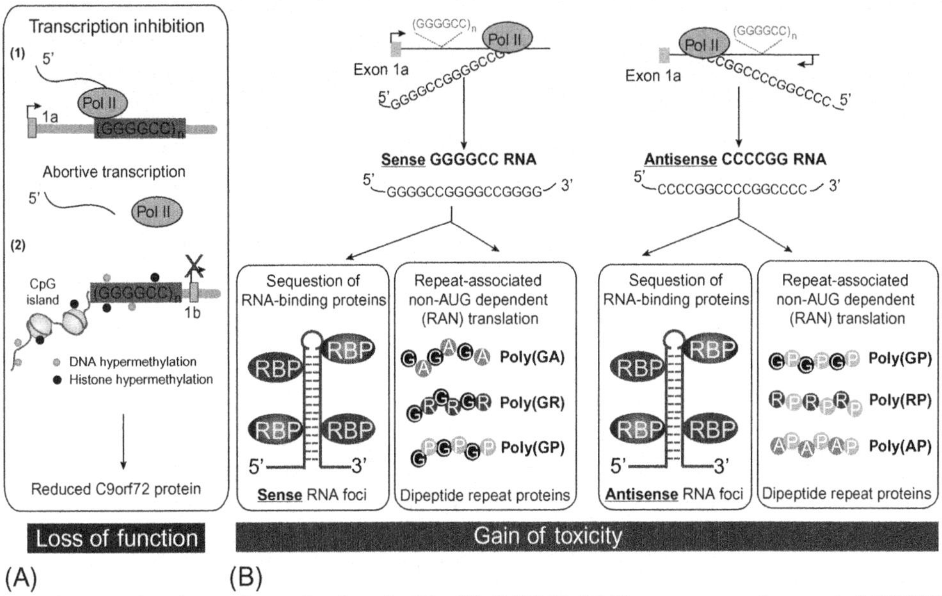

Fig. 2 Proposed pathogenic mechanisms in C9orf72 ALS/FTD. (A) The presence of expanded GGGGCC repeats potentially causes abortive transcription from exon 1a and hypermethylation of both DNA and histones to reduce C9orf72 RNA transcription, leading to a loss of C9orf72 protein function. (B) Bidirectionally transcribed repeat-containing RNAs cause neuronal toxicity by sequestration of RNA-binding proteins into RNA foci or production of at least 5 aberrant dipeptide repeat proteins [poly(GA), poly(GP), poly(GR), poly(PR), and poly(PA)] through a novel repeat-associated non-AUG-dependent translation mechanism.

pathway contributing to downregulation of C9orf72 gene expression and loss of C9orf72 protein function.

Since the initial identification of reduced C9orf72 RNA transcripts in patient lymphoblasts and frontal cortex, many independent groups have extended these findings with one or more C9ORF72 transcript variants in other cell lines such as induced pluripotent stem (iPS) neurons and patient tissues, including cerebellum.[4,5,39] Several studies also suggest that consistent with reduced RNAs, C9orf72 proteins are reduced in the frontal and temporal cortex in patients.[23,24,40] However, demonstrating reduced C9orf72 protein levels has been more challenging due to the low abundance of C9orf72 protein and a lack of antibodies for its detection.

Assuming C9orf72 has an essential role in autophagy and lysosomal functions, one could reasonably conclude that impairment of these functions is involved in ALS and FTD. A growing body of literature, including studies in cellular and animal models as well as human trials, support this. In human cell lines and iPSC-derived neurons from C9orf72 patients, knockdown of C9orf72 inhibits the induction of autophagy, leading to the accumulation of autophagy substrates, including P62.[30,33] Accumulated P62, is also observed in spleens of C9orf72 knockout mice.[41] IPS neurons from C9orf72 patients also showed increased sensitivity to autophagy inhibition, suggesting that reductions in C9orf72 levels contribute to cellular stress.[42] Recently, Burberry et al., found that an environment with a reduced abundance of immune-stimulating bacteria protects C9orf72-mutant mice from premature mortality and significantly ameliorates their underlying systemic inflammation and autoimmunity.[43] In addition, inflammatory phenotypes in mutant mice can be suppressed by reducing the microbial burden with broad-spectrum antibiotics or by transplanting gut microflora from a protective environment, suggesting C9orf72 functions to prevent microbiota from inducing a pathological inflammatory response.[43]

Studies in lower species model systems suggest that C9orf72 is critical for the development of motor systems and motor functions. In zebrafish, blocking C9orf72 translation or splicing with antisense morpholino oligonucleotides resulted in axonopathy and reduced mobility.[44] Overexpressing C9orf72 that lacks the complete DENN protein domain also mimics the morpholino experiment, suggesting the indispensability of the complete DENN protein domain.[45] Similarly, deletion of the C9orf72 orthologue in *Caenorhabditis elegans* led to age-dependent motility deficits, paralysis, and increased susceptibility to environmental osmotic stress.[46] Primary hippocampal neurons cultured from C9orf72 knockout mice have reduced dendritic arborization and spinal density, suggesting its role in neuronal morphogenesis.[47]

Despite evidence supporting the importance of C9orf72 in human cells and lower organisms, much evidence counters that loss of function is the main disease driver for c9ALS/FTD. Only one sporadic ALS patient has ever been found to carry a splice site mutation potentially causing a heterozygous loss of function, and it is unclear whether this mutation is the cause of disease or a variant of undetermined significance.[48,49] Further, one would expect that patients homozygous for C9orf72 repeat expansions (and thus with less C9orf72 than heterozygotes) will have more severe symptoms of disease, but this is not the case.[50] Although not conclusive, these human studies do not support a loss of function as the disease-driving mechanism.

Additionally, sustained reduction of mouse C9orf72 to 30%–40% of its normal level throughout the CNS using antisense oligonucleotides (ASOs) is well tolerated, producing no behavioral or pathological features characteristic of ALS/FTD.[51] Several ubiquitous or neuronal/glial-specific C9orf72 knockout mice have also been characterized and none of these mice produced any ALS/FTD-like pathological or behavioral phenotypes.[31,32,51–56] Interestingly, global C9orf72 knockout mice develop a proinflammatory/autoimmune phenotype with an enlarged spleen and lymph nodes, consistent with the observation that C9orf72 ALS/FTD patients also show an increased prevalence of autoimmune disease.[57] These findings are particularly noteworthy, given the growing appreciation of the role of the immune system in a variety of neurodegenerative diseases,[58] yet it is thought that C9orf72 loss-of-function alone is not the main ALS/FTD disease driver in mice.

Overall, C9orf72 loss of function by itself is insufficient to precipitate disease, and intriguingly, reduced expression of the mutant C9orf72 allele might actually be protective, potentially by decreasing the transcription of toxic repeat-containing C9orf72 RNAs.[59–61] Alternatively, given the role of C9orf72 in autophagy and microglial function, pathways implicated in ALS and FTD, haploinsufficiency might contribute to the disease process in synergy with the gain-of-toxicity mechanism.

Gain of function: DPR protein-mediated toxicity

Despite being in an intronic region, the hexanucleotide repeat of bidirectionally transcribed C9orf72 RNA transcripts can be translated through an unusual mechanism, RAN translation, in every reading frame to produce five different dipeptide repeat (DPR) proteins [poly(GA), poly(GP), poly(GR), poly(PR), and poly(PA)]. These might drive toxicity in c9ALS/FTD. Poly(GA) and poly(GR) are uniquely translated from the sense GGGGCC RNA; poly(PR) and poly(PA) are uniquely translated from the antisense CCCCGG RNA, and poly(GP) can be translated from both sense and antisense directions (Fig. 2B). Accumulations of DPR proteins have been detected using either immunohistochemical assays for DPR inclusions or immunoblotting and ELISA-based assays for the soluble form.[62–66] DPR inclusions are widely distributed throughout the brains of c9ALS/FTD but, interestingly, are less frequent in the spinal cord.[62,67–70] Of the five different types of DPRs, those encoded by the sense-strand, poly(GA), poly(GP), and poly(GR), form inclusions far more frequently than those encoded by the antisense-strand, poly(PA) and poly(PR), and they can co-occur in a same neuron.[65] Based on the staining patterns, they can be categorized into neuronal cytoplasmic inclusions, intranuclear inclusions, dystrophic neurites, and diffuse dendritic or cytoplasmic staining. So far, no DPR pathology has been reported in astrocytes, microglia, or oligodendrocytes. SDS-soluble poly(GA) and poly(GP) can also be detected in CSF and blood by immunoassay which could prove to be useful biomarkers specific to C9ALS/FTD for current and future clinical trials.[71–74]

Several clinicopathological studies have sought to determine whether or not DPR proteins are the primary culprit of c9ALS/FTD. Mackenzie et al. showed moderate associations of poly(GA)-positive dystrophic neurites with local degeneration in the frontal cortex[68] and in another study by Davidson et al., more abundant neural cytoplasmic inclusions of poly(GA) in cortical, hippocampal, and motor regions were associated with earlier age at symptom onset.[75] In a cohort of 40 c9ALS/FTD cases, middle frontal gyrus inclusions of poly(GR) but not poly(GA) or poly(GP) were strongly correlated with neurodegeneration, supporting that DPRs containing the arginine residue are the most toxic.[76] Recently, we found that in C9orf72 ALS, poly(GR) inclusions were increased in areas of neurodegeneration, frontal and motor cortex, and the anterior horn of the spinal cord, compared to areas that are spared from neurodegeneration (parietal and occipital cortex and the posterior horn of the spinal cord).[23] Compared to the other DPR species, poly(GR) was also unique in colocalizing with phosphorylated TDP-43, a hallmark pathology observed in ALS/FTD. Eighty percent of poly(GR) dendritic inclusions also had phosphorylated TDP-43 whereas this co-occurrence was nearly unseen for poly(GA) and poly(GP)]. Notably, DPRs exclusively encoded by the C9orf72 antisense transcript, poly(PA), and poly(PR) were rarely seen in the brain and spinal cord.

To move from pathology to pathogenesis, several groups have expressed individual DPR proteins in isolation in yeast,[77] mammalian cells,[78–85] Drosophila,[79,82,86–89] zebrafish,[90–92] and mice.[93–97] Of all the DPR proteins, poly(GR) and poly(PR) containing the arginine residue have proven to be the most toxic. Synthetic 20-mer poly(GR) and poly(PR) can cause rapid death of U2OS cells and cultured human astrocytes when added exogenously.[98] These DPR proteins also consistently show toxicity when overexpressed in several different cell lines, such as HEK293T cells, NSC-34 cells, and iPSC-derived neurons.[79,81,82] In Drosophila, targeted expression of poly(GR) or (PR) resulted in eye degeneration, motor deficits, and reduced survival.[79,82,86–89] Finally, AAV-mediated or transgenic expression of poly(GR) and poly(PR) in mice caused age-dependent neurodegeneration, brain atrophy, and motor/memory deficits.[93,95,97,99] To a lesser extent, poly(GA) has also been shown to exert toxicity. When synthetic poly(GA) is exogenously applied to human cells or primary neurons, there is decreased survivability within 2 days.[100] While the remaining DPR proteins, poly(PA) and poly(GP), are less likely to be the toxic species driving disease, poly(GA) has continually been shown to be toxic through overexpression in cultured cells and primary neurons,[80,84] zebrafish,[90,91] Drosophila,[88] and mouse brains.[94,96] These studies provide compelling evidence that certain DPR proteins are toxic. However, since individual DPR proteins were overexpressed, a key question remains: which (if any) of the DPR proteins are the main toxic species in c9ALS/FTD when expressed at the endogenous levels in patients?

While DPRs have been demonstrated to be toxic in numerous animal and cellular models, many pathological studies have raised doubts about the pathogenicity of DPR inclusions, i.e., no obvious correlation exists between the abundance of these inclusions and neurodegeneration.[68,70,75,101] Regions with the highest burden of DPR inclusions are cerebellum, hippocampus, and neocortex, where no significant loss of neurons occurs; and when DPRs are observed, most neurons lack TDP-43 pathology, the hallmark proteinopathy of C9ALS/FTD. Also, DPR inclusions do not differ in the neuroanatomical distributions between FTD and ALS cases and DPR inclusions are rarely observed in spinal cord motor neurons. Countering this paucity of evidence found in pathological studies is the possibility that key factors underlying disease remain unknown. These pathological studies only detect aggregated, insoluble proteins, and it is possible that soluble DPR species mediate toxicity. Most problematic is the lingering possibility that toxic DPR species, notably poly-PR which is all but absent in postmortem human tissues, have killed neurons, thus precluding their abundance and correlation with disease. Alternatively, CNS regions that have extensive DPR pathology but are unaffected by neurodegeneration might contain protective factors yet to be discovered. Another important issue is whether different DPR proteins, repeat RNA, and DPR proteins act synergistically, potentially in conjunction with loss of function of C9orf72, to elicit downstream effects.

Gain of function: RNA foci-mediated toxicity

RNAs containing sense GGGGCC and antisense CCCCGG repeats are bidirectionally transcribed from the C9orf72 mutation. These repeat-containing RNAs may form complicated secondary structures, such as hairpins, i-motifs, and G-quadruplexes,[102–104] which in turn sequester RBPs into intranuclear RNA foci that can be experimentally visualized using fluorescence in situ hybridization with probes against the repeats. These RNA foci lead to depletion of the normal function of the sequestered RBPs and neuronal death. The RBP sequestration model is best demonstrated in the neuromuscular disease myotonic dystrophy 1, where the muscleblind family proteins are sequestered by the CTG repeats in the DMPK gene and cause downstream splicing changes.[105] In c9ALS/FTD patients, both sense and antisense RNA foci are detected in fibroblast, lymphoblast, iPSC-derived neurons, and several CNS regions of c9ALS/FTD patients.[51,63,66,106–108] RNA foci are most abundant in neurons, but also occur in astrocytes, microglia, and oligodendrocytes.[51,63,108]

Several studies have overexpressed GGGGCC repeats with variable sizes and demonstrated deleterious phenotypes in cell culture,[82] *C. elegans,*[109] zebrafish,[92] and Drosophila.[87,88,110] Short or interrupted repeats are generally used, given the technical difficulties to obtain long and pure repeats. For example, adeno-associated viruses expressing 66 or 149 GGGGCC repeats were injected intraventricularly into newborn mice to induce somatic transgenesis. These mice accumulated nuclear RNA foci and inclusions of DPR proteins, and consequently exhibited hyperactivity, anxiety, antisocial behavior, and motor deficits.[111,112] Four different groups, including ours, also generated transgenic mice expressing a bacterial artificial chromosome (BAC) that can accommodate genomic DNA from c9ALS/FTD patients to express long repeats (up to ~1000 repeats) in the context of the human gene. All of these mouse models produced sense and antisense RNA foci, as well as DPR proteins from the sense strand. Two transgenic models failed to produce neurodegeneration,[113,114] and our mice, expressing exons 1–5 of the human C9orf72 gene and ~450 repeats, developed age-dependent increased anxiety and cognitive deficits, accompanied by the loss of hippocampal neurons, but no unusual motor phenotype.[115] In the last model expressing the full-length C9orf72 gene and ~500 repeats, a subset of female mice developed an acute motor phenotype, including paralysis and decreased survival. Other females and males showed slower progression.[116] The reasons for differences in these mouse models are unknown but they feature different genetic backgrounds, repeat sizes, transgene insertion sites, and expression levels which can modulate induced toxicity. An in-depth comparison of all BAC models especially the accumulation of RNA foci and DPR proteins will provide insights to the toxic species leading to motor and/or cortical neuron death.

Expressing C9orf72 GGGGCC repeats in primary neurons,[82] Drosophila,[110,117] and zebrafish,[92] suggests that the toxicity arises from RNA foci because no DPR proteins were detected. However, the inability to detect DPR proteins in these studies is insufficient to exclude a role for these proteins. In two important studies, researchers generated Drosophila expressing either pure GGGGCC repeats or interrupted repeats of similar size, but with stop codons inserted in each reading frame to prevent the translation of the repeats into DPR proteins.[87,88] Both pure and interrupted repeats form similar levels of RNA foci. Expression of pure repeats caused toxicity and early lethality whereas the interrupted ones had no effect. In another study, Drosophila expressing 160 GGGGCC repeats in the intron formed abundant sense RNA foci in the nucleus but produced little DPR protein and no neurodegeneration, again suggesting that nuclear sense RNA foci are not sufficient to drive neurodegeneration in this model.[118] There are, however, some caveats for interpreting these experiments, including whether the interrupted repeat RNAs have similar secondary structures to sequester the same key RBPs as the pure repeats, and whether RNA foci in humans may sequester RBPs that are not expressed in Drosophila.

A key to validating RNA foci-mediated toxicity is to identify the RBP(s) that are sequestered and thus lose function. Many proteins are proposed to interact with short, synthesized C9orf72 repeat RNAs in vitro, including hnRNPA1, hnRNPA3, hnRNP-H, hnRNP-H1/F, nucleolin, Pur-α, double-stranded RNA-specific editase B2 (ADARB2), RanGAP1, THO complex subunit 4 (ALYREF), serine/arginine-rich splicing factor 1 (SRSF1), SRSF2, and Zfp106.[107,110,119–124] Sense RNA foci have been shown to colocalize with core paraspeckle proteins SFPQ and NONO in c9ALS/FTD cerebellum.[125] In c9ALS/FTD brain tissues, sense and antisense RNA foci colocalize with hnRNP A1, hnRNP H/F, SRSF2, and ALYREF.[122,124,126] But most of these RNA-binding proteins appear to colocalize with RNA foci at a low frequency; only hnRNP H colocalizes with 70% of all sense foci detected.[122] Nevertheless, key data to support RNA foci-mediated toxicity are missing (i.e., demonstrating that loss of function of any of these RBPs leads to C9orf72 diseases and upregulation of their functions can rescue phenotypes caused by C9orf72 repeat expansions).

Only a few neuropathological studies have attempted to correlate the abundance and distribution of RNA foci with c9ALS/FTD clinical features. In a large patient population study conducted at the Mayo Clinic, patients with more antisense RNA foci in middle frontal gyrus (cortical layers 3–6) neurons showed a later age of onset.[127] However, a contradictory observation was made by Mizielinska et al. who showed that patients with more sense RNA foci had an earlier age of symptom onset in a small number of C9orf72

FTD patients, and antisense RNA foci showed a similar trend.[108] In agreement with this, antisense RNA foci are associated with nucleoli and mislocalization of TDP-43.[62,128] Perhaps the most striking evidence supporting the possibility that RNA toxicity underlies neurodegeneration in c9ALS/FTD is a unique case study in which nervous system was analyzed for pathology before and after death in an individual who had part of the frontal lobe dissected for epilepsy 5 years before the onset of FTD symptoms. Subsequently, postmortem tissue was analyzed 8 years after the symptoms onset. In this study by Vatsavayai et al., sense and antisense RNA foci (as well as DPR inclusions and neuronal loss) were found to precede both symptoms of FTD and TDP-43 pathology.[129]

In summary, expressing GGGGCC repeats in model systems clearly recapitulates both cellular pathology (RNA pathology and DPR aggregates) and neuronal degeneration seen in patients, even though unphysiologically high levels of expression were used. It is also challenging to dissect the relative contributions of RNA foci or DPR proteins to disease pathogenesis since these repeats will produce both RNA foci and DPR proteins, and the relative contributions of sense and antisense repeat-containing RNAs remain unclear.

Synergistic mechanisms

Sense and antisense RNA foci can occur in the same cells and either has been found to co-occur with DPR proteins. However, the presence of sense RNA foci does not affect the likelihood of detecting the abundant DPR species, namely poly(GP) and poly(GR) in the brain and spinal cord suggesting that formation of RNA foci do not prevent (or enhance) repeat-containing RNA transcripts from being transported to the cytoplasm and translated.[128] In c9ALS/FTD patients, pathological evidence exists for both loss of C9orf72 function and gain of toxicity. Whether loss of C9orf72 function synergizes with the gain of toxicity to exacerbate ALS/FTD disease is not known. Mice with loss of C9orf72 do not develop ALS/FTD-related behavioral deficits, but additional stressors might be required as second hits to induce neurodegeneration in these mouse models. Shi et al. reported that loss of C9orf72 in human iPS-derived motor neurons impact lysosomal biogenesis and vesicular trafficking, and exacerbate toxicity to poly(GR) and poly(PR) exposure.[130] Similarly, loss of C9orf72 protein sensitized cells to toxicity induced by expression of 30 and 60 GGGGCC repeats.[131] Shao et al. bred C9orf72 knockout mice with a BAC transgenic mouse expressing the repeat-expanded full-length human C9orf72 gene to determine the synergism.[132] However, since transgene overexpression would compensate for loss of the mouse C9orf72 endogenous protein, this casts doubt on such a strategy. But surprisingly, the transgenic mouse line does not express full-length C9orf72 proteins from the transgene and C9orf72 haploinsufficiency exacerbate motor behavior deficits in a dose-dependent manner.[132] More recently, Zhu et al. bred mice expressing a C9orf72 transgene with 450 repeats that did not encode the

C9ORF72 protein, and inactivation of one or both endogenous C9orf72 alleles inhibited autophagy and exacerbated cognitive deficits, hippocampal neuron loss, glial activation, and accumulation of DRPs.[133] It is premature to say that C9orf72 loss of function induces neurodegeneration together with toxic RNA and DPR proteins. Further investigations crossing C9orf72 KO mice with transgenic mice that do not express the full-length C9orf72 proteins or injecting AAV virus expressing GGGGCC repeats into C9orf72 KO mice are needed to address this question. In addition, poly(GA) recruits poly(GR) into inclusions and reduces poly(PR) toxicity when coexpressed in human cells and Drosophila.[83,89] Whether the interaction between individual DPR proteins enhances or suppresses their toxicity is thus a point of interest. Poly(GA) can also induce nuclear RNA foci in cells expressing 80 GGGGCC repeats and in patient fibroblast cells, suggesting a positive feedback loop between RNA foci and DPR proteins.[134] Finally, translational frameshifting has been suggested to occur in cells expressing short C9orf72 repeats[135] and other repeat expansions,[136] which might produce chimeric repeat peptides. However, it is unknown whether such a frameshift occurs in C9orf72 mutation carriers, but if it does, the contribution of chimeric repeat peptides to disease pathogenesis remains to be discerned.

Pathological consequences underlying c9ALS/FTD

Nucleocytoplasmic transport deficits

Active transport of proteins and RNAs through nuclear pore complexes, a process called nucleocytoplasmic transport (NCT), is essential for cellular functions.[137] Zhang et al. showed that RanGAP1 binds to GGGGCC repeat RNA in vitro and is sequestered into sense RNA foci. RanGAP1 activates Ran small GTPase, which plays an important role in regulating the interactions between cargo molecules and transport receptors, such as the importin and exportin family of proteins.[137] In both Drosophila overexpressing 30 GGGGCC repeats and c9ALS iPSC-derived neurons, the nucleocytoplasmic Ran gradient is decreased, and the nuclear import of proteins and nuclear export of RNAs are compromised. In addition, the repeat toxicity can be readily suppressed by increased activity of RanGAP1, suggesting the RNA foci-mediated NCT deficit is a fundamental pathway in c9ALS/FTD pathogenesis.[117] In parallel, unbiased genetic screenings for phenotypic modifiers of Drosophila expressing GGGGCC repeats[87] or poly(PR),[86] yeast,[77] and human cells expressing poly(GR) or poly(PR),[138] identified that proteins involved in NCT mitigate or exacerbate disease phenotypes. Mechanistically, poly(GR) and poly(PR) interact with nucleopore proteins (Nups), including the central channel of the nucleopore, to reduce trafficking.[139] Poly(GA) also sequesters nucleocytoplasmic transport proteins, such as HR23, and induces an abnormal distribution of RanGAP1 in mice.[96] The primary cause for the NCT dysfunction in c9ALS/FTD is unknown but a number of human studies have documented that abnormalities in the nuclear envelope are a part of c9ALS/FTD pathology.[97,140,141] NCT defects have also been reported in other neurodegenerative disorders, such as Alzheimer's[142] and Huntington's disease[143,144] and have been shown to be a pathological feature in SOD1-ALS,[145] suggesting that it may play a more global role in the neurodegenerative process.

Alteration in stress granules

Eukaryotic cells have evolved sophisticated strategies to combat unexpected cellular stress. Cytoplasmic ribonucleoprotein (RNP) granules, such as P-bodies and stress granules (SGs), assemble quickly under stress to sequester mRNAs, translation initiation factors, 40S ribosomes, and many other RNA-binding proteins. In this way, only essential proteins needed for survival are produced. These granular structures can either disassemble upon release of cellular stress or be degraded by the autophagy pathway. Dysregulated formation or clearance of SGs has been proposed as a key pathogenic mechanism for ALS/FTD.[146] In recent years, groundbreaking work from several groups showed that SG-associated RNA-binding proteins, such as FUS and hnRNP1/2, undergo liquid-liquid phase separation (LLPS) to form droplets in vitro and have the propensity to further fibrilize into irreversible hydrogel or even insoluble amyloid structures.[147–150] One characteristic of these RBPs is that they all contain prion-like, intrinsically disordered low-complexity domains (LCD). Poly(GR) and poly(PR) interact with these LCD proteins, impair LLPS, and disrupt the dynamics of stress granules.[151] Consistent with this, primary cortical neurons overexpressing poly(PR) showed a reduction in cytoplasmic P-bodies with larger sizes and an increase in stress granule formation,[82] and knockdown of several of the LCD proteins modifies the eye degeneration phenotype in Drosophila expressing poly(GR).[79] Activation of the stress granule response or exposure to either poly(GR) or poly(PR) was also sufficient to disrupt nucleocytoplasmic transport by promoting the sequestration of several nuclear pore proteins into SGs.[152] Interestingly, poly(GR) and poly(PR) themselves undergo LLPS at high concentrations, and GGGGCC repeat RNA also undergoes gel transition itself[153] and promotes the phase transition of RNA granule proteins in vitro and in cells.[154] While the transient nature of stress granule assembly may preclude their abundance in pathological tissues, a few studies have documented stress granule-like inclusions of TIA-1 and EIF3 that colocalize with TDP-43 inclusions in the spinal cord of ALS or brain of FTLD patients.[155] Similarly, a new mouse model of c9ALS/FTD with AAV-mediated expression of 149 GGGGCC repeats at birth leads to TIA-1 and EIF3 stress granule-like bodies co-localizing with pathological TDP-43 in neurons.[111] These mice undergo progressive neurodegeneration coinciding with behavioral defects and invaginations of the nuclear membrane that are similar to those found in humans.[23,111,141]

Nucleolar dysfunction

The nucleolus is the site of ribosome biogenesis. When overexpressed in cultured cells, astrocytes, primary neurons, iPSC-derived neurons, and Drosophila, poly(GR) and poly(PR) colocalize with nucleoli leading to abnormal nucleolar morphology and impaired ribosomal RNA maturation.[82,98] However, poly(GR) and poly(PR) do not localize within the nucleolus in C9orf72 patient brains. But sense RNA foci can bind nucleolin, a principle component of the nucleolus. This is supported by the finding of aberrant subcellular localization of nucleolin in C9orf72 iPSC-derived motor neurons and the patient motor cortex.[121] Contrary to this, antisense RNA foci have been found to associate more often with nucleoli than sense foci in tissue obtained from autopsy.[128] Nucleolin mislocalization was also found in C9orf72 BAC transgenic mice without significant changes in ribosomal RNA processing or splicing.[113] Interestingly, C9orf72 patient brains exhibit bidirectional nucleolar volume changes, with smaller nucleoli overall but enlarged nucleoli in neurons containing poly(GR) inclusions.[156]

Translation inhibition and ubiquitin-proteasome system

In addition to regulating stress granule dynamics and ribosomal RNA synthesis, overexpression of poly(GR) and (PR) blocked global translation in an in vitro translation assay and in cell lines.[78] This was attributed to the direct binding of poly(GR) and poly(PR) to mRNA, thereby blocking access to the translational machinery. Poly(PR) and poly(GR) also interact with translation initiation and elongation factors and ribosomal subunits in pull-down assays.[78] In addition, GGGGCC repeat-containing RNAs sequester ribosomal subunits trigger stress granule formation and inhibit translation.[154,157] Interestingly, 26S proteasome complexes, components of the ubiquitin-proteasome system (UPS), are sequestered within poly(GA) aggregates in cells as revealed by protein cryo-electron tomography, which allows 3D imaging of the cell interior in close-to-native conditions.[158] These results provide new insight into the mechanism by which poly(GA) deposition promotes UPS impairment and regulates protein homeostasis.

DNA damage

Nucleolar stress induces DNA damage, and postmitotic neurons are particularly susceptible to DNA damage. An age-dependent increase in DNA damage and oxidative stress was reported in iPSC-derived motor neurons of patients with C9orf72 repeat expansions.[159] DNA damage is also increased in spinal cord neurons of C9orf72 ALS patients.[160] Such DNA damage is partially caused by poly(GR) since the expression of poly(GR) in iPSC-derived control neurons or neuronal cell lines and in mice in vivo is sufficient to induce DNA damage and toxicity.[93] Mechanistically, poly(GR) binds to Atp5a1 and compromises mitochondrial function, leading to increased oxidative stress and an overactivated Ku80-dependent DNA repair pathway.[93,161] Introducing ectopic

Atp5a1 expression or reduction in oxidative stress by treatment with antioxidants partially rescues DNA damage in motor neurons from C9orf72 carriers and in neurons expressing poly(GR). As a result of DNA damage, the levels of phosphorylated ATM and P53 and other downstream proapoptotic proteins, such as PUMA, Bax, and cleaved caspase-3, are significantly increased in C9orf72 patient neurons. Partial loss of Ku80 function in these neurons through CRISPR/Cas9-mediated ablation or small RNA-mediated knockdown suppresses the apoptotic pathway. Thus, poly(GR)-mediated DNA damage contributes to ALS/FTD pathogenesis, and partial inhibition of the over-activated Ku80-dependent DNA repair pathway is a promising therapeutic target.

Therapeutic approaches for c9ALS/FTD

Transcript targeting

Since the pivotal discovery of repeat expansions in C9orf72 causing ALS and FTD, research has overall favored that gain of function from dipeptide repeat proteins and/or repeat expanded RNA drive toxicity. Thus, a primary route to effective therapeutic treatment is reducing the levels of repeat expanded RNA transcripts, a strategy that would have the potential to prevent toxicity from either RNA and/or simultaneously translation of DPRs. The most developed such therapeutic strategies are antisense oligonucleotide therapies (ASOs). These are synthetic oligonucleotides or oligonucleotide analogs that bind to target RNAs. Depending on their chemical modifications, ASOs selectively degrade mRNAs through endonuclease RNase H recruitment or prevent the interaction of RNAs with RBPs, thereby modulating their splicing/processing without degradation.[162] In 2016, the ASO Nusinersen (Spinraza) became the first FDA approved treatment for spinal muscular atrophy. Nuisinersen functions by modulating the splicing of SMN2 pre-mRNA to increase the expression of survival motor neuron protein.[163] ASOs are currently in Phase 1 or 2 trials for LRRK2-Parkinson's disease and Tau associated Alzheimer's disease and FTD and in Phase III trials for Huntington's disease and superoxide dismutase ALS.[164,165]

ASOs have shown promise in preclinical trials at reducing sense RNA foci, rescuing cells from glutamate excitotoxicity and correcting aberrant gene expression profiles in fibroblasts and patient-derived IPS motor neurons.[51,107,124] We and others have also demonstrated that ASOs can specifically target transcript variants 1 (NM_145005.6) and 3 (NM_001256054.2) that carry the repeat expansion without reducing variant 2 (NM_018325.5) expression that encodes the functional C9ORF72 protein, thereby targeting gain of function toxicity without exacerbating the loss of function. In this study, utilizing transgenic mice expressing 450 repeats, ASOs significantly reduced levels of sense RNA foci and DPRs as well as rescuing the behavioral and cognitive deficits observed in the treated mice.[115] Based on these and other encouraging results, a phase

I clinical trial of ASOs targeting the sense strand for C9ALS patients has been initiated by Ionis Pharmaceuticals and its partner Biogen Inc. in September 2018 (NCT03626012).

RNA interference (RNAi) is an alternative therapeutic strategy which involves short interfering RNAs (siRNAs), short hairpin RNAs (shRNAs), or micro RNAs (miRNAs) that also aim to mitigate gain of function toxicity by targeting RNA in the cytoplasm of cells.[166] If successful, this results in the degradation of the RNA transcripts, steric inhibition of RBPs binding the RNA and prevention of RNA foci forming, but their effect on in vivo models has yet to be tested.[51,167–169] Notably, a single administration of an adenoviral-mediated delivery of an RNAi could lead to a long-lasting therapeutic effect. This can be an advantage since ASOs require repeat administration, but in the former instance treatment is not "druggable," i.e., one cannot stop or lower therapy in the event of an adverse effect.[170]

Additional therapies aim to mitigate toxicity by targeting the structure of repeat-containing sense RNA. For example, the small molecule TMPyP4 binds the G-quadruplex and has been shown to prevent sequestration of RBPs by (GGGGCC)8 RNA,[171] reduce sense RNA foci in Drosophila overexpressing (GGGGCC)30 RNA, and rescue NCT defects and neurodegeneration.[117] In addition, several small molecules designed to specifically target the GGGGCC hairpin have been shown to significantly decrease sense RNA foci and poly(GP) in cells overexpressing (GGGGCC)66 RNA and in iPSC-derived neurons.[73,172] It remains to be seen whether these compounds also reduce poly(GR) and poly(PR), the two DPR proteins that have been shown to be most toxic in model systems, and whether they can reduce cellular toxicity both in vitro and in vivo.

Several studies have sought to reduce the transcription of repeat-containing RNA using CRISPR/Cas strategies.[173] As a proof of concept, genetic correction of C9orf72 repeat expansions in patient iPS cells has been achieved by using CRISPR/Cas9 to cut out the C9orf72 repeat expansions and repair the gene through nonhomologous end joining.[93,130] More recently, marked reduction in transcription of repeat-containing sense RNA transcripts was achieved by using CRISPR/Cas9 to remove the promotor region of C9orf72 variants 1 and 3.[174] In addition, Pinto et al. targeted enzymatically dead Cas9 to directly bind repeat expanded DNAs, such as CTG expansion in DM1, CTGG expansion in DM2, and GGGGCC expansion in C9ALS/FTD. Instead of cleaving DNA, dCas9 sterically impeded transcription of repeat-containing RNA, leading to a drastic decrease in the production of DPRs.[175] Apart from targeting DNA, RNA-targeting Cas9 with RNA endonuclease has been used to degrade repeat expanded RNAs, reducing RNA foci, and DPRs in a manner similar to ASOs.[176] The clinical use of CRISPR/Cas technology in c9ALS/FTD is still in its infancy. Ethical concerns, adequate methods of delivery and cell targeting, off-target and safety measures, and treatment efficacy are the main issues that need to be addressed before applying this method to patients.

Several nonredundant proteins including SUPT4H and PAF1C, have been identified that facilitate RNA polymerase II transcription of C9orf72 through the complex secondary structure formed by the GC repeat expansion.[177,178] Depleting SUPT4H and PAF1C has been shown to reduce sense and antisense C9orf72 repeat transcripts, as well as RNA foci and DPR products, and ameliorate neurodegeneration in C9orf72 iPSC-derived neurons and in Drosophila.[177,178] However, an effective therapeutic strategy would require specific action against C9orf72 repeat RNA since depleting SUPT4H results in a global reduction of RNA.[179]

DPR reduction using antibody strategies

In addition to preventing toxic DPR production by targeting RNA, another therapeutic approach is utilizing antibodies targeting DPRs to mitigate gain of function toxicity by reducing or possibly eliminating DPRs. Both active and passive immunizations have been explored to combat the prion-like propagation of misfolded proteins such as amyloid-β and tau in Alzheimer's disease, and α-synuclein in Parkinson's disease.[180] An immunization approach could also prevent the cell-to-cell transmission of DPRs but a main concern is developing antibodies that are specific to the DPRs since similar short repetitive sequences are present in the human proteome.[134,181–183] Nevertheless, antibodies against poly(GA) were recently shown to reduce poly(GA) as well as poly(GP) and poly(GR), improve behavioral deficits, decrease neuroinflammation and neurodegeneration, and increase survival in mice.[184] An alternative approach to target DPRs involves therapy utilizing the heat shock protein, HSPB8. Overexpression of HSPB8 has been shown to facilitate the autophagy-mediated clearance of misfolded proteins and reduce the levels of insoluble DPRs including poly(GR) and poly(PR).[185]

Pathway inhibition using small molecule approaches

Even further downstream, therapies are aimed at correcting cellular pathways affected by C9orf72 repeat expansion, such as stress granule assembly and nuclear-cytoplasmic transport. For example, inhibiting stress granule assembly by using ASOs targeting Ataxin 2 mitigates nucleocytoplasmic transport defects and neurodegeneration in patient-derived IPS neurons and mice.[152] Reducing Ataxin 2 also extends the lifespan of TDP-43 mice, which is promising for c9ALS/FTD and additional diseases involving stress granule assembly and NCT defects.[186] Furthermore, Zhang et al. demonstrated that reducing nuclear export factors SRSF1 or exportin 1, or using selective inhibitors of nuclear export (SINEs), a drug that blocks exportin 1, ameliorates the toxicity seen in a Drosophila model of c9ALS/FTD.[117] The therapeutic benefit of inhibiting nuclear export is perhaps an effect of reducing levels of repeat-containing RNA,[187] thus lowering levels of DPRs or reversing mislocalization of RBPs such as TDP-43 caused by NCT defects.[117] SINE compounds were also shown to be beneficial in rats overexpressing mutant TDP-43, improving primary neuron survival, and partially rescuing loss of motor function.[188] Currently, Karyopharm Therapeutics and its partner Biogen Inc. are taking one of the SINE compounds, KPT-350, into clinical trials (Fig. 3).

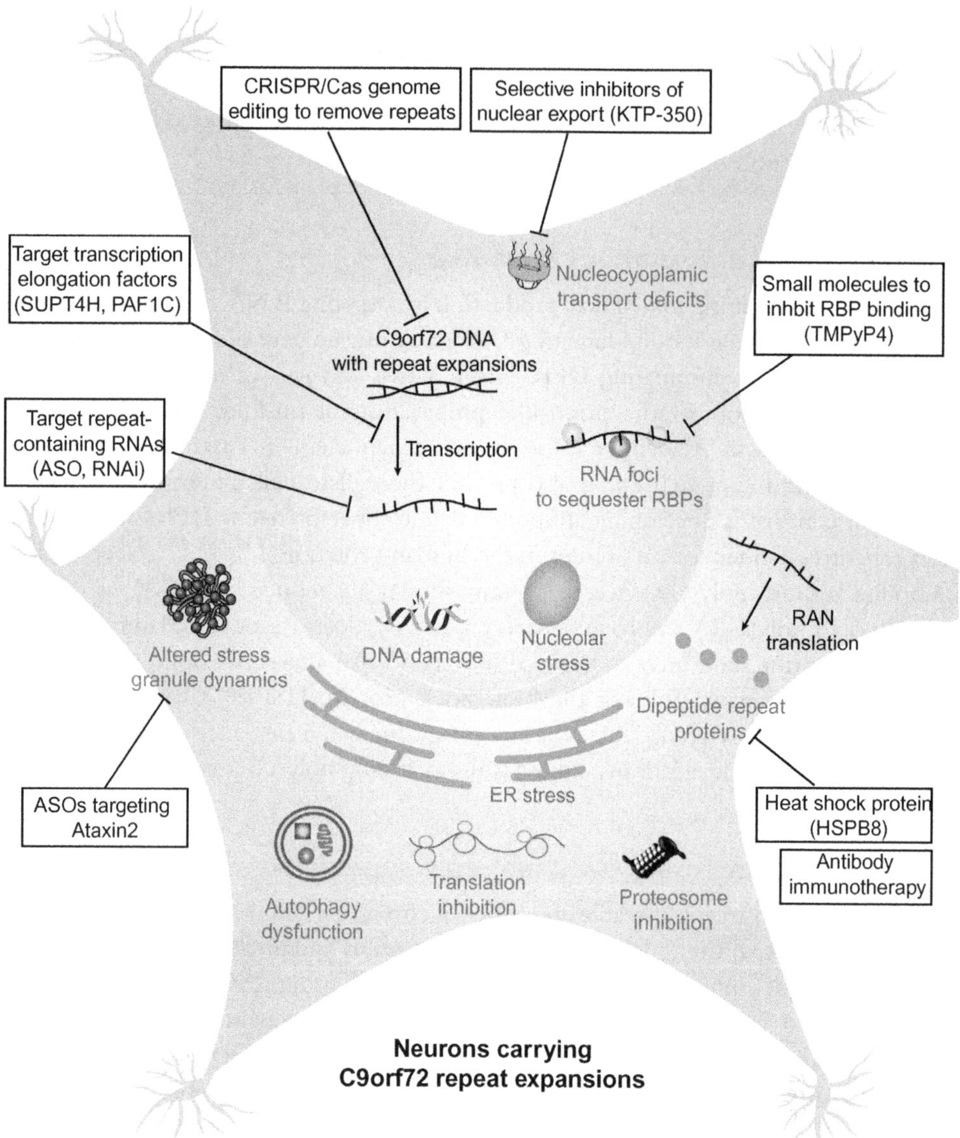

Fig. 3 Cellular processes impaired by the C9orf72 repeat expansions and potential therapeutic interventions. A wide range of cellular pathways have been implicated in c9ALS/FTD, including DNA damage, nucleolar stress, nucleocytoplasmic transport deficits, ER stress, autophagy dysfunction, translational inhibition, proteasome inhibition, and altered stress granule dynamics. Therapies targeting these deficits, as well as directly targeting repeat expanded C9orf72 DNA/RNA and DPR proteins, are also highlighted.

Conclusions and future perspectives

Since the discovery of C9orf72 repeat expansions, an impressive effort has been developed to investigate disease mechanisms. Many disease models were developed to determine loss of function and/or gain of toxicity and downstream cellular function alterations. Therapeutic approaches and biomarkers have also been explored. Excitingly, antisense oligonucleotide therapy to selectively degrade repeat-containing RNAs has gone into a Phase I clinical trial for C9orf72 ALS patients. Based on the pace of research, promising therapy for the most common form of familial ALS is on the horizon.

Acknowledgments

OAA is supported by the National Science Foundation Graduate Research Fellowship (DGE-1650112) and the Nicholas Martin Jr. Family Foundation. JJ is supported by the National Institute for Neurological Disorders and Stroke (NIH R21NS114908). JR is the recipient of grants awarded by the ALS Association (5356S3), Target ALS (20134792), and the National Institute of Neurological Diseases and Stroke (NIH R01NS088578 and NS047101). Pam Golden is acknowledged for her involvement in Biogen-sponsored clinical trials for which JR is the principal investigator.

References

1. Taylor JP, Brown RH, Cleveland DW. Decoding ALS: from genes to mechanism. *Nature.* 2016;539:197–206. https://doi.org/10.1038/nature20413.
2. Olney NT, Spina S, Miller BL. Frontotemporal dementia. *Neurol Clin.* 2017;35:339–374. https://doi.org/10.1016/j.ncl.2017.01.008.
3. Ling SC, Polymenidou M, Cleveland DW. Converging mechanisms in ALS and FTD: disrupted RNA and protein homeostasis. *Neuron.* 2013;79:416–438. https://doi.org/10.1016/j.neuron.2013.07.033.
4. Dejesus-hernandez M, Mackenzie IR, Boeve BF, et al. Expanded GGGGCC hexanucleotide repeat in noncoding region of C9ORF72 causes chromosome 9p-linked FTD and ALS. *Neuron.* 2011;72:245–256. https://doi.org/10.1016/j.neuron.2011.09.011.
5. Gijselinck I, Van Langenhove T, van der Zee J, et al. A C9orf72 promoter repeat expansion in a Flanders-Belgian cohort with disorders of the frontotemporal lobar degeneration-amyotrophic lateral sclerosis spectrum: a gene identification study. *Lancet Neurol.* 2012;11:54–65. https://doi.org/10.1016/S1474-4422(11)70261-7.
6. Renton AE, Majounie E, Waite A, et al. A hexanucleotide repeat expansion in C9ORF72 is the cause of chromosome 9p21-linked ALS-FTD. *Neuron.* 2011;72:257–268. https://doi.org/10.1016/j.neuron.2011.09.010.
7. Nguyen HP, Van Broeckhoven C, van der Zee J. ALS genes in the genomic era and their implications for FTD. *Trends Genet.* 2018;34:404–423. https://doi.org/10.1016/j.tig.2018.03.001.
8. La Spada AR, Taylor JP. Repeat expansion disease: progress and puzzles in disease pathogenesis. *Nat Rev Genet.* 2010;11:247–258. https://doi.org/10.1016/j.physbeh.2017.03.040.
9. Beck J, Poulter M, Hensman D, et al. Large C9orf72 hexanucleotide repeat expansions are seen in multiple neurodegenerative syndromes and are more frequent than expected in the UK population. *Am J Hum Genet.* 2013;92:345–353. https://doi.org/10.1016/j.ajhg.2013.01.011.
10. Buchman VL, Cooper-Knock J, Connor-Robson N, et al. Simultaneous and independent detection of C9ORF72 alleles with low and high number of GGGGCC repeats using an optimised protocol of southern blot hybridisation. *Mol Neurodegener.* 2013;8:1–6. https://doi.org/10.1186/1750-1326-8-12.
11. Dobson-Stone C, Hallupp M, Loy CT, et al. C9ORF72 repeat expansion in Australian and Spanish frontotemporal dementia patients. *PLoS ONE.* 2013;8https://doi.org/10.1371/journal.pone.0056899.

12. Hübers A, Marroquin N, Schmoll B, et al. Polymerase chain reaction and southern blot-based analysis of the C9orf72 hexanucleotide repeat in different motor neuron diseases. *Neurobiol Aging*. 2014;35:1214.e1–1214.e6. https://doi.org/10.1016/j.neurobiolaging.2013.11.034.
13. Ishiura H, Takahashi Y, Mitsui J, et al. C9ORF72 repeat expansion in amyotrophic lateral sclerosis in the Kii peninsula of Japan. *Arch Neurol*. 2012;69:1154–1158. https://doi.org/10.1001/archneurol.2012.1219.
14. Gijselinck I, Van Mossevelde S, Van Der Zee J, et al. The C9orf72 repeat size correlates with onset age of disease, DNA methylation and transcriptional downregulation of the promoter. *Mol Psychiatry*. 2016;21:1112–1124. https://doi.org/10.1038/mp.2015.159.
15. Proudfoot M, Gutowski NJ, Edbauer D, et al. Early dipeptide repeat pathology in a frontotemporal dementia kindred with C9ORF72 mutation and intellectual disability. *Acta Neuropathol*. 2014;127:451–458. https://doi.org/10.1007/s00401-014-1245-7.
16. Van Mossevelde S, Van Der Zee J, Gijselinck I, et al. Clinical evidence of disease anticipation in families segregating a C9orf72 repeat expansion. *JAMA Neurol*. 2017;74:445–452. https://doi.org/10.1001/jamaneurol.2016.4847.
17. Benussi L, Rossi G, Glionna M, et al. C9ORF72 hexanucleotide repeat number in frontotemporal lobar degeneration: a genotype-phenotype correlation study. *J Alzheimers Dis*. 2014;38:799–808. https://doi.org/10.3233/JAD-131028.
18. Van Blitterswijk M, Dejesus-hernandez M, Niemantsverdriet E, et al. Associations of repeat sizes with clinical and pathological characteristics in C9ORF72 expansion carriers (Xpansize-72): a cross-sectional cohort study. *Lancet Neurol*. 2013;12:1–18. https://doi.org/10.1016/S1474-4422(13)70210-2.Associations.
19. Dols-Icardo O, García-Redondo A, Rojas-García R, et al. Characterization of the repeat expansion size in C9orf72 in amyotrophic lateral sclerosis and frontotemporal dementia. *Hum Mol Genet*. 2014;23:749–754. https://doi.org/10.1093/hmg/ddt460.
20. Nordin A, Akimoto C, Wuolikainen A, et al. Extensive size variability of the GGGGCC expansion in C9orf72 in both neuronal and non-neuronal tissues in 18 patients with ALS or FTD. *Hum Mol Genet*. 2014;24:3133–3142. https://doi.org/10.1093/hmg/ddv064.
21. Suh E, Lee EB, Neal D, et al. Semi-automated quantification of C9orf72 expansion size reveals inverse correlation between hexanucleotide repeat number and disease duration in frontotemporal degeneration EunRan. *Neuropathology*. 2015;130:363–372. https://doi.org/10.1007/s00401-015-1445-9.Semi-automated.
22. Xiao S, MacNair L, McGoldrick P, et al. Isoform-specific antibodies reveal distinct subcellular localizations of C9orf72 in amyotrophic lateral sclerosis. *Ann Neurol*. 2015;78:568–583. https://doi.org/10.1002/ana.24469.
23. Saberi S, Stauffer JE, Jiang J, et al. Sense-encoded poly-GR dipeptide repeat proteins correlate to neurodegeneration and uniquely co-localize with TDP-43 in dendrites of repeat-expanded C9orf72 amyotrophic lateral sclerosis. *Acta Neuropathol*. 2017;1–16. https://doi.org/10.1007/s00401-017-1793-8.
24. Frick P, Sellier C, Mackenzie IRA, et al. Novel antibodies reveal presynaptic localization of C9orf72 protein and reduced protein levels in C9orf72 mutation carriers. *Acta Neuropathol Commun*. 2018;6:72. https://doi.org/10.1186/s40478-018-0579-0.
25. Levine TP, Daniels RD, Gatta AT, Wong LH, Hayes MJ. The product of C9orf72, a gene strongly implicated in neurodegeneration, is structurally related to DENN Rab-GEFs. *Bioinformatics*. 2013;29:499–503. https://doi.org/10.1093/bioinformatics/bts725.
26. Zhang D, Iyer LM, He F, Aravind L. Discovery of novel DENN proteins: implications for the evolution of eukaryotic intracellular membrane structures and human disease. *Front Genet*. 2012;3:1–10. https://doi.org/10.3389/fgene.2012.00283.
27. Amick J, Roczniak-Ferguson A, Ferguson SM. C9orf72 binds SMCR8, localizes to lysosomes, and regulates mTORC1 signaling. *Mol Biol Cell*. 2016;27:3040–3051. https://doi.org/10.1091/mbc.E16-01-0003.
28. Farg MA, Sundaramoorthy V, Sultana JM, et al. C9ORF72, implicated in amytrophic lateral sclerosis and frontotemporal dementia, regulates endosomal trafficking. *Hum Mol Genet*. 2014;23:3579–3595. https://doi.org/10.1093/hmg/ddu068.

29. Jung J, Nayak A, Schaeffer V, et al. Multiplex image-based autophagy RNAi screening identifies SMCR8 as ULK1 kinase activity and gene expression regulator. *Elife*. 2017;6:1–32. https://doi.org/10.7554/eLife.23063.
30. Sellier C, Campanari M, Julie Corbier C, et al. Loss of C9 ORF 72 impairs autophagy and synergizes with polyQ Ataxin-2 to induce motor neuron dysfunction and cell death. *EMBO J*. 2016;35:1276–1297. https://doi.org/10.15252/embj.201593350.
31. Sullivan PM, Zhou X, Robins AM, et al. The ALS/FTLD associated protein C9orf72 associates with SMCR8 and WDR41 to regulate the autophagy-lysosome pathway. *Acta Neuropathol Commun*. 2016;4:51. https://doi.org/10.1186/s40478-016-0324-5.
32. Ugolino J, Ji YJ, Conchina K, et al. Loss of C9orf72 enhances autophagic activity via deregulated mTOR and TFEB signaling. *PLoS Genet*. 2016;12:1–26. https://doi.org/10.1371/journal.pgen.1006443.
33. Webster CP, Smith EF, Bauer CS, et al. The C9orf72 protein interacts with Rab1a and the ULK1 complex to regulate initiation of autophagy. *EMBO J*. 2016;35:1656–1676. https://doi.org/10.15252/embj.201694401.
34. Yang M, Chen L, Swaminathan K, et al. A C9ORF72/SMCR8-containing complex regulates ULK1 and plays a dual role in autophagy. *Sci Adv*. 2016;2:1–17. https://doi.org/10.1126/sciadv.1601167.
35. Coffee B, Zhang F, Warren ST, Reines D. Acetylated histones are associated with FMR1 in normal but not fragile X-syndrome cells. *Nat Genet*. 1999;22:98–101. https://doi.org/10.1038/9738.
36. Xi Z, Zinman L, Moreno D, et al. Hypermethylation of the CpG island near the G4C2 repeat in ALS with a C9orf72 expansion. *Am J Hum Genet*. 2013;92:981–989. https://doi.org/10.1016/j.ajhg.2013.04.017.
37. Xi Z, Zhang M, Bruni AC, et al. The C9orf72 repeat expansion itself is methylated in ALS and FTLD patients. *Acta Neuropathol*. 2015;129:715–727. https://doi.org/10.1007/s00401-015-1401-8.
38. Belzil VV, Bauer PO, Prudencio M, et al. Reduced C9orf72 gene expression in c9FTD/ALS is caused by histone trimethylation, an epigenetic event detectable in blood. *Acta Neuropathol*. 2013;126:895–905. https://doi.org/10.1007/s00401-013-1199-1.
39. van Blitterswijk M, Gendron TF, Baker MC, et al. Novel clinical associations with specific C9ORF72 transcripts in patients with repeat expansions in C9ORF72. *Acta Neuropathol*. 2015;130:863–876. https://doi.org/10.1007/s00401-015-1480-6.
40. Waite AJ, Bäumer D, East S, et al. Reduced C9orf72 protein levels in frontal cortex of amyotrophic lateral sclerosis and frontotemporal degeneration brain with the C9ORF72 hexanucleotide repeat expansion. *Neurobiol Aging*. 2014;35:1779.e5–1779.e13. https://doi.org/10.1016/j.neurobiolaging.2014.01.016.
41. Aoki Y, Manzano R, Lee Y, et al. C9orf72 and RAB7L1 regulate vesicle trafficking in amyotrophic lateral sclerosis and frontotemporal dementia. *Brain*. 2017;140:887–897. https://doi.org/10.1093/brain/awx024.
42. Almeida S, Gascon E, Tran H, et al. Modeling key pathological features of frontotemporal dementia with C9ORF72 repeat expansion in iPSC-derived human neurons. *Acta Neuropathol*. 2013;126:385–399. https://doi.org/10.1007/s00401-013-1149-y.
43. Burberry A, Wells MF, Limone F, et al. C9orf72 suppresses systemic and neural inflammation induced by gut bacteria. *Nature*. 2020. https://doi.org/10.1038/s41586-020-2288-7.
44. Ciura S, Lattante S, Le Ber I, et al. Loss of function of C9orf72 causes motor deficits in a zebrafish model of amyotrophic lateral sclerosis. *Ann Neurol*. 2013;74:180–187. https://doi.org/10.1002/ana.23946.
45. Yeh TH, Liu HF, Li YW, et al. C9orf72 is essential for neurodevelopment and motility mediated by Cyclin G1. *Exp Neurol*. 2018;304:114–124. https://doi.org/10.1016/j.expneurol.2018.03.002.
46. Therrien M, Rouleau GA, Dion PA, Parker JA. Deletion of C9ORF72 results in motor neuron degeneration and stress sensitivity in *C. elegans*. *PLoS ONE*. 2013;8:1–10. https://doi.org/10.1371/journal.pone.0083450.
47. Ho WY, Tai YK, Chang JC, et al. The ALS-FTD-linked gene product, C9orf72, regulates neuronal morphogenesis via autophagy. *Autophagy*. 2019;15:827–842. https://doi.org/10.1080/15548627.2019.1569441.

48. Harms M, Cadya J, Zaidmana C, et al. Lack of C9ORF72 coding mutations supports a gain of function for repeat expansions in ALS. *Neurobiol Aging*. 2013;34:2234. https://doi.org/10.1016/j.immuni.2010.12.017.Two-stage.
49. Liu F, Liu Q, Lu CX, et al. Identification of a novel loss-of-function C9orf72 splice site mutation in a patient with amyotrophic lateral sclerosis. *Neurobiol Aging*. 2016;47:219.e1–219.e5. https://doi.org/10.1016/j.neurobiolaging.2016.07.027.
50. Fratta P, Poulter M, Lashley T, et al. Homozygosity for the C9orf72 GGGGCC repeat expansion in frontotemporal dementia. *Acta Neuropathol*. 2013;126:401–409. https://doi.org/10.1007/s00401-013-1147-0.
51. Lagier-Tourenne C, Baughn M, Rigo F, et al. Targeted degradation of sense and antisense C9orf72 RNA foci as therapy for ALS and frontotemporal degeneration. *Proc Natl Acad Sci*. 2013;110: E4530–E4539. https://doi.org/10.1073/pnas.1318835110.
52. Atanasio A, Decman V, White D, et al. C9orf72 ablation causes immune dysregulation characterized by leukocyte expansion, autoantibody production, and glomerulonephropathy in mice. *Sci Rep*. 2016;6:1–14. https://doi.org/10.1038/srep23204.
53. Burberry A, Suzuki N, Wang JY, et al. Loss-of-function mutations in the C9ORF72 mouse ortholog cause fatal autoimmune disease. *Sci Transl Med*. 2016;8:347ra93. https://doi.org/10.1016/j.physbeh.2017.03.040.
54. Koppers M, Blokhuis AM, Westeneng HJ, et al. C9orf72 ablation in mice does not cause motor neuron degeneration or motor deficits. *Ann Neurol*. 2015;78:426–438. https://doi.org/10.1002/ana.24453.
55. O'Rourke JG, Bogdanik L, Yáñez A, et al. C9orf72 is required for proper macrophage and microglial function in mice. *Science*. 2016;351:1324–1329. https://doi.org/10.1126/science.aaf1064.
56. Sudria-Lopez E, Koppers M, de Wit M, et al. Full ablation of C9orf72 in mice causes immune system-related pathology and neoplastic events but no motor neuron defects. *Acta Neuropathol*. 2016;132:145–147. https://doi.org/10.1007/s00401-016-1581-x.
57. Miller ZA, Sturm VE, Camsari GB, et al. Increased prevalence of autoimmune disease within C9 and FTD/MND cohorts. *Neurol Neuroimmunol Neuroinflamm*. 2016;3:1–8. https://doi.org/10.1212/NXI.0000000000000301.
58. Molteni M, Rossetti C. Neurodegenerative diseases: the immunological perspective. *J Neuroimmunol*. 2017;313:109–115. https://doi.org/10.1016/j.jneuroim.2017.11.002.
59. Liu EY, Russ J, Wu K, et al. C9orf72 hypermethylation protects against repeat expansion-associated pathology in ALS/FTD. *Acta Neuropathol*. 2014;128:525–541. https://doi.org/10.1007/s00401-014-1286-y.
60. McMillan CT, Russ J, Wood EM, et al. C9orf72 promoter hypermethylation is neuroprotective. *Neurology*. 2015;84:1622–1630. https://doi.org/10.1212/WNL.0000000000001495.
61. Russ J, Liu EY, Wu K, et al. Hypermethylation of repeat expanded C9orf72 is a clinical and molecular disease modifier. *Acta Neuropathol*. 2015;129:39–52. https://doi.org/10.1007/s00401-014-1365-0.
62. Ash PEA, Bieniek KF, Gendron TF, et al. Unconventional translation of C9ORF72 GGGGCC expansion generates insoluble polypeptides specific to c9FTD/ALS. *Neuron*. 2013;77:639–646. https://doi.org/10.1016/j.neuron.2013.02.004.
63. Gendron TF, Bieniek KF, Zhang YJ, et al. Antisense transcripts of the expanded C9ORF72 hexanucleotide repeat form nuclear RNA foci and undergo repeat-associated non-ATG translation in c9FTD/ALS. *Acta Neuropathol*. 2013;126:829–844. https://doi.org/10.1007/s00401-013-1192-8.
64. Mori K, Arzberger T, Grasser FA, et al. The C9orf72 GGGGCC repeat is translated into aggregating dipeptide-repeat proteins in FTLD/ALS. *Science*. 2013;126:1335–1338. https://doi.org/10.1126/science.1232927.
65. Mori K, Arzberger T, Grässer FA, et al. Bidirectional transcripts of the expanded C9orf72 hexanucleotide repeat are translated into aggregating dipeptide repeat proteins. *Acta Neuropathol*. 2013;126:881–893. https://doi.org/10.1007/s00401-013-1189-3.
66. Zu T, Liu Y, Banez-Coronel M, et al. RAN proteins and RNA foci from antisense transcripts in C9ORF72 ALS and frontotemporal dementia. *Proc Natl Acad Sci*. 2013;110:E4968–E4977. https://doi.org/10.1073/pnas.1315438110.

67. Gomez-Deza J, Lee YB, Troakes C, et al. Dipeptide repeat protein inclusions are rare in the spinal cord and almost absent from motor neurons in C9ORF72 mutant amyotrophic lateral sclerosis and are unlikely to cause their degeneration. *Acta Neuropathol Commun*. 2015;3:38. https://doi.org/10.1186/s40478-015-0218-y.
68. Mackenzie IRA, Frick P, Grässer FA, et al. Quantitative analysis and clinico-pathological correlations of different dipeptide repeat protein pathologies in C9ORF72 mutation carriers. *Acta Neuropathol*. 2015;130:845–861. https://doi.org/10.1007/s00401-015-1476-2.
69. Schipper LJ, Raaphorst J, Aronica E, et al. Prevalence of brain and spinal cord inclusions, including dipeptide repeat proteins, in patients with the C9ORF72 hexanucleotide repeat expansion: a systematic neuropathological review. *Neuropathol Appl Neurobiol*. 2016;42:547–560. https://doi.org/10.1111/nan.12284.
70. Schludi MH, May S, Grässer FA, et al. Distribution of dipeptide repeat proteins in cellular models and C9orf72 mutation cases suggests link to transcriptional silencing. *Acta Neuropathol*. 2015;130:537–555. https://doi.org/10.1007/s00401-015-1450-z.
71. Floeter MK, Gendron TF. Biomarkers for amyotrophic lateral sclerosis and frontotemporal dementia associated with hexanucleotide expansion mutations in C9orf72. *Front Neurol*. 2018;9:1–9. https://doi.org/10.3389/fneur.2018.01063.
72. Lehmer C, Oeckl P, Weishaupt JH, et al. Poly-GP in cerebrospinal fluid links C9orf72-associated dipeptide repeat expression to the asymptomatic phase of ALS/FTD. *EMBO Mol Med*. 2017;9:859–868. https://doi.org/10.15252/emmm.201607486.
73. Su Z, Zhang Y, Gendron TF, et al. Discovery of a biomarker and lead small molecules to target r (GGGGCC)-associated defects in c9FTD/ALS. *Neuron*. 2014;83:1043–1050. https://doi.org/10.1016/j.neuron.2014.07.041.Discovery.
74. Gendron TF, Chew J, Stankowski JN, Al E. Poly(GP) proteins are a useful pharmacodynamic marker for C9ORF72-associated amyotrophic lateral sclerosis. *Sci Transl Med*. 2017;9. https://doi.org/10.1016/j.physbeh.2017.03.040.
75. Davidson YS, Barker H, Robinson AC, et al. Brain distribution of dipeptide repeat proteins in frontotemporal lobar degeneration and motor neurone disease associated with expansions in C9ORF72. *Acta Neuropathol Commun*. 2014;2:1–13. https://doi.org/10.1186/2051-5960-2-70.
76. Sakae N, Bieniek KF, Zhang YJ, et al. Poly-GR dipeptide repeat polymers correlate with neurodegeneration and clinicopathological subtypes in C9ORF72-related brain disease. *Acta Neuropathol Commun*. 2018;6:63. https://doi.org/10.1186/s40478-018-0564-7.
77. Jovičič A, Mertens J, Boeynaems S, et al. Modifiers of C9orf72 dipeptide repeat toxicity connect nucleocytoplasmic transport defects to FTD/ALS. *Nat Neurosci*. 2015;18:1226–1229. https://doi.org/10.1038/nn.4085.
78. Kanekura K, Yagi T, Cammack AJ, et al. Poly-dipeptides encoded by the C9ORF72 repeats block global protein translation. *Hum Mol Genet*. 2016;25:1803–1813. https://doi.org/10.1093/hmg/ddw052.
79. Lee KH, Zhang P, Kim HJ, et al. C9orf72 dipeptide repeats impair the assembly, dynamics, and function of membrane-less organelles. *Cell*. 2016;167:774–788.e17. https://doi.org/10.1016/j.cell.2016.10.002.
80. May S, Hornburg D, Schludi MH, et al. C9orf72 FTLD/ALS-associated Gly-Ala dipeptide repeat proteins cause neuronal toxicity and Unc119 sequestration. *Acta Neuropathol*. 2014;128:485–503. https://doi.org/10.1007/s00401-014-1329-4.
81. Tao Z, Wang H, Xia Q, et al. Nucleolar stress and impaired stress granule formation contribute to C9orf72 RAN translation-induced cytotoxicity. *Hum Mol Genet*. 2015;24:2426–2441. https://doi.org/10.1093/hmg/ddv005.
82. Wen X, Tan W, Westergard T, et al. Antisense proline-arginine RAN dipeptides linked to C9ORF72-ALS/FTD form toxic nuclear aggregates that initiate invitro and invivo neuronal death. *Neuron*. 2014;84:1213–1225. https://doi.org/10.1016/j.neuron.2014.12.010.
83. Yamakawa M, Ito D, Honda T, et al. Characterization of the dipeptide repeat protein in the molecular pathogenesis of c9FTD/ALS. *Hum Mol Genet*. 2015;24:1630–1645. https://doi.org/10.1093/hmg/ddu576.

84. Zhang YJ, Jansen-West K, Xu YF, et al. Aggregation-prone c9FTD/ALS poly(GA) RAN-translated proteins cause neurotoxicity by inducing ER stress. *Acta Neuropathol*. 2014;128:505–524. https://doi.org/10.1007/s00401-014-1336-5.
85. Zu T, Gibbens B, Doty NS, et al. Non-ATG-initiated translation directed by microsatellite expansions. *Proc Natl Acad Sci U S A*. 2011;108:260–265. https://doi.org/10.1073/pnas.1013343108.
86. Boeynaems S, Bogaert E, Michiels E, et al. Drosophila screen connects nuclear transport genes to DPR pathology in c9ALS/FTD. *Sci Rep*. 2016;6:7–14. https://doi.org/10.1038/srep20877.
87. Freibaum BD, Lu Y, Lopez-Gonzalez R, et al. GGGGCC repeat expansion in C9orf72 compromises nucleocytoplasmic transport. *Nature*. 2015;525:129–133. https://doi.org/10.1038/nature14974.
88. Mizielinska S, Grönke S, Niccoli T, et al. C9orf72 repeat expansions cause neurodegeneration in Drosophila through arginine-rich proteins. *Science*. 2014;345:1192–1195. https://doi.org/10. 1126/science.1256800.
89. Yang D, Abdallah A, Li Z, Lu Y, Almeida S, Gao FB. FTD/ALS-associated poly(GR) protein impairs the Notch pathway and is recruited by poly(GA) into cytoplasmic inclusions. *Acta Neuropathol*. 2015;130:525–535. https://doi.org/10.1007/s00401-015-1448-6.
90. Ohki Y, Wenninger-Weinzierl A, Hruscha A, et al. Glycine-alanine dipeptide repeat protein contributes to toxicity in a zebrafish model of C9orf72 associated neurodegeneration. *Mol Neurodegener*. 2017;12:1–11. https://doi.org/10.1186/s13024-016-0146-8.
91. Swaminathan A, Bouffard M, Liao M, et al. Expression of C9orf72-related dipeptides impairs motor function in a vertebrate model. *Hum Mol Genet*. 2018;27:1754–1762. https://doi.org/10.1093/hmg/ddy083.
92. Swinnen B, Bento-Abreu A, Gendron TF, et al. A zebrafish model for C9orf72 ALS reveals RNA toxicity as a pathogenic mechanism. *Acta Neuropathol*. 2018;135:427–443. https://doi.org/10.1007/s00401-017-1796-5.
93. Choi SY, Lopez-Gonzalez R, Krishnan G, et al. C9ORF72-ALS/FTD-associated poly(GR) binds Atp 5a1 and compromises mitochondrial function in vivo. *Nat Neurosci*. 2019;22:851–862. https://doi.org/10.1038/s41593-019-0397-0.
94. Schludi MH, Becker L, Garrett L, et al. Spinal poly-GA inclusions in a C9orf72 mouse model trigger motor deficits and inflammation without neuron loss. *Acta Neuropathol*. 2017;134:241–254. https://doi.org/10.1007/s00401-017-1711-0.
95. Zhang Y-J, Gendron TF, Ebbert MTW, et al. Poly(GR) impairs protein translation and stress granule dynamics in C9orf72-associated frontotemporal dementia and amyotrophic lateral sclerosis. *J Autism Dev Disord*. 2017;47:549–562. https://doi.org/10.1097/CCM.0b013e31823da96d.Hydrogen.
96. Zhang Y, Gendron TF, Grima JC, et al. C9ORF72 poly(GA) aggregates sequester and impair HR23 and nucleocytoplasmic transport proteins. *Nat Neurosci*. 2016;19:668–677. https://doi.org/10.1038/nn.4272.C9ORF72.
97. Zhang Y, Guo L, Gonzales PK, et al. Heterochromatin anomalies and double-stranded RNA accumulation underlie C9orf72 poly(PR) toxicity. *Science*. 2019;363:1–27. https://doi.org/10.1126/science.aav2606.Heterochromatin.
98. Kwon I, Xiang S, Kato M, et al. Poly-dipeptides encoded by the C9orf72 repeats bind nucleoli, impede RNA biogenesis, and kill cells. *Science*. 2014;345:1139–1145. https://doi.org/10.1126/science.1254917.
99. Hao Z, Liu L, Tao Z, et al. Motor dysfunction and neurodegeneration in a C9orf72 mouse line expressing poly-PR. *Nat Commun*. 2019;10. https://doi.org/10.1038/s41467-019 10956-w.
100. Flores BN, Dulchavsky ME, Krans A, et al. Distinct c9orf72-associated dipeptide repeat structures correlate with neuronal toxicity. *PLoS ONE*. 2016;11:1–18. https://doi.org/10.1371/journal.pone.0165084.
101. MacKenzie IR, Arzberger T, Kremmer E, et al. Dipeptide repeat protein pathology in C9ORF72 mutation cases: clinico-pathological correlations. *Acta Neuropathol*. 2013;126:859–879. https://doi.org/10.1007/s00401-013-1181-y.
102. Fratta P, Mizielinska S, Nicoll AJ, et al. C9orf72 hexanucleotide repeat associated with amyotrophic lateral sclerosis and frontotemporal dementia forms RNA G-quadruplexes. *Sci Rep*. 2012;2:1–6. https://doi.org/10.1038/srep01016.

103. Kovanda A, Zalar M, Šket P, Plavec J, Rogelj B. Anti-sense DNA d(GGCCCC)n expansions in C9ORF72 form i-motifs and protonated hairpins. *Sci Rep.* 2015;5:3–9. https://doi.org/10.1038/srep17944.
104. Reddy K, Zamiri B, Stanley SYR, Macgregor RB, Pearson CE. The disease-associated r(GGGGCC)n repeat from the C9orf72 gene forms tract length-dependent uni- and multimolecular RNA G-quadruplex structures. *J Biol Chem.* 2013;288:9860–9866. https://doi.org/10.1074/jbc.C113.452532.
105. Rahimov F, Kunkel LM. Cellular and molecular mechanisms underlying muscular dystrophy. *J Cell Biol.* 2013;201:499–510. https://doi.org/10.1083/jcb.201212142.
106. Cooper-Knock J, Higginbottom A, Stopford MJ, et al. Antisense RNA foci in the motor neurons of C9ORF72-ALS patients are associated with TDP-43 proteinopathy. *Acta Neuropathol.* 2015; 130:63–75. https://doi.org/10.1007/s00401-015-1429-9.
107. Donnelly CJ, Zhang PW, Pham JT, et al. RNA toxicity from the ALS/FTD C9ORF72 expansion is mitigated by antisense intervention. *Neuron.* 2013;80:415–428. https://doi.org/10.1016/j.neuron.2013.10.015.
108. Mizielinska S, Lashley T, Norona FE, et al. C9orf72 frontotemporal lobar degeneration is characterised by frequent neuronal sense and antisense RNA foci. *Acta Neuropathol.* 2013;126:845–857. https://doi.org/10.1007/s00401-013-1200-z.
109. Wang X, Hao L, Saur T, et al. Forward genetic screen in Caenorhabditis elegans suggests F57A10.2 and acp-4 as suppressors of C9ORF72 related phenotypes. *Front Mol Neurosci.* 2016;9:1–10. https://doi.org/10.3389/fnmol.2016.00113.
110. Xu Z, Poidevin M, Li X, et al. Expanded GGGGCC repeat RNA associated with amyotrophic lateral sclerosis and frontotemporal dementia causes neurodegeneration. *Proc Natl Acad Sci.* 2013; 110:7778–7783. https://doi.org/10.1073/pnas.1219643110.
111. Chew J, Cook C, Gendron TF, et al. Aberrant deposition of stress granule-resident proteins linked to C9orf72-associated TDP-43 proteinopathy. *Mol Neurodegener.* 2019;14:1–15. https://doi.org/10.1186/s13024-019-0310-z.
112. Chew J, Gendron TF, Prudencio M, et al. C9ORF72 repeat expansions in mice cause TDP-43 pathology, neuronal loss, and behavioral deficits. *Science.* 2015;348:1151–1154. https://doi.org/10.1126/science.aaa9344.C9ORF72 HHS Public Access.
113. O'Rourke JG, Bogdanik L, Muhammad AKMG, et al. C9orf72 BAC transgenic mice display typical pathologic features of ALS/FTD. *Neuron.* 2015;88:892–901. https://doi.org/10.1016/j.neuron.2015.10.027.
114. Peters OM, Cabrera GT, Tran H, et al. Human C9ORF72 hexanucleotide expansion reproduces RNA foci and dipeptide repeat proteins but not neurodegeneration in BAC transgenic mice. *Neuron.* 2015;88:902–909. https://doi.org/10.1016/j.neuron.2015.11.018.
115. Jiang J, Zhu Q, Gendron TF, et al. Gain of toxicity from ALS/FTD-linked repeat expansions in C9ORF72 is alleviated by antisense oligonucleotides targeting GGGGCC-containing RNAs. *Neuron.* 2016;90:535–550. https://doi.org/10.1016/j.neuron.2016.04.006.
116. Liu Y, Pattamatta A, Zu T, et al. C9orf72 BAC mouse model with motor deficits and neurodegenerative features of ALS/FTD. *Neuron.* 2016;90:521–534. https://doi.org/10.1016/j.neuron.2016.04.005.
117. Zhang K, Donnelly CJ, Haeusler AR, et al. The C9orf72 repeat expansion disrupts nucleocytoplasmic transport. *Nature.* 2015;525:56–61. https://doi.org/10.1038/nature14973.
118. Tran H, Almeida S, Moore J, et al. Differential toxicity of nuclear RNA foci versus dipeptide repeat proteins in a drosophila model of C9ORF72 FTD/ALS. *Neuron.* 2015;87:1207–1214. https://doi.org/10.1016/j.neuron.2015.09.015.
119. Celona B, Von Dollen J, Vatsavayai SC, et al. Suppression of c9orf72 RNA repeat-induced neurotoxicity by the ALS-associated RNA-binding protein Zfp106. *Elife.* 2017;6:1–17. https://doi.org/10.7554/eLife.19032.
120. Conlon EG, Lu L, Sharma A, et al. The C9ORF72 GGGGCC expansion forms RNA G-quadruplex inclusions and sequesters hnRNP H to disrupt splicing in ALS brains. *Elife.* 2016;5:1–28. https://doi.org/10.7554/eLife.17820.

121. Haeusler AR, Donnelly CJ, Periz G, et al. C9orf72 nucleotide repeat structures initiate molecular cascades of disease. *Nature*. 2014;507:195–200. https://doi.org/10.1038/nature13124.
122. Lee YB, Chen HJ, Peres JN, et al. Hexanucleotide repeats in ALS/FTD form length-dependent RNA foci, sequester RNA binding proteins, and are neurotoxic. *Cell Rep*. 2013;5:1178–1186. https://doi.org/10.1016/j.celrep.2013.10.049.
123. Mori K, Lammich S, Mackenzie IRA, et al. HnRNP A3 binds to GGGGCC repeats and is a constituent of p 62-positive/TDP43-negative inclusions in the hippocampus of patients with C9orf72 mutations. *Acta Neuropathol*. 2013;125:413–423. https://doi.org/10.1007/s00401-013-1088-7.
124. Sareen D, O'Rourke JG, Meera P, et al. Targeting RNA foci in iPSC-derived motor neurons from ALS patients with a C9ORF72 repeat expansion. *Sci Transl Med*. 2013;5:208ra149.
125. Bajc Česnik A, Darovic S, Mihevc SP, et al. Nuclear RNA foci from C9ORF72 expansion mutation form paraspeckle-like bodies. *J Cell Sci*. 2019;132:jcs224303.
126. Cooper-Knock J, Walsh MJ, Higginbottom A, et al. Sequestration of multiple RNA recognition motif-containing proteins by C9orf72 repeat expansions. *Brain*. 2014;137:2040–2051. https://doi.org/10.1093/brain/awu120.
127. Dejesus M, Nicole H, Xue AF, et al. In-depth clinico-pathological examination of RNA foci in a large cohort of C9ORF72 expansion carriers. *Acta Neuropathol*. 2017;134:255–269. https://doi.org/10.1007/s00401-017-1725-7.
128. Aladesuyi Arogundade O, Stauffer JE, Saberi S, et al. Antisense RNA foci are associated with nucleoli and TDP-43 mislocalization in C9orf72-ALS/FTD: a quantitative study. *Acta Neuropathol*. 2019;137:527–530. https://doi.org/10.1016/j.physbeh.2017.03.040.
129. Vatsavayai SC, Yoon SJ, Gardner RC, et al. Timing and significance of pathological features in *C9orf72* expansion-associated frontotemporal dementia. *Brain*. 2016;aww250. https://doi.org/10.1093/brain/aww250.
130. Shi Y, Lin S, Staats KA, et al. Haploinsufficiency leads to neurodegeneration in C9ORF72 ALS/FTD human induced motor neurons. *Nat Med*. 2018;24:313–325. https://doi.org/10.1038/nm.4490.
131. Maharjan N, Künzli C, Buthey K, Saxena S. C9ORF72 regulates stress granule formation and its deficiency impairs stress granule assembly, hypersensitizing cells to stress. *Mol Neurobiol*. 2017;54:3062–3077. https://doi.org/10.1007/s12035-016-9850-1.
132. Shao Q, Liang C, Chang Q, Zhang W, Yang M, Chen JF. C9orf72 deficiency promotes motor deficits of a C9ALS/FTD mouse model in a dose-dependent manner. *Acta Neuropathol Commun*. 2019;7:32. https://doi.org/10.1186/s40478-019-0685-7.
133. Zhu Q, Jiang J, Gendron TF, et al. Reduced C9ORF72 function exacerbates gain of toxicity from ALS/FTD-causing repeat expansion in C9orf72. *Nat Neurosci*. 2020;23:615–624. https://doi.org/10.1038/s41593-020-0619-5.
134. Zhou Q, Lehmer C, Michaelsen M, et al. Antibodies inhibit transmission and aggregation of C9orf72 poly- GA dipeptide repeat proteins. *EMBO Mol Med*. 2017;9:687–702. https://doi.org/10.15252/emmm.201607054.
135. Tabet R, Schaeffer L, Freyermuth F, et al. CUG initiation and frameshifting enable production of dipeptide repeat proteins from ALS/FTD C9ORF72 transcripts. *Nat Commun*. 2018;9:1–14. https://doi.org/10.1038/s41467-017-02643-5.
136. Wojciechowska M, Olejniczak M, Galka-Marciniak P, Jazurek M, Krzyzosiak WJ. RAN translation and frameshifting as translational challenges at simple repeats of human neurodegenerative disorders. *Nucleic Acids Res*. 2014;42:11849–11864. https://doi.org/10.1093/nar/gku794.
137. Knockenhauer KE, Schwartz TU. The nuclear pore complex as a flexible and dynamic gate. *Cell*. 2016;164:1162–1171. https://doi.org/10.1016/j.cell.2016.01.034.
138. Kramer NJ, Haney MS, Morgens DW, et al. CRISPR-Cas9 screens in human cells and primary neurons identify modifiers of C9orf72 dipeptide repeat protein toxicity Nicholas. *Nat Genet*. 2018;50:603–612. https://doi.org/10.1038/s41588-018-0070-7.CRISPR-Cas9.
139. Shi KY, Mori E, Nizami ZF, et al. Toxic PRn poly-dipeptides encoded by the C9orf72 repeat expansion block nuclear import and export. *Proc Natl Acad Sci U S A*. 2017;114:E1111–E1117. https://doi.org/10.1073/pnas.1620293114.
140. Hutten S, Dormann D. Nucleocytoplasmic transport defects in neurodegeneration—cause or consequence? *Semin Cell Dev Biol*. 2019;1–12. https://doi.org/10.1016/j.semcdb.2019.05.020.

141. Vatsavayai SC, Nana AL, Yokoyama JS, Seeley WW. C9orf72-FTD/ALS pathogenesis: evidence from human neuropathological studies. *Acta Neuropathol.* 2018;1–26. https://doi.org/10.1007/s00401-018-1921-0.
142. Eftekharzadeh B, Daigle JG, Kapinos LE, et al. Tau protein disrupts nucleocytoplasmic transport in Alzheimer's disease. *Neuron.* 2018;99:925–940. https://doi.org/10.1016/j.neuron.2018.07. 039.Tau.
143. Gasset-rosa F, Chillon-marinas C, Goginashvili A, et al. Polyglutamine-expanded huntingtin exacerbates age-related disruption of nuclear integrity and nucleocytoplasmic transport. *Neuron.* 2017;94:48–57. https://doi.org/10.1016/j.neuron.2017.03.027.Polyglutamine-expanded.
144. Grima JC, Daigle JG, Arbez N, et al. Mutant huntingtin disrupts the nuclear pore complex. *Neuron.* 2017;94:93–107. https://doi.org/10.1016/j.neuron.2017.03.023.Mutant.
145. Shang J, Yamashita T, Nakano Y, et al. Aberrant distributions of nuclear pore complex proteins in ALS mice and ALS patients. *Neuroscience.* 2017;350:158–168. https://doi.org/10.1016/j.neuroscience.2017.03.024.
146. Li YR, King OD, Shorter J, Gitler AD. Stress granules as crucibles of ALS pathogenesis. *J Cell Biol.* 2013;201:361–372. https://doi.org/10.1083/jcb.201302044.
147. Kato M, Han T, Xie S, et al. Cell-free formation of RNA granules: low complexity sequence domains form dynamic fibers within hydrogels. *Cell.* 2012. https://doi.org/10.1016/j.physbeh.2017.03.040.
148. Molliex A, Temirov J, Lee J, et al. Phase separation by low complexity domains promotes stress granule assembly and drives pathological fibrillization. *Cell.* 2015;163:123–133. https://doi.org/10.1016/j.cell.2015.09.015.
149. Murakami T, Qamar S, Lin JQ, et al. ALS/FTD mutation-induced phase transition of FUS liquid droplets and reversible hydrogels into irreversible hydrogels impairs RNP granule function. *Neuron.* 2015;88:678–690. https://doi.org/10.1016/j.neuron.2015.10.030.
150. Patel A, Lee HO, Jawerth L, et al. A liquid-to-solid phase transition of the ALS protein FUS accelerated by disease mutation. *Cell.* 2015;162:1066–1077. https://doi.org/10.1016/j.cell.2015.07.047.
151. Boeynaems S, Bogaert E, Kovacs D, et al. Phase separation of C9orf72 dipeptide repeats perturbs stress granule dynamics. *Mol Cell.* 2017;65:1044–1055. e5. https://doi.org/10.1016/j.molcel.2017.02.013.
152. Zhang K, Daigle JG, Cunningham KM, et al. Stress granule assembly disrupts nucleocytoplasmic transport. *Cell.* 2018;173:958–959. e17. https://doi.org/10.1016/j.cell.2018.03.025.
153. Jain A, Vale RD. RNA phase transitions in repeat expansion disorders. *Nature.* 2017;546:243–247. https://doi.org/10.1038/nature22386.
154. Fay MM, Anderson PJ, Ivanov P. ALS/FTD-associated C9ORF72 repeat RNA promotes phase transitions in vitro and in cells. *Cell Rep.* 2017;21:3573–3584. https://doi.org/10.1016/j.physbeh.2017.03.040.
155. Liu-Yesucevitz L, Bilgutay A, Zhang YJ, et al. Tar DNA binding protein-43 (TDP-43) associates with stress granules: analysis of cultured cells and pathological brain tissue. *PLoS ONE.* 2010;5. https://doi.org/10.1371/journal.pone.0013250.
156. Mizielinska S, Ridler CE, Balendra R, et al. Bidirectional nucleolar dysfunction in C9orf72 frontotemporal lobar degeneration. *Acta Neuropathol Commun.* 2017,5:29. https://doi.org/10.1186/s40478-017-0432-x.
157. Green KM, Glineburg MR, Kearse MG, et al. RAN translation at C9orf72-associated repeat expansions is selectively enhanced by the integrated stress response. *Nat Commun.* 2017;8. https://doi.org/10.1038/s41467-017-02200-0.
158. Guo Q, Lehmer C, Martínez-Sánchez A, et al. In situ structure of neuronal C9ORF72 poly-GA aggregates reveals proteasome recruitment. *Cell.* 2018;172:696–705. https://doi.org/10.1016/j.physbeh.2017.03.040.
159. Lopez-Gonzalez R, Lu Y, Gendron TF, et al. Poly(GR) in C9ORF72-related ALS/FTD compromises mitochondrial function and increases oxidative stress and DNA damage in iPSC-derived motor neurons. *Neuron.* 2016;92:383–391. https://doi.org/10.1016/j.neuron.2016.09.015.
160. Farg MA, Konopka A, Soo KY, Ito D, Atkin JD. The DNA damage response (DDR) is induced by the C9orf72 repeat expansion in amyotrophic lateral sclerosis. *Hum Mol Genet.* 2017;26:2882–2896. https://doi.org/10.1093/hmg/ddx170.
161. Lopez-Gonzalez R, Yang D, Pribadi M, et al. Partial inhibition of the overactivated Ku80-dependent DNA repair pathway rescues neurodegeneration in C9ORF72-ALS/FTD. *Proc Natl Acad Sci U S A.* 2019;116:9628–9633. https://doi.org/10.1073/pnas.1901313116.

162. Bennett CF, Krainer AR, Cleveland DW. Antisense oligonucleotide therapies for neurodegenerative diseases. *Annu Rev Neurosci.* 2019;42:385–406. https://doi.org/10.1146/annurev-neuro-070918-050501.
163. Hua Y, Vickers TA, Baker BF, Bennett CF, Krainer AR. Enhancement of SMN2 exon 7 inclusion by antisense oligonucleotides targeting the exon. *PLoS Biol.* 2007;5:729–744. https://doi.org/10.1371/journal.pbio.0050073.
164. Miller T, Pestronk A, David W, et al. A phase I, randomised, first-in-human study of an antisense oligonucleotide directed against SOD1 delivered intrathecally in SOD1-familial ALS patients. *Lancet Neurol.* 2013;12:435–442. https://doi.org/10.1016/S1474-4422(13)70061-9.
165. Tabrizi SJ, Leavitt BR, Landwehrmeyer GB, et al. Targeting huntingtin expression in patients with Huntington's disease. *N Engl J Med.* 2019;380:2307–2316. https://doi.org/10.1056/NEJMoa1900907.
166. Castanotto D, Rossi JJ. The promises and pitfalls of RNA-interference-based therapeutics. *Nature.* 2009;457:426–433. https://doi.org/10.1038/nature07758.
167. Hu J, Rigo F, Prakash TP, Corey DR. Recognition of c9orf72 mMutant RNA by single-stranded silencing RNAs. *Nucleic Acid Ther.* 2017;27:87–94. https://doi.org/10.1089/nat.2016.0655.
168. Martier R, Liefhebber JM, García-osta A, et al. Targeting RNA-mediated toxicity in C9orf72 ALS and/or FTD by RNAi-based gene therapy. *Mol Ther Nucleic Acids.* 2019;16:26–37. https://doi.org/10.1016/j.omtn.2019.02.001.
169. Martier R, Liefhebber JM, Miniarikova J, et al. Artificial microRNAs targeting C9orf72 can reduce accumulation of intra-nuclear transcripts in ALS and FTD patients. *Mol Ther Nucleic Acids.* 2019;14:593–608.
170. Wang D, Tai PWL, Gao G, Systems P, Tai PWL. Adeno-associated virus vector as a platform for gene therapy delivery. *Nat Rev Drug Discov.* 2019;18:358–378. https://doi.org/10.1038/s41573-019-0012-9.Adeno-associated.
171. Zamiri B, Reddy K, Macgregor RB, Pearson CE. TMPyP4 porphyrin distorts RNA G-quadruplex structures of the disease-associated r(GGGGCC)n repeat of the C9orf72 gene and blocks interaction of RNAbinding proteins. *J Biol Chem.* 2014;289:4653–4659. https://doi.org/10.1074/jbc.C113.502336.
172. Wang Z, Ursu A, Childs-disney JL, et al. The hairpin form of r(G4C2)exp in c9ALS/FTD is repeat-associated non-ATG translated and a target for bioactive small molecules. *Cell Chem Biol.* 2020;26:179–190. https://doi.org/10.1016/j.chembiol.2018.10.018.The.
173. Heidenreich M, Zhang F. Applications of CRISPR-Cas systems in neuroscience. *Nat Rev Neurosci.* 2016;17:36–44. https://doi.org/10.1016/j.physbeh.2017.03.040.
174. Krishnan G, Zhang Y, Gu Y, Kankel MW, Gao FB, Almeida S. CRISPR deletion of the C9ORF72 promoter in ALS/FTD patient motor neurons abolishes production of dipeptide repeat proteins and rescues neurodegeneration. *Acta Neuropathol.* 2020;3:6–9. https://doi.org/10.1007/s00401-020-02154-6.
175. Pinto BS, Saxena T, Oliveira R, et al. Impeding transcription of expanded microsatellite repeats by deactivated Cas9. *Mol Cell.* 2017;68:479–490. e5https://doi.org/10.1016/j.physbeh.2017.03.040.
176. Batra R, Nelles DA, Pirie E, et al. Elimination of toxic microsatellite repeat expansion RNA by RNA-targeting Cas9. *Cell.* 2017;170:899–912. https://doi.org/10.1016/j.cell.2017.07.010.Elimination.
177. Goodman LD, Prudencio M, Kramer NJ, et al. Toxic expanded GGGGCC repeat transcription is mediated by the PAF1 complex in C9orf72-associated FTD. *Nat Neurosci.* 2019;22:863–874. https://doi.org/10.1038/s41593-019-0396-1.
178. Kramer NJ, Carlomagno Y, Zhang Y, et al. Spt4 selectively regulates the expression of C9orf72 sense and antisense mutant transcripts associated with c9FTD/ALS. *Science.* 2016;353:708–712. https://doi.org/10.1126/science.aaf7791.Spt4.
179. Naguib A, Sandmann T, Yi F, Watts RJ, Lewcock JW, Dowdle WE. SUPT4H1 depletion leads to a global reduction in RNA. *Cell Rep.* 2019;26:45–53. e4https://doi.org/10.1016/j.celrep.2018.12.004.
180. Gallardo G, Holtzman DM. Antibody therapeutics targeting Aβ and tau. *Cold Spring Harb Perspect Med.* 2017;7:1–17. https://doi.org/10.1101/cshperspect.a024331.

181. Chang YJ, Jeng US, Chiang YL, Hwang IS, Chen YR. The glycine-alanine dipeptide repeat from C9orf72 hexanucleotide expansions forms toxic amyloids possessing cell-to-cell transmission properties. *J Biol Chem*. 2016;291:4903–4911. https://doi.org/10.1074/jbc.M115.694273.
182. Hock EM, Polymenidou M. Prion-like propagation as a pathogenic principle in frontotemporal dementia. *J Neurochem*. 2016;138:163–183. https://doi.org/10.1111/jnc.13668.
183. Ravits J. Focality, stochasticity and neuroanatomic propagation in ALS pathogenesis. *Exp Neurol*. 2014;262:121–126. https://doi.org/10.1016/j.expneurol.2014.07.021.
184. Nguyen L, Montrasio F, Pattamatta A, et al. Antibody therapy targeting RAN proteins rescues C9 ALS/FTD phenotypes in C9orf72 mouse model. *Neuron*. 2020;105:645–662.e11. https://doi.org/10.1016/j.neuron.2019.11.007.
185. Cristofani R, Crippa V, Vezzoli G, et al. The small heat shock protein B8 (HSPB8) efficiently removes aggregating species of dipeptides produced in C9ORF72-related neurodegenerative diseases. *Cell Stress Chaperones*. 2018;23:1–12. https://doi.org/10.1007/s12192-017-0806-9.
186. Becker LA, Huang B, Bieri G, et al. Therapeutic reduction of ataxin 2 extends lifespan and reduces pathology in TDP-43 mice. *Nature*. 2017;544:367–371. https://doi.org/10.1038/nature22038. Therapeutic.
187. Hautbergue GM, Castelli LM, Ferraiuolo L, et al. SRSF1-dependent nuclear export inhibition of C9ORF72 repeat transcripts prevents neurodegeneration and associated motor deficits. *Nat Commun*. 2017;8:1–18. https://doi.org/10.1038/ncomms16063.
188. Archbold HC, Jackson KL, Arora A, et al. TDP43 nuclear export and neurodegeneration in models of amyotrophic lateral sclerosis and frontotemporal dementia. *Sci Rep*. 2018;8:1–18. https://doi.org/10.1038/s41598-018-22858-w.

CHAPTER 3

Translating molecular therapies for spinal muscular atrophy

Brian K. Kaspar
Milo Biotechnology, San Diego, CA, United States

Introduction

Spinal muscular atrophy (SMA) is a neuromuscular disease characterized by motor dysfunction and eventual death of the alpha motor neurons residing in the ventral horn of the spinal cord. The disease occurs with an incidence of 1 case per 10,000 live births.[1–3] Most cases exhibit rapid disease progression early in life, with the majority of incident cases being the most severe form, SMA Type 1. Historically, SMA has been one of the leading causes of genetic infant mortality.[4] SMA is classified into a spectrum of disease severity to include Type 1 through Type IV.[5] Type 1 patients are not able to sit upright and typically demonstrate rapid loss of motor function before 6 months of age and typically die from respiratory issues. Indeed 50% of children succumb to the disease with fatality or permanent mechanical ventilation by 10.5 months of age.[6,7] Type II patients can sit upright but cannot ambulate without assistance, whereas Type III patients can ambulate but typically become wheel chair bound in later years. Type IV patients may only show neuromuscular issues in late life.[8] The Melki laboratory in 1995 showed that 95% of cases of SMA, regardless of type, is caused by homozygous deletion or mutation in the 5q13 survival of the motor neuron (*SMN1*) gene.[9] Humans have two forms of the *SMN* gene, a telomeric form (*SMN1*) and a centromeric form (*SMN2*). The *SMN1* gene produces a full-length mRNA transcript that encodes the SMN protein. The *SMN2* gene is identical to the *SMN1* gene other than a C to T substitution in an exonic splicing enhancer, resulting in the exclusion of exon 7 during transcription. This results in a truncated protein that is nonfunctional and rapidly degraded. However, a small amount of total mRNA transcripts arising from the *SMN2* gene contains the exon 7 sequence which encodes the normal functional SMN protein. It is thought that each *SMN2* gene produces approximately 10%–15% of full-length mRNA transcript. Therefore, the disease is dictated by the copy number of *SMN2* genes which ultimately produces slightly more SMN protein with increasing copy number and hence results in less severe types of SMA.[10–14] Having no SMN, or SMA Type 0, is typically embryonic lethal. A detailed review of SMA

Neurotherapeutics in the Era of Translational Medicine
https://doi.org/10.1016/B978-0-12-816475-4.00017-3

pathogenesis is beyond the scope of this book chapter,[15,16] however the molecular genetics and clinical understanding of SMA led way to numerous therapeutic approaches.

A number of therapeutic modalities have focused on altering *SMN2* splicing to include exon 7 in the final mRNA transcript thereby leading to full-length SMN protein expression. Antisense oligonucleotides (ASOs) are DNA molecules built to bind to their complementary sequences within an intron or exon that can enhance or disrupt the targeted splicing event. Its first neurological application targeted a familial variant of amyotrophic lateral sclerosis.[17] ASOs can be chemically modified to increase stability. The discovery that a splice site regulator ISS-N1 in intron 7 in the SMN2 gene leads to strong exon 7 inclusion and higher levels of SMN protein expression initiated the search for an ASO that could accomplish this molecular feat.[18–21] These ASOs, developed by a collaboration with Ionis Pharmaceuticals and Biogen, were tested in clinical trials in SMA patients and led to a broad approval of Spinraza for all types of SMA in December 2016.[22]

A small-molecule approach has also been investigated in SMA. Risdiplam is an experimental drug developed by Roche, PTC Therapeutics, and the SMA Foundation to treat SMA. It is a pyridazine derivative small molecule taken orally that modifies splicing in the *SMN2* gene to create functional SMN protein.[23–25] It has been in various phase II-III clinical trials and in April 2020, Roche applied for marketing authorization which was recently approved by the FDA.

Gene therapy approaches have also gained attention as potential molecular medicines due to advances in vector design, novel viral capsids, improved production, and novel delivery devices. Conceptually, gene therapy for monogenic disease is most often based on a gene replacement strategy. Gene therapists have explored a multitude of nonviral and viral vectors, with viral vectors being the most studied. Many types of viruses have been explored as the trojan horses to deliver the recombinant payloads, including adenovirus, herpes virus, retrovirus, lentivirus, and adeno-associated viruses. This chapter will briefly introduce adeno-associated viruses (AAV) as they have emerged as the most popular viral vector for treating neurological and neuromuscular disorders.

AAV was first discovered in adenoviral preparations in the mid-1960s and subsequently isolated from human tissues.[26–28] Within 20 years, understanding of AAV DNA replication and virion assembly along with the successful cloning and sequencing of the wild-type AAV2 genome had been completed.[29] AAV is a Dependoparvovirus within the family of Parvoviridae. It requires a helper virus, such as adenovirus for its rescue, replication, and packaging. To date, AAV is considered nonpathogenic and has not been implicated in human disease. Since initial isolation, a diverse number of AAV serotypes (e.g., AAV6, AAV8, AAV9) have been isolated, each with the varied sequence in the capsid.[30] The capsid is an icosahedral protein approximately 20 nm in diameter encapsulating a single-stranded DNA genome of 4.7 kB. The genome is flanked by two inverted terminal repeats (ITRs). There are two sets of wild-type AAV genes, replication genes and capsid genes. The replication (rep) gene encodes four proteins,

Rep78, Rep 68, Rep 52, and Rep 40, all named for their molecular mass. The capsid is composed of three subunit proteins, VP1, VP2, and VP3, produced from the single cap gene. Recombinant AAV is produced by removing all of the genes between the ITRs and replacing them with a promoter of choice, your gene of interest, and a polyadenylation signal. The rep and cap genes are provided in trans by a separate plasmid and the helper functions from adenovirus are provided typically in a third plasmid. These three plasmids are introduced into a cell, such as a human embryonic kidney carcinoma cell (hek293) by transient transfection and cells harvested and lyzed with AAV purification following. Purification of AAV has greatly improved with novel ion exchange as well as affinity columns that are now available. An excellent historical review of AAV as a vector platform has recently been published by Gao and colleagues.[31] AAV vectors have been explored for the treatment of SMA.

While we have tried to be inclusive of all contributors within the field, it is without a doubt that we have missed some. Rather, this chapter intends to show the dynamic research and development that overcame significant hurdles and is meant more as a story that chronicles the remarkable achievements of hundreds of dedicated scientists and clinicians.

Effectively targeting motor neurons within the spinal cord

Most drugs and therapies do not cross the blood-brain barrier to effectively target cells within the central nervous system (CNS). Indeed, the blood-brain barrier evolved to protect the CNS from pathogens such as viral infections. The widespread distribution of therapeutic molecules in the CNS is a challenge for biologics. In the gene therapy world, the typical entry point to the brain or spinal cord was via drilling a burr hole into the skull or vertebral bones and placing a syringe needle for multiple injections to achieve the desired distribution of vector within the CNS. In the early 2000s, several groups including Gage, Kaspar, and Azzouz, Mazarakis showed the promise of using retrograde transport to deliver genetic payloads peripherally via muscle where motor neuron terminals would take up small amounts of the virus and retrogradely shuttle it back to the motor neuron cell body; the result being the cell would be effectively transduced.[32–34] While highly promising, the treatment of a widespread motor neuron disease such as SMA or amyotrophic lateral sclerosis would require hundreds, if not thousands, of injections to effectively target the entire spinal cord and motor cortex. This was not translationally feasible and further work continued, based on this encouraging first "back door entry" to the spinal cord.

The series of studies to address AAV serotypes which may effectively cross the blood-brain barrier was initiated in 2007–2008 with a postdoctoral fellow, Dr. Kevin Foust, who joined The Kaspar Laboratory at Nationwide Children's Hospital (Ohio State University), having done his doctoral training with Dr. Terry Flotte and Dr. Ron Mandel at

The University of Florida-Gainesville. Kevin's prior work attempted to test the ability of AAV 8 to cross the blood-brain barrier. Unfortunately, this resulted in sparse and infrequent labeling of lower motor neurons, leading to the conclusion that systemic injection of recombinant AAV8 was not an effective delivery route to engage the target neurons.[35] However, numerous other studies between 2004 and 2006 showed adeno-associated virus (AAV) serotypes 6 and 8 ability to cross endothelial barriers following systemic intravenous or intraperitoneal injections with impressive widespread targeting of multiple tissues such as skeletal and cardiac muscles.[31,36–38] Indeed, this approach was and currently is being applied to numerous muscle disorders, including Duchenne and limb-girdle muscular dystrophies. A one-time administration led to robust targeting of both skeletal and cardiac muscles. So, while the AAV6 or AAV8 were escaping the endothelial barrier in the instance of muscle targeting, other groups including the Kaspar lab reasoned other AAV serotypes may cross the blood-brain barrier more efficiently.

The blood-brain barrier (BBB) is a highly selective semipermeable layer of endothelial cells lining the capillary wall, astrocytic end-feet ensheathing the outside of the capillary while pericytes are residents of the vessel's basement membrane. This unique cellular architecture prevents circulating blood from crossing into the extracellular space. Most molecules, including ions, nutrients, glucose, and amino acids, are actively transported, whereas water is diffused passively through the barrier via aquaporins. Importantly, the BBB blocks the passage of most pathogens, having evolved to protect the brain. Multiple groups were attempting to transiently open the BBB by hyperosmotic shock, with some degree of success.[39] However, this approach to delivering therapeutics was met with resistance by the field, as transiently opening the BBB in neurodegenerative disease with inflammatory components invokes two risks: exacerbating inflammation and/or the risk of brain hemorrhage. Considering this, our laboratory undertook a different approach to drug delivery to the CNS. Toward this end, our laboratory screened some of the newly discovered AAV variants from the laboratory of Dr. Jim Wilson and the University of Pennsylvania with the intent of discovering treatment vectors.[30] Our initial attempts of looking at various serotypes delivered intravascularly under laboratory conditions were fruitless, with discouraging faint and feeble evidence of cells within the brain and spinal cord that had been transduced and expressed the green fluorescent gene which we had incorporated as a "payload." Frustrated, we questioned the longevity of the project due to our failure to discover a virus that was capable of traversing the blood-brain barrier. However, upon looking at the genetic tree of AAV capsids,[30] we opted, after more thought, to soldier on. Within the genetic tree, we noticed a lineage, or clade, of AAV that looked promising. That serotype was AAV9. Utilizing AAV9 as a vector, we injected neonatal mice intravenously with a green fluorescent protein (GFP), ChAT, and GFAP. Two weeks later, we stained for these markers. The results were remarkable: cells throughout the brain and spinal cord were successfully transduced, an unprecedented occurrence within the field. This result had never been seen with any other AAV

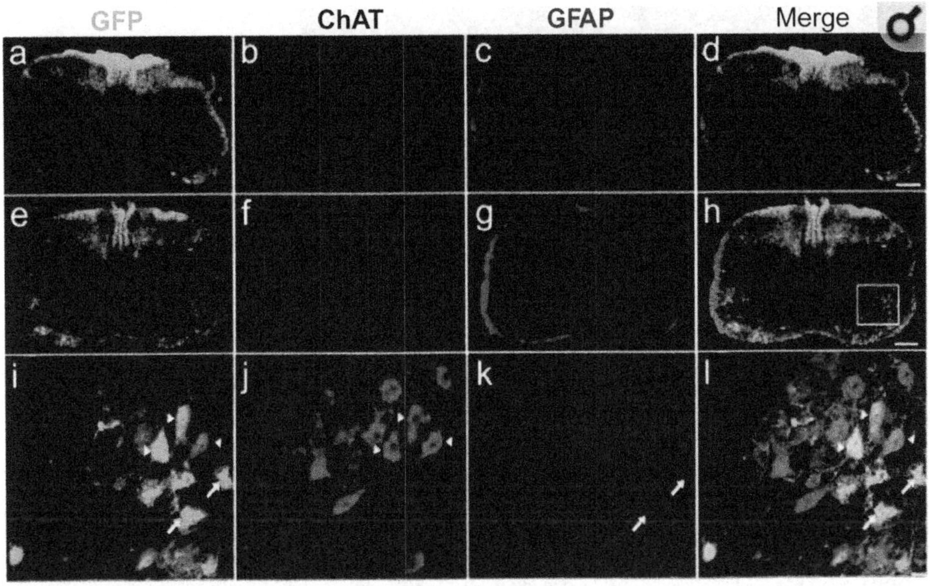

Fig. 1 Intravenous injection of AAV9-GFP leads to robust motor neuron transduction of cervical (A–D) and lumbar (E–H) spinal cord, with high power projections in (I–L).[40]

serotype. After several hours in awe over the microscope and weeks of subsequent imaging and quantification, we showed over 60% of the lumbar spinal motor neurons following a single intravascular dose of AAV9 encoded the GFP transgene (NBT, Fig. 1).[40,41]

With increasing doses, we were able to target nearly every single motor neuron in the spinal cord of mice. Not only had we effectively targeted motor neurons at all segments of the spinal cord, but it appeared that the transgene expression was persistent, potentially leading to a "one and done" gene therapy to target cells within the spinal cord. It was also gratifying to see our scientific result replicated by Dr. Martine Barkat's group in Paris, France, reporting nearly identical data of AAV9 traversing the BBB and targeting cells throughout the spinal cord, and in the case of Dr. Barkat's group, demonstrating that this ability of AAV9 occurred not only in mice but also in a larger species, i.e., a cat[42](Fig. 2).

Treatment of the SMA mouse model

Impressed with the level of motor neuron transduction, we presented the mouse data of AAV9-GFP targeting to Dr. Arthur Burghes, who had created the mouse model of spinal muscular atrophy, a disease affecting motor neurons, where we believed our gene therapy might be able to deliver the SMN transgene effectively. The scientific advisors of Fight SMA, Dr. Burghes, Dr. Chris Lorson, and Dr. Alex MacKenzie had graciously included

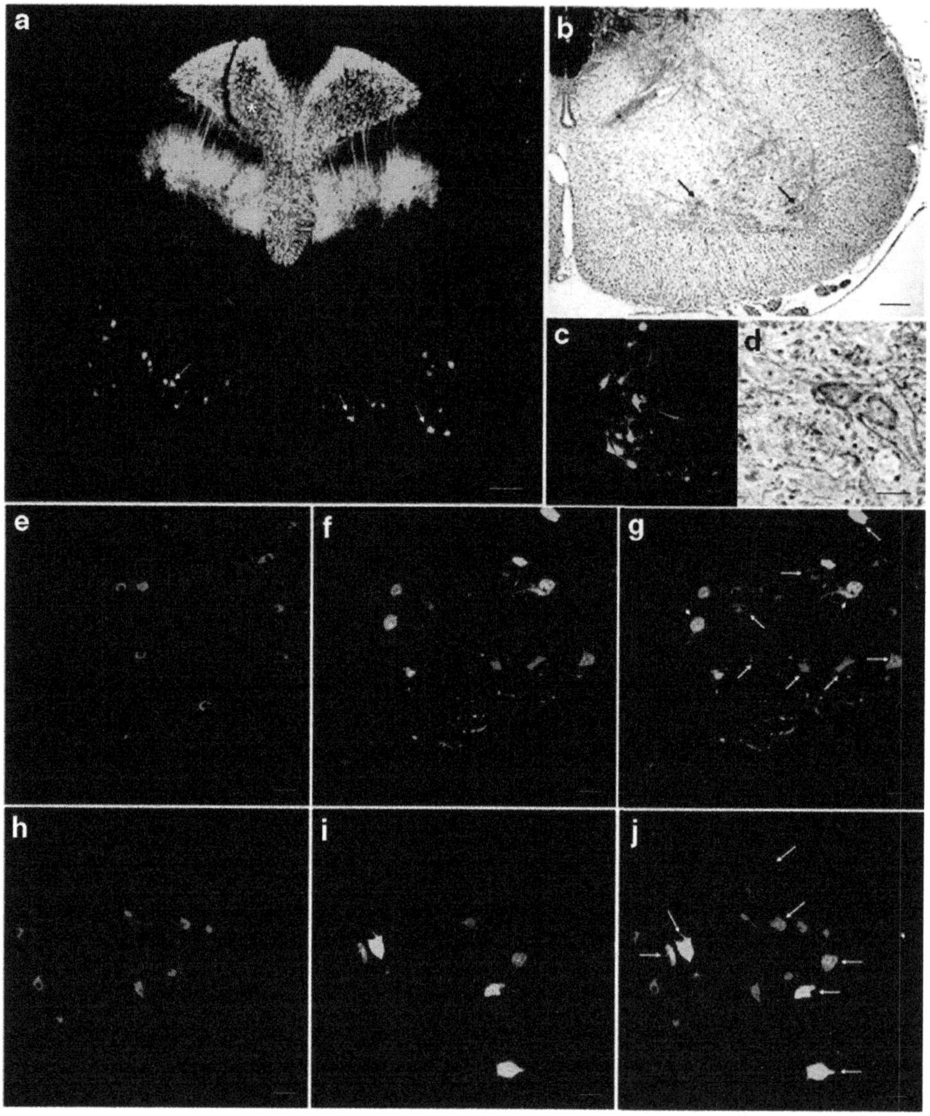

Fig. 2 Motor neuron transduction in cats following intravenous delivery of AAV9-GFP.[42] (A–D) Representative transverse sections of spinal cord from a LIX1 heterozygous cat observed by (A, C) laser scanning confocal microscopy (*arrows*: GFP-positive MNs; *star*: dorsal sensory tracts) or (B, D) treated for GFP immunohistochemistry (*arrows*: GFP-immunopositive MNs) 10 days after the injection of scAAV9-GFP into the jugular vein. (E–J) Double-labeled transverse sections of the spinal cord of both affected (E–G) and nonaffected (H–J) kitten, treated for ChAT immunofluorescence (E, H, *red*) and GFP native fluorescence (F, I, *green*). (G, J) Merge (*arrows*: transduced MNs). Bars = (A) 200 μm; (B, C) 100 μm; (D–J) 50 μm.

me for years in the scientific conferences of Fight SMA, so I was aware of the basic and translational studies that were occurring in the field and, therefore, felt comfortable thinking about SMA as an ideal target for testing our discovery as a method to deliver a therapeutic transgene. However, we needed guidance as there were many unanswered questions regarding SMA biology and disease progression. Arthur's laboratory agreed to collaborate with us, and we were armed with a seasoned SMA molecular geneticist for us to embark upon exploring gene therapy in SMA. Indeed, Arthur's laboratory had collaborated with Mazarakis and Azzouz on a lentiviral gene therapy approach back in 2004, utilizing a pseudorabies enveloped lentivirus,[43] demonstrating modest efficacy, but nonetheless, results that showed SMN expression by a viral vector could extend survival in the mouse model of SMA. We were encouraged by those results and believed the AAV9 approach would be much more efficient at targeting motor neurons. We transferred the SMA mouse model from the Burghes lab into the Kaspar laboratory and quickly learned the genotyping and husbandry procedures with the help of the Burghes lab team. Next, we thought long and hard about the specific therapeutic vector design: What promoter would we want to use? Did we want to use single-stranded AAV or self-complementary AAV vectors developed by McCarthy and Samulski? We thought about the quickness of patients succumbing to the disease and opted to utilize a pan promoter; the chicken beta-actin (CBA) promoter with widespread and strong expression in a multitude of cells. Next, we evaluated the size of the construct and believed that a self-complementary AAV vector (scAAV) would allow for the near-immediate expression of the transgene, a feature critical for halting the pace of neuron degeneration.[44] We deduced this since the virus packages as a double-stranded intermediate and is available for immediate transcription and translation. Finally, we cloned the SMN transgene into the vector and made the first batch of viruses to be injected into animals.

The SMA delta 7 mice developed by Burghes typically die at 2 weeks of age, with a median lifespan of 15 ± 2 days.[45] We treated our cohort of animals within the first day of life with an intravenous injection of the AAV9 vector carrying the SMN transgene through the facial vein of the mice. Strikingly, our treated animals were healthy at 2 weeks of age, whereas control animals were succumbing to the disease. When treated animals were 20 days of age, our initial thought was that we may have mis-genotyped them. We sent tail clippings from these animals to Dr. Burghes' laboratory in a blinded manner for re-genotyping so as to exclude the possibility that we had erringly included wild-type animals in the treatment arm of the experiment. Arthur's laboratory confirmed the SMN gene mutation, i.e., the animals included in the experiment all carried the correct genotype. Upon confirming this over the telephone, I mentioned to Arthur that the treated SMA animals were now approaching 30 days of age and appeared normal. Thirty minutes later, Arthur was in my office at Nationwide Children's to witness history in the making, both of us watching the longest-lived SMA animals in the world run around the cage like normal mice, all treated with a single intravenous infusion with our newly created gene therapy. Indeed, our vigil went on and on as animals surpassed 60, 100 and,

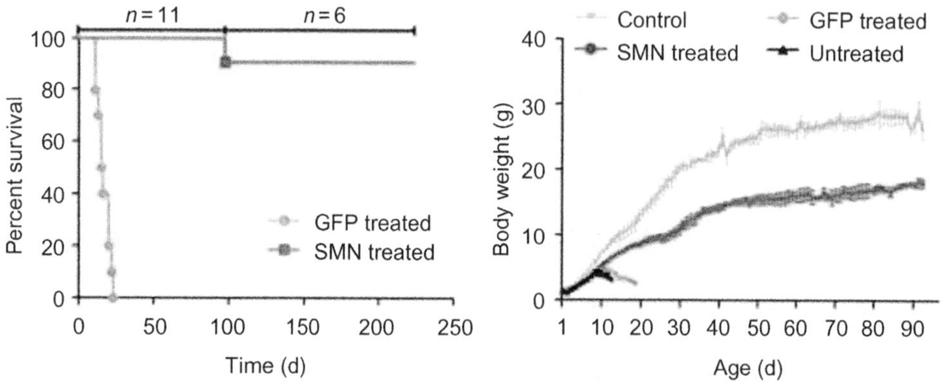

Fig. 3 Survival and body weight of SMA control and gene therapy treated animals.[46]

then 200 days of age (Fig. 3).[46] Considering that many in the field said that the mouse model of SMA was too severe to be treated, we were obligated to test it again. The result was similar upon repetition. Not only were the mice living longer, but they were healthy, running around the cage like normal animals, and at the cellular level had normal electrophysiology at the neuromuscular junction (Fig. 4).

Following the remarkable rescue of the mouse model of SMA, it was gratifying to see our work replicated in a number of laboratories including the group of Dr. Barkats in France, along with Dr. Azzouz in Sheffield, UK, and Dr. Passini's work at Genzyme.[47–49] Based on this collective experience, we were compelled to move the program forward. To do this, we elicited the support of Dr. Jerry Mendell who was the Department Chair for the Center for Gene Therapy at Nationwide Children's Hospital. Interestingly, Jerry had brought both myself and Arthur Burghes to Columbus, Ohio, hiring us as faculty. While there were human factors that led to our laboratory success—the hiring of both Arthur and I and our subsequent friendship—it was not at all certain that this would translate into a successful treatment for SMA children. Fortunately, the Nationwide Children's Hospital was able to devout their significant resources to moving our discovery from the bench to the clinic. With Jerry at the helm to guide us with his years of translational experiences,[50] we next determined to test our gene therapy approach in a larger species.

Moving to larger species and human clinical trials

We next initiated studies in nonhuman primates to determine whether systemic dosing of AAV9 resulted in widespread targeting of motor neurons in the spinal cord. Led by a graduate student in the Kaspar Laboratory, Adam Bevan, the team tested newborn, 30-day-old and 90-day-old cynomolgus macaques delivering an AAV9 encoding the green fluorescent gene. Strikingly, we found motor neurons targeted throughout the spinal cord at all time points tested (Fig. 5).[51]

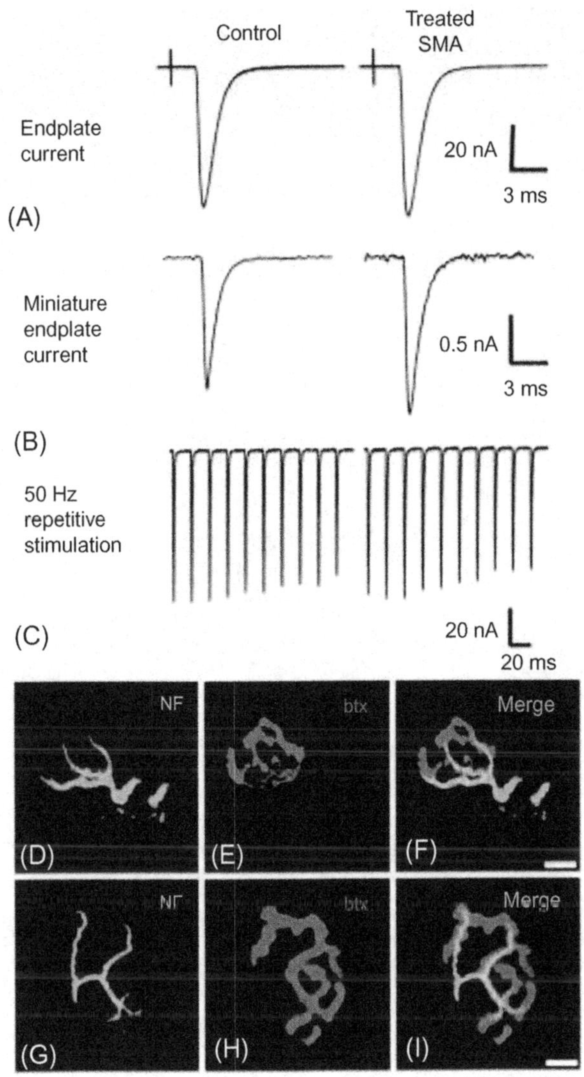

Fig. 4 Electrophysiology of treated SMA mice demonstrated that electrophysiology parameters were restored to normal and the neuromuscular junction was intact.[46] (A) scAAV9-SMN treatment restores endplate currents (EPC) in ~90-day-old SMA animals. In control mice, the mean EPC amplitude was 82.6 ± 3.5 nA, and in treated SMA mice, it was 83.4 ± 4.1 nA ($P = .89$, $n = 4$ mice for each group). (B) Affected animals treated with scAAV9-SMN had an increase in miniature endplate currents. (C) Both control and treated SMA endplate currents had a similar degree of depression during 50 Hz nerve stimulation. (D–I) Representative sections from the transverse abdominis (TVA), a proximal muscle with innervation abnormalities in SMA mice, shows normal innervation in both wild-type (D–F) and SMN-treated (G–I) animals. Scale bars, 10 μm.

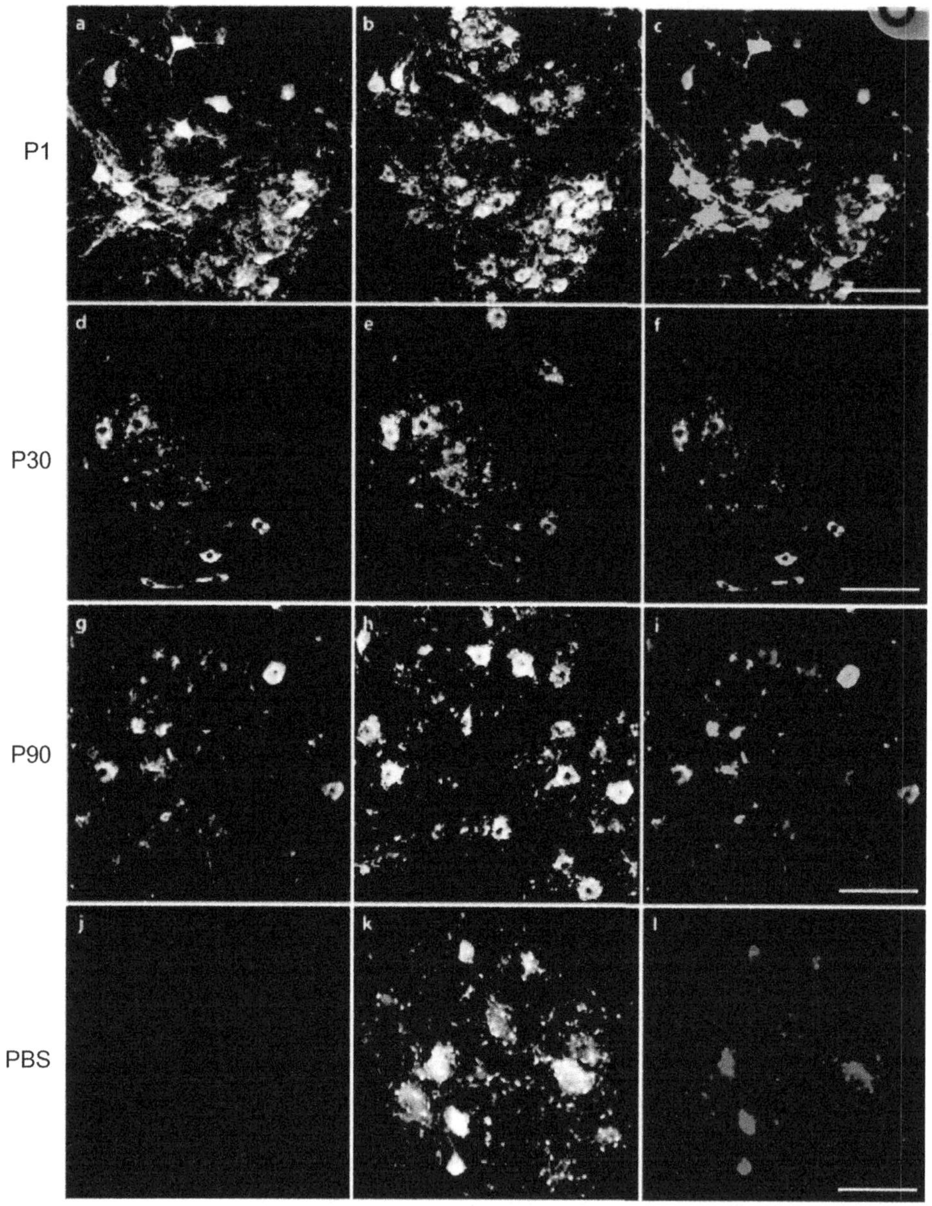

Fig. 5 Motor neuron transduction of lumbar spinal cord in nonhuman primates following systemic delivery at postnatal day 1, 30, 90 (P1, P30, P90) with AAV9-CBA-GFP. GFP staining in (A, D, G, J), Choline Acetyltransferase (ChAT) staining in (B, E, H, K) and Merged GFP (*green*) and ChAT in (C, F, I, L).[51]

Further, we characterized the robust transduction of muscles throughout the entire musculoskeletal system. Monkeys received a dosage between 1 and 3×10^{14} vg/kg in these studies. These titers were initially determined by quantitative PCR (Q-PCR) and gave us further confidence in determining the planned dosing for a first in the human clinical trial. Additional dose ranging in the SMA delta 7 mouse demonstrated a minimally effective dosage at 6.7×10^{13} vg/kg, effectively doubling median survival of a treated cohort, and a rescue dose at $2.0–3.3 \times 10^{14}$ vg/kg, again as determined by Q-PCR titer. Also, in these studies, we determined that intrathecal delivery also was effective at targeting the brain and spinal cord. In a series of studies by a postdoctoral fellow in Dr. Burghes' laboratory, Dr. Sandra Duque demonstrated robust motor neuron targeting following a single administration into the cerebrospinal fluid. This targeting was achieved at a dosage significantly less than the systemic route of administration. Further, the Burghes laboratory generated a porcine model of SMA, with the use of an AAV9- shRNA gene knockdown strategy of porcine SMN. Using immunosuppressive drugs, we were able to treat these pigs prior to or after the manifestation of hindlimb and forelimb paralysis induced by the shRNA to porcine SMN with our therapeutic construct; AAV9-SMN, encoding the human SMN. The results were clear. One could prevent further decline of weakness in animals that had already succumbed to motor neuron damage. However, if one intervened with AAV9-SMN prior to motor neuron damage and weakness, the pigs showed little to no signs of SMA.[52] This result demonstrated the urgency for early intervention in this disease. Indeed, SMA, particularly SMA Type 1 needed to be thought of as a medical emergency. Armed with preclinical efficacy and biodistribution studies in multiple species, Jerry Mendell and the Kaspar Laboratory designed a series of safety studies in mice and nonhuman primates to support an investigational new drug application (IND). With Jerry's and the Nationwide Children's translational team's help, we compiled the necessary studies to support a first in human trial for systemic dosing in SMA.

The next challenge was to make enough of the gene therapy for human clinical trials. Fortunately, Nationwide Children's Hospital had invested in a GMP manufacturing facility. The only problem was that no one to our knowledge at the time had made enough virus to support doses up to 3.3×10^{14} vg/kg. The amount of virus that was required, even for infants would be enormous. The GMP facility was led by Dr. K. Reed Clark, a faculty member with extensive experience in AAV biology and manufacturing. A new cell culture technology had just emerged from Corning, moving from 10-stack "cell hotels" to hyperstacks, a 36-layer plastic chamber for cell culture growth. Reed's group was the first to implement these hyperstacks for producing AAV and over the course of a year, developed the method and produced the GMP material to support the first in human studies. We tested his GMP virus and a single injection of this virus rescued the lifespan of the delta 7 SMA mice identical to the research-grade preparations. After productive discussions with the FDA in a pre-

IND meeting, Jerry and the team submitted the IND to the FDA, and Jerry was given the green light to initiate the first inhuman studies. The initial plans were to treat six patients with SMA Type 1, the most severe form of the disease with most patients not surviving past 20 months of age. Three patients would be dosed at the 6.7×10^{13} vg/kg dose and three patients at the 3.3×10^{14} vg/kg dose. The trial was ultimately changed to three patients being dosed at 6.7×10^{13} vg/kg and 12 patients dosed at 2.0×10^{14} vg/kg again using the QPCR method for determining titers. Subsequent studies changed the method of quantification from QPCR to a droplet digital PCR method which was more robust and reliable for determining the titer of the Phase 1 trial material and subsequent batches of the virus.

The first patient was dosed in May 2014 in what at the time was a historic event for the gene therapy field. This was the largest amount of AAV gene therapy vector that had ever been dosed to a human at the time. In fact, several prominent leaders of the field cautioned Jerry and I not to dose the patients, that it was too risky. It is here that the courage and experience of Jerry shown through. The first patient experienced liver function enzyme (LFE) alterations approximately 3 weeks following administration of the gene therapy, which Dr. Mendel mitigated by administration of prednisolone. Jerry subsequently amended the clinical protocol to include a prophylactic steroid regimen 1 day prior to gene administration for the remaining patients in the Phase I trial which mitigated most of the LFE elevations. The results of the Phase 1 trial were stunning: at the time of data cutoff, all of the patients had reached 20 months of age and did not require permanent mechanical ventilation. Natural history studies in contrast showed that only 8% of patients would not require permanent mechanical ventilation at 20 months of age. In the 2.0×10^{14} vg/kg dose, 11 of 12 patients were able to sit unassisted for at least 5 seconds. Further, these Type 1 patients were able to gain head control, attain the ability to speak, were able to roll over, crawl, and two patients were even able to stand and walk independently.[53] These motor developmental milestones are never achieved in Type 1 historical controls.

In anticipation of clinical success, I cofounded a biotechnology company, AveXis, in late 2013. The goal for AveXis would be to scale manufacturing to supply the global SMA population which includes thousands of patients, a herculean task if using hyperstacks for cell culture growth. Also, the company would have to develop new analytical assays to ensure that Batch # 01 was the same and as efficacious as Batch #33 or Batch #333 would be. Most of these technical hurdles in gene therapy clinical practice had not been done before, many technologies or concepts were untested, so we were literally building the rocket ship as we launched off the ground. The sense of urgency came not only from ourselves but also from the SMA community onto us: babies were continuing to die of this disease. AveXis built our own manufacturing division as well as our own Good Manufacturing Practice Facility to produce enough to supply the need and to meet our logistical needs, we recruited topnotch biotechnology experts in regulatory, analytical, quality, and medical fields. We also formed outstanding relationships with critical equipment and reagent suppliers that provided the bioreactors, transfection reagents, and purification columns. In late 2017, a pivotal trial using the company's GMP produced material was initiated at 16 trial sites within the United States. With the more precise analytical assays for potency and using digital droplet PCR, the dose tested in patients was recalculated as 1.1×10^{14} vg/kg determined by ddPCR. The clinical readout of the Phase 3 trial mirrored the benefits seen in the Phase 1 studies,[53,54] (Fig. 6) allowing AveXis to file a New Drug Application which was the basis for regulatory approval in May 2019 of the new drug, Zolgensma in the United States. As of this writing, Zolgensma has been approved in the United States, Japan, and Europe with over 600 patients treated to date.

Future of SMA

With Spinraza and Zolgensma on the market, as well as Risdiplam, recently approved by the FDA, the future of treating SMA is bright. Not long ago, the treatment of SMA was palliative care. The genetics underlying SMA allowed for critical molecular discoveries to be made that have resulted in the drugs listed above. The field continues to learn about these new technologies, their strengths, weaknesses, promises, and shortcomings. A landmark win in 2018 for SMA patients was the recommendation by the US Secretary of Health and Human Services that all newborns receive SMA screening. Subsequently, SMA was added to the Federal Recommended Uniform Screening Panel (RUSP). At the time of writing this chapter, 31 states currently screen for SMA with approximately 65% of newborn babies in the United States being screened. This will dramatically facilitate the treatment of patients prior to the rapid motor neuron decline that characterizes SMA.

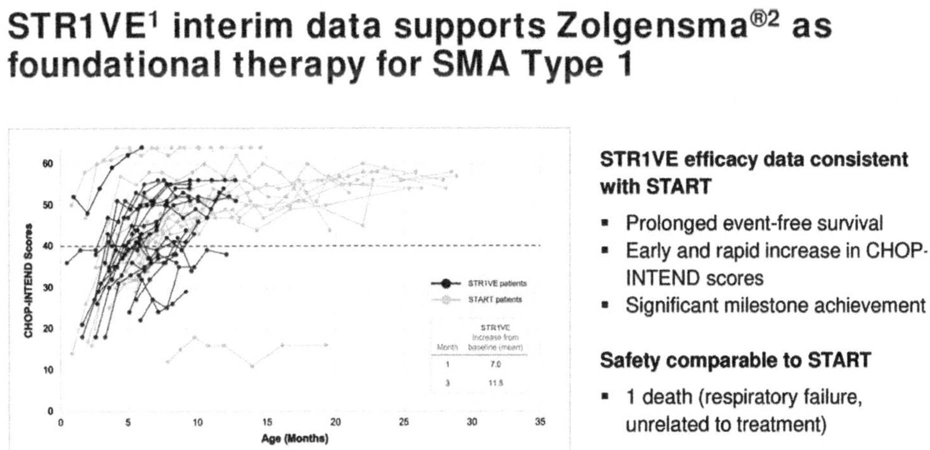

Fig. 6 Motor function improvements in Phase 1/2a (START) and Phase 3 (STR1VE) studies.[54]

In addition, neuromuscular clinics have bolstered their holistic approach to treating SMA by implementing recommendations for nutrition, pulmonary care, and physical therapy routines to enhance the care of patients with SMA. Sometime in the near future, it is likely that the advances chronicled herein will give way to major advances in science, including the application of gene editing strategies utilizing the CRISPR/CAS9 strategy along with zinc-finger nucleases. Combinatorial therapies may have a further beneficial impact for SMA, such as the addition of muscle-enhancing strategies, perhaps the use of follistatin.[55] Finally, cellular replacement of motor neurons looms in the horizon but major hurdles remain to guarantee engraftment and physiological connectivity of the graft. However, additional work including reprogramming methods have shown amazing results in early preclinical studies for Parkinson's and Huntington's Disease.[56] With the advancement of science, the future continues to look bright for new therapies for SMA.

Concluding thoughts and acknowledgments

Throughout the drug development process, we knew from the early days of studying AAV serotypes, this was only going to be accomplished by teamwork. It's difficult to name everyone and all of the entities that are required to develop a novel therapy. This is what it took: the early foundations supporting us, the National Institutes of Health, administrators at Nationwide Children's Hospital, GMP operators, clinicians, physical therapists, clinical trial coordinators, pulmonary care specialists, physicians, neurologists, and regulators around the world, upstream and downstream technologists, supply chain operators, and finally the unseen and often unacknowledged workers who labor behind the scene. Ron Bartek says it best in his prologue in this book—it's a "family affair."

Pioneering a new gene therapy approach and the drug development process is humbling. I am deeply grateful to the teams, both academic and industry who devoted their lives and stuck with the development of this gene therapy, solving hundreds of challenges along the way, with many long days and sleepless nights. Finally, we are forever thankful for the 15 pioneers who braved the first inhuman trial to pave the way for generations of patients to benefit.

References

1. Pearn J. Incidence, prevalence, and gene frequency studies of chronic childhood spinal muscular atrophy. *J Med Genet.* 1978;15(6):409–413.
2. Prior TW, et al. Newborn and carrier screening for spinal muscular atrophy. *Am J Med Genet A.* 2010;152A(7):1608–1616.
3. Sugarman EA, et al. Pan-ethnic carrier screening and prenatal diagnosis for spinal muscular atrophy: clinical laboratory analysis of >72,400 specimens. *Eur J Hum Genet.* 2012;20(1):27–32.
4. Roberts DF, Chavez J, Court SD. The genetic component in child mortality. *Arch Dis Child.* 1970;45(239):33–38.

5. Zerres K, Rudnik-Schoneborn S. Natural history in proximal spinal muscular atrophy. Clinical analysis of 445 patients and suggestions for a modification of existing classifications. *Arch Neurol.* 1995;52 (5):518–523.
6. Finkel RS, et al. Observational study of spinal muscular atrophy type I and implications for clinical trials. *Neurology.* 2014;83(9):810–817.
7. Kolb SJ, et al. Baseline results of the NeuroNEXT spinal muscular atrophy infant biomarker study. *Ann Clin Transl Neurol.* 2016;3(2):132–145.
8. Piepers S, et al. A natural history study of late onset spinal muscular atrophy types 3b and 4. *J Neurol.* 2008;255(9):1400–1404.
9. Burglen L, et al. SMN gene deletion in variant of infantile spinal muscular atrophy. *Lancet.* 1995;346 (8970):316–317.
10. Lefebvre S, et al. Correlation between severity and SMN protein level in spinal muscular atrophy. *Nat Genet.* 1997;16(3):265–269.
11. Coovert DD, et al. The survival motor neuron protein in spinal muscular atrophy. *Hum Mol Genet.* 1997;6(8):1205–1214.
12. Lorson CL, et al. A single nucleotide in the SMN gene regulates splicing and is responsible for spinal muscular atrophy. *Proc Natl Acad Sci U S A.* 1999;96(11):6307–6311.
13. Monani UR, et al. A single nucleotide difference that alters splicing patterns distinguishes the SMA gene SMN1 from the copy gene SMN2. *Hum Mol Genet.* 1999;8(7):1177–1183.
14. Burnett BG, et al. Regulation of SMN protein stability. *Mol Cell Biol.* 2009;29(5):1107–1115.
15. Arnold WD, Burghes AH. Spinal muscular atrophy: development and implementation of potential treatments. *Ann Neurol.* 2013;74(3):348–362.
16. Kolb SJ, Kissel JT. Spinal muscular atrophy: a timely review. *Arch Neurol.* 2011;68(8):979–984.
17. Smith RA, et al. Antisense oligonucleotide therapy for neurodegenerative disease. *J Clin Invest.* 2006;116(8):2290–2296.
18. Lim SR, Hertel KJ. Modulation of survival motor neuron pre-mRNA splicing by inhibition of alternative 3′ splice site pairing. *J Biol Chem.* 2001;276(48):45476–45483.
19. Singh NK, et al. Splicing of a critical exon of human survival motor neuron is regulated by a unique silencer element located in the last intron. *Mol Cell Biol.* 2006;26(4):1333–1346.
20. Hua Y, et al. Antisense masking of an hnRNP A1/A2 intronic splicing silencer corrects SMN2 splicing in transgenic mice. *Am J Hum Genet.* 2008;82(4):834–848.
21. Passini MA, et al. Antisense oligonucleotides delivered to the mouse CNS ameliorate symptoms of severe spinal muscular atrophy. *Sci Transl Med.* 2011;3(72):72ra18.
22. Finkel RS, et al. Nusinersen versus sham control in infantile-onset spinal muscular atrophy. *N Engl J Med.* 2017;377(18):1723–1732.
23. Poirier A, et al. Risdiplam distributes and increases SMN protein in both the central nervous system and peripheral organs. *Pharmacol Res Perspect.* 2018;6(6):e00447.
24. Sturm S, et al. A phase 1 healthy male volunteer single escalating dose study of the pharmacokinetics and pharmacodynamics of risdiplam (RG7916, RO7034067), a SMN2 splicing modifier. *Br J Clin Pharmacol.* 2019;85(1):181–193.
25. Ratni H, et al. Discovery of Risdiplam, a selective survival of motor neuron-2 (SMN2) gene splicing modifier for the treatment of spinal muscular atrophy (SMA). *J Med Chem.* 2018;61(15):6501–6517.
26. Atchison RW, Casto BC, Hammon WM. Adenovirus-associated defective virus particles. *Science.* 1965;149(3685):754–756.
27. Hoggan MD, Blacklow NR, Rowe WP. Studies of small DNA viruses found in various adenovirus preparations: physical, biological, and immunological characteristics. *Proc Natl Acad Sci U S A.* 1966;55(6):1467–1474.
28. Blacklow NR, Hoggan MD, Rowe WP. Isolation of adenovirus-associated viruses from man. *Proc Natl Acad Sci U S A.* 1967;58(4):1410–1415.
29. Samulski RJ, et al. Cloning of adeno-associated virus into pBR322: rescue of intact virus from the recombinant plasmid in human cells. *Proc Natl Acad Sci U S A.* 1982;79(6):2077–2081.
30. Gao G, et al. Clades of adeno-associated viruses are widely disseminated in human tissues. *J Virol.* 2004;78(12):6381–6388.

31. Wang Z, et al. Adeno-associated virus serotype 8 efficiently delivers genes to muscle and heart. *Nat Biotechnol.* 2005;23(3):321–328.
32. Kaspar BK, et al. Targeted retrograde gene delivery for neuronal protection. *Mol Ther.* 2002;5(1): 50–56.
33. Kaspar BK, et al. Retrograde viral delivery of IGF-1 prolongs survival in a mouse ALS model. *Science.* 2003;301(5634):839–842.
34. Mazarakis ND, et al. Rabies virus glycoprotein pseudotyping of lentiviral vectors enables retrograde axonal transport and access to the nervous system after peripheral delivery. *Hum Mol Genet.* 2001;10 (19):2109–2121.
35. Foust KD, et al. Neonatal intraperitoneal or intravenous injections of recombinant adeno-associated virus type 8 transduce dorsal root ganglia and lower motor neurons. *Hum Gene Ther.* 2008;19(1):61–70.
36. Blankinship MJ, et al. Efficient transduction of skeletal muscle using vectors based on adeno-associated virus serotype 6. *Mol Ther.* 2004;10(4):671–678.
37. Inagaki K, et al. Robust systemic transduction with AAV9 vectors in mice: efficient global cardiac gene transfer superior to that of AAV8. *Mol Ther.* 2006;14(1):45–53.
38. Pacak CA, et al. Recombinant adeno-associated virus serotype 9 leads to preferential cardiac transduction in vivo. *Circ Res.* 2006;99(4):e3–e9.
39. Fu H, et al. Self-complementary adeno-associated virus serotype 2 vector: global distribution and broad dispersion of AAV-mediated transgene expression in mouse brain. *Mol Ther.* 2003;8(6):911–917.
40. Foust KD, et al. Intravascular AAV9 preferentially targets neonatal neurons and adult astrocytes. *Nat Biotechnol.* 2009;27(1):59–65.
41. Foust KD, Kaspar BK. Over the barrier and through the blood: to CNS delivery we go. *Cell Cycle.* 2009;8(24):4017–4018.
42. Duque S, et al. Intravenous administration of self-complementary AAV9 enables transgene delivery to adult motor neurons. *Mol Ther.* 2009;17(7):1187–1196.
43. Azzouz M, et al. Lentivector-mediated SMN replacement in a mouse model of spinal muscular atrophy. *J Clin Invest.* 2004;114(12):1726–1731.
44. McCarty DM, Monahan PE, Samulski RJ. Self-complementary recombinant adeno-associated virus (scAAV) vectors promote efficient transduction independently of DNA synthesis. *Gene Ther.* 2001;8 (16):1248–1254.
45. Monani UR, et al. The human centromeric survival motor neuron gene (SMN2) rescues embryonic lethality in SMN(-/-) mice and results in a mouse with spinal muscular atrophy. *Hum Mol Genet.* 2000;9(3):333–339.
46. Foust KD, et al. Rescue of the spinal muscular atrophy phenotype in a mouse model by early postnatal delivery of SMN. *Nat Biotechnol.* 2010;28(3):271–274.
47. Dominguez E, et al. Intravenous scAAV9 delivery of a codon-optimized SMN1 sequence rescues SMA mice. *Hum Mol Genet.* 2011;20(4):681–693.
48. Valori CF, et al. Systemic delivery of scAAV9 expressing SMN prolongs survival in a model of spinal muscular atrophy. *Sci Transl Med.* 2010;2(35):35ra42.
49. Passini MA, et al. CNS-targeted gene therapy improves survival and motor function in a mouse model of spinal muscular atrophy. *J Clin Invest.* 2010;120(4):1253–1264.
50. Mendell JR. Translational medicine: evolution, fulfillment, and belief in gene therapy. *Hum Gene Ther.* 2020;31(1–2):14–19.
51. Bevan AK, et al. Systemic gene delivery in large species for targeting spinal cord, brain, and peripheral tissues for pediatric disorders. *Mol Ther.* 2011;19(11):1971–1980.
52. Duque SI, et al. A large animal model of spinal muscular atrophy and correction of phenotype. *Ann Neurol.* 2015;77(3):399–414.
53. Mendell JR, et al. Single-dose gene-replacement therapy for spinal muscular atrophy. *N Engl J Med.* 2017;377(18):1713–1722.
54. Relations NI. *STR1VE Interim Data Supports Zolgensma as Foundational Therapy for SMA Type 1.* Q1 2019 Results-Novartis Investor Presentation, 2019.
55. Kota J, et al. Follistatin gene delivery enhances muscle growth and strength in nonhuman primates. *Sci Transl Med.* 2009;1(6):6ra15.
56. Qian H, et al. Reversing a model of Parkinson's disease with in situ converted nigral neurons. *Nature.* 2020;582(7813):550–556.

CHAPTER 4

From uncovering the mechanism of transthyretin aggregation to the drug tafamidis for ameliorating neurodegeneration and cardiomyopathy

Evan T. Powers[a] and Jeffery W. Kelly[a,b]
[a]Department of Chemistry, The Scripps Research Institute, La Jolla, CA, United States
[b]The Skaggs Institute for Chemical Biology, The Scripps Research Institute, La Jolla, CA, United States

Introduction

Transthyretin-aggregation-based peripheral polyneuropathy

Human genetic, pharmacologic, and pathologic evidence to be covered below all indicate that the disease transthyretin (TTR) amyloid polyneuropathy results from the process of TTR aggregation, which includes the formation of numerous distinct aggregate structures, including cross-β-sheet amyloid fibrils.[1–3] In 1952, Andrade reported a progressive length-dependent axonal sensory, motor, and autonomic neuropathy, with variable involvement of other organs, linked to the presence of amyloid fibril pathology.[4–6] These aggregates were later established to be composed mostly of mutated (V30M) TTR,[7] the remainder being wild-type (WT) TTR. This autosomal dominant disease is now called hereditary TTR amyloidosis with polyneuropathy (hATTR-PN), a malady formerly called familial amyloid polyneuropathy. After about a decade of hATTR-PN progression, this malady also typically compromises the central nervous system (CNS).[8–11] The reason that CNS degeneration lags behind the peripheral neurodegeneration is likely because the transthyretin concentration is an order of magnitude lower in the CNS than in the periphery. The rate of protein aggregation is strongly concentration dependent; thus, TTR aggregation is simply slower in the brain than in the periphery. The function of the eyes, the kidneys, and other organs is often also adversely affected by the process of TTR aggregation in hATTR-PN.[12–17] If untreated, hATTR-PN causes death within 15 years from diagnosis.

Neurotherapeutics in the Era of Translational Medicine
https://doi.org/10.1016/B978-0-12-816475-4.00004-5

Transthyretin-aggregation-based brain neurodegeneration

The aggregation of a different category of TTR mutants—those that form exceptionally destabilized tetramers, like D18G and A25T—leads to a CNS degenerative disease in which patients present with dementia, and/or focal neurological episodes, and/or cerebral vascular bleeding.[8, 10, 18–20] The CNS TTR aggregate pathology is observed in the meninges and its vessels, and in the meningocortical vessels, as well as in the brain parenchyma. PET amyloid imaging studies support the notion that the dementia, focal neurological episodes, and cerebral vascular bleeding are based on TTR aggregation, including amyloid deposition in the brain.[10] The tissue tropism of this neurodegenerative disease is reasonably well understood and is discussed below.[21]

Transthyretin-aggregation-based cardiomyopathy

The most common transthyretin amyloid disease is late-onset cardiomyopathy that is caused by the aggregation of WT TTR.[22–26] This disease, wild-type TTR amyloid cardiomyopathy (wtATTR-CM), previously called senile systemic amyloidosis, leads to congestive heart failure and death in ≈7 years from diagnosis, if untreated.[23] Patients with wtATTR-CM typically also present with compromised autonomic nervous system function.[22, 23] Hereditary TTR cardiomyopathy/dysautonomia (hATTR-CM), formerly called familial amyloid cardiomyopathy, presents analogously to wtATTR-CM.[27, 28] Up to 4% of people of African ancestry carry the V122I mutation in TTR, of these allele carriers ≈44% develop hATTR-CM and congestive heart failure.[29]

Tissue tropism of the transthyretin amyloid diseases

The tissue tropism of the TTR amyloid diseases presumably results from the different structural ensembles of conformationally distinct non-native TTR aggregates (Box 1) being produced by different TTR sequences.[1] For example, in cardiomyopathy (wtATTR-CM and hATTR-CM), little non-native TTR is circulating in blood[2] and the fibrils deposited in the extracellular space of the heart comprise both an N-terminal and a C-terminal fragment of TTR.[3, 30] In polyneuropathy (hATTR-PN), the cross-β-sheet amyloid fibrils also deposited extracellularly are composed predominantly of full-length TTR[30] and notably, there is a large amount of non-native TTR circulating in the blood.[2] The vulnerability of specific postmitotic tissues to the aggregation of a given TTR sequence(s) likely also involves the composition of the extracellular matrix, including the glycosaminoglycans. Glycosaminoglycans are established to enhance protein aggregation.[31, 32] The ineffectiveness of the extracellular matrix and the glycosaminoglycans to convert non-native TTR aggregates to putatively less toxic cross-β-sheet amyloid fibrils could exacerbate proteotoxicity as one possibility.[31–33] We have a lot to learn about the structural ensemble of TTR aggregates formed (Box 1) in the various TTR amyloid diseases and their organ toxicity.[1, 34]

BOX 1

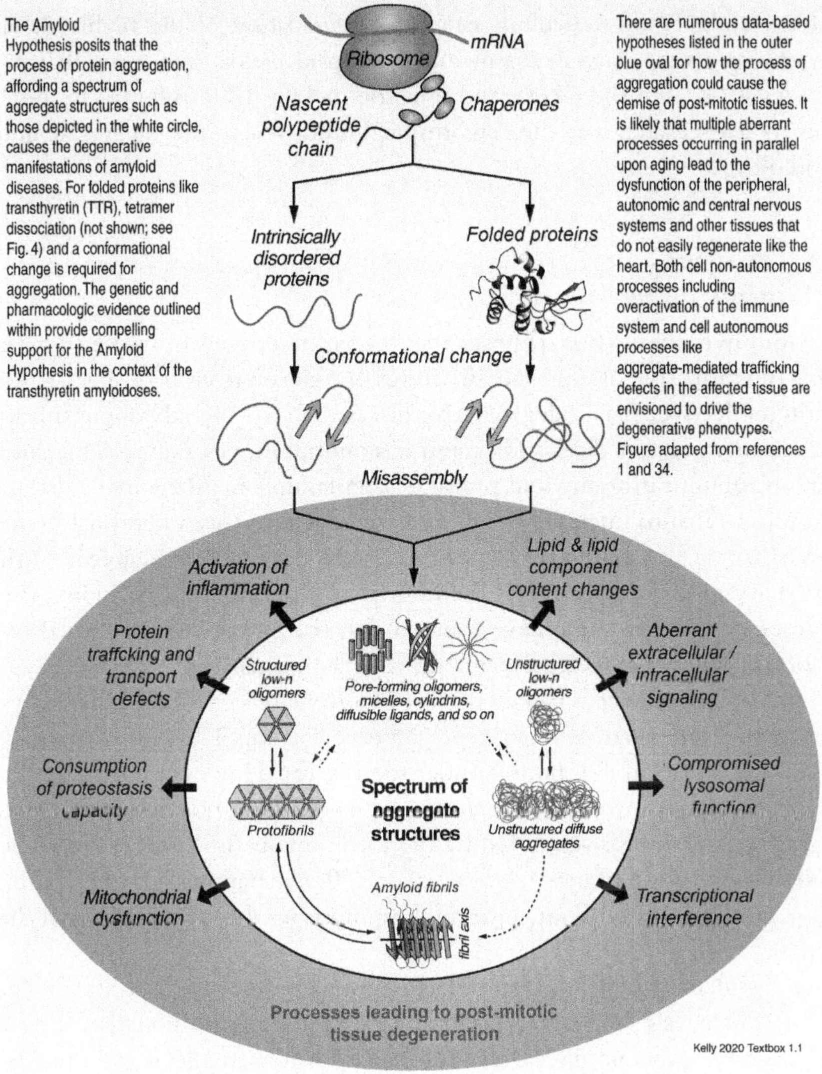

The hereditary CNS degenerative disease tissue tropism is largely the result of the ability of the brain, but not the peripheral tissues, to secrete highly destabilized TTR. In other words, highly destabilized TTR mutants are not secreted from the liver, as they are detected as being aberrantly folded in the endoplasmic reticulum of hepatocytes and are directed to degradation by the proteasome or the lysosome by the secretory pathway proteostasis network.[21, 35, 36] Thus, transthyretin cardiomyopathy or peripheral polyneuropathy is not possible with the highly destabilized TTR variants because endoplasmic reticulum quality control in the liver mostly prevents their secretion into the blood.[21] In contrast, endoplasmic reticulum-associated degradation of the highly destabilized transthyretin variants is largely absent in the choroid plexus (the tissue that produces and secretes TTR into the CNS), and thus the mutant TTR subunits in the context of tetramers get secreted into the cerebrospinal fluid, where they aggregate and cause CNS pathology.[21]

How does transthyretin aggregation compromise the functions of postmitotic tissue?

The amyloid hypothesis (Box 1) posits that the entire process of protein aggregation, including the formation of aggregate structures in addition to cross-β-sheet-based amyloid fibrils, collectively cause degenerative diseases.[34, 37–39] Doubt about this hypothesis has arisen in the medical and scientific communities because of the numerous aggregation inhibition or amyloid removal drug failures in Alzheimer's disease,[40, 41] and because it remains unclear how protein aggregation causes neuronal dysfunction and loss (Box 1).[1, 34] In other words, we still lack a clear aggregate structure-proteotoxicity relationship in all of the human amyloid diseases, including the TTR amyloidoses.[1, 34] While some have claimed that the tissue damage caused by TTR aggregation appears to stem largely from the toxicity of small diffusible TTR aggregates (Box 1),[42–44] this evidence is based on acute cytotoxicity experiments that may not be relevant to the degenerative phenotypes observed in the TTR amyloidoses, where aging is clearly a critical, but poorly understood factor. However, the hypothesis that small diffusible non-native TTR structures play a significant role in neurodegeneration and organ deterioration is supported by the cardiomyopathy clinical data, where the cessation of aggregation of newly synthesized TTR by tafamidis treatment affords significant clinical benefit without apparently diminishing the amyloid load of the heart (vide infra).

There is evidence that TTR aggregation consumes proteostasis capacity,[45, 46] and that TTR amyloidogenesis also leads to the chronic activation of inflammation (Box 1).[47] In addition, in other amyloid diseases, compromised protein trafficking,[48] mitochondrial dysfunctions,[49, 50] as well as aberrant signaling,[51] are also observed. The genetic evidence that compromised lysosomal function[52] also contributes to neurodegeneration is strong.[53] Moreover, increasing evidence indicates that lipid and lipid component content changes contribute to neurodegeneration, although little is known about etiology.[54–56] We still have a lot to learn to answer the important question of how the aggregation of TTR leads to compromised function of postmitotic tissue and ultimately the loss of certain postmitotic tissues (Box 1).[1, 34]

The role of TTR amyloid fibrils in tissue degeneration is poorly understood

There are now four drugs that ameliorate hATTR-PN (*vide infra*); however, in patients that respond well to treatment, the amyloid load as a function of the treatment period has not been systematically quantified in any of the clinical trials or clinical studies carried out to date.[57–60] Amyloid clearance does not seem to correlate with pharmacologic clinical response in hATTR-PN, but this needs to be rigorously studied. While there is some data on the substantial decrease in circulating non-native TTR levels upon tafamidis treatment using peptide probes that incorporate into circulating aggregates composed of non-native TTR, we do not understand the structures of these soluble TTR aggregates or the details of how the peptide probes bind them or the clinical importance of circulating TTR aggregates at this time.[2, 34] Nor do we understand whether amyloid clearance from the heart is a prerequisite for the response to tafamidis therapy in TTR cardiomyopathy patients, however this does not seem to be the case.[61, 62]

Transthyretin structural and cell biology

Transthyretin is a tetrameric, 127-residue β-sheet-rich protein that circulates in the bloodstream at a concentration of $\approx 3–5\,\mu M$ (Fig. 1A).[64–67] Most individuals have two WT alleles, thus they have WT TTR homotetramers circulating in their blood, their cerebrospinal fluid (CSF), and their vitreous fluid of the eye. The hATTR amyloidoses are autosomal dominant diseases, and almost all hATTR amyloidosis patients are heterozygotes. Heterozygotes have four TTR heterotetramers comprising mutant and WT subunits, as well as two homotetramers, made up of all WT subunits or all mutant subunits in their circulation (Fig. 1B).[68] The liver secretes 95% of TTR into the bloodstream,[69] whereas the choroid plexus and the retinal pigment epithelial cells secrete most of TTR into the CSF bathing the brain and the vitreous fluid within the eye, respectively.[70–72]

For the majority of the TTR mutations associated with the hATTR amyloidoses, the endoplasmic reticulum quality control system does not detect the folded mutant protein as being aberrant (beyond the 30%–40% of TTR that typically misfolds and is degraded intracellularly) and thus the mutant TTR is secreted into the blood at WT-like levels, largely as heterotetramers.[21] The established function of TTR in the blood is to transport a sub-stoichiometric quantity of *holo*-retinol binding protein and a very small quantity of thyroid hormone (T_4; $0.1\,\mu M$ concentration in blood).[73–76] Greater than 99% of the T_4 binding sites in TTR are unoccupied in the blood of humans because albumin ($620\,\mu M$; $K_{d,T4} = 1.5\,\mu M$) and thyroid-binding globulin ($0.3\,\mu M$; $K_{d,T4} = 0.1\,pM$) carry the vast majority of thyroid hormone.[77, 78] In our kinetic stabilization strategy, we take advantage of the unoccupied TTR T_4 binding sites in blood, to occupy them with a kinetic stabilizer to prevent newly synthesized TTR from aggregating in humans, which results in the slowing or stopping of TTR amyloid disease progression in most patients,

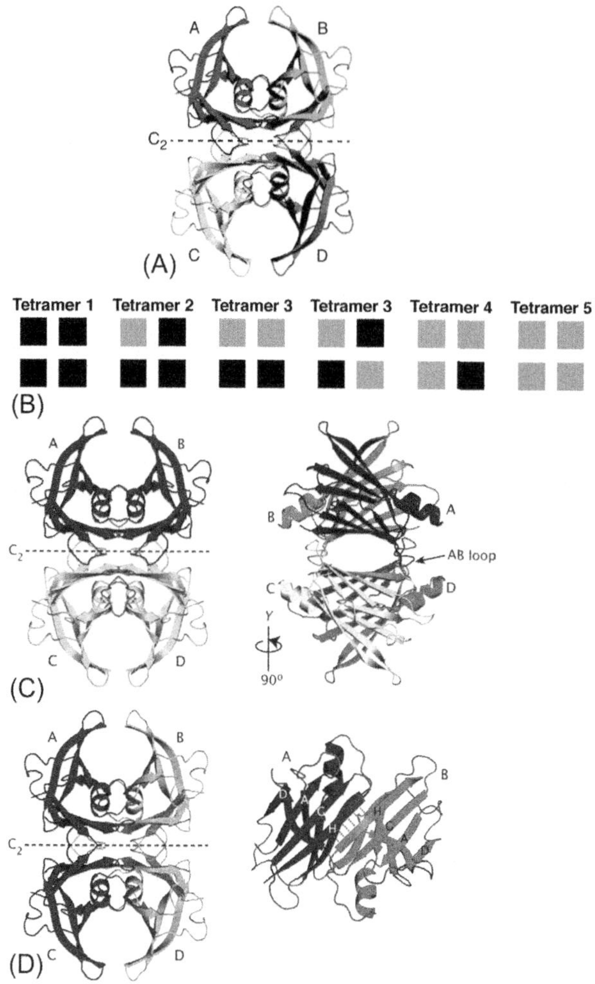

Fig. 1 Structure of transthyretin that circulates in the bloodstream of humans at a concentration of ≈3–5 μM, and in the cerebrospinal fluid[20] and vitreous fluid of the eye[63] at concentrations of ≈200–400 nM. (A) Human transthyretin adopts a tetrameric quaternary structure, made up of four identically folded 127-residue β-sheet-rich subunits colored differently so that they can be easily differentiated. (B) In TTR heterozygotes, the tetramer is generally composed of WT and/or mutant subunits allocated in a statistical distribution. The ribbon diagram representation of the subunits shown in (A) is simplified in (B) in that individual squares represent monomeric β-sheet-rich subunits in the tetramers shown. In heterozygous TTR amyloidosis patients exhibiting hereditary amyloidosis, one allele produces WT TTR subunits (represented by *black squares*) while the other allele produces mutant TTR subunits (represented by *gray squares*). The tetramers afforded in heterozygotes generally exhibit a statistical distribution of subunits; thus, Tetramer 1, Tetramer 2, Tetramers 3, Tetramer 4, and Tetramer 5 are present in a 1:4:6:4:1 ratio. In (C), we illustrate the less stable or so-called AB CD dimer-dimer interface of TTR. Subunits A and B depicted in ribbon diagram format are colored black, whereas the C and D subunits are colored gray. The less stable AB CD dimer-dimer interface of TTR, bisected by the crystallographic twofold axis (C_2), is stabilized by AB loop interactions (*right panel*; same image and coloring in *left panel* rotated by 90 degree). The AB CD dimer-dimer interface creates the two identical thyroxine (T_4) binding sites, which are also the sites that kinetic stabilizers bind to. In (D), we illustrate the more stable or so-called AC BD dimer-dimer interface. Subunits A and C depicted in ribbon diagram format are colored *black*, whereas the B and D subunits are colored *gray*. The AC BD dimer-dimer interface is the stronger of the two distinct dimer-dimer interfaces in TTR because this interface is stabilized by intersubunit antiparallel β-sheet hydrogen bonding and hydrophobic interactions (shown in *right panel*).

vide infra.[57, 58, 61, 79, 80] While TTR is the primary carrier of T_4 in CSF, the low concentration of T_4 in CSF means that the majority of the TTR (300 nM CSF concentration) T_4 binding sites are unoccupied in the CSF, like the blood,[70, 81] hence these sites can be engaged by kinetic stabilizers to inhibit TTR aggregation.[20]

Two different dimer-dimer interfaces: One binds small molecules

The TTR tetramer is characterized by two different dimer-dimer interfaces, the less stable of which is the interface that creates the two identical T_4 binding sites bisected by the crystallographic twofold axis of TTR (Fig. 1C; the so-called AB CD dimer-dimer interface).[64, 82–84] This weak dimer-dimer interface is largely mediated by the interaction of the AB loops (Fig. 1C; right panel). The strong dimer-dimer interface (Fig. 1D; the so-called AC BD dimer-dimer interface) is created by hydrogen-bonded intermolecular antiparallel β-sheet interactions. This hydrogen-bonded interface between the A and the B subunits is depicted in the right panel of Fig. 1D.

The two T_4 binding sites, created by the AB CD dimer-dimer interface (Fig. 2A), are each composed of a small inner subsite and a larger outer cavity (Fig. 2B) made up of positively (Lys-15) and negatively (Glu-54) charged TTR side chains positioned at the periphery of the T_4 binding sites. The Lys-15 and Glu-54 side chains complement the zwitterionic amino acid substructure of T_4 (Fig. 2A and B).[65–67, 84–86] The two T_4 binding sites probably do not have identical structures when only one is occupied compared to when both are occupied, as most ligands bind with negative cooperativity to TTR.[87, 88] There are three pairs of symmetric hydrophobic depressions, referred to as the halogen binding pockets (HBPs; Fig. 2B) that comprise the small inner subsite and the larger outer cavity of the T_4 binding site, so named because the iodine atoms of T_4 (Fig. 2A; bottom) are positioned at HBP 1,1′ and 3,3′ upon T_4 binding (Fig. 2B).[65, 66, 85, 86]

Conformational change hypothesis and the first TTR amyloid structure

Substantial evidence supporting the "conformational change hypothesis" as underpinning aggregation of proteins that are initially natively folded was published in the early 1990s.[89–96] These studies revealed that folded amyloidogenic proteins like TTR and immunoglobulin light chain must partially unfold in order to misassemble into a spectrum of aggregate structures, including amyloid fibrils.[89, 90, 93–100] Partial unfolding of their tertiary structures exposes stretches of largely uncharged hydrophobic amino acid residues in an extended conformation that can efficiently misassemble into the cross-β sheet amyloid structures directly[3, 101, 102] or the ensemble of extended conformers can misassemble into largely unstructured spherical aggregates that may ultimately undergo conformational conversion into cross-β sheet amyloid fibril structures (Box 1).[103–109]

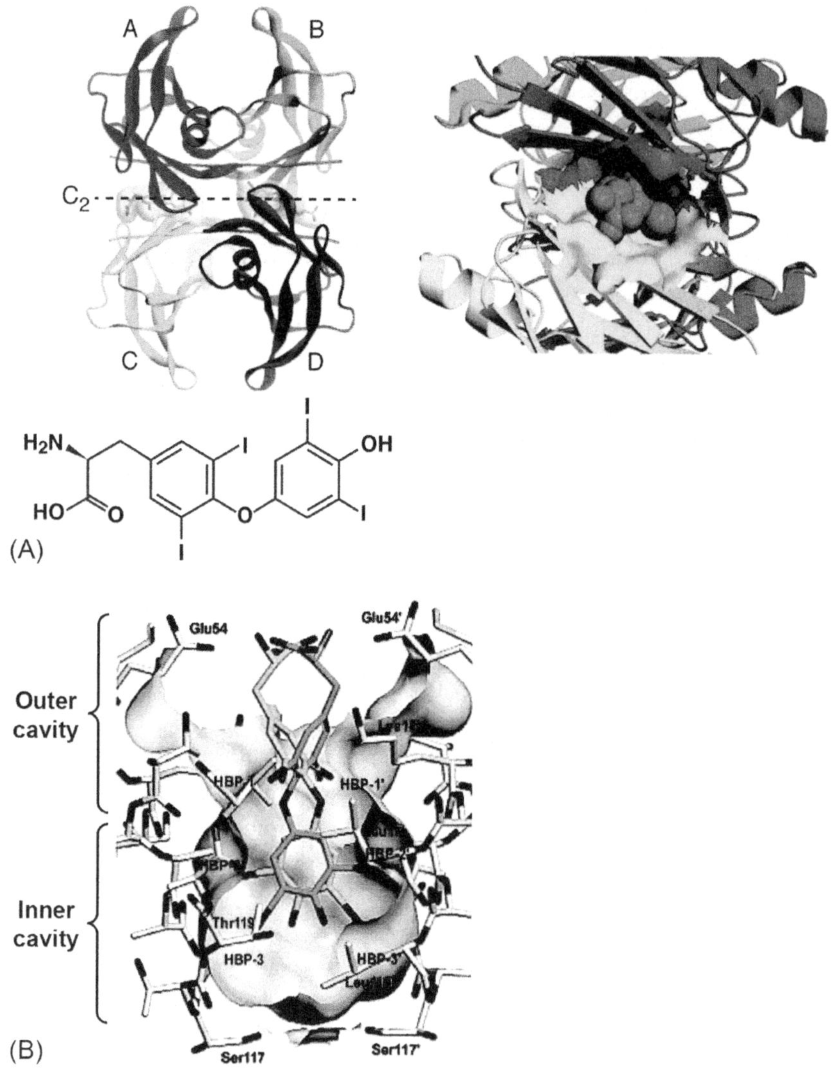

Fig. 2 X-ray crystal structure of the (thyroxine)$_2$•transthyretin complex (TTR; PDB code: 2ROX). (A) Ribbon diagram depiction of the TTR tetramer with the crystallographic twofold axis (C_2) bisecting the two occupied thyroxine (T_4) binding sites. The AB CD dimer-dimer interface of TTR, stabilized by AB loop interactions, creates the T_4 binding sites. In the right panel, a close-up view of one TTR thyroxine (T_4) binding site is shown by rotating the view in (A) by 90 degrees. The reader is looking down the C_2 axis which bisects the AB CD dimer-dimer interface. T_4 is shown as a red CPK depiction and TTR in ribbon diagram format. A line drawing of thyroxine is found in the middle of this figure. (B) Close-up view of one thyroid hormone binding site in TTR with bound T_4 in stick representation. This view shows the iodinated aromatic rings of the biaryl ether occupying symmetry-related halogen binding pockets (HBPs). Primed amino acids refer to those comprising symmetry-related TTR subunits. *(Adapted from Connelly S, Choi S, Johnson SM, Kelly JW, Wilson IA. Structure-based design of kinetic stabilizers that ameliorate the transthyretin amyloidoses. Curr Opin Struct Biol 2010;20(1):54–62).*

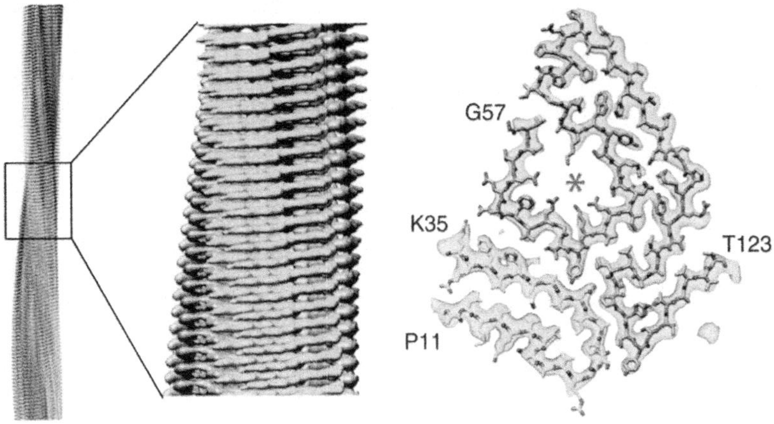

Fig. 3 Cryo-electron microscopy reconstruction of a TTR amyloid fibril, comprising fragmented TTR, extracted from the hearts of two V30M heterozygous polyneuropathy patients exhibiting cardiac involvement. Left panel is a side-view of the cross-β-sheet amyloid fibril, wherein the N-terminal TTR fragment is shown in cyan and the C-terminal TTR fragment is depicted in orange. The cross-sectional view of the two fragments of TTR comprising one layer of the cross-β-sheet amyloid fibril is shown in the right panel, with the same coloring scheme as in the left panel. *(Adapted from Schmidt M, Wiese S, Adak V, et al. Cryo-EM structure of a transthyretin-derived amyloid fibril from a patient with hereditary ATTR amyloidosis. Nat Commun 2019;10).*

The structure of a cross-β sheet amyloid fibril extracted from the hearts of late-onset V30M hATTR-PN and CM patients has recently been solved by cryo-electron microscopy.[3] In the cross-β-sheet structure depicted in Fig. 3, there is an N-terminal TTR fragment and a C-terminal TTR fragment in each layer of the cross-β-sheet amyloid fibril structure. It is abundantly clear that the quaternary and tertiary structure of TTR changes dramatically in going from a native tetramer to a cross-β-sheet amyloid structure (cf. Figs. 1–3), further demonstrating that monomer misfolding is critical for TTR aggregation.[3] Schmidt et al. argue, reasonably, that at least in these late-onset V30M patients, endoproteolysis likely occurs after TTR aggregation.[3] This is in contrast to proposals that tetramer dissociation, monomer misfolding, and aberrant endoproteolysis or aberrant endoproteolysis of the TTR tetramer enable ATTR-CM.[110–112] It may turn out that there are different TTR amyloid structures for early and late-onset hATTR-PN, and/or that there are additional aggregate structures or strains that contribute to tissue tropism and other facets of TTR amyloid disease etiology. These hypotheses require further scrutiny.

TTR tetramer dissociation is rate limiting for TTR aggregation

Tetramer dissociation is the rate-limiting step for the process of TTR aggregation (Fig. 4A).[90, 93, 113–119] The resulting monomers also have to misfold in order for TTR misassembly to occur, in accord with the conformational change hypothesis of

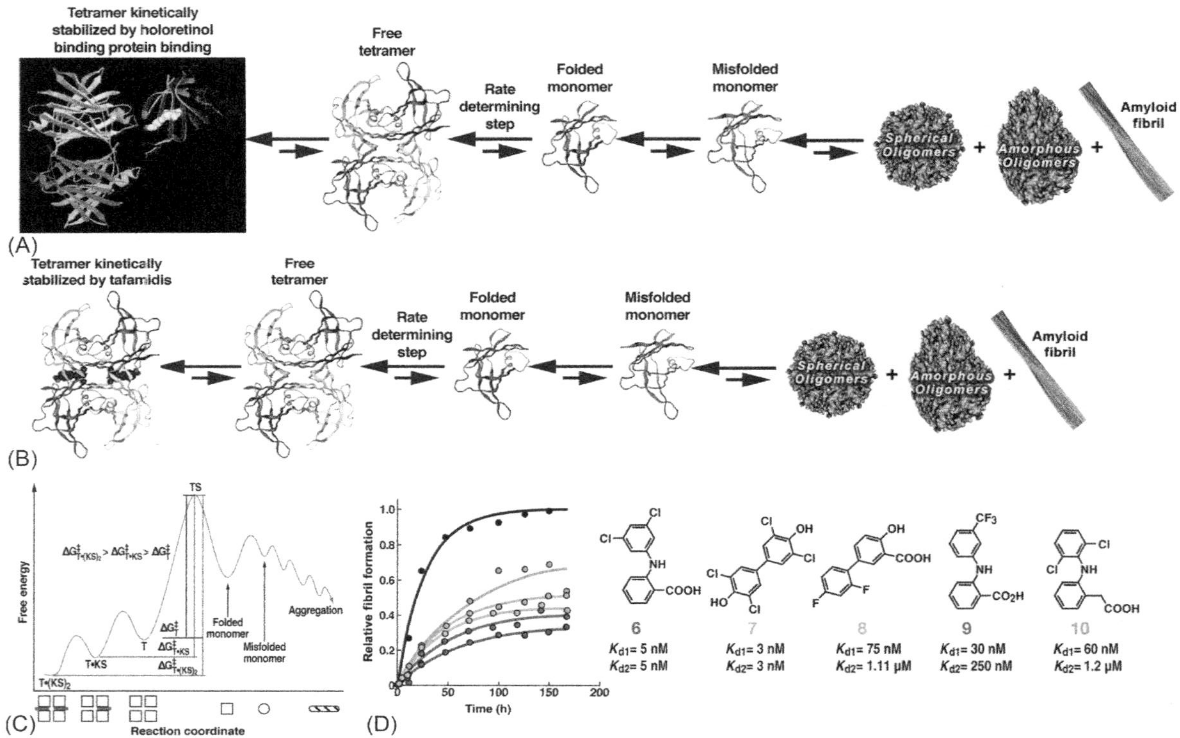

Fig. 4 Transthyretin dissociation, misfolding, and aggregation pathway. (A) Top-left figure shows that holo-retinol binding protein, owing to the approximate 0.5:1 holo-retinol binding protein to TTR stoichiometry, can bind to and stabilize about half of the TTR tetramer in blood, rendering that population of the TTR tetramer protected from aggregation. (B) The remaining ≈50% of the TTR tetramer can dissociate to the TTR monomer in a process that is rate limiting for TTR aggregation. Once the monomer misfolds it can aggregate into a spectrum of aggregate structures, including amyloid fibrils. Collectively, the aggregate structures appear to cause neurodegeneration by some of the mechanisms outlined in Box 1. Kinetic stabilizer binding to TTR, shown on the far left, kinetically stabilizes the TTR tetramer by substantially slowing tetramer dissociation. This pharmacologic strategy prevents monomeric TTR from being produced in a sufficient concentration to aggregate. (C) Free energy reaction coordinate diagram indicating that the TTR tetramer (T) dissociation to the folded monomer has the largest free energy of activation and therefore is the rate-limiting step for TTR aggregation. Kinetic stabilizers (KS) that selectively bind to the tetramer over the dissociative transition state increase the free energy of activation for tetramer dissociation, inhibiting aggregation proportional to the number of kinetic stabilizers bound. (D) Influence of kinetic stabilizers **6–10** (at a concentration of 3.6 µM) on the rate of acid-mediated (pH 4.4) WT TTR (3.6 µM tetramer) aggregation. *(Adapted from Hammarstrom P, Wiseman RL, Powers ET, Kelly JW. Prevention of transthyretin amyloid disease by changing protein misfolding energetics. Science 2003;299(5607):713–6).*

amyloidogenesis.[90, 93–95, 120] In humans with hATTR, a point mutation in one of the two TTR alleles leads to destabilized heterotetramers (Fig. 5A).[21, 113, 121, 122] Specifically, the four mutant TTR/WT TTR heterotetramers and the mutant homotetramer circulating are destabilized (Fig. 5A), allowing those tetramers to dissociate faster and to a greater extent than the wild-type TTR homotetramer. Therefore, the aggregates comprise mostly mutant TTR. Thus, the hATTR amyloidoses typically present earlier and can be more severe because the process of TTR aggregation that drives postmitotic tissue dysfunction and degeneration is more facile.

How TTR aggregates in a human is unknown

It is not known what aging-associated change enhances TTR tetramer dissociation and monomer misfolding enough to make TTR aggregation detectable in a human, especially in the case of wtATTR-CM. From biophysical experiments, we have learned that TTR aggregates efficiently under acidic conditions.[90, 93, 123, 124] We hypothesize that TTR heterotetramer dissociation and monomer misfolding could occur in the acidic vesicles that transport TTR to the cell surface during secretion, leading to aggregation.[46, 90, 93] Alternatively, TTR aggregation may occur during attempted degradation in hepatocytes, either in acidified endocytic vesicles and/or in the lysosome, and since the lysosomal function can deteriorate with aging, the hypothesis that failed lysosomal TTR degradation is the source of acid-mediated TTR aggregation upon aging merits further investigation.[90, 93, 117, 118, 123–130]

The first therapeutic strategy to treat hereditary TTR polyneuropathy

The first strategy to ameliorate hATTR-PN associated with mutant and wild-type TTR co-aggregation (Fig. 5A) was liver transplantation.[131–133] In this surgical procedure, patients who are heterozygous for a disease-associated TTR mutation have their liver replaced by a liver from a donor that is homozygous for WT TTR.[131, 133] Since TTR destined for the blood is mainly synthesized by the liver, liver transplantation reduces circulating mutant TTR levels by 90%–95%,[69] reducing TTR aggregation by largely eliminating the destabilized mutant TTR/WT TTR heterotetramers (Fig. 5A). For the ~90% of patients that survive the surgery (amyloid in the vasculature makes surgery more challenging than for a typical liver transplant), the polyneuropathy progression is slowed dramatically and lifespan is extended (*vide infra*).[69, 132–135] This surgical form of gene therapy or TTR gene replacement is more effective the earlier in the disease course the surgery is performed.[133] Liver transplantation appears to be less effective for hATTR-PN patients with TTR variants other than V30M. This may be due to these non V30M patients being older and their disease having progressed further before they are transplanted (hATTR patients with rarer TTR mutations and who lack a family history typically see numerous physicians before they are properly diagnosed).[136, 137] For

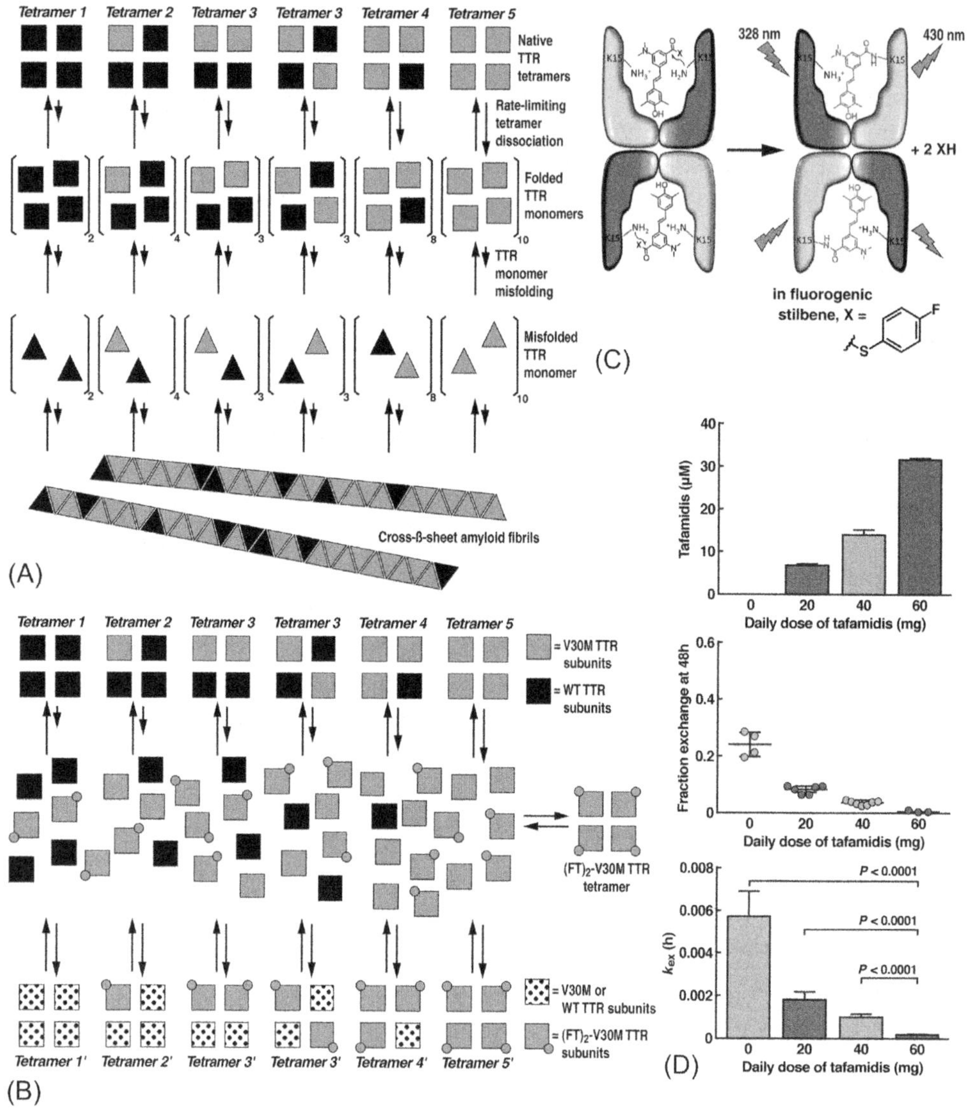

Fig. 5 Transthyretin mutant homotetramers and TTR heterotetramers comprising mutant and WT subunits are generally kinetically less stable, i.e., they dissociate faster, generating a higher concentration of folded and misfolded monomers, putting humans at greater risk for developing hereditary TTR amyloidosis. (A) In heterozygous hereditary TTR amyloidosis patients, six different tetramers are secreted into the blood. The mutant homotetramer (Tetramer 5) is generally the least kinetically stable TTR tetramer. Its subunits dissociate fastest, thus there are more mutant subunits (colored *gray*) to misfold and aggregate. Tetramer 4 comprising 3 mutant and 1 WT subunit (colored *black*) also has a fast dissociation rate, thus it will be more dissociated than Tetramers 3, Tetramer 2, and Tetramer 1, resulting in mostly mutant TTR subunits to misfold and aggregate. Tetramers 3, made up of equal quantities of mutant and WT TTR subunits, dissociate faster and thus

a subset of patients, continuation of WT TTR aggregation after about a decade posttransplantation leads to cardiomyopathy.[11, 131, 136, 138] Combined heart and liver transplantation has been employed to treat hATTR-CM patients, whereas heart transplants have been used to ameliorate wtATTR-CM.[139–141]

Tafamidis: The first pharmacologic agent to treat TTR amyloidosis

Support for the amyloid hypothesis

TTR aggregation proceeds by a so-called downhill polymerization process.[142] This means that after rate-limiting tetramer dissociation and monomer misfolding, the partially denatured TTR monomers aggregate efficiently (Fig. 4B) because the misassembled dimer is more stable than the misfolded monomer and the misassembled trimer is more

Fig. 5, cont'd have a greater concentration of monomers than dissociation from Tetramers 2 and 1. Since the mutant TTR containing tetramers dissociate and misfold to a greater degree, the resulting aggregates in hereditary TTR amyloidosis patients (only the cross-β-sheet amyloid fibrils are shown for simplicity) are composed predominantly of mutant TTR subunits. (B) Transthyretin subunit exchange experiments in plasma allow quantification of pharmacologic kinetic stabilization of TTR under physiological (non-denaturing) conditions, since tetramer dissociation is the rate-limiting step for TTR subunit exchange. The observed TTR tetramer dissociation rate is measured by the rate of subunit exchange between endogenous TTR (comprising mutant heterotetramers 2, 3, and 4 as well as WT TTR homotetramers or mutant TTR homotetramers depicted at the top of this panel) and flag-tagged kinetically unstable V30M TTR homotetramers added to plasma, shown middle, right in the panel. The observed rate constant for exchange encompasses numerous subunit exchange processes depicted diagrammatically. The six resulting tetramers that are chromatographically distinguishable are shown at the bottom of this panel. (C) The ability to detect all the TTR tetramers, comprising tagged and untagged TTR subunits, present in plasma during the TTR subunit exchange time course is challenging since there are several thousand plasma proteins. Transthyretin tetramer detection was made possible by discovering fluorogenic TTR small molecule probes that, upon binding to and reacting with the TTR tetramers, render the TTR conjugates fluorescent. This reaction allows quantification of the extent of subunit exchange as a function of time in human plasma by using ion exchange chromatography coupled to fluorescence detection. (D) The influence of orally administered tafamidis on WT TTR kinetic stability in a wtATTR-CM patient, as assessed by subunit exchange in plasma, is shown. *(Top panel)* Plasma tafamidis concentration measured in triplicate by reverse phase HPLC using fluorescence detection is shown after daily oral tafamidis doses of 20 mg versus 40 mg versus 60 mg (standard deviation reflected in the error bars after weeks of dosing). *(Middle panel)* Fraction of subunit exchange after 48 h (the approximate half-life of plasma TTR) of exchange as a function of the oral dose of tafamidis after weeks of dosing. *(Bottom panel)* The rate of tetramer dissociation (k_{ex}) was calculated from subunit exchange time courses and is expressed as the average of the rates obtained at 48 h and 96 h into the subunit exchange process. Each bar graph is an average of four or more individual technical replicates, wherein the error bars represent standard deviation. *(Adapted from Cho Y, Baranczak A, Helmke S, et al. Personalized medicine approach for optimizing the dose of tafamidis to potentially ameliorate wild-type transthyretin amyloidosis (cardiomyopathy). Amyloid 2015;22(3):175–180).*

stable than the misassembled dimer, and so on (Fig. 4B and C).[142] In other words, TTR aggregation does not require high energy multimeric nucleus formation like Aβ and Tau aggregation require—the aggregation of which is linked to Alzheimer's disease.[143] In a downhill aggregation process, the misfolded TTR monomer spontaneously aggregates as long as TTR is above the critical concentration required its aggregation.[142] The critical concentration, a thermodynamic parameter, is defined as the minimum concentration of an amyloidogenic protein required to support its aggregation.[144] Based on the clinical effectiveness of the TTR mRNA lowering agents, the effective critical concentration in humans is estimated to be around 1 μM.[59, 60] Because TTR aggregation is very efficient once the misfolded monomer state is reached (Fig. 4C), it seems unwise to try to block TTR aggregation after rate-limiting tetramer dissociation, although progress has been made on such an approach.[145]

The TTR kinetic stabilizer approach

In the kinetic stabilizer approach for slowing TTR aggregation (Fig. 4B), the idea is to fashion ligands that bind selectively and with high affinity to the native tetrameric state of TTR, stabilizing it, while not binding to the dissociative transition state linked to TTR tetramer dissociation.[66, 84, 113, 116, 146–165] This results in an increase in the free energy of activation ($\Delta G^{\ddagger}$) for tetramer dissociation, the rate-limiting step of TTR aggregation (Fig. 4C).[113, 146] In other words, small molecule binding to the properly folded TTR tetramer makes tetramer dissociation, the first and the slowest step in the TTR aggregation cascade, much slower (Fig. 4B and C).[115–118] Notably, kinetic stabilizers block the entire TTR aggregation process (Fig. 4B). We envision this to be the most conservative strategy for treating the TTR amyloidoses because we still do not know which misfolded or misfolded and aggregated TTR structures compromise the function of postmitotic tissue, ultimately leading to its loss.[42–44, 166–168] This strategy should be effective even if the misfolded monomer or small soluble TTR oligomers turn out to mediate cytotoxicity.[169–171] This pharmacologic kinetic stabilization principle was first demonstrated in vitro with thyroxine, a natural TTR ligand, and with 2,4,6-triiodophenol also known to bind to TTR, although not as strongly, each of which inhibited acid-mediated TTR aggregation.[146] Moreover, the use of kinetic stabilizer binding to TTR to prevent newly synthesized transthyretin from aggregating is also a conservative strategy in that the TTR•(kinetic stabilizer)$_2$ complex is still capable of functioning as a holo-retinol binding protein transporter (Fig. 4A),[68] a function that is compromised in the siRNA and antisense oligonucleotide-based TTR mRNA knockdown approaches.[59, 60]

From the beginning, we have sought high-affinity small molecule TTR-binding ligands that are neither thyroid agonists nor antagonists, in order to mitigate potential toxicity in treating humans with kinetic stabilizers over long treatment periods. Over a thousand novel organic small molecule TTR kinetic stabilizers exhibiting structural complementarity to the TTR T_4 binding sites were synthesized in the past 25 years.[84, 116, 125, 146–165, 172–175] The vast majority of these were envisioned using the principles of structure-based drug design; that is, we employed TTR•(small molecule kinetic stabilizer)$_2$ co-crystal structures as well as TTR stability and aggregation assays outlined below to iteratively improve the TTR kinetic stabilizers.[65, 66, 68, 116, 153, 176] Small-scale screening efforts also led to several new TTR kinetic stabilizer chemotypes, although these chemotypes were not pursued beyond their identification.[68, 149, 173, 175, 177] Small molecule TTR kinetic stabilizers are typically composed of two aromatic rings, either linked directly as a biaryl or separated by short linkers of diverse chemical structure (Fig. 6A).[66, 84, 113, 116, 146–165]

Biophysical experiments demonstrate that kinetic stabilizer (KS) binding to the first binding site in the TTR tetramer (T), affording the T•KS complex, and to the second T_4 binding site, affording the T•(KS)$_2$ complex, sequentially increase the activation barrier

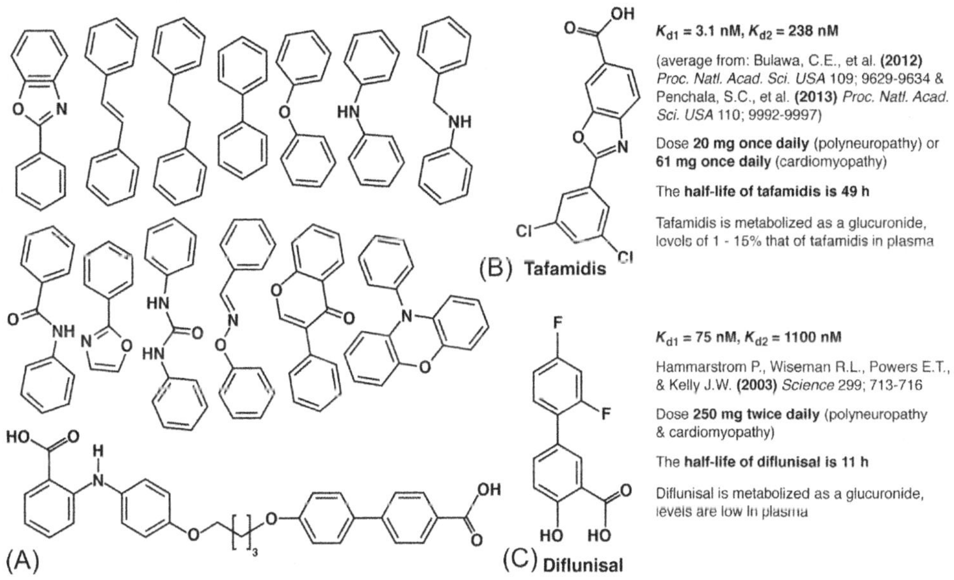

Fig. 6 The unfunctionalized core chemical structures of the kinetic stabilizer chemotypes and line drawings of tafamidis and diflunisal listing their key pharmacologic properties. (A) Shown are the unfunctionalized core chemical structures of kinetic stabilizers that exhibit structural complementarity to the thyroid hormone binding sites in TTR. Missing are the substituents that are important for making electrostatic and hydrophobic effect-based binding contributions, enabling these small molecules to bind selectively to TTR over the 4000+ additional proteins in patient plasma. (B) Shown is the protonated version of tafamidis now sold as a once-daily 61-mg tablet, bioequivalent to the 80 mg dose used in the cardiomyopathy registration trial. The remainder of the panel summarizes the key tafamidis pharmacology properties. (C) Shown is the protonated version of diflunisal, repurposed as a TTR kinetic stabilizer at a dose of 250 mg twice daily. The remainder of the panel summarizes the key diflunisal pharmacology properties.

associated with tetramer dissociation ($\Delta G^{\ddagger}$) (Fig. 4C), thereby kinetically stabilizing native TTR through the stabilization of the tetrameric ground state relative to the dissociative transition state ($\Delta G^{\ddagger}_{T\bullet KS_2} > \Delta G^{\ddagger}_{T\bullet KS} > \Delta G^{\ddagger}_{T}$, Fig. 4C).[113] The extent of kinetic stabilization is a function of the aggregation inhibitor's dissociation constants, as well as the plasma concentration of the kinetic stabilizer and the binding selectivity of the kinetic stabilizer for TTR over competitor proteins like albumin.[115, 178]

The assessment of candidate small molecule kinetic stabilizer potency in human plasma used during the TTR kinetic stabilizer development period was based on comparisons made under denaturing conditions (human plasma was acidified or urea was added to plasma).[125] After incubation of plasma under denaturing conditions, in the presence of a kinetic stabilizer or vehicle alone, the TTR was subjected to glutaraldehyde cross-linking. The TTR tetramer preserved by kinetic stabilizer binding is then quantified either using SDS-PAGE visualized by Western blotting or by immunoturbidity (the basis for the clinical laboratory test for measuring tetrameric TTR concentration in patient blood). The more tetramer that remains, the better the stabilizer performs under denaturing conditions in human plasma. Unfolded monomeric TTR is highly modified by glutaraldehyde treatment, which prevents it from being recognized in the Western blot or by immunoturbidity. Since these denaturation experiments reduce the binding affinity of the kinetic stabilizers to tetrameric TTR, and since the extent of reduction may be kinetic stabilizer structure dependent, this approach should not be used as the only means of rank-ordering compounds.

We also used inhibition of acid-mediated TTR aggregation as a function of kinetic stabilizer concentration to help rank-order our candidate kinetic stabilizers, since tetramer dissociation is rate limiting for TTR aggregation (Fig. 4B). The rate of acid-mediated (pH 4.4) WT TTR (3.6 μM tetramer) aggregation decreases at a kinetic stabilizer concentration of 3.6 μM, roughly proportional to the dissociation constants (determined at pH 7.5) of the kinetic stabilizers (Fig. 4D). The best kinetic stabilizers completely inhibit TTR aggregation at concentrations twice (7.2 μM) that of TTR.[66, 84, 93, 113, 116, 146–165]

The degree of selective kinetic stabilizer binding to TTR in human plasma was also employed for candidate kinetic stabilizer prioritization.[179] We used an immunoprecipitation approach that isolates TTR and bound kinetic stabilizer from plasma and quantifies the amount of small molecule kinetic stabilizers bound to the TTR by HPLC analysis. Wash-associated losses lead to underestimation of the true TTR binding stoichiometry.[179] This assay demonstrated that only a small subset of the kinetic stabilizers that saturate the TTR binding sites in vitro do so in human plasma. The advantage of this method is that it does not require determining or guessing which other plasma protein may be a competitor for TTR kinetic stabilizer binding.

Subunit exchange experiments in human plasma only became available at the later stages of the development of tafamidis.[115, 178] Subunit exchange experiments allow one to quantify small molecule-based kinetic stabilization of TTR under physiological (non-denaturing) conditions since tetramer dissociation is the rate-limiting step for TTR subunit exchange.[118] The TTR tetramer dissociation rate is measured by the rate of exchange of subunits between endogenous TTR (WT homotetramers or WT and/or mutant homo- and heterotetramers (Fig. 5B)) and flag-tagged TTR homotetramers added to plasma (Fig. 5B; an appropriate sequence is chosen so that dissociation of the flag-tagged TTR homotetramers is not rate limiting).[115, 117, 118, 178] The challenge in making the subunit exchange assay possible in human plasma is detecting all the TTR tetramers present, comprising tagged and untagged subunits, resulting from subunit exchange, among the 4000+ other plasma proteins that are present. Transthyretin tetramer detection was made possible by making fluorogenic TTR small molecule probes—small molecules that on binding to and reacting with the TTR tetramers render the TTR conjugates fluorescent (Fig. 5C). Detection of all TTR tetramers enables the quantification of the extent of subunit exchange as a function of time in human plasma (Fig. 5C).[115, 162, 174, 178] Subunit exchange experiments in human plasma reveal that the kinetic stabilizer dissociation constants to TTR, the plasma concentration of the kinetic stabilizer, and the kinetic stabilizer TTR binding selectivity influence the TTR tetramer dissociation rate, i.e., the kinetic stability of the TTR tetramers.[20, 115, 118, 178] This methodology demonstrates that the majority of the best TTR kinetic stabilizers at plasma concentrations between 10 and 30 μM increase the kinetic barrier of TTR dissociation such that it becomes insurmountable on a biologically relevant time scale (Fig. 5D).[115, 178]

The commercialization of TTR kinetic stabilizers and emergence of tafamidis

Because of the limitations of the academic environment for drug development, we saw the value of creating a company to perform professional drug development. The formation of FoldRx Pharmaceuticals Inc. was made possible because Christopher Mirabelli of Health Care Ventures shared our vision for the kinetic stabilizer concept to ameliorate the TTR amyloidoses. Approximately 700 small molecules were licensed to FoldRx by The Scripps Research Institute, which when combined with the late Prof. Susan Lindquist's yeast-based neurodegenerative disease screening platform,[180] allowed us to raise sufficient capital to complete a phase II/III clinical trial to demonstrate the effectiveness of the TTR kinetic stabilizer strategy. We were privileged to recruit medicinal chemist Richard Labaudinière as CEO. Richard along with Paul Anderson (serving as a consultant, retired from Senior Management at Merck & Dupont-Merck) fairly quickly after the formation of FoldRx chose 2-(3,5-diclorophenyl)-1,3-benzoxazole-6-carboxylic acid or tafamidis as it was later called[116, 153, 178] as one of the prioritized kinetic stabilizers licensed from the Kelly laboratory (Fig. 6B). The choice of tafamidis as a clinical candidate was driven by its tight binding to TTR (K_{d1} ~ 2 nM and K_{d2} ~ 200 nM[116, 181]) and good binding selectivity for TTR over albumin and the other plasma proteins, its potency to inhibit acid-mediated aggregation, combined with its ability to stabilize TTR in plasma subjected to urea and acid-mediated denaturation stresses. Late in the development process, after we knew that tafamidis exhibited high oral bioavailability, a long half-life and a favorable safety profile in humans, we also demonstrated that tafamidis was potent at preventing TTR subunit exchange in human plasma. The subunit exchange experiment revealed that the dissociation constants of tafamidis, its plasma concentration upon oral dosing, and the tafamidis TTR binding selectivity combined to stop TTR subunit exchange above a dose ≥60 mg of tafamidis once daily (QD; Fig. 5D).[178] Moreover, it was shown that tafamidis (20 mg QD) slows tetramer dissociation to ≈15% of normal and, thus, slows the process of TTR aggregation in the blood of hATTR-PN patients (Fig. 5D).[2]

Tafamidis' amelioration of the TTR amyloidoses by inhibiting TTR aggregation provides support for the amyloid hypothesis

In the hATTR-PN placebo-controlled clinical trial on the efficacy of tafamidis treatment (20 mg once daily), the coprimary endpoints were the Neuropathy Impairment Score-Lower Limbs (NIS-LL) responder analysis (<2 point change is a responder)[182] and treatment-group difference in the mean change from baseline in the Norfolk quality-of-life-diabetic neuropathy total score (TQOL) in the intent-to-treat (ITT) population. There was a higher-than-anticipated liver transplantation dropout rate during the trial because the rules changed for those on the liver transplantation waiting list. Because those hATTR-PN patients approved for transplant were considered nonresponders when analyzing the results of the tafamidis hATTR-PN trial, the trial was underpowered in the ITT population ($n=125$) regarding the NIS-LL ($P=0.068$) (Fig. 7A) and the quality-of-life assessment ($P=0.116$) (Fig. 7A).[57] Analyses of the coprimary endpoints were also performed in the efficacy-evaluable (EE) population ($n=87$) comprising ITT patients who completed the trial per protocol. Scrutinizing the EE population was prespecified because it was anticipated that the majority of patients enrolled in the trial would be on the liver transplant list and that some would undergo liver transplantation (then the standard of care) during the clinical trial if an organ became available. Significantly more tafamidis-treated patients than placebo-treated hATTR-PN patients in the EE group were NIS-LL responders (60% vs 38%; $P=0.041$) (Fig. 7A). Moreover, tafamidis-treated patients had better preserved quality of life ($P=0.045$) (Fig. 7A; right panel). In a prespecified secondary endpoint analyses in the ITT population, tafamidis patients exhibited 52% less neurological deterioration at month 18 than patients who received placebo. This amounted to a difference of three NIS-LL points (2.81 for tafamidis vs 5.83 for placebo; $P=0.027$) (Fig. 7B). Tafamidis treatment also improved autonomic neuropathy and muscle wasting, as reflected by an increase (39 ± 11.5) in modified body mass index (mBMI) from baseline in contrast to a declining mBMI in hATTR-PN patients who received placebo (-33.8 ± 11.8; $P<0.0001$) (Fig. 7C).[57] A 12-month open-label extension study evaluated the longer-term efficacy of tafamidis (20 mg QD) in 86 hATTR-PN

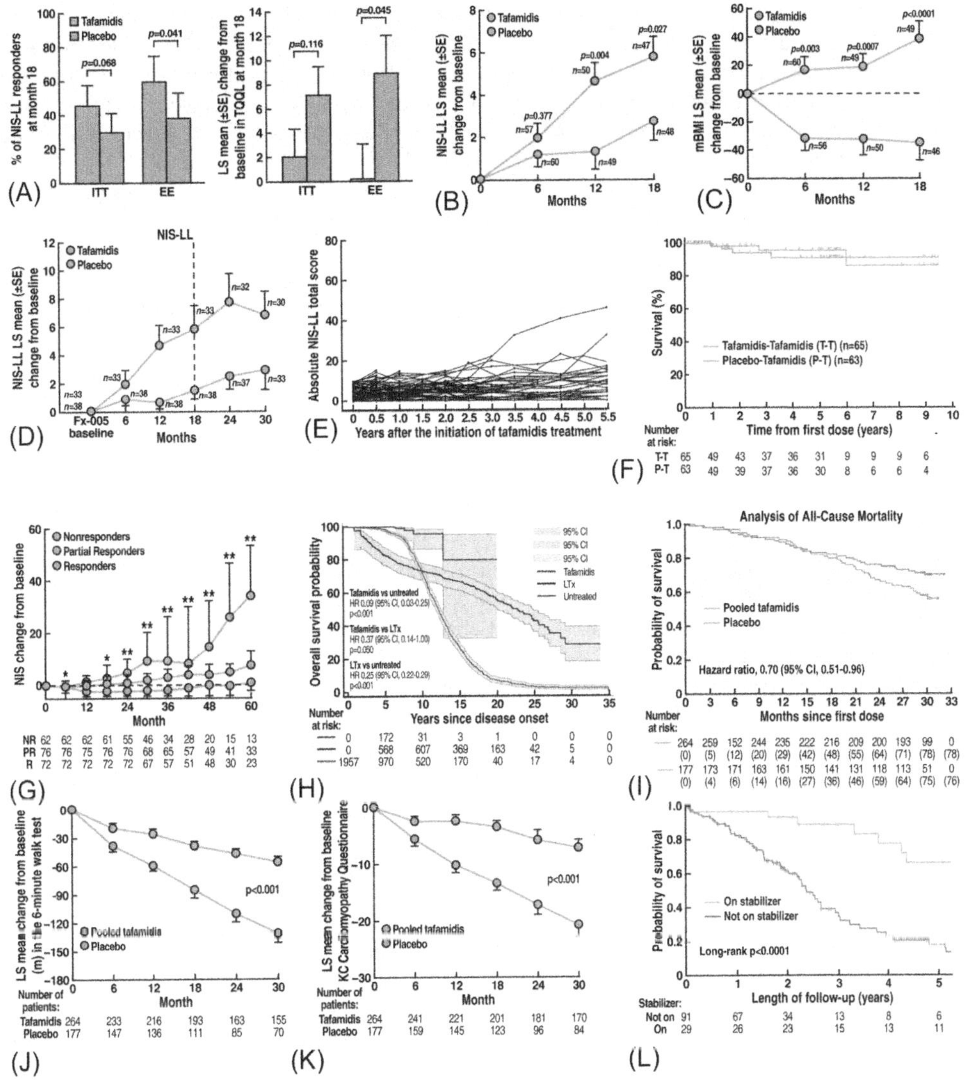

Fig. 7 Summary of tafamidis clinical trial data and data from the larger clinical studies. (A) Neuropathy Impairment Score-Lower Limbs (NIS-LL) and Norfolk quality-of-life (QOL) diabetic neuropathy questionnaire (NorfolkQOL-DN) data displayed is from the multicenter, international, 18-month duration hATTR-V30M-PN registration trial.[57] *(Left panel)* Percentage of the intent-to-treat (ITT) population and the efficacy-evaluable (EE) population (prespecified secondary analyses) in the tafamidis (20 mg QD) treatment group versus the placebo treatment group classified as NIS-LL responders after 18 months of study is displayed in bar graph format. Responders exhibit an increase of <2 points in the NIS-LL total score. *(Right panel)* Least-squares (LS) mean change from baseline in the NorfolkQOL-DN. The NorfolkQOL-DN score ranges from −2 (best QOL) to 140 (worst QOL). (B) Neuropathy Impairment Score-Lower Limbs least squares (LS) mean change from baseline ± standard error (SE) as a function of time into the 18-month hATTR-V30M-PN registration trial[57] for the tafamidis and placebo treatment cohorts. (C) Least squares (LS) mean change from baseline ± standard error (SE) in the modified body mass index (mBMI) for the tafamidis and placebo cohorts at the indicated time points during the 18-month hATTR-V30M-PN registration

(Continued)

Fig. 7, cont'd trial.[57] (D) Neuropathy Impairment Score-Lower Limbs least squares (LS) change from baseline ± standard error (SE) for the 18-month hATTR-V30M-PN registration trial[57] and for the year-long extension study in the intent-to-treat population.[80] After the 18-month registration trial, the placebo group started taking tafamidis (20 mg QD), indicated by a change in color (orange to blue) of the symbols in the placebo-tafamidis group. (E) Effects of tafamidis treatment (20 mg QD) on the cohort of patients exhibiting mild neuropathy after 5.5 years of therapy. Absolute Neuropathy Impairment Score-Lower Limbs total scores throughout tafamidis (20 mg QD) treatment over 5.5 years for each of 71 patients is shown.[183] The patients in this cohort exhibited a NIS-LL ≤ 10 before commencement of tafamidis treatment. This data come from the 18-month hATTR-V30M-PN registration trial,[57] the 12-month open-label extension study,[80] and the ongoing 10-year extension study. Notably, the data evaluated 5.5 years into this study reveals a sustained delay in neurologic progression and long-term preservation of nutritional status (data not shown) reflecting preservation of autonomic nervous system function.[183] (F) Effects of tafamidis (20 mg QD) on mortality of patients in the 18-month hATTR-V30M-PN registration trial,[57] the 12-month open-label extension study,[80] and the ongoing 10-year extension study evaluated 8.5 years into the study.[184] Kaplan-Meier curves of hATTR-V30M-PN patient survival while on tafamidis (20 mg QD) suggest that tafamidis will extend lifespan compared to panel H for perspective. (G) A distinct retrospective cohort of 210 patients with hATTR-V30M-PN (previously unstudied) treated with commercial tafamidis (20 mg QD) was studied wherein they were classified as responders (34%; $n = 72$), partial responders (36%; $n = 76$), or nonresponders (30%; $n = 62$), based on expert opinion and selected outcome measures.[79] One of the selected outcome measures depicted is NIS change from baseline. Notably, the complete NIS Score (upper and lower limbs) was used in this study because there were more advanced hATTR-V30M-PN patients in this study.[79] (H) Multi-institutional, hospital-based study of hATTR-V30M-PN patients followed until December 2016, grouped into untreated ($n = 1771$), liver transplant (LTx)-treated ($n = 975$), or tafamidis (20 mg QD)-treated ($n = 432$) cohorts.[135] Median overall survival of the untreated (11.61 years) and LTx-treated cohorts (20.9 years following LTx) demonstrate the effectiveness of LTx as a treatment strategy. Notably, median overall survival was not reached in the maximum follow-up of 10 years in the tafamidis-treated group. Kaplan-Meier survival estimates from disease onset depicted with 95% confidence interval shown (by shading) in hATTR-V30M-PN patient cohorts, allowing comparison of treatment effectiveness among these early stage patients.[135] Even though the confidence intervals after more than a decade overlap, it is clear that tafamidis treatment (20 mg QD) is superior to LTx as a strategy to treat these hATTR-V30M-PN patients.[135] (I) In a multicenter, international, 30-month duration double-blind placebo-controlled phase III registration trial, 441 wtATTR-CM and h-ATTR-CM patients were randomized at a 2:1:2 ratio to receive tafamidis (80 mg QD), tafamidis (20 mg QD), or placebo.[61] In the primary analysis, hierarchical all-cause mortality, followed by the frequency of cardiovascular-related hospitalizations according to the Finkelstein-Schoenfeld method was assessed, demonstrating that tafamidis treatment (20+80 mg QD data) was superior to placebo ($P < 0.001$).[61] Kaplan-Meier survival curves showed that tafamidis treatment resulted in a reduction in all-cause mortality after 18 months. Tafamidis-treated New York Heart Association (NYHA) class I subjects in the study exhibited a 64.4% reduction in death in contrast to the later stage class II subjects who exhibited a 39.6% reduction in death—revealing the advantage of early treatment.[61] (J) A key secondary endpoint in the wtATTR-CM and h-ATTR-CM phase III registration trial includes the change from baseline to month 30 in the distance walked during the 6-min walk test. Pooled tafamidis treatment reduced the decline in distance walked over the 6-min period in comparison to placebo ($P < 0.001$).[61] (K) In the 441-patient wtATTR-CM and h-ATTR-CM phase III registration trial, pooled tafamidis treatment also reduced the decline in the Kansas City Cardiomyopathy Questionnaire-Overall Summary score ($P < 0.001$).[61] (L) In a single center 120-patient retrospective study of hATTR-CM and wtATTR-CM patients, tafamidis (20 mg QD; $n = 16$) or diflunisal (250 mg BID, $n = 13$) administration was compared to an untreated demographically similar group ($n = 91$) by Kaplan-Meier analysis, revealing a statistically significant advantage of being treated with a kinetic stabilizer ($P < 0.0001$).[62] In concert with the registration hATTR-CM and wtATTR-CM trial data depicted in panels I, J, and K, it is clear that stabilizers slow the progression of ATTR-CM.[61, 62]

patients who earlier either received tafamidis or placebo in the aforementioned clinical trial. To assess the effects of tafamidis on disease progression during the 18-month trial and in a 12-month open-label extension study wherein everyone received tafamidis (20 mg QD), the placebo-tafamidis group was compared to the tafamidis-tafamidis group. Compared with hATTR-PN patients originally randomized to the placebo arm, neurological function (NIS-LL (Fig. 7D), muscle weakness, large, and small-fiber function) in the tafamidis-tafamidis group remained stable, and notably the improvement in the mBMI was preserved.[80] This was sufficient for regulatory agency approval of tafamidis (20 mg QD) in Europe and eventually in >40 countries to ameliorate early-stage hATTR-PN, but was insufficient evidence for FDA approval of tafamidis (20 mg QD) in the United States for this indication. The tafamidis-tafamidis and placebo-tafamidis groups (n=71) are still being studied in a second ongoing longer-term study that was evaluated at the 5½ year time point.[183] Tafamidis treatment (20 mg QD) resulted in sustained delay in neurologic progression with mean changes from baseline in NIS-LL of 5.3 over 5.5 years, or less than 1 point/year (Fig. 7E).[183] The published mean rate of V30M hATTR-PN progression is ≈4 points/year,[80] so if tafamidis were ineffective, we would have anticipated a 22-point change over 5.5 years, underscoring the long-term benefits of early intervention with tafamidis in terms of slowing disease progression.[183] Preservation of nutritional status (mBMI) over the 5.5 years was also observed, as was a favorable safety profile.[183] The influence of tafamidis on mortality in V30M hATTR-PN was evaluated in the aforementioned tafamidis-tafamidis group (n=65) and the placebo-tafamidis group (n=63) now monitored for 9 years in the ongoing 10-year extension study.[184] As depicted in Fig. 7F, 85.9% of the tafamidis-tafamidis group and 91.1% of the patients in the placebo-tafamidis group are still alive. Notably, none of the treatment related deaths were attributed to the ineffectiveness of tafamidis; the vast majority were associated with liver transplantation.[184] In a non-V30M hATTR-PN group (n=21) treated with 20 mg QD of tafamidis studied for up to 8 years, 75.9% are still alive.[184] The Kelly and Coelho groups carried out a retrospective cohort study of a different group of hATTR-PN patients taking commercial tafamidis comprising 210 patients, followed for at least 18 months and as long as 66 months (20 mg QD; these patients were not enrolled in hATTR-PN clinical trial or in the follow-on studies).[79] At the end of the study, the patients were classified by an expert (Coelho) as responders (34%, exhibiting an almost complete arrest of disease progression), partial responders (36%, having a partial to complete arrest in progression of some but not all disease phenotypes), or as nonresponders (30%, progressing despite taking 20 mg QD tafamidis). Thus, over two-thirds of the patients receiving tafamidis had at least a partial response to therapy. The expert categorization of the three groups was strongly supported by the change in the NIS, calculated as the difference between the NIS score at each time point and the NIS score at baseline (Fig. 7G), and the change in weight.[79] A 1.5-year-long phase III tafamidis (20 mg QD) open-label study in Japanese hATTR-PN patients (n=10) revealed an 80% responder rate based on a NIS-LL score change of <2/year, also demonstrating maintained quality-of-life and nutritional status.[185] A multi-institutional, retrospective study of Portuguese V30M hATTR-PN patients who were untreated ($n=1771$), who were treated by liver transplantation ($n=957$), or who were treated by tafamidis administration (20 mg QD; $n=432$) in their hospitals was carried out through December 2016.[135] Kaplan-Meier and Cox regression methodology were employed to estimate survival probabilities, derive adjusted hazard ratios, and generate 95% confidence intervals (CIs) for all-cause mortality. Median overall survival of untreated patients (followed between the years 1972 and 1991) and liver transplant-treated cohorts was 11.6 versus 20.9 years, respectively (Fig. 7H).[135] Although the median overall survival was not reached in the tafamidis-treated cohort, as the maximal follow-up period was 10 years, it is noteworthy that the observed survival rate at 10 years after treatment initiation was 93% (Fig. 7H).[135] Notably, tafamidis reduced mortality risk in V30M hATTR-PN patients <50 years old by 91% (hazard ratio 0.09) when compared to untreated patients and by 63% (hazard ratio 0.37) when compared to liver transplant patients.[135] In V30M hATTR-PN patients presenting with symptoms after 50 years of age, tafamidis reduced mortality risk by 82% compared with untreated patients (hazard ratio 0.18).[135] For each year tafamidis treatment was delayed, the risk of mortality increased by 39% (hazard ratio 1.39). The effectiveness of tafamidis for the treatment of hATTR-PN was also evaluated using data from the Transthyretin Amyloidosis Outcomes Survey (THAOS) registry.[186] Patients receiving tafamidis (20 mg QD) ($n=252$) were compared with untreated subjects in a nonrandomized matched cohort analysis based on genetic mutation, birth location, and mean treatment propensity score over a 2-year period.[186] Subjects treated with tafamidis showed significantly less NIS-LL deterioration ($P<0.001$), significantly less worsening on the Norfolk quality-of-life scale ($P<0.001$), and of note, there were no deaths in the tafamidis-treated cohort.[186]

Tafamidis also significantly slows the progression of wtATTR-CM.[61] In a phase III placebo-controlled clinical trial, 441 patients were randomized in a 2:1:2 ratio to receive 80 mg of tafamidis QD, 20 mg of tafamidis QD, or placebo for 30 months. For the primary analysis, all-cause mortality was hierarchically assessed, followed by frequency of cardiovascular-related hospitalizations employing the Finkelstein-Schoenfeld methodology. All-cause mortality and rates of cardiovascular-related hospitalizations were lower among the 264 tafamidis-treated (pooled 20 and 80 mg QD data) patients than in the placebo-treated ($n=177$) TTR-CM patients ($P<0.001$) (Fig. 7I).[61] Tafamidis also met the secondary endpoints, i.e., it was associated with a lower rate of decline in distance covered in the 6-min walk test (Fig. 7J; $P<0.001$) and was associated with a lower rate of decline in the Kansas City Cardiomyopathy Questionnaire-Overall Summary (Fig. 7K)[61] ($P<0.001$). Maurer and colleagues also carried out an open-label cardiomyopathy study assessing tafamidis (20 mg QD; $n=16$) and diflunisal (250 mg BID; $n=13$). A Kaplan-Meier analysis revealed that median time to the combined outcome of death or orthotopic heart transplant was significantly worse ($P<0.0001$) if the patients were not taking a kinetic stabilizer (Fig. 7L).[62, 187]

The tafamidis results covered in Fig. 7 provide compelling pharmacologic evidence that the process of TTR aggregation causes the degenerative phenotypes in hATTR-PN, in wtATTR-CM, as well as in hATTR-CM. In other words, the results from the two registration trials, as well as the results from the numerous subsequent clinical studies strongly support the amyloid hypothesis underlying the etiology of hATTR-PN, wtATTR-CM, and hATTR-CM, the idea that the process of TTR aggregation drives postmitotic tissue dysfunction and loss. The safety of tafamidis is exemplary: in the largest trial conducted, the safety profile of tafamidis at the 80 mg dose (the highest dose tested in a registration trial) was comparable to that of the placebo.[57, 61, 188, 189]

Will tafamidis ameliorate central nervous system degeneration?

We have recently shown that tafamidis (20 mg QD) is present in cerebrospinal fluid of hATTR-PN patients at concentrations possibly sufficient to stabilize TTR (Fig. 8),[20] and therefore potentially high enough to slow CNS degeneration.[190] That tafamidis ameliorates CNS degeneration is hinted at by case reports describing diminished focal neurological episodes after tafamidis treatment in V30M polyneuropathy patients

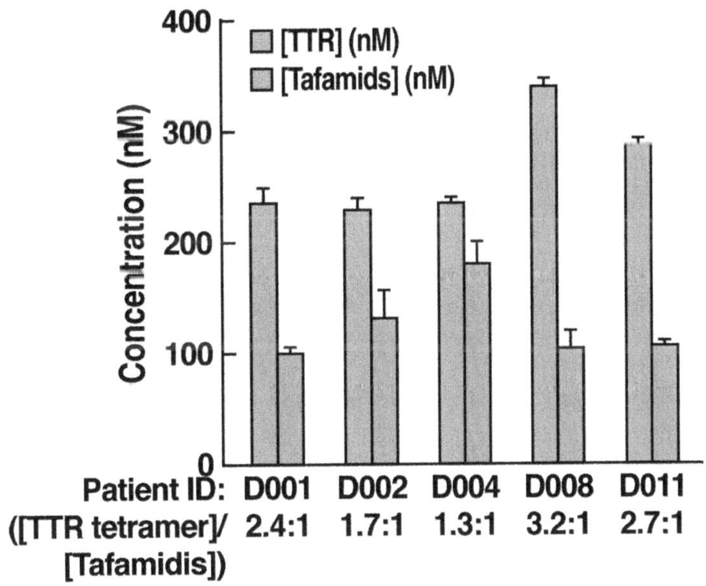

Fig. 8 Comparison of tafamidis and transthyretin tetramer concentrations in the cerebrospinal fluid (CSF) upon tafamidis oral dosing (20 mg QD) reveals that tafamidis levels in the CSF may be sufficient to ameliorate central nervous system degeneration,[20] although there is not yet clinical data to support this hypothesis. It is expected that the tafamidis CSF concentration will rise as much as fourfold upon increasing the oral dose to 80 mg or the bioequivalent 61 mg formulation (QD), however those critical measurements have not yet been made. *(Adapted from Monteiro C, da Silva AM, Ferreira N, et al. Cerebrospinal fluid and vitreous body exposure to orally administered tafamidis in hereditary ATTRV30M (p.TTRV50M) amyloidosis patients. Amyloid 2018;25(2):120–128).*

exhibiting CNS degenerative disease phenotypes.[190] Whether tafamidis protects against CNS pathology can be scrutinized in the future by comparing hATTR-PN patients who have taken tafamidis for more than a decade relative to liver transplant patients treated for the same period, and by comparison to hATTR-PN patients treated with TTR mRNA lowering agents known not to cross the blood-brain barrier. It is reasonable to expect that doses higher than 20 mg QD (80 mg QD or the bioequivalent 61 mg QD formulation approved for treating TTR cardiomyopathies) will afford higher CNS exposure, but these measurements have not yet been made.[191]

Repurposing diflunisal as a TTR kinetic stabilizer: More amyloid hypothesis support

Diflunisal (Fig. 6C), is an FDA-approved nonsteroidal anti-inflammatory drug that binds to TTR with negative cooperativity (K_{d1} = 75 nM, K_{d2} = 1100 nM).[147, 152, 192, 193] In addition to its modest binding affinity to TTR, diflunisal exhibits acceptable binding selectivity to TTR over the thousands of proteins present in human plasma. Despite its moderate binding affinity, preclinical studies conducted by us showed that diflunisal kinetically stabilizes TTR tetramers in human plasma quite effectively owing to diflunisal's excellent oral bioavailability. In other words, diflunisal achieves 100–200 μM plasma concentrations after oral dosing (250 mg BID), while exhibiting a half-life of 11 h.[192, 193] A placebo-controlled phase III clinical trial in genotypically diverse polyneuropathy patients to test the efficacy of diflunisal for the treatment of hATTR-PN demonstrated a statistically significant slowing of polyneuropathy progression (Fig. 9),[58, 194] as did a subsequent clinical study.[194] Another open-label clinical study indicated efficacy in slowing cardiomyopathy progression as well (Fig. 7L).[62] The sensitivity of some patients to the gastrointestinal, cardiac, and renal side effects from taking 0.5 g of diflunisal per day limits its applicability to those who can tolerate this dose, however many can.[58, 62, 194, 195]

Subsequent pharmacologic evidence that TTR mRNA lowering agents slow hATTR-PN

Further support for the amyloid hypothesis

Subsequent to the tafamidis and diflunisal clinical trial publications, it was shown that the TTR mRNA lowering agents, patisiran[60] and inotersen,[59] slow polyneuropathy progression in the majority of hATTR-PN patients based on double-blind placebo-controlled clinical trial results. These agents also stop the amyloid cascade at the beginning by lowering the blood TTR concentration below the TTR critical concentration.[34, 196, 197] In other words, the TTR mRNA lowering agents decrease the plasma concentration of TTR to about 1 μM or less, which stops newly synthesized TTR from aggregating. These registration trial results show that stopping the amyloid cascade by a different

Longitudinal intention-to-treat analyses of primary (NIS+7) and secondary outcomes

Outcomes	Mean (95% CI)			
	Placebo change from baseline	Diflunisal change from baseline	Difference, Placebo-Diflunisal	*P* value
NIS+7 score				
At 1 year	12.5 (8.6–16.4)	6.2 (2.8–9.6)	6.4 (1.2–11.6)	0.02
At 2 years	26.3 (20.2–32.4)	8.2 (2.9–13.6)	18.0 (9.9–26.2)	<0.001
NIS score				
At 1 year	10.1 (6.9–13.3)	4.1 (1.2–6.9)	6.0 (1.7–10.3)	0.007
At 2 years	23.2 (17.8–28.5)	6.4 (1.6–11.2)	16.8 (9.6–24.0)	<0.001
NIS LL score				
At 1 year	6.0 (3.9–8.2)	3.2 (1.3–5.2)	2.8 (−0.1–5.7)	0.06
At 2 years	12.1 (8.9–15.3	3.8 (0.9–6.6)	8.3 (4.1–12.6)	<0.001

Fig. 9 In the repurposing diflunisal phase III randomized, double-blind placebo-controlled TTR polyneuropathy clinical trial ($n=130$ hATTR-PN patients), the use of diflunisal (250 mg BID) reduced the rate of progression of neurological impairment, while preserving quality of life for those who could tolerate the NSAID activity of diflunisal.[58] In the intention-to-treat hATTR-PN population, diflunisal slowed disease progression after 2 years, based on the complete NIS score (upper and lower limbs, the NIS-lower limb score, and the NIS plus seven nerve tests (NIS+7) score, as outlined in the table.[58] Table adapted from Ref. 58. Since the repurposed kinetic stabilizer diflunisal (250 mg BID) and the fit-for-purpose kinetic stabilizer tafamidis (20 mg QD) both slowed the progression of hATTR-PN, collectively these data strongly support the amyloid hypothesis or the notion that transthyretin aggregation drives the degenerative phenotypes in transthyretin polyneuropathy, likely by one or more of the mechanisms summarized in Box 1.[34]

mechanism of action stops disease progression in hATTR-PN, further supporting the amyloid hypothesis.[59, 60]

Genetic evidence supporting the amyloid hypothesis in hATTR-PN

Just before the Kelly laboratory began developing small molecule kinetic stabilizers to slow or prevent TTR tetramer dissociation (Fig. 4) and TTR aggregation, Coelho and coworkers reported Portuguese families that appeared to exhibit either very mild hATTR-PN or no discernable pathology, despite carrying the V30M mutation.[198, 199] After sequencing the DNA of these individuals, it was revealed that they were compound heterozygotes who express the V30M mutation associated with highly penetrant hATTR-PN on one allele, and they express the T119M TTR variant from their second allele instead of WT TTR. This results in the formation of TTR heterotetramers exhibiting a statistical distribution of V30M and T119M subunits.[114, 117]

Interallelic trans-suppression by T119M TTR is achieved by kinetic stabilization

Biophysical experiments performed in the Kelly laboratory revealed that T119M subunit inclusion into tetramers otherwise composed of V30M subunits proportionately reduces the acid-mediated aggregation rate, as well as the rate of tetramer dissociation at neutral pH in urea (Fig. 10A).[114, 125] The dissociation rate of the T119M TTR homotetramer is ~25-fold slower than the dissociation rate of the WT TTR homotetramer based on urea-mediated dissociation kinetics, revealing that the T119M TTR homotetramer has a much higher dissociation barrier relative to that exhibited by the WT TTR homotetramer. Subunit exchange experiments under physiological conditions reveal that the T119M tetramer dissociation barrier is effectively insurmountable (Fig. 10B). Wild-type and T119M TTR homotetramers exhibit similar thermodynamic stabilities but differ considerably in their dissociation kinetics (Fig. 10C).[113, 114, 125] T119M TTR subunit inclusion into a tetramer otherwise composed of disease-associated subunits (V30M) raises the kinetic barrier of tetramer dissociation proportionately to the number of T119M subunits included in the tetramer (Fig. 10D). Mechanistically, inclusion of T119M subunits into a TTR tetramer destabilizes the dissociative transition state (Fig. 10C), protecting these individuals from TTR aggregation and a postmitotic tissue degenerative disease by a process referred to as interallelic trans-suppression-mediated TTR kinetic stabilization.[114, 125, 198, 199]

Timeline

One of us (JWK) commenced TTR amyloidosis research efforts at Texas A&M University in the Fall of 1989 (Fig. 11), primarily focused on understanding the requirements for TTR aggregation. At this time in history, TTR aggregation was thought to be mediated by aberrant endoproteolysis of the transthyretin protein affording peptides that aggregated. In 1989, we hypothesized that TTR dissociation and conformational changes within the monomer alone would enable TTR aggregation. The initial reaction of my Biochemistry Department colleagues at Texas A&M was that studying protein aggregation was an intractable quagmire that could "end my career" and NIH study section feedback on our first grant proposal on this topic was harsh. Nonetheless we persevered, and my colleagues at Texas A&M in both the Chemistry and Biochemistry Departments were very supportive. Moreover, I was fortunate enough to recruit Wilfredo Colon into my group to commence this project. He made enough progress to convince an NIH study section that this problem was tractable. It took at least 50 trainees, first at Texas A&M and after 1997 at the Scripps Research Institute, working for more than three decades to come up with enough biophysical information on the mechanism of TTR aggregation and sufficient structure-activity information on small molecule TTR kinetic

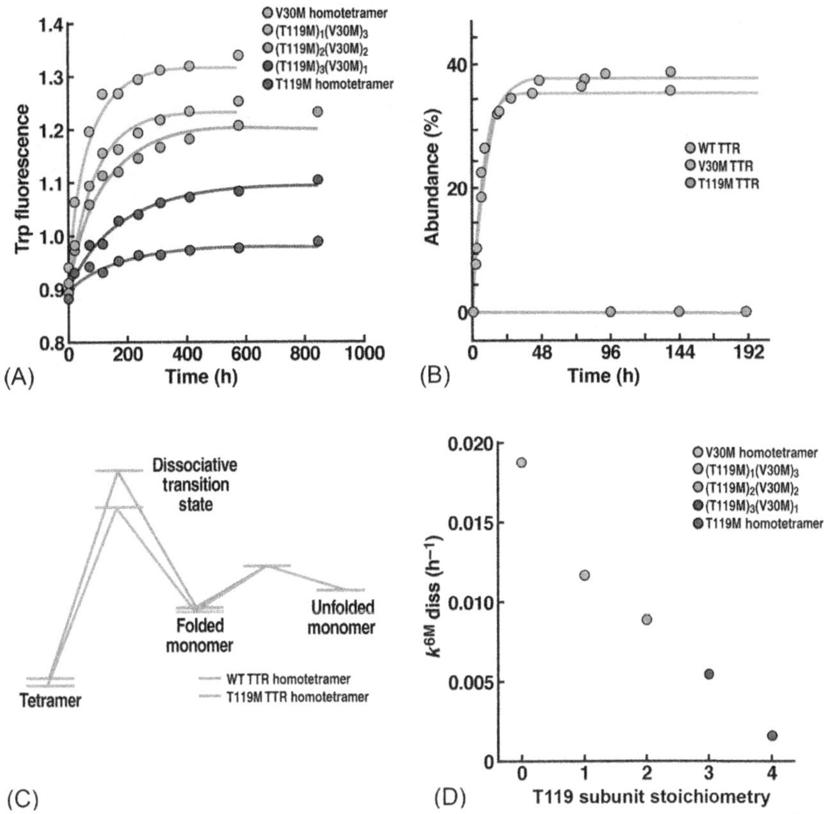

Fig. 10 Translation of the human genetic evidence discovered by the Coelho group[198, 199] into biochemical evidence by the Kelly group that collectively support the amyloid hypothesis as the basis for the postmitotic tissue degeneration in the transthyretin amyloidoses. (A) Urea-mediated tetramer dissociation time courses of the T119M TTR homotetramer, the V30M TTR homotetramer, and the heterotetramers comprising V30M and T119M subunits produced by the coexpression of the two different subunits, the stoichiometry being indicated in the panel. These data show that T119M TTR subunit incorporation into TTR tetramers otherwise composed of misfolding-prone, disease-associated TTR subunits, dramatically slows heterotetramer dissociation in the compound heterozygotes, inhibiting TTR aggregation.[113, 114] (B) Tetramer dissociation is rate limiting for subunit exchange, thus subunit self-exchange was employed to assess the kinetic stability of WT homotetramers *(blue line)*, V30M TTR homotetramers *(orange line)*, and T119M homotetramers *(green line)*. Homotetramer subunit self-exchange time courses at 4°C, reveal that the T119M homotetramer dissociates imperceptibly and is thus kinetically very stable relative to self-exchange exhibited by V30M and WT TTR homotetramers. (C) Free energy diagram illustrating the increase in activation energy required for the dissociation of the T119M homotetramer *(in orange)* relative to a V30M homotetramer *(in blue)*. (D) T119M subunit inclusion into tetramers otherwise composed of hereditary disease associated misfolding-prone V30M subunits proportionately slows dissociation, aggregation, and polyneuropathy. Dependence of the tetramer dissociation rate, k (6 M urea; data from panel A), on T119M suppressor subunit stoichiometry relative to V30M subunits is clearly demonstrated (color scheme as in panel A). The interallelic trans-suppression mechanism of hATTR-PN disease prevention or disease progression slowing discovered by Coelho et al. provide strong support for the amyloid hypothesis (Box 1), the notion that TTR aggregation causes TTR amyloidosis. T119M subunit integration into tetramers kinetically stabilizes the heterotetramers, slowing heterotetramer dissociation, misfolding and aggregation, explaining how disease onset is inhibited or progression is slowed. *(Adapted from Hammarstrom P, Schneider F, Kelly JW. Trans-suppression of misfolding in an amyloid disease. Science 2001;293(5539):2459–62 and Hammarstrom P, Wiseman RL, Powers ET, Kelly JW. Prevention of transthyretin amyloid disease by changing protein misfolding energetics. Science 2003;299(5607):713–6).*

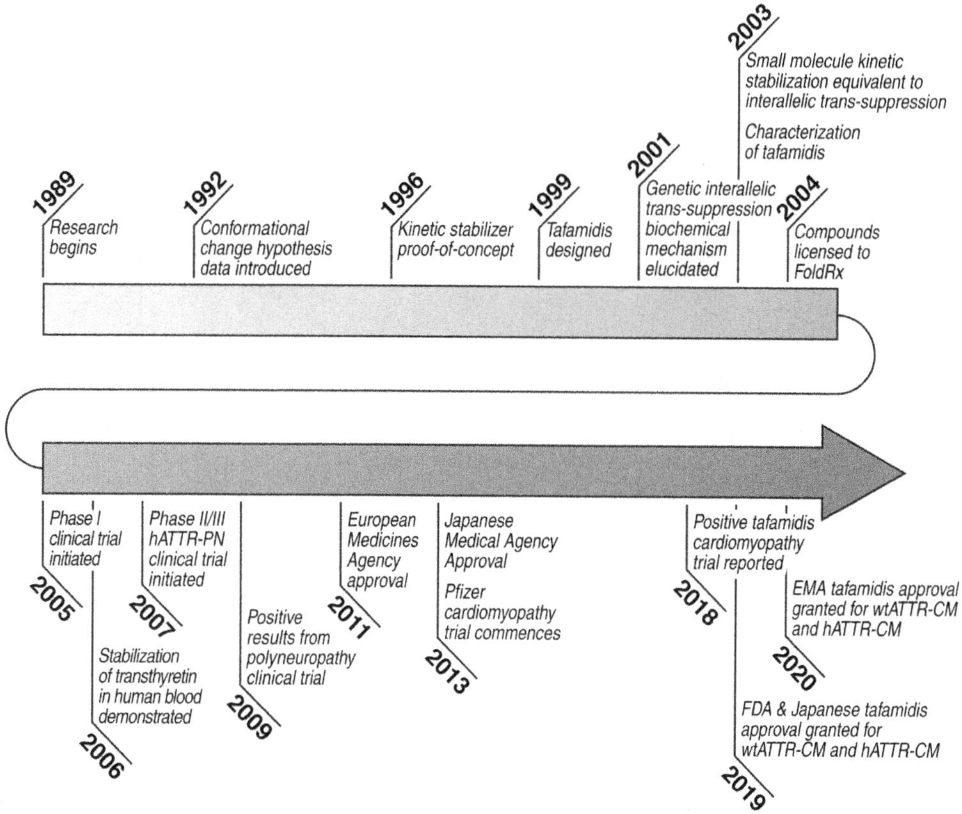

Fig. 11 The 30+ year academic, biotechnology company and pharmaceutical industry journey leading to the approval of the first-in-class drug tafamidis for the amelioration of the TTR amyloidoses is summarized by the timeline shown.

stabilizers/aggregation inhibitors to ultimately facilitate clinical trials with tafamidis. Tafamidis was discovered primarily by the efforts of one of us (ETP). Hossein Razavi made the molecule, and physician-scientist Yoshi Sekijima made tafamidis a reality by thinking through a clinical trial plan. A tafamidis registration trial for hATTR-PN, carried out by FoldRx Pharmaceuticals, led to tafamidis (20 mg QD) being approved by regulatory agencies in >40 countries throughout the world to ameliorate polyneuropathy. We are now approaching nearly 10 years of follow-up clinical studies with tafamidis, demonstrating a durable clinical response. A tafamidis registration trial for wtATTR-CM and hATTR-CM led to tafamidis (80/61 or 20 mg QD) being recently approved by the FDA, the EMA, and the Japanese regulators for use in slowing the relentless progression of TTR cardiomyopathy/dysautonomia. While tafamidis has been approved by the FDA for cardiomyopathy, it has still not been approved for polyneuropathy in the United States, despite the overwhelming amount of evidence supporting its efficacy and safety (summarized in Fig. 7). The FDA decision went against the recommendation by the FDA Peripheral and Central Nervous System Drugs Advisory Committee who voted 13-4 that the data provided substantial evidence of efficacy for a surrogate endpoint that is reasonably likely to predict clinical benefit (The Advisory Committee did not find evidence of efficacy based on the clinical endpoint). The take-home message of this paragraph is that developing first-in-class drugs for unmet medical needs is an ultramarathon. Along the way not everyone will be supportive and unforeseeable challenges will arise. By and large, we were very fortunate to have the resources and support to get tafamidis to most of the patients in need, but there are many examples of efficacious drugs that do not make it to any patients in need for myriad reasons.

Perspective and conclusions

By carefully studying the mechanism of TTR aggregation, we established that tetramer dissociation was rate limiting for monomeric TTR misfolding and misassembly into amyloid fibrils and other aggregate structures. This information allowed us to envision kinetic stabilizers: small molecules that bind selectively to the natively folded TTR tetramer over the dissociative transition state, rendering TTR tetramer dissociation, and TTR aggregation much slower. Two of these TTR kinetic stabilizers, tafamidis, discovered employing structure-based design, and diflunisal, repurposed as a TTR kinetic stabilizer, stopped, or slowed degeneration of postmitotic tissues in the autonomic and peripheral nervous systems in the majority of hATTR-PN patients by inhibiting newly synthesized TTR from aggregating. Tafamidis also slows the progression of the cardiomyopathy associated with wtATTR-CM and hATTR-CM, extending lifespan. A smaller open-label study also indicates that diflunisal slows TTR cardiomyopathy progression while extending lifespan. Collectively, the tafamidis and diflunisal data support the validity of the amyloid hypothesis as it applies to the TTR amyloidoses, in which the process of TTR aggregation drives the degeneration of postmitotic tissue in TTR polyneuropathy and cardiomyopathy. We envision a future wherein mechanistically distinct drug combinations will be used to ameliorate human amyloid diseases in a greater fraction of patients. For example, we envision small molecule TTR kinetic stabilizers to be used in combination with drugs that enhance the capacity of the proteostasis network to achieve proteome maintenance (e.g., autophagy activators),[45, 200, 201] possibly along with drugs or drug candidates that attenuate neuroinflammation.

Dedications

This chapter is dedicated to Susan Lindquist, Christopher Dobson, Corino de Andrade, and George Glenner, who made substantial contributions to understanding amyloid diseases before their untimely passing.

Acknowledgments

We are eternally grateful for the hard work, determination, and creativity exhibited by the Kelly Laboratory Graduate Students and Postdoctoral fellows over a period of nearly 30 years, who collectively produced the knowledge and TTR kinetic stabilizer candidates that tafamidis emerged from. The demonstration that tafamidis ameliorates neurodegeneration and cardiomyopathy would not have been possible without the creative clinical insights of Professors Teresa Coelho and Mat Maurer, who helped design the tafamidis polyneuropathy and cardiomyopathy clinical trials, respectively. The demonstration that diflunisal ameliorates neurodegeneration and likely cardiomyopathy would not have been possible without John Berk and Yoshi Sekijima, who conceptualized and led the completion of the diflunisal clinical trial for polyneuropathy. Tafamidis would not have been possible without sustained NIH funding of Kelly Laboratory research (DK 046335) since 1990. Supplemental financial support from the Skaggs Institute for Chemical Biology and historic support from Lita Annenberg Hazen Foundation also critically accelerated progress. The creation of FoldRx Pharmaceuticals Inc., which commercialized tafamidis, was made possible by Christopher Mirabelli's leadership and vision, along with the support of his Health Care Ventures partners. The development of tafamidis was efficiently and skillfully accomplished by Richard Labaudinière (FoldRx; President and CEO), Donna Grogan (FoldRx; Chief Medical Officer), Jeff Packman (FoldRx; Sr. Director, Drug Development Operations), and their colleagues. Paul S. Anderson (retired from Merck in 1994 to become Senior Vice-President for Chemical and Physical Sciences at Dupont-Merck Pharmaceuticals (retired)) helped the FoldRx team and the founders choose tafamidis out of the numerous appealing kinetic stabilizer candidates handed over to FoldRx by the Scripps Research Institute team. The success of the FoldRx team was substantially enabled by the comradery, clinical insights, and work ethic of Dr. Teresa Coelho, MD, PhD who recruited the majority of patients to the TTR polyneuropathy clinical trial, which was the first trial to demonstrate that pharmacologic intervention was possible. We also value the creative scientific and medical input from Giampaolo Merlini, a scholar in the amyloid disease field. We thank Colleen Fearns PhD for scientific insights and sophisticated editing of the text and Janet Hightower for professional assistance with figure creation.

Declaration of Interests

JWK and ETP discovered tafamidis, receive sales royalties and sales milestone payments, as well as related payments from Pfizer.

References

1. Eisele YS, Monteiro C, Fearns C, et al. Targeting protein aggregation for the treatment of degenerative diseases. *Nat Rev Drug Discov*. 2015;14(11):759–780.
2. Schonhoft JD, Monteiro C, Plate L, et al. Peptide probes detect misfolded transthyretin oligomers in plasma of hereditary amyloidosis patients. *Sci Transl Med*. 2017;9(407).
3. Schmidt M, Wiese S, Adak V, et al. Cryo-EM structure of a transthyretin-derived amyloid fibril from a patient with hereditary ATTR amyloidosis. *Nat Commun*. 2019;10.
4. Coelho T. Familial amyloid polyneuropathy: new developments in genetics and treatment. *Curr Opin Neurol*. 1996;9(5):355–359.
5. Andrade C. A peculiar form of peripheral neuropathy; familiar atypical generalized amyloidosis with special involvement of the peripheral nerves. *Brain*. 1952;75(3):408–427.
6. Benson MD. Familial amyloidotic polyneuropathy. *Trends Biochem Sci*. 1989;12:88–92.
7. Saraiva MJ, Costa PP, Birken S, Goodman DS. Presence of an abnormal transthyretin (prealbumin) in Portuguese patients with familial amyloidotic polyneuropathy. *Trans Assoc Am Phys*. 1983;96:261–270.

8. Maia LF, Magalhaes R, Freitas J, et al. CNS involvement in V30M transthyretin amyloidosis: clinical, neuropathological and biochemical findings. *J Neurol Neurosurg Psychiatry*. 2015;86(2):159–167.
9. da Silva AM, Cavaco S, Fernandes J, et al. Age-dependent cognitive dysfunction in untreated hereditary transthyretin amyloidosis. *J Neurol*. 2018;265(2):299–307.
10. Sekijima Y, Yazaki M, Oguchi K, et al. Cerebral amyloid angiopathy in posttransplant patients with hereditary ATTR amyloidosis. *Neurology*. 2016;87(8):773–781.
11. Wange N, Anan I, Ericzon BG, et al. Atrial fibrillation and central nervous complications in liver transplanted hereditary transthyretin amyloidosis patients. *Transplantation*. 2018;102(2):e59–e66.
12. Ciulla TA, Tolentino F, Morrow JF, Dryja TP. Vitreous amyloidosis in familial amyloidotic polyneuropathy—report of a case with the val30met transthyretin mutation. *Surv Ophthalmol*. 1995;40 (3):197–206.
13. Ezawa N, Katoh N, Oguchi K, Yoshinaga T, Yazaki M, Sekijima Y. Visualization of multiple organ amyloid involvement in systemic amyloidosis using C-11-PiB PET imaging. *Eur J Nucl Med Mol Imaging*. 2018;45(3):452–461.
14. Hund E, Linke RP, Willig F, Grau A. Transthyretin-associated neuropathic amyloidosis—pathogenesis and treatment. *Neurology*. 2001;56(4):431–435.
15. Latasiewicz M, Sala-Puigdollers A, Gonzalez-Ventosa A, Milla E, Civera AA. Multimodal retinal imaging of familial amyloid polyneuropathy. *Ophthalmic Genet*. 2019;40(5):407–420.
16. Sandgren O, Kjellgren D, Suhr OB. Ocular manifestations in liver transplant recipients with familial amyloid polyneuropathy. *Acta Ophthalmol*. 2008;86(5):520–524.
17. Buxbaum JN, Brannagan T, Buades-Reines J, et al. Transthyretin deposition in the eye in the era of effective therapy for hereditary ATTRV30M amyloidosis. *Amyloid*. 2019;26(1):10–14.
18. Uitti RJ, Donat JR, Rozdilsky B, Schneider RJ, Koeppen AH. Familial oculoleptomeningeal amyloidosis—report of a new family with unusual features. *Arch Neurol*. 1988;45(10):1118–1122.
19. Ushiyama M, Ikeda S, Yanagisawa N. Transthyretin-type cerebral amyloid angiopathy in type-I familial amyloid polyneuropathy. *Acta Neuropathol*. 1991;81(5):524–528.
20. Monteiro C, da Silva AM, Ferreira N, et al. Cerebrospinal fluid and vitreous body exposure to orally administered tafamidis in hereditary ATTRV30M (p.TTRV50M) amyloidosis patients. *Amyloid*. 2018;25(2):120–128.
21. Sekijima Y, Wiseman RL, Matteson J, et al. The biological and chemical basis for tissue-selective amyloid disease. *Cell*. 2005;121(1):73–85.
22. Ruberg FL, Grogan M, Hanna M, Kelly JW, Maurer MS. Transthyretin amyloid cardiomyopathy JACC state-of-the-art review. *J Am Coll Cardiol*. 2019;73(22):2872–2891.
23. Westermark P, Sletten K, Johansson B, Cornwell 3rd. GG. Fibril in senile systemic amyloidosis is derived from normal transthyretin. *Proc Natl Acad Sci U S A*. 1990;87(7):2843–2845.
24. Cornwell 3rd GG, Sletten K, Johansson B, Westermark P. Evidence that the amyloid fibril protein in senile systemic amyloidosis is derived from normal prealbumin. *Biochem Biophys Res Commun*. 1988;154(2):648–653.
25. Tanskanen M, Peuralinna T, Polvikoski T, et al. Senile systemic amyloidosis affects 25% of the very aged and associates with genetic variation in alpha2-macroglobulin and tau: a population-based autopsy study. *Ann Med*. 2008;40(3):232–239.
26. Ueda M, Horibata Y, Shono M, et al. Clinicopathological features of senile systemic amyloidosis: an ante- and post-mortem study. *Mod Pathol*. 2011;24(12):1533–1544.
27. Gorevic PD, Prelli FC, Wright J, Pras M, Frangione B. Systemic senile amyloidosis—identification of a new prealbumin (transthyretin) variant in cardiac tissue—immunological and biochemical similarity to one form of familial amyloidotic polyneuropathy. *J Clin Investig*. 1989;83(3):836–843.
28. Jacobson DR, Pastore RD, Yaghoubian R, et al. Variant-sequence transthyretin (isoleucine 122) in late-onset cardiac amyloidosis in black Americans. *New Engl J Med*. 1997;336(7):466–473.
29. Damrauer SM, Chaudhary K, Cho JH, et al. Association of the V122I hereditary transthyretin amyloidosis genetic variant with heart failure among individuals of African or hispanic/latino ancestry. *JAMA*. 2019;322(22):2191–2202.
30. Bergstrom J, Gustavsson A, Hellman U, et al. Amyloid deposits in transthyretin-derived amyloidosis: cleaved transthyretin is associated with distinct amyloid morphology. *J Pathol*. 2005;206(2):224–232.

31. Snow AD, Willmer J, Kisilevsky R. Sulfated glycosaminoglycans—a common constituent of all amyloids. *Lab Investig*. 1987;56(1):120–123.
32. Bourgault S, Solomon JP, Reixach N, Kelly JW. Sulfated glycosaminoglycans accelerate transthyretin amyloidogenesis by quaternary structural conversion. *Biochemistry*. 2011;50(6):1001–1015.
33. Bourgault S, Choi S, Buxbaum JN, Kelly JW, Price JL, Reixach N. Mechanisms of transthyretin cardiomyocyte toxicity inhibition by resveratrol analogs. *Biochem Biophys Res Commun*. 2011;410(4):707–713.
34. Kelly JW. Clinical trial support for the hypothesis that the process of protein aggregation causes neurodegeneration. *Sci Trans Med*. 2020; in press.
35. Hammarstrom P, Sekijima Y, White JT, et al. D18G transthyretin is monomeric, aggregation prone, and not detectable in plasma and cerebrospinal fluid: a prescription for central nervous system amyloidosis? *Biochemistry*. 2003;42(22):6656–6663.
36. Sekijima Y, Hammarstrom P, Matsumura M, et al. Energetic characteristics of the new transthyretin variant A25T may explain its atypical central nervous system pathology. *Lab Investig*. 2003;83(3):409–417.
37. Selkoe DJ, Hardy J. The amyloid hypothesis of Alzheimer's disease at 25 years. *Embo Mol Med*. 2016;8(6):595–608.
38. Hardy J, Selkoe DJ. Medicine—the amyloid hypothesis of Alzheimer's disease: progress and problems on the road to therapeutics. *Science*. 2002;297(5580):353–356.
39. Tanzi RE, Bertram L. Twenty years of the Alzheimer's disease amyloid hypothesis: a genetic perspective. *Cell*. 2005;120(4):545–555.
40. Panza F, Lozupone M, Watling M, Imbimbo BP. Do BACE inhibitor failures in Alzheimer patients challenge the amyloid hypothesis of the disease? *Expert Rev Neurother*. 2019;19(7):599–602.
41. Kametani F, Hasegawa M. Reconsideration of amyloid hypothesis and tau hypothesis in Alzheimer's disease. *Front Neurosci*. 2018;12.
42. Reixach N, Deechongkit S, Jiang X, Kelly JW, Buxbaum JN. Tissue damage in the amyloidoses: transthyretin monomers and nonnative oligomers are the major cytotoxic species in tissue culture. *Proc Natl Acad Sci U S A*. 2004;101(9):2817–2822.
43. Sousa MM, Cardoso I, Fernandes R, Guimaraes A, Saraiva MJ. Deposition of transthyretin in early stages of familial amyloidotic polyneuropathy: evidence for toxicity of nonfibrillar aggregates. *Am J Pathol*. 2001;159(6):1993–2000.
44. Lambert MP, Barlow AK, Chromy BA, et al. Diffusible, nonfibrillar ligands derived from Abeta1-42 are potent central nervous system neurotoxins. *Proc Natl Acad Sci U S A*. 1998;95(11):6448–6453.
45. Madhivanan K, Greiner ER, Alves-Ferreira M, et al. Cellular clearance of circulating transthyretin decreases cell-nonautonomous proteotoxicity in Caenorhabditis elegans. *Proc Natl Acad Sci U S A*. 2018;115(33):E7710–E7719.
46. Chen JJ, Genereux JC, Suh EH, et al. Endoplasmic reticulum proteostasis influences the oligomeric state of an amyloidogenic protein secreted from mammalian cells. *Cell Chem Biol*. 2016;23(10):1282–1293.
47. Kurian SM, Novais M, Whisenant T, et al. Peripheral blood cell gene expression diagnostic for identifying symptomatic transthyretin amyloidosis patients: male and female specific signatures. *Theranostics*. 2016;6(11):1792–1809.
48. Stokin GB, Lillo C, Falzone TL, et al. Axonopathy and transport deficits early in the pathogenesis of Alzheimer's disease. *Science*. 2005;307(5713):1282–1288.
49. Lustbader JW, Cirilli M, Lin C, et al. ABAD directly links A beta to mitochondrial toxicity in Alzheimer's disease. *Science*. 2004;304(5669):448–452.
50. Wang XL, Su B, Lee HG, et al. Impaired balance of mitochondrial fission and fusion in Alzheimer's disease. *J Neurosci*. 2009;29(28):9090–9103.
51. Olzscha H, Schermann SM, Woerner AC, et al. Amyloid-like aggregates sequester numerous metastable proteins with essential cellular functions. *Cell*. 2011;144(1):67–78.
52. Hardy J. Failures in protein clearance partly underlie late onset neurodegenerative diseases and link pathology to genetic risk. *Front Neurosci*. 2019;13.

53. Menzies FM, Fleming A, Rubinsztein DC. Compromised autophagy and neurodegenerative diseases. *Nat Rev Neurosci.* 2015;16(6):345–357.
54. Fanning S, Selkoe D, Dettmer U. Parkinson's disease: proteinopathy or lipidopathy? *NPJ Parkinsons Dis.* 2020;6(1).
55. Imberdis T, Negri J, Ramalingam N, et al. Cell models of lipid-rich alpha-synuclein aggregation validate known modifiers of alpha-synuclein biology and identify stearoyl-CoA desaturase. *Proc Natl Acad Sci U S A.* 2019;116(41):20760–20769.
56. Vincent BM, Tardiff DF, Piotrowski JS, et al. Inhibiting stearoyl-CoA desaturase ameliorates alpha-synuclein cytotoxicity. *Cell Rep.* 2018;25(10):2742.
57. Coelho T, Maia LF, da Silva AM, et al. Tafamidis for transthyretin familial amyloid polyneuropathy A randomized, controlled trial. *Neurology.* 2012;79(8):785–792.
58. Berk JL, Suhr OB, Obici L, et al. Repurposing diflunisal for familial amyloid polyneuropathy: A randomized clinical trial. *JAMA.* 2013;310(24):2658–2667.
59. Benson MD, Waddington-Cruz M, Berk JL, et al. Inotersen treatment for patients with hereditary transthyretin amyloidosis. *N Engl J Med.* 2018;379(1):22–31.
60. Adams D, Gonzalez-Duarte A, O'Riordan WD, et al. Patisiran, an RNAi therapeutic, for hereditary transthyretin amyloidosis. *N Engl J Med.* 2018;379(1):11–21.
61. Maurer MS, Schwartz JH, Gundapaneni B, et al. Tafamidis treatment for patients with transthyretin amyloid cardiomyopathy. *N Engl J Med.* 2018;379(11):1007–1016.
62. Rosenblum H, Castano A, Alvarez J, Goldsmith J, Helmke S, Maurer MS. TTR (Transthyretin) stabilizers are associated with improved survival in patients with TTR cardiac amyloidosis. *Circ Heart Fail.* 2018;11(4):e004769.
63. Van Aken E, De Letter EA, Veckeneer M, et al. Transthyretin levels in the vitreous correlate with change in visual acuity after vitrectomy. *Br J Ophthalmol.* 2009;93(11):1539–1545.
64. Blake CC, Geisow MJ, Oatley SJ, Rerat B, Rerat C. Structure of prealbumin: secondary, tertiary and quaternary interactions determined by Fourier refinement at 1.8 A. *J Mol Biol.* 1978;121(3):339–356.
65. Klabunde T, Petrassi HM, Oza VB, Raman P, Kelly JW, Sacchettini JC. Rational design of potent human transthyretin amyloid disease inhibitors. *Nat Struct Biol.* 2000;7(4):312–321.
66. Connelly S, Choi S, Johnson SM, Kelly JW, Wilson IA. Structure-based design of kinetic stabilizers that ameliorate the transthyretin amyloidoses. *Curr Opin Struct Biol.* 2010;20(1):51–62.
67. Hornberg A, Eneqvist T, Olofsson A, Lundgren E, Sauer-Eriksson AE. A comparative analysis of 23 structures of the amyloidogenic protein transthyretin. *J Mol Biol.* 2000;302(3):649–669.
68. McCammon MG, Scott DJ, Keetch CA, et al. Screening transthyretin amyloid fibril inhibitors: characterization of novel multiprotein, multiligand complexes by mass spectrometry. *Structure.* 2002;10(6):851–863.
69. Holmgren G, Steen L, Ekstedt J, et al. Biochemical effect of liver transplantation in two Swedish patients with familial amyloidotic polyneuropathy (FAP-Met30). *Clin Genet.* 1991;40(3):242–246.
70. Herbert J, Wilcox JN, Pham KT, et al. Transthyretin: a choroid plexus-specific transport protein in human brain. The 1986 S. Weir Mitchell award. *Neurology.* 1986;36(7):900–911.
71. Cavallaro T, Martone RL, Dwork AJ, Schon EA, Herbert J. The retinal-pigment epithelium is the unique site of transthyretin synthesis in the rat eye. *Invest Ophthalmol Vis Sci.* 1990;31(3):497–501.
72. Ong DE, Davis JT, Oday WT, Bok D. Synthesis and secretion of retinol-binding protein and transthyretin by cultured retinal-pigment epithelium. *Biochemistry.* 1994;33(7):1835–1842.
73. Monaco HL, Rizzi M, Coda A. Structure of a complex of two plasma proteins: transthyretin and retinol-binding protein. *Science.* 1995;268(5213):1039–1041.
74. Kopelman M, Cogan U, Mokady S, Shinitzky M. The interaction between retinol-binding proteins and prealbumins studied by fluorescence polarization. *Biochim Biophys Acta.* 1976;439(2):449–460.
75. Zanotti G, Berni R. Plasma retinol-binding protein: structure and interactions with retinol, retinoids, and transthyretin. *Vitam Horm.* 2004;69:271–295.
76. White JT, Kelly JW. Support for the multigenic hypothesis of amyloidosis: the binding stoichiometry of retinol-binding protein, vitamin A, and thyroid hormone influences transthyretin amyloidogenicity in vitro. *Proc Natl Acad Sci U S A.* 2001;98(23):13019–13024.

77. Bartalena L, Robbins J. Variations in thyroid hormone transport proteins and their clinical implications. *Thyroid*. 1992;2(3):237–245.
78. Bartalena L, Robbins J. Thyroid hormone transport proteins. *Clin Lab Med*. 1993;13(3):583–598.
79. Monteiro C, Mesgazardeh JS, Anselmo J, et al. Predictive model of response to tafamidis in hereditary ATTR polyneuropathy. *JCI Insight*. 2019;4(12):e126526.
80. Coelho T, Maia LF, da Silva AM, et al. Long-term effects of tafamidis for the treatment of transthyretin familial amyloid polyneuropathy. *J Neurol*. 2013;260(11):2802–2814.
81. Hagen GA, Elliott WJ. Transport of thyroid hormones in serum and cerebrospinal fluid. *J Clin Endocrinol Metab*. 1973;37(3):415–422.
82. Foss TR, Wiseman RL, Kelly JW. The pathway by which the tetrameric protein transthyretin dissociates. *Biochemistry*. 2005;44(47):15525–15533.
83. Foss TR, Kelker MS, Wiseman RL, Wilson IA, Kelly JW. Kinetic stabilization of the native state by protein engineering: implications for inhibition of transthyretin amyloidogenesis. *J Mol Biol*. 2005;347 (4):841–854.
84. Johnson SM, Wiseman RL, Sekijima Y, Green NS, Adamski-Werner SL, Kelly JW. Native state kinetic stabilization as a strategy to ameliorate protein misfolding diseases: a focus on the transthyretin amyloidoses. *Acc Chem Res*. 2005;38(12):911–921.
85. Wojtczak A, Cody V, Luft JR, Pangborn W. Structures of human transthyretin complexed with thyroxine at 2.0 A resolution and 3',5'-dinitro-N-acetyl-L-thyronine at 2.2 A resolution. *Acta Crystallogr D*. 1996;52(Pt 4):758–765.
86. Wojtczak A, Luft J, Cody V. Mechanism of molecular recognition. Structural aspects of 3,3'-diiodo-L-thyroxine binding to human serum transthyretin. *J Biol Chem*. 1992;267:353–357.
87. Neumann P, Cody V, Wojtczak A. Structural basis of negative cooperativity in transthyretin. *Acta Biochim Pol*. 2001;48(4):867–875.
88. Tomar D, Khan T, Singh RR, et al. Crystallographic study of novel transthyretin ligands exhibiting negative-cooperativity between two thyroxine binding sites. *PLoS One*. 2012;7(9).
89. Hurle MR, Helms LR, Li L, Chan W, Wetzel R. A role for destabilizing amino acid replacements in light-chain amyloidosis. *Proc Natl Acad Sci U S A*. 1994;91(12):5446–5450.
90. Colon W, Kelly JW. Partial denaturation of transthyretin is sufficient for amyloid fibril formation in vitro. *Biochemistry*. 1992;31(36):8654–8660.
91. Colon W, Lai Z, McCutchen SL, Miroy GJ, Strang C, Kelly JW. FAP mutations destabilize transthyretin facilitating conformational changes required for amyloid formation. *CIBA Found Symp*. 1996;199:228–238 discussion 239-42.
92. Kelly JW, Colon W, Lai Z, et al. Transthyretin quaternary and tertiary structural changes facilitate misassembly into amyloid. *Adv Protein Chem*. 1997;50:161–181.
93. Lai Z, Colon W, Kelly JW. The acid-mediated denaturation pathway of transthyretin yields a conformational intermediate that can self-assemble into amyloid. *Biochemistry*. 1996;35(20):6470–6482.
94. Kelly JW. Alternative conformations of amyloidogenic proteins govern their behavior. *Curr Opin Struct Biol*. 1996;6(1):11–17.
95. Kelly JW. The alternative conformations of amyloidogenic proteins and their multi-step assembly pathways. *Curr Opin Struct Biol*. 1998;8(1):101–106.
96. Booth DR, Sunde M, Bellotti V, et al. Instability, unfolding and aggregation of human lysozyme variants underlying amyloid fibrillogenesis. *Nature*. 1997;385(6619):787–793.
97. Guijarro JI, Sunde M, Jones JA, Campbell ID, Dobson CM. Amyloid fibril formation by an SH3 domain. *Proc Natl Acad Sci U S A*. 1998;95(8):4224–4228.
98. Wiseman RL, Powers ET, Kelly JW. Partitioning conformational intermediates between competing refolding and aggregation pathways: insights into transthyretin amyloid disease. *Biochemistry*. 2005;44 (50):16612–16623.
99. McParland VJ, Kad NM, Kalverda AP, et al. Partially unfolded states of beta(2)-microglobulin and amyloid formation in vitro. *Biochemistry*. 2000;39(30):8735–8746.
100. Hurshman Babbes AR, Powers ET, Kelly JW. Quantification of the thermodynamically linked quaternary and tertiary structural stabilities of transthyretin and its disease-associated variants: the relationship between stability and amyloidosis. *Biochemistry*. 2008;47(26):6969–6984.

101. Nelson R, Sawaya MR, Balbirnie M, et al. Structure of the cross-beta spine of amyloid-like fibrils. *Nature*. 2005;435(7043):773–778.
102. Sawaya MR, Sambashivan S, Nelson R, et al. Atomic structures of amyloid cross-beta spines reveal varied steric zippers. *Nature*. 2007;447(7143):453–457.
103. Serio TR, Cashikar AG, Kowal AS, et al. Nucleated conformational conversion and the replication of conformational information by a prion determinant. *Science*. 2000;289(5483):1317–1321.
104. Lee J, Culyba EK, Powers ET, Kelly JW. Amyloid-beta forms fibrils by nucleated conformational conversion of oligomers. *Nat Chem Biol*. 2011;7:602–609.
105. Chiti F, Stefani M, Taddei N, Ramponi G, Dobson CM. Rationalization of the effects of mutations on peptide and protein aggregation rates. *Nature*. 2003;424:805–808.
106. DuBay KF, Pawar AP, Chiti F, Zurdo J, Dobson CM, Vendruscolo M. Prediction of the absolute aggregation rates of amyloidogenic polypeptide chains. *J Mol Biol*. 2004;341:1317–1326.
107. Tartaglia GG, Pawar AP, Campioni S, Dobson CM, Chiti F, Vendruscolo M. Prediction of aggregation-prone regions in structured proteins. *J Mol Biol*. 2008;380:425–436.
108. Goldschmidt L, Teng PK, Riek R, Eisenberg D. Identifying the amylome, proteins capable of forming amyloid-like fibrils. *Proc Natl Acad Sci U S A*. 2010;107(8):3487–3492.
109. Fernandez-Escamilla A-M, Rousseau F, Schymkowitz J, Serrano L. Prediction of sequence-dependent and mutational effects on the aggregation of peptides and proteins. *Nat Biotechnol*. 2004;22:1302–1306.
110. Mangione PP, Porcari R, Gillmore JD, et al. Proteolytic cleavage of Ser52Pro variant transthyretin triggers its amyloid fibrillogenesis. *Proc Natl Acad Sci U S A*. 2014;111(4):1539–1544.
111. Mangione PP, Verona G, Corazza A, et al. Plasminogen activation triggers transthyretin amyloidogenesis in vitro. *J Biol Chem*. 2018;293(37):14192–14199.
112. Marcoux J, Mangione PP, Porcari R, et al. A novel mechano-enzymatic cleavage mechanism underlies transthyretin amyloidogenesis. *Embo Mol Med*. 2015;7(10):1337–1349.
113. Hammarstrom P, Jiang X, Hurshman AR, Powers ET, Kelly JW. Sequence-dependent denaturation energetics: a major determinant in amyloid disease diversity. *Proc Natl Acad Sci U S A*. 2002;99:16427–16432.
114. Hammarstrom P, Schneider F, Kelly JW. Trans-suppression of misfolding in an amyloid disease. *Science*. 2001;293(5539):2459–2462.
115. Rappley I, Monteiro C, Novais M, et al. Quantification of transthyretin kinetic stability in human plasma using subunit exchange. *Biochemistry*. 2014;53(12):1993–2006.
116. Bulawa CE, Connelly S, Devit M, et al. Tafamidis, a potent and selective transthyretin kinetic stabilizer that inhibits the amyloid cascade. *Proc Natl Acad Sci U S A*. 2012;109(24):9629–9634.
117. Schneider F, Hammarstrom P, Kelly JW. Transthyretin slowly exchanges subunits under physiological conditions: a convenient chromatographic method to study subunit exchange in oligomeric proteins. *Protein Sci*. 2001;10(8):1606–1613.
118. Wiseman RL, Green NS, Kelly JW. Kinetic stabilization of an oligomeric protein under physiological conditions demonstrated by a lack of subunit exchange: implications for transthyretin amyloidosis. *Biochemistry*. 2005;44(25):9265–9274.
119. Wiseman RL, Johnson SM, Kelker MS, Foss T, Wilson IA, Kelly JW. Kinetic stabilization of an oligomeric protein by a single ligand binding event. *J Am Chem Soc*. 2005;127(15):5540–5551.
120. Jiang X, Smith CS, Petrassi HM, et al. An engineered transthyretin monomer that is nonamyloidogenic, unless it is partially denatured. *Biochemistry*. 2001;40(38):11442–11452.
121. McCutchen SL, Colon W, Kelly JW. Transthyretin mutation Leu-55-Pro significantly alters tetramer stability and increases amyloidogenicity. *Biochemistry*. 1993;32(45):12119–12127.
122. McCutchen SL, Lai Z, Miroy GJ, Kelly JW, Colon W. Comparison of lethal and nonlethal transthyretin variants and their relationship to amyloid disease. *Biochemistry*. 1995;34(41):13527–13536.
123. Lashuel HA, Wurth C, Woo L, Kelly JW. The most pathogenic transthyretin variant, L55P, forms amyloid fibrils under acidic conditions and protofilaments under physiological conditions. *Biochemistry*. 1999;38(41):13560–13573.
124. Palaninathan SK, Mohamedmohaideen NN, Snee WC, Kelly JW, Sacchettini JC. Structural insight into pH-induced conformational changes within the native human transthyretin tetramer. *J Mol Biol*. 2008;382(5):1157–1167.

125. Hammarstrom P, Wiseman RL, Powers ET, Kelly JW. Prevention of transthyretin amyloid disease by changing protein misfolding energetics. *Science*. 2003;299(5607):713–716.
126. Lashuel HA, Lai Z, Kelly JW. Characterization of the transthyretin acid denaturation pathways by analytical ultracentrifugation: implications for wild-type, V30M, and L55P amyloid fibril formation. *Biochemistry*. 1998;37(51):17851–17864.
127. Quintas A, Saraiva MJ, Brito RM. The tetrameric protein transthyretin dissociates to a non-native monomer in solution. A novel model for amyloidogenesis. *J Biol Chem*. 1999;274(46):32943–32949.
128. Liu K, Cho HS, Hoyt DW, et al. Deuterium-proton exchange on the native wild-type transthyretin tetramer identifies the stable core of the individual subunits and indicates mobility at the subunit interface. *J Mol Biol*. 2000;303(4):555–565.
129. Liu K, Cho HS, Lashuel HA, Kelly JW, Wemmer DE. A glimpse of a possible amyloidogenic intermediate of transthyretin. *Nat Struct Biol*. 2000;7(9):754–757.
130. Liu K, Kelly JW, Wemmer DE. Native state hydrogen exchange study of suppressor and pathogenic variants of transthyretin. *J Mol Biol*. 2002;320(4):821–832.
131. Holmgren G, Ericzon BG, Groth CG, et al. Clinical improvement and amyloid regression after liver transplantation in hereditary transthyretin amyloidosis. *Lancet*. 1993;341(8853):1113–1116.
132. Herlenius G, Wilczek HE, Larsson M, Ericzon BG. Ten years of international experience with liver transplantation for familial amyloidotic polyneuropathy: results from the Familial Amyloidotic Polyneuropathy World Transplant Registry. *Transplantation*. 2004;77(1):64–71.
133. Ericzon BG, Wilczek HE, Larsson M, et al. Liver transplantation for hereditary transthyretin amyloidosis: after 20 years still the best therapeutic alternative? *Transplantation*. 2015;99(9):1847–1854.
134. Ando Y, Tashima K, Tanaka Y, et al. Treatment of a Japanese patient with familial amyloidotic polyneuropathy with orthotopic liver transplantation. *Intern Med*. 1994;33(11):730–732.
135. Coelho T, Ines M, Conceicao I, Soares M, de Carvalho M, Costa J. Natural history and survival in stage 1 Val30Met transthyretin familial amyloid polyneuropathy. *Neurology*. 2018;91(21):E1999–E2009.
136. Hornsten R, Wiklund U, Olofsson BO, Jensen SM, Suhr OB. Liver transplantation does not prevent the development of life-threatening arrhythmia in familial amyloidotic polyneuropathy, Portuguese-type (ATTR Val30Met) patients. *Transplantation*. 2004;78(1):112–116.
137. Liepnieks JJ, Zhang LQ, Benson MD. Progression of transthyretin amyloid neuropathy after liver transplantation. *Neurology*. 2010;75:324–327.
138. Olofsson BO, Backman C, Karp K, Suhr OB. Progression of cardiomyopathy after liver transplantation in patients with familial amyloidotic polyneuropathy. Portuguese type. *Transplantation*. 2002;73(5):745–751.
139. Dubrey SW, Burke MM, Hawkins PN, Banner NR. Cardiac transplantation for amyloid heart disease: the United Kingdom experience. *J Heart Lung Transplant*. 2004;23(10):1142–1153.
140. Falk RH. Diagnosis and management of the cardiac amyloidoses. *Circulation*. 2005;112(13):2047–2060.
141. Ruygrok PN, Gane EJ, McCall JL, Chen XZ, Haydock DA, Munn SR. Combined heart and liver transplantation for familial amyloidosis. *Intern Med J*. 2001;31(1):66–67.
142. Hurshman AR, White JT, Powers ET, Kelly JW. Transthyretin aggregation under partially denaturing conditions is a downhill polymerization. *Biochemistry*. 2004;43(23):7365–7381.
143. Ferrone F. Analysis of protein aggregation kinetics. *Methods Enzymol*. 1999;309:256–274.
144. Kroes-Nijboer A, Venema P, Bouman J, van der Linden E. The critical aggregation concentration of beta-lactoglobulin-based fibril formation. *Food Biophys*. 2009;4(2):59–63.
145. Saelices L, Chung K, Lee JH, et al. Amyloid seeding of transthyretin by ex vivo cardiac fibrils and its inhibition. *Proc Natl Acad Sci U S A*. 2018;115(29):E6741–E6750.
146. Miroy GJ, Lai Z, Lashuel HA, Peterson SA, Strang C, Kelly JW. Inhibiting transthyretin amyloid fibril formation via protein stabilization. *Proc Natl Acad Sci U S A*. 1996;93(26):15051–15056.
147. Miller SR, Sekijima Y, Kelly JW. Native state stabilization by NSAIDs inhibits transthyretin amyloidogenesis from the most common familial disease variants. *Lab Investig*. 2004;84(5):545–552.
148. Oza VB, Smith C, Raman P, et al. Synthesis, structure, and activity of diclofenac analogues as transthyretin amyloid fibril formation inhibitors. *J Med Chem*. 2002;45(2):321–332.

149. Baures PW, Peterson SA, Kelly JW. Discovering transthyretin amyloid fibril inhibitors by limited screening. *Bioorg Med Chem*. 1998;6(8):1389–1401.
150. Baures PW, Oza VB, Peterson SA, Kelly JW. Synthesis and evaluation of inhibitors of transthyretin amyloid formation based on the non-steroidal anti-inflammatory drug, flufenamic acid. *Bioorg Med Chem*. 1999;7(7):1339–1347.
151. Peterson SA, Klabunde T, Lashuel HA, Purkey H, Sacchettini JC, Kelly JW. Inhibiting transthyretin conformational changes that lead to amyloid fibril formation. *Proc Natl Acad Sci U S A*. 1998;95 (22):12956–12960.
152. Adamski-Werner SL, Palaninathan SK, Sacchettini JC, Kelly JW. Diflunisal analogues stabilize the native state of transthyretin. Potent inhibition of amyloidogenesis. *J Med Chem*. 2004;47(2):355–374.
153. Razavi H, Palaninathan SK, Powers ET, et al. Benzoxazoles as transthyretin amyloid fibril inhibitors: synthesis, evaluation, and mechanism of action. *Angew Chem Int Ed*. 2003;42(24):2758–2761.
154. Razavi H, Powers ET, Purkey HE, et al. Design, synthesis, and evaluation of oxazole transthyretin amyloidogenesis inhibitors. *Bioorg Med Chem Lett*. 2005;15(4):1075–1078.
155. Petrassi HM, Klabunde T, Sacchettini J, Kelly JW. Structure-based design of N-phenyl phenoxazine transthyretin amyloid fibril inhibitors. *J Am Chem Soc*. 2000;122:2178–2192.
156. Purkey HE, Palaninathan SK, Kent KC, et al. Hydroxylated polychlorinated biphenyls selectively bind transthyretin in blood and inhibit amyloidogenesis: rationalizing rodent PCB toxicity. *Chem Biol*. 2004;11(12):1719–1728.
157. Petrassi HM, Johnson SM, Purkey HE, et al. Potent and selective structure-based dibenzofuran inhibitors of transthyretin amyloidogenesis: kinetic stabilization of the native state. *J Am Chem Soc*. 2005;127 (18):6662–6671.
158. Johnson SM, Connelly S, Wilson IA, Kelly JW. Toward optimization of the linker substructure common to transthyretin amyloidogenesis inhibitors using biochemical and structural studies. *J Med Chem*. 2008;51(20):6348–6358.
159. Johnson SM, Connelly S, Wilson IA, Kelly JW. Biochemical and structural evaluation of highly selective 2-arylbenzoxazole-based transthyretin amyloidogenesis inhibitors. *J Med Chem*. 2008;51 (2):260–270.
160. Johnson SM, Connelly S, Wilson IA, Kelly JW. Toward optimization of the second aryl substructure common to transthyretin amyloidogenesis inhibitors using biochemical and structural studies. *J Med Chem*. 2009;52(4):1115–1125.
161. Johnson SM, Petrassi HM, Palaninathan SK, et al. Bisaryloxime ethers as potent inhibitors of transthyretin amyloid fibril formation. *J Med Chem*. 2005;48(5):1576–1587.
162. Choi S, Connelly S, Reixach N, Wilson IA, Kelly JW. Chemoselective small molecules that covalently modify one lysine in a non-enzyme protein in plasma. *Nat Chem Biol*. 2010;6:133–139.
163. Choi S, Reixach N, Connelly S, Johnson SM, Wilson IA, Kelly JW. A substructure combination strategy to create potent and selective transthyretin kinetic stabilizers that prevent amyloidogenesis and cytotoxicity. *J Am Chem Soc*. 2010;132(4):1359–1370.
164. Green NS, Foss TR, Kelly JW. Genistein, a natural product from soy, is a potent inhibitor of transthyretin amyloidosis. *Proc Natl Acad Sci U S A*. 2005;102(41):14545–14550.
165. Green NS, Palaninathan SK, Sacchettini JC, Kelly JW. Synthesis and characterization of potent bivalent amyloidosis inhibitors that bind prior to transthyretin tetramerization. *J Am Chem Soc*. 2003;125 (44):13404–13414.
166. Lacor PN, Buniel MC, Chang L, et al. Synaptic targeting by Alzheimer's-related amyloid beta oligomers. *J Neurosci*. 2004;24(45):10191–10200.
167. Lashuel HA, Hartley D, Petre BM, Walz T, Lansbury Jr. PT. Neurodegenerative disease: amyloid pores from pathogenic mutations. *Nature*. 2002;418(6895):291.
168. Sousa MM, Fernandes R, Palha JA, Taboada A, Vieira P, Saraiva MJ. Evidence for early cytotoxic aggregates in transgenic mice for human transthyretin Leu55Pro. *Am J Pathol*. 2002;161 (5):1935–1948.
169. Brenner DA, Jain M, Pimentel DR, et al. Human amyloidogenic light chains directly impair cardiomyocyte function through an increase in cellular oxidant stress. *Circ Res*. 2004;94(8):1008–1010.

170. Liao RL, Jain M, Teller P, et al. Infusion of light chains from patients with cardiac amyloidosis causes diastolic dysfunction in isolated mouse hearts. *Circulation*. 2001;104(14):1594–1597.
171. Shi JR, Guan J, Jiang BB, et al. Amyloidogenic light chains induce cardiomyocyte contractile dysfunction and apoptosis via a non-canonical p38 alpha MAPK pathway. *Proc Natl Acad Sci U S A*. 2010;107 (9):4188–4193.
172. Oza VB, Petrassi HM, Purkey HE, Kelly JW. Synthesis and evaluation of anthranilic acid-based transthyretin amyloid fibril inhibitors. *Bioorg Med Chem Lett*. 1999;9(1):1–6.
173. Reixach N, Adamski-Werner SL, Kelly JW, Koziol J, Buxbaum JN. Cell based screening of inhibitors of transthyretin aggregation. *Biochem Biophys Res Commun*. 2006;348(3):889–897.
174. Choi S, Ong DS, Kelly JW. A stilbene that binds selectively to transthyretin in cells and remains dark until it undergoes a chemoselective reaction to create a bright blue fluorescent conjugate. *J Am Chem Soc*. 2010;132(45):16043–16051.
175. Choi S, Kelly JW. A competition assay to identify amyloidogenesis inhibitors by monitoring the fluorescence emitted by the covalent attachment of a stilbene derivative to transthyretin. *Bioorg Med Chem*. 2011;19(4):1505–1514.
176. Sacchettini JC, Kelly JW. Therapeutic strategies for human amyloid diseases. *Nat Rev Drug Discov*. 2002;1(4):267–275.
177. Alhamadsheh MM, Connelly S, Cho A, et al. Potent kinetic stabilizers that prevent transthyretin-mediated cardiomyocyte proteotoxicity. *Sci Transl Med*. 2011;3(97):97ra81.
178. Cho Y, Baranczak A, Helmke S, et al. Personalized medicine approach for optimizing the dose of tafamidis to potentially ameliorate wild-type transthyretin amyloidosis (cardiomyopathy). *Amyloid*. 2015;22(3):175–180.
179. Purkey HE, Dorrell MI, Kelly JW. Evaluating the binding selectivity of transthyretin amyloid fibril inhibitors in blood plasma. *Proc Natl Acad Sci U S A*. 2001;98(10):5566–5571.
180. Outeiro TF, Lindquist S. Yeast cells provide insight into alpha-synuclein biology and pathobiology. *Science*. 2003;302(5651):1772–1775.
181. Gundapaneni BK, Sultan MB, Keohane DJ, Schwartz JH. Tafamidis delays neurological progression comparably across Val30Met and non-Val30Met genotypes in transthyretin familial amyloid polyneuropathy. *Eur J Neurol*. 2018;25(3):464–468.
182. Dyck PJ, Davies JL, Litchy WJ, O'Brien PC. Longitudinal assessment of diabetic polyneuropathy using a composite score in the Rochester Diabetic Neuropathy Study cohort. *Neurology*. 1997;49 (1):229–239.
183. Cruz MW, Amass L, Keohane D, Schwartz J, Li HH, Gundapaneni B. Early intervention with tafamidis provides long-term (5.5-year) delay of neurologic progression in transthyretin hereditary amyloid polyneuropathy. *Amyloid*. 2016;23(3):178–183.
184. Merlini G, Coelho T, Cruz MW, Li HH, Stewart M, Ebede B. Evaluation of mortality during long-term treatment with tafamidis for transthyretin amyloidosis with polyneuropathy: clinical trial results up to 8.5 years. *Neurol Ther*. 2020;9(1):105–115.
185. Yukio A, Yoshiki S, Konen O, et al. Effects of tafamidis treatment on transthyretin (TTR) stabilization, efficacy, and safety in Japanese patients with familial amyloid polyneuropathy (TTR-FAP) with Val30-Met and non-Varl30Met: a phase III, open-label study. *J Neurol Sci*. 2016;362:266–271.
186. Mundayat R, Stewart M, Alvir J, et al. Positive effectiveness of tafamidis in delaying disease progression in transthyretin familial amyloid polyneuropathy up to 2 years: an analysis from the transthyretin amyloidosis outcomes survey (THAOS). *Neurol Ther*. 2018;7(1):87–101.
187. Sultan MB, Gundapaneni B, Schumacher J, Schwartz JH. Treatment with tafamidis slows disease progression in early-stage transthyretin cardiomyopathy. *Clin Med Insights Cardiol*. 2017;11.
188. Barroso FA, Judge DP, Ebede B, et al. Long-term safety and efficacy of tafamidis for the treatment of hereditary transthyretin amyloid polyneuropathy: results up to 6 years. *Amyloid*. 2017;24(3):194–204.
189. Huber P, Flynn A, Sultan MB, et al. A comprehensive safety profile of tafamidis in patients with transthyretin amyloid polyneuropathy. *Amyloid*. 2019;26(4):203–209.
190. Romero-Imbroda J, Sagrario-Fustero T, Del Canto-Perez C. Tafamidis for a transplant patient with transthyretin amyloid polyneuropathy. *J Clin Neurol*. 2017;13(4):444–446.

191. Lockwood PA, Le VH, O'Gorman MT, et al. The bioequivalence of Tafamidis 61-mg free acid capsules and Tafamidis Meglumine 4 x 20-mg capsules in healthy volunteers. *Clin Pharmacol Drug Dev*. 2020. https://doi.org/10/1002/cpdd.789.
192. Sekijima Y, Dendle MA, Kelly JW. Orally administered diflunisal stabilizes transthyretin against dissociation required for amyloidogenesis. *Amyloid*. 2006;13(4):236–249.
193. Tojo K, Sekijima Y, Kelly JW, Ikeda S. Diflunisal stabilizes familial amyloid polyneuropathy-associated transthyretin variant tetramers in serum against dissociation required for amyloidogenesis. *Neurosci Res*. 2006;56(4):441–449.
194. Sekijima Y, Tojo K, Morita H, Koyama J, Ikeda S. Safety and efficacy of long-term diflunisal administration in hereditary transthyretin (ATTR) amyloidosis. *Amyloid*. 2015;22(2):79–83.
195. Ikram A, Donnelly JP, Sperry BW, Samaras C, Valent J, Hanna M. Diflunisal tolerability in transthyretin cardiac amyloidosis: a single center's experience. *Amyloid*. 2018;25(3):197–202.
196. Buxbaum JN. Oligonucleotide drugs for transthyretin amyloidosis. *N Engl J Med*. 2018;379(1):82–85.
197. Buxbaum JN. Treatment of hereditary and acquired forms of transthyretin amyloidosis in the era of personalized medicine: the role of randomized controlled trials. *Amyloid*. 2019;26(2):55–65.
198. Coelho T, Carvalho M, Saraiva MJ, Alves I, Almeida MR, Costa PP. A strikingly benign evolution of FAP in an individual found to be a compound heterozygote for two TTR mutations: TTR MET 30 and TTR MET 119. *J Rheumatol*. 1993;20:179.
199. Coelho T, Chorao R, Sausa A, Alves I, Torres MF, Saraiva MJ. Compound heterozygotes of transthyretin Met30 and transthyretin Met119 are protected from the devastating effects of familial amyloid polyneuropathy. *Neuromuscul Disord*. 1996;6:27.
200. Balch WE, Morimoto RI, Dillin A, Kelly JW. Adapting proteostasis for disease intervention. *Science*. 2008;319(5865):916–919.
201. Powers ET, Morimoto RI, Dillin A, Kelly JW, Balch WE. Biological and chemical approaches to diseases of proteostasis deficiency. *Annu Rev Biochem*. 2009;78:959–991.

CHAPTER 5

A novel treatment for Parkinson's disease and ALS: Combined cell and gene therapies

Soshana P. Svendsen and Clive N. Svendsen
Board of Governors Regenerative Medicine Institute, Cedars-Sinai, Los Angeles, CA, United States

Introduction

Neurological diseases such as Parkinson's disease (PD) and amyotrophic lateral sclerosis (ALS, also known as Lou Gehrig's disease) continue to challenge the medical community. Despite billions of research dollars, large international clinical trials, and major investments from just about every arm of big pharma, the drugs used today have not changed much over the past 50 years. Unlike cancer and infections where the mechanism and genesis of disease are often largely understood, we still do not know what causes the majority of Parkinson's and ALS cases. There are genetic forms that are amenable to gene therapy approaches, but these account for only about 10% of all cases, with varied prevalence depending on the population sampled.[1] The majority of remaining cases are sporadic, though a combination of unknown genetic and environmental risk factors may lead to disease. Neuropathological studies have shown that PD is associated with a build-up of α-synuclein that aggregates into Lewy bodies within dying neurons and, in a similar fashion, abnormal distribution and accumulation of TDP43 have been found in the brains and spinal cords of ALS patients.[2] This has led to the idea that both of these diseases are associated with abnormal protein aggregation, and for sporadic PD, novel methods have been developed to reduce levels of α-synuclein.[3,4] While protein accumulation may be important, there are of course other potential causes of these diseases that are the focus of many ongoing research studies including RNA homeostasis, mitochondrial function, axonal transport, and many other phenotypes seen in patient postmortem samples and animal models.[5] Thus, what triggers proteins to accumulate in the central nervous system remains to be established and they may simply be endpoints of a complex cellular process rather than the cause of the disease.

Both PD and ALS are degenerative diseases, starting with the loss of specific sets of neurons but at later stages leading to more widespread degeneration. In addition, both involve various levels of inflammation and the activation of glial cells. For PD, there have been huge successes treating one of the cardinal features of the disease—loss of dopamine

Neurotherapeutics in the Era of Translational Medicine
https://doi.org/10.1016/B978-0-12-816475-4.00016-1

neurons in the substantia nigra that largely project to the striatum. Drugs that boost the function of dopamine, either by increasing dopamine release (L-Dopa) from remaining dopamine neurons or activating dopamine receptors (dopamine agonists), can maintain critical brain function for many years and can truly be called miracle drugs.[6] However, they do not attack the root cause or stop the continual degeneration of both dopamine neurons and other neurons within the brain. Nor do they affect many other non-dopaminergic symptoms of PD such as constipation, depression, and autonomic problems.[6] For ALS, there are only two approved drugs. Riluzole has been a mainstay for patients and affects glutamate toxicity in motor neurons whereas Edaravone is an anti-oxidant.[7] However, neither has any major impact on disease progression and only add a few months to the expected lifespan of the patient. The motor neuron degeneration in both the motor cortex and spinal cord is relentless.

A significant problem is that the neurological symptoms with both of these diseases appear after a majority of neurons are lost. Interestingly, in PD, there are very early non-motor symptoms including changes in bowel function and sleep patterns that could predict disease,[8] but these symptoms often occur independently and it has been difficult to use them as a predictor.[9] If the neurons are already lost, the only remaining strategies are to either replace them or prevent the continual loss of further neurons soon after diagnosis. This would at least prevent disease progression once diagnosed.

Clinical trials using primary fetal tissue for Parkinson's disease provided the foundation for the cell therapy field

Since the cardinal feature of PD is the loss of dopamine neurons in the substantia nigra, this led investigators to consider neuronal cell replacement as a therapy. The only source of new dopamine neurons originally available was fetal brain tissue. The developing ventral mesencephalon from rodents and humans could be isolated and was shown to contain living dopamine neurons.[10,11] This tissue source allowed transplantation of dopamine neurons into the adult rodent and primate striatum in models of PD.[12] Amazingly, these cells could survive and produce dopamine neurons that could innervate local neurons in the striatum and release dopamine. Even though these dopamine neurons were in an ectopic location (the striatum rather than the substantia nigra), they were able to produce functional recovery in these animals. This was assumed to be through the local release of dopamine within the striatum that by-passed the normal circuits of dopamine neurons from the substantia nigra to the striatum.[13] Attempts to transplant the DA neurons back into their site of origin in the substantia nigra showed that the cells could survive and in some cases lead to anatomical and functional recovery in a rodent model of PD.[14] However, the extension of new axons to reform the lesioned nigrostriatal pathway appeared age dependent and not possible in the older rodent brain.[15] While attempts have been made in the primate,[16,17] dopaminergic reinnervation of the striatum by nigral grafts in a primate model of PD has not been demonstrated and remains a challenge for the field.

Through a series of elegant experiments, fetal tissue transplants for PD were taken from animal models into the clinic for patients, originally performed in Sweden.[18] These studies and trials led to an era of transplantation for PD in the 1990s that has involved over 300 patients around the world. In postmortem tissue, transplanted dopamine neurons were shown to survive over 20 years and innervate large areas of the host striatum, however, graft health and function appeared to diminish overtime.[19–21] That said, many centers around the world did report patients with both surviving grafts and functional recovery.[22] But there was still a serious problem related to the impracticalities and ethics of using aborted fetal tissue. Every patient required 4–8 aborted fetal brain samples, which was not possible at many clinical sites and was simply prohibited in some countries.

Nevertheless, these pioneering studies were the first to show that primary fetal tissue could survive in the brain, produce new dopamine neurons, and in some cases provide functional benefit in PD. A European network of researchers was formed over 25 years ago that focused on using fetal tissue for neurological diseases (The Network for European CNS Transplantation and Restoration, NECTAR, nectar-eu.com), with a similar group formed in the United States (The American Society for Neural Therapy and Repair, ANSTR, asntr.org). These societies allowed active debate regarding the outcomes of these initial trials and the preclinical animal studies on which they were based.

Following the early encouraging results from the Swedish trials, a subsequent double-blinded trial from the United States showed that there was no significant difference between patients with fetal transplants and those who were in the placebo group.[23] There were many discussions at meetings suggesting that this trial's 6-month immune suppression regime may have been insufficient to prevent rejection of the fetal dopamine neurons, compared to the Swedish trials that used a 1-year immune suppression regime. However, around the same time, a non-blinded study from the United States used no immune suppression, a slightly different way of preparing the fetal mesencephalon, and a different surgical approach and found some clinical benefit of the transplants in younger but not older patients.[24] However, they also found some patients with so-called off-medication dyskinesias.[25] Dyskinesias are uncontrollable movements often seen in PD patients in response to dopamine-activating drugs. The mechanisms underlying these graft-induced dyskinesias are not fully understood, however, animal studies suggested that dyskinesias resulted from serotonergic neurons within the transplant. Excessive serotonergic innervation of the grafted striatum was also shown in patients who developed these dyskinesias.[26]

Based on the above, it became clear that there were significant challenges for the dopamine neuron transplantation field including the ectopic delivery of dopamine neurons, ethical concerns with fetal tissue as well as tissue procurement, and side-effects of the transplant possibly from serotonergic neurons.[27,28] Given these setbacks, fetal tissue transplants for PD stopped around the world in the early 2000s. Then Dr. Roger Barker spearheaded a new fetal transplant trial for PD in Europe with EU funding called TRANSEURO (ClinicalTrials.gov Identifier: NCT01898390). This ambitious trial aimed to test in a single well-controlled study whether dopamine neuron transplants could have a significant clinical effect. However, this trial also encountered issues collecting sufficient fetal brain tissue and was only half enrolled when forced to stop. Meanwhile, a new area of innovative biology was arising that could provide a unique source of tissue for transplantation studies that would finally remove the need for aborted fetal tissue.

Transitioning to neural stem and progenitor cells—opportunities and challenges

Very early on, the field of neural transplantation realized that human fetal tissue was going to be a bottleneck and began searching for alternatives. Interestingly, leads for how to overcome this problem came from trying to expand stem or progenitor cells from the adult brain. Dr. Rusty Gage and others had begun to develop methods of expanding neural progenitors from adult rat and mouse hippocampal tissue using the mitogen fibroblast growth factor-2 (FGF-2).[29] This was one region of the adult brain where constant neurogenesis occurs in rodents. Subsequently, another group published a seminal paper showing that epidermal growth factor (EGF) could drive a different type of stem cell found in the adult subventricular zone and coined the term "neurosphere" as these cells would grow as free-floating aggregates in the petri dish.[30] Together with Dr. Ron McKay and others, the field of neural stem and progenitor cell biology began to develop.[31,32] Fetal, neonatal, and adult rodent brain tissues were also found to harbor neural stem and progenitor cells that could be expanded in culture and transplanted into the CNS. While there were reports of the generation and transplantation of dopamine neurons derived from the rat mesencephalon,[33] it became clear that following expansion of adult somatic neural cells, there were limits on both the scale-up of the process and the maintenance of stem cells that remained able to make key neurons. In particular, the ability of rodent mesencephalic progenitor cells expanded in culture with FGF-2 or EGF to produce dopamine neurons appeared to decline over time. Though interestingly, these progenitors were able to make another cell type, now believed to be a glial cell, that enhanced the survival and functional efficacy of primary embryonic dopamine neuron cocultures.[34]

Based on these findings using rodents, a number of groups focused on trying to propagate neural progenitors or stem cells from the developing human mesencephalon in the hope of catching a stem cell or dopamine progenitor cell that could be expanded indefinitely and then differentiated into a mature dopamine neuron for transplantation. This would remove the need for primary fetal tissue and perhaps cells could even be frozen and stored in large batches for clinical use. In our own studies using brain fetal samples, it was clear that expanding primary cells within the human mesencephalon using either FGF-2, EGF or a combination of both was not able to expand a progenitor or stem cell but rather a later cell type that could no longer give rise to dopamine neurons upon differentiation.[35] Expanding progenitor or stem cells from the area of the brain that made dopamine neurons was such a tantalizing idea many groups continued to try and recapitulate the "niche" in which DA neurons are born. Perhaps going to earlier fetal time points in rodents would allow the isolation of the elusive progenitor or stem cell that gave rise to dopamine neurons. But as hard as everyone tried—the strategy continued to fail. It seemed that the niche requirements for the production of dopamine neurons in vitro were just not being met.

Using embryonic stem and induced pluripotent stem cells: Opportunities and challenges

In parallel to all of the fetal and adult progenitor work, there were a number of studies focusing on an early stem cell that could be isolated from the mouse blastocyst called mouse embryonic stem cells (mESCs). These were being used to make mouse transgenics and also being used with various in vitro differentiation protocols, including the production of dopamine neurons. It was shown that mESCs could be made into functional dopamine neurons in the culture dish and that these cells could be transplanted into animal models of PD.[36,37] But ESCs from humans were not available until Dr. James Thomson from the University of Wisconsin published his seminal papers in 1990 showing that stem cells could be isolated from the human blastocyst and expanded indefinitely in culture while retaining the potential to produce any cell of the human body.[38] This amazing finding shifted the thinking toward using human ESCs (hESCs) as a source of human brain tissue. One of the first studies showing neural tissue could be made from hESCs came from the laboratory of Su Chun Zhang in collaboration with the Thomson group.[39] Meanwhile, others showed that because hESCs are not somatic neural progenitor cells but precursors to all body lineages and a true pluripotent stem cell, they retained the potential to make dopamine neurons even after extensive expansion.[40,41] Furthermore, the hESC-derived dopamine neurons could engraft and provide behavioral recovery following transplantation into rodent models of PD and survive in the parkinsonian monkey.[42] A new era of unlimited dopamine neuron production was born.

But, almost before it could take off, this vital source of human tissue was hit with resounding ethical issues. The production of hESCs requires human embryos, which were usually sourced from in vitro fertilization clinics. The Bush administration was against the destruction of embryos and banned government support for the production of new hESC lines, limiting the field to a set of early hESC lines that were placed on a National Institutes of Health (NIH) registry. During this period, many academic centers and companies were weary of using hESCs and the field's growth was severely impacted. However, with a reversal of these policies by the Obama administration and continual investment in the field, techniques developed to direct hESCs into dopamine neurons once again advanced.

In addition, a Nobel prize-winning technology was set to disrupt the field. In 2008, Shinya Yamanka and Jamie Thomson simultaneously published papers showing that human adult fibroblasts could be reprogramed back in time or "induced" to become cells that were almost identical to hESCs.[43,44] These induced pluripotent stem cells (iPSCs) invigorated the field, eliminated the ethical issues associated with hESCs, and opened up the exciting potential of autologous transplantation to avoid immune rejection complications that accompany allogeneic transplantation. A group of researchers (GForce) interested in bringing dopamine neurons to the clinic for PD using either hESCs or iPSCs had formed. As a result, a number of different approaches were used to move these more flexible and controllable cell sources toward the clinic. GForce included Drs. Roger Barker, Lorenz Studer, Malin Palmer, Jun Takahashi as the steering committee and established cGMP methods for differentiating hESCs and iPSCs into dopamine neurons for treating PD.[45] The Kyoto trial with Dr. Takahasi is already underway with one patient.[46] The remarkable thing about GForce is that they are combining the clinical protocols and outcomes from fetal dopamine neuron clinical studies with the hESC and hiPSC studies, which will allow direct comparisons between these two different but complementary approaches. This is clearly the way forward for utilizing these complex cellular therapies and will become a model system as this area of clinical research moves forward for other diseases. While these novel and exciting technologies permitted the ethical and unlimited production of human dopamine neurons, the ectopic nature of dopamine neuron replacement and possible side effects from runaway dyskinesias still remain a potential issue. Protecting the patient's own dopamine neurons to allow continual function of the entire circuit remained an attractive option.

Combining cell and gene therapy approaches for Parkinson's disease

As the use of dopamine neuronal replacement for PD continued to develop, another promising therapy for PD is the protection of remaining host dopamine neurons. This may be possible with an area of translational drug development using growth factors. The use of growth factors such as nerve growth factor (NGF) was found to prevent neuronal death in models of Alzheimer's disease and both ex vivo cell delivery and viral delivery of NGF have been used in clinical trials[47,48] (clinicaltrials.gov Identifier NCT00876863). Another growth factor glial cell line-derived neurotrophic factor (GDNF) was discovered and patented by the company Amgen (Thousand Oaks, CA). This neurotrophic factor was shown to protect dopamine neurons from dying in a number of rodent and nonhuman primate models of PD.[49–51] However, neither NGF nor GDNF is able to cross the blood–brain barrier and so delivery to the striatum or substantia nigra was a challenge to the field. To circumvent this limitation, one clinical study administered

GDNF intraventricularly to PD patients. Unfortunately, there was no clinical effect, likely because GDNF did not reach the putamen and substantia nigra.[52,53] A neurosurgeon, Steve Gill, from Bristol, UK was proposing to use a new technology called convection-enhanced delivery (CED) to administer drugs for PD. In CED, pioneered by Dr. Edward Oldfield's group at NIH,[54] the delivery cannula has a narrow outlet that causes fluid delivery under high pressure, leading to a larger area of "spraying" of the drug into the brain tissue. Based on conversations with Dr. Gill, we decided that convection-enhanced delivery would be the ideal method to directly administer GDNF to the caudate-putamen of PD patients. Therefore, an open-label Phase 1 trial in Bristol, UK was performed. It showed that GDNF could be continuously delivered to the brain and had positive effects on both dopamine levels and the overall clinical scores for all five treated patients.[55] A 2-year outcome study confirmed that GDNF continuously infused into the posterior putamen was safe and provided improvement in patients' off-medication motor function and activities of daily living, and a separate postmortem study showed that GDNF induced sprouting of dopamine neurons.[56,57] A subsequent open-label trial with 10 patients with advanced PD further showed that intraputaminal GDNF infusion was safe and resulted in significant, sustained bilateral effects.[58] The exciting results from these initial studies led to a large double-blind trial by Amgen, the company that held the patent for and provided GDNF.[59] Unfortunately, a critical technical divergence with this trial from the Bristol clinical trial was the use of a different cannula device that did not provide convection-enhanced delivery of GDNF. The Amgen trial did not meet the end points. This, along with some safety concerns from a study by Amgen delivering extremely high doses of GDNF to nonhuman primates,[60] led the company to no longer providing GDNF for PD patients. Interestingly, positive clinical effects were maintained even after GDNF delivery was halted.[61,62] Accordingly, there has been much discussion about how to move forward with successful GDNF delivery.[63] Serendipitously, expiration of the patent on GDNF has permitted the community to once again pursue GDNF as a therapeutic for PD. A recent study assessing convection-enhanced delivery of GDNF to the nonhuman primate has now confirmed the safety of GDNF, with no GDNF-related pathologic findings in the brain, spinal cord, dorsal root ganglia, or trigeminal ganglia.[64] And, a very recent clinical trial,[65] along with an extension trial of the parent trial,[66] showed that while endpoints were not met, GDNF was safe out to 80 weeks. Future testing with GDNF may require a longer treatment window and/or a higher dose.

In addition to the benefits provided by neurotrophic factors, the field also became more interested in the use of nonneuronal cells for brain repair—including the use of astrocytes that are the support cells within the CNS and are now known to be crucial for neuronal functioning.[67,68] Our own studies showed that human fetal cortical tissue could be expanded in EGF or FGF-2 as free-floating aggregates of neural progenitor cells or neurospheres[69] Upon differentiation, these progenitors could give rise to astrocytes. This led to the concept that by simply expanding cortical neural progenitors, a large bank of cells could be established that subsequently could be expanded as astrocytes in culture and that following transplantation could be utilized for clinical trials. In support of this, our early studies showed that following long-term transplantation into the rodent striatum, human neural progenitors did indeed give rise to astrocytes[70] (Fig. 1). The next logical step was to engineer neural progenitors that were capable of producing GDNF and that could be considered as "trojan horses" for GDNF delivery to the patient striatum. We felt that the synergistic effect of GDNF and astrocytes may be the most powerful way to protect dying dopamine neurons. Together with our collaborator Patrick Aebischer at EPFL (Lausanne, Switzerland), we optimized the use of integrating lentiviral constructs

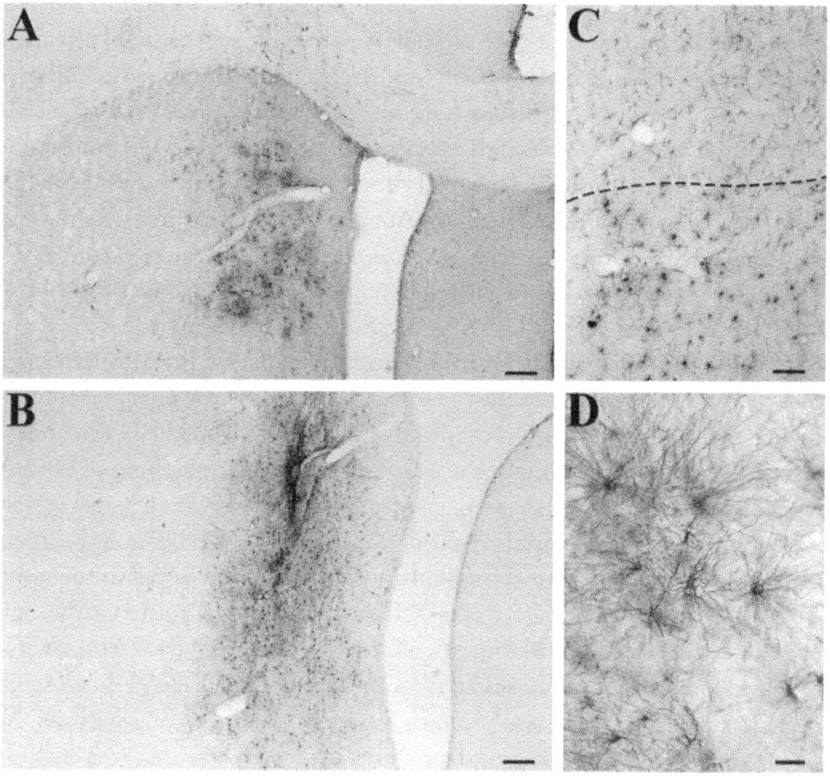

Fig. 1 Human GFAP staining in a section rostral to a human precursor cell transplant (A) and at the level of the transplant in another animal (B). Note the large number of cells migrating out from the remaining core of the graft. (C) A region of both larger, densely stained human astrocytes with many radial processes (*below dotted line*) and smaller and lighter stained astrocytes (*above dotted line*). (D) Higher power image of the densely staining astrocytes shown in (C). Scale bar, 500 μm in A and B; 250 μm in C and 100 μm in D. *(From Svendsen C.N., Caldwell M.A., Shen J., et al. Long-term survival of human central nervous system progenitor cells transplanted into a rat model of Parkinson's disease. Exp Neurol. 1997;148(1):135–146. https://doi.org/10.1006/exnr.1997.6634.)*

encoding GDNF to stably transduce human neural progenitor cells (hNPCs).[71] While our previous Phase 1 trial with Steve Gill and Amgen's subsequent trial delivered bacterial-derived GDNF, we now switched to GDNF produced by human mammalian cells, which may be more effective due to mammalian posttranslational modification.[72] Remarkably, this system worked very efficiently—hNPCs releasing GDNF (hNPC-GDNF) were able to survive transplantation into the striatum of both rodents and non-human primates and release GDNF at physiological levels for many months[50,73,74] (Fig. 2). Importantly, the hNPC-GDNF could functionally protect dopamine neurons and provide behavioral recovery in the rodent and primate models.

ALS: Different disease and different challenges for stem and progenitor cell transplantation

Following a move to Wisconsin in 2000, and after introduction to Jeffrey Alan Kaufman and The ALS Association chapter in Wisconsin, it became clear how devastating ALS could be. Jeff was an outstanding athlete, successful lawyer, amazing husband, and father to a wonderful family. But he was struck with ALS at the age of 33. Within 3 years he went from being healthy to being completely paralyzed. Inspired by the severity of this disease and the lack of any treatments, our lab decided to shift its focus to ALS.

ALS involves the death of motor neurons that control the voluntary motor system. The upper motor neuron in the motor cortex initiates the movement through a long connection to the spinal cord, where a lower motor neuron has a very lengthy projection to the appropriate muscle. While both Parkinson's disease and ALS involve the degeneration of neurons, their treatment options are very different. In PD, dopamine neurons could be transplanted ectopically into the striatum to supply dopamine to affect specific output circuits. However, it would not be possible to simply transplant new motor neurons into the muscle itself, as voluntary muscle movement requires activation of both upper and lower motor neurons. An alternative idea would be that diseased motor neurons in the spinal cord could be replaced with a healthy population. Indeed, work had shown that motor neurons could be produced from hESCs.[75] However, there is no evidence in animal models that transplanted neurons could extend axonal processes to actually reinnervate muscle—as this connection happened early in development and would be difficult to reproduce in an adult environment. So direct replacement of motor neurons, as done for dopamine neurons in PD, was not possible for ALS.

Over time, the scientific field realized that not only were motor neurons affected in ALS, but increasing evidence showed that astrocytes surrounding the motor neurons also played a crucial role.[76] While the ideal animal model would be of sporadic ALS (which represents the majority of cases)—these cannot be generated as the cause remains unknown. Therefore, the field has turned to genetic models of ALS. The most widely used of these has been rodents over-expressing mutant copper-zinc superoxide dismutase

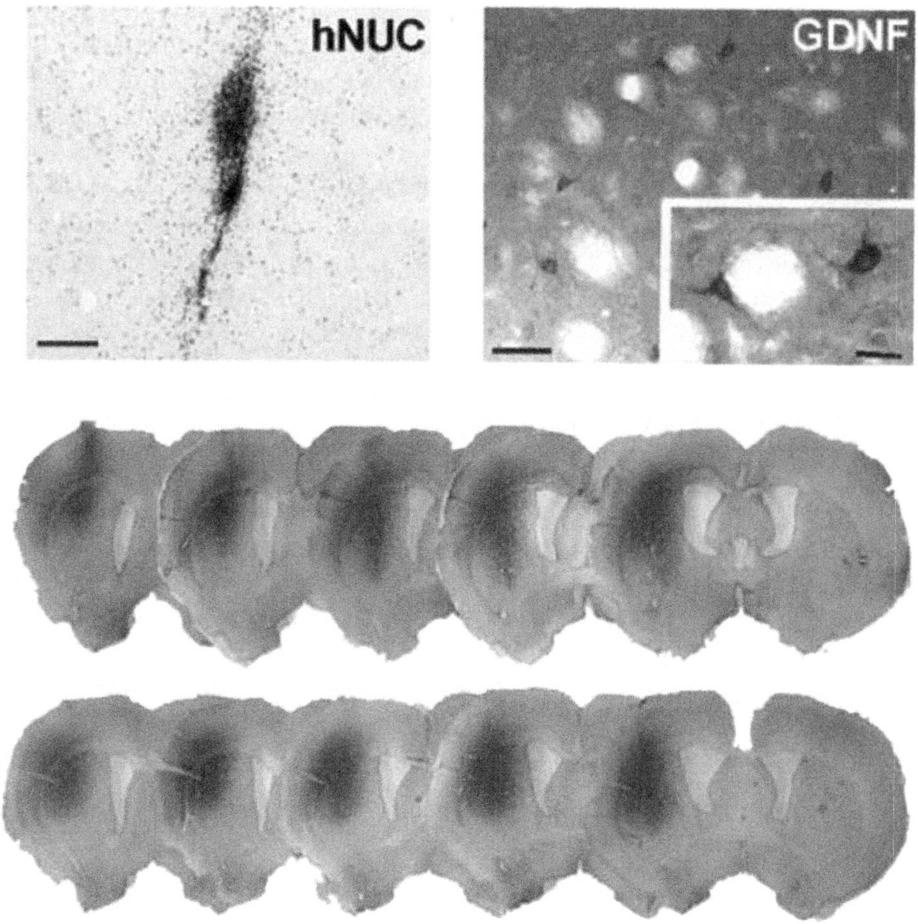

Fig. 2 hNPC-GDNF express GDNF in long term in the partially lesioned striatum and provide protection. Rats receiving hNPC-GDNF or no cells postlesion were assessed for functional recovery by amphetamine rotations (A). Rats receiving hNPC-GDNF had a trend toward fewer rotations at 5 and 6 weeks posttransplantation compared to the control group. By 8 weeks, control rats showed similar recovery from the partial lesion (B). More TH-positive neurons are present in the hNPC-GDN-transplanted rats compared to no transplant rats (C). Cells, identified with the human nuclear specific antibody hNUC, survive for at least 8 weeks following transplantation (D) and release GDNF, which is taken up by host neurons (E) and distributed to cover a large region of the striatum (F). Scale bars are 500 μm in (D), 100 μm in (E) and 30 μm in (E inset). *(From Behrstock S., Ebert A., McHugh J., et al. Human neural progenitors deliver glial cell line-derived neurotrophic factor to parkinsonian rodents and aged primates. Gene Ther. 2006;13(5):379–388. https://doi.org/10.1038/sj.gt.3302679.)*

(mSOD1).[77–79] Mutations in this gene account for about 10%–20% of familial ALS cases.[1] The mSOD1 protein accumulates in cells and both lower and upper motor neurons selectively die leading to paralysis at about 120 days and death a few weeks later. A number of animal studies suggested that astrocytes, in particular, may be important in preventing the death of motor neurons. In support of this idea, chimeric animal studies have shown that astrocytes expressing multiple copies of mSOD1 can have detrimental effects on normal, healthy motor neurons and lead to paralysis. Of most importance, the presence of healthy astrocytes leads to the survival of motor neurons carrying a SOD1 mutation.[80,81] In addition, studies showed that transplantation of astrocytes into mouse models of ALS could be protective.[82] Our own studies demonstrated that the human neural progenitor cells (hNPCs) we isolated from fetal tissue and expanded in EGF and FGF-2 could make astrocytes when transplanted into the spinal cord of the mSOD1 rat.[83,84] But these hNPC-derived astrocytes were not able to slow disease progression or protect motor neurons. Critically, as with dopamine neurons, GDNF also has a high affinity for motor neurons in the spinal cord. The next question was whether combining both astrocytes and GDNF may protect dying motor neurons. When hNPCs were engineered to stably make GDNF, in the same fashion to our approach for PD, the cells provided a major protective effect on dying motor neurons in the rat mSOD1 ALS model. This therapeutic approach was very appealing as it combined both stem cell and gene therapy to treat a disease for which there was little hope. In addition, we had our extensive work in the PD models using astrocytes secreting GDNF as a therapeutic approach (see above) to draw on for knowledge and experience with the cells.

How to move from an idea to the clinic for ALS: Obtaining an IND without a company

The evolution of the preclinical animal work using astrocytes secreting GDNF to treat either PD or ALS had been fairly linear over the years. However, to actually do a clinical trial, it was going to take an exponential increase in both funding and personnel—or the creation of a new biotech company or collaboration with the private sector. Fortunately, the initiation of the California Institute for Regenerative Medicine (CIRM) opened up the possibility to do an entire IND-enabling preclinical study and clinical trial under one funding umbrella. Funded through Prop 71 in California as a response to President Bush banning federal funding for new embryonic stem cell line production, CIRM had $3 billion over 10 years to fund innovative stem cell science and clinical trials. The Svendsen lab moved to Cedars-Sinai Medical Center in 2010 and set to work establishing a team of investigators and clinicians within the medical center to transplant fetal-derived human neural progenitor cells secreting GDNF into the spinal cord of patients with ALS.

Looking back over those early years of moving from the bench to the bedside, it is clear that basic scientists face many barriers. The first issue is the amount of time that these processes require in order to understand the workings of the FDA—time that cannot be spent generating data for the next scientific paper. There is no defined academic track for a basic scientist to do a clinical trial and no real encouragement. The general mantra is that companies provide the IND and physicians do the trials. Scientists simply produce ideas for the companies to translate. The major roadblock is that scientists are of course not trained or permitted to treat patients. If they want to be involved with drug development, the normal route would be to join a drug company and help move the product along. CIRM offered a unique opportunity to a number of basic scientists in California to obtain solid funding to carry out preclinical studies that could be part of an IND. The Svendsen laboratory was one of the first to make the most of this opportunity.

The critical requirement was the conducive environment in which the project would take place. Cedars-Sinai Medical Center had all of the parts required to do a clinical trial: pharmacy, risk management, IRB, surgery suites, clinical staff, grants management, and most importantly principal investigators to conduct the clinical trial. The clinical protocol for the project really dictates the initial enabling studies in animals and so finding a motivated clinician to partner with is crucial. If there is a surgical component to the administration of the stem cells, it is key to have a surgeon who is both motivated and has the time to help shape the clinical protocol and eventually participate in the trial.

The other crucial aspect is to hire an FDA consultant who could help navigate the deep waters of IND preparation. The choice was Sheryl Osborne—a delightful person with extensive training as an FDA consultant. Together, we mapped out how one should approach the preclinical studies and manufacturing of the hNPC-GDNF product. Unfortunately, Sheryl passed away just as we received IND approval. She will always be remembered as being instrumental in moving this project forward. Another consultant, Becky Donahue, joined the team as we began the trial and again made us realize how important a good FDA consultant is for the actual trial. Reporting of serious adverse events, death and other aspects of data capture have to be done following specific FDA guidelines. On one hand, it was daunting how much information the FDA would require and how rigorous the studies needed to be. On the other hand, the foundation of an IND application is extensively and clearly documented on the FDA website.

The process to obtain an IND involves breaking the work into different volumes: Chemistry, Manufacturing, and Controls (CMC), preclinical dose-ranging studies, tumor toxicology studies, large animal studies, and finally the preparation of an integrated summary. This is followed by the clinical protocol which outlines the exact parameters of the trial (Immune suppression, number of patient visits, outcomes, etc.). The example discussed below is a general outline of our effort to bring human neural progenitor cells releasing GDNF to the clinic for ALS.

CMC: Central to the development of any medicinal cellular product is the requirement to delineate the details underlying its manufacture. This requires a lot of attention and effort. The cellular product has to be manufactured under sterile conditions and meet specific quality control measures. Further, the methods used in the general laboratory have to be scaled for production in a good manufacturing practice (cGMP) facility. This process is often complex. Things that seem simple in an academic tissue culture laboratory take on another dimension when it is required to use fully protective clothing and to precisely follow strict protocols for an accurate and reproducible "batch record" system.

Our cellular product was the human fetal-derived cortical neural progenitor cells that were grown as neurospheres and were expanded using a unique chopping process to avoid single-cell dissociation.[85] The first task was to transfer the cells into a cGMP facility (City of Hope, Duarte, CA), which required full adventitious agent testing to prove that there were no contaminants or infectious agents in the cultures. Once the cells passed this test, they could then be propagated within the cGMP facility following strict batch record documentation. This led to a clinical-grade Master Cell Bank (MCB) of hNPCs, termed CNS010.

As this was a combined cell and gene therapy product, we had to include a lentiviral infection procedure to engineer the cells to produce GDNF. The lenti-GDNF was produced by a separate CRO (The National Gene Vector Laboratories, University of Indiana), under cGMP and FDA guidelines. The clinical-grade lenti-GDNF and accompanying documents were then transferred to the City of Hope. At a specific stage of the expansion, the CNS10 cells were transduced with lenti-GDNF and then further expanded for cryopreservation and banking of our final cellular and gene clinical product, termed CNS10-GDNF.

The cryopreservation and storage of the cellular product required extensive laboratory studies to develop reliable freezing protocols (that involved control rate freezers) and proof that upon thaw the product retained efficacy both in vitro and in vivo. The final stage was to optimize dose preparation and shipment of products to the clinical site. We developed a hibernate media (initially in collaboration with Sigma-Aldrich) that could maintain final thawed cells at 4°C for up to 4 days with no reduction in survival or efficacy following transplantation. The final part of the CMC protocol was to demonstrate that the banked CNS10-GDNF could be thawed, rinsed, and prepared at the final dose in a vial, which was then shipped from the cGMP facility to the clinical site pharmacy for storage overnight at 4°C and ultimate delivery to the patient the next day.

Preclinical dose-ranging studies: These studies are aimed at establishing the optimal dose to pursue in your IND-enabling tumorigenicity and toxicology studies. The dose-ranging studies in an animal model that shows some features of the targeted disease are generally carried out in a non-GLP laboratory. The model should include some features of the targeted disease and include a good outcome of efficacy—either cell survival or functional improvement of the animals. However, these preclinical studies are still part of the FDA application and everything that is undertaken must be documented using good laboratory practice (GLP) record keeping. It is not permitted to stop the study, disregard data, and start again. As the FDA considers any experiments with the cellular product as part of the IND and reportable to the agency, any study that fails or creates surprising results needs to be documented as a study report. The dose-ranging studies need to range from the lowest dose (where there is no effect) to the maximal feasible dose (where there is sometimes a toxic effect). Once the window of effect is established, the goal is to determine the optimal dose that is efficacious but not toxic. The dose-ranging study typically involves a log increase in cell number (e.g., 1000, 10,000, 100,000, and 1,000,000 cells). Ideally, the dose-ranging study uses the final cell product from the Master Cell Bank, although a highly equivalent product can be used in order to reduce costs and reserve cells in the Master Cell Bank for the clinical trial. In our extensive preclinical studies, we delivered these cells to the spinal cord of mutant SOD1 rats and showed efficacy based on increased survival of motor neurons[83,84] as well as to the spinal cord of immunocompromised rats and showed safety based on no observed tumors at 6 months[86] (Fig. 3). The dose-ranging study replicated our published findings and established an ideal dose to be used in the tumor tox studies described below.

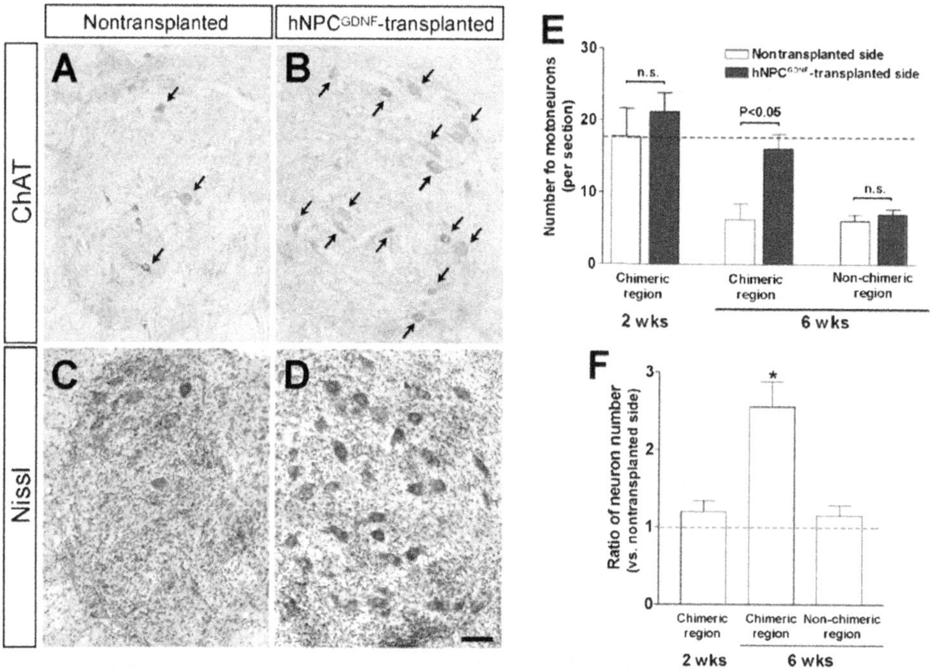

Fig. 3 Motor neurons are protected by GDNF secreting hNPC transplants in the spinal cord of $SOD1^{G93A}$ rats. The number of motor neurons was higher in hNPC-transplanted (chimeric) side (B, D) compared to the contralateral nongrafted (nonchimeric) hemisphere (A, C). Cell number reduced by 70% at 6 weeks postgrafting in sham treated $SOD1^{G93A}$ rats (E). There was almost complete protection of motor neuron loss within chimeric (transplant) regions of the spinal cord. Outside of these transplant (nonchimeric) regions motor neuron numbers dropped to less than 70%. The broken line indicates the number of motor neurons in wild-type rats. (F) The ratio of motor neuron numbers compared to the contralateral side. $^{*}P < .05$ vs. 2 weeks and 6 weeks outside groups. Scale bar: 100 μm in (9D). *(From Suzuki M., McHugh J., Tork C., et al. GDNF secreting human neural progenitor cells protect dying motor neurons, but not their projection to muscle, in a rat model of familial ALS. PLoS One. 2007;2(8):e689. https://doi.org/10.1371/journal.pone.0000689.)*

Tumor tox studies: This is the crucial part of the IND process where the product is delivered to animals and shown not to have any tumor activity or toxicity. Two things must happen at this stage of the process. First, all these studies need to use the final product from the Master Cell Bank. As such, the original calculations for how many vials to bank needs to plan for enough to both obtain the IND and complete the trial. Second, these studies have to be conducted under the auspices of a GLP-compliant CRO and performed under highly regulated GLP conditions. It is vital to visit and even audit the CRO in order to ensure that cell delivery to the animals can be properly performed. Tumor tox studies generally require 6–9 months of in vivo testing in an immunodeficient animal (normally a mouse or rat). The report back from the CRO is nothing like any scientific paper. It includes an exhaustive analysis of the findings for each treated animal in the report. Any toxic effect on any organ system and/or the presence of tumor growth is reported. Obviously, the area of the transplant is critically assessed, along with cell distribution. Our tumor tox studies for CNS10-NPC-GDNF showed that the cells survived well, did not form tumors, and were limited to the region of the transplant. We did see some swelling of the dorsal roots at the site of the spinal cord injection and formation of neuromas, which was related to the cell product and reported in the IND to the FDA.

Large animal studies and device development: This is a critical aspect in studies that involve the direct injection of cells into the brain or spinal cord. A range of large animals can be used, but pigs or nonhuman primates are the most common. The idea is to show safety using the identical targeting device and cannula that will be used in the clinical study. Part of our IND application included a new minimally invasive stereotaxic device to deliver the cells to the patient's spinal cord. This, in itself, was a major undertaking and involved a separate series of interactions with another part of the FDA—CDRH—in order to get approval for use under our main IND. We showed that the newly developed stereotactic device provided successful delivery of the cells to the minipig spinal cord with survival for up to 1 month and no long-term paralysis in any treated animal.

Clinical protocol. At the heart of any IND is the clinical protocol, which describes exactly what will occur in the clinical trial. Classically, Phase I trials enroll control populations and assess the overall safety of the product, Phase II studies enroll patient groups and assess safety (IIa) and often efficacy (IIb) and Phase III studies are large efficacy studies to demonstrate whether the product has significant effects on patients. Ours was a Phase IIa study in that its primary endpoint was safety. Obviously, the invasive approach of delivering our product to the spinal cord could not be performed in control subjects.

After all of this information was collected, everything needed to be condensed into an integrated summary that covered all the major aspects of the various reports. After the entire document was collated and indexed for the FDA, our application was 5 volumes and over 5000 pages (Fig. 4).

Fig. 4 Paper copy of the Investigational New Drug (IND) submitted to the Food and Drug Administration for the first-ever combined cell and gene therapy for ALS. This shows four copies of the submission, with each copy consisting of 4646 pages!

After IND submission: Submitting a complex IND to the FDA that includes cell therapy, gene therapy, and a new delivery device for a process that has never been attempted in humans before requires much dedication and an amazing team effort. From the collection of data via the animal studies to the reading and proofing of the reports, every member of the team is critical to making a successful application. When the submission arrives at the FDA, it is split into the various sections (CMC, dose ranging, tumor/toxicology, large animal/device development) and each is sent out for review. After the submission date, the FDA has just 30 days to respond. In our case, there were multiple phone calls about various aspects of the application, creating a number of clinical holds over a period of 6 months. Once all of the questions had been addressed, we finally received FDA approval to move forward with the trial—a very exciting day for the entire team.

Executing the trial: Once we received FDA-approval to initiate the clinical trial, we applied to CIRM, who had already funded all the preclinical work, to fund the trial itself. This grant was successfully funded and in May 2017, the first ALS patient was dosed with CNS10-NPC-GDNF (Clinicaltrials.gov, NCT02943850). This is a dose-escalation study using 2,000,000 ($n = 9$) or 5,000,000 ($n = 9$) CNS10-NPC-GDNF in 10 unilateral injections into the lumbar spinal cord of 18 subjects with moderate leg involvement. As we have shown that bilateral limbs deteriorate at a similar rate in ALS,[87] we chose unilateral injections so that the progression rate could be monitored against the contralateral nontreated side as a way to measure efficacy. The primary outcome measurement is safety as evaluated by adverse events and serious adverse events, postoperative MRI, and clinical laboratory assessments. Secondary outcome measurements include assessment of muscle strength in the treated vs nontreated legs over time.

It was a very exciting time seeing the cumulation of many years work finally come to fruition, but of course, also an anxious time until product safety was confirmed. The FDA required a 1-month interval between patients in order to make sure that there were no serious adverse events associated with this first in-man study. The first patient recovered well from the surgery and the FDA allowed us to continue. Thanks to the outstanding teamwork from a group of over 30 professionals, we completed the 18-patient trial within our targeted time frame. There were no unanticipated adverse events seen to date—demonstrating the safety of this cellular product. One of the bigger challenges of the trial was the 1-year immune suppression regime, which caused side effects in many patients, and in some cases, the suppression had to be halted. This is a frustrating requirement, as it is not certain whether allogeneic transplants to the CNS even require immune suppression after the first few weeks, once the blood–brain barrier repairs to prevent the entry of host immune cells. At the time of writing this chapter, we are still collecting patient data, but we can say that all postmortem cases to date have surviving grafts up to 2 years after transplantation that were still secreting GDNF into the spinal cord. This was very satisfying and established that CNS10-NPC-GDNF can survive in the human spinal cord, even after withdrawal of immune suppression for at least 1 year.

One IND provides parallel trials for ALS: ALS involves more than the loss of motor neurons in the spinal cord. There is also a loss of upper motor neurons in the cortex, which initiate movement. Interestingly, in recent studies, we have shown that knocking down SOD1 only in the motor cortex or transplanting CNS10-NPC-GDNF directly into the motor cortex of the mutant SOD1 rat model of ALS can delay the death of both upper and lower motor neurons.[88,89] Critically, delivering GDNF-releasing human neural progenitors to the motor cortex also delayed motor function decline and extended lifespan. Through extensive preclinical studies in rodents and nonhuman primates, we have already obtained FDA approval to transplant CNS10-NPC-GDNF directly into the motor cortex of ALS patients. Ultimately, the answer for good functional effects in patients will probably be dual transplantation into both the upper motor neuron region and the spinal cord to protect both sets of motor neurons. These studies will hopefully be carried out over the coming years.

A cellular product for many different disease indications

While performing the IND-enabling studies and advancing the use of CNS10-NPC-GDNF for the treatment of ALS, the Svendsen laboratory has also been busy using these cells in other indications. These cells had initially been developed for Parkinson's disease, in order to provide young astrocytes secreting GDNF that would protect remaining dopamine neurons and regenerate new terminals in the striatum.[73,90] To extend our published work with these cells in rodent and nonhuman primate models of PD,[50,73,74] we applied for CIRM funding to perform further IND-enabling studies to move to the clinic for this disorder as well. In collaboration with Dr. Krys Bankiewicz at UCSF, we were awarded a CIRM grant to a produce a new cGMP bank of CNS10-NPC-GDNF (using an identical process as described above for ALS) as well as to complete all of the required preclinical studies, specifically dose-ranging and tumor/tox studies in rodents and large animal studies in nonhuman primates to show that the cells survive, to produce GDNF, and are safe. These studies are ongoing at the time of writing and will be used to initiate a Phase I/IIa trial to transplant CNS10-NPC-GDNF into the striatum of PD patients.

In addition to ALS and PD that respond to CNS10-NPC-GDNF, other indications may respond to just the parent neural progenitor cell product (without GDNF modification), which we have developed and banked under cGMP (CNS10-NPC). This is because the neural progenitor cell-derived young astrocytes may release protective factors and/or take-up deleterious factors that then protects a range of cell types within the brain, spinal cord, and even retina. Wild-type cortical-derived neural progenitor cells can be delivered to the cervical spinal cord of aging rats, where they differentiate into astrocytes and protect motor neurons.[91] In addition, delivery of these wild-type neural progenitors to the cervical spinal cord of ALS rats leads to motor neuron protection and preservation of respiration, which was an unexpected finding as motor neuron protection in the lumbar spinal cord and the motor cortex required the synergistic effect from cells and GDNF.[92] When wild-type CNS10-NPC are transplanted into the rat cortex 7 days poststroke, we have shown in collaboration with Dr. Gary Steinberg at Stanford that the cells exhibited robust survival and migration toward the lesion at 5 weeks posttransplantation, leading to axonal rewiring and functional recovery. Corresponding in vitro studies demonstrated that the neural progenitor cells secreted various factors that could enhance brain repair and plasticity to aid recovery after stroke.[93] In addition, in collaboration with Matt Blurton-Jones at UCLA, we showed that transplantation of CNS10-NPC into the striatum of a mouse model of Lewy body dementia led to reduced synuclein levels and increased behavioral function.[94] Similarly, striatal transplants of cortical-derived human neural progenitors led to improved motor function in a rodent model of Huntington's Disease.[95] Finally, working with both Drs. David Gamm and Shaomei Wang, we have used the well-established Royal College of Surgeons (RCS) rat and USH2A mouse models of retinal degeneration and shown that our neural progenitor cells can survive long-term following subretinal transplantation, leading to significant photoreceptor rescue and preservation of vision.[96,97] Our extensive preclinical work, which includes dose-ranging studies in the RCS rat, GLP tumorigenicity, and toxicity studies in the RCS rat, as well as scale-up and surgical feasibility in a large animal pig model, has led to IND-approval of CNS10-NPC for the treatment of retinitis pigmentosa (RP). We applied to CIRM, who had already funded much of the preclinical work, to fund the trial itself. This grant was successfully funded and the Phase 1/2a trial expects to enroll the first RP patient in March 2020. This will be an open-label unilateral dose-escalation study ($n = 3$ low dose and $n = 3$ high dose in advanced RP and $n = 10$ best-tolerated dose in moderate RP) with safety as the primary endpoint. Secondary endpoints include comparisons of vision between treated and nontreated eyes.

Together, these studies suggest that one cell product may have protective effects in regions of ongoing degeneration in a range of different disorders (Fig. 5). The cells probably work through multiple mechanisms to support the niche around aging or sick neurons or retinal cells leading to survival and enhanced function. This promising area of neurotherapeutics is entering a very critical period and the coming years will be an exciting time for the field.

The growth and commercialization of cellular neurotherapeutics based on fetal neural progenitors and stem cells

While this chapter has been focused on our experience bringing stem cells and gene therapy to the clinic for ALS and PD, there is a broad field of cellular neurotherapeutics that has emerged from academic laboratories into commercial ventures. One of the first of these companies in the United States was called Stem Cells, Inc. founded by a group

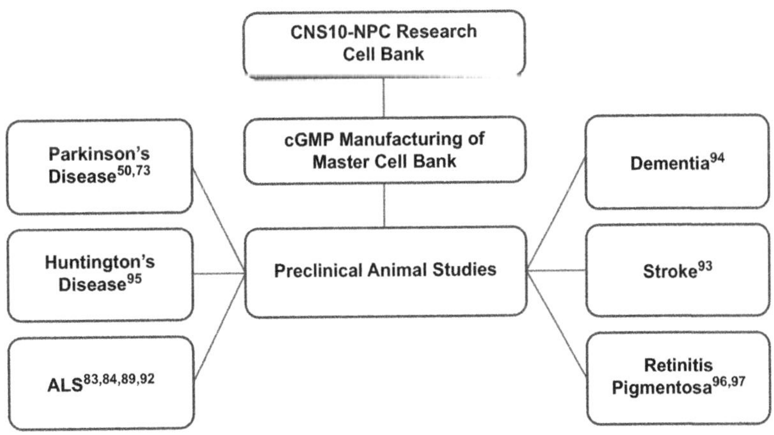

Fig. 5 One cell product has been transplanted in animal models for a range of different disorders and shown protective effects in regions of ongoing degeneration. Associated references are provided for each disorder.

of academic researchers and focused on using human fetal tissue-derived neural progenitors in the clinic for a range of neurological disorders. The company would obtain multiple fetal brain samples and expand them with EGF and FGF-2 as neurospheres, which could then give rise to neurons, astrocytes, or oligodendrocytes following transplantation. These cells (termed HuCNS-SC) received FDA approval as an investigational new drug in March 2006.[98] Promising preclinical data came from transplantation of research-grade HuCNS-SC in spinal cord injury,[99] retinitis pigmentosa,[100] a rare leukodystrophy known as Pelizaeus-Merzbacher disease (PMD)[101] and a fatal childhood neurological disorder known as infantile and late-infantile neuronal ceroid lipofuscinosis (NCLs) that are a subset of lysosomal storage diseases.[102] Numerous clinical trials were performed, including for NCLs (ClinicalTrials.gov Identifier: NCT00337636); PMD (NCT01391637); age-related macular degeneration (NCT01632527 and NCT02467634); as well as thoracic and cervical spinal cord injury (NCT01321333 and NCT02163876). One of the initial clinical trials transplanted HuCNS-SC in children with NCL and confirmed the absence of transplantation-related serious adverse events, however, observations regarding efficacy were limited by the enrollment criteria requiring that patients be in advanced stages of disease.[103]

While the number and breadth of clinical trials performed with HuCNS-SC product are impressive, the outcome was disappointing. For instance, in the Phase 2 study that enrolled about 30 adults with spinal cord injuries for the delivery of clinical-grade HuCNS-SC, the first six treated patients notched significant improvements at 6 months, but these tapered off toward baseline by 12 months. A 6-month interim analysis of the remaining participants revealed only modest benefits in the treatment group that did not reach statistical significance. Company scientists halted the trial, concluding that the gains in the larger cohort were unlikely to reach the primary endpoint.

This study outcome may have been predicted with additional preclinical studies, as indicated by separate publications comparing the performance of clinical and research-grade HuCNS-SC in mouse models of spinal cord injury and Alzheimer's disease.[104, 105] In both studies, the research-grade lines showed a treatment benefit, but the clinical-grade cells did not. These findings suggest that there may have been an issue in the clinical manufacturing process. Another potential problem was that StemCells Inc. generated cell lines from numerous different fetal samples, creating large variability between lines. This highlights the importance of solid cGMP processing and the possible complications of procuring new fetal brain tissue for each production lot. The company did not survive after August 2016, although recent papers where these cells had been used in clinical trials are still emerging, for instance, a preclinical study for retinal degeneration[106] and clinical trial results based on patients that received cells for PMD as well as cervical and thoracic spinal cord injury.[107, 108]

Another company, NeuralStem, has the largest Phase 1 and 2 trials to date for ALS using intraspinal transplantation of human spinal stem cells (termed NSI-566). Here a single fetal spinal cord sample was expanded in FGF-2 based on the pioneering work of Ron McKay.[32] These cells, which differentiate mostly into interneurons and less frequently into astrocytes and release numerous growth factors, have been shown to survive transplantation into the rodent spinal cord in models of ALS and in some cases have functional effects. Further, these cells were demonstrated to be safe in a large pig model.[109,110] Cell delivery parameters were in part optimized in collaboration with the Svendsen lab using human neural progenitor cells.[111,112] Based on preclinical studies, the company was awarded an IND to treat ALS patients. The phase 1 single-center trial (ClinicalTrials.gov NCT01348451) delivered cells to the lumbar and/or cervical spinal cord of ALS patients and met primary endpoint showing safety and tolerability of intraspinal injections.[113] This led to a Phase 2 study (ClinicalTrials.gov NCT01730716), with expansion to three study centers and dose escalation from 2 million cells via 20 injections into the cervical cord, to 16 million cells via 40 injections into both the lumbar and cervical cord.[114,115] Again, this trial met the primary endpoint of safety, though two of the highest dose patients experienced significant, nonfatal complications. Ultimately, there was no dramatic effect across patients, and without a larger clinical trial in ALS along with control groups, it is difficult to detect smaller effects on outcome measures. A phase 3 trial may be the next step, but this is not funded or initiated at this time. The company also provided NSI-566 cells to clinical trials for spinal cord injury[116,117] (ClinicalTrials.gov NCT01772810) and stroke[118] (ClinicalTrials.gov NCT03296618). But in September 2019, NeuralStem was renamed as Seneca Biopharma Inc. and, according to the company's executive chairman, the change points to the company's shift away from a prior emphasis on research related to neurological disorders.

In addition to trials in the United States, a trial in ALS using expanded human fetal-derived neural stem cells derived from the forebrain of the same two donors has also been completed in Europe (ClinicalTrials.gov Identifier. NCT01640067). It was recently reported that for up to 60 months after surgery, none of the patients manifested severe adverse effects or increased disease progression because of the treatment.[119] There was a decrease in progression of the ALS Functional Rating Scale Revised, starting within the first month after surgery and up to 4 months after transplantation, but this was transitory.

A therapeutics company from the United Kingdom called ReNeuron was founded over 30 years ago based on modifying human fetal cortical cells with the oncogene cMYC to enable large-scale stable banking.[120] The clonal cell line (named CTX0E03) can survive transplantation into models of stroke and the mechanism of action may be through paracrine trophic factors.[121] The company performed a successful Phase I safety trial termed PISCES (ClinicalTrials.gov NCT01151124),[122] which led to a Phase IIa trial termed PISCES II (ClinicalTrials.gov NCT02117635). As a result of the positive data from this trial, the FDA approved a randomized, placebo-controlled, Phase IIb trial termed PISCES III (ClinicalTrials.gov NCT03629275) for stroke patients, which is currently enrolling approximately 130 patients across up to 40 centers. Reneuron also has a current Phase I/IIa clinical trial for patients with RP using human retinal progenitor cells (hRPCs) (ClinicalTrials.gov Identifier: NCT02464436), which is based on preclinical data showing that hRPCs are safe and can preserve vision in the RCS rat.[123,124] The trial design is an open-label, dose-escalation study to evaluate the safety, tolerability, and preliminary efficacy of the hRPC therapy for patients with advanced RP. As with our trial using CNS10-NPC, the hRPCs were delivered as a single subretinal injection. While lacking published preclinical data, another company, JCyte, is also delivering human retinal progenitor cells (termed JCell) in Phase I/IIa and IIb clinical trials for retinitis pigmentosa, via intravitreal injection (ClinicalTrials.gov Identifier: NCT02320812 and NCT03073733). JCyte recently received funding for an additional trial, which is yet to commence. Thus far, both the ReNeuron and JCyte therapies have had a good safety profile and shown early signs of efficacy. ReNeuron maybe one of the most successful stem cell companies to survive the past few decades.

In addition to these companies that have active clinical trials based around human fetal brain tissue products, another promising source of cells for treating retinal degeneration comes from human embryonic stem cells (hESCs). For instance, Phase I/II open-label, multicenter clinical trials delivering hESC-derived retinal pigmented epithelial (ma09-hRPE) cells to patients with either Stargardt's Macular Dystrophy (ClinicalTrials.gov Identifier: NCT01345006) or age-related macular degeneration (ClinicalTrials.gov Identifier: NCT01344993).[125,126] Finally, there are emerging companies with a focus on human-induced pluripotent stem cell (iPSC) technology. Leading the charge is BlueRock Therapeutics (under Bayer following an acquisition), which is currently optimizing human iPSC-derived dopamine neurons as a source of dopamine neuron replacement for PD as presented earlier in this chapter.

Conclusions

In summary, the commercial efforts, as described herein, are driving the move of fetal and stem cell-derived neural progenitors into real clinical products that, upon FDA-approval, will be available for use in the general patient population. However, it is a tough road to FDA approval and traditional "Big Pharma" often finds it too risky to invest in the lengthy and stringent requirements of preclinical research for untested therapies. It will likely be efforts of basic science laboratories that expand beyond simply academic research and small start-up companies with innovative ideas and cellular neurotherapeutic products that eventually make the breakthrough and bring these exciting possibilities to patients.

References

1. Harms MB, Baloh RH. Clinical neurogenetics: amyotrophic lateral sclerosis. *Neurol Clin*. 2013;31(4):929–950. https://doi.org/10.1016/j.ncl.2013.05.003.
2. Brettschneider J, Del Tredici K, Lee VM-Y, Trojanowski JQ. Spreading of pathology in neurodegenerative diseases: a focus on human studies. *Nat Rev Neurosci*. 2015;16(2):109–120. https://doi.org/10.1038/nrn3887.
3. Alarcón-Arís D, Recasens A, Galofré M, et al. Selective α-Synuclein knockdown in monoamine neurons by intranasal oligonucleotide delivery: potential therapy for Parkinson's disease. *Mol Ther*. 2018;26(2):550–567. https://doi.org/10.1016/j.ymthe.2017.11.015.
4. Tóth G, Neumann T, Berthet A, et al. Novel small molecules targeting the intrinsically disordered structural ensemble of α-Synuclein protect against diverse α-Synuclein mediated dysfunctions. *Sci Rep*. 2019;9(1):16947. https://doi.org/10.1038/s41598-019-52598-4.
5. Taylor JP, Brown RH, Cleveland DW. Decoding ALS: from genes to mechanism. *Nature*. 2016;539(7628):197–206. https://doi.org/10.1038/nature20413.
6. Diaz NL, Waters CH. Current strategies in the treatment of Parkinson's disease and a personalized approach to management. *Expert Rev Neurother*. 2009;9(12):1781–1789. https://doi.org/10.1586/ern.09.117.
7. Jaiswal MK. Riluzole and edaravone: a tale of two amyotrophic lateral sclerosis drugs. *Med Res Rev*. 2019;39(2):733–748. https://doi.org/10.1002/med.21528.
8. Mehndiratta M, Garg RK, Pandey S. Nonmotor symptom complex of Parkinson's disease–an under-recognized entity. *J Assoc Physicians India*. 2011;59:302–308 313.
9. Bassetti CL. Nonmotor disturbances in Parkinson's disease. *Neurodegener Dis*. 2011;8(3):95–108. https://doi.org/10.1159/000316613.
10. Spenger C, Haque NS, Studer L, et al. Fetal ventral mesencephalon of human and rat origin maintained in vitro and transplanted to 6-hydroxydopamine-lesioned rats gives rise to grafts rich in dopaminergic neurons. *Exp Brain Res*. 1996;112(1):47–57. https://doi.org/10.1007/bf00227177.
11. Brundin P, Barbin G, Strecker RE, Isacson O, Prochiantz A, Björklund A. Survival and function of dissociated rat dopamine neurones grafted at different developmental stages or after being cultured in vitro. *Brain Res*. 1988;467(2):233–243. https://doi.org/10.1016/0165-3806(88)90027-2.
12. Lindvall O. Clinical application of neuronal grafts in Parkinson's disease. *J Neurol*. 1994;242(1 Suppl 1):S54–S56. https://doi.org/10.1007/bf00939243.
13. Björklund A, Stenevi U. Reconstruction of the nigrostriatal dopamine pathway by intracerebral nigral transplants. *Brain Res*. 1979;177(3):555–560. https://doi.org/10.1016/0006-8993(79)90472-4.
14. Nikkhah G, Cunningham MG, McKay R, Björklund A. Dopaminergic microtransplants into the substantia nigra of neonatal rats with bilateral 6-OHDA lesions. II. Transplant-induced behavioral recovery. *J. Neuroscience*. 1995;15(5 Pt 1):3562–3570.
15. Bentlage C, Nikkhah G, Cunningham MG, Björklund A. Reformation of the nigrostriatal pathway by fetal dopaminergic micrografts into the substantia nigra is critically dependent on the age of the host. *Exp Neurol*. 1999;159(1):177–190. https://doi.org/10.1006/exnr.1999.7110.
16. Redmond DE, Elsworth JD, Roth RH, et al. Embryonic substantia nigra grafts in the mesencephalon send neurites to the host striatum in non-human primate after overexpression of GDNF. *J Comp Neurol*. 2009;515(1):31–40. https://doi.org/10.1002/cne.22028.
17. Wakeman DR, Redmond DE, Dodiya HB, et al. Human neural stem cells survive long term in the midbrain of dopamine-depleted monkeys after GDNF overexpression and project neurites toward an appropriate target. *Stem Cells Transl Med*. 2014;3(6):692–701. https://doi.org/10.5966/sctm.2013-0208.
18. Dunnett SB, Björklund A, Lindvall O. Cell therapy in Parkinson's disease—stop or go? *Nat Rev Neurosci*. 2001;2(5):365–369. https://doi.org/10.1038/35072572.
19. Li W, Englund E, Widner H, et al. Extensive graft-derived dopaminergic innervation is maintained 24 years after transplantation in the degenerating parkinsonian brain. *Proc Natl Acad Sci U S A*. 2016;113(23):6544–6549. https://doi.org/10.1073/pnas.1605245113.

20. Kordower JH, Goetz CG, Chu Y, et al. Robust graft survival and normalized dopaminergic innervation do not obligate recovery in a Parkinson disease patient. *Ann Neurol*. 2017;81(1):46–57. https://doi.org/10.1002/ana.24820.
21. Kurowska Z, Englund E, Widner H, Lindvall O, Li J-Y, Brundin P. Signs of degeneration in 12-22-year old grafts of mesencephalic dopamine neurons in patients with Parkinson's disease. *J Parkinsons Dis*. 2011;1(1):83–92. https://doi.org/10.3233/JPD-2011-11004.
22. Björklund A, Lindvall O. Replacing dopamine neurons in Parkinson's disease: how did it happen? *J Parkinsons Dis*. 2017;7(s1):S21–S31. https://doi.org/10.3233/JPD-179002.
23. Olanow CW, Goetz CG, Kordower JH, et al. A double-blind controlled trial of bilateral fetal nigral transplantation in Parkinson's disease. *Ann Neurol*. 2003;54(3):403–414. https://doi.org/10.1002/ana.10720.
24. Freed CR, Greene PE, Breeze RE, et al. Transplantation of embryonic dopamine neurons for severe Parkinson's disease. *N Engl J Med*. 2001;344(10):710–719. https://doi.org/10.1056/NEJM200103083441002.
25. Hagell P, Piccini P, Björklund A, et al. Dyskinesias following neural transplantation in Parkinson's disease. *Nat Neurosci*. 2002;5(7):627–628. https://doi.org/10.1038/nn863.
26. Politis M, Wu K, Loane C, et al. Serotonergic neurons mediate dyskinesia side effects in Parkinson's patients with neural transplants. *Sci Transl Med*. 2010;2(38):38ra46. https://doi.org/10.1126/scitranslmed.3000976.
27. Olanow CW, Kordower JH, Lang AE, Obeso JA. Dopaminergic transplantation for Parkinson's disease: current status and future prospects. *Ann Neurol*. 2009;66(5):591–596. https://doi.org/10.1002/ana.21778.
28. Stoker TB, Blair NF, Barker RA. Neural grafting for Parkinson's disease: challenges and prospects. *Neural Regen Res*. 2017;12(3):389–392. https://doi.org/10.4103/1673-5374.202935.
29. Palmer TD, Ray J, Gage FH. FGF-2-responsive neuronal progenitors reside in proliferative and quiescent regions of the adult rodent brain. *Mol Cell Neurosci*. 1995;6(5):474–486. https://doi.org/10.1006/mcne.1995.1035.
30. Doetsch F, Petreanu L, Caille I, Garcia-Verdugo JM, Alvarez-Buylla A. EGF converts transit-amplifying neurogenic precursors in the adult brain into multipotent stem cells. *Neuron*. 2002;36(6):1021–1034. https://doi.org/10.1016/s0896-6273(02)01133-9.
31. Svendsen CN, Caldwell MA, Ostenfeld T. Human neural stem cells: isolation, expansion and transplantation. *Brain Pathol*. 1999;9(3):499–513.
32. Brüstle O, McKay RD. Neuronal progenitors as tools for cell replacement in the nervous system. *Curr Opin Neurobiol*. 1996;6(5):688–695. https://doi.org/10.1016/S0959-4388(96)80104-8.
33. Studer L, Tabar V, McKay RD. Transplantation of expanded mesencephalic precursors leads to recovery in parkinsonian rats. *Nat Neurosci*. 1998;1(4):290–295. https://doi.org/10.1038/1105.
34. Ostenfeld T, Horn P, Aardal C, Orpen I, Caldwell MA, Svendsen CN. Mouse epidermal growth factor-responsive neural precursor cells increase the survival and functional capacity of embryonic rat dopamine neurons in vitro. *Neuroreport*. 1999;10(9):1985–1992. https://doi.org/10.1097/00001756-199906230-00035.
35. Caldwell MA, He X, Wilkie N, et al. Growth factors regulate the survival and fate of cells derived from human neurospheres. *Nat Biotechnol*. 2001;19(5):475–479. https://doi.org/10.1038/88158.
36. Lee SH, Lumelsky N, Studer L, Auerbach JM, McKay RD. Efficient generation of midbrain and hindbrain neurons from mouse embryonic stem cells. *Nat Biotechnol*. 2000;18(6):675–679. https://doi.org/10.1038/76536.
37. Kim J-H, Auerbach JM, Rodríguez-Gómez JA, et al. Dopamine neurons derived from embryonic stem cells function in an animal model of Parkinson's disease. *Nature*. 2002;418(6893):50–56. https://doi.org/10.1038/nature00900.
38. Thomson JA, Itskovitz-Eldor J, Shapiro SS, et al. Embryonic stem cell lines derived from human blastocysts. *Science*. 1998;282(5391):1145–1147. https://doi.org/10.1126/science.282.5391.1145.
39. Zhang SC, Wernig M, Duncan ID, Brüstle O, Thomson JA. In vitro differentiation of transplantable neural precursors from human embryonic stem cells. *Nat Biotechnol*. 2001;19(12):1129–1133. https://doi.org/10.1038/nbt1201-1129.

40. Perrier AL, Tabar V, Barberi T, et al. Derivation of midbrain dopamine neurons from human embryonic stem cells. *Proc Natl Acad Sci U S A*. 2004;101(34):12543–12548. https://doi.org/10.1073/pnas.0404700101.
41. Yan Y, Yang D, Zarnowska ED, et al. Directed differentiation of dopaminergic neuronal subtypes from human embryonic stem cells. *Stem Cells*. 2005;23(6):781–790. https://doi.org/10.1634/stemcells.2004-0365.
42. Kriks S, Shim J-W, Piao J, et al. Dopamine neurons derived from human ES cells efficiently engraft in animal models of Parkinson's disease. *Nature*. 2011;480(7378):547–551. https://doi.org/10.1038/nature10648.
43. Yu J, Vodyanik MA, Smuga-Otto K, et al. Induced pluripotent stem cell lines derived from human somatic cells. *Science*. 2007;318(5858):1917–1920. https://doi.org/10.1126/science.1151526.
44. Takahashi K, Tanabe K, Ohnuki M, et al. Induction of pluripotent stem cells from adult human fibroblasts by defined factors. *Cell*. 2007;131(5):861–872. https://doi.org/10.1016/j.cell.2007.11.019.
45. Barker RA, Parmar M, Studer L, Takahashi J. Human trials of stem cell-derived dopamine neurons for Parkinson's disease: dawn of a new era. *Cell Stem Cell*. 2017;21(5):569–573. https://doi.org/10.1016/j.stem.2017.09.014.
46. Takahashi J. Preparing for first human trial of induced pluripotent stem cell-derived cells for Parkinson's disease: an interview with Jun Takahashi. *Regen Med*. 2019;14(2):93–95. https://doi.org/10.2217/rme-2018-0158.
47. Tuszynski MH, Thal L, Pay M, et al. A phase 1 clinical trial of nerve growth factor gene therapy for Alzheimer disease. *Nat Med*. 2005;11(5):551–555. https://doi.org/10.1038/nm1239.
48. Rafii MS, Tuszynski MH, Thomas RG, et al. Adeno-associated viral vector (serotype 2)-nerve growth factor for patients with Alzheimer disease: a randomized clinical trial. *JAMA Neurol*. 2018;75(7):834–841. https://doi.org/10.1001/jamaneurol.2018.0233.
49. Lin LF, Doherty DH, Lile JD, Bektesh S, Collins F. GDNF: a glial cell line-derived neurotrophic factor for midbrain dopaminergic neurons. *Science*. 1993;260(5111):1130–1132. https://doi.org/10.1126/science.8493557.
50. Emborg ME, Ebert AD, Moirano J, et al. GDNF-secreting human neural progenitor cells increase tyrosine hydroxylase and VMAT2 expression in MPTP-treated cynomolgus monkeys. *Cell Transplant*. 2008;17(4):383–395.
51. Björklund A, Rosenblad C, Winkler C, Kirik D. Studies on neuroprotective and regenerative effects of GDNF in a partial lesion model of Parkinson's disease. *Neurobiol Dis*. 1997;4(3–4):186–200. https://doi.org/10.1006/nbdi.1997.0151.
52. Kirik D, Rosenblad C, Björklund A. Preservation of a functional nigrostriatal dopamine pathway by GDNF in the intrastriatal 6-OHDA lesion model depends on the site of administration of the trophic factor. *Eur J Neurosci*. 2000;12(11):3871–3882. https://doi.org/10.1046/j.1460-9568.2000.00274.x.
53. Nutt JG, Burchiel KJ, Comella CL, et al. Randomized, double-blind trial of glial cell line-derived neurotrophic factor (GDNF) in PD. *Neurology*. 2003;60(1):69–73. https://doi.org/10.1212/wnl.60.1.69.
54. Bobo RH, Laske DW, Akbasak A, Morrison PF, Dedrick RL, Oldfield EH. Convection-enhanced delivery of macromolecules in the brain. *Proc Natl Acad Sci U S A*. 1994;91(6):2076–2080. https://doi.org/10.1073/pnas.91.6.2076.
55. Gill SS, Patel NK, Hotton GR, et al. Direct brain infusion of glial cell line-derived neurotrophic factor in Parkinson disease. *Nat Med*. 2003;9(5):589–595. https://doi.org/10.1038/nm850.
56. Patel NK, Bunnage M, Plaha P, Svendsen CN, Heywood P, Gill SS. Intraputamenal infusion of glial cell line-derived neurotrophic factor in PD: a two-year outcome study. *Ann Neurol*. 2005;57(2):298–302. https://doi.org/10.1002/ana.20374.
57. Love S, Plaha P, Patel NK, Hotton GR, Brooks DJ, Gill SS. Glial cell line-derived neurotrophic factor induces neuronal sprouting in human brain. *Nat Med*. 2005;11(7):703–704. https://doi.org/10.1038/nm0705-703.
58. Slevin JT, Gerhardt GA, Smith CD, Gash DM, Kryscio R, Young B. Improvement of bilateral motor functions in patients with Parkinson disease through the unilateral intraputaminal infusion of glial cell

line-derived neurotrophic factor. *J Neurosurg.* 2005;102(2):216–222. https://doi.org/10.3171/jns.2005.102.2.0216.

59. Lang AE, Gill S, Patel NK, et al. Randomized controlled trial of intraputamenal glial cell line-derived neurotrophic factor infusion in Parkinson disease. *Ann Neurol.* 2006;59(3):459–466. https://doi.org/10.1002/ana.20737.
60. Hovland DN, Boyd RB, Butt MT, et al. Six-month continuous intraputamenal infusion toxicity study of recombinant methionyl human glial cell line-derived neurotrophic factor (r-metHuGDNF) in rhesus monkeys. *Toxicol Pathol.* 2007;35(5):676–692. https://doi.org/10.1080/01926230701481899.
61. Slevin JT, Gash DM, Smith CD, et al. Unilateral intraputaminal glial cell line-derived neurotrophic factor in patients with Parkinson disease: response to 1 year each of treatment and withdrawal. *Neurosurg Focus.* 2006;20(5):E1https://doi.org/10.3171/foc.2006.20.5.2.
62. Patel NK, Pavese N, Javed S, Hotton GR, Brooks DJ, Gill SS. Benefits of putaminal GDNF infusion in Parkinson disease are maintained after GDNF cessation. *Neurology.* 2013;81(13):1176–1178. https://doi.org/10.1212/WNL.0b013e3182a55ea5.
63. Sherer TB, Fiske BK, Svendsen CN, Lang AE, Langston JW. Crossroads in GDNF therapy for Parkinson's disease. *Mov Disord.* 2006;21(2):136–141. https://doi.org/10.1002/mds.20861.
64. Luz M, Allen PC, Bringas J, et al. Intermittent convection-enhanced delivery of GDNF into rhesus monkey putamen: absence of local or cerebellar toxicity. *Arch Toxicol.* 2018;92(7):2353–2367. https://doi.org/10.1007/s00204-018-2222-z.
65. Whone A, Luz M, Boca M, et al. Randomized trial of intermittent intraputamenal glial cell line-derived neurotrophic factor in Parkinson's disease. *Brain.* 2019;142(3):512–525. https://doi.org/10.1093/brain/awz023.
66. Whone AL, Boca M, Luz M, et al. Extended treatment with glial cell line-derived neurotrophic factor in Parkinson's disease. *J Parkinsons Dis.* 2019;9(2):301–313. https://doi.org/10.3233/JPD-191576.
67. Booth HDE, Hirst WD, Wade-Martins R. The role of astrocyte dysfunction in parkinson's disease pathogenesis. *Trends Neurosci.* 2017;40(6):358–370. https://doi.org/10.1016/j.tins.2017.04.001.
68. de Majo M, Koontz M, Rowitch D, Ullian EM. An update on human astrocytes and their role in development and disease. *Glia.* 2020. https://doi.org/10.1002/glia.23771.
69. Svendsen CN, ter Borg MG, Armstrong RJ, et al. A new method for the rapid and long term growth of human neural precursor cells. *J Neurosci Methods.* 1998;85(2):141–152. https://doi.org/10.1016/s0165-0270(98)00126-5.
70. Svendsen CN, Caldwell MA, Shen J, et al. Long-term survival of human central nervous system progenitor cells transplanted into a rat model of Parkinson's disease. *Exp Neurol.* 1997;148(1):135–146. https://doi.org/10.1006/exnr.1997.6634.
71. Capowski EE, Schneider BL, Ebert AD, et al. Lentiviral vector-mediated genetic modification of human neural progenitor cells for ex vivo gene therapy. *J Neurosci Methods.* 2007;163(2):338–349. https://doi.org/10.1016/j.jneumeth.2007.02.022.
72. Piccinini E, Kalkkinen N, Saarma M, Runeberg-Roos P. Glial cell line-derived neurotrophic factor: characterization of mammalian posttranslational modifications. *Ann Med.* 2013;45(1):66–73. https://doi.org/10.3109/07853890.2012.663927.
73. Behrstock S, Ebert A, McHugh J, et al. Human neural progenitors deliver glial cell line-derived neurotrophic factor to parkinsonian rodents and aged primates. *Gene Ther.* 2006;13(5):379–388. https://doi.org/10.1038/sj.gt.3302679.
74. Ebert AD, Beres AJ, Barber AE, Svendsen CN. Human neural progenitor cells over-expressing IGF-1 protect dopamine neurons and restore function in a rat model of Parkinson's disease. *Exp Neurol.* 2008;209(1):213–223. https://doi.org/10.1016/j.expneurol.2007.09.022.
75. Li X-J, Du Z-W, Zarnowska ED, et al. Specification of motoneurons from human embryonic stem cells. *Nat Biotechnol.* 2005;23(2):215–221. https://doi.org/10.1038/nbt1063.
76. Bruijn LI, Becher MW, Lee MK, et al. ALS-linked SOD1 mutant G85R mediates damage to astrocytes and promotes rapidly progressive disease with SOD1-containing inclusions. *Neuron.* 1997;18(2):327–338. https://doi.org/10.1016/s0896-6273(00)80272-x.
77. Gurney ME, Pu H, Chiu AY, et al. Motor neuron degeneration in mice that express a human Cu,Zn superoxide dismutase mutation. *Science.* 1994;264(5166):1772–1775. https://doi.org/10.1126/science.8209258.

78. Ripps ME, Huntley GW, Hof PR, Morrison JH, Gordon JW. Transgenic mice expressing an altered murine superoxide dismutase gene provide an animal model of amyotrophic lateral sclerosis. *Proc Natl Acad Sci U S A*. 1995;92(3):689–693. https://doi.org/10.1073/pnas.92.3.689.
79. Howland DS, Liu J, She Y, et al. Focal loss of the glutamate transporter EAAT2 in a transgenic rat model of SOD1 mutant-mediated amyotrophic lateral sclerosis (ALS). *Proc Natl Acad Sci U S A*. 2002;99(3):1604–1609. https://doi.org/10.1073/pnas.032539299.
80. Clement AM, Nguyen MD, Roberts EA, et al. Wild-type nonneuronal cells extend survival of SOD1 mutant motor neurons in ALS mice. *Science*. 2003;302(5642):113–117. https://doi.org/10.1126/science.1086071.
81. Yamanaka K, Chun SJ, Boillee S, et al. Astrocytes as determinants of disease progression in inherited amyotrophic lateral sclerosis. *Nat Neurosci*. 2008;11(3):251–253. https://doi.org/10.1038/nn2047.
82. Lepore AC, Rauck B, Dejea C, et al. Focal transplantation-based astrocyte replacement is neuroprotective in a model of motor neuron disease. *Nat Neurosci*. 2008;11(11):1294–1301. https://doi.org/10.1038/nn.2210.
83. Klein SM, Behrstock S, McHugh J, et al. GDNF delivery using human neural progenitor cells in a rat model of ALS. *Hum Gene Ther*. 2005;16(4):509–521. https://doi.org/10.1089/hum.2005.16.509.
84. Suzuki M, McHugh J, Tork C, et al. GDNF secreting human neural progenitor cells protect dying motor neurons, but not their projection to muscle, in a rat model of familial ALS. *PLoS One*. 2007;2(8):e689https://doi.org/10.1371/journal.pone.0000689.
85. Shelley BC, Gowing G, Svendsen CN. A cGMP-applicable expansion method for aggregates of human neural stem and progenitor cells derived from pluripotent stem cells or fetal brain tissue. *J Vis Exp*. 2014;88. https://doi.org/10.3791/51219.
86. Gowing G, Shelley B, Staggenborg K, et al. Glial cell line-derived neurotrophic factor-secreting human neural progenitors show long-term survival, maturation into astrocytes, and no tumor formation following transplantation into the spinal cord of immunocompromised rats. *Neuroreport*. 2014;25(6):367–372. https://doi.org/10.1097/WNR.0000000000000092.
87. Rushton DJ, Andres PL, Allred P, Baloh RH, Svendsen CN. Patients with ALS show highly correlated progression rates in left and right limb muscles. *Neurology*. 2017;89(2):196–206. https://doi.org/10.1212/WNL.0000000000004105.
88. Thomsen GM, Gowing G, Latter J, et al. Delayed disease onset and extended survival in the SOD1G93A rat model of amyotrophic lateral sclerosis after suppression of mutant SOD1 in the motor cortex. *J Neurosci*. 2014;34(47):15587–15600. https://doi.org/10.1523/JNEUROSCI.2037-14.2014.
89. Thomsen GM, Avalos P, Ma AA, et al. Transplantation of neural progenitor cells expressing glial cell line-derived neurotrophic factor into the motor cortex as a strategy to treat amyotrophic lateral sclerosis. *Stem Cells*. 2018;36(7):1122–1131. https://doi.org/10.1002/stem.2825.
90. Rosenblad C, Martinez-Serrano A, Björklund A. Intrastriatal glial cell line-derived neurotrophic factor promotes sprouting of spared nigrostriatal dopaminergic afferents and induces recovery of function in a rat model of Parkinson's disease. *Neuroscience*. 1998;82(1):129–137. https://doi.org/10.1016/s0306-4522(97)00269-8.
91. Das MM, Avalos P, Suezaki P, et al. Human neural progenitors differentiate into astrocytes and protect motor neurons in aging rats. *Exp Neurol*. 2016;280:41–49. https://doi.org/10.1016/j.expneurol.2016.03.023.
92. Nichols NL, Gowing G, Satriotomo I, et al. Intermittent hypoxia and stem cell implants preserve breathing capacity in a rodent model of amyotrophic lateral sclerosis. *Am J Respir Crit Care Med*. 2013;187(5):535–542. https://doi.org/10.1164/rccm.201206-1072OC.
93. Andres RH, Horie N, Slikker W, et al. Human neural stem cells enhance structural plasticity and axonal transport in the ischaemic brain. *Brain*. 2011;134(Pt 6):1777–1789. https://doi.org/10.1093/brain/awr094.
94. Goldberg NRS, Marsh SE, Ochaba J, et al. Human neural progenitor transplantation rescues behavior and reduces α-Synuclein in a transgenic model of dementia with Lewy bodies. *Stem Cells Transl Med*. 2017;6(6):1477–1490. https://doi.org/10.1002/sctm.16-0362.
95. McBride J. Kordower. Human neural stem cell transplants improve motor function in a rat model of Huntington's disease. *J Comp Neurol*. 2004;475(2):211–219. https://doi.org/10.1002/cne.20176.

96. Wang S, Girman S, Lu B, et al. Long-term vision rescue by human neural progenitors in a rat model of photoreceptor degeneration. *Invest Ophthalmol Vis Sci*. 2008;49(7):3201–3206. https://doi.org/10.1167/iovs.08-1831.
97. Lu B, Wang S, Francis PJ, et al. Cell transplantation to arrest early changes in an ush2a animal model. *Invest Ophthalmol Vis Sci*. 2010;51(4):2269–2276. https://doi.org/10.1167/iovs.09-4526.
98. Taupin P. HuCNS-SC (StemCells). *Curr Opin Mol Ther*. 2006;8(2):156–163.
99. Piltti KM, Salazar DL, Uchida N, Cummings BJ, Anderson AJ. Safety of human neural stem cell transplantation in chronic spinal cord injury. *Stem Cells Transl Med*. 2013;2(12):961–974. https://doi.org/10.5966/sctm.2013-0064.
100. McGill TJ, Cottam B, Lu B, et al. Transplantation of human central nervous system stem cells - neuroprotection in retinal degeneration. *Eur J Neurosci*. 2012;35(3):468–477. https://doi.org/10.1111/j.1460-9568.2011.07970.x.
101. Uchida N, Chen K, Dohse M, et al. Human neural stem cells induce functional myelination in mice with severe dysmyelination. *Sci Transl Med*. 2012;4(155):155ra136https://doi.org/10.1126/scitranslmed.3004371.
102. Tamaki SJ, Jacobs Y, Dohse M, et al. Neuroprotection of host cells by human central nervous system stem cells in a mouse model of infantile neuronal ceroid lipofuscinosis. *Cell Stem Cell*. 2009;5(3):310–319. https://doi.org/10.1016/j.stem.2009.05.022.
103. Selden NR, Al-Uzri A, Huhn SL, et al. Central nervous system stem cell transplantation for children with neuronal ceroid lipofuscinosis. *J Neurosurg Pediatr*. 2013;11(6):643–652. https://doi.org/10.3171/2013.3.PEDS12397.
104. Anderson AJ, Piltti KM, Hooshmand MJ, Nishi RA, Cummings BJ. Preclinical efficacy failure of human CNS-derived stem cells for use in the pathway study of cervical spinal cord injury. *Stem Cell Rep*. 2017;8(2):249–263. https://doi.org/10.1016/j.stemcr.2016.12.018.
105. Marsh SE, Yeung ST, Torres M, et al. HuCNS-SC human NSCs fail to differentiate, form ectopic clusters, and provide no cognitive benefits in a transgenic model of Alzheimer's disease. *Stem Cell Rep*. 2017;8(2):235–248. https://doi.org/10.1016/j.stemcr.2016.12.019.
106. McGill TJ, Osborne L, Lu B, et al. Subretinal transplantation of human central nervous system stem cells stimulates controlled proliferation of endogenous retinal pigment epithelium. *Transl Vis Sci Technol*. 2019;8(3):43. https://doi.org/10.1167/tvst.8.3.43.
107. Gupta N, Henry RG, Kang S-M, et al. Long-term safety, immunologic response, and imaging outcomes following neural stem cell transplantation for Pelizaeus-Merzbacher disease. *Stem Cell Rep*. 2019;13(2):254–261. https://doi.org/10.1016/j.stemcr.2019.07.002.
108. Levi AD, Okonkwo DO, Park P, et al. Emerging safety of intramedullary transplantation of human neural stem cells in chronic cervical and thoracic spinal cord injury. *Neurosurgery*. 2018;82(4):562–575. https://doi.org/10.1093/neuros/nyx250.
109. Usvald D, Vodicka P, Hlucilova J, et al. Analysis of dosing regimen and reproducibility of intraspinal grafting of human spinal stem cells in immunosuppressed minipigs. *Cell Transplant*. 2010;19(9):1103–1122. https://doi.org/10.3727/096368910X503406.
110. Hefferan MP, Galik J, Kakinohana O, et al. Human neural stem cell replacement therapy for amyotrophic lateral sclerosis by spinal transplantation. *PLoS One*. 2012;7(8):e42614. https://doi.org/10.1371/journal.pone.0042614.
111. Riley J, Federici T, Park J, et al. Cervical spinal cord therapeutics delivery: preclinical safety validation of a stabilized microinjection platform. *Neurosurgery*. 2009;65(4):754–761. discussion 761. https://doi.org/10.1227/01.NEU.0000343524.45387.9E.
112. Gutierrez J, Lamanna JJ, Grin N, et al. Preclinical validation of multilevel intraparenchymal stem cell therapy in the porcine spinal cord. *Neurosurgery*. 2015;77(4):604–612. discussion 612. https://doi.org/10.1227/NEU.0000000000000882.
113. Feldman EL, Boulis NM, Hur J, et al. Intraspinal neural stem cell transplantation in amyotrophic lateral sclerosis: phase 1 trial outcomes. *Ann Neurol*. 2014;75(3):363–373. https://doi.org/10.1002/ana.24113.
114. Glass JD, Hertzberg VS, Boulis NM, et al. Transplantation of spinal cord-derived neural stem cells for ALS: analysis of phase 1 and 2 trials. *Neurology*. 2016;87(4):392–400. https://doi.org/10.1212/WNL.0000000000002889.

115. Goutman SA, Brown MB, Glass JD, et al. Long-term phase 1/2 intraspinal stem cell transplantation outcomes in ALS. *Ann Clin Transl Neurol*. 2018;5(6):730–740. https://doi.org/10.1002/acn3.567.
116. van Gorp S, Leerink M, Kakinohana O, et al. Amelioration of motor/sensory dysfunction and spasticity in a rat model of acute lumbar spinal cord injury by human neural stem cell transplantation. *Stem Cell Res Ther*. 2013;4(3):57. https://doi.org/10.1186/scrt209.
117. Curtis E, Martin JR, Gabel B, et al. A first-in-human, phase I study of neural stem cell transplantation for chronic spinal cord injury. *Cell Stem Cell*. 2018;22(6):941–950.e6. https://doi.org/10.1016/j.stem.2018.05.014.
118. Zhang G, Li Y, Reuss JL, et al. Stable intracerebral transplantation of neural stem cells for the treatment of paralysis due to ischemic stroke. *Stem Cells Transl Med*. 2019;8(10):999–1007. https://doi.org/10.1002/sctm.18-0220.
119. Mazzini L, Gelati M, Profico DC, et al. Results from phase I clinical trial with Intraspinal injection of neural stem cells in amyotrophic lateral sclerosis: a long-term outcome. *Stem Cells Transl Med*. 2019; 8(9):887–897. https://doi.org/10.1002/sctm.18-0154.
120. Pollock K, Stroemer P, Patel S, et al. A conditionally immortal clonal stem cell line from human cortical neuroepithelium for the treatment of ischemic stroke. *Exp Neurol*. 2006;199(1):143–155. https://doi.org/10.1016/j.expneurol.2005.12.011.
121. Stroemer P, Patel S, Hope A, Oliveira C, Pollock K, Sinden J. The neural stem cell line CTX0E03 promotes behavioral recovery and endogenous neurogenesis after experimental stroke in a dose-dependent fashion. *Neurorehabil Neural Repair*. 2009;23(9):895–909. https://doi.org/10.1177/1545968309335978.
122. Kalladka D, Sinden J, Pollock K, et al. Human neural stem cells in patients with chronic ischaemic stroke (PISCES): a phase 1, first-in-man study. *Lancet*. 2016;388(10046):787–796. https://doi.org/10.1016/S0140-6736(16)30513-X.
123. Luo J, Baranov P, Patel S, et al. Human retinal progenitor cell transplantation preserves vision. *J Biol Chem*. 2014;289(10):6362–6371. https://doi.org/10.1074/jbc.M113.513713.
124. Semo M, Haamedi N, Stevanato L, et al. Efficacy and safety of human retinal progenitor cells. *Transl Vis Sci Technol*. 2016;5(4):6. https://doi.org/10.1167/tvst.5.4.6.
125. Schwartz SD, Regillo CD, Lam BL, et al. Human embryonic stem cell-derived retinal pigment epithelium in patients with age-related macular degeneration and Stargardt's macular dystrophy: follow-up of two open-label phase 1/2 studies. *Lancet*. 2015;385(9967):509–516. https://doi.org/10.1016/S0140-6736(14)61376-3.
126. Mehat MS, Sundaram V, Ripamonti C, et al. Transplantation of human embryonic stem cell-derived retinal pigment epithelial cells in macular degeneration. *Ophthalmology*. 2018;125(11):1765–1775. https://doi.org/10.1016/j.ophtha.2018.04.037.

CHAPTER 6

Parkinson's disease: Genetic-driven therapeutic approaches

S. Pablo Sardi[a] and Marco A.S. Baptista[b]
[a]Rare and Neurological Diseases Therapeutic Area, Sanofi, Framingham, MA, United States
[b]The Michael J. Fox Foundation for Parkinson's Research, New York, NY, United States

Introduction

Parkinson's disease (PD; OMIM #168600) is the most common neurodegenerative movement disorder and the second most common neurodegenerative disorder that afflicts approximately 1 million people in the United States and over 6 million worldwide. PD affects one in 100 people over the age of 60 and there are 60,000 new cases each year in the United States. Neurological disorders are now the leading source of disability globally, and the fastest-growing neurological disorder in the world is PD.[1] This increase is driven principally by aging and some projections estimate that the number of PD patients could double again to over 12 million by 2040.

The impact of PD on patients, caregivers, and payers is devastating on many levels. According to a comprehensive economic burden study conducted and published by The Michael J. Fox Foundation for Parkinson's Research with support from organizations including Parkinson's Foundation, American Parkinson's disease Association, The Parkinson Alliance, and industry partners, the total cost of PD is $51.9 billion every year. That figure includes $25.4 billion attributable to direct medical costs and $26.5 billion in nonmedical costs such as missed work, lost wages, early forced retirement, and family caregiver time.[2] The study also revealed that the federal government spends nearly $25 billion annually caring for people with PD, with about $2 billion covered by Social Security and the remaining in Medicare costs.[2]

The British physician Dr. James Parkinson initially described a series of six cases of a "Shaking Palsy" (*paralysis agitans*) in 1817.[3] This landmark essay defined the clinical diagnosis of PD for nearly two centuries. The classical disease is clinically manifested by a triad of cardinal motor symptoms—rigidity, bradykinesia, and resting tremor—due to loss of dopaminergic neurons in the substantia nigra pars compacta of the midbrain. This historical definition has evolved over the last 15 years and many nonmotor symptoms are now recognized as central to the disease. These symptoms include anxiety, apathy, dementia, depression, fatigue, gastrointestinal dysfunction, hallucinations, orthostatic

Neurotherapeutics in the Era of Translational Medicine
https://doi.org/10.1016/B978-0-12-816475-4.00008-2

hypotension, and others. The current medical management of PD aims at controlling signs and symptoms for as long as possible while minimizing adverse effects. Dopaminergic replacement therapies are effective at controlling the dopaminergic deficits; however, the current therapies have no effect on the progression of the disease. Therapies directed at controlling disease progression are desperately needed.[4,5]

The development of therapeutics for PD is progressing at an unprecedented pace. Recent advances have led to the initiation of human studies that test novel therapeutic targets, with the hope of slowing progression of symptoms. This momentum is the result of several genetic discoveries, decades of academic and pharmaceutical research, patient-volunteerism in natural history and biomarker studies, the development of novel biosensor technologies, and funding and partnerships including government, foundations, and the private sector. These advances provide an ideal springboard to evaluate a new class of therapies aimed at slowing PD progression[6,7] (Table 1).

This chapter summarizes how genetic discoveries originated over the last two decades have shaped the current clinical development of novel therapeutic approaches aimed at slowing PD progression. These initial genetic findings have enabled targeting therapies for α-synuclein (*SNCA*), glucocerebrosidase (*GBA*), and leucine-rich repeat kinase (*LRRK2*), which are the most advanced in clinical development.

Parkinson's disease genetics: Laying the foundation for genetic-driven therapeutic approaches

Over the last two decades, genetic studies have revolutionized our understanding of the etiology of PD.[8–10] Initially believed to be a pure sporadic disease, Parkinson's is now widely accepted to have a substantial genetic component.[11] The α-synuclein gene (*SNCA*) was the first to be unequivocally associated with PD.[12] Although monogenic, inherited forms of PD are rare and constitute around 5%–10% of all cases, 20% of patients with PD report having one affected first or second-degree relative.[9] The initial discoveries of α-synuclein gene point mutations and multiplications as a cause of PD were followed by the identification of numerous genes linked to PD pathogenesis. Since then, many other genes have been associated with PD, either as causative genes and/or risk factors for developing the disease.[10]

Initial studies identified rare disease-causing genetic variants through linkage analysis in large families with PD. The advent of higher throughput methods such as next-generation sequencing techniques and increased worldwide collaborations has led to the identification of common genetic components of PD using genome-wide association studies (GWAS). To date, at least 90 independent risk signals have been identified from 7.8 million single nucleotide polymorphisms evaluated in more than 37,000 patients, 17,000 "proxy cases" (i.e., individuals with a parent with PD), and 1.4 million controls.[13,14]

Table 1 Genetic-based targeted therapies currently being tested for Parkinson's disease.[a]

Mechanism	Therapeutic	Sponsor	ClinicalTrials.gov Identifier	Stage I	Stage II	Stage III	Stage Registration
Vaccine against α-synuclein	UB-312	United Neuroscience	NCT04075318				
	AFFITOPE PDC1A	AFFiRiS	NCT01568099				
	AFFITOPE PDC3A		NCT02267434				
Antibody therapy targeting α-synuclein	MEDI-1341	AstraZeneca/Takeda	NCT03272165				
	Lu-AF82422	Lundbeck	NCT03611569				
	ABBV-0805	AbbVie	NCT04127695				
	BIIB054	Biogen	NCT02459886				
	Prasinezumab	Prothena/Roche	NCT03100149				
Small molecule α-synuclein disaggregator	NPT200-11 (aka UCB0599)	Neuropore/UCB	NCT02606682				
LRRK2 kinase inhibitor	DNL201	Denali	NCT03710707				
	DNL151	Denali	NCT04056689				
LRRK2 antisense oligonucleotide	BIIB-094	Biogen	NCT03976349				
GBA gene therapy	PR-C01	Prevail Therapeutics	NCT04127578				
Glucocerebrosidase enhancer	Ambroxol	Lawson Health Research Institute	NCT02914366				
	Ambroxol	University College London	NCT02941822				
Glucosylceramide synthase inhibitor	Venglustat	Sanofi Genzyme	NCT02906020				

These genetic signals provide a biological context for the development of the disease. Evaluation of these, in the context of pathway, tissue, and cellular functionality, revealed that genes near PD risk variants are enriched for brain expression, and almost exclusively in neuronal cell types.[14] This contrasts with Alzheimer's disease hits, which tend to concentrate on microglia.[15] Mechanistically, PD-associated genes are enriched in chemical signaling pathways and pathways involving the response to stressors, which might help increase the translatability of disease models. In addition, mechanistic studies highlight a connection between synaptic, mitochondrial, and lysosomal function, suggesting that defects in these pathways may conspire to cause neuronal demise in PD.[16]

Genetic information is valuable for identifying pathogenic mechanisms of PD. However, genetic testing for PD is not yet routinely ordered in clinical practice.[17] There are several obstacles that need to be addressed before it becomes an everyday tool for neurologists. Population-wide genetic screening for individuals who are likely to develop PD is currently not feasible due to an extremely high false positive rate. Additional data sources (e.g., smell tests, age, sex, family history) would need to be incorporated to develop algorithms with increased predictive value for population screenings.[18] These multimodal predictive models will be extremely relevant to select prodromal populations for premotor clinical trials. Genetic information is being employed to select populations with a higher likelihood to benefit from specific therapeutic interventions, and clinical trials are underway for GBA-related PD (ClinicalTrials.gov Identifier: NCT02906020) and LRRK2-related PD (ClinicalTrials.gov Identifiers: NCT03710707, NCT03976349). Genetic testing can also inform the progression of the disease and potentially inform medical or personal lifestyle decisions.[13,19,20] For these reasons, there is an urgent need to increase knowledge and reduce practical barriers to genetic counseling and testing in PD.[17] Nonprofit patient organizations have recently launched large genetic screening efforts. The "*PD GENEration*" by the Parkinson's Foundation is a national initiative that offers free genetic testing for clinically relevant PD-related genes and free genetic counseling to help participants better understand their results.[21] "*Fox Insight*," an online clinical study sponsored by The Michael J. Fox Foundation, continues to provide no-cost genetic testing and counseling to people with PD and plans to expand testing for *LRRK2* and *GBA* mutations and counseling services to qualified individuals in 2020.[22]

SNCA-related therapeutic approaches

Multiple lines of evidence support a central role for α-synuclein in PD pathogenesis.[23] α-Synuclein (encoded by *SNCA*) is a highly soluble 14-kDa protein that is enriched at presynaptic terminals, where it binds lipids and regulates the release of synaptic vesicles.[24,25] In neurons, α-synuclein participates in the regulation of neurotransmitter release from the presynaptic terminals.[24] α-Synuclein is also the major component of abnormal neuronal aggregates known as Lewy bodies, the pathological hallmark of PD.[26]

Point mutations and multiplications in *SNCA* cause familial PD.[9,12] Genome-wide association studies (GWAS) have expanded this notion by consistently identifying common *SNCA* genetic variants that increase the risk for idiopathic PD.[13,27] Among these risk variants, a noncoding distal enhancer element of *SNCA* was shown to increase α-synuclein expression, further implicating increased α-synuclein levels in the development of PD.[28]

Increased levels of α-synuclein can cause profound cellular dysfunction, including synaptic dysfunction, mitochondrial impairment, defective endoplasmic reticulum function, and deficits in the proteosomal and autophagy-lysosomal degradation pathways.[29] Formation of α-synuclein oligomers and fibrils that are ineffectively cleared by lysosomal or ubiquitin proteasome systems results in aggregation and Lewy body pathology.[30] Misfolded α-synuclein is allegedly responsible for its toxic effects, although the precise noxious entity is under debate and subject to intense investigations.

Another interesting feature of aggregated misfolded α-synuclein is its ability to propagate between cells. This cell-to-cell spreading was initially suggested when investigators found Lewy bodies in fetal dopaminergic grafts implanted in the brains of patients with PD who had survived more than 10 years after surgery.[31,32] These findings lead to a series of studies demonstrating direct transmission via endocytosis to neighboring neurons in cell culture and animal models[33,34] and recruitment of endogenous α-synuclein to form Lewy body-like pathology upon exogenous administration of α-synuclein fibrils in animal models.[35] These studies favored a pathogenic hypothesis involving α-synuclein's prion like properties, consistent with the staged propagation of α-synuclein pathology along the nervous system described by Braak and colleagues.[36,37]

The presence of aggregated α-synuclein in specific brain regions is a hallmark of PD and suggests a central role for this protein in the sporadic disorder. In addition, the observation of abnormal α-synuclein seeding activity in PD patient cerebrospinal fluid (CSF) provides additional support to this hypothesis.[38,39] A wide range of approaches is being developed in attempts to prevent α-synuclein accumulation, aggregation, and cell-to-cell propagation of misfolded α-synuclein species with the aim of slowing PD progression (Fig. 1). Several α-synuclein-directed therapeutics are now entering the clinical arena. The majority of these efforts focus on reducing α-synuclein levels and/or its propensity to propagate.

Gene expression knockdown is being pursued as a strategy to reduce α-synuclein (Fig. 1, Panel 1). Potential approaches include antisense oligonucleotides or viral-mediated delivery of short hairpin RNAs.[40,41] The advent of these mRNA approaches has dramatically increased the druggable space, increasing the options to modulate specific targets. Antisense oligonucleotides have already entered the clinical space for other neurodegenerative diseases, demonstrating the successful distribution of oligonucleotides in a phase III trial in spinal muscular atrophy[42] and decreased mutant huntingtin in Huntington disease.[43]

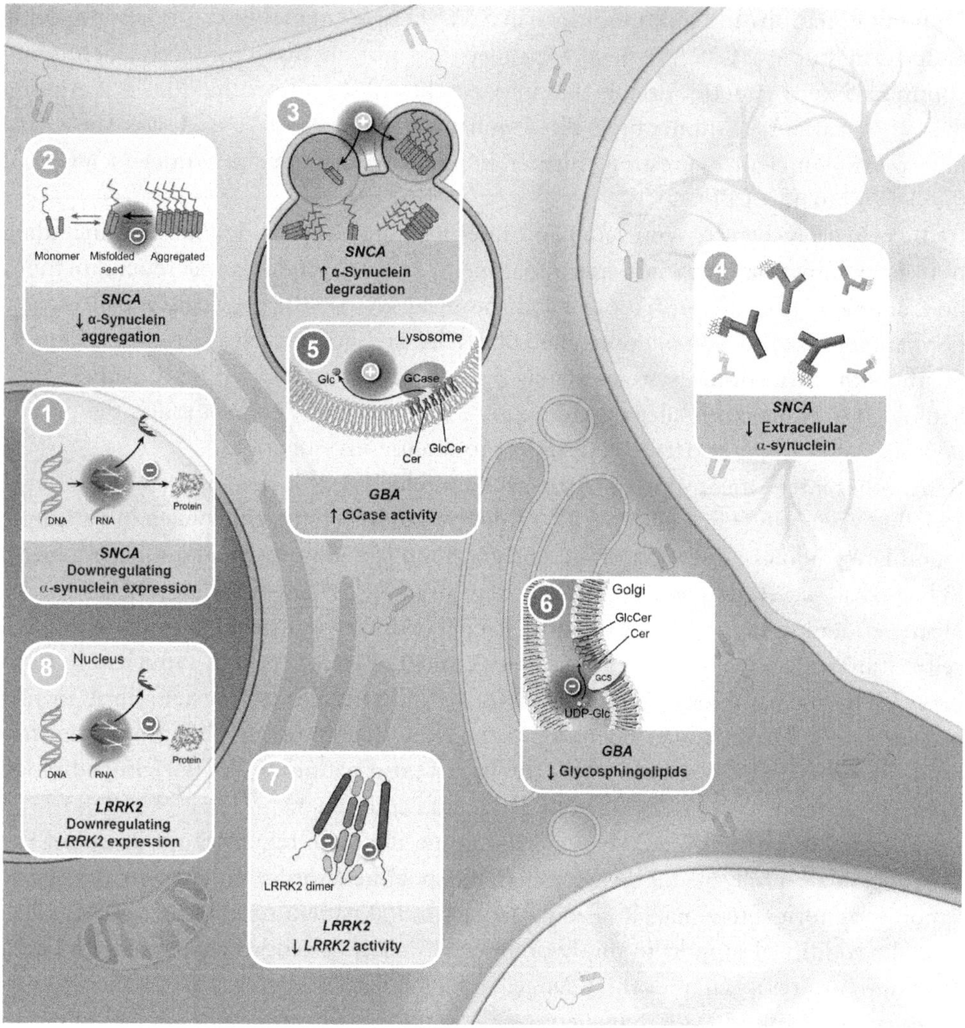

Fig. 1 Potential mechanisms for genetic-based investigational therapies for Parkinson's disease targeting α-synuclein (*SNCA*), glucocerebrosidase (*GBA*), and leucine-rich repeat kinase (*LRRK2*). Targeting *SNCA*: (1) reducing α-synuclein production, (2) decreasing α-synuclein aggregation, (3) enhancing autophagy, and (4) reducing the availability of extracellular aggregates. Targeting the *GBA* pathway: (5) increasing glucocerebrosidase activity and (6) modulating *GBA*-related glycosphingolipids. Targeting *LRRK2*: (7) inhibiting *LRRK2* kinase activity and (8) reducing *LRRK2* synthesis.

Several approaches to reduce α-synuclein expression using viral vector-mediated production of small interfering RNA (siRNA) against α-synuclein in the substantia nigra have been successfully tested in animal models.[41] Antisense oligonucleotides designed to promote RNAse-H mediated degradation of α-synuclein messenger RNA have been shown to safely lower levels of α-synuclein and to protect nigral dopaminergic neurons in rodent experiments.[40] One potential caveat is that in some of the animal models a marked downregulation of α-synuclein was associated with nigral degeneration (i.e., the opposite of the predicted effect).[44] Current studies are trying to clarify if this toxicity is related to the degree of α-synuclein knockdown or specific properties of the RNA lowering mechanism and if the system can be fine-tuned so that the targeted neurons maintain levels of α-synuclein needed for survival.[45]

β2-adrenoreceptor (β2AR) ligands have been reported to modulate *SNCA* transcription through histone 3 lysine 27 acetylation of its promoter and enhancers.[46] Examination of the Norwegian Prescription Database validated these results as prescription of the antiasthmatic β2AR agonist salbutamol correlated with a decreased risk of PD (risk ratio 0.66), whereas propranolol, a β2AR antagonist prescribed for cardiovascular diseases, associated to an increased risk of PD (risk ratio 2.20). Of note, a recent epidemiological study of United States Medicare beneficiaries failed to replicate the association between β2-adrenergic agonists or antagonists and PD risk.[47] These findings warrant further studies in additional population cohorts and in animal models of α-synucleinopathies.

The reduction of α-synuclein's tendency to aggregate, an alternative approach, has entered clinical testing (Fig. 1, Panel 2). These programs are founded on the premise that aggregated α-synuclein is the pathological hallmark of synucleinopathies and the more recent discovery of abnormal α-synuclein seeding activity in CSF from patients.[38,39] For example, the small molecule NPT200-11 has been shown to target α-synuclein and stabilize its physiological structure, leaving it incapable of assembling into toxic oligomers. NPT200-11 prevented the formation of pathogenic α-synuclein oligomeric aggregates, leading to improvements in neuropathological and biochemical outcomes in α-synuclein transgenic murine models.[48] A small phase I clinical trial was completed to evaluate NPT200-11's safety, tolerability, and plasma exposure (ClinicalTrials.gov Identifier: NCT02606682). However, data from this study is not available, and further plans for the development of this compound have not been announced. Another small molecule, ANLE138b, has been shown to cross the blood-brain barrier and reduce aggregation of α-synuclein as well as block the aggregation of pathogenic forms of prion protein in cell culture and mouse models.[49]

Autophagy is a collective term for different molecular pathways by which cells remove unnecessary or dysfunctional organelles and proteins, including misfolded and aggregated α-synuclein.[50] After the autophagosome—a cytoplasmic organelle with a double membrane structure surrounding a lumen—has formed around its target, it can merge with a lysosome to form an autophagolysosome where lysosomal enzymes degrade the cargo. When autophagy is inhibited, aggregated α-synuclein accumulates intracellularly and also undergoes secretion via the autophagy-secretory pathway.[51] Not only can intracellular accumulation of aggregated α-synuclein potentially damage or disable the host neuron, but the secretion of aggregated α-synuclein into the extracellular space may promote cell-to-cell propagation of α-synuclein pathology to neighboring cells.[52]

Autophagy is believed to play a major role in the degradation of intracellular α-synuclein aggregates.[53,54] The proposed approach aims to enhance the autophagic processes to increase the clearance of pathological α-synuclein (Fig. 1, Panel 3). Rapamycin and its analogues act via the mammalian target of rapamycin (mTOR) to increase macro-autophagy function. This class of drugs has been shown in numerous studies to reduce α-synuclein aggregation and toxicity in overexpression-based cellular and animal models.[45] However, lack of specificity (i.e., effects on other essential pathways) and side-effects (e.g., immunosuppression) have limited the potential use of such drugs for PD when long-term treatment would be needed.[45] Potential alternative candidates that alter autophagy might be repurposed from other indications. MSDC-0160 is an inhibitor of the mitochondrial pyruvate carrier originally developed for Type II diabetes that can also affect mTOR activity.[55] MSDC-0160 displays a favorable safety profile, suitable brain penetrance, and reduction of neuronal α-synuclein toxicity in a *Caenorhabditis elegans* model.[55]

Repurposing of c-Abl (ABL1, Abelson tyrosine kinase) inhibitors is another approach that might result in increased autophagy of misfolded α-synuclein. c-Abl inhibitors are currently approved as treatments for various forms of leukemia. c-Abl is a member of Abl family of nonreceptor tyrosine kinases and affects multiple targets that are involved in a wide variety of functions ranging from cell growth, shape, and migration, and in the nervous system, it appears to influence normal processes such as neurogenesis and neurite extension[56] c-Abl inhibitors reportedly inhibit protein aggregation and neurodegeneration in PD models.[56,57] While the precise mechanisms of action are still debated, the preclinical evidence for efficacy was sufficiently compelling to warrant a small safety trial of the c-Abl inhibitor nilotinib. In a nonplacebo, open-label safety trial, patients with dementia with Lewy bodies and PD and dementia were reported to exhibit some improvement in motor function within 6 months of starting treatment.[58] Despite the significant caveats of this trial,[59] these results triggered further trials with nilotinib and alternative c-Abl inhibitors in PD (ClinicalTrials.gov Identifiers: NCT03205488, NCT03655236). Investigators from the NILO-PD study (NCT03205488) indicated in a press release that nilotinib was safe and tolerable in 76 participants with moderate to advanced PD and without heart issues or other health challenges, but did not show a clinically meaningful benefit or biological effect to benefit those with PD. The complete report of this study is forthcoming.[60]

The notion that α-synuclein could act as a prion-like protein[31,32] provided the foundation for a fundamentally different therapeutic approach, targeting extracellular α-synuclein by passive or active immunization (Fig. 1, Panel 4). This approach assumes that intraneuronal α-synuclein aggregates need to be secreted into the extracellular space and taken up by neighboring neurons to seed aggregation of endogenous, natively unfolded α-synuclein in the recipient cells. This process has been demonstrated in cell culture and experimental animal models.[45] To be targeted by antibody therapy, pathogenic seeding-prone α-synuclein must be present in the extracellular space. Recently, structural variances of α-synuclein seeding kinetics and products have been observed in different synucleinopathies, which raises the possibility for the future development of strain-specific therapies.[61] There are now several clinical programs that use either active (immunizing the patient with modified α-synuclein) or passive (injecting antibodies targeting α-synuclein at regular intervals) immunotherapy approaches to target different conformers of α-synuclein aggregates (ClinicalTrials.gov Identifiers: NCT03100149, NCT03272165, NCT02267434, NCT02459886, NCT03611569, NCT04127695). The first reports for PRX002/RG7935 and BIIB054 show that the approach is generally safe and well tolerated, with no serious or severe drug-related treatment-emergent adverse events (TEAEs).[62,63] PRX002/RG7935 showed a clear reduction in free serum α-synuclein levels, whereas BIIB054 showed a paradoxical increase in plasma α-synuclein. This puzzling increase in α-synuclein levels was speculated to correspond to a prolongation of the protein half-life due to its binding to the antibody within the intravascular compartment, where the KD for monomeric α-synuclein is exceeded thus decreasing its clearance.[63] Both studies showed that these antibodies were present in the cerebrospinal fluid at concentrations from 0.1% to 0.5% relative to the serum. The evidence of biologic activity supported the dose selection and continuous investigation of these therapies in phase II studies.

Active vaccination to induce an antibody response, rather than passive antibody delivery, is also under investigation to reduce extracellular α-synuclein. The advantages of such an approach include the potential calibration for immune response specificity and selectivity, subcutaneous rather than intravenous administration, and the possibility of boosting immune response if antibody response is not achieved therapeutically. The potential pitfalls are lack of accurate prespecified level of individual immune response and related potential safety concerns, although none have been reported so far in the early programs. Two active immunotherapy programs have advanced to clinical testing: AFFITOPE PD01A and PD03A (ClinicalTrials.gov Identifiers: NCT01568099, NCT02267434). These synthetically produced vaccines contain α-synuclein-mimicking peptides that elicit an α-synuclein antibody response. These investigational vaccines were administered to patients with early PD. AFFITOPE PD01A and PD03A were well tolerated and no drug-related serious adverse events or reactions were reported. The majority of adverse events were mild and local reactions. AFFITOPE PD03A reportedly exhibited a dose- and time-dependent immune response against the peptide itself and cross-reactivity against α-synuclein, as well as antibody reactivation upon booster immunization.[64,65]

One general concern about immunotherapy for PD, be it active or passive, is whether antibodies will have sufficient brain penetration to achieve adequate target engagement. To date, none of the immunotherapy programs has reported an effect on levels of α-synuclein in CSF. Importantly, the reduction of amyloid-beta plaques in Alzheimer's disease by the antibody aducanumab provides evidence that effective CNS antibody concentrations can be achieved.[66]

A common limitation of all α-synuclein-directed trials is the lack of biomarkers of α-synuclein related pathology for patient selection and stratification, target engagement, and proof of biological activity. This is a high priority for The Michael J. Foundation for Parkinson's Research, which has launched several initiatives to develop biomarkers of α-synuclein. These include positron emission tomography (PET) tracers of pathology; biochemical assays to evaluate α-synuclein in native, oligomeric, and aggregated states in biofluids and tissues; assays for specific α-synuclein posttranslational modifications; and tests to evaluate α-synuclein prion-like seeding activity.[7,38,39,67–69] Importantly, these assays need independent replication with blinded samples and rigorous analytical validation of the methods before widespread application. The development of such biomarkers will prove essential for the successful translation and advancement of α-synuclein experimental therapies.

Human testing of the α-synuclein-centric hypothesis for PD has been highly anticipated. However, we should not underestimate the challenges ahead as our understanding of the underlying pathogenic mechanisms is incomplete, and the path to approval for disease-modifying therapies is still being charted.[6] Abundant evidence from experimental animal models suggests that α-synuclein aggregates trigger neuronal dysfunction and death. However, it remains unclear which molecular species of α-synuclein is the appropriate target and is the "toxic" species of this protein. In addition, the presence of α-synuclein pathology in most PD cases does not prove that Lewy body pathology is *per se* toxic. There remains the possibility that these aggregates represent an innocuous epiphenomenon or perhaps a protective response whereby the cell sequesters toxic misfolded protein.[70] Future precision medicine efforts for molecular targeting of neurodegenerative diseases may require analyses not anchored on current clinicopathologic criteria but instead on biological signals generated from large, deeply phenotyped populations or from smaller but well-defined genetic-molecular cohorts.[7,70]

GBA-related therapeutic approaches

The discovery of *GBA* mutations as a risk factor for PD started with clinical observations of parkinsonism in patients with Gaucher disease, a lipid storage disorder.[71] These observations were extended by reports of increased frequency of parkinsonism in family members of Gaucher patients and several small genetic studies, suggesting a role for heterozygous *GBA* mutations in PD.[72,73] This association was confirmed by a collaborative consortium that analyzed *GBA* mutations in a large cohort by sequencing the entire coding region.[74] Numerous genetic studies have substantiated the increased frequency of mutations in *GBA* in patients with PD and dementia with Lewy bodies (DLB).[75] Heterozygous carriers of *GBA* mutations have an increased frequency of

PD; however, most *GBA* mutation carriers will not develop the disease, suggesting the presence of additional genetic modifiers.[76]

The *GBA* gene encodes a lysosomal hydrolase, glucocerebrosidase, which hydrolyzes glucosylceramide into ceramide and glucose. Glucocerebrosidase deficiency, as a result of homozygous or compound heterozygous mutations, causes Gaucher disease through the accumulation of undegraded substrates.[77] Partial enzymatic deficiency due to heterozygous *GBA* mutations increases the risk of developing PD and other synucleinopathies.[74,76]

GBA mutations are the most frequent genetic risk factor for PD identified to date; with 7%–10% of PD patients carrying one of approximately 300 reported *GBA* mutations.[74,76] As mentioned, most subjects carrying *GBA* mutations will not develop PD.[78] The utility of *GBA* genetic testing in routine clinical practice is limited by the incomplete penetrance of the mutations and lack of impact on practical treatment decisions. This practice will necessarily evolve as we update the definition of PD for an era of personalized medicine.[7,17] Accounting for genetic variability will be a useful factor in understanding disease course and in minimizing heterogeneity in clinical trials.[79]

GBA-associated PD is clinically, pharmacologically, and pathologically indistinguishable from sporadic PD. However, evidence from longitudinal clinical cohorts indicates that PD patients with *GBA* mutations present an accelerated disease course, higher risk for nonmotor symptoms, and reduced survival.[19,20,80,81] Patients carrying a *GBA* mutation exhibit an earlier and more rapid cognitive decline compared to noncarriers and the risk for dementia is strongly modulated by the type of mutation and its residual enzymatic activity. Severe *GBA* mutations yielding significantly reduced residual enzymatic activity correlated with a more severe phenotype.[19,20] Dementia is a serious complication in late-stage idiopathic PD and greatly affects quality of life and survival. The increased risk for dementia in *GBA*-related PD is consistent with studies demonstrating the increased frequency of *GBA* mutations in patients with PD dementia (PDD) and dementia with Lewy bodies.[75,82]

The basic mechanisms underlying the association between *GBA* mutations and synucleinopathies remain elusive. The leading hypothesis posits that *GBA*-mediated loss of function causes an abnormal glycosphingolipid environment leading to cellular protein mishandling (proteinopathy) and neuronal dysfunction.[83] In animal models of disease, decreased glucocerebrosidase activity results in increased CNS α-synuclein, ubiquitin, and tau aggregates and associated cognitive and motor deficits.[84,85] A growing body of epidemiological, clinical, and basic science studies supports this concept.[86,87]

Heterozygous *GBA* mutations can initiate a pathogenic cascade by decreasing glucocerebrosidase activity and initiating a self-perpetuating loop. Decreased glucocerebrosidase activity leads to substrate glucosylceramide and glucosylsphingosine accumulation in specific neurons.[88–90] Alterations in glycosphingolipid homeostasis that affect membrane composition can impair vesicular transport, lysosomal/endosomal function, and mitochondrial activity and promote α-synuclein aggregation, all of which can result in synaptic dysfunction and selective neuronal demise.[86,91–95] These changes in α-synuclein levels in addition to the lysosomal damage lead to further reduced glucocerebrosidase activity and continuous propagation of this pathogenic forward loop.

Brains from PD patients carrying nonmutated *GBA* alleles show decreased glucocerebrosidase activity, suggesting a role for this feedback loop in sporadic disease.[96] These studies were rapidly followed by many others demonstrating reduced enzymatic activity in CSF and blood.[97,98] The involvement of glucocerebrosidase in sporadic PD is also supported by studies showing decreased brain enzymatic activity due to aging, another critical risk factor for PD and other synucleinopathies.[99] Mechanistically, mitochondrial oxidant stress leading to the accumulation of oxidized dopamine quinones might be responsible for decreasing glucocerebrosidase activity by directly reacting with cysteine residues in the enzyme catalytic site.[100] These studies support a pathogenic hypothesis in which the accumulation of oxidative stress in aging postmitotic neurons leads to decreased glucocerebrosidase activity and subsequent accumulation of substrates. They also provide a rational explanation for the higher susceptibility of specific neurons in PD, where aging and dopaminergic synthesis colludes with a high metabolic rate to generate toxic dopamine adducts that impair key cellular processes.[101]

The availability of successful therapies for Gaucher disease provided a springboard to accelerate the development of therapeutics targeting the *GBA* pathway. However, novel therapeutics had to be developed as the currently approved therapies for Gaucher disease do not affect glycosphingolipid accumulation in the CNS. Viral gene therapy-mediated overexpression of exogenous glucocerebrosidase in the CNS provided the initial evidence of amelioration (and even reversion) of pathological and behavioral aberrations in animal models overexpressing α-synuclein or displaying *GBA*-associated synucleinopathy.[84,85] These initial observations were substantiated by multiple studies overexpressing glucocerebrosidase using viral vectors or small molecule compounds to modulate the endogenous enzymatic activity.[102–106] These studies provided strong support for glucocerebrosidase augmentation as a therapy for *GBA*-associated PD and conceivably for certain forms of sporadic PD. As a consequence, several therapeutic strategies aimed at increasing glucocerebrosidase activity in the CNS through gene therapy or small molecules are under active investigation.

Adeno-associated virus (AAV) vector platforms provide a safe, persistent expression of biologically active proteins to target structures in the human brain. During the last decade, various gene therapy clinical trials for PD have been initiated and completed.[107] Preclinical studies have demonstrated that AAV-mediated glucocerebrosidase augmentation can reverse cognitive deficits and reduce pathological protein aggregates in a mouse model of Gaucher-related synucleinopathy and modulate α-synuclein levels in an A53T-SNCA mouse model.[85] A significant increase in glucocerebrosidase activity in the CNS would be required for clinical translation of the therapeutic benefit. A human phase 1/2 clinical trial is investigating the safety of intracisternal PR001A, (AAV9-GBA) administration in patients with moderate to severe PD with at least one pathogenic GBA mutation (ClinicalTrials.gov Identifier: NCT04127578). Two escalating dose cohorts are planned (low dose and high dose, with a sham procedure as control). The duration of the study is 5 years. During the first year, patients will be evaluated for the effect of PR001A on safety, tolerability, immunogenicity, biomarkers, and clinical efficacy measures. Patients will continue to be followed for an additional 4 years to continue to monitor safety as well as selected biomarker and efficacy measures.

An alternative approach for increasing glucocerebrosidase activity is the use of small molecules. Brain-penetrant small molecules have the potential to overcome the limited distribution of gene therapy approaches.[86] Brain-penetrant small molecular chaperones capable of increasing lysosomal glucocerebrosidase activity are being developed for PD and related synucleinopathies. Several noninhibitory chaperones have been shown to reduce lipid substrate accumulation, increase glucocerebrosidase lysosomal translocation and activity, and reverse α-synuclein accumulation in cellular and animal models of Gaucher disease or PD.[108–110] Advancing these chemophores to the clinical stages will require further medicinal chemistry and toxicological evaluation to produce compounds with appropriate physicochemical and pharmacological properties.

Ambroxol, an EMA-approved mucolytic, is another chaperone being investigated for PD. At high concentrations in mouse models of synucleinopathy, ambroxol reportedly displayed glucocerebrosidase chaperone activity improving lysosomal transport and reducing α synuclein and S129-phosphorylated α-synuclein protein levels (implicated in the pathogenesis of PD).[105] Based on these results, the safety, tolerability, and efficacy of ambroxol in PD is currently being tested in Canada and the United Kingdom (ClinicalTrials.gov Identifiers: NCT02914366, NCT02941822).

An alternative approach aims at modulating the abnormal glycosphingolipid accumulation by antagonizing the enzyme responsible for its synthesis, namely glucosylceramide synthase (GCS). This approach is referred to as substrate reduction therapy and does not target the mutant enzyme. Substrate reduction therapy offsets the buildup of lipid substrates, equilibrating biosynthesis with the reduced breakdown due to *GBA* mutations. The viability of GCS inhibition as a potential approach for GBA-associated PD is supported by the approval of eliglustat, a nonbrain penetrant GCS inhibitor, for systemic Gaucher disease.[111] In PD cell models with and without *GBA* mutations, GCS inhibitors were shown to restore glycosphingolipid balance and reduce α-synuclein pathological conformers.[92,93] Importantly, a novel brain-penetrant GCS inhibitor improved α-synuclein processing and behavioral abnormalities in mouse models of *GBA*-related synucleinopathy and α-synuclein overexpression.[90] These results have prompted the initiation of a multinational double-blind, placebo-controlled study to assess the efficacy and safety of a GCS inhibitor in PD patients carrying a *GBA* mutation (ClinicalTrials.gov Identifier: NCT02906020).

Increasing glucocerebrosidase activity and/or modulating glycosphingolipid levels are potential approaches currently being pursued for PD and other synucleinopathies. The main challenge for the development of these therapeutics is the incomplete understanding of the underlying mechanisms governing the role of *GBA* mutations in the progression of synucleinopathies, which could inform the appropriate timing of the interventions, levels of modulation required, and reversibility of the pathological changes. Additional challenges for the development of glucocerebrosidase-related therapies include the current lack of indicators of biological activity (e.g., α-synuclein aggregation) and understanding the precise level of pathway modulation required for a therapeutic effect.

LRRK2-related therapeutic approaches

Leucine-rich repeat kinase 2 (LRRK2) is another key therapeutic target for PD borne from disease genetics that is now in the clinic. There are several reasons why this gene is the target of intense research both in academia and the pharmaceutical industry. First, *LRRK2* mutations are associated with inherited PD and the genetic locus containing the *LRRK2* gene contains a risk factor for sporadic PD.[13,27,112,113] Second, the most common mutation, G2019S, is located on the kinase domain and increases kinase activity by two- to threefold.[114] The field of kinase inhibition has seen remarkable advances in recent decades, particularly in oncology. Thus, LRRK2 is considered a "druggable" target. Third, mounting evidence indicates that the LRRK2 pathway might play a role in sporadic disease, which is significant as the cause of the majority of PD cases is unknown (i.e., idiopathic).[115] LRRK2 kinase inhibitors provide an opportunity to develop a medicine that may help more than just the fraction of patients carrying an *LRRK2* mutation. Recent studies have uncovered common physiological mechanisms linking PD genetic pathways and dysfunctional cellular systems in PD. As the current treatments for PD only affect the symptoms of the disease, the potential for an LRRK2-associated therapeutic to be disease modifying is heartening.

Our understanding of LRRK2 genetics and its biological functions in health and disease has exponentially expanded in the last 15 years. Since the initial discovery of LRRK2 gene mutations causing PD in 2004, numerous studies have revealed a variety of genetic ties to PD across *LRRK2*.[112] These include common alleles with broad effects on risk, likely through alterations at the protein sequence and expression.[116] Researchers have functionally characterized LRRK2 and its disease-causing mutations, and considerable progress has been made linking increased LRRK2 activity to both genetic and sporadic PD. Investigators are exploring the kinase activity of LRRK2 and its potential as a therapeutic target. Viable small molecule candidates and antisense oligonucleotides are already in clinical testing.

The LRRK2 gene encodes a large multidomain protein (2527 amino acids, 286 kDa) containing an ankyrin repeat (ANK), leucine-rich repeat (LRR) domain, and a C-terminal WD-40 domain. LRRK2 belongs to the Roco family of proteins that contains two catalytic domains: a Ras of complex (Roc) G-domain followed by dimerization domain COR (C-terminal of Roc) but uniquely alongside a kinase domain.[117] Most of the mutations that increase the risk of PD are located on these catalytic domains with the most common mutation, G2019S, located on the kinase domain.[118] All mutations, including G2019S, are gain-of-function mutations that increase the kinase activity of LRRK2, supporting LRRK2 kinase inhibition as a therapeutic approach for PD.[114] This premise is further sustained by the similar frequencies of *LRRK1* and *LRRK2* loss-of-function variants present in PD cases and controls.[119] The working hypothesis is that the aberrant increase in LRRK2 kinase activity induces downstream metabolic alterations that are deleterious to neuronal survival.[120]

One indicator of how intriguing LRRK2 is as a drug target for PD is the number of LRRK2 kinase inhibitor drug programs initiated without knowledge of its substrate and the signaling pathways that are activated or inhibited by kinases. This is fundamental knowledge that forms the basis for how kinase activity is measured, which is important for understanding how effective kinase inhibitors are in suppressing activity and downstream pathways. The discovery that LRRK2 mutations increase the risk of PD occurred in 2004 and the discovery of the first substrate only in 2016. The reason for this long search was mainly due to relatively low levels of LRRK2 expression and kinase activity. Although many publications claimed to identify LRRK2 substrates, further scrutiny failed to validate them. And yet, the field continued to heavily invest in researching LRRK2 biology and developing LRRK2 kinase inhibitors. In 2012, The Michael J. Fox Foundation initiated an international public-private research consortium to tackle this field-wide challenge. The foundation brought together industry and academic experts in LRRK2, kinase biochemistry, and mass spectrometry to set up and execute a comprehensive plan to unequivocally identify LRRK2 substrate/s. This challenge required a multidisciplinary team that could contribute key research tools to find such an elusive target. Two workstreams were used to create the proper experimental controls and internal validation necessary for this discovery. The first workstream focused on testing two LRRK2 kinase inhibitors in a transgenic LRRK2 G2019S mouse. The second workstream tested a novel potent and selective LRRK2 kinase inhibitor (MLi-2)[121] in a wild-type mouse and a transgenic mouse (A2016T) that was designed to be resistant to the kinase inhibitor. These parallel workstreams were designed to use multiple LRRK2 kinase inhibitors and specific drug-resistant mice to precisely identify nonspecific effects. These communal efforts paid off and in 2016, the specific residue in the Rab10 GTPase (T73) was identified as an LRRK2 substrate.[122]

The small Rab GTPases are master regulators of membrane trafficking in eukaryotic cells,[123,124] and the LRRK2 phosphorylation site sits in a critical switch 2 regions of the Rab protein. This region is the exact location that binds to cytosolic GDP dissociation inhibitors (GDI) that regulate the insertion of Rabs into cellular membranes and where GEFs catalyze the exchange of GDP to the activating GTP state. LRRK2 mutations increase the phosphorylation of Rabs on this switch 2 regions and thus prevent them from binding to the GDI in the cytosol and from being activated. The end result is a cellular traffic jam where inactivated Rabs are improperly inserted into cellular membranes. This alters the equilibrium between membrane-bound and cytosolic Rabs that may contribute to the pathophysiology of LRRK2 PD. Interestingly, specific Rabs have been identified as genetic risk variants for PD. Among the five genes in the Park16 locus is Rab7L1 (or also known as Rab29) which increases the risk of nonfamilial PD.[125] Also, Rab39b at the Park21 locus is a risk factor for PD.[126,127]

The discovery of Rab10 as a substrate of LRRK2 led to the systematic evaluation of additional family members and the discovery of a subset of 14 Rab proteins (Rab3A/B/C/D, Rab5A/B/C, Rab8A/B, Rab10, Rab12, Rab29, Rab35, and Rab43) that are direct substrates of LRRK2.[128] The next logical step is to identify the upstream and downstream regulators of the LRRK2-Rab pathway and the dysfunctional consequences associated with LRRK2 mutations. While Rab29 (aka Rab7L1) is a direct substrate of LRRK2, it also recruits mutated LRRK2 to the trans-Golgi network and stimulates LRRK2 activity as measured by phosphorylation of constitutive sites on the N-terminal region of LRRK2 (S910, S935, S955, and S973) and downstream Rab8A and Rab10.[129] This mechanism of Rab29 activation via the LRRK2 ankyrin domain is consistent with the current structural model of full-length dimeric LRRK2 that has been generated based on homology models, chemical cross-linking, negative-stain electron microscopy, and small-angle X-ray scattering.[130] The model suggests that the N-terminal ankyrin domain of LRRK2 lies in close proximity to the kinase domain and thus would allow for the activation of the LRRK2 kinase domain upon Rab29 binding. Unfortunately, a high-resolution atomic structure of LRRK2 is still missing. Multiple efforts are underway using diverse approaches including microgravity experiments on the International Space Station.[131] The idea behind the project is that under microgravity conditions, lack of sedimentation and convection currents may improve the conditions for protein crystal growth.

The N-terminal biomarkers sites are consistently dephosphorylated by every known LRRK2 kinase inhibitor suggesting that LRRK2 itself is regulating these sites through some unknown mechanism.[129] Hence, both the N-terminal sites and the Rabs can be utilized as target engagement markers. While the former can be used as target engagement readouts, there are efforts underway to understand if the phosphorylation of Rabs can become markers of disease progression. For example, a Michael J. Fox Foundation-led consortium tested different ELISA platforms and immune cell isolation protocols to determine phosphorylation levels at pS935 in healthy individuals or Parkinson's Disease patients carrying wild-type or mutant LRRK2. Surprisingly, the consortium found that the dephosphorylation on the S935 site was greatest in LRRK2 carriers manifesting the disease.[132] Additional studies will help consolidate the relationship between these N-terminal sites and the LRRK2 substrates on the pathophysiology of PD.

The functional consequences of the LRRK2 recruitment to the trans-Golgi network and activation of downstream Rabs have also been extensively investigated. There is strong evidence that LRRK2 inhibition plays a role in autophagy and lysosomal biology.[115,133,134] A recent report suggests that LRRK2 also plays a critical role in ciliogenesis.[135] LRRK2 mutations may disrupt ciliogenesis by altering the phosphorylation of both Rab8A and Rab10 and increasing binding of RILP1 (Rab interacting lysosomal protein-like 1). The loss of cilia might affect the ability of striatal cholinergic cells to sense sonic hedgehog and other neurotrophic factors.[136]

Another mechanism that may explain the role of LRRK2 in PD is inflammation. LRRK2 is highly expressed in immune cells (e.g., macrophages, monocytes, and neutrophils), suggesting it might play a role in the cellular immune response against intracellular pathogens.[133,137] Human genetic evidence supports this hypothesis, as there are protective and risk LRRK2 alleles linked to Crohn's disease[138] and single-nucleotide polymorphisms associated with increased susceptibility to leprosy infection.[139] In mice, LRRK2 is required for immunity against *Listeria monocytogenes* and *Salmonella typhimurium* during bacterial infection.[139,140] LRRK2 reportedly interacts and phosphorylates the inflammasome protein NLRC4 (NLR family CARD domain-containing protein 4) required for inflammasome formation. This leads to the activation of caspase 1, which produces the pro-inflammatory interleukin-1β. While early pro-inflammatory activation may be beneficial, prolonged LRRK2-related immune activation may increase the risk of PD.[133] It remains to be determined if LRRK2 inhibitors will have any potential safety liabilities given LRRK2's role in the immune system. For example, mice lacking LRRK2 or treated with an LRRK2 kinase inhibitor have an impaired ability to clear *Salmonella typhimurium* infection.[140]

LRRK2 knockout rodents exhibit a kidney and lung phenotype described as an increase in secondary lysosomes and increased lamellar bodies, respectively.[141,142] Furthermore, LRRK2 kinase inhibitors administered to nonhuman primates induce a similar phenotype in the lung.[143] The lamellar bodies are secretory organelles in type II pneumocytes containing surfactant, needed to lower surface tension. However, in the nonhuman primate study, it was not clear if LRRK2 kinase inhibitors had any off-target effects that drove the phenotype. Furthermore, it was unclear if the LRRK2 kinase inhibitor-induced lung effect was reversible, and if there were any functional deficits. A precompetitive consortium was assembled and led by The Michael J. Fox Foundation to find answers to these critical questions. This LRRK2 Safety Initiative (LSI) was an unprecedented precompetitive private-public collaboration, with pharmaceutical companies providing novel LRRK2 kinase inhibitors with different off-target selectivity to determine if these histopathological findings were target-related, reversible, and functionally deleterious. Results from these studies demonstrated that three different LRRK2 kinase inhibitors induced the lung phenotype, which was completely reversible after a 2-week withdrawal in nonhuman primates. In addition, no lung functional deficits were observed despite the high-dosing regimen.[144] A recent study analyzed lung, kidney, and other phenotypes in individuals carrying LRRK2 loss-of-function variants and found no association with any specific phenotype or disease state.[145] Although the study was not powered to detect subtle or infrequent clinical phenotypes arising from LRRK2 haploinsufficiency, it suggests that any clinical phenotype associated with partial reduction of LRRK2 will be substantially more benign than early-onset PD.

There are currently three LRRK2 therapeutics in the clinic from two pharmaceutical companies. In a healthy volunteer phase 1 study (ClinicalTrials.gov Identifier: NCT03710707), DNL201, a small molecule LRRK2 inhibitor, achieved on average greater than 90% inhibition of LRRK2 kinase activity at peak and greater than 50% inhibition at trough drug levels at the highest multiple-dose tested in blood as measured by LRRK2 S935 phosphorylation and phosphorylation of the substrate Rab10. DNL201 also achieved appropriate central nervous system penetration as determined by measurement of the compound in CSF. DNL201 was generally well tolerated in healthy young volunteers and doses up to 80 mg BID being tolerated in elderly subjects. Importantly, there were no clinically meaningful changes or dose-related trends in all functional pulmonary readouts tested.[146] DNL151, another oral LRRK2 inhibitor, is also undergoing phase 1b testing in subjects with PD (ClinicalTrials.gov Identifier: NCT04056689).

Another approach to reducing LRRK2 activity targets the messenger RNA using antisense oligonucleotides. These studies were prompted by the remarkable success of oligonucleotide delivery in spinal muscular atrophy (see Chapter 3 of this book) and preclinical studies demonstrating that reduction of endogenous LRRK2 levels decreased α-synuclein pathology and increased the number of tyrosine hydroxylase-positive neurons.[147] The direct administration of the oligonucleotide therapy into the CNS represents an invasive approach; however, it provides a viable therapeutic alternative. The LRRK2 knockdown might be restricted to the CNS and provide a stable, long-lasting reduction of protein throughout the brain without causing potential adverse side effects observed with LRRK2 inhibition in peripheral tissues. A clinical study to evaluate the safety, tolerability, and pharmacokinetics of the LRRK2 antisense oligonucleotide BIIB094 in adults with PD was recently initiated (ClinicalTrials.gov Identifier: NCT03976349).

Concluding remarks

The development of therapeutics for PD and other neurodegenerative diseases is undergoing an unprecedented transformation. Halting or slowing disease progression is now an achievable goal. Advances are emerging from genetics, basic and clinical sciences, digital technologies, and innovative public-private alliances. Despite our hope and optimism, history tells us most therapeutic candidates will not be approved.[148] There are many challenges for drug development in PD including erratic placebo response, absence of established biomarkers to enable informed go/no go decisions, limited understanding of disease progression and nonmotor manifestations of the disease, scarcity of diagnostic tools, and lack of outcome measures to objectively measure clinically meaningful changes.[149] Observational data, new molecular, biomarker, and clinical insights' global collaborations, and emerging technologies will enable more efficient and productive clinical trials.

We have highlighted three genetic discoveries that have prompted the initiation of clinical trials for therapies targeting the α-synuclein, GBA, and LRRK2 pathways. Although progress over the last two decades has been remarkable, the mechanistic picture is still developing. Well designed and powered basic and clinical studies will be vital to advance understanding of underlying disease mechanisms. We remain confident that the speed in the development of PD therapeutics will continue to accelerate and lead to the first treatment that slows or stops the progression of this disease.

Acknowledgments

The authors would like to thank Alison E. Schoeder and Brian DelGiudice (Sanofi Biomedical Media Services) for figure artwork, and Krishna Knabe (The Michael J. Fox Foundation) and Jennifer Matthews (Sanofi) for language editing.

Relevant conflicts of interest/financial disclosures

S.P.S. is an employee and stockholder of Sanofi SA. M.A.S.B. is an employee of The Michael J. Fox Foundation for Parkinson's Research.

References

1. Dorsey ER, Sherer T, Okun MS, Bloem BR. The emerging evidence of the parkinson pandemic. *J Parkinsons Dis*. 2018;8:S3–S8.
2. Yang W, Hamilton JL, Kopil C, et al. Current and projected future economic burden of Parkinson's disease in the U.S. *J Parkinsons Dis*. 2020;6.
3. Parkinson J. An essay on the shaking palsy. 1817. *J Neuropsychiatry Clin Neurosci*. 2002;14:223–236 discussion 222.
4. Poewe W, Seppi K, Tanner CM, et al. Parkinson disease. *Nat Rev Dis Primers*. 2017;3:17013.
5. Lang AE, Espay AJ. Which therapeutic developments are more likely to modify progression in Parkinson's disease? *Mov Disord*. 2018.
6. Sardi SP, Cedarbaum JM, Brundin P. Targeted therapies for Parkinson's disease: from genetics to the clinic. *Mov Disord*. 2018;33:684–696.
7. Chen-Plotkin AS, Zetterberg H. Updating our definitions of Parkinson's disease for a molecular age. *J Parkinsons Dis*. 2018;8:S53–S57.

8. Hardy J. Genetic analysis of pathways to Parkinson disease. *Neuron.* 2010;68:201–206.
9. Domingo A, Klein C. Genetics of Parkinson disease. *Handb Clin Neurol.* 2018;147:211–227.
10. Blauwendraat C, Nalls MA, Singleton AB. The genetic architecture of Parkinson's disease. *Lancet Neurol.* 2019;19:170–178.
11. Goedert M, Spillantini MG, Del Tredici K, Braak H. 100 years of Lewy pathology. *Nat Rev Neurol.* 2013;9:13–24.
12. Polymeropoulos MH, Lavedan C, Leroy E, et al. Mutation in the alpha-synuclein gene identified in families with Parkinson's disease. *Science.* 1997;276:2045–2047.
13. Chang D, Nalls MA, Hallgrimsdottir IB, et al. A meta-analysis of genome-wide association studies identifies 17 new Parkinson's disease risk loci. *Nat Genet.* 2017;49:1511–1516.
14. Nalls MA, Blauwendraat C, Vallerga CL, et al. Identification of novel risk loci, causal insights, and heritable risk for Parkinson's disease: a meta-analysis of genome-wide association studies. *Lancet Neurol.* 2019;18:1091–1102.
15. Nott A, Holtman IR, Coufal NG, et al. Brain cell type-specific enhancer-promoter interactome maps and disease-risk association. *Science.* 2019;366:1134–1139.
16. Nguyen M, Wong YC, Ysselstein D, Severino A, Krainc D. Synaptic, mitochondrial, and lysosomal dysfunction in Parkinson's disease. *Trends Neurosci.* 2019;42:140–149.
17. Alcalay RN, Kehoe C, Shorr E, et al. Genetic testing for Parkinson disease: current practice, knowledge, and attitudes among US and Canadian movement disorders specialists. *Genet Med.* 2019;22:574–580.
18. Nalls MA, McLean CY, Rick J, et al. Diagnosis of Parkinson's disease on the basis of clinical and genetic classification: a population-based modelling study. *Lancet Neurol.* 2015;14:1002–1009.
19. Liu G, Boot B, Locascio JJ, et al. Specifically neuropathic Gaucher's mutations accelerate cognitive decline in Parkinson's. *Ann Neurol.* 2016;80:674–685.
20. Cilia R, Tunesi S, Marotta G, et al. Survival and dementia in GBA-associated Parkinson disease: the mutation matters. *Ann Neurol.* 2016. https://doi.org/10.1002/ana.24777.
21. *Parkinson's Foundation Announces New Genetic Initiative Connecting Parkinson's Genetic Data with Clinical Care.*
22. *Fox Insight is an Online Clinical Study Building a Large, Diverse Cohort of People With Parkinson's and Age-Matched Control Volunteers to Provide the Research Community With Critical Insight Into the Lived Experience, Genetics and Variability of Parkinson's Disease.*
23. Goedert M, Jakes R, Spillantini MG. The synucleinopathies: twenty years on. *J Parkinsons Dis.* 2017;7:S53–S71.
24. Burre J. The synaptic function of alpha-synuclein. *J Parkinsons Dis.* 2015;5:699–713.
25. Bendor JT, Logan TP, Edwards RH. The function of alpha-synuclein. *Neuron.* 2013;79:1044–1066.
26. Spillantini MG, Schmidt ML, Lee VM, Trojanowski JQ, Jakes R, Goedert M. Alpha-synuclein in Lewy bodies. *Nature.* 1997;388:839–840.
27. Nalls MA, Pankratz N, Lill CM, et al. Large-scale meta-analysis of genome-wide association data identifies six new risk loci for Parkinson's disease. *Nat Genet.* 2014;46:989–993.
28. Soldner F, Stelzer Y, Shivalila CS, et al. Parkinson-associated risk variant in distal enhancer of alpha-synuclein modulates target gene expression. *Nature.* 2016;533:95–99.
29. Wong YC, Krainc D. Alpha-synuclein toxicity in neurodegeneration: mechanism and therapeutic strategies. *Nat Med.* 2017;23:1–13.
30. Dehay B, Bourdenx M, Gorry P, et al. Targeting alpha-synuclein for treatment of Parkinson's disease: mechanistic and therapeutic considerations. *Lancet Neurol.* 2015;14:855–866.
31. Li JY, Englund E, Holton JL, et al. Lewy bodies in grafted neurons in subjects with Parkinson's disease suggest host-to-graft disease propagation. *Nat Med.* 2008;14:501–503.
32. Kordower JH, Chu Y, Hauser RA, Freeman TB, Olanow CW. Lewy body-like pathology in long-term embryonic nigral transplants in Parkinson's disease. *Nat Med.* 2008;14:504–506.
33. Desplats P, Lee HJ, Bae EJ, et al. Inclusion formation and neuronal cell death through neuron-to-neuron transmission of alpha-synuclein. *Proc Natl Acad Sci U S A.* 2009;106:13010–13015.
34. Lee SJ, Desplats P, Sigurdson C, Tsigelny I, Masliah E. Cell-to-cell transmission of non-prion protein aggregates. *Nat Rev Neurol.* 2010;6:702–706.
35. Luk KC, Kehm V, Carroll J, et al. Pathological alpha-synuclein transmission initiates Parkinson-like neurodegeneration in nontransgenic mice. *Science.* 2012;338:949–953.

36. Braak H, Braak E. Pathoanatomy of Parkinson's disease. *J Neurol*. 2000;247(Suppl 2):II3–10.
37. Braak H, Del Tredici K. Neuropathological staging of brain pathology in sporadic Parkinson's disease: separating the wheat from the chaff. *J Parkinsons Dis*. 2017;7:S73–S87.
38. Fairfoul G, McGuire LI, Pal S, et al. Alpha-synuclein RT-QuIC in the CSF of patients with alpha-synucleinopathies. *Ann Clin Transl Neurol*. 2016;3:812–818.
39. Shahnawaz M, Tokuda T, Waragai M, et al. Development of a biochemical diagnosis of Parkinson disease by detection of alpha-synuclein misfolded aggregates in cerebrospinal fluid. *JAMA Neurol*. 2017;74:163–172.
40. Cole T, Paumier K, Zhao H, Weihofen A, Kordasiewicz HB, Swayze EE. Snca targeted antisense oligonucleotides mediate progression of pathological deposition in alpha synuclein rodent transmission models of Parkinson's disease. *Neurology*. 2016;86(16 Supplement):P6.239.
41. Zharikov AD, Cannon JR, Tapias V, et al. shRNA targeting alpha-synuclein prevents neurodegeneration in a Parkinson's disease model. *J Clin Invest*. 2015;125:2721–2735.
42. Mercuri E, Darras BT, Chiriboga CA, et al. Nusinersen versus sham control in later-onset spinal muscular atrophy. *N Engl J Med*. 2018;378:625–635.
43. Wild EJ, Tabrizi SJ. Therapies targeting DNA and RNA in Huntington's disease. *Lancet Neurol*. 2017;16:837–847.
44. Gorbatyuk OS, Li S, Nash K, et al. In vivo RNAi-mediated alpha-synuclein silencing induces nigrostriatal degeneration. *Mol Ther*. 2010;18:1450–1457.
45. Brundin P, Dave KD, Kordower JH. Therapeutic approaches to target alpha-synuclein pathology. *Exp Neurol*. 2017;298:225–235.
46. Mittal S, Bjornevik K, Im DS, et al. Beta2-Adrenoreceptor is a regulator of the alpha-synuclein gene driving risk of Parkinson's disease. *Science*. 2017;357:891–898.
47. Searles Nielsen S, Gross A, Camacho-Soto A, Willis AW, Racette BA. Beta2-adrenoreceptor medications and risk of Parkinson disease. *Ann Neurol*. 2018;84:683–693.
48. Price DL, Koike MA, Khan A, et al. The small molecule alpha-synuclein misfolding inhibitor, NPT200-11, produces multiple benefits in an animal model of Parkinson's disease. *Sci Rep*. 2018;8:16165.
49. Levin J, Schmidt F, Boehm C, et al. The oligomer modulator anle138b inhibits disease progression in a Parkinson mouse model even with treatment started after disease onset. *Acta Neuropathol*. 2014;127:779–780.
50. Xilouri M, Brekk OR, Stefanis L. Alpha-synuclein and protein degradation systems: a reciprocal relationship. *Mol Neurobiol*. 2013;47:537–551.
51. Poehler AM, Xiang W, Spitzer P, et al. Autophagy modulates SNCA/alpha-synuclein release, thereby generating a hostile microenvironment. *Autophagy*. 2014;10:2171–2192.
52. Lopes da Fonseca T, Villar-Pique A, Outeiro TF. The interplay between alpha-synuclein clearance and spreading. *Biomolecules*. 2015;5:435–471.
53. Decressac M, Mattsson B, Weikop P, Lundblad M, Jakobsson J, Bjorklund A. TFEB-mediated autophagy rescues midbrain dopamine neurons from alpha-synuclein toxicity. *Proc Natl Acad Sci U S A*. 2013;110:E1817–E1826.
54. Spencer B, Potkar R, Trejo M, et al. Beclin 1 gene transfer activates autophagy and ameliorates the neurodegenerative pathology in alpha-synuclein models of Parkinson's and Lewy body diseases. *J Neurosci*. 2009;29:13578–13588.
55. Ghosh A, Tyson T, George S, et al. Mitochondrial pyruvate carrier regulates autophagy, inflammation, and neurodegeneration in experimental models of Parkinson's disease. *Sci Transl Med*. 2016;8:368ra174.
56. Brahmachari S, Karuppagounder SS, Ge P, et al. c-Abl and Parkinson's disease: mechanisms and therapeutic potential. *J Parkinsons Dis*. 2017;7:589–601.
57. Karuppagounder SS, Brahmachari S, Lee Y, Dawson VL, Dawson TM, Ko HS. The c-Abl inhibitor, nilotinib, protects dopaminergic neurons in a preclinical animal model of Parkinson's disease. *Sci Rep*. 2014;4:4874.
58. Pagan F, Hebron M, Valadez EH, et al. Nilotinib effects in Parkinson's disease and Dementia with Lewy bodies. *J Parkinsons Dis*. 2016;6:503–517.
59. Wyse RK, Brundin P, Sherer TB. Nilotinib—differentiating the hope from the hype. *J Parkinsons Dis*. 2016;6:519–522.
60. *Though Safe, Nilotinib Does not Show Promise for Benefit for Parkinson's Disease*.

61. Candelise N, Schmitz M, Llorens F, et al. Seeding variability of different alpha synuclein strains in synucleinopathies. *Ann Neurol*. 2019;85:691–703.
62. Jankovic J, Goodman I, Safirstein B, et al. Safety and tolerability of multiple ascending doses of PRX002/RG7935, an anti-alpha-synuclein monoclonal antibody, in patients with Parkinson disease: a randomized clinical trial. *JAMA Neurol*. 2018;75:1206–1214.
63. Brys M, Fanning L, Hung S, et al. Randomized phase I clinical trial of anti-alpha-synuclein antibody BIIB054. *Mov Disord*. 2019;34:1154–1163.
64. AFFiRiS_Website. AFFiRiS Announces Top Line Results of First-in-Human Clinical Study Using AFFITOPE® PD03A, Confirming Immunogenicity and Safety Profile in Parkinson's Disease Patients. http://wwwaffiriscom/news/affiris-announces-top-line-results-of-first-in-human-clinical-study-using-affitope/.
65. Zella SMA, Metzdorf J, Ciftci E, et al. Emerging immunotherapies for Parkinson disease. *Neurol Ther*. 2019;8:29–44.
66. Sevigny J, Chiao P, Bussiere T, et al. The antibody aducanumab reduces Abeta plaques in Alzheimer's disease. *Nature*. 2016;537:50–56.
67. Eberling JL, Dave KD, Frasier MA. Alpha-synuclein imaging: a critical need for Parkinson's disease research. *J Parkinsons Dis*. 2013;3:565–567.
68. Mollenhauer B, Batrla R, El-Agnaf O, et al. A user's guide for alpha-synuclein biomarker studies in biological fluids: perianalytical considerations. *Mov Disord*. 2017;32:1117–1130.
69. Majbour NK, Vaikath NN, Eusebi P, et al. Longitudinal changes in CSF alpha-synuclein species reflect Parkinson's disease progression. *Mov Disord*. 2016;31:1535–1542.
70. Espay AJ, Vizcarra JA, Marsili L, et al. Revisiting protein aggregation as pathogenic in sporadic Parkinson and Alzheimer diseases. *Neurology*. 2019;92:329–337.
71. Neudorfer O, Giladi N, Elstein D, et al. Occurrence of Parkinson's syndrome in type I Gaucher disease. *QJM*. 1996;89:691–694.
72. Goker-Alpan O, Schiffmann R, LaMarca ME, Nussbaum RL, McInerney-Leo A, Sidransky E. Parkinsonism among Gaucher disease carriers. *J Med Genet*. 2004;41:937–940.
73. Aharon-Peretz J, Rosenbaum H, Gershoni-Baruch R. Mutations in the glucocerebrosidase gene and Parkinson's disease in Ashkenazi Jews. *N Engl J Med*. 2004;351:1972–1977.
74. Sidransky E, Nalls MA, Aasly JO, et al. Multicenter analysis of glucocerebrosidase mutations in Parkinson's disease. *N Engl J Med*. 2009;361:1651–1661.
75. Nalls MA, Duran R, Lopez G, et al. A multicenter study of glucocerebrosidase mutations in dementia with Lewy bodies. *JAMA Neurol*. 2013;70:727–735.
76. Beavan MS, Schapira AH. Glucocerebrosidase mutations and the pathogenesis of Parkinson disease. *Ann Med*. 2013;45:511–521.
77. Grabowski GA. Phenotype, diagnosis, and treatment of Gaucher's disease. *Lancet*. 2008;372: 1263–1271.
78. Alcalay RN, Dinur T, Quinn T, et al. Comparison of Parkinson risk in Ashkenazi Jewish patients with Gaucher disease and GBA heterozygotes. *JAMA Neurol*. 2014;71:752–757.
79. Iwaki H, Blauwendraat C, Leonard HL, et al. Genetic risk of Parkinson disease and progression: an analysis of 13 longitudinal cohorts. *Neurol Genet*. 2019;5:e348.
80. Davis MY, Johnson CO, Leverenz JB, et al. Association of GBA mutations and the E326K polymorphism with motor and cognitive progression in Parkinson disease. *JAMA Neurol*. 2016;73: 1217–1224.
81. Zhang Y, Shu L, Zhou X, et al. A meta-analysis of GBA-related clinical symptoms in Parkinson's disease. *Parkinsons Dis*. 2018;2018:3136415.
82. Tsuang D, Leverenz JB, Lopez OL, et al. GBA mutations increase risk for Lewy body disease with and without Alzheimer disease pathology. *Neurology*. 2012;79:1944–1950.
83. Mazzulli JR, Xu YH, Sun Y, et al. Gaucher disease glucocerebrosidase and alpha-synuclein form a bidirectional pathogenic loop in synucleinopathies. *Cell*. 2011;146:37–52.
84. Sardi SP, Clarke J, Kinnecom C, et al. CNS expression of glucocerebrosidase corrects {alpha}-synuclein pathology and memory in a mouse model of Gaucher-related synucleinopathy. *Proc Natl Acad Sci U S A*. 2011;108:12101–12106.

85. Sardi SP, Clarke J, Viel C, et al. Augmenting CNS glucocerebrosidase activity as a therapeutic strategy for parkinsonism and other Gaucher-related synucleinopathies. *Proc Natl Acad Sci U S A*. 2013;110: 3537–3542.
86. Sardi SP, Cheng SH, Shihabuddin LS. Gaucher-related synucleinopathies: the examination of sporadic neurodegeneration from a rare (disease) angle. *Prog Neurobiol*. 2015;125C:47–62.
87. Aflaki E, Westbroek W, Sidransky E. The complicated relationship between Gaucher disease and parkinsonism: insights from a rare disease. *Neuron*. 2017;93:737–746.
88. Schondorf DC, Aureli M, McAllister FE, et al. iPSC-derived neurons from GBA1-associated Parkinson's disease patients show autophagic defects and impaired calcium homeostasis. *Nat Commun*. 2014;5:4028.
89. Taguchi YV, Liu J, Ruan J, et al. Glucosylsphingosine promotes alpha-synuclein pathology in mutant gba-associated Parkinson's disease. *J Neurosci*. 2017;37:9617–9631.
90. Sardi SP, Viel C, Clarke J, et al. Glucosylceramide synthase inhibition alleviates aberrations in synucleinopathy models. *Proc Natl Acad Sci U S A*. 2017;114:2699–2704.
91. Galvagnion C. The role of lipids interacting with alpha-synuclein in the pathogenesis of Parkinson's disease. *J Parkinsons Dis*. 2017;7:433–450.
92. Zunke F, Moise AC, Belur NR, et al. Reversible conformational conversion of alpha-synuclein into toxic assemblies by glucosylceramide. *Neuron*. 2017;97:92–107.
93. Kim S, Yun SP, Lee S, et al. GBA1 deficiency negatively affects physiological alpha-synuclein tetramers and related multimers. *Proc Natl Acad Sci U S A*. 2018;115:798–803.
94. Papadopoulos VE, Nikolopoulou G, Antoniadou I, et al. Modulation of beta-glucocerebrosidase increases alpha-synuclein secretion and exosome release in mouse models of Parkinson's disease. *Hum Mol Genet*. 2018;27:1696–1710.
95. Li H, Ham A, Ma TC, et al. Mitochondrial dysfunction and mitophagy defect triggered by heterozygous GBA mutations. *Autophagy*. 2019;15:113–130.
96. Gegg ME, Burke D, Heales SJ, et al. Glucocerebrosidase deficiency in substantia nigra of parkinson disease brains. *Ann Neurol*. 2012;72:455–463.
97. Parnetti L, Chiasserini D, Persichetti E, et al. Cerebrospinal fluid lysosomal enzymes and alpha-synuclein in Parkinson's disease. *Mov Disord*. 2014;29:1019–1027.
98. Alcalay RN, Levy OA, Waters CC, et al. Glucocerebrosidase activity in Parkinson's disease with and without GBA mutations. *Brain*. 2015;138:2648–2658.
99. Hallett PJ, Huebecker M, Brekk OR, et al. Glycosphingolipid levels and glucocerebrosidase activity are altered in normal aging of the mouse brain. *Neurobiol Aging*. 2018;67:189–200.
100. Burbulla LF, Song P, Mazzulli JR, et al. Dopamine oxidation mediates mitochondrial and lysosomal dysfunction in Parkinson's disease. *Science*. 2017;357:1255–1261.
101. Vila M, Laguna A, Carballo-Carbajal I. Intracellular crowding by age-dependent neuromelanin accumulation disrupts neuronal proteostasis and triggers Parkinson disease pathology. *Autophagy*. 2019; 15:2028–2030.
102. Rocha EM, Smith GA, Park E, et al. Glucocerebrosidase gene therapy prevents alpha-synucleinopathy of midbrain dopamine neurons. *Neurobiol Dis*. 2015;82:495–503.
103. Rockenstein E, Clarke J, Viel C, et al. Glucocerebrosidase modulates cognitive and motor activities in murine models of Parkinson's disease. *Hum Mol Genet*. 2016. https://doi.org/10.1093/hmg/ddw124.
104. Morabito G, Giannelli SG, Ordazzo G, et al. AAV-PHP.B-mediated global-scale expression in the mouse nervous system enables GBA1 gene therapy for wide protection from synucleinopathy. *Mol Ther*. 2017;25:2727–2742.
105. Migdalska-Richards A, Daly L, Bezard E, Schapira AH. Ambroxol effects in glucocerebrosidase and alpha-synuclein transgenic mice. *Ann Neurol*. 2016;80:766–775.
106. Magalhaes J, Gegg ME, Migdalska-Richards A, Schapira AH. Effects of ambroxol on the autophagy-lysosome pathway and mitochondria in primary cortical neurons. *Sci Rep*. 2018;8:1385.
107. Bartus RT, Weinberg MS, Samulski RJ. Parkinson's disease gene therapy: success by design meets failure by efficacy. *Mol Ther*. 2014;22:487–497.
108. Mazzulli JR, Zunke F, Tsunemi T, et al. Activation of beta-glucocerebrosidase reduces pathological alpha-synuclein and restores lysosomal function in Parkinson's patient midbrain neurons. *J Neurosci*. 2016;36:7693–7706.

109. Aflaki E, Borger DK, Moaven N, et al. A new glucocerebrosidase chaperone reduces alpha-synuclein and glycolipid levels in iPSC-derived dopaminergic neurons from patients with Gaucher disease and parkinsonism. *J Neurosci*. 2016;36:7441–7452.
110. Burbulla LF, Jeon S, Zheng J, Song P, Silverman RB, Krainc D. A modulator of wild-type glucocerebrosidase improves pathogenic phenotypes in dopaminergic neuronal models of Parkinson's disease. *Sci Transl Med*. 2019;11:eaau6870.
111. Poole RM. Eliglustat: first global approval. *Drugs*. 2014;74:1829–1836.
112. Paisan-Ruiz C, Jain S, Evans EW, et al. Cloning of the gene containing mutations that cause PARK8-linked Parkinson's disease. *Neuron*. 2004;44:595–600.
113. Zimprich A, Biskup S, Leitner P, et al. Mutations in LRRK2 cause autosomal-dominant parkinsonism with pleomorphic pathology. *Neuron*. 2004;44(4):601–607.
114. West AB, Moore DJ, Biskup S, et al. Parkinson's disease-associated mutations in leucine-rich repeat kinase 2 augment kinase activity. *Proc Natl Acad Sci U S A*. 2005;102:16842–16847.
115. Di Maio R, Hoffman EK, Rocha EM, et al. LRRK2 activation in idiopathic Parkinson's disease. *Sci Transl Med*. 2018;10(451):eaar5429.
116. Paisan-Ruiz C, Lewis PA, Singleton AB. LRRK2: cause, risk, and mechanism. *J Parkinsons Dis*. 2013;3:85–103.
117. Gilsbach BK, Kortholt A. Structural biology of the LRRK2 GTPase and kinase domains: implications for regulation. *Front Mol Neurosci*. 2014;7:32.
118. Mata IF, Wedemeyer WJ, Farrer MJ, Taylor JP, Gallo KA. LRRK2 in Parkinson's disease: protein domains and functional insights. *Trends Neurosci*. 2006;29:286–293.
119. Blauwendraat C, Reed X, Kia DA, et al. Frequency of loss of function variants in LRRK2 in Parkinson disease. *JAMA Neurol*. 2018;75:1416–1422.
120. Greggio E, Jain S, Kingsbury A, et al. Kinase activity is required for the toxic effects of mutant LRRK2/dardarin. *Neurobiol Dis*. 2006;23:329–341.
121. Scott JD, DeMong DE, Greshock TJ, et al. Discovery of a 3-(4-pyrimidinyl) indazole (MLi-2), an orally available and selective leucine-rich repeat kinase 2 (LRRK2) inhibitor that reduces brain kinase activity. *J Med Chem*. 2017;60:2983–2992.
122. Steger M, Tonelli F, Ito G, et al. Phosphoproteomics reveals that Parkinson's disease kinase LRRK2 regulates a subset of Rab GTPases. *Elife*. 2016;5:e12813.
123. Wandinger-Ness A, Zerial M. Rab proteins and the compartmentalization of the endosomal system. *Cold Spring Harb Perspect Biol*. 2014;6:a022616.
124. Pfeffer SR. Rab GTPases: master regulators that establish the secretory and endocytic pathways. *Mol Biol Cell*. 2017;28:712–715.
125. MacLeod DA, Rhinn H, Kuwahara T, et al. RAB7L1 interacts with LRRK2 to modify intraneuronal protein sorting and Parkinson's disease risk. *Neuron*. 2013;77:425–439.
126. Wilson GR, Sim JC, McLean C, et al. Mutations in RAB39B cause X-linked intellectual disability and early-onset Parkinson disease with alpha-synuclein pathology. *Am J Hum Genet*. 2014;95:729–735.
127. Mata IF, Jang Y, Kim CH, et al. The RAB39B p.G192R mutation causes X-linked dominant Parkinson's disease. *Mol Neurodegener*. 2015;10:50.
128. Steger M, Diez F, Dhekne HS, et al. Systematic proteomic analysis of LRRK2-mediated Rab GTPase phosphorylation establishes a connection to ciliogenesis. *Elife*. 2017;6:e31012.
129. Purlyte E, Dhekne HS, Sarhan AR, et al. Rab29 activation of the Parkinson's disease-associated LRRK2 kinase. *EMBO J*. 2018;37:1–18.
130. Guaitoli G, Gilsbach BK, Raimondi F, Gloeckner CJ. First model of dimeric LRRK2: the challenge of unrevealing the structure of a multidomain Parkinson's-associated protein. *Biochem Soc Trans*. 2016;44:1635–1641.
131. Mathea S, Baptista M, Reichert P, et al. Crystallizing the Parkinson's disease protein LRRK2 under microgravity conditions. *bioRxiv*. 2018, https://doi.org/10.1101/259655.
132. Padmanabhan S, Lanz TA, Gorman D, et al. An assessment of LRRK2 serine 935 phosphorylation in human peripheral blood mononuclear cells in idiopathic Parkinson's disease and G2019S LRRK2 cohorts. *J Parkinsons Dis*. 2020;10(2):623–629.
133. Alessi DR, Sammler E. LRRK2 kinase in Parkinson's disease. *Science*. 2018;360:36–37.

134. Ysselstein D, Nguyen M, Young TJ, et al. LRRK2 kinase activity regulates lysosomal glucocerebrosidase in neurons derived from Parkinson's disease patients. *Nat Commun*. 2019;10:5570.
135. Dhekne HS, Yanatori I, Gomez RC, et al. A pathway for Parkinson's disease LRRK2 kinase to block primary cilia and sonic hedgehog signaling in the brain. *Elife*. 2018;7:e40202.
136. Gonzalez-Reyes LE, Verbitsky M, Blesa J, et al. Sonic hedgehog maintains cellular and neurochemical homeostasis in the adult nigrostriatal circuit. *Neuron*. 2012;75:306–319.
137. Shutinoski B, Hakimi M, Harmsen IE, et al. Lrrk2 alleles modulate inflammation during microbial infection of mice in a sex-dependent manner. *Sci Transl Med*. 2019;11:eaas9292.
138. Hui KY, Fernandez-Hernandez H, Hu J, et al. Functional variants in the LRRK2 gene confer shared effects on risk for Crohn's disease and Parkinson's disease. *Sci Transl Med*. 2018;10:eaai7795.
139. Dzamko NL. LRRK2 and the immune system. *Adv Neurobiol*. 2017;14:123–143.
140. Liu W, Liu X, Li Y, et al. LRRK2 promotes the activation of NLRC4 inflammasome during Salmonella Typhimurium infection. *J Exp Med*. 2017;214:3051–3066.
141. Baptista MA, Dave KD, Frasier MA, et al. Loss of leucine-rich repeat kinase 2 (LRRK2) in rats leads to progressive abnormal phenotypes in peripheral organs. *PLoS One*. 2013;8:e80705.
142. Ness D, Ren Z, Gardai S, et al. Leucine-rich repeat kinase 2 (LRRK2)-deficient rats exhibit renal tubule injury and perturbations in metabolic and immunological homeostasis. *PLoS One*. 2013;8: e66164.
143. Fuji RN, Flagella M, Baca M, et al. Effect of selective LRRK2 kinase inhibition on nonhuman primate lung. *Sci Transl Med*. 2015;7:273ra215.
144. Baptista MAS, Merchant K, Barrett T, et al. LRRK2 kinase inhibitors induce a reversible effect in the lungs of non-human primates with no measurable pulmonary deficits. *Sci Transl Med*. 2020;12(540): eaav0820.
145. Whiffin N, Armean IM, Kleinman A, et al. The effect of LRRK2 loss-of-function variants in humans. *Nat Med*. 2020;26:869–877.
146. Denali_Website. Denali Therapeutics Announces Advancement and Expansion of Its LRRK2 Inhibitor Clinical Program for Parkinson's Disease. http://investorsdenalitherapeuticscom/news-releases/news-release-details/denali-therapeutics-announces-advancement-and-expansion-its#ir-pages; 2018.
147. Zhao HT, John N, Delic V, et al. LRRK2 antisense oligonucleotides ameliorate alpha-synuclein inclusion formation in a Parkinson's disease mouse model. *Mol Ther Nucleic Acids*. 2017;8:508–519.
148. Kesselheim AS, Hwang TJ, Franklin JM. Two decades of new drug development for central nervous system disorders. *Nat Rev Drug Discov*. 2015;14:815–816.
149. Shihabuddin LS, Brundin P, Greenamyre JT, Stephenson D, Sardi SP. New frontiers in parkinson's disease: from genetics to the clinic. *J Neurosci*. 2018;38:9375–9382.

CHAPTER 7

Neural stem cell-mediated brain tumor therapy

Rachael Mooney[a], Hoi Wa Ngai[a], Jana Portnow[b], and Karen Aboody[a]
[a]Department of Developmental and Stem Cell Biology, Beckman Research Institute of City of Hope, Duarte, CA, United States
[b]Department of PS Medical Oncology, Beckman Research Institute of City of Hope, Duarte, CA, United States

Introduction

Glioblastoma (GBM) is the most common and aggressive diffuse malignant primary brain tumor in adults. Making up 60% of all gliomas, the incidence of GBM in the United States is 3.6 per 100,000 adults.[1] Despite aggressive multimodal therapy consisting of surgery, radiation, chemotherapy with temozolomide as well as alternating electric fields, patients affected by primary GBMs generally die 15–20 months after diagnosis, whereas patients with secondary GBMs that have transformed from low-grade astrocytomas die within 31 months.[2] Only about 10% of patients will be alive 5 years after they are diagnosed with GBM.[3] This sobering clinical scenario underscores the critical need for novel delivery strategies that will safely increase anticancer agent concentrations at invasive tumor foci while sparing normal brain cells.

Treatment failure is largely attributed to the tumor's highly invasive and heterogeneous nature leading to further recurrences and development of chemo- and radioresistant populations of tumor cells.[4] Major obstacles to effective treatment include suboptimal passage of chemotherapy and other therapeutics across the blood-brain-barrier and (BBB), and poor biodistribution through tumor sites.[5] The toxicity of current chemo- and radiation therapies limits the doses that can be used. The same limitations exist for the treatment of brain metastases from melanoma, breast, and lung cancer. The development of more effective brain tumor-targeted therapies is critical to improving clinical outcomes.

One strategy to achieve more effective tumor targeting is to couple therapeutics with cells that inherently migrate to tumors. The tumor microenvironment is a rich source of a myriad of signals (i.e., angiogenesis, extracellular matrix production, hypoxia, inflammation, tumor-derived cytokines) that attract migratory cell types including immune cells, stem cells, and other tumor cells.[6–10]

Stem cells began to dominate the "carrier cell" field because they had more appealing safety and immunogenicity profiles. Patient-derived stem cells (hematopoietic, mesenchymal, and induced pluripotent: HSCs, MSCs, and iPSCs) have been heavily explored

Neurotherapeutics in the Era of Translational Medicine
https://doi.org/10.1016/B978-0-12-816475-4.00007-0

for the delivery of antitumor treatments to cancer.[11–14] These studies have shown substantial benefits including more effective distribution through tumor foci, limited off-target distribution, and superior therapeutic efficacy.[15–17]

While the use of autologous cells reduces the risk of immune rejection, we believe there are several limitations to this cell source, as discussed below. This chapter describes an alternative cell source, the immortalized human neural stem cell (NSC) line (HB1.F3.CD21), which is intended as an allogeneic, "off-the-shelf" cell line available for large-scale treatments. The use of a clinically relevant allogeneic cell line is an important development in several key respects:

1. *Cell supply and time*: A major bottleneck in the clinical translation of cellular therapies is clinical scale-up compliant with current good manufacturing practices (cGMP). After isolation from the patient, autologous cells can only be expanded in culture for a limited number of passages (~5–6 passages, or 30 population doublings) before they lose their tumor-homing properties, accumulate chromosomal and functional instabilities, and senesce.[17] The first attempt to use patient-derived MSCs to deliver an oncolytic virus to ovarian cancer metastases required 2 weeks post-isolation to generate enough MSCs for patient treatment, and 20% showed abnormal karyotypes.[17] In contrast, we can produce a cGMP-grade lot of up to 2.4×10^{10} clinically viable NSCs in just 9 days.[18]
2. *Standardization and reproducibility*: Autologous treatments require that a new source of primary tissue be obtained for each production cycle, which prevents the thorough characterization of cells pretransplantation. Furthermore, each patient receives a different product, and the amount and quality of cells that can be isolated depend on patient age and current health status, confounding treatment analysis. This variability results in unpredictable yields, rates of proliferation, and *ex vivo* transduction capacities. Conversely, our immortalized HB1.F3.CD21 NSCs can be reliably expanded, characterized, and banked as an "off-the-shelf" product to obtain reproducible results across patients and treatment sites.
3. *Cost*: Analysts have estimated that cell therapies involving modified autologous cells are likely to cost up to $100,000 per patient, making them among the most expensive therapies currently available.[19] The lower production costs of allogeneic alternatives will likely expand the reach and increase the feasibility and potential impact of stem cell-based therapies.

We have brought two NSC-mediated enzyme/prodrug gene therapies from "bench to bedside" with the first-in-human phase I dose-escalation trials for recurrent high-grade glioma patients. A third strategy developed in collaboration with Maciej Lesniak, MD at Northwestern University using NSCs to deliver oncolytic virotherapy has demonstrated safety in a phase I trial for newly diagnosed high-grade glioma patients when given as an adjunct to radiation and temozolomide. Here, we describe different NSC-mediated therapeutic brain tumor strategies, our preclinical pipeline, our path to the clinic, and trial results to date.

Immortalized, allogeneic HB1.F3.CD21 neural stem cells (NSCs)

NSCs hold great promise for brain tumor therapy due to their inherent tumor-tropic properties, making them an ideal vehicle to deliver therapeutic payloads selectively to infiltrative CNS tumor foci. Our lab has established a well-characterized, v-*myc* immortalized, clonal NSC line which is genetically and functionally stable over time and passage, and HLA Class 2 negative (Fig. 1).[11] NSCs initially derived from normal XX karyotype 15-week telencephalon were stably transduced with replication-incompetent retrovirus that inserted 1 copy of the avian v-*myc* gene into chromosome 19. Avian v-*myc* was chosen because it maintains the cells' stem-like state and spontaneously downregulates after removal of mitogenic stimulation (i.e., upon transplantation), a necessary safety feature for clinical applications.[20] A second stable transduction with replication-incompetent retrovirus inserted 1 copy of the *E-coli* cytosine deaminase (CD) gene into chromosome 9.[11] The CD served as both a suicide gene and a prodrug-converting enzyme as described below (Section "NSC-mediated enzyme/prodrug gene therapy"). Clones were screened for tumor-tropism.

Clone 21 (HB1.F3.CD21) was shown to migrate well to invasive primary and secondary brain tumor sites in orthotopic preclinical models, whether delivered intracerebrally into the same or opposite hemisphere, the lateral ventricle, or intravenously—demonstrating the ability of NSCs to cross the BBB (Fig. 2).[21] Microenvironmental factors contributing to this tumor-tropism include high VEGF concentrations, hypoxia, inflammatory cytokines, and PI3K signaling.[7,9,22,23]

NSCs have a natural ability to migrate to locations where it is notoriously difficult to achieve adequate coverage when delivering either free-drugs, nanoparticle-encapsulated

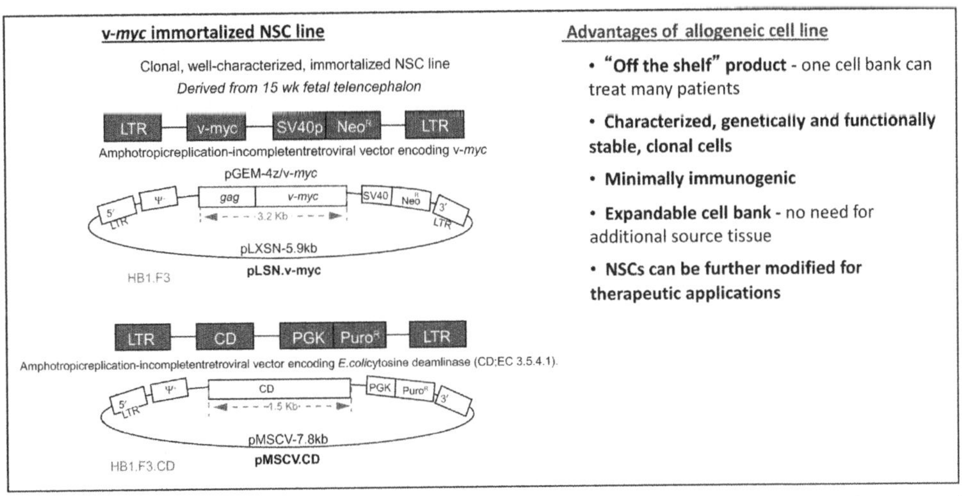

Fig. 1 Schematic representation showing genetic modification of HB1.F3 NSCs to create the parental HB1.F3.CD allogeneic cell line.

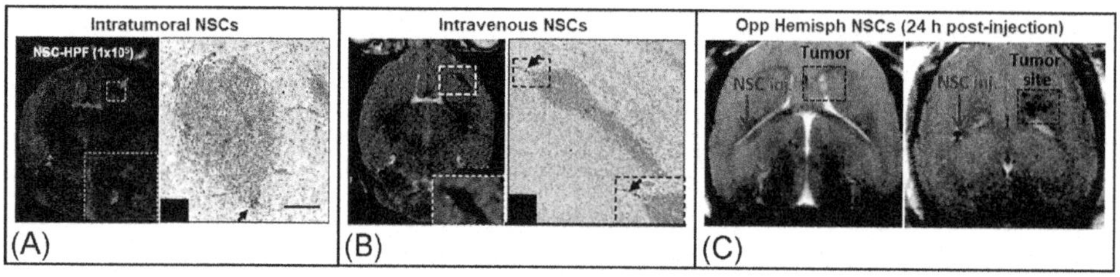

Fig. 2 Intracerebrally injected feraheme-labeled NSCs migrate and distribute to orthotopic U251T.eGFP.ffluc gliomas. Intracerebral xenografts in mice were established by stereotactic implantation of 2×10^5 U251T.eGFP.ffluc human glioma cells into the right frontal hemisphere. Three days later, feraheme-labeled NSCs were injected caudolateral to the tumor. T2-weighted magnetic resonance images of mouse brains are shown. Arrows indicate hypointense *(black)* signals associated with feraheme-labeled NSCs. Boxed areas are magnified in the Prussian blue-stained and pararosaniline-counterstained brain tumor sections with NSCs *(black arrows)*. HPF-labeled NSCs that migrated to tumor are stained blue (Scale bars = 200 μm). *(A) and (B) adapted from Gutova M, Frank JA, D'Apuzzo M, et al. Magnetic resonance imaging tracking of ferumoxytol-labeled human neural stem cells: studies leading to clinical use. Stem Cells Transl Med 2013;2(10):766–775.*

drugs, gene therapy expression vectors, or oncolytic viruses. For example, NSCs administered intracerebrally or intravenously are able to penetrate the main tumor mass, migrate to hypoxic areas, and disseminated micrometastases throughout the brain.[9,21,24] NSCs can, therefore, serve as a platform delivery technology to improve the distribution and efficacy of various anticancer therapies to tumor sites. Our vision is to shift the current paradigm of passive drug delivery to a more active, NSC-mediated delivery in an effort to improve the therapeutic index of existing and novel therapeutics for solid tumors. NSCs can increase the concentration of anticancer agents specifically at tumor foci while sparing toxicity to normal tissues—potentially improving patient quality of life.

NSC-mediated brain tumor therapies

As described, the parental NSC line (HB1.F3.CD21) has demonstrated clinical safety, migration to distant tumor sites, non-tumorigenicity, and non-immunogenicity in a first-in-human, single-dose study for recurrent glioma patients.[25] We also demonstrated proof-of-concept for brain tumor localized chemotherapy production by the NSCs. We have further modified this parental line with different therapeutic cargo, demonstrating the capacity of this NSC delivery platform for improving therapeutic distribution, retention, and efficacy (Fig. 3). Here, we describe our efforts focused on NSCs and NSC-derived exosomes improving the delivery of (1) enzyme/prodrug gene therapy, (2) oncolytic virotherapy, (3) therapeutic antibodies, (4) oligonucleotide therapeutics (ONTs), and (5) therapeutic nanoparticles (NPs).

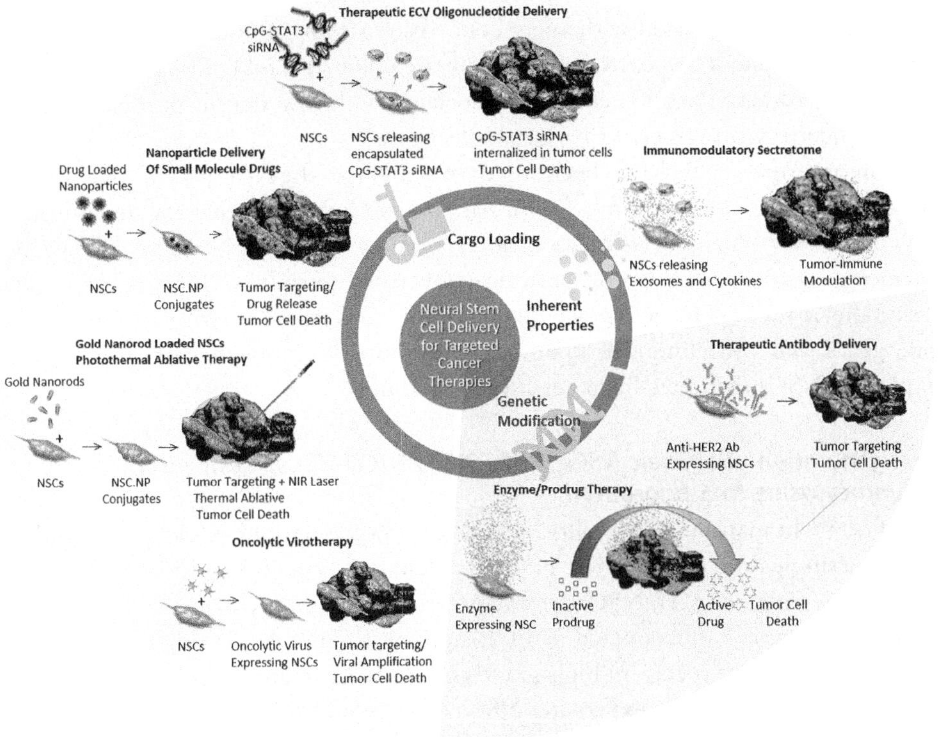

Fig. 3 Engineering NSCs to deliver targeted anticancer payloads. Tumor-tropic NSCs can be engineered to deliver anticancer agents selectively to tumor foci. Approaches explored include enzyme/prodrug gene therapy, oncolytic virotherapy, therapeutic protein delivery (antibody delivery shown as an example), sustained-release or stimuli-responsive nanoparticle delivery, and extracellular vesicle oligonucleotide delivery. *NSC*, neural stem cell; *NP*, nanoparticle; *Ab*, antibody.

(1) NSC-mediated enzyme/prodrug gene therapy

Enzyme/prodrug gene therapy involves local expression of enzymes that convert an inactive prodrug into a tumor-toxic effector drug. Over 50 different enzyme/prodrug combinations have been developed over the past 2 decades.[26] Once generated, each effector molecule can kill multiple surrounding tumor cells via a "bystander effect", the size of which is dependent on the effector drug produced.[27–30] Enzyme/prodrug treatments have such potent bystander effects that only a small percentage (~2%–10%) of the tumor needs to express the prodrug-converting enzyme.[31]

Despite the potent bystander effect, it is a major challenge to attain sufficient, sustained therapeutic enzyme expression in tumors using traditional viral or nonviral vectors. Viral vectors, such as retrovirus, lentivirus, adenovirus, and adeno-associated virus, are usually associated with safety concerns regarding immunogenicity and mutagenesis; while nonviral vectors, such as synthetic polymers and lipids, amino acid-based vectors and bacteria-based vectors, are associated with low transfection efficiency and safety concerns with respect to bacterial infections.[32]

Tumor-tropic cells have been used to enhance the efficacy of five different enzyme-prodrug combinations.[33] Our lab has performed preclinical investigational new drug (IND)-enabling efficacy and safety/toxicity studies for NSCs which have been engineered to express two prodrug-converting enzymes for tumor-localized chemotherapy.[34,35] The genes encoding prodrug-converting enzymes also serve as suicide genes that will eliminate, upon administration of complementary prodrugs, any implanted NSCs that have been genetically modified.[36]

First-generation allogeneic NSCs: HB1.F3.CD21 (CD-NSCs), converting 5-fluorocytosine to 5-fluorourail

Our first-in-human pilot feasibility study assessed the safety of intracerebrally administering a single dose of allogeneic NSCs that had been genetically modified to express cytosine deaminase (CD-NSCs), for converting 5-fluorocytosine (5-FC) into the chemotherapeutic 5-fluorouracil (5-FU), a potent thymidylate synthetase inhibitor.[25] Fifteen patients with recurrent high-grade gliomas were treated in cohorts of three with an increasing dose of CD-expressing NSCs (CD-NSCs) injected into the brain at the time of resection or biopsy and followed 3 days later by a 7-day course of oral 5-FC. Results demonstrated safety and feasibility for brain tumor localized conversion of 5-FC to 5-FU by the CD-NSCs (Fig. 4).[25] There were no significant toxicities associated with the CD-NSCs. Autopsy data demonstrated NSC migration to distant tumor sites, even in the opposite hemisphere, as well as non-tumorigenicity of the NSCs, confirming the shutdown of v-*myc* (likely due to epigenetic factors). Intracerebral microdialysis data showed detectable levels of 5-FU in the brain for the 7-day duration of 5-FC administration, indicating that CD continued to be expressed. Furthermore, minimal or nondetectable levels of 5-FU in the patients' plasma indicated that the conversion to 5-FU resulted from the CD-NSCs, rather than from CD produced by gut *E. coli* converting 5-FC to 5-FU systemically and then the 5-FU crossing into the brain. We thus demonstrated proof-of-concept for tumor-localized production of an anticancer agent.

We have recently completed a phase I study of administering multiple doses of CD-NSCs in combination with 5-FC and leucovorin to patients with recurrent high-grade gliomas (NCT02015819). In this study, a Rickham reservoir/catheter was placed into the tumor resection cavity for use in administering the CD-NSCs intracerebrally every 2 weeks, followed each time by a 7-day course of oral 5-FC. This manuscript has been submitted for publication.

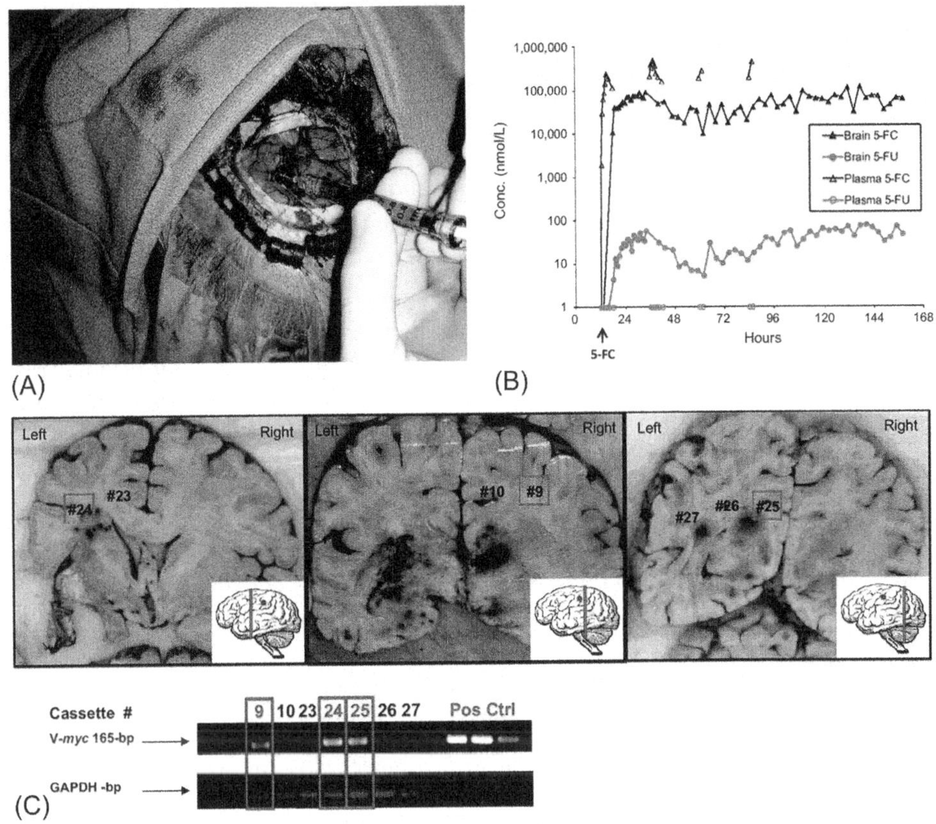

Fig. 4 First-in-human safety study established proof-of-principle prodrug conversion and NSC migration after intracranial administration of CD-NSCs. (A) Representative photograph of NSC injection into the tumor resection cavity. (B) Intracerebral microdialysis data: brain interstitial 5-FC and 5-FU concentrations from a single patient on dose level 1 demonstrating the ability of the NSCs to convert 5-FC to 5-FU in the brain throughout the 7-day course of oral 5-FC. In this patient, no 5-FU was detected in the plasma, indicating that all of the 5-FU measured in the brain came from NSC-mediated conversion of 5-FC to 5-FU. (C) Brain autopsy data from a male patient. Representative coronal sections, depicted from rostral to occipital (inset shows the plane of section, *red line*; and the plane of NSC injection, *red dot*), show the classic variegated appearance of involvement by a high-grade infiltrating glioma, including diffuse cerebral edema with mass effect (right side larger than left side), necrosis, and hemorrhage. Numbers on the coronal sections represent areas from which tissue samples were obtained (sampling from additional coronal sections not shown). The presence of NSCs *(red boxes)* was confirmed by nested PCR analysis in the right frontoparietal white matter, just occipital to the NSC injection site (cassette #9; middle) as well as at distant sites, including the left internal capsule and corona radiata (cassette #24, left) and left occipital region (cassette #25, right). The presence of infiltrating glioblastoma was confirmed histologically at these three sampled sites. (D) Representative nested PCR results. *Red text and boxes*, v-*myc*-positive bands. GAPDH was used as a DNA integrity control, and normal brain tissue spiked with NSC DNA was used as a positive control for v-*myc*. The assay was repeated 3 times per tissue block sample, and the detection of v-*myc* in 2 of 3 PCR assays was considered positive for NSCs. *Adapted from Portnow J, Synold TW, Badie B, et al., Neural stem cell-based anticancer gene therapy: a first-in-human study in recurrent high-grade glioma patients. Clin Cancer Res 2017;23 (12):2951–2960.*

Second-generation NSCs: HB1.F3.CD21.hCE1m6 (CE-NSCs), converting prodrug irinotecan to SN-38

The first-generation NSCs were further modified ex vivo by replication-incompetent adenovirus to secrete a modified human carboxylesterase (CE-NSCs) that converts the chemotherapy irinotecan (IRN; CPT-11) into its more active metabolite SN-38, a potent topoisomerase I inhibitor.[34] Our preclinical in vivo microdialysis studies of intravenously administered IRN demonstrated higher concentrations of both IRN and SN-38 in the plasma of brain tumor-bearing rats, which match the pharmacokinetic property of plasma protein binding of both IRN and SN-38.[34,37] Despite the low permeability of IRN through the BBB, we found an increase in IRN conversion to SN-38 in the brain due to the presence of CE-NSCs compared to controls, but no increase in plasma SN-38, demonstrating the tumor-selective conversion of IRN to SN-38 by CE-NSCs.[34] A phase I clinical trial of serial doses of CE-NSCs administered intracerebrally and in combination with intravenous IRN in recurrent high glioma patients, is ongoing (NCT02192359).

The primary objectives of the CD-NSC and CE-NSC phase I studies are to define the phase II recommended dose for these NSC-based treatments. Secondary study objectives include pharmacokinetics, immunogenicity, and cell fate correlative studies.

(2) NSC-mediated oncolytic virotherapy

Oncolytic viruses can induce cancer cell death irrespective of radio- or chemoresistance, and they can stimulate immune recognition of cancer cells, due to exposure of tumor antigens upon lysis.[38] Recent early phase trials with several of these viruses have demonstrated that subsets of recurrent glioblastoma/glioma patients may significantly benefit from oncolytic virotherapy and survive >3 years after treatment.[39] These encouraging preliminary clinical data raise a glimmer of hope in fighting glioblastoma after many years of intensive search for better treatments. One reason that positive responses were not observed in the majority of patients may be that the efficacy of the virus is limited by delivery hurdles such as rapid immune inactivation, poor viral penetration of tumors, and an inability of the viruses to reach invasive foci that are separated from the main tumor mass by normal tissue.[40]

In collaboration with Dr. Lesniak's group, we further modified the HB1.F3.CD21 NSC line to produce and deliver a conditionally replicative adenovirus, CRAd-Survivin-pk7, modified with poly-L-lysine incorporation into the C terminus of the fiber knob protein (pk7) and driven by the *survivin* promoter (Fig. 5).[41] The pk7 fiber knob proteins interact with heparin sulfate proteoglycans, which enhance the adenoviral infection of NSCs and glioma cells. NSCs provide a safe home for adequate viral amplification and protection from rapid immune clearance before reaching tumor foci to release the CRAd-Survivin-pk7 upon cell burst.[42] The NSCs migrate across normal brain tissue, allowing seeding of the virus to reach distant invasive tumor foci (Fig. 6).[43] The virus

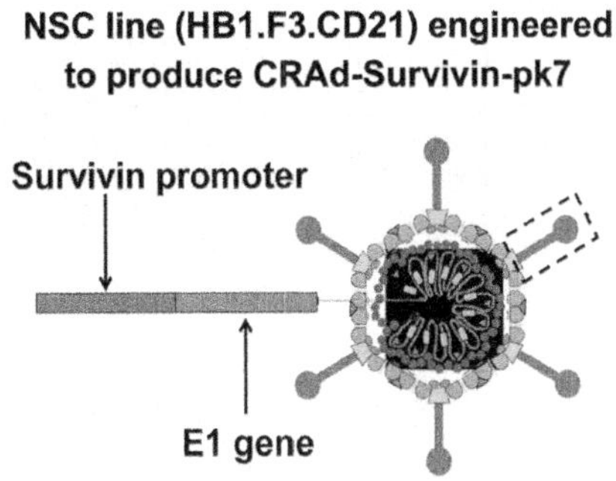

Viral element	Effect
Survivin	Restricts viral replication in tumor cells
E1 gene	Essential gene for viral replication
Pk7 fiber	Enhances infectivity via integrin binding

Fig. 5 Schematic representation showing engineered CRAd-S-pk7 virus. In constructing adenovirus CRAd-S-pk7, the native E1a promoter of Ad5 has been replaced with a portion of the survivin promoter to direct tumor-selective virus replication. An additional modification to the vector is the incorporation of a short poly-lysine (pk7)-encoding sequence into the fiber knob gene. This modification improves the transduction efficiency of the virus in tumor cells with limited coxsackie-adenovirus receptors (CAR).

then replicates specifically in cells that overexpress *survivin*, which is highly expressed in both newly diagnosed and recurrent glioma cells (upregulated by radiation), but not in normal differentiated cells.[44] In other words, once seeded into the tumor, the conditionally replicative adenovirus will continue to reproduce only in tumor cells, amplifying the effect to neighboring tumor cells until normal tissue is reached and it ceases.

We recently completed a phase 1 study demonstrating the safety and feasibility of administering a single dose of the NSC.CRAd-S-pk7 to newly diagnosed patients with high-grade glioma (NCT03072134) who were then treated with radiation and temozolomide. Due to volume limitations, the highest dose administered intracerebrally was 1.5×10^7 NSCs. Future plans include studying repeated administrations of the NSC. CRAd-s-pk7, in both newly diagnosed and recurrent glioma patients whose tumors overexpress *survivin*. We hypothesize multiple doses will both increase viral coverage of tumors and stimulate a secondary immune response without the development of neutralizing antibodies.

(3) NSC-mediated antibody delivery

The treatment of human epidermal growth factor receptor 2 (HER2)-overexpressing breast cancer has been revolutionized by trastuzumab (Herceptin). However, increased survival time now predisposes these patients to develop brain metastases, as the

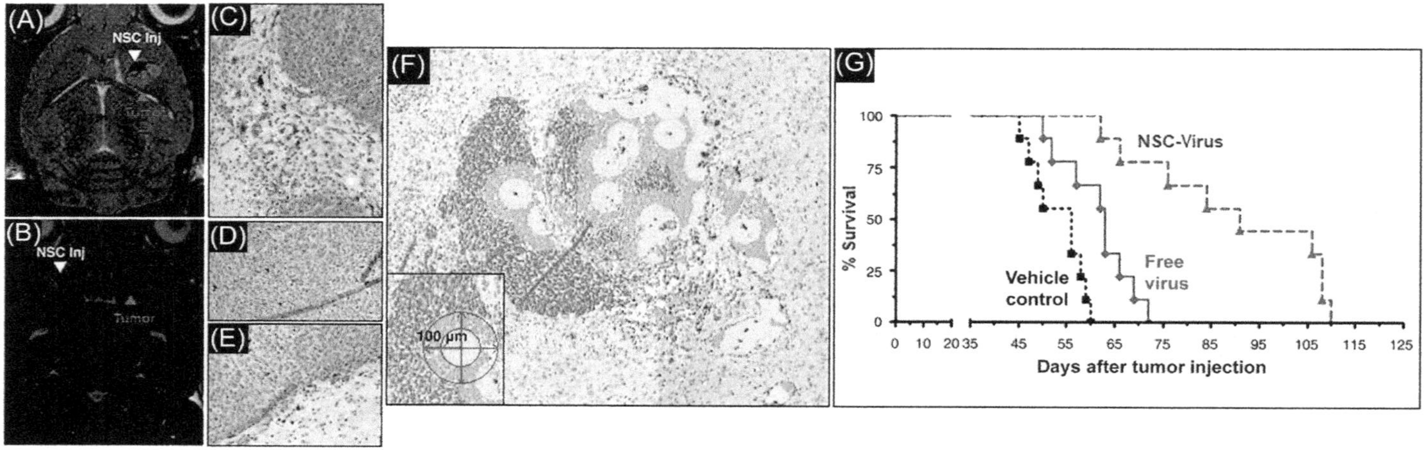

Fig. 6 Oncolytic virus-loaded neural stem cells (NSCs) inhibit xenograft growth and prolong survival of mice with orthotopic glioblastoma. (A–B) Magnetic resonance imaging (MRI) visualization of NSC-CRAd-S-pk7 migration. NSC-CRAd-S-pk7 were administered either in the contralateral cortical hemisphere (a) or adjacent to (b) established GBM43 tumors. T2-weighted MRI scans and brain tissue sections were acquired on day 4. (C–D) Prussian blue staining of tissue sections from these same animals were carried out to confirm the presence of ferumoxytol-labeled stem cells. (C–E) Determination of tumor coverage by NSC and NSC-CRAd-S-pk7. (C–E) Visual representation of tumor coverage analysis. NSCs were identified by histology based on Prussian blue staining. (F) A therapeutic diameter of 100 μm was used in the calculation of tumor coverage. (F) 2.5×10^5 U87 cells were injected stereotactically into the right hemisphere of the brains of 9- to 12-week-old *nu/nu* mice followed by intratumoral (IT) injection of CRAd-S-pk7 alone (2.5×10^7 I.U./animal) or loaded into NSCs (5×10^5 NSC/animal loaded with 50 I.U./NSC, total 2.5×10^7 I.U. viral dose) at 5 days after tumor implantation. *Reproduced from Morshed RA, Gutova M, Juliano J, et al., Analysis of glioblastoma tumor coverage by oncolytic virus-loaded neural stem cells using MRI-based tracking and histological reconstruction. Cancer Gene Ther 2015;22 (1): 55–61 and Ahmed AU, Thaci B, Alexiades NG, et al. Neural stem cell-based cell carriers enhance therapeutic efficacy of an oncolytic adenovirus in an orthotopic mouse model of human glioblastoma. Mol Ther 2011;19(9): 1714–1726.*

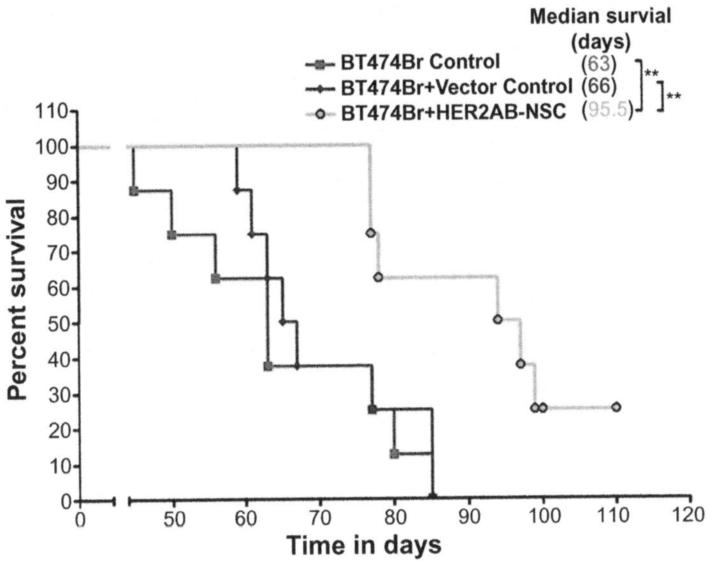

Fig. 7 Survival analysis using the intracranial injection of HER2Ab-NSCs. Kaplan Meier survival curve of PBS, vector control, and HER2Ab-NSCs injected. Mice were monitored weekly and sacrificed when moribund. The log rank test was applied to compare mice survival ($n=8$). *Adapted from Kanojia D, Balyasnikova IV, Morshed RA, et al., Neural stem cells secreting anti-HER2 antibody improve survival in a preclinical model of HER2 overexpressing breast cancer brain metastases. Stem Cells 2015;33(10): 2985–2994, with permission.*

therapeutic antibodies cannot effectively cross the BBB. We, therefore, modified our NSC line through adenoviral and lentiviral transduction to test the respective transient and stable secretion of the full-length anti-HER2 antibody, functionally equivalent to trastuzumab.[45] In vivo experiments using a breast cancer brain metastasis mouse model demonstrate that intracerebral injection of HER2Ab-NSCs significantly improves survival (Fig. 7).[46] In effect, these NSCs provide tumor localized and sustained production of HER2Ab, minimizing potential off target side effects.

(4) NSC-derived exosomes as carriers for oligonucleotide therapeutics

Oligonucleotides are short synthetic DNA or RNA polymers, which modulate gene expression through Watson-Crick base pairing with targeted nucleic acids.[47,48] ONTs can overcome the targeting limitations of small molecule inhibitors and antibodies, especially against undruggable molecular targets such as oncogenic transcription factors. ONT delivery, however, is challenging due to short circulatory half-lives and limited intratumoral penetration within the GBM microenvironment. To overcome this problem, synthetic ONTs can be packaged into NSC-derived exosomes.[49] Preliminary studies suggest NSCs rapidly internalize exosome-encapsulated oligonucleotides for up to

3 days. Compared to free ONT delivery, NSC delivered exosome-encapsulated ONTs improved the number of tumor-associated myeloid cells present within glioma to generate direct cytotoxic and immune-mediated antitumor effects.

(5) NSC-mediated nanoparticle delivery

Because intravenously administered chemotherapies do not efficiently penetrate the BBB or reach hypoxic tumor regions, intratumoral infusions of therapeutic NPs have been combined with convection-enhanced delivery for the treatment of brain tumors.[50] Obstacles to successful treatment include poor NP tumor penetration and distribution, as well as poor NP retention at the tumor site. NSCs are appealing for use as NP carriers in order to overcome these biodistribution challenges. We have shown that NSCs maintain their tumor tropism when either surface bound or internalized and are exploring two NSC-NP treatment strategies for brain tumors.[51]

NSC-NPs for small-molecule drug delivery

TMZ is one of <2% of small-molecule drugs that can cross the BBB; however, brain concentrations only reach ~20% of plasma levels (Portnow et al., Clin Cancer Res 2009) and drug toxicity limits dosing.[52,53] The combination of NSCs and NPs offers a platform for local delivery of anticancer drugs, where NSCs can migrate to and retain NPs at invasive tumor sites for triggered or slow drug release. We have shown that NPs can be either surface-coupled or internalized by NSCs, and subsequently deliver NP cargo to ipsilateral and contralateral glioma lesions (Fig. 8).[51,54] The NPs used were ~fivefold larger than those typically used for cancer therapies, which translates into a >100-fold increase in the loading potential of a potent GBM chemotherapeutic, doxorubicin.

NSC-AuNRs+NIR for thermal ablative therapy

Photothermal ablative therapy utilizes inert gold nanorods as tumor-localized antennas that convert near-infrared light (NIR) into heat to essentially 'burn' the tumor tissue—irrespective of chemoresistance or molecular phenotype. Advances in image-guided procedures and nanotechnology enhance synergy with other treatments, but in the instance when the entire tumor mass has been heated, adequate distribution and retention at the tumor site are required to be effective and to avoid collateral damage to surrounding healthy tissue. NSCs can internalize AuNRs and maintain their viability and tumor homing properties. NSC-AuNRs improve distribution and retention selectively at tumor sites, as compared to free AuNRs, and can improve localized heating upon exposure to NIR to kill surrounding tumor cells while minimizing damage to non-tumor tissue.[55] Thermal-ablative therapy has not yet been clinically applied to brain tumors, however, preclinical studies continue.

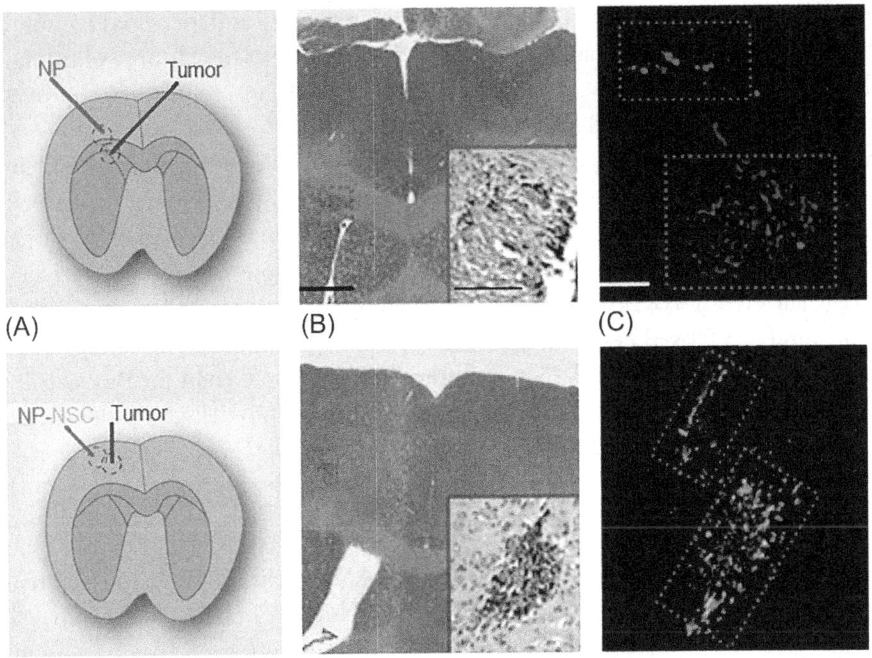

(A) (B) (C)

(D) (E) (F)

Fig. 8 Advantages of using migratory NSCs to selectively target NPs to glioma. (A, B) Schematic representation of injection paradigm for upper and lower panels. (C, D) Four days after NP injection, brains were sectioned throughout the entire injection and tumor site. Every fifth section was DAPI stained and imaged using fluorescence microscopy. The slice that contained the greatest surface area of NPs was identified and adjacent sections stained with hematoxylin and eosin (C, D) (Scale bar = 500 μm). Anti-ffLuc immunohistochemistry was also performed (insets in C, D) (Scale bar = 100 μm) to confirm the presence of tumor cells. (E, F) Representative merged fluorescence images of the tumor *(blue dotted box)* and injection sites *(red dotted box)* (NPs, *red/orange*; NSCs, *green*; nuclei, *blue*). Tumors are visible as dense clusters of DAPI-stained nuclei *(blue)* (Scale bar = 100 μm). *Adapted with permission from Mooney R, Weng Y, Tirughana-Sambandan R, et al., Neural stem cells improve intracranial nanoparticle retention and tumor-selective distribution. Future Oncol 2014;10(3): 401–415.*

Neural stem cell manufacturing

Cell source

Tumor-tropic NSCs are a platform technology that could improve the delivery of a wide repertoire of therapeutics, to a wide variety of malignant tumors. One of the most critical issues considered early in the development process was the source and ability to expand a sufficient supply of NSCs and to characterize their stability over time and passage. Primary NSC cultures generated from developing brain can be expanded by repeated subculture of polyclonal neurospheres. However, progressive passages lead to a decreased capacity for cellular self-renewal and differentiation potential and increased accumulation of chromosomal and functional instabilities.[56] This requires a new source of primary tissue for each production cycle, which makes process scale-up, regulatory approval, and clinical translation substantially more difficult and costly.[33]

We took the approach of genetic modification to generate immortalized NSC lines by retroviral transduction with a v-*myc* gene of early gestational NSC pools and clonal selection.[57] Our NSC therapies have utilized a well-characterized, normal female karyotype clonal line, HB1.F3.CD21, that demonstrates genetic stability, tumor-tropism, non-tumorigenicity, and stem cell characteristics over time and passage, thereby enabling product standardization and scale-up. It is HLA Class II negative, minimizing potential immune responses. This cell line is established as a Master Cell Bank (MCB) at the City of Hope Center for Biomedicine and Genetics GMP facility. This MCB can be further engineered to deliver various therapeutic cargo to generate multiple "off the shelf" allogeneic cell banks for patient treatment.

Scale-up GMP manufacturing

As cell-based therapies move from phase I to phase II-III clinical trials and commercialization, methods for scale-up manufacture and production of GMP-grade stem cells in a cost-effective, safe, and reproducible manner need to be addressed. We have used both multilayer cell factories and hollow fiber bioreactors for our adherent NSCs, with each clinical bank release tested for sterility, mycoplasma, endotoxin, identity, viability, therapeutic cargo levels, and karyotype. We are now able to reproducibly expand the NSCs from initial seeding of 4×10^7 cells to production of $1.5–3 \times 10^9$ genetically and functionally normal cells within 7–9 days.[18] This would not be possible using an autologous cell source.[33] Such stringent quality control is necessary to ensure patient safety when using an allogeneic cell line. The FDA has created guidance documents that address the various controls and safeguards starting with donor eligibility, initial collection of the source tissue under current good tissue practice (cGTP), and subsequent manufacturing steps under current good manufacturing practice (cGMP).[58]

Improving NSC administration

Poor NSC survival, due to transplantation into a hostile microenvironment post-resection is a major roadblock that limits the therapeutic potential of cell-mediated therapeutic delivery. Following transplantation into the resection cavity wall, NSCs must migrate and localize to residual sites of glioma located around the resection margin and to distant tumor sites along conductive flow routes (e.g., white matter tracts, perivascular spaces). Because 80%–90% of GBM recurrences are located within 2 cm of the primary resection cavity, we are investigating ways to support NSC survival and function within this environment.[59] We have demonstrated that the baseline level of NSC apoptosis within the resection cavity can be delayed by overexpressing Bcl-2, an antiapoptotic protein known to reduce implantation-associated cell death.[60] Another approach is to deliver NSCs within a hydrogel that acts as a "reservoir," that prevents post-injection reflux and provides the tumor margin with a continuous supply of therapeutic NSCs that can migrate from the gel toward the invasive GBM.[61,62] In an effort to provide better tumor coverage by NSCs to all micro tumor foci throughout the brain, we have also explored the possibility of delivering NSCs, intraventricularly.[33,63] More easily administered routes of NSC delivery are being investigated, including intranasal and intravenous delivery.[64–66]

Conclusions

NSC-mediated therapies have the potential to improve the treatment of brain cancer as well as possibly improve quality of life since treatment is localized to tumor sites, minimizing off-target toxicities. Further avenues for development include optimizing NSC delivery, maximizing NSC viability, tumor coverage, and improving ease of administration. In addition, we are actively exploring delivery of other therapeutic payloads to different tumor types, and combinatorial regimens, including coupling NSC-mediated treatments with immunomodulatory strategies. There is much work still needed to realize the full potential of NSC-mediated cancer treatments, particularly to augment endogenous and/or engineered cytokine expression and exosome production.

Bench-to-bedside: Perspective and lessons learned

While drug development has typically been the purview of industry, academic researchers face challenges in translating their research into clinical products. These limitations include lack of familiarity with venture capital funding, GMP manufacturing, and FDA regulatory requirements.[67] Translation, by which we mean advancing scientific discoveries from the laboratory to the clinic for patient benefit, i.e., "bench to bedside," requires a comprehensive collaborative team approach: research scientists and clinicians must work closely with regulatory agencies, patient advocacy groups, ethics bodies, cell manufacturing facilities, and industry to achieve the quality of studies and necessary funding to ensure success. This requires new partnership models for research in which traditional silos are broken down, translational teams are created, and new mechanisms for effective hand-off from nonprofit to for-profit are generated.

Despite these difficulties, steady progress toward this goal is being made, spearheaded by industry, academic institutions, and nonprofit foundations in conjunction with a recent focus by the National Institutes of Health and the Food and Drug Administration in the United States on both translational research and regenerative medicine. Recognizing the 'valley of death' where progress is halted by lack of funding, several private foundations target and support translational research for specific neurological diseases. NIH has created the National Center for Advancing Translational Sciences and an NIH Center for Regenerative Medicine. The FDA has issued guidelines for preclinical safety evaluation and assessment of investigational cellular and gene therapy products. Human cellular products such as stem and progenitor cells have unique requirements for characterization, manufacturing, and testing that are regulated by the Center for Biologics Evaluation and Research and its Office of Cellular, Tissue, and Gene Therapies. In addition to the demonstration of preclinical proof of concept, efficacy/safety/toxicity studies are required for filing an Investigational New Drug Application (IND) with the FDA. Further, Institutional Review Board (IRB), Institutional Biosafety Committee, and Stem Cell Research Oversight Committee approvals are required to initiate a clinical trial.

Acknowledgments

Financial support was provided by the Ben and Catherine Ivy Foundation, the Daphna and Richard Ziman Family Foundation, The Rosalinde and Arthur Gilbert Foundation, The Alvarez Family Foundation, The Anthony F. & Susan M. Markel Foundation, City of Hope, and National Cancer Institute grants R01 CA198076, R01 FD004816-A1, U01NS082328-01 R43 CA86768, R44 CA8678, and P30 CA033572. COI Statement: KSA is on the Board of TheraBiologics Inc.

References

1. Mesfin FB, Al-Dhahir MA. Gliomas. In: *StatPearls*. Treasure Island (FL): StatPearls Publishing; 2020.
2. Silantyev AS, Falzone L, Libra M, et al. Current and future trends on diagnosis and prognosis of glioblastoma: from molecular biology to proteomics. *Cells*. 2019;8(8).
3. Guzauskas GF, Pollom EL, Stieber VW, Wang BCM, Garrison LP. Tumor treating fields and maintenance temozolomide for newly-diagnosed glioblastoma: a cost-effectiveness study. *J Med Econ*. 2019; 22(10):1006–1013.
4. Nicholas MK, Lukas RV, Chmura S, Yamini B, Lesniak M, Pytel P. Molecular heterogeneity in glioblastoma: therapeutic opportunities and challenges. *Semin Oncol*. 2011;38(2):243–253.
5. Halle B, Mongelard K, Poulsen FR. Convection-enhanced drug delivery for glioblastoma: a systematic review focused on methodological differences in the use of the convection-enhanced delivery method. *Asian J Neurosurg*. 2019;14(1):5–14.
6. Gutova M, Najbauer J, Frank RT, et al. Urokinase plasminogen activator and urokinase plasminogen activator receptor mediate human stem cell tropism to malignant solid tumors. *Stem Cells*. 2008; 26(6):1406–1413.
7. Kendall SE, Najbauer J, Johnston HF, et al. Neural stem cell targeting of glioma is dependent on phosphoinositide 3-kinase signaling. *Stem Cells*. 2008;26(6):1575–1586.
8. Schmidt NO, Przylecki W, Yang W, et al. Brain tumor tropism of transplanted human neural stem cells is induced by vascular endothelial growth factor. *Neoplasia*. 2005;7(6):623–629.
9. Zhao D, Najbauer J, Garcia E, et al. Neural stem cell tropism to glioma: critical role of tumor hypoxia. *Mol Cancer Res*. 2008;6(12):1819–1829.
10. Ziu M, Schmidt NO, Cargioli TG, Aboody KS, Black PM, Carroll RS. Glioma-produced extracellular matrix influences brain tumor tropism of human neural stem cells. *J Neuro-Oncol*. 2006;79(2):125–133.
11. Aboody KS, Najbauer J, Metz MZ, et al. Neural stem cell-mediated enzyme/prodrug therapy for glioma: preclinical studies. *Sci Transl Med*. 2013;5(184):184ra59.

12. Nouri FS, Banerjee D, Hatefi A. Practical issues with the use of stem cells for cancer gene therapy. *Stem Cell Rev.* 2015;11(5):688–698.
13. Papapetrou EP. Patient-derived induced pluripotent stem cells in cancer research and precision oncology. *Nat Med.* 2016;22(12):1392–1401.
14. Xiang B, Chen L, Wang X, Xiang C. Mesenchymal stem cells as therapeutic agents and in gene delivery for the treatment of glioma. *J Zhejiang Univ Sci B.* 2017;18(9):737–746.
15. Huang B, Abraham WD, Zheng Y, Bustamante López SC, Luo SS, Irvine DJ. Active targeting of chemotherapy to disseminated tumors using nanoparticle-carrying T cells. *Sci Transl Med.* 2015; 7(291):291ra94.
16. Li L, Guan Y, Liu H, et al. Silica nanorattle-doxorubicin-anchored mesenchymal stem cells for tumor-tropic therapy. *ACS Nano.* 2011;5(9):7462–7470.
17. Mader EK, Butler G, Dowdy SC, et al. Optimizing patient derived mesenchymal stem cells as virus carriers for a phase I clinical trial in ovarian cancer. *J Transl Med.* 2013;11:20.
18. Tirughana R, Metz MZ, Li Z, et al. GMP production and scale-up of adherent neural stem cells with a quantum cell expansion system. *Mol Ther Methods Clin Dev.* 2018;10:48–56.
19. Lopes AG, Sinclair A, Frohlich B. Cost analysis of cell therapy manufacture: autologous cell therapies, part 1. *Bioprocess Int.* 2018;16:S3–S8.
20. Flax JD, Aurora S, Yang C, et al. Engraftable human neural stem cells respond to developmental cues, replace neurons, and express foreign genes. *Nat Biotechnol.* 1998;16(11):1033–1039.
21. Aboody KS, Brown A, Rainov NG, et al. Neural stem cells display extensive tropism for pathology in adult brain: evidence from intracranial gliomas. *Proc Natl Acad Sci U S A.* 2000;97(23):12846–12851.
22. Schmidt NO, Koeder D, Messing M, et al. Vascular endothelial growth factor-stimulated cerebral microvascular endothelial cells mediate the recruitment of neural stem cells to the neurovascular niche. *Brain Res.* 2009;1268:24–37.
23. Zhao D, Najbauer J, Annala AJ, et al. Human neural stem cell tropism to metastatic breast cancer. *Stem Cells.* 2012;30(2):314–325.
24. Aboody KS, Najbauer J, Schmidt NO, et al. Targeting of melanoma brain metastases using engineered NSPCs. *Neuro-Oncol.* 2006;8(2):119–126.
25. Portnow J, Synold TW, Badie B, et al. Neural stem cell-based anticancer gene therapy: a first-in-human study in recurrent high-grade glioma patients. *Clin Cancer Res.* 2017;23(12):2951–2960.
26. Greco O, Dachs GU. Gene directed enzyme/prodrug therapy of cancer: historical appraisal and future prospectives. *J Cell Physiol.* 2001;187(1):22–36.
27. Agard C, Ligeza C, Dupas B, et al. Immune-dependent distant bystander effect after adenovirus-mediated suicide gene transfer in a rat model of liver colorectal metastasis. *Cancer Gene Ther.* 2001; 8(2):128–136.
28. Kuriyama S, Tsujinoue H, Yoshiji H. Immune response to suicide gene therapy. *Methods Mol Med.* 2004;90:353–369.
29. Moolten FL. Tumor chemosensitivity conferred by inserted herpes thymidine kinase genes: paradigm for a prospective cancer control strategy. *Cancer Res.* 1986;46(10):5276–5281.
30. Rainov NG. A phase III clinical evaluation of herpes simplex virus type 1 thymidine kinase and ganciclovir gene therapy as an adjuvant to surgical resection and radiation in adults with previously untreated glioblastoma multiforme. *Hum Gene Ther.* 2000;11(17):2389–2401.
31. Huber BE, Austin EA, Richards CA, Davis ST, Good SS. Metabolism of 5-fluorocytosine to 5-fluorouracil in human colorectal tumor cells transduced with the cytosine deaminase gene: significant antitumor effects when only a small percentage of tumor cells express cytosine deaminase. *Proc Natl Acad Sci U S A.* 1994;91(17):8302–8306.
32. Malekshah OM, Chen X, Nomani A, Sarkar S, Hatefi A. Enzyme/prodrug systems for cancer gene therapy. *Curr Pharmacol Rep.* 2016;2(6):299–308.
33. Mooney R, Abdul Majid A, Batalla J, Annala AJ, Aboody KS. Cell-mediated enzyme prodrug cancer therapies. *Adv Drug Deliv Rev.* 2017;118:35–51.
34. Metz MZ, Gutova M, Lacey SF, et al. Neural stem cell-mediated delivery of irinotecan-activating carboxylesterases to glioma: implications for clinical use. *Stem Cells Transl Med.* 2013;2(12):983–992.

35. Mooney R, Hammad M, Batalla-Covello J, Abdul Majid A, Aboody KS. Concise review: neural stem cell-mediated targeted cancer therapies. *Stem Cells Transl Med*. 2018;7(10):740–747.
36. Li W, Xiang AP. Safeguarding clinical translation of pluripotent stem cells with suicide genes. *Organogenesis*. 2013;9(1):34–39.
37. Mathijssen RH, van Alphen RJ, Verweij J, et al. Clinical pharmacokinetics and metabolism of irinotecan (CPT-11). *Clin Cancer Res*. 2001;7(8):2182–2194.
38. Ding J. Oncolytic virus as a cancer stem cell killer: progress and challenges. *Stem Cell Invest*. 2014;1:22.
39. Lang FF, Conrad C, Gomez-Manzano C, et al. Phase I study of DNX-2401 (delta-24-RGD) oncolytic adenovirus: replication and immunotherapeutic effects in recurrent malignant glioma. *J Clin Oncol*. 2018;36(14):1419–1427.
40. Ahmed AU, Thaci B, Tobias AL, et al. A preclinical evaluation of neural stem cell-based cell carrier for targeted antiglioma oncolytic virotherapy. *J Natl Cancer Inst*. 2013;105(13):968–977.
41. Tyler MA, Ulasov IV, Sonabend AM, et al. Neural stem cells target intracranial glioma to deliver an oncolytic adenovirus in vivo. *Gene Ther*. 2009;16(2):262–278.
42. Ahmed AU, Thaci B, Alexiades NG, et al. Neural stem cell-based cell carriers enhance therapeutic efficacy of an oncolytic adenovirus in an orthotopic mouse model of human glioblastoma. *Mol Ther*. 2011;19(9):1714–1726.
43. Morshed RA, Gutova M, Juliano J, et al. Analysis of glioblastoma tumor coverage by oncolytic virus-loaded neural stem cells using MRI-based tracking and histological reconstruction. *Cancer Gene Ther*. 2015;22(1):55–61.
44. Tobias AL, Thaci B, Auffinger B, et al. The timing of neural stem cell-based virotherapy is critical for optimal therapeutic efficacy when applied with radiation and chemotherapy for the treatment of glioblastoma. *Stem Cells Transl Med*. 2013;2(9):655–666.
45. Frank RT, Edmiston M, Kendall SE, et al. Neural stem cells as a novel platform for tumor-specific delivery of therapeutic antibodies. *PLoS One*. 2009;4(12):e8314.
46. Kanojia D, Balyasnikova IV, Morshed RA, et al. Neural stem cells secreting anti-HER2 antibody improve survival in a preclinical model of HER2 overexpressing breast cancer brain metastases. *Stem Cells*. 2015;33(10):2985–2994.
47. Krichevsky AM, Uhlmann EJ. Oligonucleotide therapeutics as a new class of drugs for malignant brain tumors: targeting mRNAs, regulatory RNAs, mutations, combinations, and Beyond. *Neurotherapeutics*. 2019;16(2):319–347.
48. Takakura K, Kawamura A, Torisu Y, Koido S, Yahagi N, Saruta M. The Clinical potential of oligonucleotide therapeutics against pancreatic cancer. *Int J Mol Sci*. 2019;20(13).
49. Kooijmans SAA, Stremersch S, Braeckmans K, et al. Electroporation-induced siRNA precipitation obscures the efficiency of siRNA loading into extracellular vesicles. *J Control Release*. 2013; 172(1):229–238.
50. Allard E, Passirani C, Benoit J-P. Convection-enhanced delivery of nanocarriers for the treatment of brain tumors. *Biomaterials*. 2009;30(12):2302–2318.
51. Mooney R, Weng Y, Tirughana-Sambandan R, et al. Neural stem cells improve intracranial nanoparticle retention and tumor-selective distribution. *Future Oncol*. 2014;10(3):401–415.
52. Ferber D. Bridging the blood-brain barrier: new methods improve the odds of getting drugs to the brain cells that need them. *PLoS Biol*. 2007;5(6):e169.
53. Tentori L, Graziani G. Recent approaches to improve the antitumor efficacy of temozolomide. *Curr Med Chem*. 2009;16(2):245–257.
54. Cheng Y, Morshed R, Cheng S-H, et al. Nanoparticle-programmed self-destructive neural stem cells for glioblastoma targeting and therapy. *Small (Germany)*. 2013;9(24):4123–4129.
55. Mooney R, Roma L, Zhao D, et al. Neural stem cell-mediated intratumoral delivery of gold nanorods improves photothermal therapy. *ACS Nano*. 2014;8(12):12450–12460.
56. Kallos MS, Behie LA. Inoculation and growth conditions for high-cell-density expansion of mammalian neural stem cells in suspension bioreactors. *J Biochem Microbiol Technol Eng*. 1999;63(4):473–483.
57. Kim SU, Nagai A, Nakagawa E, et al. Production and characterization of immortal human neural stem cell line with multipotent differentiation property. *Methods Mol Biol*. 2008;438:103–121.

58. Burger SR. Current regulatory issues in cell and tissue therapy. *Cytotherapy*. 2003;5(4):289–298.
59. Wallner KE, Galicich JH, Krol G, Arbit E, Malkin MG. Patterns of failure following treatment for glioblastoma multiforme and anaplastic astrocytoma. *Int J Radiat Oncol Biol Phys*. 1989;16(6):1405–1409.
60. Mooney R, Majid AA, Mota D, et al. Bcl-2 Overexpression improves survival and efficacy of neural stem cell-mediated enzyme prodrug therapy. *Stem Cells Int*. 2018;2018:7047496.
61. Kauer TM, Figueiredo JL, Hingtgen S, Shah K. Encapsulated therapeutic stem cells implanted in the tumor resection cavity induce cell death in gliomas. *Nat Neurosci*. 2011;15(2):197–204.
62. Sheets KT, Ewend MG, Mohiti-Asli M, et al. Developing implantable scaffolds to enhance neural stem cell therapy for post-operative glioblastoma. *Mol Ther*. 2020;28(4):1056–1067.
63. Gutova M, Flores L, Adhikarla V, et al. Quantitative evaluation of intraventricular delivery of therapeutic neural stem cells to orthotopic glioma. *Front Oncol*. 2019;9:68.
64. Dey M, Yu D, Kanojia D, et al. Intranasal oncolytic virotherapy with CXCR4-enhanced stem cells extends survival in mouse model of glioma. *Stem Cell Rep*. 2016;7(3):471–482.
65. Reekmans KP, Praet J, De Vocht N, et al. Clinical potential of intravenous neural stem cell delivery for treatment of neuroinflammatory disease in mice? *Cell Transplant*. 2011;20(6):851–869.
66. Reitz M, Demestre M, Sedlacik J, et al. Intranasal delivery of neural stem/progenitor cells: a noninvasive passage to target intracerebral glioma. *Stem Cells Transl Med*. 2012;1(12):866–873.
67. Aboody K, Capela A, Niazi N, Stern JH, Temple S. Translating stem cell studies to the clinic for CNS repair: current state of the art and the need for a Rosetta stone. *Neuron*. 2011;70(4):597–613.

CHAPTER 8

Advances in migraine therapeutics: The role of calcitonin gene-related peptide

Nada Hindiyeh[a] and Hanna Hovren[b]
[a]Department of Neurology, Stanford University, Stanford, CA, United States
[b]Center for Neurologic Study, La Jolla, CA, United States

Introduction

Migraine is a common, recurrent disabling hereditary neurological disorder that remains underdiagnosed and undertreated. The Global Burden of Disease (GBD) metrics ranks migraine as the sixth most prevalent and second most disabling medical condition worldwide, as well as the leading cause of years lived with disability in people aged 15–49 globally.[1,2] The 1-year worldwide prevalence is estimated at 14.7% among both genders, affecting over 1 billion people.[3] Migraine is 2–3 times more common in women and, at some point in their lives, will affect 20% of women and 10% of men.[4,5]

Despite these statistics, migraine is often misdiagnosed and suboptimally treated. The 2004 Landmark study highlighted the importance of considering migraine as a diagnosis for any patient visiting their primary doctor with a complaint of headache. This prospective, open-label study evaluated 1203 participants who visited their physician (93% primary care) with a chief complaint of headache. Participant self-diagnosis and physician diagnoses were then compared to an expert panel diagnosis. Results indicated that physician diagnosis of migraine was correct 98% of the time, but when physician diagnosis was reported as nonmigraine, they were wrong 82% of the time, and the diagnosis was later determined to be migraine by the expert panel. Similarly, a patient self-diagnosis of migraine was correct 99.5% of the time and when patient self-diagnosis was nonmigraine, 86% of the time this was later concluded to be a migraine. Thus, for any recurring headache significant enough to be brought to the attention of a physician, the default diagnosis should be migraine.[6]

Correct diagnosis of migraine and its subtypes is the key first step toward optimizing appropriate therapy and minimizing migraine disability. The International Classification of Headache Disorders (ICHD-3) is widely used to aid in the diagnosis of headache types.[7] Here, migraine is categorized as either with or without aura and is separately categorized as chronic migraine if headaches occur more days than not. Migraine without

Neurotherapeutics in the Era of Translational Medicine
https://doi.org/10.1016/B978-0-12-816475-4.00019-7

aura by definition is a recurrent headache disorder with attacks lasting 4–72h characterized by 2 of the 4 clinical features: unilateral location, pulsating quality, moderate or severe intensity, or aggravation by routine physical activity and associated with nausea and/or photophobia and phonophobia. Thus, migraine can be diagnosed in someone with a moderate intensity headache associated with nausea and intensified by activity, such as walking upstairs. Accordingly, the absence of unilateral location or throbbing pain does not exclude migraine as a diagnosis and migraine can often be bilateral, mild, or associated with neck pain, often the basis for misdiagnosis of tension-type headache (TTH) (Fig. 1).

Only 25%–30% of people experience aura with their migraine, usually occurring shortly before headache onset. The aura consists of sensory disturbances most commonly visual (wavy or zig-zag lines, bright or dark spots), language dysfunction, or sensory changes such as numbness or tingling that develop gradually and typically last for 5–60min. Most people who do have migraine with aura also have migraine without aura.

About 2.5% of people with episodic migraine (migraine with and without aura) transform to chronic migraine each year. And this typically happens gradually over months to years.[8] Headache occurs on more days than not and often appear phenotypically as a tension-type headache, but, by definition, eight or more days a month typically have migrainous features. In fact, if a patient meets the diagnostic criteria for chronic migraine (CM) this excludes them from a diagnosis of chronic TTH. TTH by definition is not bothersome, with patients unlikely to pursue medical attention.

Migraine phases

A migraine attack can be categorized into four clinical phases: premonitory, aura, headache, and postdrome with an interictal period between attacks characterized by complete resolution of symptoms (Fig. 2). Although these phases may occur in successive order, there is often considerable overlap between phases and not each phase may be clinically present during each migraine attack.

Premonitory phase

The premonitory phase, defined by nonpainful symptomatology, can precede headache by up to 72h. Premonitory symptoms can be subdivided into three different categories: fatigue and cognitive changes, homeostatic alterations, and sensory sensitivities.[9] Cognitive changes include concentration difficulties, changes in mood such as depression, elation, or irritability among others. Homeostatic changes include thirst, food cravings, yawning, and frequent urination. Sensory disturbances include neck stiffness, photophobia, phonophobia, and nausea. Thus, many factors thought to be triggers for migraine such as bright light, hunger, or sleep deprivation, may actually be premonitory symptoms with increasing evidence implicating the hypothalamus and other brainstem nuclei in the pathophysiological process.[10]

<table>
<tr><th colspan="2">ICHD-3 diagnostic criteria</th></tr>
<tr><td>Migraine without aura

A. At least five attacks fulfilling criteria B–D
B. Headache attacks lasting 4–72 h (when untreated or unsuccessfully treated)
C. Headache has at least two of the following four characteristics:
1. unilateral location
2. pulsating quality
3. moderate or severe pain intensity
4. aggravation by or causing avoidance of routine physical activity (e.g. walking or climbing stairs)
D. During headache at least one of the following:
1. nausea and/or vomiting
2. photophobia and phonophobia
E. Not better accounted for by another ICHD-3 diagnosis</td><td>Migraine with aura

A. At least two attacks fulfilling criteria B and C
B. One or more of the following fully reversible aura symptoms:
1. visual
2. sensory
3. speech and/or language
4. motor
5. brainstem
6. retinal
C. At least three of the following six characteristics:
1. at least one aura symptom spreads gradually over 5 min
2. two or more aura symptoms occur in succession
3. each individual aura symptom lasts 5–60 min
4. at least one aura symptom is unilateral
5. at least one aura symptom is positive
6. the aura is accompanied, or followed within 60 min, by headache
D. Not better accounted for by another ICHD-3 diagnosis</td></tr>
<tr><td>Chronic migraine

A. Headache (migraine-like or tension type-like[1]) on ≥15 days/month for >3 months, and fulfilling criteria B and C
B. Occurring in a patient who has had at least five attacks fulfilling criteria B-D for 1.1 Migraine without aura and/or criteria B and C for 1.2 Migraine with aura
C. On ≥8 days/month for >3 months, fulfilling any of the following:
1.criteria C and D for 1.1 Migraine without aura
2. criteria B and C for 1.2 Migraine with aura
3. believed by the patient to be migraine at onset and relieved by a triptan or ergot derivative
4. Not better accounted for by another ICHD-3 diagnosis</td><td>Frequent episodic tension-type headache

1. At least 10 episodes of headache occurring on 1–4 days/month on average for >3 months (≥12 and <180 days/year) and fulfilling criteria B-D
2. Lasting from 30 min to 7 days
3. At least two of the following four characteristics:
1. bilateral location
2. pressing or tightening (nonpulsating) quality
3. mild or moderate intensity
4. not aggravated by routine physical activity such as walking or climbing stairs
4. Both of the following:
1. no nausea or vomiting
2. no more than one of photophobia or phonophobia
5. Not better accounted for by another ICHD-3 diagnosis</td></tr>
</table>

Fig. 1 ICHD-3 diagnostic criteria.

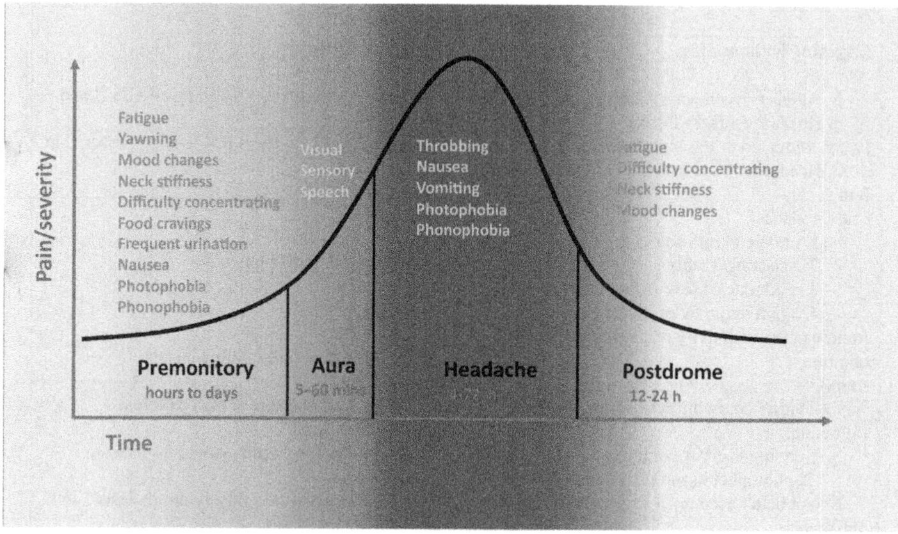

Fig. 2 Phases of migraine.

Few studies have evaluated the prevalence of premonitory symptoms in migraine: three retrospective and one small prospective study. In a large cross-sectional retrospective study, a questionnaire identified 2223 people with episodic migraines among which 77% reported premonitory symptoms. Of the 14 predefined premonitory symptoms, yawning was identified as the most common in 34% of the study population, closely followed by mood change and lethargy.[11] In a small prospective study of 97 people who felt they could predict migraines from their premonitory symptoms, 72% did accurately predict the onset of headache. The most common premonitory symptoms were fatigue (72%), difficulty concentrating (51%), and a stiff neck (50%).[12] More recently, presented in abstract form, a large migraine trial prospectively evaluated 438 participants for premonitory symptoms during the run-in phase. A total of 75.6% of participants experienced one or more premonitory symptoms with excessive fatigue being the most frequently reported (21.5% of subject-days), followed by yawning and mood changes.[13]

Aura

As mentioned above, aura only occurs in about a third of people with migraine and, as defined by the ICHD-3, presents as brief, unilateral, and fully reversible attacks manifested by visual, sensory, or other central nervous system symptoms that usually develop gradually and are usually followed by a headache. Most people who have migraine with aura also have migraine without aura. The clinical features of aura vary widely not only from person to person but also from attack to attack in an individual. They are further categorized as typical or brainstem aura. Visual aura is the most common typical symptom, occurring in over 90% of patients, followed by sensory or language changes, and when sensory or language symptoms do occur, they often occur in conjunction with visual symptoms and can follow visual symptoms in a reliable temporal progression.

The most commonly reported visual aura is flashing light, and often isolated scotomas, but other visual distortions are reported. When it does occur, the classic scintillating scotoma, typically begins at the center of the visual field and expands outward and may take the form of a zigzag pattern known as a fortification spectrum. The clinical progression of the fortification spectra correlates with the pathophysiological process of cortical spreading depolarization also known as cortical spreading depression (CSD): a 2–6 mm/min propagating wave of neuronal depolarization followed by inhibition of cortical brain activity. The most common sensory disturbances are paresthesia of the hand and face, although this can spread to involve the trunk and limbs. The most common speech symptoms are reported as word-finding difficulties but slowed speech and difficulty with articulation are also reported.

Dizziness and vertigo, often described in association with migraine, can also be symptoms of aura. Vertigo, dysarthria, tinnitus, hypacusis, diplopia, ataxia, and decreased level of consciousness are included as part of the diagnosis of migraine with brainstem aura (previously referred to as basilar migraine). At least two of these seven symptoms are required to meet diagnostic criteria in addition to typical aura symptoms during the attack.

Aura can be a frightening experience since focal neurological deficits can often be mistaken for a stroke. A key distinguishing characteristic is that aura has a gradual onset and progression as opposed to the sudden or abrupt onset of stroke symptoms. Furthermore, unlike a stroke, with aura, positive symptoms such as flashing lights and paresthesia are more common than negative ones (vision loss and numbness).[14–16]

Headache phase

The headache phase is easily the most recognizable of the migraine attack and is thought to be a clinical manifestation of the activation of the trigeminovascular system. Headaches can occur at any time, but often occur during sleep, upon awakening, or shortly thereafter in the morning.

The average duration of a headache is 24 h but ranges between 4 and 72 h with a median time to peak intensity after 1 h. Sixty percent of people with migraines report unilateral location, 50% a throbbing nature, 90% report aggravation by routine physical activity or head movement, and 75% report neck pain. The pain can involve any part of the head, neck, or face and often leads to misdiagnosis as TTH or sinus headache because of this. Pain can be on a spectrum from unilateral to holocephalic and can vary with different attacks.

Nausea and/or vomiting occur during migraine in 73% of patients and photophobia and phonophobia in 80%. Cranial autonomic features such as lacrimation, nasal congestion, and rhinorrhea are not uncommon during a migraine and can often lead to a misdiagnosis of a trigeminal autonomic cephalalgia such as cluster headache.

Postdrome

Postdrome is the most newly recognized and least studied phase of migraine. It is recognized as the time between the resolution of a headache and when a person feels completely normal.[17] The postdrome is characterized by nonheadache symptoms often associated with significant disability. A prospective electronic diary study of 120 people with migraine showed that 81% of participants reported at least one nonheadache symptom during the postdrome. The most common symptom was feeling tired/weary in 88%, difficulty with concentration in 56% and neck stiffness (42%). Symptoms have been categorized broadly as follows: neuropsychiatric (e.g., mood changes and sleep disturbances); sensory (e.g., head soreness and photophobia); gastrointestinal (e.g., nausea and food cravings); and general symptoms (e.g., tiredness).[18] The duration has been reported in several studies to range from 18 to 25.5h with 93% exhibiting resolution within 24h after pain has resolved.[17]

Barriers to migraine care

Although proper diagnosis of migraine remains a significant barrier to appropriate care, its not the only limitation. Epidemiological studies suggest people with migraine often do not seek consultation from healthcare providers, and the few that do receive suboptimal migraine treatment. In a study of 775 participants with episodic migraine, only 45.5% had consulted their healthcare provider in the previous year; a total of 39.5% of these received an accurate diagnosis of migraine, and only 26.3% of the total cohort received appropriate migraine treatment.[19] Similarly, in a cohort of 1254 participants with chronic migraine, only 40.8% consulted a healthcare provider, with only 10% diagnosed accurately and 4.5% of the total cohort receiving appropriate acute and preventive chronic migraine treatment.[20] Thus, education empowering both patients and healthcare providers are key to overcoming these hurdles.

Even for patients who are able to access treatment, inadequate medication and/or tolerability, and patient adherence remain barriers to care. Commonly used preventative oral medications include many that are nonmigraine-specific: antidepressants, beta-blockers, antiepileptics, etc. Most of these are dosed daily or multiple times a day but may have undesirable side effects at optimal dosing. However, in the instance of migraine-specific acute medications (i.e., triptans), patients with cardiovascular risk factors are at risk for complications, and these medications may have limited tolerability and effectiveness in some and lead to medication overuse headache in others.

Pathophysiology of migraine

The understanding of migraine pathophysiology continues to evolve and has substantially progressed over the last several decades; from what was initially thought to be a disorder of cerebral vasodilation to a now accepted theory of trigeminovascular system activation and sensitization and brainstem and diencephalic nuclei activation. It is suggested that a genetic predisposition to a hyperexcitable brain along with dysregulation of sensory processing is the common denominator in persons with migraine.

Dr. Harold Wolff, considered to be a pioneer in headache research, dedicated much of his work to the question "How do the cerebral blood vessels, many of them covered with a network of nerve fibers, participate in migraine headache?"[21] Alongside his colleagues, they observed that in awake patients undergoing craniotomies mechanical stimulation of brain parenchyma did not produce pain, but the stimulation of meninges and cerebral and meningeal blood vessels generated severe ipsilateral headache.[22,23] They found pain-sensitive components in the meninges, dura matter, extensively in the vasculature (venous sinuses, carotid, vertebral, basilar arteries, etc.), and further in neuronal structures including the trigeminal cranial nerves, upper cervical nerves, and the glossopharyngeal and vagus nerve.[21] In a further milestone, Graham and Wolff demonstrated that relief of the pain associated with migraine was associated with decreased amplitude of arterial pulsations. These observations were made following the administration of IV ergotamine, a potent vasoconstrictor, to migraine patients. This led to the vasogenic theory of migraine, a theory that dominated thinking about the pathophysiology of the disease for years.[24]

Shortly after these publications, Lashley published data from his personal observations of his scintillating scotomas in 1941. He illustrated his own visual auras temporally, leading him to determine that the wave of excitation propagated at a rate of 3 mm/min across the visual cortex.[25–27] Later, Leão observed cortical spreading depression (CSD) in rabbit brains following electrical stimulation.[26,28] The slowly propagating wave of neuronal depolarization at 2–6 mm/min was followed by a suppression of neural activity, an observation that was somewhat parallel to the findings Lashley described. It is believed that the depolarization of glial and neuronal cell membranes disrupts ionic gradients, causing a rise in extracellular potassium and glutamate release with corresponding blood flow changes of transient hyperemia followed by prolonged oligemia.[14,29]

It was not until 1979 that the trigeminovascular hypothesis of migraine was introduced by Moskowitz and colleagues suggesting that vascular changes were only an epiphenomenon in migraine. They highlighted the role of the trigeminal nerve and neuropeptides as essential components in migraine pathogenesis and developed a rodent model of neurogenic-induced dural inflammation. Electrical stimulation of the trigeminal ganglion in rats triggered plasma protein extravasation, vasodilation, increased mast cell degranulation, and release of neuropeptides. They hypothesized that the "headache phase of migraine may develop as a result of an abnormal interaction and perhaps abnormal release of vasoactive neurotransmitters from terminals of the trigeminal nerve with surrounding large intracranial and extracranial blood vessels."[30,31] In 1983, Moskowitz and colleagues coined the term "trigeminovascular system," that underscores the prevailing theory of migraine pathogenesis today.

Later that year, the vasoactive neuropeptide substance P (SP), first discovered in 1931, was found to be released from trigeminovascular afferents suggesting a role as a neuromediator in the meninges.[32] CGRP, a significantly more potent vasodilator than SP was the second neuropeptide to be implicated in the trigeminovascular system a year later.[33, 34] In 1988, pituitary adenylate cyclase-activating peptide (PACAP) was identified in trigeminal nerve fibers around cerebral blood vessels as well as the trigeminal ganglia, implicating it as a potential important neuropeptide in migraine pathophysiology. Immunoreactive studies of the human trigeminal ganglia axonal projections to the meninges found that CGRP predominates in 40% of cells; a small number of neurons, 18%, contain substance P, while 20% contain PACAP and 15% contain nitric oxide synthase.[35] These findings have helped pave the way for therapeutic targets of migraines as discussed below.

The initiation of a migraine attack and head pain have been longstanding topics of debate. Many people with migraines report external and internal triggers such as stress, skipped meals, hormonal fluctuations, sleep disturbances, or increased sensory stimuli as triggers of a migraine attack. However, this notion is being questioned as evidence suggests the premonitory phase of migraine can start hours to days before the onset of a headache with features overlapping those previously reported as triggers.[10,36,37] These symptoms have been interpreted as evidence that central brain areas, such as the hypothalamus or cortex, are possible initiators of the classic premonitory or aura symptoms.[38] The subsequent activation of the trigeminal vascular system and meningeal nociceptors is thought to initiate the headache phase of migraine.

The trigeminovascular system acts as a pathway for nociceptive transmission of information from the dura to central brain areas and the cortex. These primary afferent neurons in the dura and intracranial vasculature are comprised of C and A-delta fibers largely arising from the ophthalmic division of the trigeminal nerve from the trigeminal ganglion and the upper cervical roots. Activation of these dural afferents stimulates the release of vasoactive peptides such as CGRP, PACAP-38, and substance P. Afferents from the trigeminal ganglion converge with C1-C2 innervated tissue including pericranial and paraspinal muscle to synapse on second-order neurons in the trigeminal cervical complex (TCC) which is comprised of the trigeminal nucleus caudalis (TNC) and the dorsal horns of C1 and C2. This convergence likely accounts for symptoms of neck pain, occipital, and periorbital pain often reported in migraine.

Second-order neurons from the TCC then project to neurons in multiple brainstems, thalamic, hypothalamic, and basal ganglia nuclei. These in turn project to various cortical areas including visual, auditory, and olfactory cortices that could account for characteristic migrainous features of photophobia, phonophobia, and osmophobia, while projections to somatosensory, insular, motor, parietal association, and retrosplenial cortical areas likely accounts for the sensory-discriminative, emotional, and cognitive aspects of migraine[38,39] (Fig. 3).

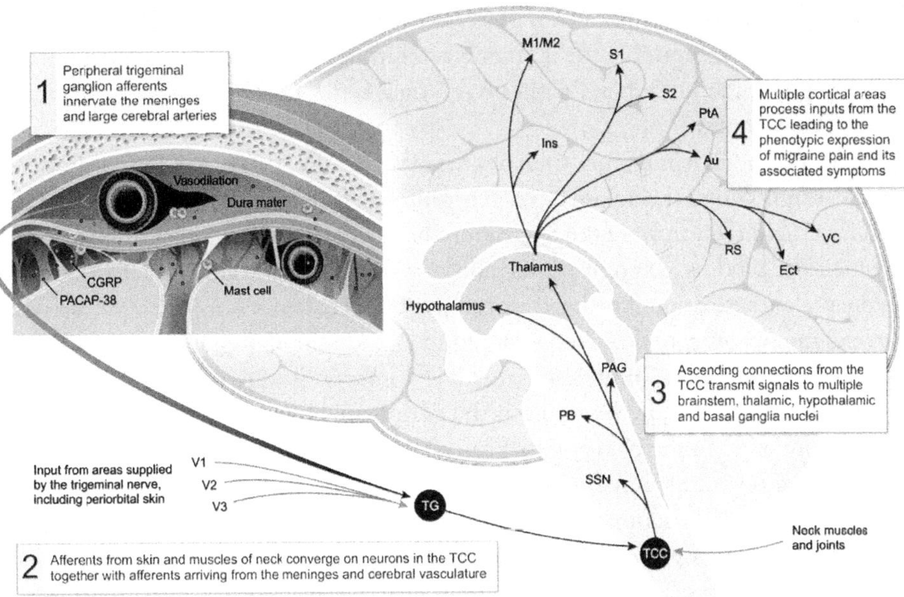

Fig. 3 The triggering of pain via the trigeminovascular pathway. *Au*, auditory; *Ect*, ectorhinal; *Ins*, insula; *M1/M2*, motor cortices; *PAG*, periaqueductal gray; *PB*, parabrachial nucleus; *PtA*, parietal association; *RS*, retrosplenial; *S1/S2*, somatosensory cortices; *SSN*, superior salivatory nucleus; *TCC*, trigeminal cervical complex; *TG*, trigeminal ganglion; *VC*, visual cortices; *V1*, ophthalmic branch of trigeminal nerve; *V2*, maxillary branch of trigeminal nerve; *V3*, mandibular branch of the trigeminal nerve. *(From Dodick DW. A phase-by-phase review of migraine pathophysiology.* Headache *2018;58(Suppl. 1):4–16.)*

Targets in migraine therapeutics

As the understanding of migraine pathophysiology continues to advance, so has the development of migraine therapeutics and mechanism-based targeted therapy. Initial migraine therapies were thought to work in part because of their role as vasoconstrictors based on the vascular theory of headache whereby changes in vessel diameter trigger pain. But currently, our understanding of migraine has dramatically shifted over the last several decades, highlighting the importance of neuropeptides such as CGRP.

The use of ergot to treat migraine was undertaken by Rothlin in 1925. Within quick succession, 1934 and 1937, Lennox conducted controlled migraine studies utilizing ergotamine, and Graham and Wolff determined it to be a vasoconstrictor.[40,41]

Over the ensuing years, however, publications on the importance of neuropeptides in the trigeminovascular system in migraine pathogenesis argued against the predominant theory of vascular tone. About 10 years later, researchers at the Mayo Clinic initiated treatment with dihydroergotamine (DHE), which was as efficacious as ergotamine in the treatment of migraine but exerted less of a vasoconstrictor effect. Based on the surmounting data, consideration was given to the idea that mechanisms other than vasoconstriction were responsible for migraine resolution.[40] Thus, the long-held belief that vasodilation was the underlying pathophysiologic factor in the genesis of migraine was further cast in doubt.

The renaissance of migraine therapeutics dates to the 1960s and primarily owes its birth to insights into the nuances of serotonin pharmacology. The clues that drove research in this direction were the realization that ergot alkaloids bind to serotonin receptors, albeit the binding was nonspecific and responsible for the adverse effects associated with the use of these drugs. Further, serotonin, known to be a vasoconstrictor, was shown by several investigators to be effective in reducing the pain of migraine, although its clinical application was again limited due to adverse effects. Studies suggested that the therapeutic effects of ergots are the result of the peripheral blockade of neurogenic-induced neuroinflammation in the dura and that the effect was mediated by 5HTIB/ID receptors localized to sensory trigeminal neurons.[42]

The breakthrough, seminal studies undertaken by Humphrey et al. at Glaxo Pharmaceuticals, was the development of a 5-HT1B/1D receptor agonist (Sumatriptan), a first in the class of "triptans" to enter the migraine pharmacopeia.[43] In contrast to ergotamine, Sumatriptan is well-tolerated, but its limited bioavailability and short plasma half-life led to the development of second- and third-generation products. These are now available in various forms, including those administered nasally, orally, subcutaneously, etc.[44]

Thus, what started as a search for vascular effects had culminated with a new understanding of migraine pathophysiology and the implications of the trigeminovascular system in triptan and ergot pharmacology. However, the fact that these novel migraine drugs had undesirable vasoconstrictive effects lead researches down the path in search of antimigraine drugs without vasoconstriction. While triptans remain commonly prescribed for acute migraine treatment, the side effect profile for their vasoconstrictor effects makes them contraindicated for patients with cardiovascular and cerebrovascular disease.

Substance P antagonists were the first to be studied in a translational model that was focused on neuropeptide targets of the trigeminovascular system. Substance P specifically binds with the high affinity to the tachykinin, neurokinin-1 (NK-1) receptors in the peripheral and central nervous system. In a rat model of neurogenic inflammation, NK-1 receptor antagonists were shown to block plasma protein extravasation in a dose-dependent manner suggesting a possible translational target in migraine treatment.[34] However, human studies in the mid-1990s were not as successful as hoped. A double-blind, placebo-controlled study of the oral NK-1 antagonist RPR100893 demonstrated no effect over placebo in migraine.[45] Similarly, the NK-1 antagonist lanepitant was not effective in acute migraine treatment over placebo in a three-way crossover study.[46] It was argued that these oral drugs might not be lipophilic enough to induce antimigraine effects due to poor CNS penetration. Subsequently, two intravenous NK-1 antagonists were studied, but, again, neither was found to be therapeutically useful.[47]

Nevertheless, many studies demonstrated that substance P does not act alone and is released in conjunction with other neuropeptides such as CGRP. What further dictated the new era in migraine therapeutics was a landmark study by Goadsby and colleagues in 1990. Via radioimmunoassay, they measured neuropeptide levels including CGRP, VIP, SP, and NPY from the jugular veins of humans during migraines. CGRP was the only neuropeptide found to be significantly elevated during a migraine attack.[48] These findings laid the foundation for CGRP models of migraine treatment that have transformed the current landscape of migraine therapeutics.

Calcitonin gene-related peptide

Shortly after its discovery in 1982, CGRP was recognized as a key mediator in the trigeminovascular system. It is a potent vasodilator and is implicated in sensory neural transmission. CGRP is a 37 amino acid neuropeptide and member of the calcitonin family which also includes calcitonin, amylin, and adrenomedullin. Two distinct isoforms of CGRP, encoded by different genes exist: CGRPalpha and beta. CGRPβ or CGRP2 has not been as extensively studied but is expressed primarily in the enteric nervous system. CGRPα also known as CGRP1 is transcribed by the calcitonin-CGRP gene located on chromosome 11 and is the most relevant for migraine. It is abundantly found in neurons of the peripheral nervous system and various sites throughout the central nervous system. After cleavage from a pro-peptide precursor, CGRP is then processed through the Golgi apparatus and packaged into dense-core vesicles for transport to axon terminals where it is stored and later released. CGRP is expressed in C-fibers and A-delta fibers of the trigeminal ganglion and when stimulated, CGRP is released from its vesicles via calcium-dependent exocytosis.[49]

CGRP receptors are found in many areas implicated in migraine pathophysiology including the cortex, limbic system thalamus, brainstem, dura, trigeminal ganglion, dorsal root ganglion, and trigeminocervical complex.[50] The work of McLatchie and colleagues illuminated the complexity of the CGRP receptor complex. The main subunit of this receptor is a G-protein coupled receptor, the G-protein calcitonin-like receptor (CLR). CLR can only function when linked to a single transmembrane protein known as receptor activity-modifying protein (RAMP). Three RAMPs exist (RAMP1, RAMP2, and RAMP3) and RAMP1 localized with CLR gives rise to a functional CGRP receptor.[51]

Several studies have paved the way for the translational research that has focused on CGRP as a target for migraine therapy. As previously mentioned, CGRP is elevated during a migraine attack in humans and, conversely, declines in interictal periods.[48,50] Furthermore, when infused with CGRP, participants with migraine developed a headache meeting ICHD criteria for migraine over the following 11 h.[52] CGRP receptor activation has been shown to induce the cascade of events initiating cortical spreading depolarization or depression that is known to be linked to the aura of migraine. Likewise, CGRP release in the meninges triggers vasodilation, mast cell degranulation, and neurogenic inflammation; all components of migraine pain.[50] Moreover, triptans, have been shown to interrupt this cascade and prevent the release of CGRP. Triptans simultaneously prevented the increase in plasma CGRP and aborted the headache attack. Thus, the thrust to develop drugs targeting CGRP as a headache therapeutic ensued.

Early drug targets of CGRP

The CGRP antagonists, collectively termed "gepants," were the first to be developed for migraine treatment. CGRP antagonists are small molecules, typically 0.2–1 kilodalton (kDa), that are typically administered orally. Their short half-lives, on the order of hours, allow for dosing daily or more frequently and are thus more suited for acute rather than preventive therapy. They are metabolized mainly by the hepatic CYP system which renders them susceptible to drug-drug interactions.

BIBN 4096 BS, or olcegepant, an intravenous CGRP antagonist, was the first to be developed by Boehringer Ingelheim GmbH, Germany, for the acute treatment of migraine. Doods and colleagues established that BIBN 4096 BS olcegepant is a selective CGRP antagonist with high affinity and a potent inhibitor of neurogenic vasodilation.[53] While the drug asserted a significant, palliative effect in a treatment trial, development was discontinued due to the difficulty of developing an oral formulation. Subsequently, Merck Research Laboratories developed an orally available CGRP antagonist MK-0974 or Telcagepant. Again, in spite of a successful treatment outcome, the development was suspended, in this instance due to hepatic toxicity. Until recently, efforts by several companies to develop a gepant were unsuccessful.

CGRP antibodies

Monoclonal antibodies have provided an alternative option to small-molecule antagonists in the modulation of CGRP for migraine treatment. As large molecules, monoclonal antibodies (mAbs) do not typically cross the blood-brain barrier unless it has been compromised. The size of a mAb, around 150 kDa is in stark contrast to a gepant which typically ranges from 0.2 to 1 kDa. A parallel has been drawn to the differences in size between a truck and a grain of rice.[50] Due to their large size, mAbs must be administered by injection, either subcutaneously or intravenously whereas gepants can be administered orally. Although this can be seen as a drawback to those who are averse to needles, many other features highlight the advantages of mAbs as a prophylactic migraine treatment.

In the late 1980s, animal studies were first conducted showing that intrathecal injection of CGRP antisera could block pain induced by noxious stimuli.[49,54,55] In the 1990s,

Tan and colleagues at Merck developed an anti-CGRP monoclonal antibody that was shown to block experimentally induced vasodilation in an animal model. These studies were the first to show that a CGRP antibody could block vasodilation induced by sensory nerve stimulation.[56]

In the ensuing years, it became evident that monoclonal antibodies were in fact successful in preventing migraines in humans.

Considering the issue of liver toxicity associated with the administration of gepants, it was essential to demonstrate that a monoclonal antibody targeting the CGRP ligand was safe. This would be of utmost concern for the development of peripherally acting CGRP mAbs.[50] However, to date, none of the developed CGRP mAbs have exhibited liver toxicity. Four mAbs are currently FDA approved in the prevention of migraine, with clinical trials in episodic and chronic migraine: Erenumab (Aimovig), Galcanezumab (Emgality), Fremanezumab (Ajovy), and Eptinezumab (Vyepti).

Erenumab

Erenumab or Aimovig was the first of the mAbs to be FDA approved (2018) for the prevention of episodic and chronic migraine. It also distinguishes itself from the others as the only mAb to target the CGRP receptor rather than the ligand itself and as the only fully human mAb. It is an IgG2 initially synthesized by Dr. Cen Xu and colleagues at Amgen which announced its collaboration with Novartis in 2017 and together marketed and launched the drug as Aimovig. It is FDA approved at 70 mg monthly with a caveat that some people may benefit from a higher dose (140 mg/month).

Several episodic and chronic migraine studies have been done to date evaluating erenumab in the prevention of migraine. The primary outcome measure was the mean change in monthly migraine days (MMD) from baseline, which has become the FDA-approved standard primary endpoint for successive mAb migraine prevention trials. Comparing the last 4 weeks of a treatment trial of patients with episodic migraine, it was established that a 70 mg dose exhibited a modest reduction (P=.021) in monthly migraine days compared to a placebo group in a phase 2 trial.[57] Furthermore, a secondary endpoint (the percentage of patients achieving at least a 50% reduction in MMD) was found to be significant at 46% of patients receiving the drug (P=.011). In an open-label extension trial, 65% of participants achieved a ≥50% reduction in MMD, and a quarter of the patients enjoyed 100% relief of their headaches.[58]

Two large pivotal phase 3 trials reinforced the earlier findings in episodic migraine. The ARISE trial again confirmed 70 mg as an efficacious dose against placebo in MMD reduction and decreased monthly migraine days by half in 40% of patients (P=.010). Whereas the STRIVE trial also evaluated a 140 mg dose of erenumab in comparison to 70 mg and placebo. Both doses proved to be effective in the reduction of MMD and showed statistically significant responder rates, i.e., 43.3% in the 70 mg group and 50% in the 140 mg group.[59,60]

Somewhat similar results were obtained when the drug was tested in patients with chronic migraine at both doses with a majority of participants enrolled in the trial having previously failed one preventive.[61]

Fremanezumab

Fremanezumab, or Ajovy, ultimately developed by Teva Pharmaceuticals, is unique compared to erenumab as it targets the CGRP molecule directly. It is a fully humanized IgG2 mAb and has a half-life of 39–48 days.[49] As a result, it can be administered subcutaneously monthly or quarterly at a higher dose.

Phase 2 trials established doses of both 225 and 675 mg monthly as effective compared to placebo in the reduction of MMD in episodic migraine. In the chronic migraine trial, drug dosages varied and a primary endpoint of reduction of headache hours was utilized but similarly, a significant difference was established over placebo.[62,63]

Phase 3 episodic migraine trial, HALO, evaluated the current dosing regimen for fremanezumab of 225 mg monthly or 675 quarterly versus placebo with both drug doses successfully reaching the primary end point of MMD reduction and 50% responder rates. Likewise, the chronic migraine HALO trial successfully met its targeted endpoints.[50,64,65]

Galcanezumab

Galcanezumab or Emgality is an IgG4 humanized mAb that is marketed by Eli Lilly. It too is a mAb targeting the CGRP ligand itself with a half-life of 28 days.[49] It is unique among the CGRP mAbs as the only treatment FDA approved for the prevention of episodic cluster headache as well as episodic and chronic migraine.

Galcanezumab was initially studied in a phase 2 trial of episodic migraine with subcutaneous dosing every 2 weeks with statistically significant improvement in MMD reduction compared to placebo. Furthermore, a significant 50% responder rate was reached by 70% of participants, a greater than 75% response was achieved by almost half, and 32% had complete elimination of migraine.[66] A subsequent study evaluating monthly dosing was also successful.[67]

The phase 3 episodic migraine trials EVOLVE-1 and EVOLVE-2 varied only in geography but were both similarly successful in achieving MMD reduction compared to placebo. Moreover, up to 60% of patients reached a 50% response rate, slightly over 1/3 attained a 75% response, and approximately 13% achieved migraine freedom.[68,69]

The phase 3 REGAIN chronic migraine trial was equally successful in reaching MMD reduction in both active groups, also showing a 50% responder rate in 27% of participants. Its open-label extension continued to show MMD reduction in active groups.[70,71]

Eptinezumab

Finally, the most recently FDA approved mAb for episodic and chronic migraine prevention eptinezumab, or Vyepti, distinguishes itself from the others as the only mAb therapy approved for quarterly intravenous administration. It is a mAb of the IgG1 subtype targeting the CGRP ligand with a half-life of 31 days. Surprisingly, it has been noted that the drug can reduce headache soon after administration, suggesting that eptinezumab will be useful for both migraine prevention and acute treatment.

A phase 2 trial evaluated the safety and efficacy of 1000-mg IV eptinezumab versus placebo in participants with episodic migraine. Migraine days were significantly reduced in the eptinezumab group compared to placebo ($P = .0306$).[72] Furthermore, 41% of patients reported complete headache freedom.

Two pivotal phase 3 trials, PROMISE-1 and PROMISE-2, evaluated eptinezumab in the prevention of episodic and chronic migraine, respectively, over 3 months duration. Unique to these studies is a secondary endpoint investigating the probability of a migraine attack at 1-day postinfusion. Both studies not only exhibited significant reductions in MMD in the active groups but also showed a reduction of the likelihood of migraine on day 1 postinfusion.[73,74]

Based on these collective results, the new CGRP mAbs for migraine prevention have revolutionized the prevention of migraine. All 4 mAbs have proven efficacy in episodic and chronic migraine and are well tolerated. However, the long-term safety data of this class of drugs will require vigilance (Figs. 4 and 5).

CGRP antagonists: New developments

Despite setbacks in the previous development of CGRP antagonists, two new gepants have been FDA approved for the acute treatment of migraine: ubrogepant in late 2019 and rimegepant early in 2020. Both are available in oral form. No hepatic abnormalities have been reported in this generation of gepants. Most of these studies evaluated pain freedom 2h after gepant administration for the treatment of a single headache episode.

CGRP mAb	Target	Indication	Administration route	Dosing	Side effects
Erenumab	CGRP receptor	EM, CM	SC	70 or 140 mg monthly	Injection site reactions, constipation, hypertension
Galcanezumab	CGRP molecule	EM, CM, eCH	SC	240 mg loading dose, followed by 120 mg monthly	Injection site reactions
Fremanezumab	CGRP molecule	EM, CM	SC	225 mg monthly or 675 mg quarterly	Injection site reactions
Eptinezumab	CGRP molecule	EM, CM	IV	100 or 300 mg quarterly	Nasopharyngitis, hypersensitivity

Fig. 4 Theraputic monoclonal antibodies for migraine prevention.

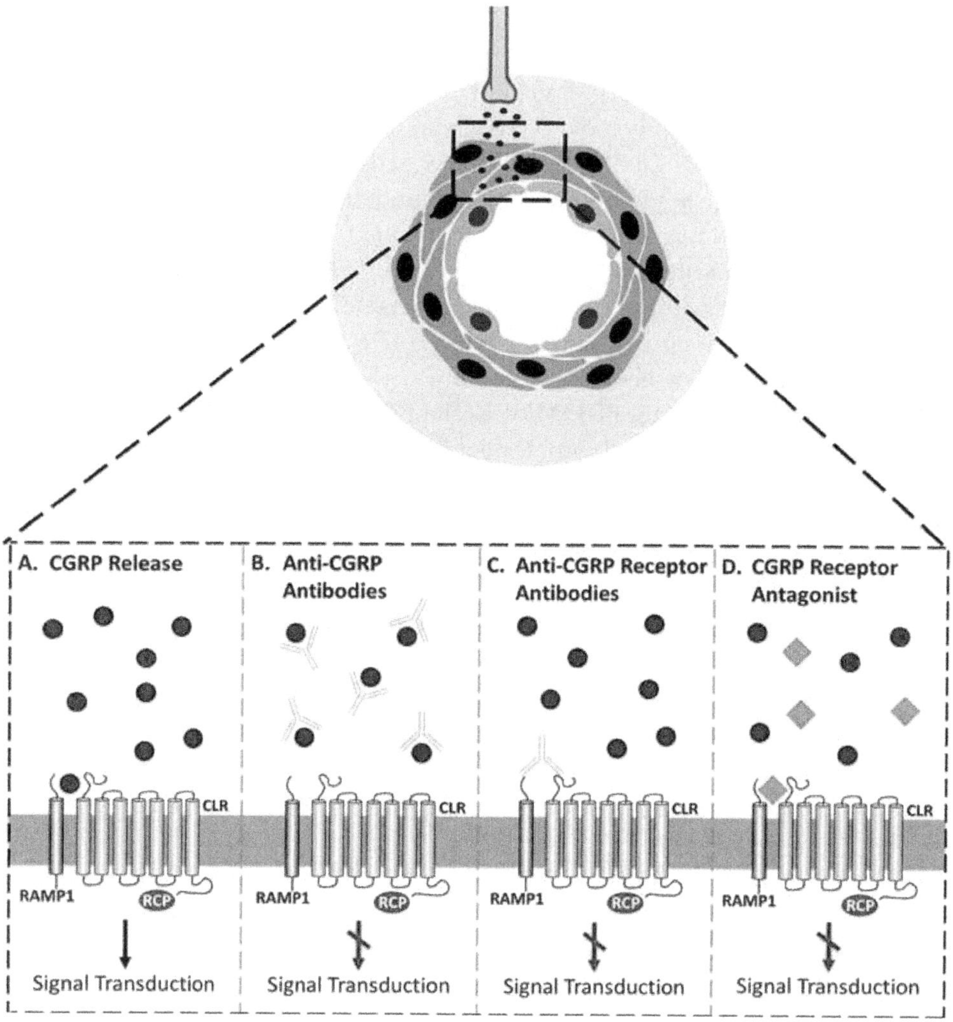

Fig. 5 Possible mechanisms to block calcitonin gene-related peptide (CGRP) signaling. (A) The neuropeptide, CGRP, is released from sensory nerves and acts through a heterodimeric receptor complex, composed of the calcitonin receptor-like receptor (CLR) and receptor activity-modifying protein (RAMP1), along with an intracellular receptor component protein (RCP). Upon binding, signal transduction occurs, leading to vasodilation and other cardioprotective mechanisms. (B) One possible method of blocking CGRP is the use of anti-CGRP antibodies targeting the ligand CGRP itself. (C) Antibodies can also be raised against the CGRP receptor complex to block signaling. (D) CGRP receptor antagonists, such as the small molecule gepants may also be used to block CGRP signaling. *(From Brain SD, Geppetti P.* Calcitonin Gene-Related Peptide (CGRP) Mechanisms: Focus on Migraine. Handbook of Experimental Pharmacology (eBook). *Switzerland: Springer Nature; 2019.)*

Ubrogepant

A phase 2 dose-defining study published in 2016 established 100 mg of ubrogepant as efficacious when compared to placebo for the acute treatment of migraine (25.5% vs 8.9%).[75] The phase 3 trials, ACHIEVE-1 and ACHIEVE-2 further established efficacy in acute migraine.[76] In both studies, the most common adverse events were nausea and dizziness, and most importantly no hepatic abnormalities were identified. Participants with cardiovascular risk factors such as hypertension were allowed in the studies, however, those with cardiovascular and cerebrovascular disease were excluded. Because the mechanism of action varies from that of triptans, this may be a safe new option for patients with vascular disease.[77]

Rimegepant

Rimegepant's first phase 2 dose-ranging study established that doses of 75, 150, and 300 mg were superior to placebo in the acute treatment of migraine. In two subsequent randomized controlled studies (301 and 302) of rimegepant (75 mg dose), participants with a single migraine attack were treated, and pain freedom at 2 h was the primary endpoint. Similar results were attained in both studies.[78] In the registration study (303), participants received either 75 mg ODT rimegepant or placebo. The results, again, were significant in that pain freedom at 2 h in the 75 mg Rimegepant limb compared favorably to placebo (21% vs 11% for placebo—$P < .0001$). Consistent with previous studies, the most common adverse events were nausea and urinary tract infection. No hepatic abnormalities reported.[79] Furthermore, in a long-term safety study of rimegepant (75 mg dose), reduction of migraine frequency without signs of medication overuse headache or liver toxicity was seen in patients who experienced greater than 14 migraine days a month.

These results are important, leading to a therapeutic alternative for people who cannot tolerate triptans. In short, rimegepant may find utility as a treatment for both acute migraine as well as prevention. Further, this benefit comes without the risk of MOH (Fig. 6).

Gepant	Clinical phase	Indication	Administration route	Dosing	Side effects/ comments
Ubrogepant	FDA approved	Acute migraine attack	Oral	50 or 100mg (may take 2nd dose ≥2 h later; not to exceed 200 mg in 24 h)	Nausea, somnolence
Rimegepant	FDA approved	Acute migraine attack	Oral disintegrating tablet	75mg (not to exceed 1 dose in 24 h)	Nausea New data suggesting daily use can prevent migraine without causing MOH
Atogepant	Phase 2b/3 clinical trial	Preventive treatment	Oral	Varying 10mg 4 times a day to 60 mg BID	Nausea, fatigue, constipation, nasopharyngitis, UTI First gepantto be developed solely for migraine prevention
Vazegepant	Phase 2/3 clinical trial	Acute migraine attack	Intranasal	10–20 mg	Dysgeusia, nasal discomfort

Fig. 6 Gepants for migraine treatment.

Conclusion

Translational medicine, as applied to migraine, has led to a bounty of therapeutic choices. Nonetheless, a "cure" for migraine, the holy grail of therapy, has eluded us thus far. However, one can be optimistic about the future since it is certain that genetic risk factors will be identified that will provide clues to therapy, and it would not be surprising if serendipity proves to be a factor as well.

References

1. Steiner TJ, Stovner LJ, Vos T. GBD 2015: migraine is the third cause of disability in under 50s. *J Headache Pain*. 2016;17(1):104.
2. GBD 2016 Disease and Injury Incidence and Prevalence Collaborators. Global, regional, and national incidence, prevalence, and years lived with disability for 328 diseases and injuries for 195 countries, 1990–2016: a systematic analysis for the Global Burden of Disease Study 2016. *Lancet*. 2017;390(10100):1211–1259.
3. GBD 2016 Headache Collaborators. Global, regional, and national burden of migraine and tension-type headache, 1990–2016: a systematic analysis for the Global Burden of Disease Study 2016. *Lancet Neurol*. 2018;17(11):954–976.
4. Weatherall MW. The diagnosis and treatment of chronic migraine. *Ther Adv Chronic Dis*. 2015;6(3):115–123.
5. Vetvik KG, MacGregor EA. Sex differences in the epidemiology, clinical features, and pathophysiology of migraine. *Lancet Neurol*. 2017;16(1):76–87.
6. Tepper SJ, Dahlöf CG, Dowson A, et al. Prevalence and diagnosis of migraine in patients consulting their physician with a complaint of headache: data from the landmark study. *Headache*. 2004;44(9):856–864.
7. Headache Classification Committee of the International Headache Society (IHS). The international classification of headache disorders, 3rd edition. *Cephalalgia*. 2018;38:211.
8. Bigal ME, Lipton RB. Concepts and mechanisms of migraine chronification. *Headache*. 2008;48(1):7–15.
9. Karsan N, Bose P, Goadsby PJ. The migraine premonitory phase. *Continuum (Minneap Minn)*. 2018;24(4, Headache):996–1008.
10. Hoffmann J, Recober A. Migraine and triggers: post hoc ergo propter hoc? *Curr Pain Headache Rep*. 2013;17(10):370.
11. Laurell K, Artto V, Bendtsen L, et al. Premonitory symptoms in migraine: a cross-sectional study in 2714 persons. *Cephalalgia*. 2016;36(10):951–959.
12. Giffin NJ, Ruggiero L, Lipton RB, et al. Premonitory symptoms in migraine: an electronic diary study. *Neurology*. 2003;60(6):935–940.
13. Hindiyeh N, Kellerman D, Schmidt P. Characteristics of headaches and premonitory symptoms during the run-in phase of a large migraine trial. In: *61st Annual Scientific Meeting of the American Headache Society; Philadelphia, PA*; 2019.
14. Charles A. The migraine aura. *Continuum (Minneap Minn)*. 2018;24(4, Headache):1009–1022.
15. Dodick DW. Migraine. *Lancet*. 2018;391(10127):1315–1330.
16. Dodick DW. A phase-by-phase review of migraine pathophysiology. *Headache*. 2018;58(Suppl. 1):4–16.
17. Giffin NJ, Lipton RB, Silberstein SD, Olesen J, Goadsby PJ. The migraine postdrome: an electronic diary study. *Neurology*. 2016;87(3):309–313.
18. Bose P, Goadsby PJ. The migraine postdrome. *Curr Opin Neurol*. 2016;29(3):299–301.
19. Lipton RB, Serrano D, Holland S, Fanning KM, Reed ML, Buse DC. Barriers to the diagnosis and treatment of migraine: effects of sex, income, and headache features. *Headache*. 2013;53(1):81–92.

20. Dodick DW, Loder EW, Manack Adams A, et al. Assessing barriers to chronic migraine consultation, diagnosis, and treatment: results from the chronic migraine epidemiology and outcomes (CaMEO) study. *Headache*. 2016;56(5):821–834.
21. Silberstein SD, Lipton RB, Dalessio DJ. *Wolff's Headache and Other Head Pain*. 7th ed. Oxford University Press; 2001.
22. Penfield W. A contribution to the mechanism of intracranial pain. *Research Nerv Ment Dis Proc*. 1935;15:399.
23. Ray B, Wolff H. Experimental studies on headache. Pain-sensitive structures of the head and their significance in headache. *Arch Surg*. 1940;41:813–856.
24. Graham J, Wolff H. Mechanism of migraine headache and action of ergotamine tartrate. *Arch Neurol Psychiatry*. 1938;39(4):737–763.
25. Lashley K. Patterns of cerebral integration indicated by the scotoma of migraine. *Arch Neurol Psychiatry*. 1941;42:259–264.
26. Tfelt-Hansen PC, Koehler PJ. One hundred years of migraine research: major clinical and scientific observations from 1910 to 2010. *Headache*. 2011;51(5):752–778.
27. Tfelt-Hansen PC. History of migraine with aura and cortical spreading depression from 1941 and onwards. *Cephalalgia*. 2010;30(7):780–792.
28. Leao A. Pial circulation and spreading depression of activity in cerebral cortex. *J Neurophysiol*. 1944;7:391–396.
29. Charles AC, Baca SM. Cortical spreading depression and migraine. *Nat Rev Neurol*. 2013;9 (11):637–644.
30. Moskowitz MA, Reinhard Jr. JF, Romero J, Melamed E, Pettibone DJ. Neurotransmitters and the fifth cranial nerve: is there a relation to the headache phase of migraine? *Lancet*. 1979;2 (8148):883–885.
31. Moskowitz MA. Pathophysiology of headache—past and present. *Headache*. 2007;47(Suppl. 1): S58–S63.
32. Moskowitz MA, Brody M, Liu-Chen LY. In vitro release of immunoreactive substance P from putative afferent nerve endings in bovine pia arachnoid. *Neuroscience*. 1983;9(4):809–814.
33. Mason RT, Peterfreund RA, Sawchenko PE, Corrigan AZ, Rivier JE, Vale WW. Release of the predicted calcitonin gene-related peptide from cultured rat trigeminal ganglion cells. *Nature*. 1984;308 (5960):653–655.
34. Ashina M, Hansen JM, Do TP, Melo-Carrillo A, Burstein R, Moskowitz MA. Migraine and the trigeminovascular system-40 years and counting. *Lancet Neurol*. 2019;18(8):795–804.
35. Edvinsson L. Sensory nerves in man and their role in primary headaches. *Cephalalgia*. 2001;21(7): 761–764.
36. Goadsby PJ, Holland PR, Martins-Oliveira M, Hoffmann J, Schankin C, Akerman S. Pathophysiology of migraine: a disorder of sensory processing. *Physiol Rev*. 2017;97(2):553–622.
37. Goadsby PJ, Holland PR. An update: pathophysiology of migraine. *Neurol Clin*. 2019;37(4):651–671.
38. Burstein R, Noseda R, Borsook D. Migraine: multiple processes, complex pathophysiology. *J Neurosci*. 2015;35(17):6619–6629.
39. Noseda R, Jakubowski M, Kainz V, Borsook D, Burstein R. Cortical projections of functionally identified thalamic trigeminovascular neurons: implications for migraine headache and its associated symptoms. *J Neurosci*. 2011;31(40):14204–14217.
40. Silberstein SD, Hargreaves RJ. The history and pharmacology of ergotamine and dihydroergotamine. In: Diener HC, ed. *Drug Treatment of Migraine and Other Headaches*. *1*. Basel, Switzerland: Karger; 2000:52–65.
41. Silberstein SD, McCrory DC. Ergotamine and dihydroergotamine: history, pharmacology, and efficacy. *Headache*. 2003;43(2):144–166.
42. Buzzi MG, Moskowitz MA, Peroutka SJ, Byun B. Further characterization of the putative 5-HT receptor which mediates blockade of neurogenic plasma extravasation in rat dura mater. *Br J Pharmacol*. 1991;103(2):1421–1428.
43. Humphrey PP. The discovery and development of the triptans, a major therapeutic breakthrough. *Headache*. 2008;48(5):685–687.

44. Antonaci F, Ghiotto N, Wu S, Pucci E, Costa A. Recent advances in migraine therapy. *Springerplus.* 2016;5:637.
45. Diener HC. RPR100893, a substance-P antagonist, is not effective in the treatment of migraine attacks. *Cephalalgia.* 2003;23(3):183–185.
46. Goldstein DJ, Wang O, Saper JR, Stoltz R, Silberstein SD, Mathew NT. Ineffectiveness of neurokinin-1 antagonist in acute migraine: a crossover study. *Cephalalgia.* 1997;17(7):785–790.
47. May A, Goadsby PJ. Substance P receptor antagonists in the therapy of migraine. *Expert Opin Investig Drugs.* 2001;10(4):673–678.
48. Goadsby PJ, Edvinsson L, Ekman R. Vasoactive peptide release in the extracerebral circulation of humans during migraine headache. *Ann Neurol.* 1990;28(2):183–187.
49. Brain SD, Geppetti P. *Calcitonin Gene-Related Peptide (CGRP) Mechanisms: Focus on Migraine. Handbook of Experimental Pharmacology (eBook).* Switzerland: Springer Nature; 2019.
50. Tepper SJ. History and review of anti-calcitonin gene-related peptide (CGRP) therapies: from translational research to treatment. *Headache.* 2018;58(Suppl. 3):238–275.
51. McLatchie LM, Fraser NJ, Main MJ, et al. RAMPs regulate the transport and ligand specificity of the calcitonin-receptor-like receptor. *Nature.* 1998;393(6683):333–339.
52. Lassen LH, Haderslev PA, Jacobsen VB, Iversen HK, Sperling B, Olesen J. CGRP may play a causative role in migraine. *Cephalalgia.* 2002;22(1):54–61.
53. Doods H, Hallermayer G, Wu D, et al. Pharmacological profile of BIBN4096BS, the first selective small molecule CGRP antagonist. *Br J Pharmacol.* 2000;129(3):420–423.
54. Kuraishi Y, Nanayama T, Ohno H, Minami M, Satoh M. Antinociception induced in rats by intrathecal administration of antiserum against calcitonin gene-related peptide. *Neurosci Lett.* 1988;92(3):325–329.
55. Kurosawa M, Messlinger K, Pawlak M, Schmidt RF. Increase of meningeal blood flow after electrical stimulation of rat dura mater encephali: mediation by calcitonin gene-related peptide. *Br J Pharmacol.* 1995;114(7):1397–1402.
56. Tan KK, Brown MJ, Longmore J, Plumpton C, Hill RG. Demonstration of the neurotransmitter role of calcitonin gene-related peptides (CGRP) by immunoblockade with anti-CGRP monoclonal antibodies. *Br J Pharmacol.* 1994;111(3):703–710.
57. Sun H, Dodick DW, Silberstein S, et al. Safety and efficacy of AMG 334 for prevention of episodic migraine: a randomised, double-blind, placebo-controlled, phase 2 trial. *Lancet Neurol.* 2016;15(4):382–390.
58. Ashina M, Dodick D, Goadsby PJ, et al. Erenumab (AMG 334) in episodic migraine: interim analysis of an ongoing open-label study. *Neurology.* 2017;89(12):1237–1243.
59. Goadsby PJ, Reuter U, Hallström Y, et al. A controlled trial of Erenumab for episodic migraine. *N Engl J Med.* 2017;377(22):2123–2132.
60. Dodick DW, Ashina M, Brandes JL, et al. ARISE: a phase 3 randomized trial of erenumab for episodic migraine. *Cephalalgia.* 2018;38(6):1026–1037.
61. Tepper S, Ashina M, Reuter U, et al. Safety and efficacy of erenumab for preventive treatment of chronic migraine: a randomised, double-blind, placebo-controlled phase 2 trial. *Lancet Neurol.* 2017;16(6):425–434.
62. Bigal ME, Dodick DW, Rapoport AM, et al. Safety, tolerability, and efficacy of TEV-48125 for preventive treatment of high-frequency episodic migraine: a multicentre, randomised, double-blind, placebo-controlled, phase 2b study. *Lancet Neurol.* 2015;14(11):1081–1090.
63. Bigal ME, Edvinsson L, Rapoport AM, et al. Safety, tolerability, and efficacy of TEV-48125 for preventive treatment of chronic migraine: a multicentre, randomised, double-blind, placebo-controlled, phase 2b study. *Lancet Neurol.* 2015;14(11):1091–1100.
64. Dodick DW, Silberstein SD, Bigal ME, et al. Effect of Fremanezumab compared with placebo for prevention of episodic migraine: a randomized clinical trial. *JAMA.* 2018;319(19):1999–2008.
65. Silberstein SD, Dodick DW, Bigal ME, et al. Fremanezumab for the preventive treatment of chronic migraine. *N Engl J Med.* 2017;377(22):2113–2122.
66. Dodick DW, Goadsby PJ, Spierings EL, Scherer JC, Sweeney SP, Grayzel DS. Safety and efficacy of LY2951742, a monoclonal antibody to calcitonin gene-related peptide, for the prevention of migraine: a phase 2, randomised, double-blind, placebo-controlled study. *Lancet Neurol.* 2014;13(9):885–892.

67. Skljarevski V, Oakes TM, Zhang Q, et al. Effect of different doses of Galcanezumab vs placebo for episodic migraine prevention: a randomized clinical trial. *JAMA Neurol.* 2018;75(2):187–193.
68. Stauffer V, Zhang Q, Skljarevski V, et al. Phase 3 study (EVOLVE-1) of galcanezumab in episodic migraine. *Headache.* 2017;57:1336 abstract: PS88LB.
69. Skljarevski V, Zhang Q, Detke H, et al. Phase 3 study (EVOLVE-2) of galcanezumab in episodic migraine. *Headache.* 2017;57:1312 abstract:IOR-12LB.
70. Detke H, Wang S, Skljarevski V. A phase 3 placebo-controlled study of galcanezumab in patients with chronic migraine: results from the 3-month double-blind treatment phase of the REGAIN study. *Headache.* 2017;57:1336–1337 (abstract PS89LB).
71. Detke H, Pozo-Rosich P, Reuter U, et al. One-year treatment with galcanezumab in patients with chronic migraine: results from the open-label phase of the REGAIN study (P2.10–010). *Neurology.* 2019;92.
72. Dodick DW, Goadsby PJ, Silberstein SD, et al. Safety and efficacy of ALD403, an antibody to calcitonin gene-related peptide, for the prevention of frequent episodic migraine: a randomised, double-blind, placebo-controlled, exploratory phase 2 trial. *Lancet Neurol.* 2014;13(11):1100–1107.
73. Silberstein S, McAlliser P, Berman G, et al. Eptinezumab reduced migraine frequency, duration, and pain intensity through week 24: results from the phase 3 PROMISE-1 trial. *Neurology.* 2018;90:4.091.
74. Kudrow D, Diamond M, McGill L, et al. Eptinezumab achieved meaningful reductions in migraine activity as early as day 1 that were sustained through week 12: results from PROMISE-2. *Headache.* 2018;275:P4–470.
75. Voss T, Lipton RB, Dodick DW, et al. A phase IIb randomized, double-blind, placebo-controlled trial of ubrogepant for the acute treatment of migraine. *Cephalalgia.* 2016;36(9):887–898.
76. Lipton RB, Dodick DW, Ailani J, et al. Effect of Ubrogepant vs placebo on pain and the most bothersome associated symptom in the acute treatment of migraine: the ACHIEVE II randomized clinical trial. *JAMA.* 2019;322(19):1887–1898.
77. Moreno-Ajona D, Pérez-Rodríguez A, Goadsby PJ. Gepants, calcitonin-gene-related peptide receptor antagonists: what could be their role in migraine treatment? *Curr Opin Neurol.* 2020;33(3):309–315.
78. Lipton RB, Croop R, Stock EG, et al. Rimegepant, an oral calcitonin gene-related peptide receptor antagonist, for migraine. *N Engl J Med.* 2019;381(2):142–149.
79. Croop R, Goadsby PJ, Stock DA, et al. Efficacy, safety, and tolerability of rimegepant orally disintegrating tablet for the acute treatment of migraine: a randomised, phase 3, double-blind, placebo-controlled trial. *Lancet.* 2019;394(10200):737–745.

CHAPTER 9

The impact of translational research on the development of therapeutic agents for multiple sclerosis

Stanley L. Cohan[a], Elisabeth B. Lucassen[a], Kyle E. Smoot[a], Kiren Kresa-Reahl[b], Meghan Romba[c], and Stefanie N. Linch[d]
[a]Providence Multiple Sclerosis Center, Providence Brain and Spine Institute, Portland, OR, United States
[b]Atara Biotherapeutics Inc., Thousand Oaks, CA, United States
[c]Department of Neurology, School of Medicine, University of Washington, Seattle, WA, United States
[d]Cancer Early Detection Advanced Research Center (CEDAR), Knight Cancer Institute, Oregon Health and Science University, Portland, OR, United States

Challenges with preclinical models of multiple sclerosis

The relative inaccessibility of the central nervous system (CNS) is a major obstacle in biological research of neurological diseases—be that access to live tissue, imaging of the brain dynamically in vivo, monitoring or probing real-time biochemical and physiological events, and within the current context, the relatively limited ability to monitor the impact of pharmacological interventions beyond clinical assessment and imaging parameters. The management of certain diseases, such as epilepsy, have benefitted greatly from advances in surface, meningo-cortical and deep brain electrophysiological monitoring, magnetic resonance imaging (MRI), and functional MRI and positron emission tomography. Furthermore, the development of both ex vivo and in vivo animal models has yielded important insights into the functional abnormalities characterizing this group of diseases.

Multiple sclerosis (MS) research, on the other hand, has been held back by imperfect animal models, very limited and infrequent access to CNS samples from living patients, the lack of biomarker specificity for MS, and a paucity of relevant in vivo physiological functional monitoring technologies. In order to discern important dynamic biological events and assess cytological and biochemical alterations within the CNS, researchers have typically had to sample peripheral sources such as cerebrospinal fluid and blood to evaluate individual responses to pharmacological interventions. Although important insights have been obtained in this manner, they cannot begin to provide the depth of information that direct examination of brain and spinal cord tissue in vivo could provide. Even MRI scanning—which in many important ways has revolutionized our understanding of MS pathogenesis, prognosis, and response to therapy—is notably

Neurotherapeutics in the Era of Translational Medicine
https://doi.org/10.1016/B978-0-12-816475-4.00010-0

lacking in that it provides static, cross-sectional data and only very crude biochemical information through the use of MR spectroscopy. Furthermore, these advances still provide only limited actionable insights into the noninflammatory drivers of CNS degeneration, particularly in patients with nonrelapsing, progressive forms of MS.

Although the cause(s) of MS remain unknown with any scientific certainty, postmortem examination of patients with MS strongly suggests that it is in large measure a cell-driven, autoimmune inflammatory disease, with a multifactorial pathogenesis. Genetic predisposition, early childhood viral infection, and exogenous environmental factors—such as sunlight exposure, vitamin D levels, tobacco consumption, obesity, possibly exogenous toxins, and the microbiome—are all increasingly the focus of investigation as significant contributory risk factors for MS.

MS is not known to occur spontaneously in any nonhuman species. Despite these challenges, the progress made in the development of MS disease-modifying therapies since the early 1990s has been impressive by any standard. One of the mainstays of this effort has been the animal model experimental allergic encephalomyelitis (EAE), of which there are varying designs in the species employed and autoimmune-inducing strategies or regimens. Interspecies differences in immune system pathways have caused some incongruous findings. Nevertheless, EAE has been an invaluable platform for experiments designed to widen our understanding of MS disease pathogenesis and to test novel pharmacological agents for efficacy.

Typically studied in rodents, and rarely in larger mammals and subhuman primates, EAE is an autoimmune disorder of the CNS, but differs significantly from MS. It is most commonly induced by repeatedly injecting myelin epitopes, such as myelin basic protein (MBP) or myelin oligodendroglial glycoprotein (MOG) peptides, with an accompanying adjuvant to produce an autoimmune, typically cell-mediated attack on CNS myelin. Important functional and mechanistic differences in the immune system between species, as well as the process of EAE induction, have led to some observations that have not been reproduced in MS patients or have failed to sufficiently replicate important fundamental biological features of MS pathogenesis. Due to the design of some EAE models, the important roles of certain immune cell subsets—such as B cells, microglial cells, myeloid cells, natural killer cells, and other innate immune system components—were not sufficiently appreciated in the past. These models also failed to recapitulate the potential pathogenic contributions of immunoglobulins, which are characteristically observed in the cerebrospinal fluid (CSF) and brain samples of the majority of MS patients. In many cases, the importance of certain cell subsets was only revealed by studying human CNS tissue, employing ex vivo experiments to analyze human immune cells, or investigating the mechanisms of action for certain therapeutic agents, which uncovered heretofore underappreciated or unknown immune pathways relevant to MS pathogenesis. Nevertheless, EAE has been a crucial tool in the translational research leading to MS disease-modifying therapies.

Immune modulators in the treatment of multiple sclerosis

Although numerous novel, clinically efficacious disease-modifying therapies have emerged in the past 25 years and are in clinical use for MS treatment, we have selected seven agents, two of which have not yet been approved for use in humans and are still in various stages of clinical development: (1) glatiramer acetate, (2) fingolimod, (3) siponimod, (4) ozanimod, (5) ponesimod, (6) natalizumab, (7) elezanumab, and (8) opicinumab. The stories of preclinical development of these agents are interesting or instructive from the perspective of their preclinical investigations and the critical role that preclinical translational research results played in their development as disease-modifying therapies. For each of these agents, we will discuss the preclinical development and, if published, the results of their pivotal human trials.

Glatiramer acetate (copolymer 1)

The development of glatiramer acetate came about quite serendipitously, when researchers in Israel in the 1960s and 1970s tried to synthesize MBP mimics capable of inducing EAE in rodents.[1,2] Instead, they had created a class of molecules that could prevent or treat EAE in a variety of animal models. One of the synthesized molecules was copolymer 1 (Cop1), later named glatiramer acetate, a polypeptide mixture, synthesized by polymerizing L-glutamic acid, L-lysine, L-alanine, and L-tyrosine in very specific ratios, designed to simulate their relative abundance in MBP.[1,3] This section outlines some of the critical preclinical translational research that led to the development, characterization, and clinical use of glatiramer acetate.

Mechanistic studies of glatiramer acetate

Preclinical studies of glatiramer acetate demonstrated efficacy in preventing or suppressing EAE in several different species.[1,4] In subhuman primates, where MBP-induced EAE is highly lethal, the administration of glatiramer acetate reversed the disease even after symptoms had appeared.[5] Repeated successes in these primate models led to the introduction of glatiramer acetate into human clinical trials. Subsequently, it was approved in 1996 for the treatment of relapsing forms of MS.[6]

Multiple mechanisms of action have been proposed to account for the therapeutic effects of glatiramer acetate in EAE and MS. Admittedly, much of this data resulted from preclinical research subsequent to glatiramer acetate approval for the treatment of MS. Early investigators focused primarily on the impact of glatiramer acetate on the adaptive immune system, primarily upon alterations in lymphocyte function, with more recent work demonstrating the impact of glatiramer acetate on the innate immune system.

Proposed therapeutic mechanisms of glatiramer acetate included the following:

1. "Promiscuous" binding and thus competition with MBP for binding to major histocompatibility complex II (MHCII) molecules, thus interfering with antigen presentation and lymphocyte activation.
2. Competition of the glatiramer acetate/MHC with that of MBP/MHC for binding to cognate T cell receptors.
3. Partial activation and induction of tolerance in MBP-specific T cells.
4. Generation of glatiramer acetate-reactive T helper 2 (Th2) regulatory cells.[7–10]
5. B cell immunomodulation.[11,12]
6. Generation of tolerogenic antigen presenting cells, including conversion of monocytes and macrophages to tolerogenic type II subsets.[8,13–18]
7. Generation of antigen presenting cell-dependent glatiramer acetate-reactive CD8+ T cells.[19,20]

Glatiramer acetate binds to the surface of MHCII molecules, suggesting that it can compete with MBP-derived auto-antigens for MHCII binding sites, thus inhibiting T cell activation by preventing auto-antigen presentation to autoreactive T cells.[21–23] However, substitution of a glatiramer acetate composed of D-amino acids avidly bound to MHCII, but did not prevent or ameliorate EAE in mice, calling into question the first two of the proposed mechanisms mentioned above.[24]

Glatiramer acetate treatment was associated with the induction of auto-tolerance in MBP-specific T cells and the expansion of glatiramer acetate-reactive Th2 lymphocytes, consistent with increased antiinflammatory cytokine production observed following exposure to MBP antigens.[25] In studies of glatiramer acetate-treated patients with MS there was a shift from pro-inflammatory Th1 toward antiinflammatory Th2 cells, possibly contributing to glatiramer acetate efficacy.[7,9,26,27] Because glatiramer acetate entrance into the CNS is probably very limited, glatiramer acetate actions on T cells probably occur primarily in peripheral lymphoid organs, after which the glatiramer acetate-reactive T cells enter the CNS where they exert their tolerogenic actions following exposure to myelin auto-antigens.[4,10,28–30]

There is evidence that glatiramer acetate exposure leads to the expansion of regulatory lymphocytes and promotes their antiinflammatory activity. Human $CD4^+$ $CD25^+$ $FoxP3^+$ T regulatory (Treg) cells stimulated in vitro with glatiramer acetate exhibited increased production of interleukin-10 (IL-10), which limits pro-inflammatory activity and increases immune tolerance.[31,32] Glatiramer acetate treatment also promoted the expansion of regulatory B (Breg) cells in mice, increasing production of IL-10 and reducing the release of the pro-inflammatory cytokines IL-6, IL-12, and tumor necrosis factor-alpha (TNFα).[11,12]

Innate immunity pathways

There is growing evidence that the innate immune system is a potential therapeutic target of glatiramer acetate through the regulation of antigen presenting cells.[8,14,16,33,34] In glatiramer acetate-treated patients with MS, in vitro stimulation of their monocytes was associated with reduced production of activation markers, reduced production of pro-inflammatory IL-12 and TNFα, and increased antiinflammatory IL-10 production, when compared to monocytes taken from untreated MS patients.[34,35] Glatiramer acetate treatment of EAE in mice also resulted in the expansion of type II tolerogenic monocytes, as well as Th2 polarization of T cells, increased CD4+CD25+FoxP3+ Treg cells, and reduced pro-inflammatory Th1 and Th17 cells.[36] Glatiramer acetate exposure also leads to the expansion of glatiramer acetate-reactive cytolytic CD8+ T cells, which appears to be dependent on binding to monocytes, macrophages, or other innate antigen presenting cells expressing MHCI.[19,20] This in turn would result in cytolytic suppression of pro-inflammatory effector T cells.[37,38] Once exposed to glatiramer acetate-treated antigen presenting cells, CD8+ T cells are antiinflammatory, but in MHCI, interferon-gamma (IFNγ), and perforin-null mice, antigen presenting cell exposure to glatiramer acetate does not produce an antiinflammatory response by CD8+ T cells. Further, in the absence of CD8+ T cells, glatiramer acetate-treatment of antigen presenting cells does not result in a therapeutic response in EAE.[20] Thus it appears that antigen presenting cells of the innate system may be critically important glatiramer acetate targets, but the major therapeutic effector cell in glatiramer acetate-treated patients is the CD8+ T cell.[13,14,20,34]

Direct neuroprotective effects

Lastly, glatiramer acetate may be directly neuroprotective—possibly related to an increase in the secretion of neurotrophic factors such as insulin-like growth factor (IGF-1) and brain-derived neurotrophic factor (BDNF), by Th2-like regulatory cells within the CNS.[39–41] The neurotrophic factor BDNF promotes differentiation and survival of neurons and glial cells; is expressed in activated T cells, B cells, and monocytes in inflammatory CNS lesions; and BDNF is upregulated in glatiramer acetate-treated T cells ex vivo.[42–47] In EAE models, glatiramer acetate treatment also upregulated BDNF, reduced neuronal and axonal loss and demyelination, promoted migration and differentiation of progenitor cells, and led to an increase in oligodendroglial progenitor cells.[48–51]

In summary, glatiramer acetate promotes a cascade of antiinflammatory, tolerogenic pathways, and potentially promotes or activates neuroprotective pathways. An integrated concept of its mechanisms of action that reconciles the wide-ranging observations presented would place antigen presenting cells of the innate immune system as the primary glatiramer acetate target, promoting their evolution into antiinflammatory, tolerizing type II cells. The tolerizing effector cells, however, are glatiramer acetate-reactive CD8+ cytolytic T cells, induced by MHCI-bearing antigen presenting cells. These glatiramer acetate-reactive CD8+ cells suppress pro-inflammatory T cells and possibly B cells, and induce expansion of tolerogenic Treg and Breg clones, thus reducing pro-inflammatory cytokine secretion, increasing antiinflammatory cytokines, and potentially explaining the efficacy observed in clinical trials of glatiramer acetate in patients with relapsing MS.

Clinical trials of glatiramer acetate

The results of clinical investigations of glatiramer acetate are consistent with rodent and subhuman primate studies. The earliest clinical use of glatiramer acetate was in three patients with acute disseminated encephalomyelitis and in four patients in the terminal stages of MS; soon thereafter formal trials were initiated in the United States.[26] In a double-blind, randomized, placebo-controlled pilot trial, 50 patients with relapsing-remitting MS were treated either with 20 mg of glatiramer acetate or placebo subcutaneously daily for 2 years.[52] Fourteen of 25 patients in the glatiramer acetate group versus six of the remaining 23 patients in the placebo group had no exacerbations ($P=.045$). Over 2 years, patients taking glatiramer acetate improved an average of 0.5 Kurtzke Disability Scale points, while those taking placebo worsened an average of 1.2 units. Based on these results, a phase III, multicenter, double-blind, placebo-controlled trial was initiated.[3] A total of 251 patients were randomized to receive glatiramer acetate 20 mg or placebo by daily subcutaneous injection for 2 years. The primary end point was a difference in MS relapse rate. The annualized relapse rate at 2 years was 1.19 ± 0.13 for patients receiving glatiramer acetate and 1.68 ± 0.13 for placebo, resulting in a 29% relative reduction in relapse rate in favor of glatiramer acetate. Significantly more patients receiving glatiramer acetate were found to have improved Expanded Disability Status Scale scores and more receiving placebo worsened ($P=.037$).[53] In both trials, the most common adverse effects included injection site reactions, and, less commonly, a transient self-limited systemic reaction following the injection, characterized by flushing, chest tightness with palpitations, anxiety, or dyspnea lasting up to 30 min. No laboratory abnormalities were observed.[3,52]

In another double-blind, multinational study, a total of 239 patients with relapsing MS were randomized to either glatiramer acetate or placebo. MRI parameters were the primary endpoint. Inclusion criteria were similar to the prior phase III trial, with the additional requirement that a screening MRI had to reveal at least one gadolinium-enhancing lesion. Patients were followed for 9 months, during which they had monthly MRI scans. The primary endpoint was the cumulative number of T1 gadolinium-enhancing lesions over the 9-month study period. The median number of cumulative T1 gadolinium-enhancing lesions for glatiramer acetate and placebo were 25.96 and 36.8, respectively ($P<.0030$).[54]

An open-label extension study of the initial 2-year phase III clinical trial reported continued clinical benefit of glatiramer acetate regarding relapse rate, neurologic disability, tolerance, and safety profile.[53] However, this latter study was impacted by potential selection bias, given the large number of patients who did not enter or complete the open-label study, and the acknowledgement by the authors that a disproportionate number of poor responders in the placebo-controlled trial did not enter the open-label extension. Nevertheless, based on these favorable clinical trial results glatiramer acetate was approved for use in relapsing MS in 1996, being one of the first disease-modifying therapies to demonstrate significant efficacy in the treatment of relapsing MS.

Despite its modest efficacy, particularly when measured against more recently approved disease-modifying therapies for relapsing MS, glatiramer acetate has withstood the test of time, and as of 2017, the latest year for which data were available, approximately 600,000 MS patients worldwide had been treated with this agent.[55] Indeed, Cop1, later named glatiramer acetate, was a very fortunate, serendipitous discovery by researchers who were wise enough to recognize that while their experiments did not produce the results they had anticipated, they were a success that ultimately led to the development of an important therapeutic agent for relapsing MS that has been in continuous use for more than 20 years.

Fingolimod, siponimod, ozanimod, and ponesimod (sphingosine-1-phosphate receptor ligands)

The agents with perhaps the broadest body of preclinical evidence thus far to support their benefit in MS are those that bind to sphingosine-1-phosphate (S1P) receptors. This section details the research on S1P and S1P receptors (S1PR), and the multifaceted impact of S1P-signaling inhibition by fingolimod, as well as evidence to support siponimod, ozanimod, and ponesimod in the treatment of MS.

Sphingosine-1-phosphate and its receptors

Sphingosine is a lysophospholipid, long thought to be an "inert" cell membrane structural component that arises primarily from the turnover of membrane sphingolipids. It is phosphorylated to S1P by sphingosine kinase (SphK), which is ubiquitously expressed, including peripheral lymphoid organs and the CNS.[56–58] Platelets and endothelial cells are the major sources of S1P in the blood.[59] To date, the most important identified biological role of S1P is as a ligand for a group of five cell membrane surface-bound G-protein coupled receptors, subtypes 1 through 5 (S1PR1–5).[60–62] The physiological roles of each of these receptor subtypes varies with cell type, distribution, and maturational state.

Neurotrophic effects

Regulation of cell growth, proliferation, and survival during embryonic development and maturation, as well as tissue repair, are among key functional roles identified for S1P and its receptors, which are actively expressed during neurogenesis.[63–66] S1PR-linked G-protein stimulation promotes neural progenitor cell proliferation via extracellular signal-regulated kinase (ERK) pathways and promotes changes in morphology and neural cell aggregation via Rho kinase (Rho-K) in progenitor cells.[65,67] In addition, S1P may suppress programmed cell death, including the inhibition of mitochondrial intrinsic death pathways/apoptosis.[68,69]

Immune regulatory effects

S1P also regulates lymphocyte kinetics through S1PRs. Although both S1PR1 and S1PR4 are expressed on the surface of lymphocytes, S1PR1 expression appears to be critical for T cell mobilization from the thymus, and mobilization of both T cells and B cells from peripheral lymphoid organs and the spleen.[70,71] The functional role of S1PR4 on lymphoid tissue is currently unknown. Mobilization of lymphocytes from peripheral lymphoid tissue is regulated by the gradient of free S1P.[70–72] The absence of S1PR1 expression on the lymphocyte surface of knockout mice renders mature T cells and B cells incapable of leaving the thymus and peripheral lymphoid organs, markedly lowering or eliminating their presence in peripheral blood.[71]

S1P signaling in the CNS

Although S1PR1, 3, and 5 are prominently expressed in the CNS, to date S1PR1 and 5 appear to be functionally very important.[73–77] S1P readily crosses the blood-brain barrier to enter and be preferentially sequestered within the CNS, where it may encounter its receptors on astrocytes, oligodendroglia, oligodendroglial progenitor cells, microglia, neurons, and neural stem cells.[78–81] S1P may promote the growth, protection, and survival of oligodendroglial progenitor cells via S1PR1, which is necessary for myelination or remyelination, and protect mature oligodendroglia via S1PR5 stimulation.[73,82,83] S1P signaling also promotes neural stem cell migration toward areas of CNS injury and neural repair.[63,84–86] Moreover, S1P signaling stimulates astrocyte proliferation, primarily through the ERK pathway.[87–89] However, S1PR1 activation may also promote astrogliosis following injury and impair structural and functional recovery through the inhibition of oligodendroglial progenitor cell migration and remyelination.[90]

Inhibition of sphingosine-1-phosphate receptor signaling with fingolimod, siponimod, ozanimod, and ponesimod

Fingolimod

Fingolimod, derived from the fungal metabolite myriocin, is a high-affinity ligand for S1PR1, 3, 4, and 5, with 1000 times the affinity for S1PR relative to S1P.[69,91–94] It exerts pharmacologic actions on many organ systems including the heart and blood vessels, pulmonary airway, endothelial cells, and lymphocytes, as well as the CNS. The active form, fingolimod, requires phosphorylation by SphK2.[95–97] It actively competes with S1P for the modulation of S1P receptors. Although its interaction with these receptors leads to a transient phase of activation, this is followed by receptor inactivation, internalization, and degradation, rendering the cell incapable of responding to S1P.[71,72,98,99]

Immune activity

Activated fingolimod inhibits naïve and central but not effector memory T and B cells from exiting peripheral lymphoid organs, greatly reducing their numbers in peripheral circulation and thus reducing their pro-inflammatory potential in MS.[71,100–104] Importantly, fingolimod exposure reduces the number of circulating Th17 cells, a highly pro-inflammatory lymphocyte subset responsible for CNS inflammation and destruction.[105,106] Recognition of the immune modulating potential of fingolimod resulted in its assessment as an agent to reduce the burden of autoimmunity in various disease models—organ transplant rejection, graft versus host disease, type 1 diabetes, rheumatoid arthritis, and MS.[107–109] A number of critical studies employing a variety of rodent EAE models demonstrated that the use of fingolimod could prevent the onset of EAE or suppress it in animals that already had EAE symptoms.[100,101,110,111] Furthermore, fingolimod also reduced mortality from EAE even at very low doses.[112]

Largely based on the above observations in S1P and S1PR1 knockout mice, it had been widely assumed that the beneficial effect of fingolimod in EAE or MS was via inhibiting autoreactive central memory and naïve T cells and B cells from exiting lymphoid organs, traversing the blood-brain barrier and entering the CNS.[111] In support of this assumption, circulating lymphocyte counts were reduced by 70%–90% within 6h of a single dose of fingolimod, which was associated with reduced inflammatory cell accumulation in brain and spinal cord in EAE models.[112,113] However, studies examining the CNS activity of fingolimod suggest a more complex mechanism(s) than solely peripheral lymphocyte sequestration.

CNS activity

While fingolimod reduces the number of circulating, pro-inflammatory lymphocytes, fingolimod also readily crosses the blood-brain barrier, where it can be phosphorylated by SphK2 to its pharmacologically active form.[95,112] As noted earlier, it has been demonstrated that S1PRs are expressed on all of the cellular elements of the CNS, raising the possibility that some of the therapeutic effects of fingolimod are mediated by CNS S1PR and that fingolimod may be directly protective of CNS cells during inflammation.[61,76,77,114,115] S1PR signaling by fingolimod may promote oligodendroglial progenitor cell survival and prevent oligodendroglia apoptosis (through activation of the aforementioned G-protein-linked ERK1/2 and Akt survival pathways), promote remyelination, protect neurons, and preserve neuronal function.[76,77,85,113,114,116–119] Oligodendroglia express S1PR1, 3, and 5, but S1PR1 and S1PR5 appear to be functionally most important.[67,76,120–122] Fingolimod modulates oligodendroglia S1PR1 and S1PR5 mRNA expression and appears to promote oligodendroglia survival.[74,123,124] Fingolimod treatment of dissociated human oligodendroglia resulted in membrane process extension in vitro, upregulated S1PR5 mRNA, and prevented apoptosis via S1PR5 interaction.[121] Fingolimod also protected and promoted the survival of oligodendroglial progenitor cells, cells necessary for the replacement of lost oligodendroglia, via S1PR1.[76,125] In vitro studies using oligodendroglial progenitor cell cultures demonstrated that fingolimod increased the number of oligodendroglial progenitor cells and oligodendroglia and increased the number of mature, MBP-positive oligodendroglia.[114] Microglia activated during inflammation may reduce oligodendroglial progenitor cell survival through the release of IL-1β, nitric oxide metabolites, and TNFα, as well as activate proapoptotic caspase-3. Fingolimod inhibited this increase in pro-inflammatory cytokine release, reduced caspase-3 production/activation, and promoted increased oligodendroglial progenitor cell survival.[77,123,126]

Fingolimod may also impact the function of astrocytes, by reducing astrogliosis and astrocyte-driven inflammation.[84,88,127,128] During CNS inflammation, astrocytes may produce pro-inflammatory cytokines IL-1β, IL-6, TNFα, and IL-17, and may attract lymphocytes by producing the chemokines CCL2, CCL5, CXCL8, and CXCL 10—a role for which astrocytes may be especially suited given the abutment of their end-foot processes on blood vessel walls.[129,130] Astrocytes may also activate pro-inflammatory Th1 and Th17 lymphocytes via IL-12 and IL-23 secretion, and promote lymphocyte migration into the CNS through the expression of vascular adhesion molecules.[131–135] Once in the CNS, T cells may be targeted to CNS epitopes by astrocytes acting as the antigen presenting cells.[136,137]

Deletion of CNS S1PR1 in rodents resulted in the loss of a protective effect of fingolimod on EAE.[138] In further support of a direct CNS protective role for fingolimod, its direct intraventricular administration to animals with EAE reduced disease activity without lowering circulating lymphocyte counts, that is, peripheral sequestration of lymphocytes was not necessary for a therapeutic response.[139] In an EAE model using mice that were null for astrocyte S1PR1, fingolimod was not protective or therapeutic, despite lymphocyte depletion from the blood, further supporting a direct protective effect of fingolimod on cellular elements within the CNS, independent of its effects on lymphocyte dynamics.[140] Collectively, these data provide compelling evidence to support the direct, neuroprotective effects of fingolimod in MS.

Siponimod, ozanimod, and ponesimod

Two second-generation molecules have been developed and approved for clinical use, siponimod and ozanimod, which are more specific in that they solely modulate S1PR1 and S1PR5; and ponesimod, which solely regulates S1PR1. Each of these molecules is highly lipophilic, readily enters the CNS, and does not require phosphorylation to be activated.

The volume of preclinical research with these compounds is far more limited than that for fingolimod, but because siponimod and ozanimod have recently been approved and ponesimod is awaiting regulatory approval for the treatment of MS, they are included in this discussion. Consistent with results obtained with fingolimod, siponimod, ozanimod, and ponesimod rapidly reduced circulating T and B cell counts, and significantly reduced disease activity in a rat model of EAE.[141] In astrocytes harvested from mouse and human cerebellar slices, siponimod promoted ERK and Akt phosphorylation, thus regulating these critical trophic signaling pathways.[142] This study also demonstrated that siponimod attenuated nonimmune lysolecithin-induced demyelination via its interaction with S1PR5.

Ozanimod treatment using models of EAE and cuprizone-induced demyelination resulted in reduced plasma levels of the axonal injury marker neurofilament light chain, improved EAE scores, reduced neuronal swelling and transections, reduced spinal cord demyelination, and reduced inflammation.[143] In addition, ozanimod-treatment of cuprizone-treated mice resulted in the activation of ERK and Akt pathways and reduced astrocyte release of the pro-inflammatory cytokines IL-1β, IL-6, and TNFα following LPS-induced activation.[144] Ozanimod-treated mice with EAE had reduced Th1, Th17, and monocyte infiltration of the spinal cord and a reduction in microglial expansion.[145] As in the case of fingolimod, it is likely that both siponimod and ozanimod reduce inflammatory damage to the CNS via both peripheral immune cell sequestration and direct cytoprotective effects within the CNS.

Ponesimod has more than 10 times the affinity for S1PR1 than any of the currently approved agents, and as a result, blocks and subsequently leads to the internalization and degradation of the S1PR1 receptor on the surface of lymphocytes. This results in a rapid reduction in circulating B and T cells by up to 70% of baseline values and was associated with improved clinical scores in mice with EAE.[146–150]

As in the case of fingolimod, it is likely that siponimod, ozanimod, and ponesimod reduce inflammatory damage to the CNS via peripheral immune cell sequestration and by direct cytoprotective effects within the CNS.

Clinical trials of fingolimod, siponimod, ozanimod, and ponesimod

Fingolimod clinical trials

Fingolimod was the first oral medication approved in 2010 for relapsing MS. As previously noted, fingolimod is a functional antagonist of S1PR, exhibiting a much higher receptor affinity than S1P, with fingolimod binding resulting in receptor internalization and degradation. Fingolimod is prescribed as 0.5 mg once daily for adults and children 10 years or older with relapsing and active secondary progressive MS. Active MS is defined as patients having had a relapse in the prior 2 years or evidence of an enhancing MRI lesion. The therapeutic efficacy of fingolimod for relapsing MS was established in two pivotal, phase III, double-blind, randomized clinical trials comparing fingolimod to placebo (FREEDOMS) and intramuscular IFN-beta-1a (IFNβ-1a; TRANSFORMS). This was then followed by the phase III extension studies FREEDOMS II and extended TRANSFORMS.

FREEDOMS

FREEDOMS was a phase III multicenter, double-blind, placebo-controlled trial with 1272 patients that demonstrated superior efficacy of fingolimod in relapsing MS when compared to placebo.[151] Inclusion criteria included a diagnosis of relapsing MS,

Expanded Disability Status Scale score ≤ 5.5, one or more clinical relapses in the past year, or two or more relapses in the past 24 months. The median age was 37 years, median disease duration was 6.7 years, and median Expanded Disability Status Scale score at baseline was 2.0. Patients were randomized in a 1:1:1 ratio to receive placebo, fingolimod 0.5 mg, or fingolimod 1.25 mg once daily. The primary endpoint was annualized relapse rate and the main secondary endpoint was time to 3-month sustained disability worsening. Other secondary endpoints included: time to first relapse, change in Expanded Disability Status Scale score and the MS functional composite z-score at baseline compared to 24-months, MRI measurements, tolerability, and safety.

Annualized relapse rate was 0.18 with 0.5 mg fingolimod, 0.16 with 1.25 mg of fingolimod, and 0.40 with placebo ($P < .0001$ for both fingolimod doses vs. placebo), representing a relative reduction in annualized relapse rate of 54% and 60% for fingolimod 0.5 and 1.25 mg, respectively. In the fingolimod groups, time to first relapse was significantly longer than that in placebo-treated patients and more patients remained relapse free in the fingolimod groups over 24 months compared to placebo.

The key secondary endpoint in FREEDOMS was time to confirmed disability progression over 24 months. Treatment with fingolimod significantly reduced the risk of disability progression at 3 months at both fingolimod doses compared to placebo (hazard ratios, 0.70 for 0.5 mg dose and 0.68 for 1.25 mg dose). Cumulative probability of disability progression after 3 months was 17.7% for fingolimod 0.5 mg, 16.6% for fingolimod 1.25 mg, and 24.1% for placebo. During the 24-month study period, Expanded Disability Status Scale score and MS functional composite z-scores remained stable in the fingolimod groups and slightly worsened in the placebo group. Fingolimod was also superior to placebo in a variety of MRI measurements ($P < .001$ for all comparisons at 24 months), including the number of new or enlarging T2 lesions (relative reduction of 74% with 0.5 mg fingolimod), gadolinium-enhancing lesions (relative reduction of 79% with 0.5 mg fingolimod), and brain volume loss (relative reduction of 36% with 0.5 mg fingolimod).[151,152]

TRANSFORMS

TRANSFORMS was the second pivotal phase III clinical trial, enrolling 1292 patients and demonstrating superior efficacy of fingolimod in a 12-month, multicenter, randomized, double-dummy, parallel group study compared to intramuscular IFNβ-1a in patients with relapsing MS.[153] Inclusion criteria included a diagnosis of relapsing MS, Expanded Disability Status Scale score ≤ 5.5, one or more documented relapses in the past 12 months, or two or more in the previous 2 years. Patients were randomized in a 1:1:1 ratio to receive fingolimod 0.5 mg once daily, fingolimod 1.25 mg once daily, or IFNβ-1a 30 μg once weekly. The primary endpoint once again was annualized relapse rate and key secondary endpoints included number of new or enlarging lesions on T2-weighted MRIs at 1 year and sustained disability progression for at least 3 months. At 12 months, annualized relapse rate was 0.16 with 0.5 mg fingolimod, 0.20 with

1.25 mg fingolimod, and 0.33 with IFNβ-1a ($P<.001$ for both fingolimod doses vs. IFNβ-1a), representing a relative reduction in annualized relapse rate of 52% for fingolimod. In the fingolimod arms versus IFNβ-1a, time to first relapse was longer and more patients remained relapse free in the fingolimod groups. MRI outcomes revealed a reduction in the mean number of new and enlarged T2 lesions in the fingolimod groups versus IFNβ-1a (1.7 with 0.5 mg fingolimod, 1.5 with 1.25 mg fingolimod, and 2.6 with IFNβ-1a with $P<.01$ for both doses vs. interferon). In comparison to IFNβ-1a, 0.5 mg of fingolimod demonstrated a 54% relative reduction in gadolinium-enhancing lesions, a relative reduction of 25% in new or newly enlarged T2 lesions, and a 31% relative reduction in percent change in brain volume from baseline.[152,153] In contrast, there was no significant difference observed among the treatment arms with respect to disability progression.

FREEDOMS II

FREEDOMS II was an additional phase III trial to further evaluate the efficacy and safety of fingolimod in relapsing MS.[154] It was a 24-month, randomized, double-blind, placebo-controlled, multicenter study evaluating the efficacy of fingolimod compared to placebo in 1083 patients. Inclusion criteria included: a diagnosis of relapsing MS, one or more documented relapses in the previous year, or two or more relapses in the previous 2 years, an Expanded Disability Status Scale score ≤ 5.5, and no relapse or steroid treatment within 30 days of enrolling into the study. Patients were randomized in a 1:1:1 ratio to receive fingolimod 0.5 mg, fingolimod 1.25 mg, or placebo once daily for 24 months. In the course of the trial, patients randomized to fingolimod 1.25 mg a day were switched to 0.5 mg a day at the recommendation of the data and safety monitoring board. The primary endpoint of the trial was annualized relapse rate, comparing fingolimod 0.5 mg to placebo. Key secondary endpoints were the effect of fingolimod 0.5 mg on brain volume change, time to 3-month sustained disability progression, change in MS functional composite score, and effect on MRI measurements in comparison to placebo.

The annualized relapse rate over 24 months was 0.21 with 0.5 mg of fingolimod, 0.20 with 1.25 mg of fingolimod, and 0.40 with placebo ($P<.0001$ for both fingolimod doses vs. placebo), a relative reduction in annualized relapse rate of 48% for those receiving fingolimod 0.5 mg and 50% for those receiving fingolimod 1.25 mg. The key secondary endpoint, percent change in brain volume from baseline to 24 months, was 0.9% for 0.5 mg fingolimod, 0.6% for 1.25 mg fingolimod, and 1.3% for placebo, a relative reduction of 31% when comparing fingolimod 0.5 mg to placebo ($P<.001$). The mean number of new and enlarged T2 lesions at 24 months was 2.3 with 0.5 mg fingolimod, 1.6 with 1.25 mg fingolimod, and 8.9 with placebo, and a relative reduction of 70% in gadolinium-enhancing lesions and 74% in new or newly enlarged T2 lesions. There was no significant difference between fingolimod and placebo in confirmed disability progression at 3 or 6 months. The time to first confirmed relapse was longer and more patients remained relapse free in both fingolimod groups compared to placebo-treated patients at 24 months. MFSC scores were also improved at month 24 in the fingolimod arms compared to the placebo-treated arm.

Extended TRANSFORMS

Extended TRANSFORMS reported long-term results (up to 4.5 years) for the core TRANSFORMS patients.[155] Of note, 92% (n = 1027) of patients completed the core study and entered this extension phase, with 75.2% completing the extension study. Patients randomized to fingolimod 0.5 mg or fingolimod 1.25 mg in the core study continued at the same dose in the extension study, whereas patients receiving IFNβ-1a were re-randomized 1:1 to either fingolimod 0.5 mg or fingolimod 1.25 mg daily. Following the sponsor's decision to discontinue fingolimod 1.25 mg in 2009, all patients subsequently received fingolimod 0.5 mg until completion of the extension phase.

The primary endpoint, annualized relapse rate at 4.5 years, was significantly reduced in patients who initiated fingolimod treatment during the core study compared to those patients switching from core study IFNβ-1a to fingolimod during the Extended TRANSFORMS: (0.17 vs. 0.27), with a 35% relative reduction in risk of relapse ($P < .001$). After switching to fingolimod, patients initially treated with IFNβ-1a subsequently experienced a 50% reduction in annualized relapse rate by the end of the extension study. MRI outcomes in the extended TRANSFORMS revealed a 63% reduction in new or newly enlarged T2 lesion count in the group initially treated with IFNβ-1a. The proportion of patients with no evidence of disease activity (no relapses, no MRI worsening, and no disability progression) increased by 50% in the first year after switching from IFNβ-1a to fingolimod (44%–66%). During the extension phase, there was no statistical difference in the number of patients in either group who remained without gadolinium-enhancing lesions. The relative reduction observed in fingolimod-treated patients in percent change in brain volume from baseline during the core study was maintained through the extension study, whereas the patients switched from IFNβ to fingolimod experienced a reduction in the rate of brain volume loss during the extension phase.

Siponimod clinical trials

As previously indicated, siponimod is a selective S1P1R and S1P5R modulator that readily crosses the blood-brain barrier. Preclinical studies suggested that it may prevent synaptic neurodegeneration and promote remyelination in the CNS. Given the results of these preclinical studies, and because of the need to develop disease-modifying therapies that might slow disability worsening in patients with secondary progressive MS, it was decided to study siponimod in this patient population.

EXPAND

EXPAND was a double-blind, randomized, multicenter, multinational phase III study of 1645 adults aged 18–60 years with secondary progressive MS and an Expanded Disability Status Scale score of 3.0–6.5.[156] Patients were randomly assigned (2:1) to once daily oral siponimod 2 mg or placebo for up to 3 years or until the occurrence of a prespecified number of confirmed disability progression events. The primary endpoint was time to 3-month confirmed disability progression. At baseline, the mean time since first MS symptoms was 16.8 years and time since conversion to secondary progressive MS was 3.8 years. Of note, 64% of patients ($n=1055$) had not had a relapse in the previous 2 years; 56% ($n=918$) needed walking assistance. For the primary endpoint, 26% (288/1096) of patients receiving siponimod and 32% (173/545) receiving placebo had 3-month confirmed disability progression (hazard ratio 0.79, 95% CI 0.65–0.95) for a relative risk reduction of 21% ($P=.013$). Of the key secondary endpoints, there was no significant difference between siponimod- and placebo-treated patients in the time to 3-month confirmed worsening of at least 20% in Timed 25 Foot Walk. The increase in T2 lesion volume from baseline was significantly lower in patients treated with siponimod compared to placebo, with a between group difference of $-695\,mm^3$ (95% CI -877.3 to -513.3; $P<.0001$). Additionally, more patients receiving siponimod than placebo were free from gadolinium-enhancing lesions (89% vs. 67%) and from new or enlarging T2 lesions (57% vs. 37%).

Ozanimod clinical trials

Ozanimod, a S1PR1 and S1PR5 modulator, has been recently approved for the treatment of relapsing forms of MS, including active secondary progressive MS. Ozanimod has been studied in two phase III clinical trials, RADIANCE Part B and SUNBEAM. Each of these trials was a multicenter, randomized, double-blind, double-dummy, parallel group, and active-controlled study, evaluating the efficacy and safety of daily oral ozanimod 0.5 or 1 mg versus weekly IFNβ-1a 30 μg intramuscular injection in patients with relapsing-remitting MS.

RADIANCE part B

RADIANCE Part B included 1313 patients with relapsing–remitting MS who were randomized to receive ozanimod 0.5 mg ($n=439$) or ozanimod 1 mg ($n=433$) versus weekly IFN β-1a 30 μg intramuscular injection ($n=441$). Inclusion criteria were ages 18–55 years, Expanded Disability Status Scale score 0.0–5.0, diagnosis of relapsing-remitting MS, clinical stability without relapse or steroid treatment within 1 month of screening, one or more documented relapses in the prior year, one or more documented relapses in the past 2 years, and one or more gadolinium-enhancing lesions in the past year. Exclusion criteria included specified cardiac conditions, such as recent myocardial infarction, stroke, prolonged QTcF interval, resting heart rate <55 bpm, and Type I diabetes, or uncontrolled Type II diabetes with Hgb A1c $>7\%$.

The primary endpoint in RADIANCE Part B was annualized relapse rate for each ozanimod dose versus interferon β-1a over 2 years of treatment. Key secondary endpoints included number of new or enlarging T2 brain lesions, number of gadolinium-enhancing brain lesions, and time to 3-month confirmed disability progression. Other secondary endpoints included relative rate of whole brain volume loss; exploratory outcomes included changes in cortical gray matter and thalamic volumes over the 2 years of observation. Annualized relapse rate at 2 years was 0.218 for ozanimod 0.5 mg, 0.172 for ozanimod 1 mg, and 0.276 for IFN β-1a. The reduction in annualized relapse rate over 2 years relative to treatment with IFNβ-1a was 21% for ozanimod 0.5 mg ($P=.0167$) and was 38% for ozanimod 1 mg ($P<.0001$). Both ozanimod 0.5 mg (34% reduction; $P=.0001$) and 1 mg (42% reduction, $P<.001$) treatment arms had a significant reduction in the number of new or enlarging T2 brain lesions over 2 years compared to IFN β-1a. Gadolinium-enhancing brain lesions at 2 years were also significantly reduced in the ozanimod-treated patients by 47% ($P=.003$) for ozanimod 0.5 mg and 53% ($P=.0006$) for ozanimod 1 mg compared to IFNβ-1a. There was no statistically significant difference between ozanimod dose and IFNβ-1a in time to 3-month confirmed disability progression. Brain volume loss at 2 years compared to IFNβ-1a was reduced by 25% in patients receiving 0.5 mg ozanimod ($P<.0001$), and by 27% in patients receiving 1.0 mg ozanimod ($P<.0001$). Cortical gray volume loss was reduced by 57% in patients receiving ozanimod 0.5 mg ($P<.0001$) and by 58% ($P<.0001$) in patients treated with ozanimod 1.0 mg compared to patients receiving IFNβ-1a. Thalamic volume loss at 2 years was reduced by 30% ($P=.0012$) in patients receiving 0.5 mg ozanimod, and by 32% ($P<.0001$) in patients receiving 1.0 mg ozanimod.

SUNBEAM

The SUNBEAM phase III study included 1346 patients with relapsing-remitting MS, randomized to daily oral ozanimod 0.5 mg ($n=451$) or 1 mg ($n=447$) versus weekly intramuscular 30 μg IFNβ-1a ($n=448$). Inclusion criteria were ages 18–55 years, Expanded Disability Status Scale score 0.0–5.0, diagnosis of relapsing-remitting MS, with the same remaining inclusion and exclusion criteria as RADIANCE Part B.

The primary endpoint in SUNBEAM was annualized relapse rate for each ozanimod dose versus IFNβ-1a, but unlike RADIANCE B, over only 1 year of observation. Key secondary endpoints included number of new or enlarging T2 brain lesions, number of gadolinium-enhancing brain lesions at 1 year, and time to 3-month confirmed disability progression. Other secondary endpoints included whole brain volume loss. Exploratory endpoints were cortical gray matter and thalamic volume changes at 1 year. Annualized relapse rate at 1 year was 0.241 for ozanimod 0.5 mg, 0.181 for ozanimod 1 mg, and 0.35 for IFNβ-1a. The reduction in annualized relapse rate over 1 year was 31% ($P=.0013$) for ozanimod 0.5 mg and 48% ($P<.0001$) for ozanimod 1 mg relative to IFNβ-1a 30-μg group. Both ozanimod at 0.5 mg (25%; $P=.00032$) and 1 mg (48%; $P<.001$) doses demonstrated a significant reduction in the number of new or enlarging T2 brain lesions over 1 year compared to IFNβ-1a. The number of gadolinium-enhancing brain lesions at 1 year were significantly reduced with ozanimod 0.5 mg (34%; $P=.0182$) and 1 mg (63%; $P<.0001$) compared to IFNβ-1a. There was no statistically significant difference between ozanimod dose and IFNβ-1a for time to 3-month confirmed disability progression. For the secondary endpoint, change in whole brain volume at 1 year, there was a 12% relative reduction ($P=.0615$) at the 0.5 mg ozanimod dose and a 33% relative reduction ($P<.0001$) for 1.0 mg ozanimod compared to patients treated with IFNβ-1a. There was a 61% ($P<.0001$) and 84% ($P<.0001$) relative reduction in cortical gray matter loss at 0.5 and 1.0 mg doses of ozanimod, respectively, and a 34% ($P<.0001$) and 39% ($P<.0001$) relative reduction in thalamic volume loss at 0.5 and 1.0 mg ozanimod doses, respectively, compared to IFNβ-1a.

Thus, the results of these two pivotal studies of daily oral ozanimod 0.5 or 1 mg indicate superior efficacy compared to intramuscular IFNβ-1a for the treatment of relapsing-remitting MS. Only the 1 mg ozanimod dose achieved a statistically significant improvement for the primary outcome, annualized relapse rate. Although both doses demonstrated significant superiority in several key MRI measures of MS disease activity, it was evident that the 1 mg ozanimod dose was numerically superior, and it is currently the approved dose. Although neither dose of ozanimod was superior to IFNβ-1a in reducing rate of whole brain atrophy over either the 1- or 2-year duration trials, significant reduction in cortical gray matter, and thalamic volume loss was observed at both doses. Future studies focusing on cognitive functioning and patient reported outcome status may further clarify the importance of reduced loss of cortical gray and thalamic volume in patients treated with ozanimod.

Ponesimod clinical trials

Ponesimod has been studied in one phase III clinical trial, OPTIMUM, a 108-week multicenter, randomized, double-blind, parallel group, and active controlled study, evaluating the efficacy, safety, and tolerability of daily oral ponesimod 20 mg versus daily oral teriflunomide 14 mg in adults with relapsing-remitting MS.[157] OPTIMUM enrolled 1133 patients with relapsing-remitting MS, randomized to receive ponesimod 20 mg or teriflunomide 14 mg. Inclusion criteria included ages 18–55 years, Expanded Disability Status Scale 0.0–5.0, and active disease in the 24 months prior to screening. The primary endpoint was annualized relapse rate over 108 weeks. Secondary endpoints included change from baseline to week 108 in fatigue-related symptoms as measured by the symptoms domain of the Fatigue Symptoms and Impacts Questionnaire-Relapsing MS (FSIQ-RMS), and the mean number of combined unique active MRI lesions per year, defined as new gadolinium-enhancing lesions, and any new or enlarging T2 lesions. An additional key secondary outcome was time to 12-week and 24-week confirmed disability progression. Sixty-five percent of the patients were female with an average age of 36.7 years, and 97% were Caucasian. Baseline characteristics were well matched in the two groups: Expanded Disability Status Scale and presence of gadolinium-enhancing lesions on MRI.

There was a 30.5% reduction ($P = .0003$) in annualized relapse rate in ponesimod-treated patients (0.202 vs. 0.290) compared to teriflunomide-treated patients. Ponesimod also demonstrated superior improvement in fatigue scores at week 108 compared to teriflunomide as measured by the FSIQ-RMS weekly symptom score (mean difference − 3.57; $P = .0019$). There was a 56% relative reduction in combined unique active lesions on MRI per year (1.405 vs. 3.164) in patients treated with ponesimod. No significant differences between ponesimod and teriflunomide were seen for 12-week or 24-week confirmed disability progression. Ponesimod is awaiting approval as of this writing.

Natalizumab (anti-α4-integrin IgG)

Natalizumab, a highly efficacious therapeutic agent with a novel and well-described mechanism of action, was approved in 2006 for the treatment of relapsing MS. Natalizumab is a recombinant, humanized monoclonal antibody produced in murine myeloma cells, which is a selective adhesion-molecule inhibitor.[158] Natalizumab is specifically directed against the α4 subunit of the α4β1 (also known as very late antigen-4; VLA-4) and α4β7 integrins. This section presents some basic biology of the α4-integrin and the clinical development of this therapeutic agent for relapsing MS.

α4β1-integrin biology

Integrins are part of a large family of cell adhesion molecules expressed widely by different cell types.[159,160] Initial discovery of their function emerged from research in the 1980s and early 1990s demonstrating their role in embryogenesis, organogenesis, wound healing, and tumor invasion.[161,162] Integrins are heterodimer transmembrane glycoproteins comprising single α and β subunits that direct specific cell-cell or cell-matrix binding.[163] To date, 18 distinct α and eight distinct β subunits have been identified that combine to form 18 different heterodimers identified.[164] In their inactivated resting state, integrin subunits lie bent on the cell surface, but following activation by chemokines, they undergo a conformational change that converts them to a functioning adhesion molecule that will couple to its respective counter-receptor on the surface of endothelial cells.[165–168] Mononuclear leukocytes express α4β1 integrin, and it is this integrin type that promotes lymphocyte and monocyte migration across postcapillary venule walls during CNS inflammation.[163,169–171] During inflammation, cell adhesion is initiated by a slowing and rolling of leukocytes over the endothelial surface, mediated by selectins E and P expressed on the endothelial cell surface. Activation of leukocytes by both E- and P-selectin and chemokine release from endothelial cells leads to increased L-selectin expression on the leukocyte surface, which in turn results in progressive slowing of leukocyte rolling velocity due to binding of the selectins to their respective counter-receptors.[172–176] Pro-inflammatory cytokines, expressed by endothelial cells, activate integrins to produce tight binding of leukocytes and subsequent crawling over the surface of endothelial cells.[177,178] Using direct intravital microscopy, it has been demonstrated that P-selectin and α4 integrin have a direct synergistic action in promoting leukocyte adhesion to the endothelial luminal surface wall.[179] Furthermore, it is the α4β1-vascular cell adhesion molecule (VCAM-1) interaction that directly leads to leukocyte adhesion to the endothelial cell surface.[180,181] As noted previously, the counter-receptor for α4β1 is VCAM-1, and is expressed on the luminal endothelial surface.[182] Once tight binding is achieved, the cell migrates over the endothelial surface, under β1 guidance, to sites, usually intercellular junctions, which are then degraded by the action of inducible matrix metalloproteinases, facilitating leukocyte migration via extravasation into the CNS parenchyma.[160–162,176,183–185]

Utilizing an in vitro cell adhesion assay on tissue sections of blood vessels from the brains of mice with EAE, only antibodies directed against α4β1 integrin succeeded in blocking the binding of lymphocytes and monocytes to the blood vessel walls.[186] In mouse EAE models, only antigen-specific lymphocytes expressing α4β1 were capable of entering the CNS, and α4-integrin blocking antibodies prevented the accumulation of lymphocytes into the CNS of animals with EAE and reduced the classical symptoms of EAE.[186–190] Natalizumab binds to the α4 subunit of the α4β1 integrin, blocking its access to its endothelial adhesion molecule VCAM-1, thus inhibiting immune cell trafficking into the CNS in EAE.[186,191]

While the inhibition of lymphocyte migration may be widely regarded as the major mechanism by which natalizumab reduces cell-mediated inflammatory damage and prevents new and/or enlarging CNS perivenular plaque formation in MS, natalizumab may further reduce CNS inflammation by inhibiting osteopontin adherence to the α4β1 site on lymphocytes, monocytes, natural killer (NK) cells, and macrophages. Although osteopontin biology and its possible inhibition by natalizumab was not incorporated into the body of preclinical translational research leading to its clinical development, natalizumab-osteopontin interaction may prove to be such an important mechanism of natalizumab efficacy in MS that we would be remiss by not incorporating it into this chapter. Osteopontin is an extracellular matrix protein, which acts as a pro-inflammatory cytokine, expressed in activated T cells, dendritic cells (DC), macrophages, and natural killer (NK) cells and promotes an increase in IL-12, IL-17, and IFNγ.[192–194] Osteopontin is modulated by thrombin-induced cleavage into N- and C-terminal fragments.[195] The N-terminal fragment expresses an α4β1 integrin-binding domain. Osteopontin blocks activated T cell apoptosis via its actions on the transcription factor NFκB, and promotes pro-inflammatory T cell differentiation toward Th1 and Th17 polarization while increasing production of IL-12 and IFNγ.[193,196–202] Consistent with the key role of osteopontin cleavage by thrombin, pretreatment with a thrombin inhibitor, hirudin, reduces disease severity, demyelination, and Th1/Th17 cytokine expression in EAE.[203,204] In MS and EAE lesions, there is a perivenular deposition of osteopontin, potentially helping to recruit lymphocytes and monocytes into areas of CNS inflammation via its interaction with cell surface α4β1.[205] In EAE, osteopontin expression is increased on neurons and microglia in inflammatory lesions. Furthermore, in osteopontin knockout animals or following treatment with antiosteopontin IgG, there is suppression of EAE. Patients with all clinical subtypes of MS (except clinically isolated syndrome) exhibited increased osteopontin concentration in CSF, and to a more variable extent in plasma particularly in patients with relapsing active disease.[194,206–211] Elevated CSF osteopontin is correlated with increased Expanded Disability Status Scale scores and increased gadolinium-enhancing brain lesions, as well as increased relapse risk.[193,210,212]

Natalizumab treatment reduces CSF osteopontin concentration, and plausibly, may compete with osteopontin for binding to α4β1 integrin expressed on circulating pro-inflammatory cells, not only blocking cell migration into the CNS, but also blocking osteopontin-induced activation of lymphocytes, monocytes, dendritic cells, macrophages, and NK cells.[213,214] Thus, the reduction of disease activity and expression of pro-inflammatory markers within the CNS may be the result of a binary effect of natalizumab, both blocking the traditionally recognized migration of pro-inflammatory cells into the CNS, and also by impairing osteopontin-integrin interaction thereby reducing the overall pro-inflammatory cascade driven by those cells that do enter the CNS.

Clinical trials of natalizumab

In an exploratory clinical study, 72 patients with active relapsing MS or secondary progressive MS received two intravenous (IV) infusions of natalizumab or placebo 4 weeks apart, and followed for 24 weeks.[215] The primary outcome was accumulated brain MRI lesion count. The natalizumab-treated group showed fewer gadolinium-enhancing lesions compared to placebo-treated patients (mean 1.6 vs. 3.3 per patient; $P=.017$) during first 12 weeks, but no difference was seen between the two groups at weeks 12–24. Fewer new or active T2 lesions were also seen in the natalizumab-treated patients (1.8 vs. 3.6 per patient; $P=.042$) over the first 12 weeks. The study was not designed or powered to address clinical efficacy.[12]

A randomized, double-blind, phase II study tested natalizumab in 213 patients with relapsing MS or relapsing secondary progressive MS.[158] Patients were randomized to receive natalizumab at 3 mg/kg, 6 mg/kg, or placebo IV every 28 days. The primary outcome measure was the number of new gadolinium-enhancing MRI lesions per patient at 6 months. Secondary clinical outcomes included relapse number and the patient-reported Visual Analog Scale (VAS) of Degree of Well-Being. There was a significant reduction in the mean number of new gadolinium-enhancing lesions in both natalizumab treatment groups (9.6 per patient in the placebo group, vs. 0.7 in the 3 mg/kg and 1.1 in the 6 mg/kg groups; $P = .001$ for both). There was a reduction in number of relapses in both natalizumab treatment groups (13 in the 3 mg/kg and 14 in the 6 mg/kg group) compared to placebo (27; $P = .02$ for both). When the relapses were analyzed solely by the more rigorous, objectively documented changes on Expanded Disability Status Scale subscores, there were 18 relapses in the placebo group compared to three in the 3 mg/kg group and eight in the 6 mg/kg group ($P = .004$). The number of relapses requiring steroid treatment also differed significantly (81% of placebo subject relapses, vs. 38% of natalizumab-treated patient relapses; $P = .002$). On the 100-point VAS, subjects given placebo had a mean worsening of 1.38 points, whereas both natalizumab groups reported a mean improved score—9.49 in the 3 mg/kg group, and 6.21 in the 6 mg/kg group ($P = .04$ and .03, respectively). No significant changes between treatment groups were seen in mean Expanded Disability Status Scale scores.

A randomized, double-blind, phase III, placebo-controlled trial enrolled 942 patients (627 on natalizumab 300 mg every 28 days vs. 315 on placebo for 2 years). The clinical relapse rate was assessed at 1 year and the rate of sustained progression of Expanded Disability Status Scale-rated disability assessed at 2 years.[215] There was a 68% relative reduction in the relapse rate in natalizumab-treated patients at 1 year compared to placebo ($P < .001$), and a reduction of sustained Expanded Disability Status Scale progression risk by 42% at 2 years in the natalizumab-treated group ($P < .001$). MRI outcomes demonstrated a 92% relative reduction in gadolinium-enhancing lesions at 1 and 2 years in natalizumab-treated patients compared to placebo, and reduced accumulation of T2 lesions by 83% over 2 years ($P < .001$). Based on these efficacy results and a favorable side effects profile, natalizumab was approved in 2006 as the first therapy in its class for patients with relapsing MS. The unexpected occurrence of three cases of progressive multifocal leukoencephalopathy (PML) led to its temporary withdrawal only 3 months after its approval. The long and complex tale of PML and natalizumab in patients with MS is outside the scope of this chapter, and is reviewed extensively elsewhere.[216,217]

Natalizumab is an excellent example of the power of translational research, evolving directly from preclinical research to clinical trials to approval. Basic mechanistic studies revealed important insights into lymphocyte trafficking and adhesion through cell-surface integrins. Through the systematic in vitro screening of EAE tissue, natalizumab emerged from a host of other monoclonal antibodies as the most effective inhibitor of lymphocyte adhesion through blockade of α4-integrin. Furthermore, as in the case of both glatiramer acetate and the S1PR modulators, postapproval basic research has expanded and enriched our understanding of potential multiple therapeutic mechanisms that were not readily evident during the crucial translational research demonstrating the therapeutic potential of these agents.

Elezanumab (repulsive guidance molecule A inhibitor)

Nerve-axon regeneration through the inhibition of repulsive guidance molecule A-neogenin signaling

Despite the remarkable growth in the number of immune-modifying and immune-suppressing agents that have been developed and have significantly improved the prognosis for MS patients over the past 25 years, the majority of patients still experience residual fixed neurological impairment, which has important functional consequences for their quality of life. Although poorly understood, and only partially attributable to acute and chronic inflammatory injury, progressive brain and spinal cord atrophy due to the loss of neurons, axons, and synaptic relations is a fundamental feature of MS. This is true even for patients in whom, by conventional measures, there is seeming early control of the inflammatory features of the disease. Thus, even "successfully" treated patients are typically left with significant and permanent structural impairment that may worsen over time. The need to remyelinate axons in order to restore their functional and physiological integrity, and more fundamentally, the need to regenerate axonal projections and synaptic integration, remain critical unaddressed needs for MS patients.

Repulsive guidance molecule A, neogenin, and axonal regeneration

In 1990, Stahl reported the presence of highly expressed repulsive proteins in the embryonic chick tectum.[218] One of the first molecules identified that appeared to impact axonal regeneration was repulsive guidance molecule a (RGMa), both a soluble and membrane-bound glycoprotein found on numerous cell types.[219,220] First described in 2002, RGMa promotes topographic directional guidance of retinal ganglion cells to the proper site in the tectum of chick embryos.[220] Whereas RGMa appears to promote neuronal differentiation in the embryo, RGMa also governs axonal guidance, via repulsion, according to its concentration gradient within the tectum.[221] However, RGMa interaction with its membrane-bound receptor, neogenin, also prevents axonal fiber growth by causing collapse of the neuronal growth cone.[221,222] Consistent with these findings, blocking RGMa gene expression promotes axonal projection and growth, demonstrating that the distribution of RGMa and its receptors selectively and topographically determines where axons proliferate, and promotes growth into areas lacking RGMa expression.

In the adult rat, neogenin is expressed on neurons and oligodendroglia in the spinal cord and RGMa is upregulated following spinal cord injury. This evidence combined with prior research suggested that RGMa-neogenin interaction might prevent axonal regeneration.[223–225] Within hours following a spinal cord injury there is expression of RGMa at the lesion site, in both oligodendroglia and microglia.[225] Since RGMa had been shown to inhibit neurite outgrowth, it was reasoned that an anti-RGMa antibody might inhibit RGMa and promote axonal regeneration after injury. In cultured neurons, RGMa antibodies partially reversed the inhibitory effects of myelin-containing RGMa upon neurite growth. In rats made paraplegic by focal thoracic spinal cord lesions, functional recovery was improved by in vivo administration of RGMa antibodies. Postmortem examination of spinal cords 10 weeks after initiation of RGMa antibody treatment revealed increased nerve fiber counts, both proximal and distal to the lesion site. In contrast, animals not treated with RGMa antibody had no fibers present distal to the lesion. Furthermore, in RGMa antibody-treated rats, axons regenerated into and through the injury site scar.[225]

It has been demonstrated that lesion-infiltrating microglia inhibit injury-induced axonal regeneration, and that microglia infiltrating sites of CNS injury express RGMa.[225–227] Neurons cocultured with activated, RGMa-expressing microglia demonstrated direct contact-induced collapse of growth cones and reduced neurite growth. This inhibition was reversed by coculture with RGMa antibody, by blocking RGMa gene expression, or by blocking microglial activation by pretreatment with minocycline.[228–230] Inhibition of neogenin expression using siRNA also reversed the inhibitory effects of activated microglia on neurite sprouting and growth cones, demonstrating that RGMa-neogenin signaling in microglia was required for the inhibition of axonal regeneration following injury.[228]

Preclinical and translational research supporting repulsive guidance molecule A inhibition in EAE

Because RGMa and neogenin have an inhibitory effect on axonal regeneration, they have been investigated in several preclinical models of MS. In a MOG-induced mouse EAE model, neogenin was upregulated in pro-inflammatory CD4+ T cells. Treatment with anti-RGMa antibody reduced T cell proliferative responses, reduced the production of the pro-inflammatory cytokines IFNγ, IL-2, IL-4, and IL-17, and reduced neurological impairment as measured by EAE score,[231] suggesting a potentially important role for RGMa signaling in CD4+ T cells in the pathogenesis of MS. In the nonobese diabetic mouse EAE model, secondary progressive neurological impairment was significantly reduced by IV RGMa antibody treatment and was associated with reduced axonal degeneration and reduced demyelination in the spinal cord.[232] In a focal spinal cord model of EAE, IV RGMa antibody administration reduced the size of the inflammatory lesion and was associated with axonal preservation, reduced demyelination, increased axonal sprouting, and improved motor recovery compared to control mice.[232,233] Collectively, these preclinical studies provide evidence that inhibition of RGMa may potentially provide a therapeutic strategy for the restoration of neuronal structure and function in MS.

In human autopsy specimens of progressive MS, RGMa expression was identified in MS plaques, as well as in normal appearing white matter and normal appearing gray matter. RGMa expression was found on macrophages, microglia, CD4+ T cells, and brain parenchyma, and soluble RGMa was identified in cerebrospinal fluid.[234] In a small number of patients with progressive MS treated with intrathecal corticosteroids, a reduction in soluble RGMa in CSF was associated with a reduction in disability level. Furthermore, a recent study identified two single nucleotide polymorphisms in the disease-associated region of the human RGMa gene that were positively associated with increased expression of the pro-inflammatory cytokine IFNγ.[235] Evidence from in vivo studies in two murine MOG-induced EAE models demonstrated that the use of ABT-555 (later renamed elezanumab) to block RGMa signaling reduced inflammation in the spinal cord and promoted axon regeneration.[236]

Given the wealth of preclinical evidence suggesting an important potential etiological role of RGMa and neogenin signaling in preventing neurite growth, remyelination, and possibly promoting apoptosis, combined with the findings of RGMa expression in MS patient lesions and CSF, RGMa regulation would appear to be a plausible pathway for MS treatment, both from the perspective of reducing disease activity and promoting CNS repair.

Clinical trials with elezanumab

Elezanumab is a fully humanized inhibitory monoclonal antibody directed against RGMa. To date, two phase I clinical trials have been conducted evaluating safety. The first of these was a randomized, double-blind, placebo-controlled trial, including 47 healthy controls in the United States.[237] This trial employed a single IV elezanumab infusion of 50, 150, 450, 1000, or 1600 mg or 150 mg subcutaneously (SC). The most common side effects include headache, infusion pain, and nausea. Two serious adverse events were reported that were considered unrelated to the study medication: a 34-year-old male who had received 150 mg elezanumab IV was found dead at home 118 days later and had a positive drug screen for amphetamines, methocarbamol, cyclobenzaprine, fentanyl, and 5-MIO-MiPT. The second death occurred of a 27-year-old male 22 days after receiving a dose of 1600 mg and was attributed to cocaine and oxycodone overdose.

The second phase I trial, which evaluated safety and CSF biomarkers, was a 20-patient, randomized, double-blind, and placebo-controlled examination of escalating doses of elezanumab.[238] There were three elezanumab treatment groups and one placebo group with five patients per group, 18 of whom had relapsing MS and 2 had secondary progressive MS. Patients on elezanumab treatment received 300, 1800, or 3600 mg, half of which was given as a loading dose with the remainder given in four equal doses, once every 4 weeks. The most common adverse event reported was headache. The only serious adverse event reported was an MS relapse, occurring 66 days after the last dose of study drug, deemed unrelated to elezanumab treatment. No deaths occurred in this trial and no adverse event led to discontinuation of the study medication.

In all, 13 patients had CSF samples obtained at baseline and at day 113. There was an elezanumab dose-dependent reduction in free RGMa in the CSF—the placebo group had 22.71 ng/mL compared to 11.40 ng/mL for the patients receiving 1800 mg of elezanumab. The antiinflammatory cytokine IL-10 was increased in the CSF of elezanumab-treated patients, with a 63% increase in patients receiving the 1800 mg dose compared to their baseline levels. No significant changes were reported in Expanded Disability Status Scale scores, not unexpected given the short duration of the trial and the small patient population. No clinically significant laboratory abnormalities were observed in either study, but transient increases in blood pressure during infusions were experienced by some patients.

Given the favorable safety data of the two phase I clinical trials, two phase II clinical trials are currently enrolling patients. One clinical trial, with a target enrollment of 90 adults with the progressive forms of MS, will be randomized 1:1:1 to receive an IV infusion of 400 mg, 1800 mg, or placebo every 4 weeks for 52 weeks (NCT03737812). The second phase II trial seeks to enroll 165 patients with relapsing MS, and will employ the same study design as the progressive MS study (NCT03737851). The primary endpoint of both studies is the mean overall response score at week 52, a composite score that consists of four components: Expanded Disability Status Scale, Timed 25 Foot Walk, and the 9-Hole Peg Test in both the dominant and nondominant hands. The secondary endpoints are 12-week and 24-week confirmed disability improvement response rate on the Expanded Disability Status Scale, Overall Response Score at weeks 12, 24, and 36, and time to first improvement on 12-week confirmed Expanded Disability Status Scale, Timed 25 Foot Walk, or 9-Hole Peg Test (either hand). MRI endpoints include T2 lesion counts and volumes, atrophy measures, magnetization transfer ratio to assess myelination, and diffusion tensor imaging to assess nerve tract integrity.

Opicinumab (anti-LINGO-1 IgG)

Demyelination of the CNS is a hallmark of MS, and as previously noted, is driven by inflammatory damage to oligodendroglia-derived myelin sheaths and largely due to the actions of autoreactive lymphocytes, monocytes, and microglia, as well as proinflammatory cytokines such as TNFα, IFNγ, IL-17, and nitric oxide species.[239] Although there may be some myelin repair after cessation of an inflammatory attack, it is typically incomplete.[240] Remyelination depends on the growth, maturation, and differentiation of the oligodendroglial progenitor cell into mature oligodendroglia.[241] Despite an abundance of oligodendroglial progenitor cells in regions of demyelination, they appear to be negatively regulated, and thus inhibited from maturating into myelin-producing oligodendroglia.[242] Although a number of molecules, including myelin breakdown products such as myelin-associated glycoprotein (MAG) and MBP may upregulate inhibitory mechanisms, a great deal of interest has been generated in the neurite outgrowth inhibitor (Nogo)-leucine-rich repeat and immunoglobulin 1 domain

(LINGO-1) receptor complex.[243–245] This section details the preclinical and translational research on LINGO-1 in MS.

LINGO-1 expression

Although the full NoGo-LINGO-1 co-receptor system is expressed on neurons, the LINGO-1 domain is expressed only on oligodendroglial progenitor cells/oligodendroglia as a transmembrane glycoprotein, which becomes upregulated in a variety of CNS demyelinating diseases and injuries including MS.[243–246] LINGO-1 negatively regulates differentiation and maturation of oligodendroglial progenitor cells into myelin-producing oligodendroglia, and its expression diminishes oligodendroglial progenitor cell potential to reach maturation.[243–247] Although LINGO-1 has a transmembrane segment and a cytoplasmic tail, recent evidence indicates that its role in negatively regulating oligodendroglial progenitor cells/oligodendroglia is solely accomplished by the extracellular surface domain of the molecule.[247] Oligodendroglial progenitor cells expressing only the extracellular portion of LINGO-1 blocked oligodendroglial progenitor cell differentiation and maturation and the addition of extracellular soluble extracellular LINGO-1 domain to oligodendroglial progenitor cells was sufficient to block their maturation into MBP-producing oligodendroglial progenitor cells. Furthermore, transgenic astrocytes expressing only the extracellular portion of LINGO-1 were able to inhibit pro-oligodendroglia maturation and myelin formation.[247] These observations support the proposal that cell-to-cell trans-presentation of the extracellular LINGO-1 domain is sufficient to negatively regulate the maturation of oligodendroglial progenitor cells into myelin-producing cells. The Rho-GTPase pathway negatively regulates oligodendroglial progenitor cell maturation and remyelination, and LINGO-1, by trans-presentation or self-presentation, may activate the RhoA-linked system of GTPases, thus inhibiting oligodendroglial progenitor cell maturation and remyelination.[243, 246, 248, 249]

Axonal remyelination through LINGO-1 inhibition

There is growing preclinical evidence that LINGO-1 inhibition, in various in vivo and in vitro inflammatory and noninflammatory models of demyelination, may promote and partially restore myelination, and improve neural function. The discussion that follows will present this data.

In vitro inhibition

Using purified cell cultures, LINGO-1 was identified on rat oligodendroglia, and LINGO-1 mRNA was expressed during all stages of maturation of oligodendroglial progenitor cells and oligodendroglia.[243] Addition of LINGO-1 inhibitory RNA or dominant-negative LINGO-1 DNA resulted in the enhanced differentiation and maturation of oligodendroglia. Blocking the LINGO-1 receptor complex with a

LINGO-1-Fc fusion protein also promoted the maturation of oligodendroglial progenitor cells, while increasing myelin production and reducing RhoA-GTP production.[243]

Overexpression of LINGO-1 resulted in marked reduction of oligodendroglia maturation. However, the addition of an anti-LINGO-1 antibody to dorsal ganglion and oligodendroglial progenitor cells, corpus callosum, and cerebellar slice oligodendroglia cultures resulted in increased myelination, expression of MBP, and axonal internode MBP expression, as well as a marked increase in the number of mature oligodendroglia and MBP clusters without promoting oligodendroglial progenitor cell proliferation.[245, 250]

Inflammatory demyelination-EAE models

Anti-LINGO-1 antibody-treated EAE rats had increased remyelination of optic nerves, reduced clinical severity of disease, and improved EAE scores.[245] In LINGO-1 knockout mice, improved EAE scores correlated with the degree of remyelination, increased number and density of myelinated fibers.[244] Although CNS penetration of anti-LINGO-1 was limited (AUC 0.15% in CNS vs. blood), anti-LINGO-1 did penetrate the CNS and the degree of CNS binding correlated with the degree of remyelination.[251] Postmortem MRI diffusion tensor imaging of spinal cords from anti-LINGO-1 treated mice with EAE demonstrated better axonal preservation and reduced amyloid β-precursor protein content relative to control mice, which would indicate better axonal health.[250]

To demonstrate that LINGO-1 inhibition strategies did not operate via the immune system in EAE models, one study examined the function of T cells harvested from LINGO-1 knockout and wild-type mice with EAE.[244] The T cells from either animal group proliferated in response to antigen stimulation with MOG peptides, accompanied by pro-inflammatory cytokine production. Furthermore, T cells transferred from either knockout or wild-type EAE mice into naïve mice resulted in the development of EAE. This demonstrates that the reduction in EAE and restoration of myelination observed with LINGO-1 inhibition was not due to alterations in peripheral immune function.[244]

Noninflammatory demyelination model

Using a noninflammatory lysolecithin model of spinal cord demyelination, treatment with anti-LINGO-1 immunoglobulin or LINGO-1 germline knockout resulted in enhanced CNS functional recovery as measured by improved amplitude and reduced latency of transcranial magnetic motor-evoked potentials, when compared to the use of a control immunoglobulin or wild-type mice.[252] Histological examination revealed reduced area of demyelinating lesions, thicker myelin sheaths, a larger number of myelinated axons, and increased myelinated axon densities in the knockout and anti-LINGO-1 treated animals.[252, 253] Conversely, upregulation of LINGO-1 inhibited oligodendroglial progenitor cell differentiation, and reduced the number of MBP+ cells, indicating reduced myelination.[250] In lysophosphatidylcholine- and cuprizone-induced nonautoimmune models of demyelination, there was no increase in LINGO-1-expressing oligodendroglial progenitor cells, but anti-LINGO-1 antibody treatment resulted in a reduced number of LINGO-1-positive oligodendroglial progenitor cells and increased mature, MBP-expressing oligodendroglia in the spinal cord.[245] Morphological examination demonstrated that anti-LINGO-1 treatment greatly reduced spinal cord lesion size, accompanied by more MBP deposition and more normal-appearing myelinated axons in the lesions.

Clinical trials with opicinumab

RENEW

Encouraged by the results of LINGO-1 inhibition in vitro and in vivo models of rodent optic neuritis and EAE, a fully human monoclonal anti-LINGO-1 antibody, opicinumab, has been used in a clinical trial of new onset optic neuritis.[254] In this randomized, double-blind, placebo-controlled, parallel-group, phase II RENEW trial, 82 patients received opicinumab 100 mg/kg or placebo IV every 4 weeks for six infusions. Full-field visual evoked potential latency measurement was the primary outcome measure. Secondary outcomes were retinal nerve fiber thickness, measured by spectral domain orbital coherence tomography, and low contrast letter visual acuity. Although nerve conduction improved in opicinumab-treated relative to placebo-treated patients, as measured by reduced latency, there was neither improvement in low contrast letter visual acuity nor an increase in retinal nerve fiber layer thickness as measured by orbital coherence tomography.

SYNERGY

The phase II, randomized, placebo-controlled, double-blind SYNERGY trial examined opicinumab at 3–100 mg/kg IV every 4 weeks for up to 72 weeks as add-on therapy to IFNβ-1a in 466 patients with active relapsing MS or secondary progressive MS.[255] A total of 334 patients completed the study. The primary outcome, a 3-month confirmed improvement in a functional composite, the Overall Response Score—comprising the Expanded Disability Status Scale, Timed 25 Foot Walk, the 9-Hole Peg Test with both dominant and nondominant hands, and 3-s Paced Auditory Serial Addition Test—compared to placebo-treated patients, was not achieved. The study also failed to meet its secondary outcome, 3-month reduction in confirmed disability progression. However, exploratory analysis of brain MRI appeared to show that 10 mg/kg opicinumab-treated patients had lower diffusion tensor imaging values relative to the population median, which could indicate biological efficacy.[256]

AFFINITY

The placebo-controlled phase II trial (AFFINITY) is evaluating IV opicinumab 750 mg versus placebo every 4 weeks, as add-on therapy for patients with relapsing MS currently receiving a disease-modifying therapy (IFNβ, glatiramer acetate, dimethylfumarate, or natalizumab). The primary objective of this study is to evaluate the efficacy of opicinumab versus placebo over 72 weeks as measured by sustained disability improvement, employing the Overall Response Score. Secondary outcomes will include additional clinical functional measures, a large number of patient-reported outcomes, traditional MRI measures, as well as the use of magnetization transfer ratio to assess myelin repair.

There are currently no agents that have proven efficacy in CNS repair and/or regeneration, yet, along with control of pathogenesis, these are crucial challenges to the future of MS management. Extensive preclinical research has served as the translational basis for bringing both elezanumab and opicinumab to clinical efficacy trials. Hopefully they are just the beginning of a broadening attack on the issues of CNS regeneration and repair. A deepening understanding of the molecular pathways that lie at the heart of nerve growth and differentiation and deeper understanding of the bases of the biological responses to injury will be required if the scientific and clinical research community, and most importantly their patients, are to experiences the necessary successes to complement the progress thus far achieved for attacking MS pathogenesis.

Summary

Great progress has been made since the first therapy, IFNβ-1b, was approved for the chronic treatment of MS more than 25 years ago. There are now 16 disease-modifying therapies that have demonstrated significant efficacy in the treatment of MS, primarily the relapsing form of the disease. The increase in number and relative efficacy of these agents has largely been the result of growing knowledge of the cells, messenger molecules, receptors, and downstream effects of ligand-receptor interactions that drive immune system disequilibrium that is responsible for many, but by no means all, of the pathogenic pathways responsible for much of the clinical profile of MS. Growing knowledge of genetic regulation, the impact of nucleotide polymorphisms upon determining risk, and the burgeoning wealth of data on disease biomarkers, are among the key areas in which important progress will be forthcoming. However, the basis of therapy development repeatedly evolves from intense, persistent, and informed research that exploits in vivo disease models developed in nonhuman animals, as well as in vitro and ex vivo experimentation. The majority of MS therapeutics presented in this chapter all depend heavily on key insights provided by carefully designed preclinical studies and meticulous data analysis, to provide critical information about future therapeutic agents and expand our understanding of agents already in clinical use. There was a notable absence of a number of MS therapeutic agents covered in this chapter, which should not be interpreted as being indicative of inferior efficacy, but simply because there was not a substantial body of prior translational research, related to MS, that fostered their development as therapeutic agents. Although there have been significant benefits in MS treatment arising from repurposing medications previously developed for other diseases, such as cladribine and alemtuzumab, we should continue to advocate for the promotion of the broadest, multidisciplinary preclinical research—biochemical, physiological, pharmacological, toxicological, and genetic—married to the most effective computer-assisted data search engines to assure continuing rational strategies for therapeutic agent development.

Disclosures

SLC has received research support, has served as a paid consultant on advisory boards, or has received speaking honoraria from AbbVie, Adamas, Alithios, Biogen, Celgene, MedDay, NARCRMS Registry, Novartis, Pear Therapeutics, Roche Genentech, Sage Therapeutics, and Sanofi Genzyme.

EBL has received research support and/or has served as a paid consultant on advisory boards or lecture honoraria from Biogen, EMD Serono, Novartis, Roche Genentech, and Sanofi Genzyme.

KES has received research support and/or has served as a paid consultant on advisory boards or lecture honoraria from Biogen, EMD Serono, Novartis, Roche Genentech, and Sanofi Genzyme.

KKR is a medical director and employee of Atara Biotherapeutics Inc.

MR and SNL have no potential conflicts to disclose.

References

1. Teitelbaum D, Meshorer A, Hirshfeld T, Arnon R, Sela M. Suppression of experimental allergic encephalomyelitis by a synthetic polypeptide. *Eur J Immunol.* 1971;1:242–248.
2. Teitelbaum D, Webb C, Meshorer A, Arnon R, Sela M. Suppression by several synthetic polypeptides of experimental allergic encephalomyelitis induced in guinea pigs and rabbits with bovine and human basic encephalitogen. *Eur J Immunol.* 1973;3:273–279.
3. Johnson KP, et al. Copolymer 1 reduces relapse rate and improves disability in relapsing-remitting multiple sclerosis: results of a phase III multicenter, double-blind placebo-controlled trial. The copolymer 1 multiple sclerosis study group. *Neurology.* 1995;45:1268–1276.
4. Lisak RP, Zweiman B, Blanchard N, Rorke LB. Effect of treatment with copolymer 1 (Cop-1) on the in vivo and in vitro manifestations of experimental allergic encephalomyelitis (EAE). *J Neurol Sci.* 1983;62:281–293.
5. Teitelbaum D, et al. Suppression of experimental allergic encephalomyelitis in Rhesus monkeys by a synthetic basic copolymer. *Clin Immunol Immunopathol.* 1974;3:256–262.
6. Lalive PH, et al. Glatiramer acetate in the treatment of multiple sclerosis: emerging concepts regarding its mechanism of action. *CNS Drugs.* 2011;25:401–414.
7. Neuhaus O, Farina C, Wekerle H, Hohlfeld R. Mechanisms of action of glatiramer acetate in multiple sclerosis. *Neurology.* 2001;56:702–708.
8. Vieira PL, Heystek HC, Wormmeester J, Wierenga EA, Kapsenberg ML. Glatiramer acetate (copolymer-1, copaxone) promotes Th2 cell development and increased IL-10 production through modulation of dendritic cells. *J Immunol.* 2003;170:4483–4488.
9. Duda PW, Schmied MC, Cook SL, Krieger JI, Hafler DA. Glatiramer acetate (Copaxone®) induces degenerate, Th2-polarized immune responses in patients with multiple sclerosis. *J Clin Invest.* 2000;105:967–976.
10. Neuhaus O, et al. Multiple sclerosis: comparison of copolymer-1- reactive T cell lines from treated and untreated subjects reveals cytokine shift from T helper 1 to T helper 2 cells. *Proc Natl Acad Sci U S A.* 2000;97:7452–7457.
11. Begum-Haque S, et al. Increased expression of B cell-associated regulatory cytokines by glatiramer acetate in mice with experimental autoimmune encephalomyelitis. *J Neuroimmunol.* 2010;219:47–53.

12. Kala M, et al. B cells from glatiramer acetate-treated mice suppress experimental autoimmune encephalomyelitis. *Exp Neurol.* 2010;221:136–145.
13. Weber MS, Hohlfeld R, Zamvil SS. Mechanism of action of glatiramer acetate in treatment of multiple sclerosis. *Neurotherapeutics.* 2007;4:647–653.
14. Weber MS, et al. Multiple sclerosis: glatiramer acetate inhibits monocyte reactivity in vitro and in vivo. *Brain.* 2004;127:1370–1378.
15. Jung S, et al. Induction of IL-10 in rat peritoneal macrophages and dendritic cells by glatiramer acetate. *J Neuroimmunol.* 2004;148:63–73.
16. Burger D, et al. Glatiramer acetate increases IL-1 receptor antagonist but decreases T cell-induced IL-1 in human monocytes and multiple sclerosis. *Proc Natl Acad Sci U S A.* 2009;106:4355–4359.
17. Toker A, Slaney CY, Bäckström BT, Harper JL. Glatiramer acetate treatment directly targets CD11b +Ly6G- monocytes and enhances the suppression of autoreactive T cells in experimental autoimmune encephalomyelitis. *Scand J Immunol.* 2011;74:235–243.
18. Sellebjerg F, et al. Glatiramer acetate antibodies, gene expression and disease activity in multiple sclerosis. *Mult Scler J.* 2012;18:305–313.
19. Karandikar NJ, et al. Glatiramer acetate (Copaxone) therapy induces CD8 + T cell responses in patients with multiple sclerosis. *J Clin Invest.* 2002;109:641–649.
20. Tyler AF, Mendoza JP, Firan M, Karandikar NJ. CD8 + T cells are required for glatiramer acetate therapy in autoimmune demyelinating disease. *PLoS One.* 2013;8:e66772.
21. Teitelbaum D, Milo R, Arnon R, Sela M. Synthetic copolymer 1 inhibits human T-cell lines specific for myelin basic protein. *Proc Natl Acad Sci U S A.* 1992;89:137–141.
22. Fridkis-Hareli M, Strominger JL. Promiscuous binding of synthetic copolymer 1 to purified HLA-DR molecules. *J Immunol.* 1998;160:4386–4397.
23. Fridkis-Hareli M, et al. Direct binding of myelin basic protein and synthetic copolymer 1 to class II major histocompatibility complex molecules on living antigen-presenting cells—specificity and promiscuity. *Proc Natl Acad Sci U S A.* 1994;91:4872–4876.
24. Aharoni R, et al. Studies on the mechanism and specificity of the effect of the synthetic random copolymer GLAT on graft-versus-host disease. *Immunol Lett.* 1997;58:79–87.
25. Freud AG, et al. A human CD34(+) subset resides in lymph nodes and differentiates into CD56bright natural killer cells. *Immunity.* 2005;22:295–304.
26. Tselis A, Khan O, Lisak RP. Glatiramer acetate in the treatment of multiple sclerosis. *Neuropsychiatr Dis Treat.* 2007;3:259–267.
27. Wiesemann E, et al. Glatiramer acetate (GA) induces IL-13/IL-5 secretion in naive T cells. *J Neuroimmunol.* 2001;119:137–144.
28. Aharoni R, Teitelbaum D, Sela M, Arnon R. Copolymer 1 induces T cells of the T helper type 2 that crossreact with myelin basic protein and suppress experimental autoimmune encephalomyelitis. *Proc Natl Acad Sci U S A.* 1997;94:10821–10826.
29. Aharoni R, Teitelbaum D, Sela M, Arnon R. Bystander suppression of experimental autoimmune encephalomyelitis by T cell lines and clones of the Th2 type induced by copolymer 1. *J Neuroimmunol.* 1998;91:135–146.
30. Chen M, et al. Glatiramer acetate induces a Th2-biased response and crossreactivity with myelin basic protein in patients with MS. *Mult Scler.* 2001;7:209–219.
31. Putheti P, Soderstrom M, Link H, Huang YM. Effect of glatiramer acetate (copaxone) on CD4 +CD25 high T regulatory cells and their IL-10 production in multiple sclerosis. *J Neuroimmunol.* 2003. https://doi.org/10.1016/j.jneuroim.2003.08.001.
32. Hong J, Li N, Zhang X, Zheng B, Zhang JZ. Induction of CD4+CD25+ regulatory T cells by copolymer-I through activation of transcription factor Foxp3. *Proc Natl Acad Sci.* 2005;102:6449–6454.
33. Hussien Y, Sanna A, Söderström M, Link H, Huang YM. Glatiramer acetate and IFN-β act on dendritic cells in multiple sclerosis. *J Neuroimmunol.* 2001;121:102–110.
34. Kim HJ, et al. Type 2 monocyte and microglia differentiation mediated by glatiramer acetate therapy in patients with multiple sclerosis. *J Immunol.* 2004;172:7144–7153.
35. Trinchieri G. Interleukin-12 and the regulation of innate resistance and adaptive immunity. *Nat Rev Immunol.* 2003;3:133–146.

36. Weber MS, et al. Type II monocytes modulate T cell-mediated central nervous system autoimmune disease. *Nat Med.* 2007;13:935–943.
37. Biegler BW, et al. Glatiramer acetate (GA) therapy induces a focused, oligoclonal CD8+ T-cell repertoire in multiple sclerosis. *J Neuroimmunol.* 2006;180:159–171.
38. Tennakoon DK, et al. Therapeutic induction of regulatory, cytotoxic CD8+ T cells in multiple sclerosis. *J Immunol.* 2006;176:7119–7129.
39. Blanco Y, et al. Effect of glatiramer acetate (Copaxone®) on the immunophenotypic and cytokine profile and BDNF production in multiple sclerosis: a longitudinal study. *Neurosci Lett.* 2006. https://doi.org/10.1016/j.neulet.2006.07.043.
40. Sarchielli P, et al. Production of brain-derived neurotrophic factor by mononuclear cells of patients with multiple sclerosis treated with glatiramer acetate, interferon-β 1a, and high doses of immunoglobulins. *Mult Scler.* 2007. https://doi.org/10.1177/1352458506070146.
41. Skihar V, et al. Promoting oligodendrogenesis and myelin repair using the multiple sclerosis medication glatiramer acetate. *Proc Natl Acad Sci U S A.* 2009;106:17992–17997.
42. Riley CP, Cope TC, Buck CR. CNS neurotrophins are biologically active and expressed by multiple cell types. *J Mol Histol.* 2004. https://doi.org/10.1007/s10735-004-0778-9.
43. Kerschensteiner M, et al. Activated human T cells, B cells, and monocytes produce brain-derived neurotrophic factor in vitro and in inflammatory brain lesions: a neuroprotective role of inflammation? *J Exp Med.* 1999;189:865–870.
44. Ziemssen T, Kümpfel T, Klinkert WEF, Neuhaus O, Hohlfeld R. Glatiramer acetate-specific T-helper 1- and 2-type cell lines produce BDNF: implications for multiple sclerosis therapy. Brain-derived neurotrophic factor. *Brain.* 2002;125:2381–2391.
45. Chen M, Valenzuela RM, Dhib-Jalbut S. Glatiramer acetate-reactive T cells produce brain-derived neurotrophic factor. *J Neurol Sci.* 2003;215:37–44.
46. Kipnis J, et al. T cell immunity to copolymer 1 confers neuroprotection on the damaged optic nerve: possible therapy for optic neuropathies. *Proc Natl Acad Sci U S A.* 2000;97:7446–7451.
47. Aharoni R, et al. The immunomodulator glatiramer acetate augments the expression of neurotrophic factors in brains of experimental autoimmune encephalomyelitis mice. *Proc Natl Acad Sci U S A.* 2005;102:19045–19050.
48. Aharoni R, et al. Demyelination arrest and remyelination induced by glatiramer acetate treatment of experimental autoimmune encephalomyelitis. *Proc Natl Acad Sci U S A.* 2008;105:11358–11363.
49. Aharoni R, Arnon R, Eilam R. Neurogenesis and neuroprotection induced by peripheral immunomodulatory treatment of experimental autoimmune encephalomyelitis. *J Neurosci.* 2005;25:8217–8228.
50. Gilgun-Sherki Y, Panet H, Holdengreber V, Mosberg-Galili R, Offen D. Axonal damage is reduced following glatiramer acetate treatment in C57/bl mice with chronic-induced experimental autoimmune encephalomyelitis. *Neurosci Res.* 2003;47:201–207.
51. Aharoni R, Kayhan B, Eilam R, Sela M, Arnon R. Glatiramer acetate-specific T cells in the brain express T helper 2/3 cytokines and brain-derived neurotrophic factor in situ. *Proc Natl Acad Sci U S A.* 2003;100:14157–14162.
52. Bornstein MB, et al. A pilot trial of Cop 1 in exacerbating–remitting multiple sclerosis. *N Engl J Med.* 2010;317:408–414.
53. Johnson KP, et al. Extended use of glatiramer acetate (Copaxone) is well tolerated and maintains its clinical effect on multiple sclerosis relapse rate and degree of disability. Copolymer 1 multiple sclerosis study group. *Neurology.* 1998;50:701–708.
54. Comi G, Filippi M, Wolinsky JS. European/Canadian multicenter, double-blind, randomized, placebo-controlled study of the effects of glatiramer acetate on magnetic resonance imaging-measured disease activity and burden in patients with relapsing multiple sclerosis. *Ann Neurol.* 2001;49:290–297.
55. Teva Pharmaceuticals. n.d., Personal Communication.
56. Kohama T, et al. Molecular cloning and functional characterization of murine sphingosine kinase. *J Biol Chem.* 1998;273:23722–23728.
57. Liu H, et al. Molecular cloning and functional characterization of a novel mammalian sphingosine kinase type 2 isoform. *J Biol Chem.* 2000;275:19513–19520.

58. Graeler M, Goetzl EJ. Activation-regulated expression and chemotactic function of sphingosine 1-phosphate receptors in mouse splenic T cells. *FASEB J*. 2002;16:1874–1878.
59. Pyne S, Pyne N. Sphingosine 1-phosphate signalling via the endothelial differentiation gene family of G-protein-coupled receptors. *Pharmacol Ther*. 2000;88:115–131.
60. Anliker B, Chun J. Lysophospholipid G protein-coupled receptors. *J Biol Chem*. 2004;279: 20555–20558.
61. Ishii I, Fukushima N, Ye X, Chun J. Lysophospholipid receptors: signaling and biology. *Annu Rev Biochem*. 2004;73:321–354.
62. Chun J. Immunology: the sources of a lipid conundrum. *Science*. 2007;316:208–210.
63. Edsall LC, Pirianov GG, Spiegel S. Involvement of sphingosine 1-phosphate in nerve growth factor-mediated neuronal survival and differentiation. *J Neurosci*. 1997;17:6952–6960.
64. MacLennan AJ, et al. An essential role for the H218/AGR16/Edg-5/LPB2 sphingosine 1-phosphate receptor in neuronal excitability. *Eur J Neurosci*. 2001;14:203–209.
65. Toman RE, et al. Differential transactivation of sphingosine-1-phosphate receptors modulates NGF-induced neurite extension. *J Cell Biol*. 2004;166:381–392.
66. Mizugishi K, et al. Essential role for sphingosine kinases in neural and vascular development. *Mol Cell Biol*. 2005;25:11113–11121.
67. Yu N, Lariosa-Willingham KD, Lin FF, Webb M, Rao TS. Characterization of lysophosphatidic acid and sphingosine-1-phosphate-mediated signal transduction in rat cortical oligodendrocytes. *Glia*. 2004;45:17–27.
68. Cuvillier O, et al. Suppression of ceramide-mediated programmed cell death by sphingosine-1-phosphate. *Nature*. 1996;381:800–803.
69. Mandala SM, et al. Molecular cloning and characterization of a lipid phosphohydrolase that degrades sphingosine-1-phosphate and induces cell death. *Proc Natl Acad Sci U S A*. 2000;97:7859–7864.
70. Graeler M, Shankar G, Goetzl EJ. Cutting edge: suppression of t cell chemotaxis by sphingosine 1-phosphate. *J Immunol*. 2002;169:4084–4087.
71. Matloubian M, et al. Lymphocyte egress from thymus and peripheral lymphoid organs is dependent on S1P receptor 1. *Nature*. 2004;427:355–360.
72. Brinkmann V, Cyster JG, Hla T. FTY720: sphingosine 1-phosphate receptor-1 in the control of lymphocyte egress and endothelial barrier function. *Am J Transplant*. 2004;4:1019–1025.
73. Jaillard C. Edg8/S1P5: an oligodendroglial receptor with dual function on process retraction and cell survival. *J Neurosci*. 2005;25:1459–1469.
74. Jung CG, et al. Functional consequences of S1P receptor modulation in rat oligodendroglial lineage cells. *Glia*. 2007;55:1656–1667.
75. Tham CS, Lin FF, Rao TS, Yu N, Webb M. Microglial activation state and lysophospholipid acid receptor expression. *Int J Dev Neurosci*. 2003;21:431–443.
76. Miron VE, et al. FTY720 modulates human oligodendrocyte progenitor process extension and survival. *Ann Neurol*. 2008;63:61–71.
77. Coelho RP, Payne SG, Bittman R, Spiegel S, Sato-Bigbee C. The immunomodulator FTY720 has a direct cytoprotective effect in oligodendrocyte progenitors. *J Pharmacol Exp Ther*. 2007;323:626–635.
78. Chun J, Contos JJA, Munroe D. A growing family of receptor genes for lysophosphatidic acid (LPA) and other lysophospholipids (LPs). *Cell Biochem Biophys*. 1999;30:213–242.
79. McGiffert C, Contos JJA, Friedman B, Chun J. Embryonic brain expression analysis of lysophospholipid receptor genes suggests roles for s1p1 in neurogenesis and s1p1-3 in angiogenesis. *FEBS Lett*. 2002;531:103–108.
80. Gardell SE, Dubin AE, Chun J. Emerging medicinal roles for lysophospholipid signaling. *Trends Mol Med*. 2006;12:65–75.
81. Birgbauer E, Chun J. New developments in the biological functions of lysophospholipids. *Cell Mol Life Sci*. 2006;63:2695–2701.
82. Gensert JAM, Goldman JE. Endogenous progenitors remyelinate demyelinated axons in the adult CNS. *Neuron*. 1997;19:197–203.
83. Saini HS, et al. Novel role of sphingosine kinase 1 as a mediator of neurotrophin-3 action in oligodendrocyte progenitors. *J Neurochem*. 2005;95:1298–1310.

84. Kimura A, et al. Essential roles of sphingosine 1-phosphate/S1P 1 receptor axis in the migration of neural stem cells toward a site of spinal cord injury. *Stem Cells*. 2007;25:115–124.
85. Shinpo K, Kikuchi S, Moriwaka F, Tashiro K. Protective effects of the TNF-ceramide pathway against glutamate neurotoxicity on cultured mesencephalic neurons. *Brain Res*. 1999;819:170–173.
86. Chin T, Hwang H, Chueh S. Distinct effects of different calcium-mobilizing agents on cell death in NG108-15 neuroblastoma X glioma cells. *Mol Pharmacol*. 2002;61:486–494.
87. Pébay A, et al. Antiproliferative properties of sphingosine-1-phosphate in human hepatic myofibroblasts. *Eur J Neurosci*. 2001;13:2067–2076.
88. Sorensen SD, et al. Common signaling pathways link activation of murine PAR-1, LPA, and S1P receptors to proliferation of astrocytes. *Mol Pharmacol*. 2003;64:1199–1209.
89. Bassi R, et al. Sphingosine-1-phosphate is released by cerebellar astrocytes in response to bFGF and induces astrocyte proliferation through G i -protein-coupled receptors. *Glia*. 2006;53:621–630.
90. Eng LF, Reier PJ, Houle JD. Astrocyte activation and fibrous gliosis: glial fibrillary acidic protein immunostaining of astrocytes following intraspinal cord grafting of fetal CNS tissue. *Prog Brain Res*. 1987;71:439–455.
91. Brinkmann V, et al. The immune modulator FTY720 targets sphingosine 1-phosphate receptors. *J Biol Chem*. 2002;277:21453–21457.
92. Paugh SW, Payne SG, Barbour SE, Milstien S, Spiegel S. The immunosuppressant FTY720 is phosphorylated by sphingosine kinase type 2. *FEBS Lett*. 2003;554:189–193.
93. Albert R, et al. Novel immunomodulator FTY720 is phosphorylated in rats and humans to form a single stereoisomer. Identification, chemical proof, and biological characterization of the biologically active species and its enantiomer. *J Med Chem*. 2005;48:5373–5377.
94. Mullershausen F, et al. Phosphorylated FTY720 promotes astrocyte migration through sphingosine-1-phosphate receptors. *J Neurochem*. 2007;102:1151–1161.
95. Billich A, et al. Phosphorylation of the immunomodulatory drug FTY720 by sphingosine kinases. *J Biol Chem*. 2003;278:47408–47415.
96. Kharel Y, et al. Sphingosine kinase 2 is required for modulation of lymphocyte traffic by FTY720. *J Biol Chem*. 2005;280:36865–36872.
97. Zemann B, et al. Sphingosine kinase type 2 is essential for lymphopenia induced by the immunomodulatory drug FTY720. *Blood*. 2006;107:1454–1458.
98. Myat LO, et al. Immunosuppressive and anti-angiogenic sphingosine 1-phosphate receptor-1 agonists induce ubiquitinylation and proteasomal degradation of the receptor. *J Biol Chem*. 2007;282: 9082–9089.
99. Gräler MH, Goetzl EJ. The immunosuppressant FTY720 down-regulates sphingosine 1-phosphate G-protein-coupled receptors. *FASEB J*. 2004;18:551–553.
100. Fujino M. Amelioration of experimental autoimmune encephalomyelitis in Lewis rats by FTY720 treatment. *J Pharmacol Exp Ther*. 2003;305:70–77.
101. Webb M, et al. Sphingosine 1-phosphate receptor agonists attenuate relapsing-remitting experimental autoimmune encephalitis in SJL mice. *J Neuroimmunol*. 2004;153:108–121.
102. Forrest M. Immune cell regulation and cardiovascular effects of sphingosine 1-phosphate receptor agonists in rodents are mediated via distinct receptor subtypes. *J Pharmacol Exp Ther*. 2004;309:758–768.
103. Sanna MG, et al. Sphingosine 1-phosphate (S1P) receptor subtypes S1P1 and S1P3, respectively, regulate lymphocyte recirculation and heart rate. *J Biol Chem*. 2004;279:13839–13848.
104. Yagi H, et al. Immunosuppressant FTY720 inhibits thymocyte emigration. *Eur J Immunol*. 2000;30:1435–1444.
105. Mehling M, et al. FTY720 therapy exerts differential effects on T cell subsets in multiple sclerosis. *Neurology*. 2008;71:1261–1267.
106. Kebir H, et al. Human TH17 lymphocytes promote blood-brain barrier disruption and central nervous system inflammation. *Nat Med*. 2007;13:1173–1175.
107. Suzuki S, et al. A novel immunosuppressant, FTY720, with a unique mechanism of action, induces long-term graft acceptance in rat and dog allotransplantation. *Transplantation*. 1996;61:200–205.
108. Matsuura M, Imayoshi T, Chiba K, Okumoto T. Effect of FTY720, a novel immunosuppressant, on adjuvant-induced arthritis in rats. *Inflamm Res*. 2000;49:404–410.

109. Yang Z, et al. The immune modulator FYT720 prevents autoimmune diabetes in nonobese diabetic mice. *Clin Immunol*. 2003;107:30–35.
110. Foster CA, et al. FTY720 rescue therapy in the dark agouti rat model of experimental autoimmune encephalomyelitis: expression of central nervous system genes and reversal of blood-brain-barrier damage. *Brain Pathol*. 2009;19:254–266.
111. Kataoka H, et al. FTY720, sphingosine 1-phosphate receptor modulator, ameliorates experimental autoimmune encephalomyelitis by inhibition of T cell infiltration. *Cell Mol Immunol*. 2005;2:439–448.
112. Foster CA, et al. Brain penetration of the oral immunomodulatory drug FTY720 and its phosphorylation in the central nervous system during experimental autoimmune encephalomyelitis: consequences for mode of action in multiple sclerosis. *J Pharmacol Exp Ther*. 2007;323:469–475.
113. Balatoni B, et al. FTY720 sustains and restores neuronal function in the DA rat model of MOG-induced experimental autoimmune encephalomyelitis. *Brain Res Bull*. 2007;74:307–316.
114. Barske C, Osinde M, Mir AK. FTY720 fingolimod enhances the number of immature and mature oligodendrocytes. *Mult Scler*. 2007;13:S148.
115. Baumruker T, Billich A, Brinkmann V. FTY720, an immunomodulatory sphingolipid mimetic: translation of a novel mechanism into clinical benefit in multiple sclerosis. *Expert Opin Investig Drugs*. 2007; https://doi.org/10.1517/13543784.16.3.283.
116. Deogracias R, et al. FTY720 induces BDNF production in neuronal cell cultures. *Neurology*. 2008;70: A373.
117. Ebner S, Dunbar M, McKinnon RD. Distinct roles for PI3K in proliferation and survival of oligodendrocyte progenitor cells. *J Neurosci Res*. 2000;62:336–345.
118. Fragoso G, et al. Developmental differences in H_2O_2-induced oligodendrocyte cell death: role of glutathione, mitogen-activated protein kinases and caspase 3. *J Neurochem*. 2004;90:392–404.
119. Cui QL, Fogle E, Almazan G. Muscarinic acetylcholine receptors mediate oligodendrocyte progenitor survival through Src-like tyrosine kinases and PI3K/Akt pathways. *Neurochem Int*. 2006;48:383–393.
120. Miron VE, Schubart A, Antel JP. Central nervous system-directed effects of FTY720 (fingolimod). *J Neurol Sci*. 2008;274:13–17.
121. Miron VE, et al. Fingolimod (FTY720) enhances remyelination following demyelination of organotypic cerebellar slices. *Am J Pathol*. 2010;176:2682–2694.
122. Novgorodov AS, El-Alwani M, Bielawski J, Obeid LM, Gudz TI. Activation of sphingosine-1-phosphate receptor S1P5 inhibits oligodendrocyte progenitor migration. *FASEB J*. 2007;21: 1503–1514.
123. Jackson SJ, Giovannoni G, Baker D. Fingolimod modulates microglial activation to augment markers of remyelination. *J Neuroinflammation*. 2011;8.
124. Ishibashi T, et al. Astrocytes promote myelination in response to electrical impulses. *Neuron*. 2006;49: 823–832.
125. Zhao C, Fancy SPJ, Kotter MR, Li WW, Franklin RJM. Mechanisms of CNS remyelination—the key to therapeutic advances. *J Neurol Sci*. 2005;233:87–91.
126. Noda H, Takeuchi H, Mizuno T, Suzumura A. Fingolimod phosphate promotes the neuroprotective effects of microglia. *J Neuroimmunol*. 2013;256:13–18.
127. Nair A, Frederick TJ, Miller SD. Astrocytes in multiple sclerosis: a product of their environment. *Cell Mol Life Sci*. 2008;65:2702–2720.
128. Wu YP, Mizugishi K, Bektas M, Sandhoff R, Proia RL. Sphingosine kinase 1/S1P receptor signaling axis controls glial proliferation in mice with Sandhoff disease. *Hum Mol Genet*. 2008;17:2257–2264.
129. Dong Y, Benveniste EN. Immune function of astrocytes. *Glia*. 2001;36:180–190.
130. Tzartos JS, et al. Interleukin-17 production in central nervous system-infiltrating T cells and glial cells is associated with active disease in multiple sclerosis. *Am J Pathol*. 2008;172:146–155.
131. Constantinescu CS, et al. Astrocytes as antigen-presenting cells: expression of IL-12/IL-23. *J Neurochem*. 2005;95:331–340.
132. Langrish CL, et al. IL-23 drives a pathogenic T cell population that induces autoimmune inflammation. *J Exp Med*. 2005;201:233–240.
133. Miljkovic D, et al. Astrocytes stimulate interleukin-17 and interferon-γ production in vitro. *J Neurosci Res*. 2007;85:3598–3606.

134. Gimenez MAT, Sim JE, Russell JH. TNFR1-dependent VCAM-1 expression by astrocytes exposes the CNS to destructive inflammation. *J Neuroimmunol.* 2004;151:116–125.
135. Barcia C, et al. In vivo mature immunological synapses forming SMACs mediate clearance of virally infected astrocytes from the brain. *J Exp Med.* 2006;203:2095–2107.
136. Tan LJ, Vanderlugt CL, McRae BL, Miller SD. Regulation of the effector stages of experimental autoimmune encephalomyelitis via neuroantigen-specific tolerance induction. III A role for anergy/deletion. *Autoimmunity.* 1998;27:13–28.
137. Kort JJ, Kawamura K, Fugger L, Weissert R, Forsthuber TG. Efficient presentation of myelin oligodendrocyte glycoprotein peptides but not protein by astrocytes from HLA-DR2 and HLA-DR4 transgenic mice. *J Neuroimmunol.* 2006;173:23–34.
138. Gardell S, et al. Evidence for neural S1P receptor signaling in EAE and FTY720 efficacy. *Mult Scler.* 2007;13:S70.
139. Schubart A, et al. CNS mediated effects of FTY720 (fingolimod). *Neurology.* 2007;68:A315.
140. Choi JW, et al. FTY720 (fingolimod) efficacy in an animal model of multiple sclerosis requires astrocyte sphingosine 1-phosphate receptor 1 (S1P1) modulation. *Proc Natl Acad Sci U S A.* 2011;108: 751–756.
141. Gergely P, et al. The selective sphingosine 1-phosphate receptor modulator BAF312 redirects lymphocyte distribution and has species-specific effects on heart rate. *Br J Pharmacol.* 2012;167:1035–1047.
142. O'Sullivan C, Schubart A, Mir AK, Dev KK. The dual S1PR1/S1PR5 drug BAF312 (Siponimod) attenuates demyelination in organotypic slice cultures. *J Neuroinflammation.* 2016;13:31.
143. Meadows KT, et al. *Ozanimod (RPC1063) reduces the plasma biomarker neurofilament light chain in preclinical rodent models of multiple sclerosis.* European Committee for Treatment and Research in Multiple Sclerosis (ECTRIMS); 2017:EP1609.
144. Meadows KT, et al. *Ozanimod (RPC1063) is potentially neuroprotective through direct CNS effects.* European Committee for Treatment and Research in Multiple Sclerosis (ECTRIMS); 2017:P1183.
145. Guimond D, et al. *Ozanimod (RPC1063) is potentially neuroprotective through direct activity on Th1 and Th17 T cell expansion and migration, monocyte migration and microglia expansion.* European Committee for Treatment and Research in Multiple Sclerosis (ECTRIMS); 2017:P1181.
146. Bolli MH, et al. 2-Imino-thiazolidin-4-one derivatives as potent, orally active S1P 1 receptor agonists. *J Med Chem.* 2010;53:4198–4211.
147. You S, et al. Therapeutic use of a selective S1P1 receptor modulator ponesimod in autoimmune diabetes. *PLoS One.* 2013;8:e77296.
148. Piali L, et al. The selective sphingosine 1-phosphate receptor 1 agonist ponesimod protects against lymphocyte-mediated tissue inflammation. *J Pharmacol Exp Ther.* 2011;337:547–556.
149. Brossard P, et al. Pharmacokinetics and pharmacodynamics of ponesimod, a selective S1P1 receptor modulator, in the first-in-human study. *Br J Clin Pharmacol.* 2013;76:888–896.
150. D'Ambrosio D, Freedman MS, Prinz J. Ponesimod, a selective S1P1 receptor modulator: a potential treatment for multiple sclerosis and other immune-mediated diseases. *Ther Adv Chronic Dis.* 2016;7:18–33.
151. Kappos L, et al. A placebo-controlled trial of oral fingolimod in relapsing multiple sclerosis. *N Engl J Med.* 2010;362:387–401.
152. Khatri BO. Fingolimod in the treatment of relapsing-remitting multiple sclerosis: long-term experience and an update on the clinical evidence. *Ther Adv Neurol Disord.* 2016;9:130–147.
153. Cohen J, et al. Oral fingolimod or intramuscular interferon for relapsing multiple sclerosis. *N Engl J Med.* 2010;362:402–415.
154. Calabresi PA, et al. Safety and efficacy of fingolimod in patients with relapsing-remitting multiple sclerosis (FREEDOMS II): a double-blind, randomised, placebo-controlled, phase 3 trial. *Lancet Neurol.* 2014;13:545–556.
155. Cohen JA, et al. Long-term (up to 4.5 years) treatment with fingolimod in multiple sclerosis: results from the extension of the randomised TRANSFORMS study. *J Neurol Neurosurg Psychiatry.* 2016;87:468–475.
156. Kappos L, et al. Siponimod versus placebo in secondary progressive multiple sclerosis (EXPAND): a double-blind, randomised, phase 3 study. *Lancet.* 2018;391:1263–1273.

157. Kappos L, et al. *Efficacy and safety of ponesimod compared to teriflunomide in patients with relapsing multiple sclerosis: results of the randomized, active-controlled, double-blind, parallel-group phase 3 OPTIMUM study.* European Committee for Treatment and Research in Multiple Sclerosis (ECTRIMS); 2019.
158. Miller DH, et al. A controlled trial of natalizumab for relapsing multiple sclerosis. *N Engl J Med.* 2003;348:15–23.
159. Gumbiner BM. Cell adhesion: the molecular basis of tissue architecture and morphogenesis. *Cell.* 1996;84:345–357.
160. Hynes R, Integrins O. A family of cell surface receptors. *Cell.* 1987;48:549–554.
161. Frenette PS, Wagner DD. Adhesion molecules—part I. *N Engl J Med.* 1996;334:1526–1529.
162. Frenette PS, Wagner DD. Adhesion molecules—part II blood vessels and blood cells. *N Engl J Med.* 1996;335:43–45.
163. Moser M, Legate KR, Zent R, Fässler R. The tail of integrins, talin, and kindlins. *Science.* 2009;324: 895–899.
164. Herter J, Zarbock A. Integrin regulation during leukocyte recruitment. *J Immunol.* 2013;190: 4451–4457.
165. Askari JA, Buckley PA, Mould AP, Humphries MJ. Linking integrin conformation to function. *J Cell Sci.* 2009;122:165–170.
166. Hogg N, Patzak I, Willenbrock F. The insider's guide to leukocyte integrin signalling and function. *Nat Rev Immunol.* 2011;11:416–426.
167. Tiwari S, Askari JA, Humphries MJ, Bulleid NJ. Divalent cations regulate the folding and activation status of integrins during their intracellular trafficking. *J Cell Sci.* 2011;124:1672–1680.
168. Wegener KL, et al. Structural basis of integrin activation by talin. *Cell.* 2007;128:171–182.
169. Chan JR, Hyduk SJ, Cybulsky MI. Chemoattractants induce a rapid and transient upregulation of monocyte α4 integrin affinity for vascular cell adhesion molecule 1 which mediates arrest: an early step in the process of emigration. *J Exp Med.* 2001;193:1149–1158.
170. Chavakis T. Leucocyte recruitment in inflammation and novel endogenous negative regulators thereof. *Eur J Clin Invest.* 2012;42:686–691.
171. Hyun Y-M, Chung H-L, McGrath JL, Waugh RE, Kim M. Activated integrin VLA-4 localizes to the lamellipodia and mediates T cell migration on VCAM-1. *J Immunol.* 2009;183:359–369.
172. Kishimoto TK, Jutila MA, Berg EL, Butcher EC. Neutrophil Mac 1 and MEL-14 adhesion proteins inversely regulated by chemotactic factors. *Science.* 1989;245:1238–1241.
173. Kansas GS. Selectins and their ligands: current concepts and controversies. *Blood.* 1996;88:3259–3287.
174. Kuwano Y, Spelten O, Zhang H, Ley K, Zarbock A. Rolling on E- or P-selectin induces the extended but not high-affinity conformation of LFA-1 in neutrophils. *Blood.* 2010;116:617–624.
175. Hafezi-Moghadam A, Klaus L. Relevance of L-selectin shedding for leukocyte rolling in vivo. *J Exp Med.* 1999;189:939–947.
176. Alon R, et al. The integrin VLA-4 supports tethering and rolling in flow on VCAM-1. *J Cell Biol.* 1995;128:1243–1253.
177. Zarbock A, Ley K, McEver RP, Hidalgo A. Leukocyte ligands for endothelial selectins: specialized glycoconjugates that mediate rolling and signaling under flow. *Blood.* 2011;118:6743–6751.
178. Yago T, et al. E-selectin engages PSGL-1 and CD44 through a common signaling pathway to induce integrin αLβ2-mediated slow leukocyte rolling. *Blood.* 2010;116:485–494.
179. Kerfoot SM, Kubes P. Overlapping roles of P-selectin and α 4 integrin to recruit leukocytes to the central nervous system in experimental autoimmune encephalomyelitis. *J Immunol.* 2002;169: 1000–1006.
180. Vajkoczy P, Laschinger M, Engelhardt B. α4-integrin-VCAM-1 binding mediates G protein-independent capture of encephalitogenic T cell blasts to CNS white matter microvessels. *J Clin Invest.* 2001;108:557–565.
181. Laschinger M, Engelhardt B. Interaction of α4-integrin with VCAM-1 is involved in adhesion of encephalitogenic T cell blasts to brain endothelium but not in their transendothelial migration in vitro. *J Neuroimmunol.* 2000;102:32–43.
182. Gahmberg CG, et al. Regulation of integrin activity and signalling. *Biochim Biophys Acta, Gen Subj.* 2009;1790:431–444.

183. Springer TA. Traffic signals for lymphocyte recirculation and leukocyte emigration: the multistep paradigm. *Cell.* 1994;76:301–314.
184. Schenkel AR, Mamdouh Z, Muller WA. Locomotion of monocytes on endothelium is a critical step during extravasation. *Nat Immunol.* 2004;5:393–400.
185. Nourshargh S, Hordijk PL, Sixt M. Breaching multiple barriers: leukocyte motility through venular walls and the interstitium. *Nat Rev Mol Cell Biol.* 2010;11:366–378.
186. Yednock TA, et al. Prevention of experimental autoimmune encephalomyelitis by antibodies against α4β1 integrin. *Nature.* 1992;356:63–66.
187. Kent SJ, Karlik SJ, Rice GPA, Horner HC. A monoclonal antibody to α4-integrin reverses the MR-detectable signs of experimental allergic encephalomyelitis in the Guinea pig. *J Magn Reson Imaging.* 1995;5:535–540.
188. Kent SJ, et al. A monoclonal antibody to α4 integrin suppresses and reverses active experimental allergic encephalomyelitis. *J Neuroimmunol.* 1995;58:1–10.
189. Baron JL, Madri JA, Ruddle NH, Hashim G, Janeway CA. Surface expression of α4 integrin by CD4 T cells is required for their entry into brain parenehyma. *J Exp Med.* 1993;177:57–68.
190. Coisne C, Mao W, Engelhardt B. Cutting edge: natalizumab blocks adhesion but not initial contact of human T cells to the blood-brain barrier in vivo in an animal model of multiple sclerosis. *J Immunol.* 2009;182:5909–5913.
191. Brocke S, Piercy C, Steinman L, Weissman IL, Veromaa T. Antibodies to CD44 and integrin 4, but not L-selectin, prevent central nervous system inflammation and experimental encephalomyelitis by blocking secondary leukocyte recruitment. *Proc Natl Acad Sci U S A.* 1999;96:6896–6901.
192. Rittling SR, Singh R. Osteopontin in immune-mediated diseases. *J Dent Res.* 2015;94:1638–1645.
193. Hur EM, et al. Osteopontin-induced relapse and progression of autoimmune brain disease through enhanced survival of activated T cells. *Nat Immunol.* 2007;8:74–83.
194. Vogt MHJ, Lopatinskaya L, Smits M, Polman CH, Nagelkerken L. Elevated osteopontin levels in active relapsing-remitting multiple sclerosis. *Ann Neurol.* 2003;53:819–822.
195. Grassinger J, et al. Thrombin-cleaved osteopontin regulates hemopoietic stem and progenitor cell functions through interactions with α9β1and α4β1 integrins. *Blood.* 2009;114:49–59.
196. Chen G, et al. Role of osteopontin in synovial Th17 differentiation in rheumatoid arthritis. *Arthritis Rheum.* 2010;62:2900–2908.
197. Santamaría MH, Corral RS. Osteopontin-dependent regulation of Th1 and Th17 cytokine responses in trypanosoma cruzi-infected C57BL/6 mice. *Cytokine.* 2013;61:491–498.
198. Ashkar S, et al. Eta-1 (osteopontin): an early component of type-1 (cell-mediated) immunity. *Science.* 2000;287:860–864.
199. O'Regan AW, Hayden JM, Berman JS. Osteopontin augments CD3-mediated interferon-γ and CD40 ligand expression by T cells, which results in IL-12 production from peripheral blood mononuclear cells. *J Leukoc Biol.* 2000;68:495–502.
200. Shinohara ML, Kim JH, Garcia VA, Cantor H. Engagement of the type I interferon receptor on dendritic cells inhibits T helper 17 cell development: role of intracellular osteopontin. *Immunity.* 2008;29:68–78.
201. Murugaiyan G, Mittal A, Weiner HL. Increased osteopontin expression in dendritic cells amplifies IL-17 production by CD4+ T cells in experimental autoimmune encephalomyelitis and in multiple sclerosis. *J Immunol.* 2008;181:7480–7488.
202. Jones AP, et al. Circulating immune cells in multiple sclerosis. *Clin Exp Immunol.* 2017;187:193–203.
203. Davalos D, et al. Early detection of thrombin activity in neuroinflammatory disease. *Ann Neurol.* 2014;75:303–308.
204. Han MH, et al. Proteomic analysis of active multiple sclerosis lesions reveals therapeutic targets. *Nature.* 2008;451:1076–1081.
205. Boggio E, et al. Thrombin cleavage of osteopontin modulates its activities in human cells in vitro and mouse experimental autoimmune encephalomyelitis in vivo. *J Immunol Res.* 2016;2016(13).
206. Altintaş A, Saruhan-Direskeneli G, Benbir G, Demir M, Purisa S. The role of osteopontin: a shared pathway in the pathogenesis of multiple sclerosis and osteoporosis? *J Neurol Sci.* 2009;276:41–44.

207. Chowdhury SA, Lin J, Sadiq SA. Specificity and correlation with disease activity of cerebrospinal fluid osteopontin levels in patients with multiple sclerosis. *Arch Neurol.* 2008;65:232–235.
208. Braitch M, Nunan R, Niepel G, Edwards LJ, Constantinescu CS. Increased osteopontin levels in the cerebrospinal fluid of patients with multiple sclerosis. *Arch Neurol.* 2008;65:633–635.
209. Agah E, et al. Osteopontin (OPN) as a CSF and blood biomarker for multiple sclerosis: a systematic review and meta-analysis. *PLoS ONE.* 2018;13:e0190252.
210. Vogt MHJ, et al. Osteopontin levels and increased disease activity in relapsing-remitting multiple sclerosis patients. *J Neuroimmunol.* 2004;155:155–160.
211. Comabella M, et al. Plasma osteopontin levels in multiple sclerosis. *J Neuroimmunol.* 2005;158: 231–239.
212. Börnsen L, Khademi M, Olsson T, Sørensen PS, Sellebjerg F. Osteopontin concentrations are increased in cerebrospinal fluid during attacks of multiple sclerosis. *Mult Scler J.* 2011;17:32–42.
213. Khademi M, et al. The effects of natalizumab on inflammatory mediators in multiple sclerosis: prospects for treatment-sensitive biomarkers. *Eur J Neurol.* 2009;16:528–536.
214. Iaffaldano P, Ruggieri M, Viterbo RG, Mastrapasqua M, Trojano M. The improvement of cognitive functions is associated with a decrease of plasma Osteopontin levels in Natalizumab treated relapsing multiple sclerosis. *Brain Behav Immun.* 2014;35:96–101.
215. Tubridy N, et al. The effect of anti-4 integrin antibody on brain lesion activity in MS. *Neurology.* 1999;53:466–472.
216. Clifford DB, et al. Natalizumab-associated progressive multifocal leukoencephalopathy in patients with multiple sclerosis: lessons from 28 cases. *Lancet Neurol.* 2010;9:438–446.
217. Schwab N, Schneider-Hohendorf T, Melzer N, Cutter G, Wiendl H. Natalizumab-associated PML. *Neurology.* 2017;88:1197–1205.
218. Stahl B, Müller B, von Boxberg Y, Cox EC, Bonhoeffer F. Biochemical characterization of a putative axonal guidance molecule of the chick visual system. *Neuron.* 1990;5:735–743.
219. Tassew NG, et al. Modifying lipid rafts promotes regeneration and functional recovery. *Cell Rep.* 2014;8:1146–1159.
220. Monnier PP, et al. RGM is a repulsive guidance molecule for retinal axons. *Nature.* 2002;419:392–395.
221. Matsunaga E, et al. RGM and its receptor neogenin regulate neuronal survival. *Nat Cell Biol.* 2004;6:749–755.
222. Rajagopalan S, et al. Neogenin mediates the action of repulsive guidance molecule. *Nat Cell Biol.* 2004;6:756–762.
223. Manitt C, Thompson KM, Kennedy TE. Developmental shift in expression of netrin receptors in the rat spinal cord: predominance of UNC-5 homologues in adulthood. *J Neurosci Res.* 2004;77:690–700.
224. Schwab JM, et al. Spinal cord injury-induced lesional expression of the repulsive guidance molecule (RGM). *Eur J Neurosci.* 2005;21:1569–1576.
225. Hata K, et al. RGMa inhibition promotes axonal growth and recovery after spinal cord injury. *J Cell Biol.* 2006;173:47–58.
226. Stirling D, et al. Minocycline treatment reduces delayed oligodendrocyte death, attenuates axonal dieback, and improves functional outcome after spinal cord injury. *J Neurosci.* 2004;24:2182–2190.
227. Busch SA, Horn KP, Silver DJ, Silver J. Overcoming macrophage-mediated axonal dieback following CNS injury. *J Neurosci.* 2009;29:9967–9976.
228. Kitayama M, Ueno M, Itakura T, Yamashita T. Activated microglia inhibit axonal growth through RGMa. *PLoS ONE.* 2011;6:e25234.
229. Tikka TM, Koistinaho JE. Minocycline provides neuroprotection against N -methyl-d-aspartate neurotoxicity by inhibiting microglia. *J Immunol.* 2001;166:7527–7533.
230. Tikka T, Fiebich BL, Goldsteins G, Keinänen R, Koistinaho J. Minocycline, a tetracycline derivative, is neuroprotective against excitotoxicity by inhibiting activation and proliferation of microglia. *J Neurosci.* 2001;21:2580–2588.
231. Muramatsu R, et al. RGMa modulates T cell responses and is involved in autoimmune encephalomyelitis. *Nat Med.* 2011;17:488–494.
232. Tanabe S, Fujita Y, Ikuma K, Yamashita T. Inhibiting repulsive guidance molecule-a suppresses secondary progression in mouse models of multiple sclerosis. *Cell Death Dis.* 2018;9:1061.

233. Kerschensteiner M, et al. Remodeling of axonal connections contributes to recovery in an animal model of multiple sclerosis. *J Exp Med*. 2004;200:1027–1038.
234. Demicheva E, et al. Targeting repulsive guidance molecule A to promote regeneration and neuroprotection in multiple sclerosis. *Cell Rep*. 2015;10:1887–1898.
235. Nohra R, et al. RGMA and IL21R show association with experimental inflammation and multiple sclerosis. *Genes Immun*. 2010;11:279–293.
236. Mueller B, Cui Y, Huang L, Greenberg S, Schoemaker H. ABT-555, a human anti-RGMa monoclonal antibody promotes axon regeneration and neuroprotection in multiple sclerosis models. *Mult Scler J*. 2015;21:275. P582.
237. Cree B, et al. ABT-555, a potential neurorestorative treatment in multiple sclerosis: pharmacokinetic, pharmacodynamic and safety results from phase 1 studies in healthy volunteers and multiple sclerosis patients. *Mult Scler J*. 2016;22:633–634. P1209.
238. Cree B, Rosebraugh M, Barger B, Ziemann A. A phase 1, multiple-dose study of elezanumab (ABT-555) in patients with relapsing forms of multiple sclerosis. *European Committee for Treatment and Research in Multiple Sclerosis*. 2018; P 899.
239. Hauser SL, Oksenberg JR. The neurobiology of multiple sclerosis: genes, inflammation, and neurodegeneration. *Neuron*. 2006;52:61–76.
240. Miller RH, Mi S. Dissecting demyelination. *Nat Neurosci*. 2007;10:1351–1354.
241. Franklin RJM. Why does remyelination fail in multiple sclerosis? *Nat Rev Neurosci*. 2002;3:705–714.
242. Chang A, Tourtellotte WW, Rudick R, Trapp BD. Premyelinating oligodendrocytes in chronic lesions of multiple sclerosis. *N Engl J Med*. 2002;346:165–173.
243. Mi S, et al. LINGO-1 negatively regulates myelination by oligodendrocytes. *Nat Neurosci*. 2005;8:745–751.
244. Mi S, et al. LINGO-1 antagonist promotes spinal cord remyelination and axonal integrity in MOG-induced experimental autoimmune encephalomyelitis. *Nat Med*. 2007;13:1228–1233.
245. Mi S, et al. Promotion of central nervous system remyelination by induced differentiation of oligodendrocyte precursor cells. *Ann Neurol*. 2009;65:304–315.
246. Mi S, et al. LINGO-1 is a component of the Nogo-66 receptor/p75 signaling complex. *Nat Neurosci*. 2004;7:221–228.
247. Jepson S, et al. LINGO-1, a transmembrane signaling protein, inhibits oligodendrocyte differentiation and myelination through intercellular self-interactions. *J Biol Chem*. 2012;287:22184–22195.
248. Liang X, Draghi N, Resh M. Signaling from integrins to Fyn to Rho family GTPases regulates morphologic differentiation of oligodendrocytes. *J Neurosci*. 2004;24:7140–7149.
249. Baer AS, et al. Myelin-mediated inhibition of oligodendrocyte precursor differentiation can be overcome by pharmacological modulation of Fyn-RhoA and protein kinase C signalling. *Brain*. 2009;132:465–481.
250. Lee X, et al. NGF regulates the expression of axonal LINGO-1 to inhibit oligodendrocyte differentiation and myelination. *J Neurosci*. 2007;27:220–225.
251. Mi S, Blake Pepinsky R, Cadavid D. Blocking LINGO-1 as a therapy to promote CNS repair: from concept to the clinic. *CNS Drugs*. 2013;27:493–503.
252. Zhang Y, et al. Inhibition of LINGO-1 promotes functional recovery after experimental spinal cord demyelination. *Exp Neurol*. 2015;266:68–73.
253. Pepinsky RB, et al. Exposure levels of anti-LINGO-1 Li81 antibody in the central nervous system and dose-efficacy relationships in rat spinal cord remyelination models after systemic administration. *J Pharmacol Exp Ther*. 2011;339:519–529.
254. Cadavid D, et al. Safety and efficacy of opicinumab in acute optic neuritis (RENEW): a randomised, placebo-controlled, phase 2 trial. *Lancet Neurol*. 2017;16:189–199.
255. Mellion M, et al. Efficacy results from the phase 2b SYNERGY study: treatment of disabling multiple sclerosis with the anti-LINGO-1 monoclonal antibody opicinumab (S33.004). *Neurology*. 2017;88: S33.007.
256. Evans KC, et al. Exploratory MRI biomarkers of opicinumab (anti-LINGO-1) show stabilization of pre-existing T2 lesions in relapsing multiple sclerosis: results from the phase 2b SYNERGY trial (P2.100). *Neurology*. 2017;88:P2.100.

CHAPTER 10

Advances in treating myasthenia gravis

Ikjae Lee[a] and Gil I. Wolfe[b]
[a]Department of Neurology, University of Alabama at Birmingham, Birmingham, AL, United States
[b]Department of Neurology, Jacobs School of Medicine and Biomedical Sciences, University in Buffalo, Buffalo, NY, United States

Introduction

Myasthenia gravis (MG) is a prototype antibody-mediated autoimmune disorder that impairs neuromuscular transmission causing fatigable muscle weakness. The estimated prevalence ranges between 15 and 179 per 1,000,000 and the estimated incidence rate is around 30 per 1,000,000.[1,2] There is a bimodal age of onset, as female cases predominate in the second and third decades and a fairly even gender distribution is observed after the sixth decade. The diagnosis of MG can be suspected in patients presenting with fluctuating weakness that typically worsens after activity and as the day progresses. Ocular muscles are affected most commonly, producing fluctuating ptosis and diplopia. Generalized symptoms develop in the majority of the patients (50%–80%) usually within 3 years of onset leading to bulbar, axial, limb, and respiratory muscle weakness.[3–6] Myasthenic crisis, rapid worsening of the disease causing respiratory insufficiency, occurs in 15%–20% of the patients and can be life threatening.[7]

The diagnosis of MG can be established with appropriate history and examination along with one or more positive results in the following three ancillary tests: serologic testing, repetitive nerve stimulation, and single fiber electromyography.[8]

Antibodies to the nicotinic skeletal muscle acetylcholine receptor (AChR) are detected in approximately 80% of MG patients, more frequently found in generalized disease compared to pure ocular MG. Direct antibody-mediated receptor modulation/blocking and complement-mediated destruction of the postsynaptic junction are both thought to be underlying pathophysiologic mechanisms in AChR MG.[9] The thymus gland has been recognized to play an important role in AChR MG supported by high rates of thymoma and thymic hyperplasia among patients. In addition, thymectomy improves the disease.[10–14] Antibodies to muscle-specific tyrosine kinase (MuSK) have been found in 1%–10% of MG patients who do not have AChR antibodies. MuSK is a receptor tyrosine kinase at the postsynaptic membrane that plays an important role in the formation and maintenance of the neuromuscular junction by inducing AChR clustering and differentiation of nerve terminals. Disruption of this pathway by autoantibody binding to MuSK leads to neuromuscular transmission defects.[15] The majority of MuSK autoantibodies

Neurotherapeutics in the Era of Translational Medicine
https://doi.org/10.1016/B978-0-12-816475-4.00003-3

belong to IgG4 subtype, which cannot activate complement, and thus are believed to directly interfere with agrin-/LRP4-dependent MuSK signaling pathway.[15,16] A pathogenic link between MuSK MG and thymus is questionable.[17] Thymic hyperplasia is rarely observed in MuSK MG patients and occurrence of thymoma has been reported as a rare exception.[18,19] While ptosis and double vision are the most common symptoms in AChR MG, these are somewhat less common in MuSK MG. Bulbar and respiratory dysfunction are more common in MuSK MG and often leads to crisis. Compared to AChR MG, patients with MuSK MG are more predominantly female, have earlier disease onset in the third or fourth decade, and are more frequently African American.[20–22] Response to treatment is reported to be different between AChR and MuSK MG. Pyridostigmine appears to be less effective in MuSK MG or might even worsen the symptoms.[20,23,24] IVIG has a variable response in MuSK MG while plasma exchange has shown consistent efficacy based on retrospective case reviews.[22,24,25] Thymectomy in general has not been associated with clinical benefits in MuSK MG.[26–28] Rituximab has risen as an efficacious treatment and is discussed in detail (see below). 3,4-Diaminopyridine (3,4-DAP) has shown promising results in a recent phase II trial of MuSK MG and is also discussed.

Around 10%–15% of MG patients whose diagnosis is confirmed by clinical features and electrodiagnostic studies remain seronegative after serologic testing through commercially available radioimmunoprecipitation assays. They represent a heterogeneous group of patients with some harboring AChR or MuSK antibodies that can be detected only by cell-based assay; others may have antibodies against new antigenic targets such as LRP4, agrin, or cortactin.[29–36] Thus, truly "seronegative MG" might be a vanishing entity with advancements in antibody detection methodology.[37] Serologic confirmation aids in diagnosis and provides useful guidance for treatment decisions. Antibody-specific treatments are increasingly being developed in the MG field.

Repetitive nerve stimulation (RNS) may demonstrate more than a 10% decrement of CMAP amplitude with low frequency stimulation (typically 2–3 Hz). Diagnostic sensitivity of repetitive nerve stimulation is higher in more severe disease reaching over 90% in myasthenic crisis while it is lower in ocular MG with reported sensitivity less than 50%.[23,38–40] Single fiber electromyography (SFEMG) is a sensitive but nonspecific test, usually spared for MG-suspected patients who have negative serologic and repetitive nerve stimulation tests.[41]

All patients need imaging of the chest with CT or MRI to evaluate the thymus for thymic hyperplasia (28%–92%)[42] and thymoma (10%–15%).[43]

Determining the severity of MG is important as it affects treatment decisions. In 2000, the MG Foundation of America (MGFA) adopted a classification of different grades of myasthenia that has been used widely in clinical and research settings. This classification system is based on the distribution (I, ocular only vs II–V generalized; (a) predominantly affecting limb or axial muscles vs (b) predominantly affecting bulbar muscles) and severity

(II, mild; III, moderate; IV, severe; V, crisis requiring intubation). While this classification provides a rough estimate of disease severity, it does not quantify the disease burden in detail and is subject to the evaluator's judgment. Quantitative myasthenia gravis (QMG) and Myasthenia Gravis Activity of Daily living (MG-ADL) scales are the two most commonly used quantitative outcome measures in recent clinical trials.[44–47] The QMG is a 13-item scale that includes physician assessment of ocular, cranial, respiratory, axial, and limb muscle function relevant to MG.[48,49] The MG-ADL is an eight-item questionnaire that asks the patients to report common symptoms.[50] Both scales were validated and shown to have high interrater reliability, reproducibility, and correlate with clinicians' impression of disease improvement.[49,51,52]

Ocular MG is considered the mildest form of MG rather than a separate entity. However, symptoms can still be disabling and negatively impact quality of life.[53] Treatment usually starts with an acetylcholinesterase inhibitor such as pyridostigmine, typically at doses of 30–60 mg three times a day, increased up to 90–120 mg every 3–4 hours based on tolerability and clinical response. Abdominal cramps, nausea, diarrhea, muscle twitching, and muscle cramps are the common side effects that might limit pyridostigmine usage. Selective muscarinic antagonists such as glycopyrrolate or propantheline can alleviate those symptoms if used concurrently. Cholinesterase inhibitors often do not resolve all symptoms, forcing the consideration of corticosteroid treatment such as prednisone.[54] In a small randomized trial by Benatar et al., low-dose prednisone started at 10 mg every other day gradually titrated every 2 weeks to 40 mg per day over 16 weeks was shown to be highly effective.[55] Retrospective studies suggest that ocular MG patients treated with corticosteroids have a higher incidence of remission and lower risk of generalization of the disease.[56,57] Some patients are refractory to corticosteroids or cannot tolerate them due to adverse effects. In this setting, steroid sparing immunosuppressive agents such as azathioprine or mycophenolate mofetil may be used as a monotherapy or with low-dose steroids.[58,59] Other immunosuppressive agents such as tacrolimus and cyclosporine can be considered in select cases. IVIG did not show significant benefit for ocular MG in one randomized trial[60] and plasma exchange has not been investigated. Thymectomy is indicated if the patient has evidence of thymoma on chest imaging. Otherwise, thymectomy is not routinely recommended in patients with pure ocular symptoms, although there is a new effort to study this in a formal trial.

Treatment must be individualized based on disease severity and comorbidities in order to achieve the goal defined by an international consensus statement as minimal manifestation status (MMS) or better with no more than grade 1 common terminology criteria for adverse events (CTCAE).[61] The majority of patients with generalized disease typically need both pyridostigmine and prednisone at the time of presentation. The use of pyridostigmine is similar to that of ocular MG, while higher doses of prednisone are typically used to treat generalized disease, especially in the early active phase. Typically, anything over 40–60 mg of daily prednisone is considered "high dose." Various dosing methods can be utilized, either starting low or starting high. There is currently no consensus guideline or data on which approach is superior. Initial worsening of MG symptoms can occur in patients treated with high-dose prednisone within 5–10 days of treatment onset. In patients with severe disease at onset, or in those that worsen once a prednisone taper has commenced, initiation of a steroid-sparing agent such as azathioprine or mycophenolate mofetil is often considered. There are no evidence-based guidelines as to when it is appropriate to begin to taper pyridostigmine or steroid-sparing agents, however it is reasonable to attempt a weaning of these medications if patients are asymptomatic or minimally symptomatic on these steroid-sparing medications for at least 1 year. In general, tapering of steroid-sparing agents should proceed no faster than a dose change every 6 months to reduce the risk of disease recurrence. IVIG and plasma exchange are both effective in the treatment of MG crisis[62]; factors such as availability, comorbidities, and local experience affect selection of one modality versus the other. IVIG was shown to be effective in a placebo-controlled trial for patients with more severe disease, and can be given in inpatient or outpatient settings as a maintenance treatment.[60] Plasma exchange is significantly more difficult to use for maintenance treatment as it often requires large-bore catheters and an inpatient stay. Further detailed information regarding the benefit and use of pyridostigmine, prednisone, nonsteroid immunosuppressants, IVIG, and plasma exchange can be obtained from review articles.[63] Again, the treatment approach should be individualized and an updated treatment algorithm is provided (see Fig. 1).

In cases of thymomatous MG, thymectomy is a requisite component of early intervention. A thymoma is discovered in 10%–15% of patients with MG. Conversely, around 30% of patients with a thymoma develop MG. Almost all patients with thymomatous MG have AChR antibodies. Thymoma in the MuSK antibody or seronegative cases is exceedingly rare.[17,64,65] Thymoma-associated MG is mainly a disease of patients beyond 50 years of age but can occur from adolescence onward.[66] Thymomas are neoplasms of thymic epithelial cells. The tumor itself is asymptomatic in 90% of patients and is detected on routine radiological examination; 10% of the patients have symptoms including chest pain, cough, or breathing difficulties. Thymomas grow very slowly over many years but may follow a malignant course in some patients. The standard therapeutic approach for thymoma remains surgery. This strategy is based on retrospective case series and reviews as prospective randomized trials are lacking due to rarity of this condition.[67,68] Completeness of thymoma resection is the most commonly cited prognostic factor, and complete removal of early-stage thymic tumors has been associated with good long-term survival.[69–71] Patients with advanced stages of thymoma or who cannot undergo complete resection might require adjuvant chemotherapy and/or radiation therapy.[72]

For many years, MG was considered an independent risk factor for postoperative mortality and morbidity in thymoma[73,74] although more recent study argues against it.[68]

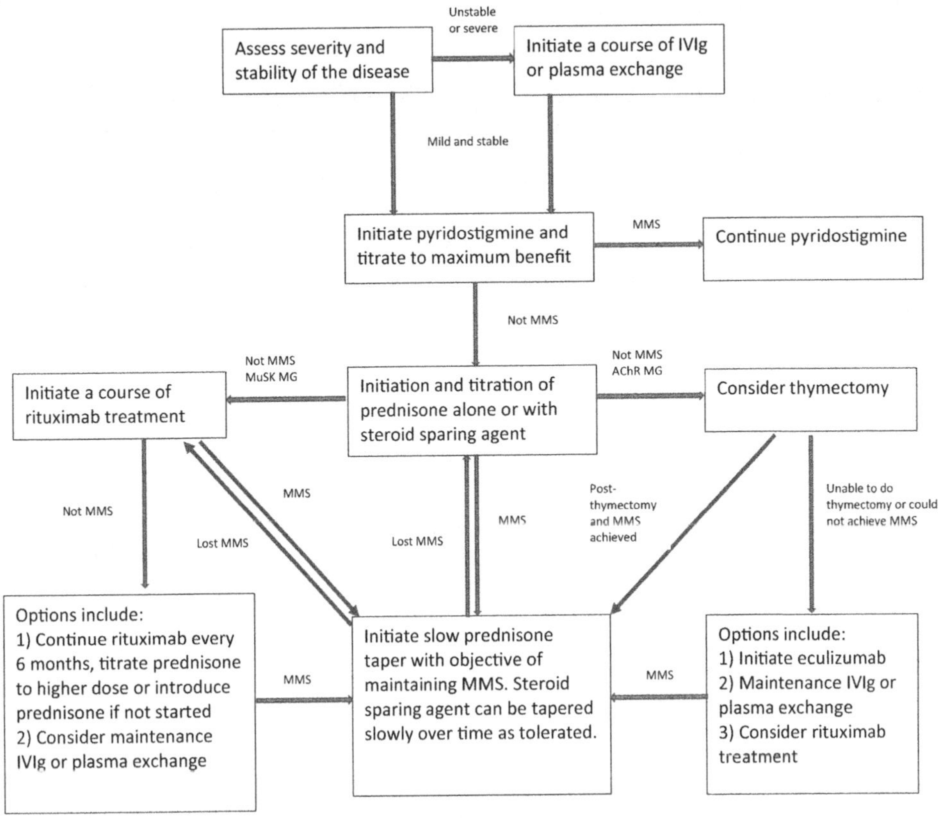

Fig. 1 Updated treatment algorithm with recent advances.

In a series of 317 thymectomy cases at a single institution where extended radical thymectomy was performed via sternotomy (83.6%), posterolateral thoracotomy (10.7%), and hemi-clamshell approach (5.7%), perioperative mortality associated with thymectomy was 1.6% and morbidity was 7.5%. In this population, MG was present at diagnosis in 276 (87.1%), and there was no difference in mortality or morbidity in patients with or without MG.[68] The authors speculate this difference is due to an improvement in anesthesia procedures and pharmacologic control of MG before and after surgery. The presence of MG has been related to a neutral or better prognosis for thymoma from 1980 onward.[69, 75–78] This might be from earlier detection of thymoma[78] related to MG symptoms. Otherwise, the tumor can be asymptomatic for prolonged periods of time until it invades adjacent structure or grows large enough to cause direct compression. From an MG standpoint, the presence of thymoma is considered a poor prognostic factor with greater severity of symptoms, poorer response to treatment, and a higher mortality rate.[6, 79–81]

Recent advances in treatment

Thymectomy

In 1936, Blalock removed a cystic thymic tumor from a 19-year-old girl resulting in a dramatic improvement in her MG. His reports on the serial experiences with thymectomy in MG had a catalytic effect on this operative treatment for MG patients.[10] Since then, thymectomy was routinely performed as a treatment for MG based on anecdotal experience and retrospective nonrandomized trials.[14] While thymectomy is clearly indicated when thymoma is present, it had been debated whether it should be performed in MG patients without tumors. Introduction of corticosteroids and other immunosuppressants amplified this debate as patients had expanding medical options for the treatment of MG. The frequency of thymectomy decreased precipitously between 2000 and 2005 in the United States when hospital admissions related to MG were examined.[82] While thymectomy has its own risk of complications and a low but potential mortality, long-term corticosteroid, and immunosuppressant use is not free from adverse effects either. The randomized trial of thymectomy in myasthenia gravis (MGTX) was the first prospective, rater-blinded controlled trial in non-thymomatous MG. MGTX compared thymectomy plus prednisone with prednisone alone and demonstrated a clinical benefit and steroid sparing effect of thymectomy in AChR antibody-positive generalized MG patients.[45, 47]

MGTX was an international, multicenter, 3-year trial followed by a voluntary 2-year extension for patients with AChR antibody-positive MG without thymoma. A total of 6958 participants were screened and 126 participants were enrolled between 2006 and 2012 at 36 sites worldwide. Patients were included if they have confirmed diagnosis of MG with positive AChR titer, disease duration less than 5 years, age between 18 and 65 years, and not on immunosuppressants other than prednisone. Further detailed inclusion and exclusion criteria can be found elsewhere.[45] Among the 126 participants, 66 participants were randomized to thymectomy plus prednisone and 60 participants to prednisone alone. Per protocol, thymectomy was performed within 30 days for those randomized to the surgery group. Prednisone naïve patients were started at 10 mg on alternate days which was increased incrementally (10 mg steps) to a maximum dose of 100 mg on alternate days or to 1.5 mg/kg body weight, whichever was lower. Patients who were taking prednisone at baseline were switched if needed to an equivalent alternate day dosing and followed the above titration. Patients who did not achieve MMS were allowed to titrate the prednisone dose up to 120 mg on alternate days. Once MMS was achieved, prednisone was tapered down based on the predefined protocol as long as MMS was maintained. Plasma exchange or IVIG were permitted at the discretion of the unblinded neurologist in patients whose condition was unstable, but they were not permitted to maintain MMS. Patients who did not achieve MMS at 12 months or who had unacceptable side effects from prednisone could take azathioprine at a dose of 2.5 mg/kg per day or another immunosuppressant such as cyclosporine if azathioprine caused side effects.[45]

The primary outcomes were the time-weighted average QMG score and time-weighted average required dose of prednisone over a 3-year period. Primary analysis demonstrated that the time-weighted average QMG score over the 3-year period was lower (better) in the thymectomy plus prednisone group with an estimated difference of 2.85 points (99.5% CI 0.47–5.22; $P<0.001$) and the time-weighted average alternate-day prednisone dose was 22 mg lower (95% CI 7–25; $P<0.001$) versus the prednisone alone group. No significant difference was found between treatment associated complications based on patient surveys ($P=0.73$), although a symptom survey indicated fewer treatment associated symptoms in the thymectomy plus prednisone group ($P<0.001$). Significantly fewer patients were hospitalized in the thymectomy plus prednisone group (9% vs 37%, $P<0.001$). The thymectomy plus prednisone group demonstrated a significantly lower frequency of azathioprine use (17% vs 48%, $P<0.001$), a higher percentage of patients reaching minimal-manifestation status at 3 years (67% vs 47%, $P=0.03$), and lower scores on the MG Activities of Daily Living scale (2.24 vs 3.41, $P=0.008$). Based on these results, the multinational group of authors concluded that thymectomy is beneficial for patients with generalized non-thymomatous MG with respect to improved clinical outcomes, reduced requirements for prednisone and other immunosuppressive therapies, decreased number of symptoms and their distress level related to medical therapies, and a lower need for hospitalization to manage disease exacerbations.

Subjects who completed the MGTX trial were given the option to enroll in the extension study.[47] A total of 35 patients in the thymectomy plus prednisone group and 33 patients in the prednisone alone group participated. Of these, 26 patients in the thymectomy plus prednisone group and 24 patients in the prednisone alone group completed the 60-month visit. At 60 months, patients in the thymectomy plus prednisone group had significantly lower (better) time-weighted mean QMG scores from month 0 to 60 compared with prednisone group (5.47 vs 9.34; $P=0.0007$). Time-weighted mean alternate-day prednisone dose was also significantly lower in the thymectomy plus prednisone group than the prednisone alone group (24 vs 48 mg; $P=0.0002$). The extension study continued to demonstrate other benefits of thymectomy through month 60 seen in the main 3-year MGTX trial, including reduced requirements for immunosuppressive agents and hospitalizations.[47]

While MGTX provides solid evidence that thymectomy is beneficial in treating generalized non-thymomatous MG, conclusions from the trial cannot be applied to the entire MG population as it only enrolled patients with AChR antibodies. The role of thymectomy in the treatment of MuSK and seronegative MG remains unclear. The MGTX trial did not enroll patients over 65 years of age and most of the participants were in younger groups under 50 years of age. Thymus undergoes fatty replacement in middle age and beyond, and antigen-specific T cells are uncommonly seen in atrophied, fat-replaced thymus.[8] Of note, greater intrathymic fat content correlated with higher time-weighted prednisone requirements in MGTX.[83] Treatment decisions surrounding thymectomy must be individualized in elderly MG patients and those who do not have AChR antibodies.[84]

Other important questions regarding thymectomy remain unanswered. One of them is the timing of thymectomy. The MGTX trial enrolled participants up to 5 years after symptom onset. We do not know whether thymectomy after a longer duration of symptoms would still be beneficial over medical therapy alone. One argument to support early thymectomy is that surgery decreases the requirement for corticosteroid and other immunosuppressant use. The lower cumulative doses of these agents over time are almost certainly of benefit to patients.[85]

While not an issue in the MGTX trial, MG exacerbation and crisis after thymectomy is known to occur both in thymomatous and nonthymomatous MG patients.[86–88] Japanese investigators studied MG patients who had thymectomy to find characteristics that help predict myasthenic crisis after thymectomy.[87] They first reviewed 275 patients with AChR antibody who underwent thymectomy to identify baseline and clinical variables associated with the risk of postoperative myasthenic crisis. A scoring system model was developed that fit best with outcomes using these identified variables and was validated in 118 patients with AChR antibody who underwent thymectomy at four centers in Japan. Factors significantly associated with the occurrence of postoperative crisis included (1) the time between MG onset and thymectomy (particularly less than 3 months), (2) MG severity using the MGFA classification, (3) thymoma, (4) bulbar symptoms, (5) neck weakness, and (6) forced vital capacity (FVC, a measure of lung function). The Postoperative Myasthenic Crisis Predictive Score was formulated by giving 1 point for preoperative bulbar symptoms, 2 points for duration ≤ 3 months, and 3 points if preoperative FVC was less than 80% predicted. The results demonstrated that only 1% of the low-score (<2) group experienced postoperative crisis compared to 22% of the high-score (≥ 3) group. The study suggests that stabilization of MG patients before thymectomy would help prevent postoperative crisis in patients with higher risk scores.

Another important question is whether other approaches including minimally invasive thymectomy would have similar beneficial effect compared to the extended transsternal thymectomy used in MGTX. Surgical approaches for thymectomy include transsternal, transcervical, minimally invasive video-assisted thoracoscopy, and related robotic techniques. Transsternal and transcervical approaches are open surgery, sometimes used in combination. Extended transsternal thymectomy involves the removal of all thymic tissue in the mediastinum, including the two cervical lobes of thymic tissue extending into the neck. It also includes resection of all fatty tissue in the anterosuperior mediastinum between the phrenic nerves from the diaphragm to the bottom of the thyroid gland.[89] Extended transsternal approach was used in the trial as surgeons across centers were all familial with this approach, and *en bloc* resection of all mediastinal tissue could be confidently accomplished this way. During MGTX, the question arose whether the trial should permit the increasingly utilized minimally invasive thoracoscopic approaches to facilitate recruitment. This was ultimately rejected as a trial result showing no benefit for thymectomy would quickly raise concern as to whether the resection of thymic tissue had been adequate. Extended transsternal thymectomy proved to be well tolerated in the trial with only one patient experiencing a surgical complication. That being said, minimally invasive approaches are now standard at most centers that participated in MGTX due to benefits seen in shorter postoperative recovery, reduced blood loss and pain, and improved cosmetics.[90–93]

The goal of thymectomy in MG is to remove the thymus gland as completely as possible. This is not a simple task due to the anatomy of thymus, which is amorphous, noncontiguous, and interspersed with fat in the neck and mediastinum. Studies have demonstrated that resected specimen weights were similar between extended transsternal and video-assisted thoracoscopic approaches[94] and that complete resection was comparable between minimally invasive and open approaches for thymoma resection.[95] These study results do argue that a minimally invasive approach can accomplish the primary purpose of thymectomy in MG. More prospective longitudinal follow-up studies should further bolster confidence in these newer, less invasive approaches. Whether a randomized trial comparing different thymectomy approaches is feasible at this point remains in question.

Eculizumab

Complement is important in both innate and adaptive immunity. Activation of the complement system protects the host against invading pathogens via cell lysis, opsonization with complement fragments, chemotaxis of inflammatory cells, and formation of the membrane attack complex (MAC).[96, 97] In the adaptive immune response, complement is the effector system for the primary and secondary antibody responses of B cells. Complement activation is regulated by a series of approximately 30 plasma and membrane proteins participating in classical, alternative, and lectin pathways. The classical pathway is activated by immune complexes containing antigen and IgM or complement fixating IgG. The alternative pathway is activated by foreign pathogens and polymeric IgA. The lectin pathway is initiated by binding of mannose-binding lectin to microbial pathogens. All three pathways converge at the cleavage of C3 by specific convertase enzymes followed by the generation of the MAC. In the context of MG, most AChR antibodies are complement fixating IgG1 and IgG3. Complement activation has been shown to be one of the major pathogenic mechanisms in AChR antibody-positive MG and experimental animal model of MG.[98, 99] It was logical to consider complement inhibition as a novel therapeutic approach in the treatment of MG. A therapeutic effect was clearly demonstrated in experimental animal models of the MG when an anti-C5 antibody was administered to Lewis rats.[100]

Eculizumab is a humanized murine monoclonal antibody that blocks activation of complement by uniquely binding to C5 and preventing enzymatic cleavage to C5a and C5b. This blocks activation of the terminal complement cascade and formation of the MAC. Eculizumab was previously approved in the treatment of atypical hemolytic uremic syndrome and paroxysmal nocturnal hemoglobinuria, both complemented-mediated disorders,[101, 102] and has been used in the clinical setting since 2007. The role of eculizumab in MG was initially explored in a pilot phase II crossover trial enrolling 14 patients with severe, refractory generalized MG. During the 16-week study periods, 86% of patients treated with eculizumab improved by at least three QMG points, and QMG improvement was significantly better in the eculizumab versus placebo group ($P=0.01$). Eculizumab was well tolerated with comparable adverse effects to placebo-treated patients.[103] These results prompted REGAIN, a larger phase 3 clinical trial.

The REGAIN study was a randomized, double blinded, placebo-controlled trial involving 76 sites in 17 countries.[46] Eligible patients were adults with refractory AChR antibody-positive generalized MG with an MG-ADL score of 6 or more despite treatment for at least 1 year with two or more immunosuppressant or immunotherapeutic agents (chronic IVIG or plasma exchange could be included as one of the therapies). Patients were excluded if they had thymic tumor, thymectomy in the past year, use of IVIG or plasma exchange in the last 4 weeks, or use of rituximab in the prior 6 months. All participants were required to have been vaccinated against *Neisseria meningitides* due to increased risk of severe meningococcal infection with complement inhibition. The study enrolled 126 subjects: 63 received eculizumab and 63 received placebo. Eculizumab was given as intravenous infusions, 900 mg weekly for the first month followed by 1200 mg every other week to 26 weeks. The authors evaluated disease severity by using patient-reported MG-ADL and MG-QOL15, the QMG and MG composite. Primary worst-rank analysis did not reveal a significant difference in the MG-ADL score change between the groups. Additional analyses, however, did show benefit for eculizumab in QMG score change and a higher proportion of patients (60% vs 40%, $P=0.0229$) achieving the clinically meaningful response of an MG-ADL improvement of 3 points or more. Overall, eculizumab was well tolerated with a safety profile similar to placebo group. In both groups, headache and upper respiratory infections were the most common adverse events. There was no increase in serious infections or reports of meningococcal infection, although in the open label phase of the study, one uncomplicated case of *N. meningitidis* did occur. Participants treated with eculizumab had fewer MG exacerbations, required less rescue medication, and experienced fewer hospital admissions. Eculizumab was approved by the FDA for treatment of AChR antibody-positive generalized MG in 2017. It is given as a 900 mg intravenous infusion weekly for the first 4 weeks followed by 1200 mg every 2 weeks.

Meningococcal vaccination at least 2 weeks prior is required before eculizumab treatment, and it is contraindicated in those who had previous meningococcal infections.

N. meningitidis is a Gram-negative bacterium that colonizes the nasopharynx and can cause meningitis and meningococcemia, a life-threatening sepsis. While meningococcal infections are very rare (approximately 1/100,000 per year) and children under 5 years of age are at highest risk followed by young adults, the use of eculizumab has been associated with a 1000- to 2000-fold increased incidence of meningococcal complications. The bacterium is classified into at least 13 distinct serogroups based on immunological reactivity and structure of the capsule's polysaccharide. Only six serogroups (A, B, C, W-135, X, Y) cause life-threatening disease.[104] Currently, two types of vaccination are available, serogroup B meningococcal vaccines and conjugate vaccines (group A, C, W, Y). After the initial vaccination, subsequent doses should follow Centers for Disease Control and Prevention (CDC) guidelines which are often updated. It is important to recognize that vaccination does not completely eliminate the risk for infection. Heightened awareness and rapid treatment of any symptoms consistent with meningococcal disease are essential for all patients receiving eculizumab, regardless of meningococcal vaccination or antimicrobial prophylaxis status.

Rituximab

The disease mechanism in MG is primarily defined by pathogenic autoantibodies targeting postsynaptic proteins integral to maintenance of transmission at the neuromuscular junction including AChR, MuSK, and more recently LRP4. Therefore, B cell targeted therapies that would interfere with production of these autoantibodies have been a recent focus of investigation.[105] Rituximab is a chimeric monoclonal antibody that targets the CD20 antigen found on subsets of the B cell lineage. CD20 is expressed by all mature B cells but not on pre-B or differentiated plasmablasts or plasma cells. Rituximab has been used as part of the standard therapy for non-Hodgkin lymphoma and also emerged as an effective tool in the management of various autoimmune diseases.[106] Interest in its use for MG began after a report of a favorable response in a patient with both lymphoma and MG.[107] Retrospective case reviews reported beneficial effects of rituximab in treatment refractory MG in terms of clinical improvement and subsequent tapering of other immunotherapies.[108–113] Although beneficial effects are often reported as durable, post-rituximab relapses are also reported.[113] While it seems intuitive that B cell depletion is bound to diminish pathogenic autoantibody production leading to disease improvement, further inspection reveals a story that is more complicated. Most circulating immunoglobulins are produced by long-lived plasma cells that do not express CD20. This is supported by unchanged immunization-generated and plasma cell-dependent antibody titers such as those against tetanus following rituximab treatment. Furthermore, the pathogenesis of AChR antibody-mediated MG, at least, is further complicated by complement-mediated destruction. A recent phase II randomized clinical trial (BeatMG) failed to demonstrate a significant prednisone-sparing benefit of rituximab versus placebo in AChR MG at 12 months (publication pending).

While the autoantibodies directed against AChR are predominantly complement fixating IgG1 and IgG3, MuSK autoantibodies are predominantly IgG4 and do not effectively activate complement. Thus, the pathogenesis in MuSK MG is likely dominated by antibody-mediated disruption of MuSK. This makes B cell targeted treatment an attractive option for MuSK MG patients, many of whom do not respond adequately to conventional therapies. Small retrospective case studies indeed demonstrate striking clinical benefit in MuSK MG that often produces a prolonged remission.[108–112, 114–116] Considering that MuSK MG is quite rare, randomized clinical trials are difficult. To overcome this limitation, Heir et al. performed a prospective chart review study to evaluate the efficacy and safety of rituximab by comparing MuSK MG patients treated with rituximab with those without such treatment.[117]

In this multicenter, blinded, prospective review, authors screened 119 MuSK MG patients across 10 sites and included 55 patients, 24 being treated with rituximab and 31 without such treatment. The primary outcome of this blinded review was to compare the achievement of a Myasthenia Gravis Status and Treatment Intensity (MGSTI) score of 2 or better. The MGSTI reflects both attainment of favorable clinical status and the immunotherapy requirements to achieve such status. There are seven different levels of MGSTI ranging from a score 0 being complete stable remission without immunotherapy to a score of 6 being symptomatic and requiring recent hospitalization for IVIG or plasma exchange treatment. The primary outcome target of a score ≤ 2 (level 2 reflects achievement of MMS on relatively modest doses of immunosuppressive agents such as mycophenolate <2000 mg/day or prednisone <20 mg/day) was reached significantly more frequently in the rituximab treated group (58%) than in controls (16%). In addition, only 29% of rituximab-treated patients were taking prednisone at the last visit, significantly less than the 74% in the control group. This study provides class IV evidence that rituximab treatment increases the probability of a favorable outcome in MuSK MG. The initial dose of rituximab in this prospective review was 375 mg/m^2 weekly for 4 weeks, the dose derived from the cancer literature. Various other regimens have been used successfully in MG.[118–120] After the first cycle of treatment, a second cycle can be repeated at 6 months or based on the provider's clinical judgment. Elevated MuSK antibody titers and recovery of B lymphocyte counts have been associated with disease recurrence after rituximab treatment and might be useful predictors for repeat dosing.[118, 121, 122] Further studies are needed to define the optimal treatment regimens for rituximab in MG.

Rituximab treatment has been associated with infusion reactions, mucocutaneous infections, reactivation of hepatitis C, tuberculosis and herpes zoster, and rare cases of progressive multifocal leukoencephalopathy. Complete blood count with differential, immunoglobulin measurements, liver function, basic metabolic panel, and hepatitis C, tuberculosis, and HIV testing are recommended prior to its administration. There are inadequate human data to guide rituximab use in pregnancy.

Other ongoing investigations

FcRn inhibitor

The neonatal Fc receptor (FcRn) plays a central role in IgG homeostasis by protecting IgG from lysosomal degradation, accounting for the long half-lives of IgG compared to other Ig isotypes.[123] It has been shown that IgG has an abnormally short half-life in FcRn-deficient mice and that FcRn blockade significantly increases the catabolism of serum IgG. In experimental autoimmune MG, anti-rat FcRn monoclonal antibody treatment resulted in dose-dependent amelioration of disease symptoms and a significant reduction in concentrations of pathogenic antibodies.[124] These findings support investigations of antihuman FcRn antagonists to decrease pathologic IgG antibodies and improve clinical status in MG patients.

Efgartigimod is an engineered molecule consisting of an IgG1 Fc portion that has been mutated at five residues to increase its FcRn affinity at both physiologic and acidic pH.[125] In a first-in-human study, a single administration of efgartigimod reduced total IgG by about 50%, while repeated dosing further lowered the levels by approximately 75%.[126] In a very recent randomized phase 2 study, efgartigimod was safe, well-tolerated, and demonstrated clinical benefit in 75% of participants based on the MG-ADL after 4 doses of treatment over a 3-week period, followed to week 11.[127] A larger randomized multicenter phase 3 trial is underway to confirm these favorable phase 2 findings.

Rozanolixizumab is an antihuman FcRn monoclonal antibody. In cynomolgus monkeys, rozanolixizumab was well tolerated and reduced IgG to a maximum 75%–90% in about 10 days.[128] In humans, rozanolixizumab demonstrated an acceptable safety profile after single administration of doses up to 4 mg/kg IV and 7 mg/kg SC. At 7 mg/kg IV, severe headache and back pain were reported. There were sustained dose-dependent reductions in serum IgG concentrations with both means of administration. A phase 2 randomized, double-blinded study to evaluate the efficacy and safety of rozanolixizumab in MG has been completed with some indication of a positive response, although a full report of the data is pending.

Subcutaneous immunoglobulin

Subcutaneous administration of immunoglobulin (SCIg) has gained interest in recent years as a simpler means than intravenous delivery, with a potential to increase patient autonomy and reduce health-care costs.[129–131] More stable serum IgG levels are achieved with SCIg, with a potentially reduced wearing-off effect at the end of treatment cycles and improved side effect profile.[132, 133] Comparisons between IVIG and SCIg have been recently reviewed.[131] SCIg has been successful in patients with immune-mediated neuropathies including chronic inflammatory demyelinating polyradiculoneuropathy,[134–137] and multifocal motor neuropathy.[138–141] In MG, a few retrospective studies[142, 143] show that SCIg is well tolerated and safe. More recently, SCIg was tested in a prospective open-label phase 3 trial in 23 patients with mild-to-moderate MG exacerbation.[144] Participants received SCIg 2 g/kg, self-administered over 4 weeks, infused 2–4 times per week with each infusion distributed over 4 anatomical sites. The primary outcome measure was the change in QMG score from baseline to the end of study at 6 weeks. After 4 weeks of SCIg treatment, QMG scores declined (improved) significantly at 6 weeks compared to baseline (9.8 vs 14.9, $P<0.0001$). Most common adverse events were headache and injection site reaction (17 and 14 out of 23). The authors concluded that SCIg is tolerable, safe, and effective in patients with mild-to-moderate MG exacerbation. One of the potential concerns of using SCIg in the treatment of acute MG exacerbation is that serum IgG levels rise more gradually than with intravenous delivery. In this study, IgG levels gradually increased over the 4-week treatment period with the major rise in level after 2 weeks. This suggests that IVIG could be considered in more severe exacerbation cases and crisis for a prompter clinical benefit. Efforts are ongoing to define the role of SCIg in the treatment of MG.

Subcutaneous complement inhibitor

Along similar lines, subcutaneous delivery of complement inhibitors is being studied. Zilucoplan is a subcutaneously administered macrocyclic peptide that binds to C5 and inhibits its cleavage into C5a and C5b, preventing the production of MAC. The results of a phase 2 randomized, placebo-controlled trial were recently presented.[145] In all, 44 patients were randomized 1:1:1 to placebo, zilucoplan 0.1 mg/kg, or zilucoplan 0.3 mg/kg subcutaneously daily over 12 weeks. Zilucoplan dosed at 0.3 mg/kg achieved a mean reduction of QMG from baseline of 6.0 points (placebo-corrected −2.8, $P=0.05$). None of 14 participants receiving zilucoplan 0.3 mg/kg needed rescue therapy while 3 out of 15 participants in the placebo arm did. Zilucoplan was observed to have a favorable safety and tolerability profile. These results, if confirmed in a planned phase III trial, would provide another therapeutic modality to achieve complement inhibition for treatment of MG.

Longer lasting complement inhibitor

A strategy to avoid frequent administration of intravenous medication is by using agents with longer half-lives. Ravulizumab is a humanized monoclonal antibody that inhibits C5, designed via targeted substitution of four amino acids in the complementary binding and neonatal Fc regions of the eculizumab backbone. These modifications result in a half-life 4 times longer than eculizumab. In a noninferiority trial of ravulizumab versus eculizumab in adults with paroxysmal nocturnal hemoglobinuria, ravulizumab treatment every 8 weeks was comparable to eculizumab given every 2 weeks in terms of efficacy and.[146, 147] Studies of ravulizumab in MG are currently in progress.

3,4-DAP for MuSK MG

MuSK MG has distinct clinical features when compared to other populations of MG, including resistance to cholinesterase inhibitors and even worsening of symptoms with these agents. Alternatively, an experimental mouse model suggested that 3,4-DAP might play a positive role in the symptomatic treatment of MuSK MG.[148] Amifampiridine phosphate (3,4-DAP) was recently approved for Lambert-Eaton myasthenic syndrome, a presynaptic autoimmune disease affecting neuromuscular transmission. The safety, tolerability, and efficacy of amifampiridine phosphate was evaluated in a randomized, double-blind, placebo-controlled crossover study of MuSK-MG patients.[149] Adult patients with generalized disease without history of epilepsy, uncontrolled asthma, or long QT syndrome (3,4-DAP can exacerbate these conditions) were enrolled. Seven patients were randomized, with four were assigned to the placebo-amifampiridine-placebo arm and three to the amifampiridine-placebo-amifampiridine arm. When patients were crossed over from amifampiridine to placebo there was a significant worsening in outcomes. For example, QMG scores were 6.9 points higher (worse) on placebo and this worsening resolved during periods on active drug. Amifampiridine phosphate was well tolerated without serious adverse events; the most common side effect was intermittent tingling, commonly seen with this agent. A larger placebo-controlled trial is underway to confirm results of this pilot trial.

Summary

The therapeutic approach to MG has undergone unprecedented change in the past decade. The longstanding questioned efficacy of thymectomy in non-thymomatous MG was finally addressed in a randomized, medication-controlled clinical trial that demonstrated numerous benefits for the procedure. Investigations into novel, targeted therapies are underway and gaining momentum to supplement or replace traditional approaches that have largely utilized nonselective immunosuppression. Complement inhibition has proven to be efficacious in AChR antibody-positive MG and is now approved for use in patients with generalized disease. Several FcRn inhibitors that facilitate the degradation of pathologic IgG are entering phase 3 trials across MG subpopulations. The B-cell therapy rituximab still remains a viable option for MuSK MG and perhaps other forms of refractory MG based on large case series. As the field advances, verification of antibody status will become even more crucial in designing an effective therapeutic plan for patients with MG.

We have updated a treatment algorithm based on evidence from the literature as a guide to clinicians that can be applied to many patients with non-thymomatous MG. The algorithm was not generated by formal consensus of MG specialists and should not be viewed as a strict guideline for all MG patients. It will certainly require modification as results from current and future trials become available.

References

1. Carr AS, Cardwell CR, McCarron PO, McConville J. A systematic review of population based epidemiological studies in Myasthenia Gravis. *BMC Neurol.* 2010;10:46.
2. McGrogan A, Sneddon S, de Vries CS. The incidence of myasthenia gravis: a systematic literature review. *Neuroepidemiology.* 2010;34(3):171–183.
3. Bever Jr. CT, Aquino AV, Penn AS, Lovelace RE, Rowland LP. Prognosis of ocular myasthenia. *Ann Neurol.* 1983;14(5):516–519.
4. Evoli A, Tonali P, Bartoccioni E, Lo Monaco M. Ocular myasthenia: diagnostic and therapeutic problems. *Acta Neurol Scand.* 1988;77(1):31–35.
5. Kupersmith MJ, Latkany R, Homel P. Development of generalized disease at 2 years in patients with ocular myasthenia gravis. *Arch Neurol.* 2003;60(2):243–248.
6. Grob D, Brunner N, Namba T, Pagala M. Lifetime course of myasthenia gravis. *Muscle Nerve.* 2008;37 (2):141–149.
7. Wendell LC, Levine JM. Myasthenic crisis. *Neurohospitalist.* 2011;1(1):16–22.
8. Gilhus NE. Myasthenia Gravis. *N Engl J Med.* 2016;375(26):2570–2581.
9. Huijbers MG, Lipka AF, Plomp JJ, Niks EH, van der Maarel SM, Verschuuren JJ. Pathogenic immune mechanisms at the neuromuscular synapse: the role of specific antibody-binding epitopes in myasthenia gravis. *J Intern Med.* 2014;275(1):12–26.
10. Blalock A, Mason MF, Morgan HJ, Riven SS. Myasthenia gravis and tumors of the thymic region: report of a case in which the tumor was removed. *Ann Surg.* 1939;110(4):544–561.
11. Blalock A. Gordon Wilson lectures: shock or peripheral circulatory failure. *Trans Am Clin Climatol Assoc.* 1941;57:1–12.
12. Lanska DJ. Indications for thymectomy in myasthenia gravis. *Neurology.* 1990;40(12):1828–1829.
13. McQuillen MP, Leone MG. A treatment carol: thymectomy revisited. *Neurology.* 1977;27 (12):1103–1106.
14. Gronseth GS, Barohn RJ. Practice parameter: thymectomy for autoimmune myasthenia gravis (an evidence-based review): report of the Quality Standards Subcommittee of the American Academy of Neurology. *Neurology.* 2000;55(1):7–15.
15. Hubbard SR, Gnanasambandan K. Structure and activation of MuSK, a receptor tyrosine kinase central to neuromuscular junction formation. *Biochim Biophys Acta.* 2013;1834(10):2166–2169.
16. Hoch W, McConville J, Helms S, Newsom-Davis J, Melms A, Vincent A. Auto-antibodies to the receptor tyrosine kinase MuSK in patients with myasthenia gravis without acetylcholine receptor antibodies. *Nat Med.* 2001;7(3):365–368.
17. Marx A, Pfister F, Schalke B, Saruhan-Direskeneli G, Melms A, Strobel P. The different roles of the thymus in the pathogenesis of the various myasthenia gravis subtypes. *Autoimmun Rev.* 2013;12 (9):875–884.
18. Lauriola L, Ranelletti F, Maggiano N, et al. Thymus changes in anti-MuSK-positive and -negative myasthenia gravis. *Neurology.* 2005;64(3):536–538.
19. Saka E, Topcuoglu MA, Akkaya B, Galati A, Onal MZ, Vincent A. Thymus changes in anti-MuSK-positive and -negative myasthenia gravis. *Neurology.* 2005;65(5):782–783 author reply. 782-783.
20. Wolfe GI, Oh SJ. Clinical phenotype of muscle-specific tyrosine kinase-antibody-positive myasthenia gravis. *Ann N Y Acad Sci.* 2008;1132:71–75.
21. Oh SJ, Morgan MB, Lu L, et al. Racial differences in myasthenia gravis in Alabama. *Muscle Nerve* 2009;39(3):328–332.
22. Oh SJ, Morgan MB, Lu L, et al. Different characteristic phenotypes according to antibody in myasthenia gravis. *J Clin Neuromuscul Dis.* 2012;14(2):57–65.
23. Evoli A, Tonali PA, Padua L, et al. Clinical correlates with anti-MuSK antibodies in generalized seronegative myasthenia gravis. *Brain.* 2003;126(Pt 10):2304–2311.
24. Sanders DB, El-Salem K, Massey JM, McConville J, Vincent A. Clinical aspects of MuSK antibody positive seronegative MG. *Neurology.* 2003;60(12):1978–1980.
25. Takahashi H, Kawaguchi N, Nemoto Y, Hattori T. High-dose intravenous immunoglobulin for the treatment of MuSK antibody-positive seronegative myasthenia gravis. *J Neurol Sci.* 2006;247 (2):239–241.

26. Ponseti JM, Caritg N, Gamez J, Lopez-Cano M, Vilallonga R, Armengol M. A comparison of long-term post-thymectomy outcome of anti-AChR-positive, anti-AChR-negative and anti-MuSK-positive patients with non-thymomatous myasthenia gravis. *Expert Opin Biol Ther*. 2009;9(1):1–8.
27. Clifford KM, Hobson-Webb LD, Benatar M, et al. Thymectomy may not be associated with clinical improvement in MuSK myasthenia gravis. *Muscle Nerve*. 2019;59(4):404–410.
28. Kushlaf H, Li Y. The evidence is stacked against thymectomy in MuSK myasthenia gravis. *Muscle Nerve*. 2019;59(4):393–394.
29. Leite MI, Jacob S, Viegas S, et al. IgG1 antibodies to acetylcholine receptors in 'seronegative' myasthenia gravis. *Brain*. 2008;131(Pt 7):1940–1952.
30. Rodriguez Cruz PM, Al-Hajjar M, Huda S, et al. Clinical features and diagnostic usefulness of antibodies to clustered acetylcholine receptors in the diagnosis of seronegative myasthenia gravis. *JAMA Neurol*. 2015;72(6):642–649.
31. Tsonis AI, Zisimopoulou P, Lazaridis K, et al. MuSK autoantibodies in myasthenia gravis detected by cell based assay—a multinational study. *J Neuroimmunol*. 2015;284:10–17.
32. Zhang B, Tzartos JS, Belimezi M, et al. Autoantibodies to lipoprotein-related protein 4 in patients with double-seronegative myasthenia gravis. *Arch Neurol*. 2012;69(4):445–451.
33. Zhang B, Shen C, Bealmear B, et al. Autoantibodies to agrin in myasthenia gravis patients. *PLoS One*. 2014;9(3):e91816.
34. Barik A, Lu Y, Sathyamurthy A, et al. LRP4 is critical for neuromuscular junction maintenance. *J Neurosci*. 2014;34(42):13892–13905.
35. Cortes-Vicente E, Gallardo E, Martinez MA, et al. Clinical characteristics of patients with double-seronegative myasthenia gravis and antibodies to cortactin. *JAMA Neurol*. 2016;73(9):1099–1104.
36. Yan M, Xing GL, Xiong WC, Mei L. Agrin and LRP4 antibodies as new biomarkers of myasthenia gravis. *Ann N Y Acad Sci*. 2018;1413(1):126–135.
37. Kaminski HJ. Seronegative myasthenia gravis—a vanishing disorder? *JAMA Neurol*. 2016;73(9):1055–1056.
38. Padua L, Stalberg E, LoMonaco M, Evoli A, Batocchi A, Tonali P. SFEMG in ocular myasthenia gravis diagnosis. *Clin Neurophysiol*. 2000;111(7):1203–1207.
39. Mittal MK, Barohn RJ, Pasnoor M, et al. Ocular myasthenia gravis in an academic neuro-ophthalmology clinic: clinical features and therapeutic response. *J Clin Neuromuscul Dis*. 2011;13(1):46–52.
40. Oh SJ, Jeong D, Lee I, Alsharabati M. Repetitive nerve stimulation test in myasthenic crisis. *Muscle Nerve*. 2019;59(5):544–548.
41. AAEM Quality Assurance Committee, American Association of Electrodiagnostic Medicine. Practice parameter for repetitive nerve stimulation and single fiber EMG evaluation of adults with suspected myasthenia gravis or Lambert-Eaton myasthenic syndrome: summary statement. *Muscle Nerve*. 2001;24(9):1236–1238.
42. Truffault F, de Montpreville V, Eymard B, Sharshar T, Le Panse R, Berrih-Aknin S. Thymic germinal centers and corticosteroids in myasthenia gravis: an immunopathological study in 1035 cases and a critical review. *Clin Rev Allergy Immunol*. 2017;52(1):108–124.
43. Mao ZF, Mo XA, Qin C, Lai YR, Hackett ML. Incidence of thymoma in myasthenia gravis: a systematic review. *J Clin Neurol*. 2012;8(3):161–169.
44. Sanders DB, Hart IK, Mantegazza R, et al. An international, phase III, randomized trial of mycophenolate mofetil in myasthenia gravis. *Neurology*. 2008;71(6):400–406.
45. Wolfe GI, Kaminski HJ, Aban IB, et al. Randomized trial of thymectomy in myasthenia gravis. *N Engl J Med*. 2016;375(6):511–522.
46. Howard Jr. JF, Utsugisawa K, Benatar M, et al. Safety and efficacy of eculizumab in anti-acetylcholine receptor antibody-positive refractory generalised myasthenia gravis (REGAIN): a phase 3, randomised, double-blind, placebo-controlled, multicentre study. *Lancet Neurol*. 2017;16(12):976–986.
47. Wolfe GI, Kaminski HJ, Aban IB, et al. Long-term effect of thymectomy plus prednisone versus prednisone alone in patients with non-thymomatous myasthenia gravis: 2-year extension of the MGTX randomised trial. *Lancet Neurol*. 2019;18(3):259–268.
48. Jaretzki 3rd A, Barohn RJ, Ernstoff RM, et al. Myasthenia gravis: recommendations for clinical research standards. Task Force of the Medical Scientific Advisory Board of the Myasthenia Gravis Foundation of America. *Ann Thorac Surg*. 2000;70(1):327–334.

49. Sharshar T, Chevret S, Mazighi M, et al. Validity and reliability of two muscle strength scores commonly used as endpoints in assessing treatment of myasthenia gravis. *J Neurol.* 2000;247(4):286–290.
50. Wolfe GI, Herbelin L, Nations SP, Foster B, Bryan WW, Barohn RJ. Myasthenia gravis activities of daily living profile. *Neurology.* 1999;52(7):1487–1489.
51. Bedlack RS, Simel DL, Bosworth H, Samsa G, Tucker-Lipscomb B, Sanders DB. Quantitative myasthenia gravis score: assessment of responsiveness and longitudinal validity. *Neurology.* 2005;64(11):1968–1970.
52. Muppidi S. The myasthenia gravis–specific activities of daily living profile. *Ann N Y Acad Sci.* 2012;1274:114–119.
53. Suzuki S, Murai H, Imai T, et al. Quality of life in purely ocular myasthenia in Japan. *BMC Neurol.* 2014;14:142.
54. Kupersmith MJ, Ying G. Ocular motor dysfunction and ptosis in ocular myasthenia gravis: effects of treatment. *Br J Ophthalmol.* 2005;89(10):1330–1334.
55. Benatar M, McDermott MP, Sanders DB, et al. Efficacy of prednisone for the treatment of ocular myasthenia (EPITOME): a randomized, controlled trial. *Muscle Nerve.* 2016;53(3):363–369.
56. Grob D, Arsura EL, Brunner NG, Namba T. The course of myasthenia gravis and therapies affecting outcome. *Ann N Y Acad Sci.* 1987;505:472–499.
57. Kupersmith MJ. Does early immunotherapy reduce the conversion of ocular myasthenia gravis to generalized myasthenia gravis? *J Neuroophthalmol.* 2003;23(4):249–250.
58. Chan JW. Mycophenolate mofetil for ocular myasthenia. *J Neurol.* 2008;255(4):510–513.
59. Sommer N, Sigg B, Melms A, et al. Ocular myasthenia gravis: response to long-term immunosuppressive treatment. *J Neurol Neurosurg Psychiatry.* 1997;62(2):156–162.
60. Zinman L, Ng E, Bril V. IV immunoglobulin in patients with myasthenia gravis: a randomized controlled trial. *Neurology.* 2007;68(11):837–841.
61. Sanders DB, Wolfe GI, Benatar M, et al. International consensus guidance for management of myasthenia gravis: executive summary. *Neurology.* 2016;87(4):419–425.
62. Gajdos P, Chevret S, Clair B, Tranchant C, Chastang C. Clinical trial of plasma exchange and high-dose intravenous immunoglobulin in myasthenia gravis. Myasthenia Gravis Clinical Study Group. *Ann Neurol.* 1997;41(6):789–796.
63. Silvestri NJ, Wolfe GI. Myasthenia gravis. *Semin Neurol.* 2012;32(3):215–226.
64. Strobel P, Bauer A, Puppe B, et al. Tumor recurrence and survival in patients treated for thymomas and thymic squamous cell carcinomas: a retrospective analysis. *J Clin Oncol.* 2004;22(8):1501–1509.
65. Maggi L, Andreetta F, Antozzi C, et al. Thymoma-associated myasthenia gravis: outcome, clinical and pathological correlations in 197 patients on a 20-year experience. *J Neuroimmunol.* 2008;201-202:237–244.
66. Marx A, Willcox N, Leite MI, et al. Thymoma and paraneoplastic myasthenia gravis. *Autoimmunity.* 2010;43(5-6):413–427.
67. Davenport E, Malthaner RA. The role of surgery in the management of thymoma: a systematic review. *Ann Thorac Surg.* 2008;86(2):673–684.
68. Margaritora S, Cesario A, Cusumano G, et al. Thirty-five-year follow-up analysis of clinical and pathologic outcomes of thymoma surgery. *Ann Thorac Surg.* 2010;89(1):245–252. discussion 252.
69. Maggi G, Casadio C, Cavallo A, Cianci R, Molinatti M, Ruffini E. Thymoma: results of 241 operated cases. *Ann Thorac Surg.* 1991;51(1):152–156.
70. Regnard JF, Zinzindohoue F, Magdeleinat P, Guibert L, Spaggiari L, Levasseur P. Results of re-resection for recurrent thymomas. *Ann Thorac Surg.* 1997;64(6):1593–1598.
71. Johnson SB, Eng TY, Giaccone G, Thomas Jr. CR. Thymoma: update for the new millennium. *Oncologist.* 2001;6(3):239–246.
72. Hejna M, Haberl I, Raderer M. Nonsurgical management of malignant thymoma. *Cancer.* 1999;85(9):1871–1884.
73. Wilkins Jr. EW, Edmunds Jr. LH, Castleman B. Cases of thymoma at the Massachusetts General Hospital. *J Thorac Cardiovasc Surg.* 1966;52(3):322–330.
74. Batata MA, Martini N, Huvos AG, Aguilar RI, Beattie Jr. EJ. Thymomas: clinicopathologic features, therapy, and prognosis. *Cancer.* 1974;34(2):389–396.

75. Monden Y, Uyama T, Taniki T, et al. The characteristics of thymoma with myasthenia gravis: a 28-year experience. *J Surg Oncol.* 1988;38(3):151–154.
76. Wilkins KB, Sheikh E, Green R, et al. Clinical and pathologic predictors of survival in patients with thymoma. *Ann Surg.* 1999;230(4):562–572. discussion 572-564.
77. Okumura M, Miyoshi S, Takeuchi Y, et al. Results of surgical treatment of thymomas with special reference to the involved organs. *J Thorac Cardiovasc Surg.* 1999;117(3):605–613.
78. Kondo K, Monden Y. Thymoma and myasthenia gravis: a clinical study of 1,089 patients from Japan. *Ann Thorac Surg.* 2005;79(1):219–224.
79. Maggi G, Casadio C, Cavallo A, Cianci R, Molinatti M, Ruffini E. Thymectomy in myasthenia gravis. Results of 662 cases operated upon in 15 years. *Eur J Cardiothorac Surg.* 1989;3(6):504–509. discussion 510-501.
80. Lucchi M, Ricciardi R, Melfi F, et al. Association of thymoma and myasthenia gravis: oncological and neurological results of the surgical treatment. *Eur J Cardiothorac Surg.* 2009;35(5):812–816. discussion 816.
81. Vachlas K, Zisis C, Rontogianni D, Tavernarakis A, Psevdi A, Bellenis I. Thymoma and myasthenia gravis: clinical aspects and prognosis. *Asian Cardiovasc Thorac Ann.* 2012;20(1):48–52.
82. Omorodion JO, Pines JM, Kaminski HJ. Inpatient cost analysis for treatment of myasthenia gravis. *Muscle Nerve.* 2017;56(6):1114–1118.
83. Weis CA, Aban IB, Cutter G, et al. Histopathology of thymectomy specimens from the MGTX-trial: entropy analysis as strategy to quantify spatial heterogeneity of lymphoid follicle and fat distribution. *PLoS One.* 2018;13(6):e0197435.
84. Mi W, Silvestri NJ, Wolfe GI. A neurologist's perspective on thymectomy for myasthenia gravis: current perspective and future trials. *Thorac Surg Clin.* 2019;29(2):143–150.
85. Curtis JR, Westfall AO, Allison J, et al. Population-based assessment of adverse events associated with long-term glucocorticoid use. *Arthritis Rheum.* 2006;55(3):420–426.
86. Nam TS, Lee SH, Kim BC, et al. Clinical characteristics and predictive factors of myasthenic crisis after thymectomy. *J Clin Neurosci.* 2011;18(9):1185–1188.
87. Kanai T, Uzawa A, Sato Y, et al. A clinical predictive score for postoperative myasthenic crisis. *Ann Neurol.* 2017;82(5):841–849.
88. Lee HS, Lee HS, Lee HE, et al. Predictive factors for myasthenic crisis after videoscopic thymectomy in patients with myasthenia gravis. *Muscle Nerve.* 2015;52(2):216–220.
89. Daniel VC, Wright CD. Extended transsternal thymectomy. *Thorac Surg Clin.* 2010;20(2):245–252.
90. Toker A, Eroglu O, Ziyade S, et al. Comparison of early postoperative results of thymectomy: partial sternotomy vs. videothoracoscopy. *Thorac Cardiovasc Surg.* 2005;53(2):110–113.
91. Cheng YJ, Kao EL, Chou SH. Videothoracoscopic resection of stage II thymoma: prospective comparison of the results between thoracoscopy and open methods. *Chest.* 2005;128(4):3010–3012.
92. Wagner AJ, Cortes RA, Strober J, et al. Long-term follow-up after thymectomy for myasthenia gravis: thoracoscopic vs open. *J Pediatr Surg.* 2006;41(1):50–54. discussion 50-54.
93. Jurado J, Javidfar J, Newmark A, et al. Minimally invasive thymectomy and open thymectomy: outcome analysis of 263 patients. *Ann Thorac Surg.* 2012;94(3):974–981. discussion 981-972.
94. Lee CY, Kim DJ, Lee JG, Park IK, Bae MK, Chung KY. Bilateral video-assisted thoracoscopic thymectomy has a surgical extent similar to that of transsternal extended thymectomy with more favorable early surgical outcomes for myasthenia gravis patients. *Surg Endosc.* 2011;25(3):849–854.
95. Burt BM, Yao X, Shrager J, et al. Determinants of complete resection of thymoma by minimally invasive and open thymectomy: analysis of an International Registry. *J Thorac Oncol.* 2017;12(1):129–136.
96. Walport MJ. Complement. Second of two parts. *N Engl J Med.* 2001;344(15):1140–1144.
97. Walport MJ. Complement. First of two parts. *N Engl J Med.* 2001;344(14):1058–1066.
98. Sahashi K, Engel AG, Lambert EH, Howard Jr. FM. Ultrastructural localization of the terminal and lytic ninth complement component (C9) at the motor end-plate in myasthenia gravis. *J Neuropathol Exp Neurol.* 1980;39(2):160–172.
99. Kusner LL, Kaminski HJ, Soltys J. Effect of complement and its regulation on myasthenia gravis pathogenesis. *Expert Rev Clin Immunol.* 2008;4(1):43–52.

100. Zhou Y, Gong B, Lin F, Rother RP, Medof ME, Kaminski HJ. Anti-C5 antibody treatment ameliorates weakness in experimentally acquired myasthenia gravis. *J Immunol.* 2007;179(12):8562–8567.
101. Hillmen P, Hall C, Marsh JC, et al. Effect of eculizumab on hemolysis and transfusion requirements in patients with paroxysmal nocturnal hemoglobinuria. *N Engl J Med.* 2004;350(6):552–559.
102. Legendre CM, Licht C, Muus P, et al. Terminal complement inhibitor eculizumab in atypical hemolytic-uremic syndrome. *N Engl J Med.* 2013;368(23):2169–2181.
103. Howard Jr. JF, Barohn RJ, Cutter GR, et al. A randomized, double-blind, placebo-controlled phase II study of eculizumab in patients with refractory generalized myasthenia gravis. *Muscle Nerve.* 2013;48(1):76–84.
104. Rouphael NG, Stephens DS. Neisseria meningitidis: biology, microbiology, and epidemiology. *Methods Mol Biol.* 2012;799:1–20.
105. Yi JS, Guptill JT, Stathopoulos P, Nowak RJ, O'Connor KC. B cells in the pathophysiology of myasthenia gravis. *Muscle Nerve.* 2018;57(2):172–184.
106. Dorner T, Isenberg D, Jayne D, et al. Current status on B-cell depletion therapy in autoimmune diseases other than rheumatoid arthritis. *Autoimmun Rev.* 2009;9(2):82–89.
107. Gajra A, Vajpayee N, Grethlein SJ. Response of myasthenia gravis to rituximab in a patient with non-Hodgkin lymphoma. *Am J Hematol.* 2004;77(2):196–197.
108. Illa I, Diaz-Manera J, Rojas-Garcia R, et al. Sustained response to Rituximab in anti-AChR and anti-MuSK positive Myasthenia Gravis patients. *J Neuroimmunol.* 2008;201-202:90–94.
109. Evoli A, Bianchi MR, Riso R, et al. Response to therapy in myasthenia gravis with anti-MuSK antibodies. *Ann N Y Acad Sci.* 2008;1132:76–83.
110. Lebrun C, Bourg V, Tieulie N, Thomas P. Successful treatment of refractory generalized myasthenia gravis with rituximab. *Eur J Neurol.* 2009;16(2):246–250.
111. Zebardast N, Patwa HS, Novella SP, Goldstein JM. Rituximab in the management of refractory myasthenia gravis. *Muscle Nerve.* 2010;41(3):375–378.
112. Diaz-Manera J, Martinez-Hernandez E, Querol L, et al. Long-lasting treatment effect of rituximab in MuSK myasthenia. *Neurology.* 2012;78(3):189–193.
113. Robeson KR, Kumar A, Keung B, et al. Durability of the rituximab response in acetylcholine receptor autoantibody-positive myasthenia gravis. *JAMA Neurol.* 2017;74(1):60–66.
114. Hain B, Jordan K, Deschauer M, Zierz S. Successful treatment of MuSK antibody-positive myasthenia gravis with rituximab. *Muscle Nerve.* 2006;33(4):575–580.
115. Baek WS, Bashey A, Sheean GL. Complete remission induced by rituximab in refractory, seronegative, muscle-specific, kinase-positive myasthenia gravis. *J Neurol Neurosurg Psychiatry.* 2007;78(7):771.
116. Keung B, Robeson KR, DiCapua DB, et al. Long-term benefit of rituximab in MuSK autoantibody myasthenia gravis patients. *J Neurol Neurosurg Psychiatry.* 2013;84(12):1407–1409.
117. Hehir MK, Hobson-Webb LD, Benatar M, et al. Rituximab as treatment for anti-MuSK myasthenia gravis: multicenter blinded prospective review. *Neurology.* 2017;89(10):1069–1077.
118. Blum S, Gillis D, Brown H, et al. Use and monitoring of low dose rituximab in myasthenia gravis. *J Neurol Neurosurg Psychiatry.* 2011;82(6):659–663.
119. Jing Y, Liu X, Yu L, Li R. Outcomes of preoperative and postoperative corticosteroid therapies in myasthenia gravis. *Eur Neurol.* 2017;78(1-2):86–92.
120. Chan F, Swayne A, Gillis D, et al. Long-term follow-up of patients with myasthenia gravis treated with low-dose rituximab. *J Neurol Neurosurg Psychiatry.* 2019;90:955–956.
121. Bartoccioni E, Scuderi F, Minicuci GM, Marino M, Ciaraffa F, Evoli A. Anti-MuSK antibodies: correlation with myasthenia gravis severity. *Neurology.* 2006;67(3):505–507.
122. Lebrun C, Bourg V, Bresch S, et al. Therapeutic target of memory B cells depletion helps to tailor administration frequency of rituximab in myasthenia gravis. *J Neuroimmunol.* 2016;298:79–81.
123. Roopenian DC, Akilesh S. FcRn: the neonatal Fc receptor comes of age. *Nat Rev Immunol.* 2007;7(9):715–725.
124. Liu X, Ye L, Christianson GJ, Yang JQ, Roopenian DC, Zhu X. NF-kappaB signaling regulates functional expression of the MHC class I-related neonatal Fc receptor for IgG via intronic binding sequences. *J Immunol.* 2007;179(5):2999–3011.

125. Vaccaro C, Zhou J, Ober RJ, Ward ES. Engineering the Fc region of immunoglobulin G to modulate in vivo antibody levels. *Nat Biotechnol*. 2005;23(10):1283–1288.
126. Ulrichts P, Guglietta A, Dreier T, et al. Neonatal Fc receptor antagonist efgartigimod safely and sustainably reduces IgGs in humans. *J Clin Invest*. 2018;128(10):4372–4386.
127. Howard Jr. JF, Bril V, Burns TM, et al. Randomized phase 2 study of FcRn antagonist efgartigimod in generalized myasthenia gravis. *Neurology*. 2019;92(23):e2661–e2673.
128. Kiessling P, Lledo-Garcia R, Watanabe S, et al. The FcRn inhibitor rozanolixizumab reduces human serum IgG concentration: a randomized phase 1 study. *Sci Transl Med*. 2017;9(414).
129. Igarashi A, Kanegane H, Kobayashi M, Miyawaki T, Tsutani K. Cost-minimization analysis of IgPro20, a subcutaneous immunoglobulin, in Japanese patients with primary immunodeficiency. *Clin Ther*. 2014;36(11):1616–1624.
130. Vultaggio A, Azzari C, Milito C, et al. Subcutaneous immunoglobulin replacement therapy in patients with primary immunodeficiency in routine clinical practice: the VISPO prospective multicenter study. *Clin Drug Invest*. 2015;35(3):179–185.
131. Bittner B, Richter W, Schmidt J. Subcutaneous administration of biotherapeutics: an overview of current challenges and opportunities. *BioDrugs*. 2018;32(5):425–440.
132. Hagan JB, Fasano MB, Spector S, et al. Efficacy and safety of a new 20% immunoglobulin preparation for subcutaneous administration, IgPro20, in patients with primary immunodeficiency. *J Clin Immunol*. 2010;30(5):734–745.
133. Wasserman RL. Hizentra for the treatment of primary immunodeficiency. *Expert Rev Clin Immunol*. 2014;10(10):1293–1307.
134. Markvardsen LH, Debost JC, Harbo T, et al. Subcutaneous immunoglobulin in responders to intravenous therapy with chronic inflammatory demyelinating polyradiculoneuropathy. *Eur J Neurol*. 2013;20(5):836–842.
135. Bayas A, Gold R, Naumann M. Long-term treatment of Lewis-Sumner syndrome with subcutaneous immunoglobulin infusions. *J Neurol Sci*. 2013;324(1-2):53–56.
136. Cocito D, Serra G, Falcone Y, Paolasso I. The efficacy of subcutaneous immunoglobulin administration in chronic inflammatory demyelinating polyneuropathy responders to intravenous immunoglobulin. *J Peripher Nerv Syst*. 2011;16(2):150–152.
137. Lee DH, Linker RA, Paulus W, Schneider-Gold C, Chan A, Gold R. Subcutaneous immunoglobulin infusion: a new therapeutic option in chronic inflammatory demyelinating polyneuropathy. *Muscle Nerve*. 2008;37(3):406–409.
138. Eftimov F, Vermeulen M, de Haan RJ, van den Berg LH, van Schaik IN. Subcutaneous immunoglobulin therapy for multifocal motor neuropathy. *J Peripher Nerv Syst*. 2009;14(2):93–100.
139. Harbo T, Andersen H, Hess A, Hansen K, Sindrup SH, Jakobsen J. Subcutaneous versus intravenous immunoglobulin in multifocal motor neuropathy: a randomized, single-blinded cross-over trial. *Eur J Neurol*. 2009;16(5):631–638.
140. Harbo T, Andersen H, Jakobsen J. Long-term therapy with high doses of subcutaneous immunoglobulin in multifocal motor neuropathy. *Neurology*. 2010;75(15):1377–1380.
141. Misbah SA, Baumann A, Fazio R, et al. A smooth transition protocol for patients with multifocal motor neuropathy going from intravenous to subcutaneous immunoglobulin therapy: an open-label proof-of-concept study. *J Peripher Nerv Syst*. 2011;16(2):92–97.
142. Yoon MS, Gold R, Kerasnoudis A. Subcutaneous immunoglobulin in treating inflammatory neuromuscular disorders. *Ther Adv Neurol Disord*. 2015;8(4):153–159.
143. Bourque PR, Pringle CE, Cameron W, Cowan J, Chardon JW. Subcutaneous immunoglobulin therapy in the chronic management of myasthenia gravis: a retrospective cohort study. *PLoS One*. 2016;11(8):e0159993.
144. Beecher G, Anderson D, Siddiqi ZA. Subcutaneous immunoglobulin in myasthenia gravis exacerbation: a prospective, open-label trial. *Neurology*. 2017;89(11):1135–1141.
145. Howard JF, Nowak R, Wolfe GI, et al. Zilucoplan, a subcutaneously self-administered peptide inhibitor of complement component 5 (C5), for the treatment of generalized myasthenia gravis: results of a phase 2 randomised, double-blind, placebo-controlled trial and open-label long-term extension. In: *Paper Presented at American Academy of Neurology Annual Meeting, Philadelphia*; 2019.

146. Kulasekararaj AG, Hill A, Rottinghaus ST, et al. Ravulizumab (ALXN1210) vs eculizumab in C5-inhibitor-experienced adult patients with PNH: the 302 study. *Blood*. 2019;133(6):540–549.
147. Lee JW, Sicre de Fontbrune F, Wong Lee L, et al. Ravulizumab (ALXN1210) vs eculizumab in adult patients with PNH naive to complement inhibitors: the 301 study. *Blood*. 2019;133(6):530–539.
148. Mori S, Kishi M, Kubo S, et al. 3,4-Diaminopyridine improves neuromuscular transmission in a MuSK antibody-induced mouse model of myasthenia gravis. *J Neuroimmunol*. 2012;245(1-2):75–78.
149. Bonanno S, Pasanisi MB, Frangiamore R, et al. Amifampridine phosphate in the treatment of muscle-specific kinase myasthenia gravis: a phase IIb, randomized, double-blind, placebo-controlled, double crossover study. *SAGE Open Med*. 2018;6:2050312118819013.

CHAPTER 11

The endocannabinoid system as a target for the treatment of neurological disorders

Jim DeMesa[a], Giovanni Appendino[b], Gaetano Morello[c], Michael Murray[c], and Eduardo Muñoz[a,d,e,f]

[a]Emerald Health Pharmaceuticals, San Diego, CA, United States
[b]Department of Pharmaceutical Sciences, University of Eastern Piedmont, Novara, Italy
[c]Emerald Health Naturals, Vancouver, BC, Canada
[d]Maimonides Biomedical Research Institute of Córdoba (IMIBIC), Córdoba, Spain
[e]Department of Cellular Biology, Physiology and Immunology, University of Córdoba, Córdoba, Spain
[f]University Hospital Reina Sofía, Córdoba, Spain

Overview of the endocannabinoid system

The discovery, in 1964, of Δ^9-tetrahydrocannabinol (Δ^9-THC) as the main psychoactive molecule in *Cannabis sativa* by the Israeli researcher Dr. Raphael Mechoulam led to a worldwide search for the molecular underpinnings of its action. In time, as the vastness of the cannabinoid network was appreciated, its ecosystem became known as the ECS, i.e., the endocannabinoid system.[1] Early in the 1990s, two cannabinoid receptors were identified and cloned, cannabinoid receptor type 1 (CB_1) and cannabinoid receptor type 2 (CB_2), and two endocannabinoids (i.e., those produced by our body) were identified: anandamide (*N*-arachidonoylethanolamide or AEA) and 2-arachydonoyl glycerol (2-AG).[2–5] The system was given the name endocannabinoid since "endo" means within the body and "cannabinoid" highlights the role the cannabis plant played in the discovery. Considering what is now known about the ECS, it is assumed to be critical to health and has a direct effect on how we feel, mentally and physically.

To put the magnitude of this discovery into perspective, an ECS exists in all mammals, birds, reptiles, and other vertebrates. In humans, the ECS is believed to have more cellular receptor sites than any other receptor system in the body. Based on its role in maintaining physiologic homeostasis, the ECS impacts the immune, digestive, and other systems. At the level of the CNS, it influences motivation, reward, social behavior, pain, and inflammation.[6]

Despite the relatively recent discovery of the ECS, over 20,000 publications in the past 20 years contain the word "cannabinoid," representing an average of more than two scientific publications per day.

Neurotherapeutics in the Era of Translational Medicine
https://doi.org/10.1016/B978-0-12-816475-4.00002-1

Molecular and functional aspects of the endocannabinoid system

The ECS is a modulatory system of neural pathways consisting of five structurally distinct elements:

1. The 7-transmembrane G protein-coupled CB_1 receptor, found predominately in the brain, and CB_2 receptors, expressed predominately in the immune system.
2. Endocannabinoids, small molecules from the arachidonolate lipid family which includes AEA and 2-AG.
3. Enzymes that regulate the synthesis and degradation of endocannabinoids, for example, fatty acid amide hydrolase (FAAH) and monoacylglycerol lipase (MAGL)
4. Genes related to the synthesis of cannabinoid receptors and the enzymes associated with the turnover of endocannabinoids.
5. Non-canonical compounds, including additional GPCRs such as the orphan receptors GPR18, GPR55, and GPR119 and other endogenous compounds, such as amides of fatty acids with biogenic amines like ethanolamine, taurine, and dopamine, the arachidonoylated ether (noladin ether), and the reverse-anandamide virodhamine.[7,8] The pharmacology of these compounds is, however, less advanced than the canonical compounds, and their contribution to the ECS is less clear.

Endocannabinoids are produced in the brain and other organs on demand, their synthesis in the brain typically triggered by increased intracellular Ca^{2+} at postsynaptic sites in response to sustained synaptic activity.[9] AEA and 2-AG are rapidly deactivated by reuptake mechanisms and degrading enzymes such as FAAH and MAGL, respectively.[10] These ligands are orthosteric agonists for both cannabinoid receptors, but bind CB_1 receptors with higher affinity (AEA $Ki = 89$ and 321 nM for CB_1 and CB_2 receptors, respectively; 2-AG $Ki = 472$ and 1400 nM for CB_1 and CB_2 receptors, respectively).[11] In addition to the classical endocannabinoids, other allosteric endogenous ligands, including pregnenolone and lipoxin A4, have been identified and can modulate CB_1 receptor signaling with potential therapeutic value.[12,13] Further research is still needed, especially for allosteric modulators, to clearly elucidate their physiological roles and modes of action. Nevertheless, the pharmacological manipulation of endocannabinoid levels, or their actions by allosteric modulators, could provide alternative opportunities to regulate the ECS (see below).

Upon ligand binding and receptor activation, CB_1 receptors are primarily coupled to pertussis toxin (PTX)-sensitive Gi/o type G proteins which leads to a rapid decrease in levels of cyclic adenosine monophosphate (cAMP) by inhibiting adenylate cyclase activity.[10] Coupling to other G proteins, including Gs and Gq/11 proteins, can also modulate adenylate cyclase activity.[14,15] The reduction in cAMP production downregulates protein kinase A (PKA), which in turn suppresses PKA-mediated signaling events. On the other hand, the dissociated $\beta\gamma$ subunits stimulate the phosphatidylinositide 3-kinase and protein kinase B pathways which induce the phosphorylation of mitogen-activated protein kinases (MAPKs). Along with the G-protein-mediated signal transduction, the CB_1 receptor also interacts with a variety of non-G proteins, including β-arrestins and the cannabinoid receptor-interacting protein 1a (CRIP1a), to regulate its downstream effectors that include mitogen-activated protein kinases (MAPKs), multiple receptor tyrosine kinases, and extracellular signal-regulated kinases (ERK).[16,17] Thus, CB_1-interacting proteins could play a critical role in the cell-specific and ligand-specific modulation of cannabinoid signaling. Indeed, CB_1 receptors can form homodimers [18] and also heterodimers with other GPCRs.[19] These potential physical interactions are cell-type-specific and can modify the signaling of different pools of CB_1 receptors.

CB_1 receptors exhibit constitutive activity indicative of G-protein activation in the absence of agonists to mediate endocannabinoid tone.[20] The basal activity associated with this state is reversed by treatment with inverse CB_1 receptor agonists such as SR141716A. The model for GPCR activation has been adapted to include basal and ligand-induced states [21,22] with distinguishing biochemical pathways, including the extent and selectivity of G protein coupling and intracellular signaling.[23]

Like many other GPCRs, CB_1 receptors are primarily localized in the cell membrane. However, predominant intracellular localization of CB_1 receptors has been observed in diverse types of cells.[24] Aside from the constitutive and agonist-induced internalized CB_1 receptor, accumulated evidence suggests a distinct pool of intracellularly localized CB_1 receptors, with differential functionalities from their plasma membrane-localized counterparts. These intracellular CB_1 receptors reside in acid-filled endo/lysosomes, and increase the release of calcium from the endoplasmic reticulum and lysosomes upon activation by intracellular AEA.[25] Another subpopulation of CB_1 receptors, as suggested by several lines of evidence, is expressed in mitochondria where they can play a role in cellular respiration. This may be associated with cannabinoid-induced feeding behavior, learning, and memory and neuroprotection after cerebral ischemia/reperfusion injury.[26–29]

Endocannabinoids are retrograde neurotransmitters. They are not accumulated in synaptic vesicles but are generated on demand in a process triggered by the binding of a neurotransmitter to its postsynaptic neuron. Their production and release reduce synaptic input. The net result of endocannabinoid signaling is that excitation of the presynaptic neuron leads to a reduced neurotransmitter release in the synaptic cleft, a hallmark of the inhibition of transmitter release as a result cannabinoid receptor activation.[30]

Endocannabinoids are promiscuous in that they can exert opposing effects. Their effect on calcium concentration upon stimulation of the TRPV-1 receptor (a subfamily V member 1 cation channel) is opposite to their effect upon activation of the CB1 receptor.[31] In some instances, the overall response to cannabinoids at the receptor depends not only upon which is produced but upon their local concentrations. In short, they do not act in the classical manner of simple agonist/antagonist transmitters.[32]

The orthosteric ligand binding sites of cannabinoid receptors are rather large, explaining why endocannabinoids and Δ^9-THC, a potent cannabinoid, activate both CB_1 and CB_2 receptors, and also why structurally different compounds such as heterocyclic molecules (pyrazole, indole) structurally unrelated to endocannabinoids can also interact with cannabinoid receptors.[33,34] Remarkably, with the exception of Δ^9-THC and related cannabinoids, very few high-affinity ligands for these receptors have been discovered within natural products, in sharp contrast to the many structurally unsophisticated and totally synthetic heterocyclic biological analogues of phytocannabinoids. Apart from phytocannabinoids, the most important natural product agonist of the ECS is the sesquiterpene β-caryophyllene, a selective CB_2 agonist.[35]

The synthesis, release, and degradation of endocannabinoids present potential opportunities for pharmacological intervention. Postsynaptic activation stimulates the enzymatic conversion and *N*-acylation of phosphatidylethanolamine into *N*-acylphosphatidylethanolamine (NAPE), next cleaved by phospholipase D into anandamide, and a similar process presumably takes place for the production of 2-AG, using a 2-arachidonoylated phospholipid as its precursor.

Glial cells are mainly responsible for the inactivation of endocannabinoids in the brain. The enzymes responsible for their degradation, FAAH and MAGL, are cytoplasmatic, and thus endocannabinoids need to be transported across the cell membrane of glial cells by specific carriers to be inactivated. Inhibition of these enzymes and the endocannabinoid transport system is a major strategy to boost the activity of the endocannabinoid system.

The inherent complexity of the ECS and its different expression in rodents and humans makes it challenging to translate cellular and animal studies into a clinical context. But studies with $CB_1^{-/-}$ knockout mice may be of great relevance. These mice are anhedonic, anxious, and prone to depression, with severe deficits in learning and an increased incidence of seizures, suggesting that a potentiation of endocannabinoid signaling could beneficially affect these conditions.

In terms of clinical translation, activation of cannabinoid receptors by Δ^9-THC seems to have a rather narrow therapeutic index, especially related to non-neuropathic acute pain. It is therefore not surprising that few activators of CB_1 receptors are under clinical development as analgesic drugs. In contrast, inflammation induced in different neurological diseases upregulate CB_2 receptors in neurons and microglia, and the possibility exists that this upregulation could be selectively targeted by small molecules.[36] Additionally, since CB_1 receptors are also expressed in primary afferent neurons, CB_1 agonists with limited brain penetration could reduce the psychotropic side-effects associated with agonists that cross the blood-brain barrier. In this regard, preliminary evidence exists for the clinical efficacy of cannabinoid receptor agonists to treat chronic pain of peripheral origin.[37]

The role of the endocannabinoid system in health and disease

The ECS is affected by compounds that our body produces as well as other substances in food, plants, and cannabinoids that are synthetically produced.[38,39] The widespread distribution of CB_1 and CB_2 receptors throughout the body affirms the importance of the ECS to overall bodily function and health.[40,41] Because of this, it has been hypothesized that a lack of adequate endocannabinoid activity could be a key contributing factor to many debilitating conditions, including neurological disorders. This theoretical condition was termed "clinical endocannabinoid deficiency syndrome" (CEDS) by Dr. Ethan Russo.[42] Whether CEDS is the result of poor diet, lack of exercise, drug abuse, environmental toxins, genetic factors, a combination of these, or something else, endocannabinoid deficits can be associated with a reduced ability to adapt to chronic stress. Prolonged exposure to stress, both mental and physical, also depletes endocannabinoid tone, and this, in turn, may have an adverse impact on many physiological processes.[43] Conversely, scientific evidence indicates that enhancement of the ECS may lead to the prevention of these conditions and perhaps can be of therapeutic value as well.

The endocannabinoid system in the central nervous system

The distribution and function of the components of the ECS in the CNS have attracted significant research attention, leading to the recent achievement of some major milestones. The ECS exerts a modulatory effect on important functions such as neurotransmission, synaptic plasticity, glial activation, and oxidative stress. As described previously, CB_1 receptors were first identified in the brain and found to be among the most widely expressed and polymorphic receptor proteins from the GPCR family in this organ.[44,45] Regions with highest levels of CB_1 expression include the hippocampus, basal ganglia, olfactory bulb, and cerebellum.[46] Moderate expression of CB_1 receptors has been detected in the cerebral cortex, amygdala, dorsal horn septum, hypothalamus, and spinal cord.[46] The expression of this receptor on presynaptic terminals mediates retrograde signaling of endocannabinoids, which exert the majority of their known actions by directly targeting CB_1 receptors located on presynaptic nerve terminals. Acting as a feedback loop, AEA and 2-AG produced at the postsynaptic level, traverse the synaptic cleft and bind to presynaptic CB_1 receptors to decrease the probability of neurotransmitter release.[47] However, CB_1 receptors are also expressed at postsynaptic sites and functional studies demonstrate self-inhibition of cortical neurons by endocannabinoids.[48] In addition to neurons, CB_1 receptors are also expressed on microglia, astrocytes, and oligodendrocytes, where they have been shown to modulate synaptic transmission.[48,49]

In the brain, CB_2 receptor expression has been detected in microglia and resident macrophages and is associated with protection against neuroinflammation.[50,51] This selective localization, together with the modulatory effect of the CB_2 receptor on microglial function, is particularly relevant since microglial cells have been implicated in the pathogenesis of numerous neurologic disorders.[52,53] Although limited in their expression, CNS CB_2 receptors appear to play an active role in neurological activities, such as nociception, drug addiction, and neuroinflammation. Moreover, the intracellular presence of these receptors has been shown in prefrontal cortical pyramidal neurons where they modulate neuronal excitability through the regulation of Ca^{2+}-activated Cl^- channels.[54]

Endogenous allosteric modulation of cannabinoid type 1 receptors

In general, drug discovery efforts aimed at CB_1 receptors have focused on ligands targeting the orthosteric sites to which the endogenous cannabinoids bind. However, these receptors exert psychotropic side effects which have hampered drug discovery efforts aimed at regulating CB_1 receptors for therapeutic benefits.[55] Recently, endogenous allosteric modulators targeting CB_1 receptors, such as Pepcan-12, Lipoxin A_4, and pregnenolone, have been discovered and the results are paving the way for the development of novel positive and negative allosteric small molecules such as Org27569, PSNCBAM-1, and ZCZ011.[55]

For instance, the endogenous anti-inflammatory lipid lipoxin A4 has been shown to be a positive allosteric modulator (PAM) of CB_1 receptor signaling in the brain. In particular, the presence of lipoxin A4 is able to increase the affinity of AEA at CB_1 receptors, thereby potentiating the signaling and behavioral effects of this endocannabinoid. This mechanism likely modulates CB_1 receptor signaling[13] and might have important behavioral consequences. Even more intriguing, recent studies have shown that the neurosteroid pregnenolone is a negative allosteric modulator (NAM) of CB_1 receptors, able to protect the brain from excessive cannabinoid intoxication.[12] It is important to note that pregnenolone does not inhibit all signaling pathways triggered by CB_1 receptor activation as in the instance of an orthosteric antagonist. Whereas the decrease of cAMP by cannabinoids is not altered by pregnenolone, this endogenous NAM fully inhibits the CB_1-dependent modulation of the MAPK–ERK pathway and mitochondrial function.[12] Preclinical studies focused on these areas have shown interesting results, leading to clinical trials that are testing the efficacy of a pregnenolone derivative (AEF0117) to moderate abuse of cannabis (ClinicalTrials.gov: NCT03717272).

The gut-brain axis and the endocannabinoid system

Like the ECS, the human intestinal ("gut") microbiome and its relationship to overall general health has experienced unprecedented attention in the past 20 years.

Interestingly, the ECS influences the physiology and function of the gut, including the character of the gut microbiome, its permeability, etc., both locally within the gut, as well as outside it.[56,57]

Both CB_1 and CB_2 receptor activity modulates motility, hunger signaling, inflammation, interaction with gut microbiota, and epithelial gastrointestinal permeability.[56] Beyond the gut, however, it has become clear that the gut microbiome plays a major role in chronic inflammation throughout the body. A "gut-brain axis" has been identified. This refers to the biochemical nexuses that occur between the CNS and the gastrointestinal tract.[58,59]

Although the connection between the gut microbiome and the brain has long been known, it is only recently that the mechanisms of action have begun to be understood. Under certain conditions, for example, if gut permeability is increased (e.g., leaky gut, which has become very common), microbe-associated molecular patterns (MAMPs) can translocate and spread to various sites throughout the body, including the brain and its compartments. Further, it is surmised that the CNS constantly interacts with MAMPS and other microbial metabolites, which in some circumstances, improves the functioning of the CNS, but under different conditions, causes consequential negative CNS effects, including neuroinflammation and neurodegeneration.[60] Based on all of this, it is clear we need to better understand the interplay of antigen trafficking through the gut barrier and the mechanisms involved in maintaining the gut-brain homeostatic balance. This may shed further light on the etiology of neurodegenerative disease.[59]

Cannabinoids and other substances that affect the endocannabinoid system

Endocannabinoids (the "bliss molecules")

After the cannabinoid receptors were cloned and identified as the targets of Δ^9-THC, the search for endogenous ligands able to activate such receptors led to the discovery of AEA, a downstream metabolite of the essential fatty acid linoleic acid. It was given this name because its root comes from the Sanskrit word *Ananda*, which means bliss. It is a fitting name as anandamide is a key "bliss" molecule in the brain, providing a sense of happiness and mental well-being.[4] 2-AG, the second major endocannabinoid, was later discovered.[5]

Research on AEA has shown that it does much more than produce a state of increased happiness. Its additional effects relate to where it is produced in the brain, such as key areas important to memory, motivation, appetite, perception of pain, etc.

A quick look at medical marijuana and Δ^9-tetrahydrocannabinol

When marijuana is consumed, Δ^9-THC quickly affects the ECS by attaching to cannabinoid receptors throughout the body, primarily to the CB_1 receptors in the brain. The effect produced by Δ^9-THC is significantly greater than the effect of AEA. Because of this more profound effect, it overpowers the endocannabinoid system and eliminates the ability of AEA to fine-tune communication between brain cells.

While the psychoactive effects of Δ^9-THC are popularly associated with cannabis, Δ^9-THC affects virtually every cell of the body. With regard to the psychoactive effects, it is well known that Δ^9-THC can impair mental functions, such as reaction time, memory, and judgment. But the "positive" effects likely redound to the areas of the brain associated with reward, such as the striatum and prefrontal cortex. The resulting emotional state, in the vernacular, is a "high."

Further "positives" include muscle relaxation and a reduction in intraocular pressure.[61] Along with this, there are some negatives over time since Δ^9-THC can change how the ECS works in key brain areas linked to memory, motivation, and mood. Short-term physical effects include reddening of the eyes, dry mouth, and a sensation of cold or hot hands and feet.

Based on these and other benefits, the medical use of marijuana has generated considerable interest. Indications for its use include the following conditions: glaucoma, nausea and vomiting associated with cancer chemotherapy, anorexia and weight loss due to HIV/AIDS, chronic pain, spasticity associated with multiple sclerosis and spinal cord injury, epilepsy, insomnia, and depression. Unfortunately, there is currently insufficient data supported by well-controlled clinical trials to draw solid conclusions in most of these areas.

With regard to relieving pain, Δ^9-THC reduces its perception, but is not as potent as an opioid such as oxycodone (Oxycontin). However, cannabis and opioids are co-agonists, each of them magnifying the effect of the other.[62,63] This sparing effect allows for people in severe pain to use lower doses of opioids without a loss of effectiveness. In fact, statistics show that opioid use and deaths related to opioid use, are lower in areas where medical marijuana is legal. Importantly, CB_1 receptors are not present in the part of the brain that regulates heart rate and respiration, so unlike narcotics, there is no known lethal dosage threshold for Δ^9-THC. In short, the therapeutic index for Δ^9-THC is not a limiting factor in its use.[62,63]

A closer look at cannabidiol

Cannabidiol (CBD) is a nonpsychoactive cannabinoid. It does not bind to CB_1 receptors in the same manner as Δ^9-THC, nor does it exert psychotropic effects. In fact, CBD reduces the euphoria produced by Δ^9-THC. The putative health benefits are summarized in the following figure.

Potential applications of CBD

Activity	Reference
Antiemetic	64
Anticonvulsant	65–67
Antipsychotic	68
Anti-inflammatory	69–71
Antioxidant	72,73
Antitumor/anticancer	69,74,75
Mood disorders	76–78

Other cannabinoids and the entourage effect

While Δ^9-THC and CBD get the most attention, more than 120 cannabinoid molecules have been identified in the cannabis plant, as well as many other non-cannabinoid active compounds.[79] It is assumed that these molecules work in concert to produce what has been called an "entourage effect".[80] In support of this, anecdotal evidence suggests that marinol—a synthetic version of Δ^9-THC—is not as potent as cannabis according to some users. This has led researchers to investigate the potential therapeutic use of products incorporating a mixture of cannabinoids or other active compounds found in cannabis or foods including herbs and spices. However, evidence for the entourage effect has been disputed.[81]

Cannabimimetics and nutraceuticals that affect the endocannabinoid system

While much of the focus on supporting the ECS has been on cannabis compounds, the ECS can be affected by non-cannabis substances as well. Studies regarding the impact of diet, lifestyle, and various other factors that influence the ECS are at an early stage. What is known is that steps can be taken to:

1. Support the body's natural production of endocannabinoids.
2. Utilize natural products to modulate the breakdown of endocannabinoids like anandamide to retain and prolong their effectiveness in the body.
3. Utilize non-cannabis-derived plant-based compounds known to effectively activate CB_1 and CB_2 receptor sites.

Diet, the ECS, and effects on neurological disorders

A diet rich in the right type of fats is critical to the proper functioning of the ECS. What makes a fat "bad" or "good" in the instance of cellular membranes is determined by dietary intake. In short, without the right type of fats, cells simply do not function properly, and neither does the ECS.[82,83]

A dietary fat profile that has been linked to diseases, including neurological ones, incorporates an abundance of omega-6 fatty acids, saturated and trans fatty acids (hydrogenated vegetable oils), and a lack of monounsaturated and omega-3 (ω-3) fatty acids. Not surprisingly this same dietary pattern is also associated with impaired function of the ECS.

Although anandamide is derived from arachidonic acid, an omega-6 fatty acid found in meat and dairy, studies indicate that too much arachidonic acid in the diet is not only associated with greater inflammation, but also lower levels of endocannabinoids in the brain. Both monounsaturated fats and omega-3 fatty acids are required for proper endocannabinoid signaling.[84] A deficiency of omega-3 fatty acids is particularly harmful to the ECS[85] and is associated with an increased risk for heart disease and about 60 other conditions including cancer, arthritis, stroke, high blood pressure, skin diseases, diabetes, and neurological disorders.[86,87]

Interestingly, the health benefits of ω-3 fatty acids are mediated, in part, through metabolic conversion to the bioactive epoxides EPA and DHA, which, after a second metabolic step, generate endocannabinoid epoxides that exert anti-inflammatory activity through the CB_2 and PPARγ receptors.[88] Hemp seed oil is considered the best source of fatty acids in terms of its omega-3 to omega-6 ratio.

Natural products that modulate the endocannabinoid system

A number of plant extracts that exert novel effects impact the ECS.[89,90] Examples include clove oil, *Echinacea purpurea*, zanthoxylum, ginger, peony, and magnolia. These extracts are extremely rich in pharmacological activity. What researchers are excited about is how these additional benefits are supported by the ECS and vice versa.

- Clove oil (concentrated for β-caryophyllene)—the pain-relieving aspects of clove oil have been utilized over the years to soothe the irritated gums of teething infants. Clove oil contains several analgesic ingredients, but the one most researched for its effects on the ECS is β-caryophyllene. This compound is also found in many other herbs and spices. It is able to bind with great affinity to CB_2 receptors, which is thought to be one of the primary mechanisms by which it palliates pain.[35]
- *E. purpurea*—alkylamides from Echinacea species bind to CB_2 with great affinity and assist CB_2 in modulating the immune system.[91] They also prolong the effects of anandamide.[92] Clinical studies with an alkylamide-rich extract have shown positive effects in relieving stress and anxiety.
- Zanthoxylum or Sichuan pepper—contains alkylamides that display very strong affinity to CB_1 and CB_2 receptors.[93] These compounds also influence what are known as transient receptor potential channels, part of the endovanilloid system that

interacts closely with the ECS. The two systems overlap and are intricately interwoven in helping cells communicate with one another.
- Peony—contains paeoniflorin, which exerts physiologic effects similar to zanthoxylum. In a rat model, it has been shown to be neuroprotective.[94]
- Ginger—contains terpene, a phytochemical that mediates pain, inflammation, etc. based on its interaction with the ECS and transient receptor potential channels.[95]
- Magnolia—contains magnolol, a CB_1 and CB_2 receptor agonist.[96,97] Extracts exert psychotropic effects.

Palmitoylethanolamide (PEA)

Palmitoylethanolamide (PEA) was first identified over 60 years ago as an active anti-inflammatory agent. It was initially isolated from extracts of brain, liver, and muscle (rats and guinea pigs), but later found to be contained in chicken egg yolk, olive oil, safflower and soy lecithin, and peanut meal. Preclinical and human studies have investigated its effects in a number of disorders: depression, cognitive decline, autism, multiple sclerosis, obesity, and metabolic syndrome.

PEA, a cannabinoid-related compound, is an important component of the ECS. Its mechanism of action (MoA) is complex. Like other recognized cannabinoid molecules, this includes direct and/or indirect interaction with CB1, TRPV1, PPAR, GPR55, and GPR18, among other receptors.[98,99] As an agonist, PEA exhibits high affinity for GPR55, indicating a possible key MoA.[100] In regards to indirect actions on the ECS, one of the key aspects of PEA is its ability to modulate CB1 and CB2 receptors and enhance the effects of AEA, resulting in changes in cellular function, including gene expression and electrical excitability of cell membranes. Non-ECS effects include those on glial and mast cells, NF-κB, and nuclear receptor peroxisome proliferator-activated receptors (PPARs).[101] In glial and mast cell mediated models of neurogenic inflammation and neuropathic pain, PEA has demonstrated beneficial results.[102] It has also been shown to counteract key pathologic features in models of Alzheimer's and Parkinson's disease.[103] Moreover, PEA is an inhibitor of NF-κB (nuclear factor kappa-light-chain-enhancer of activated B cells), a protein complex that is part of a family of inducible transcription factors that regulate a large array of genes involved in immune and inflammatory responses.[104] An antidepressant effect has been demonstrated in animal models of depression as well as in a human double-blind, placebo-controlled trial.[105,106]

Synthetic cannabinoid analogues, prodrugs and derivatives

In addition to the natural substances that affect the ECS, synthetic cannabinoids with similar properties and potential therapeutic effects are under development. Utilizing a bioengineering strategy, novel molecules have been developed, including analogues, prodrugs, and derivatives of various cannabinoid molecules. These, many of which are new chemical entities (NCEs), can provide expanded biological activity, and/or provide improved pharmacokinetic and pharmacodynamic properties compared to the natural molecules.

Novel NCEs and other synthetic molecules that bind to CB_1 and/or CB_2 receptors in the ECS as well as affect other receptors and physiologic pathways germane to multiple sclerosis, Huntington's disease, epilepsy, etc., are viewed as commercially useful products. One example is VCE-004.8, under development by Emerald Health Pharmaceuticals. It is a bioengineered aminoquinone derivative of CBD which binds to CB_2 receptors in the ECS but has no effect on CB1 receptors, thus no psychoactive effects. Its multifaceted MoA includes anti-inflammatory, antioxidative, and neuroprotective properties, offering the promise that a single molecule could be beneficial for the treatment of inflammatory, fibrotic, and autoimmune disorders as well as neurological ones (see Multiple sclerosis section).[107,108]

Potential benefits of cannabinoids and manipulation of the endocannabinoid system in specific neurological disorders

As indicated throughout this chapter, the ECS provides an important regulatory process in the central nervous system through the regulation of neurotransmission and synaptic plasticity and plays an important role in emotional and cognitive function.[109] Further, pharmacological modulation of endocannabinoid signaling has the ability to alter common traits in neurodegeneration, thus emerging as a potential palliative or disease-modifying therapy.

Alzheimer's disease

Alzheimer's disease (AD) is an irreversible, progressive brain disorder that slowly destroys memory and thinking and, eventually, the ability to carry out simple tasks. Approximately 36 million people worldwide suffer from Alzheimer's disease, and with an increasingly aged global population, that number is estimated to triple by 2050. The accumulation of misfolded proteins in the brain results in chronic activation of the immune response, oxidative damage, and excitotoxicity leading to neurodegeneration. Although the etiopathology of AD is not well understood, the ECS is emerging as a potential druggable system of particular relevance.[110,111]

A significant increase in the expression of CB_2 receptors in microglial cells in contact with the senile plaques of the βA protein has been described and suggested as a compensatory mechanism in response to the neurodegenerative and neuroinflammatory processes initiated in AD.[112,113] On the other hand, as a consequence of the neuronal death associated with the progression of AD, a significant reduction in the expression of CB_1 receptors in the hippocampus and basal ganglia has been observed.[53] Moreover, the intracerebroventricular administration of the synthetic cannabinoid WIN55,212-2 to animals alleviated βA-induced microglial activation, cognitive impairment, and loss of neuronal markers. Other cannabinoids such as HU-210 and JWH-133 also blocked βA-induced activation of cultured microglial cells through a CB_2 receptor pathway.[53] Utilizing a different strategy, the pathogenic role of the miR-139 microRNA in modulating the neuroprotective and anti-inflammatory activity of CB_2 receptors in the hippocampus has been demonstrated in an AD model.[114]

Furthermore, changes in the levels of endocannabinoids have been demonstrated in AD preclinical models. Specifically relevant are increases in the levels of 2-AG (the most abundant endocannabinoid in the CNS) in the hippocampus in response to the appearance of neurodegeneration and gliosis.[115] With respect to the enzymatic machinery that regulates the synthesis (DAGL) and degradation (MAGL) of 2-AG, the changes in the expression levels of these enzymes are germane since the DAGL/MAGL axis modulates the increase in neuroinflammation associated with the experimental disease. In this context, MALG inhibitors have shown potential therapeutic benefit by reducing the levels of arachidonic acid and pro-inflammatory prostaglandins in the rodent brain.[116,117]

In summary, preclinical in vitro and in vivo studies of AD have suggested a potential therapeutic effect for CBD.[118] These promising results obtained in experimental models of AD require confirmation in controlled clinical trials. Additionally, the use of synthetic or natural Δ^9-THC to alleviate neuropsychiatric symptoms in AD is a subject for further study.[119,120]

Parkinson's disease

Parkinson's disease (PD) is a progressive neurological disease that results in the degeneration of dopaminergic neurons in the brain. The ECS, highly represented in the CNS, has been shown to be dysfunctional in many neurological disorders, including PD. Accordingly, the development of therapeutics targeting the ECS in the brain holds promise for the treatment of PD.[121]

The interactions between cannabinoids and dopamine in the basal ganglia involve both CB_1 and CB_2 receptors. Included in these interactions is the modulation of various neurotransmitters, such as γ-aminobutyric acid, glutamate, opioids, and certain peptides. Current research supports targeting the ECS as a potential therapy for motor symptoms, such as levodopa-induced dyskinesias (LIDs).[122] LIDs are the consequence of progressive sensitization of dopamine receptors in the striatum.[123] However, the clinical relevance of the cannabinoid-dopamine interactions of motor symptoms in PD needs to be further investigated.

Regarding specific cannabinoids, it has been shown that Δ^9-THC induces the release of dopamine (DA) in the neostriatum of the brain in rats.[124] Additionally, studies have also shown the upregulation of glial CB_2 receptors in animal models of PD. The activation of these receptors may play a role in dopaminergic neuronal survival.[125]

In addition to THC, other cannabinoids (e.g., CBD and CBG), as well as some novel bioengineered cannabinoid derivatives, have been shown to possess neuroprotective, anti-inflammatory, and antioxidant properties, which could form the basis for the development of therapeutics targeting neurodegenerative diseases, including PD.[126,127]

Multiple sclerosis

Cannabinoid-based interventions are being explored for neuroinflammatory and demyelinating diseases such as multiple sclerosis (MS). Both CBD and Δ^9-THC, separately and in combination, have shown beneficial effects in preclinical models of demyelination.[128–131] In addition, endocannabinoids and synthetic cannabinoid bioengineered derivatives also alleviate neuroinflammation and symptomatology in murine models of MS.[108,132–134] At the molecular level, these improvements are generally linked to the activation of both CB_1 receptors and CB_2 receptors, resulting in dual anti-inflammatory and neuroprotective effects.[135,136] These effects included the reduction of cytotoxic factors such as nitric oxide, reactive oxygen species, and pro-inflammatory cytokines in microglia.[136] The precise mechanisms by which receptors exert their neuroprotective activity might include activation of the mTOR1 pathway and brain-derived neurotrophic factor.[137]

To date, the only cannabinoid-based drug on the market for MS is nabiximols (Sativex, GW Pharmaceuticals), an oral spray containing CBD/Δ^9-THC (1:1). Sativex is indicated as a second-line treatment for moderate to severe spasticity in MS patients who have not responded adequately to other anti-spasticity medication.

As MS involves the dysregulation of myelin integrity and/or remyelination,[138] it is important to consider effects of the endocannabinoid system on oligodendrocytes and their precursors. It has been shown that cannabinoid receptor ligands, as well as modulators of the ECS, may enhance precursor and mature oligodendrocyte characteristics including effects on myelination and survival.[139] Agonist stimulation of oligodendrocyte precursor cells (OPCs) at both CB_1 and CB_2 receptors counter apoptotic processes via Akt/PI3K and promote proliferation via Akt/mTOR and ERK pathways.[140] CB_2 receptors promote OPC migration in neonatal development and these cells produce 2-arachidonoylglycerol (2-AG), stimulating CB_1-mediated ERK pathways responsible for the differentiation of myelin basic protein (MBP)-producing oligodendrocytes. In cell culture models of excitotoxicity, accompanied by increased reactive oxygen species and depolarization-dependent calcium influx, CB_1 agonists improved viability of oligodendrocytes. In several models of rodent encephalomyelitis, chronic treatment with cannabinoid agonists ameliorated the damage by promoting OPC survival and oligodendrocyte function.[132,141] Because of these findings, pharmacotherapeutic strategies based on ECS and oligodendrocyte production and survival are warranted for the potential management and treatment of MS and other demyelinating diseases.[139] VCE-004.8 (EHP-101), a novel bioengineered CBD-derivative in clinical development (Emerald Health Pharmaceuticals), has been shown to have a unique multimodal mechanism of action, affecting CB_2, PPARγ, and the hypoxia inducible factor (HIF) pathway, all of which are validated targets for MS. In several animal models of MS using this molecule, inflammation and demyelination were decreased, neuroprotection was demonstrated, and remyelination occurred. Based on these results, VCE-004.8 has advanced to human studies, providing the potential to be a disease-modifying therapy for MS in the future.[107,108]

Epilepsy

Early documented uses of cannabis to treat seizures include a Sumerian text from 2900 BC and an Arabian document from the 12th century.

Epilepsy affects 65 million people worldwide and is characterized by recurrent seizures. The condition includes neuroinflammation as a fundamental feature and encompasses a variety of etiologies: genetic factors, stroke, infection, and traumatic brain injury.[142]

While epilepsy affects people of all ages, it most commonly affects children and elderly adults. In 2018, the FDA approved Epidiolex for the treatment of an intractable childhood form of epilepsy, a milestone in the development of CBDs as neurotherapeutics. As is often the case, this discovery had its origin in the effort of a desperate layperson to treat an afflicted family member. The 2013 CNN documentary "Weed" chronicles the use of "Charlotte's Web," an extract of a high CBD, low Δ^9-THC cannabis strain marketed as medical cannabis under Colorado state law. The extract was named after Charlotte Figi, a young girl suffering from Dravet syndrome, a genetic disorder characterized by recurrent seizures. At the age of 5, after her first dose of medical marijuana, she experienced a dramatic reduction in seizures.

The exact mechanism of action of CBD as an antiepileptic drug is not completely understood.[143] CBD is a pleiotropic cannabinoid with antioxidant activity and the ability to modulate different G-protein-coupled receptors, nuclear receptors, and ionotropic receptors.[144] A recent study has shown that the beneficial effects of CBD in a preclinical model of Dravet syndrome is mediated by antagonizing GPR55 receptor signaling and restoring the excitability of inhibitory interneurons in the hippocampal dentate gyrus.[145]

Huntington's disease

The pathogenic hallmark of Huntington's disease (HD) is the degeneration of striatal medium-sized spiny neurons induced by mutant huntingtin protein. The mutation consists of an excessive number of CAG repeats in the first exon, leading to an excess of polyglutamine repeats in the amino terminal end of the protein (huntingtin).[146] The length of the repeat dictates the time of onset of the disease which is characterized by chorea, behavioral disturbances, and dementia. Neurodegeneration primarily affects the striatum and, subsequently, the cortex, as well as other neuronal subpopulations.[147]

Numerous changes in the ECS, including cannabinoid receptor expression, have been described in postmortem brain samples from HD patients and in animal models of HD.[148] A key observation from these studies is a defect in striatal CB_1 receptor expression and signaling.[148] Administration of a CB_1 cannabinoid agonist attenuated neurotoxicity in a quinolinic acid HD model, suggesting that this receptor constitutes a promising opportunity to treat early and late HD, as well as other neurodegenerative conditions, utilizing a neuroprotective therapeutic.[149–151] Also, CB_1 receptor activation with Δ^9-THC in the genetic model of the disease (R6/2 mice) attenuated striatal neuron loss, whereas striatal degeneration was aggravated in these mice with a genetic deficiency in CB_1 receptors.[152]

The expression of CB_2 receptors in astrocytes and microglia cells is also altered in patients with HD.[153] Selective CB_2 receptor agonists are effective in experimental HD, acting preferentially on inflammatory events and microglial activation in animal models of the disease.[34,154] Antioxidant nonpsychoactive phytocannabinoids such as CBD and CBG have also been investigated in experimental models of HD, although their effects are independent of cannabinoid receptors. CBD has been found to be neuroprotective in rats given 3-nitropropionate, a model characterized by mitochondrial damage and oxidative stress.[155] CBG was also neuroprotective in the R6/2 and the 3-nitropropionate mouse models.[156] The effect of CBD and CBG might be mediated by the activation of nonclassical endocannabinoid targets such as PPARs or by other mechanisms. Pursuing this lead, Emerald Health Pharmaceuticals bioengineered VCE-003.2 (a CBG quinone derivative) which activates PPARγ. In an experimental HD model, this protected neurons and enhanced neurogenesis caused by mutant huntingtin and 3-nitropropionate.[157,158]

The first clinical trials with cannabinoids in HD were conducted with either CBD[159] or nabilone, a Δ^9-THC synthetic analog.[160] However, aggravation of choreic movements, characteristic of HD, was noted in both cases, but another study showed that nabilone palliated chorea and behavioral disturbances.[161,162] Positive results with a combination of CBD and Δ^9-THC in preclinical models of HD led to clinical investigations using Sativex for patients with HD. Although Sativex was safe and well tolerated, it did not slow disease progression.[163] This may have been due to the trial design, which was limited to 12 weeks.

Stroke

Stroke is a significant cause of morbidity and mortality worldwide, therefore, the development of effective therapies fulfills an unmet medical need. Preclinical evidence suggests that the ECS can play an important role in the pathophysiology of stroke and opens the possibility that pharmacological manipulation of the ECS could be an opportunity to develop both short-term and long-term treatments.

It has been shown that both CB_1 agonism and antagonism and CB_2 agonism are neuroprotective in stroke.[164] A recent meta-analysis of 34 preclinical studies examining the effect of cannabinoids in poststroke neuroprotection concluded that this class of compounds was able to reduce infarct volume and improve functional outcomes in experimental stroke models.[165]

Despite severe disabilities and neurological impairments during the early poststroke period, most stroke patients achieve some degree of recovery over time. While improvement in lower limb function is observed in ~65% of patients with initial motor deficits,[166] the probability of normal recovery in the upper limbs is very low (<15%).[167] In general, the rate of clinical recovery is relatively rapid during the first few weeks after a stroke but then slows considerably between 1 and 6 months.[168] In this regard, it has been suggested that targeting the ECS represents a viable opportunity for the development of pharmacological therapies aimed to enhance functional recovery after a stroke.[169,170] This is supported by the finding that treatment of experimental stroke with WIN 55,212,2, a dual CB_1/CB_2 agonist, is neuroprotective and associated with reduction in infarct size.[171,172]

Myelin repair after stroke might occur as myelin plasticity or as remyelination through the recruitment and differentiation of oligodendrocyte precursor cells (OPCs), which become myelin-forming oligodendrocytes (OLs).[173] Because of their unique ability to replace damaged OLs, OPCs represent a potential therapeutic target for myelin repair in the context of stroke. Oligodendrocytes are vulnerable to ischemic conditions and their replacement via OPCs has been found to aid recovery in an animal model.[174] In an experimental setting, WIN 55,212,2 treatment increased OPC proliferation within the ipsilateral subventricular zone (SVZ) and decreased OPC expression of tau-1, an oligodendrocyte marker of ischemic stress. Both effects were mainly mediated through CB_1 receptor activation.[175] Although these results appear promising, it may be difficult to replicate them in a clinical trial.[176]

Traumatic brain injury and concussion

Traumatic brain injury (TBI) results in many deaths and cases of permanent disability throughout the world. A TBI can be caused by any type of physical blow, a jar to the head, or a penetrating head injury. A concussion is a type of TBI in which the brain is traumatized as a result of an impact, unaccompanied by overt neurologic deficits, i.e. paralysis, sensory loss, etc. The ECS is thought to play a neuroprotective role during TBI and concussion and may be part of the brain's natural repair mechanism in many types of neurodegenerative processes.[177]

In animal models, 2-AG levels in the brain increase after TBI.[178] This result suggests a compensatory mechanism to reduce cerebral edema, inhibit the production of inflammatory cytokines, and increase migration and differentiation of stem cells.[179] In a mouse model of TBI, both CB_1 and CB_2 receptor antagonists reverse minocycline-induced neuroprotection, supporting the notion that CB receptors are a potential therapeutic target in the instance of traumatic brain injury.[180] Clinical evidence further supports this: patients given medical cannabis after a concussion experienced significantly less post-concussive effects, such as less headache, dizziness, mood changes, and sleep disturbances.[181]

These studies and others provide promising evidence for the pursuit of pharmacologic agents to help prevent, or even reverse, the pathology associated with TBI and concussion. However, further clinical studies are needed to support this conclusion.

Mood disorders

Mood disorders refer to a group of conditions in which a person's emotional state is altered. Examples of this include the occurrence of anxiety, mania, and depression. It isn't a stretch to assume that the homeostatic function of the ECS is integral to the maintenance of emotional well-being.[182] In support of this, it is known that deficits in the ECS can result in depressive and anxiogenic behavior while augmentation of the ECS can be antidepressive and anxiolytic.[183] In the instance of mood disorders in which endocannabinoid activity is thought to be dysfunctional, synthetic inhibitors of endocannabinoid degradation may represent a therapeutic opportunity.[78] Alternatively, agonists of CB_1 and CB_2 receptors may be useful for the treatment of a wide variety of psychiatric disorders manifested as anxiety and disturbances of mood, based on their effects on memory and plasticity as demonstrated in an experimental model.[184]

Summary and conclusions

Based on the information presented in this chapter, it is clear that the endocannabinoid system (the ECS) regulates numerous physiologic functions, including those in the central and peripheral nervous systems. For this reason, the ECS is being actively explored as a key target for the prevention, management, and possible treatment of a variety of ailments, anxiety and depression in the instance of mood disorders, and Parkinson's and Alzheimer's in the instance of neurodegenerative diseases. Indications for which treatments have been approved by the FDA include spasticity and seizures. While it will take more time, it is certain that the list of treatment indications will expand, logarithmically if cannabinoids are found to be disease modifying in the instance of currently untreatable neurologic conditions.

References

1. Gaoni Y, Mechoulam R. Isolation, structure, and partial synthesis of an active constituent of hashish. *J Am Chem Soc.* 1964;86(8):1646–1647.

2. Matsuda LA, Lolait SJ, Brownstein MJ, Young AC, Bonner TI. Structure of a cannabinoid receptor and functional expression of the cloned cDNA. *Nature*. 1990;346(6284):561–564.
3. Brown SM, Wager-Miller J, Mackie K. Cloning and molecular characterization of the rat CB2 cannabinoid receptor. *Biochim Biophys Acta*. 2002;1576(3):255–264.
4. Devane WA, Hanus L, Breuer A, et al. Isolation and structure of a brain constituent that binds to the cannabinoid receptor. *Science*. 1992;258(5090):1946–1949.
5. Stella N, Schweitzer P, Piomelli D. A second endogenous cannabinoid that modulates long-term potentiation. *Nature*. 1997;388(6644):773–778.
6. Parsons LH, Hurd YL. Endocannabinoid signalling in reward and addiction. *Nat Rev Neurosci*. 2015;16 (10):579–594.
7. Alexander SP, Christopoulos A, Davenport AP, et al. THE CONCISE GUIDE TO PHARMACOLOGY 2017/18: G protein-coupled receptors. *Br J Pharmacol*. 2017;174(Suppl 1):S17–S129.
8. Bradshaw HB, Walker JM. The expanding field of cannabimimetic and related lipid mediators. *Br J Pharmacol*. 2005;144(4):459–465.
9. Chevaleyre V, Takahashi KA, Castillo PE. Endocannabinoid-mediated synaptic plasticity in the CNS. *Annu Rev Neurosci*. 2006;29:37–76.
10. Howlett AC, Breivogel CS, Childers SR, Deadwyler SA, Hampson RE, Porrino LJ. Cannabinoid physiology and pharmacology: 30 years of progress. *Neuropharmacology*. 2004;47(Suppl 1):345–358.
11. Pertwee RG, Howlett AC, Abood ME, et al. International Union of Basic and Clinical Pharmacology. LXXIX. Cannabinoid receptors and their ligands: beyond CB(1) and CB(2). *Pharmacol Rev*. 2010;62 (4):588–631.
12. Vallee M, Vitiello S, Bellocchio L, et al. Pregnenolone can protect the brain from cannabis intoxication. *Science*. 2014;343(6166):94–98.
13. Pamplona FA, Ferreira J, Menezes de Lima Jr O, et al. Anti-inflammatory lipoxin A4 is an endogenous allosteric enhancer of CB1 cannabinoid receptor. *Proc Natl Acad Sci U S A*. 2012;109(51): 21134–21139.
14. Cachope R. Functional diversity on synaptic plasticity mediated by endocannabinoids. *Philos Trans R Soc Lond Ser B Biol Sci*. 2012;367(1607):3242–3253.
15. Zygmunt PM, Petersson J, Andersson DA, et al. Vanilloid receptors on sensory nerves mediate the vasodilator action of anandamide. *Nature*. 1999,400(6743):452–457.
16. Bouaboula M, Hilairet S, Marchand J, Fajas L, Le Fur G, Casellas P. Anandamide induced PPAR-gamma transcriptional activation and 3T3-L1 preadipocyte differentiation. *Eur J Pharmacol*. 2005;517(3):174–181.
17. Smith TH, Sim-Selley LJ, Selley DE. Cannabinoid CB1 receptor-interacting proteins: novel targets for central nervous system drug discovery? *Br J Pharmacol*. 2010;160(3):451–466.
18. Wager-Miller J, Westenbroek R, Mackie K. Dimerization of G protein-coupled receptors: CB1 cannabinoid receptors as an example. *Chem Phys Lipids*. 2002;121(1–2):83–89.
19. Hudson BD, Hebert TE, Kelly ME. Ligand- and heterodimer-directed signaling of the CB(1) cannabinoid receptor. *Mol Pharmacol*. 2010;77:1):1–9.
20. Nie J, Lewis DL. Structural domains of the CB1 cannabinoid receptor that contribute to constitutive activity and G-protein sequestration. *J Neurosci*. 2001;21(22):8758–8764.
21. Perez DM, Karnik SS. Multiple signaling states of G-protein-coupled receptors. *Pharmacol Rev*. 2005;57(2):147–161.
22. Park PS, Lodowski DT, Palczewski K. Activation of G protein-coupled receptors: beyond two-state models and tertiary conformational changes. *Annu Rev Pharmacol Toxicol*. 2008;48:107–141.
23. Mukhopadhyay S, Howlett AC. CB1 receptor-G protein association. Subtype selectivity is determined by distinct intracellular domains. *Eur J Biochem*. 2001;268(3):499–505.
24. Rozenfeld R. Type I cannabinoid receptor trafficking: all roads lead to lysosome. *Traffic*. 2011;12 (1):12–18.
25. Brailoiu GC, Oprea TI, Zhao P, Abood ME, Brailoiu E. Intracellular cannabinoid type 1 (CB1) receptors are activated by anandamide. *J Biol Chem*. 2011;286(33):29166–29174.
26. Benard G, Massa F, Puente N, et al. Mitochondrial CB(1) receptors regulate neuronal energy metabolism. *Nat Neurosci*. 2012;15(4):558–564.
27. Hebert-Chatelain E, Reguero L, Puente N, et al. Cannabinoid control of brain bioenergetics: exploring the subcellular localization of the CB1 receptor. *Mol Metab*. 2014;3(4):495–504.

28. Ma L, Jia J, Niu W, et al. Mitochondrial CB1 receptor is involved in ACEA-induced protective effects on neurons and mitochondrial functions. *Sci Rep*. 2015;5:12440.
29. Hebert-Chatelain E, Desprez T, Serrat R, et al. A cannabinoid link between mitochondria and memory. *Nature*. 2016;539(7630):555–559.
30. Varga EV, Georgieva T, Tumati S, et al. Functional selectivity in cannabinoid signaling. *Curr Mol Pharmacol*. 2008;1(3):273–284.
31. Bosier B, Muccioli GG, Hermans E, Lambert DM. Functionally selective cannabinoid receptor signalling: therapeutic implications and opportunities. *Biochem Pharmacol*. 2010;80(1):1–12.
32. Turu G, Hunyady L. Signal transduction of the CB1 cannabinoid receptor. *J Mol Endocrinol*. 2010;44(2):75–85.
33. Jung SW, Cho AE, Yu W. Exploring the ligand efficacy of cannabinoid receptor 1 (CB1) using molecular dynamics simulations. *Sci Rep*. 2018;8(1):13787.
34. Hua T, Vemuri K, Pu M, et al. Crystal structure of the human cannabinoid receptor CB1. *Cell*. 2016;167(3):750–762 e714.
35. Gertsch J, Leonti M, Raduner S, et al. Beta-caryophyllene is a dietary cannabinoid. *Proc Natl Acad Sci U S A*. 2008;105(26):9099–9104.
36. Bie B, Wu J, Foss JF, Naguib M. An overview of the cannabinoid type 2 receptor system and its therapeutic potential. *Curr Opin Anaesthesiol*. 2018;31(4):407–414.
37. Uhelski ML, Khasabova I, Simone DA. Modulation of pain by endocannabinoids in the periphery. In: *Recent Advances in Cannabinoid Research*. Rijeka: IntechOpen; 2018.
38. Pandey P, Roy KK, Liu H, et al. Structure-based identification of potent natural product chemotypes as cannabinoid receptor 1 inverse agonists. *Molecules*. 2018;23(10):2630.
39. Badal S, Smith KN, Rajnarayanan R. Analysis of natural product regulation of cannabinoid receptors in the treatment of human disease. *Pharmacol Ther*. 2017;180:24–48.
40. Chanda D, Neumann D, Glatz JFC. The endocannabinoid system: overview of an emerging multifaceted therapeutic target. *Prostaglandins Leukot Essent Fat Acids*. 2019;140:51–56.
41. Chianese R, Meccariello R. The endocannabinoid system in human physiology. In: *Cannabinoids in Health and Disease*. Rijeka: IntechOpen; 2016.
42. Russo EB. Clinical endocannabinoid deficiency reconsidered: current research supports the theory in migraine, fibromyalgia, irritable bowel, and other treatment-resistant syndromes. *Cannabis Cannabinoid Res*. 2016;1(1):154–165.
43. Morena M, Patel S, Bains JS, Hill MN. Neurobiological interactions between stress and the endocannabinoid system. *Neuropsychopharmacology*. 2016;41(1):80–102.
44. Kano M, Ohno-Shosaku T, Hashimotodani Y, Uchigashima M, Watanabe M. Endocannabinoid-mediated control of synaptic transmission. *Physiol Rev*. 2009;89(1):309–380.
45. Hauser AS, Chavali S, Masuho I, et al. Pharmacogenomics of GPCR drug targets. *Cell*. 2018;172(1–2):41–54 e19.
46. Mackie K. Distribution of cannabinoid receptors in the central and peripheral nervous system. *Handb Exp Pharmacol*. 2005;168:299–325.
47. Katona I, Sperlagh B, Sik A, et al. Presynaptically located CB1 cannabinoid receptors regulate GABA release from axon terminals of specific hippocampal interneurons. *J Neurosci*. 1999;19(11):4544–4558.
48. Castillo PE, Younts TJ, Chavez AE, Hashimotodani Y. Endocannabinoid signaling and synaptic function. *Neuron*. 2012;76(1):70–81.
49. Stella N. Endocannabinoid signaling in microglial cells. *Neuropharmacology*. 2009;56(Suppl 1):244–253.
50. Palazuelos J, Aguado T, Pazos MR, et al. Microglial CB2 cannabinoid receptors are neuroprotective in Huntington's disease excitotoxicity. *Brain*. 2009;132(Pt 11):3152–3164.
51. Mackie K. Cannabinoid receptors: where they are and what they do. *J Neuroendocrinol*. 2008;20(Suppl 1):10–14.
52. Sagredo O, Garcia-Arencibia M, de Lago E, Finetti S, Decio A, Fernandez-Ruiz J. Cannabinoids and neuroprotection in basal ganglia disorders. *Mol Neurobiol*. 2007;36(1):82–91.
53. Ramirez BG, Blazquez C, Gomez del Pulgar T, Guzman M, de Ceballos ML. Prevention of Alzheimer's disease pathology by cannabinoids: neuroprotection mediated by blockade of microglial activation. *J Neurosci*. 2005;25(8):1904–1913.

54. den Boon FS, Chameau P, Schaafsma-Zhao Q, et al. Excitability of prefrontal cortical pyramidal neurons is modulated by activation of intracellular type-2 cannabinoid receptors. *Proc Natl Acad Sci U S A*. 2012;109(9):3534–3539.
55. Lu D, Immadi SS, Wu Z, Kendall DA. Translational potential of allosteric modulators targeting the cannabinoid CB1 receptor. *Acta Pharmacol Sin*. 2019;40(3):324–335.
56. Mestre L, Carrillo-Salinas FJ, Mecha M, Feliu A, Guaza C. Gut microbiota, cannabinoid system and neuroimmune interactions: new perspectives in multiple sclerosis. *Biochem Pharmacol*. 2018;157:51–66.
57. Di Marzo V. New approaches and challenges to targeting the endocannabinoid system. *Nat Rev Drug Discov*. 2018;17(9):623–639.
58. Obrenovich M, Sankar Chittoor Mana T, Rai H, et al. Recent findings within the microbiota–gut–brain–endocrine metabolic interactome. *Pathol Lab Med Int*. 2017;9:21–30.
59. Russo R, Cristiano C, Avagliano C, et al. Gut-brain axis: role of lipids in the regulation of inflammation, pain and CNS diseases. *Curr Med Chem*. 2018;25(32):3930–3952.
60. Ma Q, Xing C, Long W, Wang HY, Liu Q, Wang RF. Impact of microbiota on central nervous system and neurological diseases: the gut-brain axis. *J Neuroinflammation*. 2019;16(1):53.
61. Hingorani T, Gul W, Elsohly M, Repka MA, Majumdar S. Effect of ion pairing on in vitro transcorneal permeability of a Delta(9)-tetrahydrocannabinol prodrug: potential in glaucoma therapy. *J Pharm Sci*. 2012;101(2):616–626.
62. Cooper ZD, Bedi G, Ramesh D, Balter R, Comer SD, Haney M. Impact of co-administration of oxycodone and smoked cannabis on analgesia and abuse liability. *Neuropsychopharmacology*. 2018;43(10):2046–2055.
63. Hutchison KE, Hagerty SL, Galinkin J, Bryan AD, Bidwell LC. Cannabinoids, pain, and opioid use reduction: the importance of distilling and disseminating existing data. *Cannabis Cannabinoid Res*. 2019;4(3):158–164.
64. Kwiatkowska M, Parker LA, Burton P, Mechoulam R. A comparative analysis of the potential of cannabinoids and ondansetron to suppress cisplatin-induced emesis in the Suncus murinus (house musk shrew). *Psychopharmacology*. 2004;174(2):254–259.
65. Perucca E. Cannabinoids in the treatment of epilepsy: hard evidence at last? *J Epilepsy Res*. 2017;7(2):61–76.
66. Zaheer S, Kumar D, Khan MT, Giyanwani PR, Kiran F. Epilepsy and cannabis: a literature review. *Cureus*. 2018;10(9):e3278.
67. Silvestro S, Mammana S, Cavalli E, Bramanti P, Mazzon E. Use of cannabidiol in the treatment of epilepsy: efficacy and security in clinical trials. *Molecules*. 2019;24(8):1459.
68. Zuardi AW, Crippa JA, Hallak JE, et al. A critical review of the antipsychotic effects of cannabidiol: 30 years of a translational investigation. *Curr Pharm Des*. 2012;18(32):5131–5140.
69. Pellati F, Borgonetti V, Brighenti V, Biagi M, Benvenuti S, Corsi L. *Cannabis sativa* L. and nonpsychoactive cannabinoids: their chemistry and role against oxidative stress, inflammation, and cancer. *Biomed Res Int*. 2018;2018, 1691428.
70. Burstein S. Cannabidiol (CBD) and its analogs: a review of their effects on inflammation. *Bioorg Med Chem*. 2015;23(7):1377–1385.
71. Wang Y, Mukhopadhyay P, Cao Z, et al. Cannabidiol attenuates alcohol-induced liver steatosis, metabolic dysregulation, inflammation and neutrophil-mediated injury. *Sci Rep*. 2017;7(1):12064.
72. Hacke ACM, Lima D, de Costa F, et al. Probing the antioxidant activity of Delta(9)-tetrahydrocannabinol and cannabidiol in *Cannabis sativa* extracts. *Analyst*. 2019;144(16):4952–4961.
73. Tura M, Mandrioli M, Gallina Toschi T. Preliminary study: comparison of antioxidant activity of cannabidiol (CBD) and alpha-tocopherol added to refined olive and sunflower oils. *Molecules*. 2019;24(19):3485.
74. Pacher P, Kogan NM, Mechoulam R. Beyond THC and endocannabinoids. *Annu Rev Pharmacol Toxicol*. 2020;60:637–659.
75. Kiskova T, Mungenast F, Suvakova M, Jager W, Thalhammer T. Future aspects for cannabinoids in breast cancer therapy. *Int J Mol Sci*. 2019;20(7):1673.
76. Bonaccorso S, Ricciardi A, Zangani C, Chiappini S, Schifano F. Cannabidiol (CBD) use in psychiatric disorders: a systematic review. *Neurotoxicology*. 2019;74:282–298.

77. Lucatch AM, Coles AS, Hill KP, George TP. Cannabis and mood disorders. *Curr Addict Rep*. 2018;5(3):336–345.
78. Micale V, Di Marzo V, Sulcova A, Wotjak CT, Drago F. Endocannabinoid system and mood disorders: priming a target for new therapies. *Pharmacol Ther*. 2013;138(1):18–37.
79. Hanus LO, Meyer SM, Munoz E, Taglialatela-Scafati O, Appendino G. Phytocannabinoids: a unified critical inventory. *Nat Prod Rep*. 2016;33(12):1357–1392.
80. Russo EB. The case for the entourage effect and conventional breeding of clinical cannabis: no "strain," no gain. *Front Plant Sci*. 2018;9:1969.
81. Santiago M, Sachdev S, Arnold JC, McGregor IS, Connor M. Absence of entourage: terpenoids commonly found in Cannabis sativa do not modulate the functional activity of Delta(9)-THC at human CB1 and CB2 receptors. *Cannabis Cannabinoid Res*. 2019;4(3):165–176.
82. Watkins BA. Diet, endocannabinoids, and health. *Nutr Res*. 2019;70:32–39.
83. Watkins BA. Endocannabinoids, exercise, pain, and a path to health with aging. *Mol Asp Med*. 2018;64:68–78.
84. Dyall SC. Interplay between n-3 and n-6 long-chain polyunsaturated fatty acids and the endocannabinoid system in brain protection and repair. *Lipids*. 2017;52(11):885–900.
85. Bosch-Bouju C, Layé S. Dietary omega-6/omega-3 and endocannabinoids: implications for brain health and diseases. In: *Cannabinoids in Health and Disease*. Rijeka: IntechOpen; 2016.
86. Healy-Stoffel M, Levant B. N-3 (omega-3) fatty acids: effects on brain dopamine systems and potential role in the etiology and treatment of neuropsychiatric disorders. *CNS Neurol Disord Drug Targets*. 2018;17(3):216–232.
87. Michael-Titus A. Omega-3 fatty acids: their neuroprotective and regenerative potential in traumatic neurological injury. *Clin Lipidol*. 2009;4(3):343–353.
88. McDougle DR, Watson JE, Abdeen AA, et al. Anti-inflammatory omega-3 endocannabinoid epoxides. *Proc Natl Acad Sci U S A*. 2017;114(30):E6034–E6043.
89. Russo EB. Beyond cannabis: plants and the endocannabinoid system. *Trends Pharmacol Sci*. 2016;37(7):594–605.
90. Sharma C, Sadek B, Goyal SN, Sinha S, Kamal MA, Ojha S. Small molecules from nature targeting G-protein coupled cannabinoid receptors: potential leads for drug discovery and development. *Evid Based Complement Alternat Med*. 2015;2015, 238482.
91. Woelkart K, Xu W, Pei Y, Makriyannis A, Picone RP, Bauer R. The endocannabinoid system as a target for alkamides from *Echinacea angustifolia* roots. *Planta Med*. 2005;71(8):701–705.
92. Gertsch J. Immunomodulatory lipids in plants: plant fatty acid amides and the human endocannabinoid system. *Planta Med*. 2008;74(6):638–650.
93. Dossou KS, Devkota KP, Morton C, et al. Identification of CB1/CB2 ligands from *Zanthoxylum bungeanum*. *J Nat Prod*. 2013;76(11):2060–2064.
94. Luo XQ, Li A, Yang X, et al. Paeoniflorin exerts neuroprotective effects by modulating the M1/M2 subset polarization of microglia/macrophages in the hippocampal CA1 region of vascular dementia rats via cannabinoid receptor 2. *Chin Med*. 2018;13:14.
95. Janero DR, Makriyannis A. Terpenes and lipids of the endocannabinoid and transient-receptor-potential-channel biosignaling systems. *ACS Chem Neurosci*. 2014;5(11):1097–1106.
96. Rempel V, Fuchs A, Hinz S, et al. Magnolia extract, magnolol, and metabolites: activation of cannabinoid CB2 receptors and blockade of the related GPR55. *ACS Med Chem Lett*. 2013;4(1):41–45.
97. Fuchs A, Rempel V, Muller CE. The natural product magnolol as a lead structure for the development of potent cannabinoid receptor agonists. *PLoS ONE*. 2013;8(10):e77739.
98. Okine BN, Madasu MK, McGowan F, et al. N-palmitoylethanolamide in the anterior cingulate cortex attenuates inflammatory pain behaviour indirectly via a CB1 receptor-mediated mechanism. *Pain*. 2016;157(12):2687–2696.
99. Karwad MA, Macpherson T, Wang B, et al. Oleoylethanolamine and palmitoylethanolamine modulate intestinal permeability in vitro via TRPV1 and PPARalpha. *FASEB J*. 2017;31(2):469–481.
100. Pertwee RG. GPR55: a new member of the cannabinoid receptor clan? *Br J Pharmacol*. 2007;152(7):984–986.
101. Petrosino S, Di Marzo V. The pharmacology of palmitoylethanolamide and first data on the therapeutic efficacy of some of its new formulations. *Br J Pharmacol*. 2017;174(11):1349–1365.

102. Skaper SD, Facci L, Giusti P. Glia and mast cells as targets for palmitoylethanolamide, an anti-inflammatory and neuroprotective lipid mediator. *Mol Neurobiol.* 2013;48(2):340–352.
103. Scuderi C, Bronzuoli MR, Facchinetti R, et al. Ultramicronized palmitoylethanolamide rescues learning and memory impairments in a triple transgenic mouse model of Alzheimer's disease by exerting anti-inflammatory and neuroprotective effects. *Transl Psychiatry.* 2018;8(1):32.
104. D'Agostino G, La Rana G, Russo R, et al. Central administration of palmitoylethanolamide reduces hyperalgesia in mice via inhibition of NF-kappaB nuclear signalling in dorsal root ganglia. *Eur J Pharmacol.* 2009;613(1–3):54–59.
105. Boccella S, Cristiano C, Romano R, et al. Ultra-micronized palmitoylethanolamide rescues the cognitive decline-associated loss of neural plasticity in the neuropathic mouse entorhinal cortex-dentate gyrus pathway. *Neurobiol Dis.* 2019;121:106–119.
106. Ghazizadeh-Hashemi M, Ghajar A, Shalbafan MR, et al. Palmitoylethanolamide as adjunctive therapy in major depressive disorder: a double-blind, randomized and placebo-controlled trial. *J Affect Disord.* 2018;232:127–133.
107. del Rio C, Navarrete C, Collado JA, et al. The cannabinoid quinol VCE-004.8 alleviates bleomycin-induced scleroderma and exerts potent antifibrotic effects through peroxisome proliferator-activated receptor-gamma and CB2 pathways. *Sci Rep.* 2016;6:21703.
108. Navarrete C, Carrillo-Salinas F, Palomares B, et al. Hypoxia mimetic activity of VCE-004.8, a cannabidiol quinone derivative: implications for multiple sclerosis therapy. *J Neuroinflammation.* 2018;15(1):64.
109. Zou M, Li D, Li L, Wu L, Sun C. Role of the endocannabinoid system in neurological disorders. *Int J Dev Neurosci.* 2019;76:95–102.
110. Fagan SG, Campbell VA. Chapter 2: Endocannabinoids and Alzheimer's disease. In: *Cannabinoids in Neurologic and Mental Disease.* Academic Press; 2015:15–33.
111. Graham WV, Bonito-Oliva A, Sakmar TP. Update on Alzheimer's disease therapy and prevention strategies. *Annu Rev Med.* 2017;68:413–430.
112. Aso E, Ferrer I. CB2 cannabinoid receptor as potential target against Alzheimer's disease. *Front Neurosci.* 2016;10:243.
113. Benito C, Nunez E, Tolon RM, et al. Cannabinoid CB2 receptors and fatty acid amide hydrolase are selectively overexpressed in neuritic plaque-associated glia in Alzheimer's disease brains. *J Neurosci.* 2003;23(35):11136–11141.
114. Tang Y, Bao JS, Su JH, Huang W. MicroRNA-139 modulates Alzheimer's-associated pathogenesis in SAMP8 mice by targeting cannabinoid receptor type 2. *Genet Mol Res.* 2017;16(1). https://doi.org/10.4238/gmr16019166.
115. van der Stelt M, Mazzola C, Esposito G, et al. Endocannabinoids and beta amyloid-induced neurotoxicity in vivo: effect of pharmacological elevation of endocannabinoid levels. *Cell Mol Life Sci.* 2006;63(12):1410–1424.
116. Nomura DK, Morrison BE, Blankman JL, et al. Endocannabinoid hydrolysis generates brain prostaglandins that promote neuroinflammation. *Science.* 2011;334(6057):809–813.
117. Chen R, Zhang J, Wu Y, et al. Monoacylglycerol lipase is a therapeutic target for Alzheimer's disease. *Cell Rep.* 2012;2(5):1329–1339.
118. Watt G, Karl T. In vivo evidence for therapeutic properties of cannabidiol (CBD) for Alzheimer's disease. *Front Pharmacol.* 2017;8:20.
119. Ruthirakuhan MT, Herrmann N, Gallagher D, et al. Investigating the safety and efficacy of nabilone for the treatment of agitation in patients with moderate-to-severe Alzheimer's disease: study protocol for a cross-over randomized controlled trial. *Contemp Clin Trials Commun.* 2019;15, 100385.
120. van den Elsen GAH, Ahmed AIA, Verkes RJ, Feuth T, van der Marck MA, Olde Rikkert MGM. Tetrahydrocannabinol in behavioral disturbances in dementia: a crossover randomized controlled trial. *Am J Geriatr Psychiatry.* 2015;23(12):1214–1224.
121. Di Filippo M, Picconi B, Tozzi A, Ghiglieri V, Rossi A, Calabresi P. The endocannabinoid system in Parkinson's disease. *Curr Pharm Des.* 2008;14(23):2337–2347.
122. Kluger B, Triolo P, Jones W, Jankovic J. The therapeutic potential of cannabinoids for movement disorders. *Mov Disord.* 2015;30(3):313–327.
123. Picconi B, Centonze D, Hakansson K, et al. Loss of bidirectional striatal synaptic plasticity in L-DOPA-induced dyskinesia. *Nat Neurosci.* 2003;6(5):501–506.

124. Ng Cheong Ton JM, Gerhardt GA, Friedemann M, et al. The effects of delta 9-tetrahydrocannabinol on potassium-evoked release of dopamine in the rat caudate nucleus: an in vivo electrochemical and in vivo microdialysis study. *Brain Res*. 1988;451(1–2):59–68.
125. Fernández-Ruiz J, Romero J, Ramos JA. Endocannabinoids and neurodegenerative disorders: Parkinson's disease, Huntington's chorea, Alzheimer's disease, and others. In: Endocannabinoids IPRe, ed. *Handbook of Experimental Pharmacology*. vol. 231. Cham: Springer; 2015.
126. Stampanoni Bassi M, Sancesario A, Morace R, Centonze D, Iezzi E. Cannabinoids in Parkinson's disease. *Cannabis Cannabinoid Res*. 2017;2(1):21–29.
127. Garcia C, Gomez-Canas M, Burgaz S, et al. Benefits of VCE-003.2, a cannabigerol quinone derivative, against inflammation-driven neuronal deterioration in experimental Parkinson's disease: possible involvement of different binding sites at the PPARgamma receptor. *J Neuroinflammation*. 2018;15(1):19.
128. Moreno-Martet M, Feliu A, Espejo-Porras F, et al. The disease-modifying effects of a Sativex-like combination of phytocannabinoids in mice with experimental autoimmune encephalomyelitis are preferentially due to Delta9-tetrahydrocannabinol acting through CB1 receptors. *Mult Scler Relat Disord*. 2015;4(6):505–511.
129. Mecha M, Carrillo-Salinas FJ, Feliu A, Mestre L, Guaza C. Microglia activation states and cannabinoid system: therapeutic implications. *Pharmacol Ther*. 2016;166:40–55.
130. Feliu A, Moreno-Martet M, Mecha M, et al. A Sativex((R))-like combination of phytocannabinoids as a disease-modifying therapy in a viral model of multiple sclerosis. *Br J Pharmacol*. 2015;172(14):3579–3595.
131. Mecha M, Feliu A, Inigo PM, Mestre L, Carrillo-Salinas FJ, Guaza C. Cannabidiol provides long-lasting protection against the deleterious effects of inflammation in a viral model of multiple sclerosis: a role for A2A receptors. *Neurobiol Dis*. 2013;59:141–150.
132. Feliu A, Bonilla Del Rio I, Carrillo-Salinas FJ, et al. 2-Arachidonoylglycerol reduces proteoglycans and enhances remyelination in a progressive model of demyelination. *J Neurosci*. 2017;37(35):8385–8398.
133. Carrillo-Salinas FJ, Navarrete C, Mecha M, et al. A cannabigerol derivative suppresses immune responses and protects mice from experimental autoimmune encephalomyelitis. *PLoS ONE*. 2014;9(4):e94733.
134. Granja AG, Carrillo-Salinas F, Pagani A, et al. A cannabigerol quinone alleviates neuroinflammation in a chronic model of multiple sclerosis. *J NeuroImmune Pharmacol*. 2012;7(4):1002–1016.
135. Baker D, Pryce G, Croxford JL, et al. Cannabinoids control spasticity and tremor in a multiple sclerosis model. *Nature*. 2000;404(6773):84–87.
136. Maresz K, Pryce G, Ponomarev ED, et al. Direct suppression of CNS autoimmune inflammation via the cannabinoid receptor CB1 on neurons and CB2 on autoreactive T cells. *Nat Med*. 2007;13(4):492–497.
137. Kendall DA, Yudowski GA. Cannabinoid receptors in the central nervous system: their signaling and roles in disease. *Front Cell Neurosci*. 2016;10:294.
138. Kutzelnigg A, Lassmann H. Pathology of multiple sclerosis and related inflammatory demyelinating diseases. *Handb Clin Neurol*. 2014;122:15–58.
139. Ilyasov AA, Milligan CE, Pharr EP, Howlett AC. The endocannabinoid system and oligodendrocytes in health and disease. *Front Neurosci*. 2018;12:733.
140. Gomez O, Sanchez-Rodriguez MA, Ortega-Gutierrez S, et al. A basal tone of 2-arachidonoylglycerol contributes to early Oligodendrocyte progenitor proliferation by activating phosphatidylinositol 3-kinase (PI3K)/AKT and the mammalian target of rapamycin (MTOR) pathways. *J NeuroImmune Pharmacol*. 2015;10(2):309–317.
141. Mecha M, Yanguas-Casas N, Feliu A, et al. The endocannabinoid 2-AG enhances spontaneous remyelination by targeting microglia. *Brain Behav Immun*. 2019;77:110–126.
142. Fisher RS, van Emde Boas W, Blume W, et al. Epileptic seizures and epilepsy: definitions proposed by the international league against epilepsy (ILAE) and the International Bureau for Epilepsy (IBE). *Epilepsia*. 2005;46(4):470–472.
143. Rosenberg EC, Tsien RW, Whalley BJ, Devinsky O. Cannabinoids and epilepsy. *Neurotherapeutics*. 2015;12(4):747–768.
144. Ibeas Bih C, Chen T, Nunn AV, Bazelot M, Dallas M, Whalley BJ. Molecular targets of cannabidiol in neurological disorders. *Neurotherapeutics*. 2015;12(4):699–730.

145. Kaplan JS, Stella N, Catterall WA, Westenbroek RE. Cannabidiol attenuates seizures and social deficits in a mouse model of Dravet syndrome. *Proc Natl Acad Sci U S A*. 2017;114(42):11229–11234.
146. Wanker EE, Ast A, Schindler F, Trepte P, Schnoegl S. The pathobiology of perturbed mutant huntingtin protein-protein interactions in Huntington's disease. *J Neurochem*. 2019;151(4):507–519.
147. Smith-Dijak AI, Sepers MD, Raymond LA. Alterations in synaptic function and plasticity in Huntington disease. *J Neurochem*. 2019;150(4):346–365.
148. Aymerich MS, Aso E, Abellanas MA, et al. Cannabinoid pharmacology/therapeutics in chronic degenerative disorders affecting the central nervous system. *Biochem Pharmacol*. 2018;157:67–84.
149. Pintor A, Tebano MT, Martire A, et al. The cannabinoid receptor agonist WIN 55,212-2 attenuates the effects induced by quinolinic acid in the rat striatum. *Neuropharmacology*. 2006;51 (5):1004–1012.
150. Chiodi V, Uchigashima M, Beggiato S, et al. Unbalance of CB1 receptors expressed in GABAergic and glutamatergic neurons in a transgenic mouse model of Huntington's disease. *Neurobiol Dis*. 2012;45 (3):983–991.
151. Fernandez-Ruiz J. The endocannabinoid system as a target for the treatment of motor dysfunction. *Br J Pharmacol*. 2009;156(7):1029–1040.
152. Blazquez C, Chiarlone A, Sagredo O, et al. Loss of striatal type 1 cannabinoid receptors is a key pathogenic factor in Huntington's disease. *Brain*. 2011;134(Pt 1):119–136.
153. Dowie MJ, Grimsey NL, Hoffman T, Faull RL, Glass M. Cannabinoid receptor CB2 is expressed on vascular cells, but not astroglial cells in the post-mortem human Huntington's disease brain. *J Chem Neuroanat*. 2014;59–60:62–71.
154. Sagredo O, Gonzalez S, Aroyo I, et al. Cannabinoid CB2 receptor agonists protect the striatum against malonate toxicity: relevance for Huntington's disease. *Glia*. 2009;57(11):1154–1167.
155. Sagredo O, Ramos JA, Decio A, Mechoulam R, Fernandez-Ruiz J. Cannabidiol reduced the striatal atrophy caused 3-nitropropionic acid in vivo by mechanisms independent of the activation of cannabinoid, vanilloid TRPV1 and adenosine A2A receptors. *Eur J Neurosci*. 2007;26(4):843–851.
156. Valdeolivas S, Navarrete C, Cantarero I, Bellido ML, Munoz E, Sagredo O. Neuroprotective properties of cannabigerol in Huntington's disease: studies in R6/2 mice and 3-nitropropionate-lesioned mice. *Neurotherapeutics*. 2015;12(1):185–199.
157. Diaz-Alonso J, Paraiso-Luna J, Navarrete C, et al. VCE-003.2, a novel cannabigerol derivative, enhances neuronal progenitor cell survival and alleviates symptomatology in murine models of Huntington's disease. *Sci Rep*. 2016;6, 29789.
158. Aguareles J, Paraiso-Luna J, Palomares B, et al. Oral administration of the cannabigerol derivative VCE-003.2 promotes subventricular zone neurogenesis and protects against mutant huntingtin-induced neurodegeneration. *Transl Neurodegener*. 2019;8:9.
159. Consroe P, Laguna J, Allender J, et al. Controlled clinical trial of cannabidiol in Huntington's disease. *Pharmacol Biochem Behav*. 1991;40(3):701–708.
160. Muller-Vahl KR, Schneider U, Emrich HM. Nabilone increases choreatic movements in Huntington's disease. *Mov Disord*. 1999;14(6):1038–1040.
161. Curtis A, Rickards H. Nabilone could treat chorea and irritability in Huntington's disease. *J Neuropsychiatr Clin Neurosci*. 2006;18(4):553–554.
162. Curtis A, Mitchell I, Patel S, Ives N, Rickards H. A pilot study using nabilone for symptomatic treatment in Huntington's disease. *Mov Disord*. 2009;24(15):2254–2259.
163. Lopez-Sendon Moreno JL, Garcia Caldentey J, Trigo Cubillo P, et al. A double-blind, randomized, cross-over, placebo-controlled, pilot trial with Sativex in Huntington's disease. *J Neurol*. 2016;263 (7):1390–1400.
164. Kolb B, Saber H, Fadel H, Rajah G. The endocannabinoid system and stroke: a focused review. *Brain Circ*. 2019;5(1):1–7.
165. England TJ, Hind WH, Rasid NA, O'Sullivan SE. Cannabinoids in experimental stroke: a systematic review and meta-analysis. *J Cereb Blood Flow Metab*. 2015;35(3):348–358.
166. Hendricks HT, van Limbeek J, Geurts AC, Zwarts MJ. Motor recovery after stroke: a systematic review of the literature. *Arch Phys Med Rehabil*. 2002;83(11):1629–1637.

167. Cauraugh JH, Summers JJ. Neural plasticity and bilateral movements: a rehabilitation approach for chronic stroke. *Prog Neurobiol*. 2005;75(5):309–320.
168. Duncan PW, Goldstein LB, Matchar D, Divine GW, Feussner J. Measurement of motor recovery after stroke. Outcome assessment and sample size requirements. *Stroke*. 1992;23(8):1084–1089.
169. Scharf EL. Translating endocannabinoid biology into clinical practice: cannabidiol for stroke prevention. *Cannabis Cannabinoid Res*. 2017;2(1):259–264.
170. Zhou J, Noori H, Burkovskiy I, Lafreniere JD, Kelly MEM, Lehmann C. Modulation of the endocannabinoid system following central nervous system injury. *Int J Mol Sci*. 2019;20(2):388.
171. Sun J, Fang Y, Chen T, et al. WIN55, 212-2 promotes differentiation of oligodendrocyte precursor cells and improve remyelination through regulation of the phosphorylation level of the ERK 1/2 via cannabinoid receptor 1 after stroke-induced demyelination. *Brain Res*. 2013;1491:225–235.
172. Fernandez-Lopez D, Pradillo JM, Garcia-Yebenes I, Martinez-Orgado JA, Moro MA, Lizasoain I. The cannabinoid WIN55212-2 promotes neural repair after neonatal hypoxia-ischemia. *Stroke*. 2010;41(12):2956–2964.
173. Jia W, Kamen Y, Pivonkova H, Karadottir RT. Neuronal activity-dependent myelin repair after stroke. *Neurosci Lett*. 2019;703:139–144.
174. Dewar D, Underhill SM, Goldberg MP. Oligodendrocytes and ischemic brain injury. *J Cereb Blood Flow Metab*. 2003;23(3):263–274.
175. Sun J, Fang YQ, Ren H, et al. WIN55,212-2 protects oligodendrocyte precursor cells in stroke penumbra following permanent focal cerebral ischemia in rats. *Acta Pharmacol Sin*. 2013;34(1):119–128.
176. Kahle MP, Bix GJ. Neuronal restoration following ischemic stroke: influences, barriers, and therapeutic potential. *Neurorehabil Neural Repair*. 2013;27(5):469–478.
177. Klein TW. Cannabinoid-based drugs as anti-inflammatory therapeutics. *Nat Rev Immunol*. 2005;5(5):400–411.
178. Panikashvili D, Simeonidou C, Ben-Shabat S, et al. An endogenous cannabinoid (2-AG) is neuroprotective after brain injury. *Nature*. 2001;413(6855):527–531.
179. Gruenbaum SE, Zlotnik A, Gruenbaum BF, Hersey D, Bilotta F. Pharmacologic neuroprotection for functional outcomes after traumatic brain injury: a systematic review of the clinical literature. *CNS Drugs*. 2016;30(9):791–806.
180. Lopez-Rodriguez AB, Siopi E, Finn DP, et al. CB1 and CB2 cannabinoid receptor antagonists prevent minocycline-induced neuroprotection following traumatic brain injury in mice. *Cereb Cortex*. 2015;25(1):35–45.
181. McVige J, Bargnes VH, Shukri S, Mechtler L. Cannabis, concussion, and chronic pain: an ongoing retrospective analysis at dent neurologic Institute in Buffalo, NY. *Neurology*. 2018;91:18–19.
182. Hillard CJ, Liu QS. Endocannabinoid signaling in the etiology and treatment of major depressive illness. *Curr Pharm Des*. 2014;20(23):3795–3811.
183. Hill MN, Gorzalka BB. The endocannabinoid system and the treatment of mood and anxiety disorders. *CNS Neurol Disord Drug Targets*. 2009;8(6):451–458.
184. Akirav I. The role of cannabinoids in modulating emotional and non-emotional memory processes in the hippocampus. *Front Behav Neurosci*. 2011;5:34.

CHAPTER 12

Identification of biomarkers for diagnosing and monitoring therapy in the treatment of neurologic disorders

Mohammad Shahnawaz and Claudio Soto
Mitchell Center for Alzheimer's disease and Related Brain Disorders, Department of Neurology, University of Texas McGovern Medical School, Houston, TX, United States

Introduction

Ideal biomarkers for neurologic disorders

The diagnosis of neurologic disorders such as Alzheimer's disease (AD) and Parkinson's disease (PD) mostly relies on clinical symptoms.[1–4] By the time patients show the first clinical sign massive neuronal loss has already occurred. In PD, more than 70% of dopaminergic neurons are already damaged before the appearance of clinical signs.[5] The clinical symptoms of AD and PD overlap with other brain disorders which further decreases the accuracy of clinical diagnosis. In AD, only 75%–80% of clinical diagnoses are confirmed after autopsy. However, the accuracy of the clinical diagnosis of PD varies from 46% to 90%.[2] An enormous effort has been made to identify an accurate and reliable biomarker to diagnose AD and PD at the preclinical stage at which time therapeutic intervention might be more effective. The term "*biomarker*" is an abbreviated form of "biochemical markers," used first in the instance of cancer. A biomarker is defined as an indicator of a normal biologic or pathologic process or response to a therapeutic intervention which can be reliably and accurately measured. An ideal biomarker is a diagnostic aide that allows for the stratification of normal and at-risk populations and accurately interrogates disease progression and/or the effect of the treatment. It should be able to differentiate diseased from healthy individuals with high specificity and sensitivity, even in the instance of neurologic disorders with shared clinical symptoms. It should be reproducible, reliable, noninvasive, inexpensive, and available as a routine laboratory test.[6–8]

In search of reliable central nervous system (CNS) biomarkers, CSF has long been preferred as a biofluid of choice. Due to its proximity to the extracellular space of the brain, the CSF, usually obtained by lumbar puncture, reflects ongoing biochemical and pathologic changes in the CNS. Accordingly, by tradition, examination of spinal fluid has been utilized as a standard diagnostic tool in the evaluation of patients with neurologic disorders.[9–12] However, the collection of CSFs via lumbar puncture is invasive

Neurotherapeutics in the Era of Translational Medicine
https://doi.org/10.1016/B978-0-12-816475-4.00020-3

and not ideal when repeated sampling is required to monitor disease progression. In this case, plasma is considered as the most convenient source since the collection of blood is minimally invasive, inexpensive, and readily available compared to CSF. However, the detection of a putative biomarker in plasma is a challenge since it is expected to be present in miniscule amounts compared to that in the CSF. Another complicating factor is that the high content of plasma proteins may mask detection of the potential biomarker. Despite these complexities, a plethora of efforts are being made to identify a biomarker in plasma.

Common neurologic disorders

Neurologic disorders are considered as major causes of disability and death worldwide. Over the past two decades, the burden due to neurologic disorders has increased substantially. As of 2016, these were the leading cause of disability-adjusted life years (DALYs), approximately 276 million, and the second largest cause of death, 9 million. Alzheimer's disease and Parkinson's disease are the two most common of these disorders.

Alzheimer's disease

Alzheimer's disease (AD) is the most common type of dementia, characterized by progressive loss of memory, behavioral problems, and cognitive disabilities, resulting in a devastating impact on the daily life of patients as well as caregivers. In 2019, nearly 50 million people were living with AD and related dementias worldwide. By 2050, the number of persons afflicted is expected to reach 135 million. In the United States, AD is the sixth leading cause of death. Currently, more than 5 million persons are affected, an amount that is estimated to increase threefold in the next 30 years. Collectively, AD and dementia care in the United States cost billions of dollars, an amount that will approach over one trillion dollars by 2050.

The diagnosis of AD is based on clinical criteria such as medical history, physical examination, mental status exams, laboratory tests, and imaging.[4] However, a definite diagnosis can only be achieved postmortem by identifying the signature hallmarks of AD, the presence of amyloid plaques and neurofibrillary tangles. Amyloid-β (Aβ), a 4-kDa peptide of 39–42 amino acid residues, is the major component of the amyloid plaque. It is produced from the sequential proteolysis of amyloid precursor protein (APP) by two proteases, β- and γ-secretases.[13] The most prevalent alloforms of Aβ present in the amyloid plaques are Aβ(1–40) and Aβ(1–42). The longer form containing two additional amino acids at the C-terminus is more prone to aggregation than Aβ(1–40), thus the major component of mature plaque.[14,15]

Long before disease onset, misfolding, aggregation, and extracellular deposition of Aβ in the form of amyloid plaque are thought to occur. Its aggregation is nucleation-dependent and involves the formation of several intermediate species, ranging from low-molecular-weight (LMW) oligomers to amyloid-β-derived diffusible ligands

(ADDLs) to high-molecular-weight (HMW) oligomers, to protofibrils, and then to mature fibrils. Although identification of amyloid plaque is one of the predictors of AD, compelling evidence suggests, among the amyloid family, that Aβ oligomers are the main mediator of toxicity in AD rather than the plaque itself.[16–21]

Parkinson's disease

Parkinson's disease (PD) is the second most common neurodegenerative disorder after Alzheimer's disease.[22] Nearly 1 million people have PD in the United States, which will likely rise to 1.2 million by 2030. Over 10 million people are living with PD worldwide. Its financial burden approaches $52 billion per year in the United States alone.

The diagnosis of PD is primarily based on the identification of motor symptoms that include resting tremor, limb rigidity, bradykinesia, and postural instability.[23–26] Additionally, PD patients experience nonmotor symptoms: cognitive decline, mood changes, sleep disturbances, autonomic disturbances, etc.[27,28] Some of these nonmotor symptoms appear before the motor ones and worsen as the disease progresses.[29–31]

The most prominent pathologic change that occurs in the PD brain is the progressive degeneration of dopamine-producing neurons (dopaminergic neurons) in the substantia nigra. Degeneration of these neurons leads to dysfunction of neuronal circuits in the motor cortex and the basal ganglia, the pathologic substrates that are ultimately expressed as PD.[24,30]

Another pathological hallmark of PD is the cytoplasmic inclusion of amyloid-like aggregates in neurons, i.e., Lewy bodies, the main component of which is alpha-synuclein (α-Syn).[30,32–36] α-Syn is genetically and pathologically linked to PD.[37,38] It is a small protein mainly present in the presynaptic terminal of neurons.[39,40] Although the physiological function of α-Syn is largely unknown, it is believed to play an important role in vesicle trafficking, neurotransmitter release, synaptic plasticity, and lipid metabolism.[41–44] The 14-kDa (140 amino acids) protein is composed of three structurally distinct motifs: the N-terminal domain (1–60 residues) contains seven conserved hexameric motifs similar to that found in the amphipathic α-helix of apolipoprotein; the central domain (61–95 residues), a nonamyloid component (NAC), is a hydrophobic region of the protein which plays an important role in amyloid formation and toxicity; and the C-terminal domain (96–140 residues), rich in acidic residues that are responsible for the intrinsically disordered nature of α-Syn.[45] Under physiological conditions, α-Syn is considered to be a predominately, natively unfolded monomer.[46,47] However, multiple factors facilitate the misfolding of α-Syn: genetic mutation and posttranslational modifications of the protein, i.e., phosphorylation, nitration, truncation, and oxidation, as well as oxidative stress, proteolysis, and interaction with phospholipids and metal ions.[48–55] A point mutation in the *SNCA* (A53T) α-Syn gene, via a gain of function, is the cause of autosomal dominant PD. Further, multiplication of *SNCA* genes such as duplication and triplication has been shown to cause early-onset and more severe forms of PD.[37,38,56–64]

Compelling evidence suggests that the central event in PD is misfolding and aggregation of α-Syn. Similar to other pathogenic proteins involved in protein misfolding disorders (PMDs) including prions and Aβ, the misfolding and aggregation of α-Syn are initiated by a nucleation-dependent mechanism which results in the formation of intermediate species including soluble α-Syn oligomers or α-Syn aggregates (nonfibrillar) and protofibrils which may mature into insoluble fibrillar aggregates. Recent findings have shown that soluble α-Syn oligomers, rather than insoluble fibrillar aggregates, are primarily responsible for neurodegeneration in PD.[32,65–67]

Misfolded oligomers as surrogate biomarkers

The misfolding and oligomerization of Aβ and α-Syn are the early molecular events that precede the formation of amyloid plaques and Lewy body inclusions in AD and PD which are believed to occur years or decades before the appearance of the first notable clinical sign.[30,65,66,68–70] Initially, the accumulation of insoluble fibrillar aggregates in amyloid plaques and Lewy bodies was considered the main initiator of cellular events leading to the pathogenesis of AD and PD. However, recent findings suggest that soluble oligomers of Aβ and α-Syn are the most toxic species and the main drivers of disease pathology.[20,21,53,67,68,71–73] The amount of brain-derived Aβ oligomer was shown to correlate better with disease severity as compared to amyloid plaque burden in AD.[18,74,75] Similarly, in PD patients, the load of Lewy bodies poorly correlated with the severity of disease.[76,77] These findings indicate that insoluble fibrillar aggregates are less toxic than oligomeric species. Indeed, several in vitro and in vivo studies have shown that oligomers are the most toxic species.[78–87] Further, learning and memory deficits in an AD mouse model that overexpresses oligomers were rescued by accelerating the formation of fibrils.[88] Similarly, disease symptoms have been observed in an animal model overexpressing Aβ oligomers but, on pathologic exam of the brain, there was no evidence for accumulation of amyloid plaques.[89,90] As might be expected from these animal studies, higher concentrations of Aβ oligomers were found in CSF samples from patients with AD and mild cognitive impairment (MCI) in comparison to healthy individuals.[91] Additionally, mutations such as H6R and D7N of the Aβ sequence have been shown to alter Aβ aggregation in the earliest stages of familial Alzheimer's disease (FAD), ostensibly due to the formation of oligomers of higher order. These are presumed to be highly toxic to neural cells as compared to wild-type oligomers, implicating the role of oligomers in the pathogenesis of FAD.[92,93]

Experimentally, different forms of Aβ oligomers ranging from dimer to 20-mers, low molecular weight to high molecular weight, have been reported in vitro utilizing synthetic and natural Aβ peptide as reagents. Likewise, a wide range of soluble Aβ oligomers from low molecular weight to higher molecular weight (>100 kDa) have been detected in AD brain. These diverse forms of oligomers may account for the heterogeneity of AD.[94–97]

Small oligomers present in a soluble fraction of AD brains or extracted from the cerebral cortex of AD patients have been shown to inhibit long-term potentiation (LTP) and enhance long-term depression (LTD).[21]

In the interest of brevity, the following list summarizes further evidence for the role of soluble oligomers in the instance of AD and PD:

1. Small Aβ oligomers secreted from cultured cells at nanomolar concentrations are highly toxic to cells.[98,99]
2. Insoluble extracts from amyloid plaques do not impair LTD.
3. Oligomers extracted from brain are more toxic than synthetic oligomers.[100]
4. Aβ oligomers at femtomolar concentrations inhibit LTP and adversely affect the growth of dendritic spines in hippocampal brain slices.[68,84,101,102]
5. Aβ oligomers, nonamers, and dodecamers extracted from AD brain impair memory in a mouse model.[74]
6. Aβ-derived diffusible ligands (ADDLs) elicit neurophysiologic and pathologic effects.[95,103–106]
7. Evidence for the pathogenesis of α-Syn oligomers in PD, including their subcellular effects on mitochondria, endoplasmic reticulum, etc.[79,107–114]
8. In vitro experiments demonstrating exogenously applied α-Syn oligomers are toxic.[78,79,81,115]
9. Effect of in vivo expression of E57K and E35K α-Syn variants on dopaminergic neurons.[67]
10. Detection of α-Syn oligomers in postmortem CSF and PD brains.[55,116]

As summarized, current evidence reinforces the role of oligomers as the primary driver of the neurodegenerative process in the instance of both PD and AD; thus, the detection of these circulating misfolded oligomers holds promise as potential diagnostic biomarkers. Compelling evidence for this derives from the development of novel, highly sensitive methods that are able to detect miniscule amounts of oligomers in biological fluids (Fig. 1A and B). In this chapter, the term "oligomers (seeds)" refers to all forms of soluble nonfibrillar aggregates.

Methods for the detection of oligomers in biological fluids

Direct seed detection assay

Enzyme-linked immunosorbent assay (ELISA): ELISA has long been used not only for the detection of Aβ and α-Syn oligomers in the body fluids but also for their quantification using a known concentration of oligomers as a standard. For the sensitive measurement of these oligomers, two different approaches have been described.

ELISAs using conformational antibodies: In this approach, conformational antibodies directed against nonlinear epitopes of protein that recognize selectively a subset of

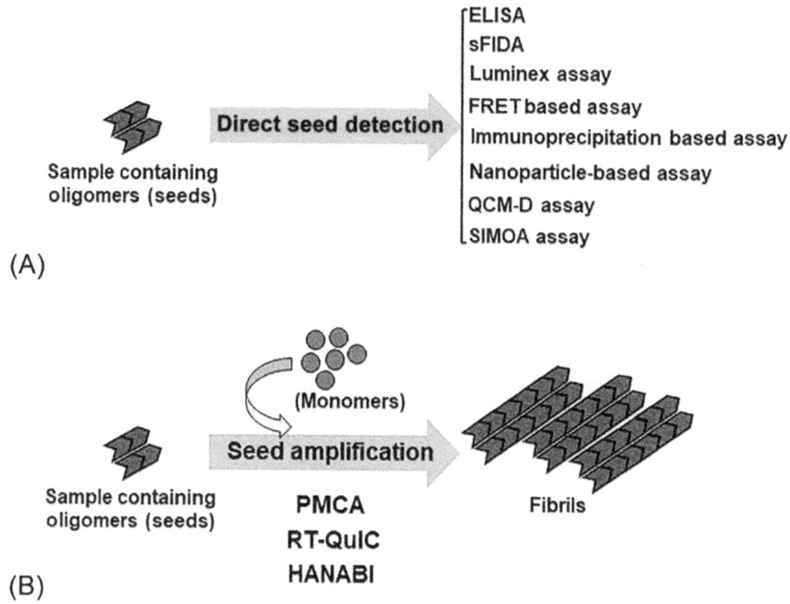

Fig. 1 Methods for the detection of misfolded oligomers in biological fluids. (A) *Direct seed detection assay.* Assays described here are used to detect oligomers in the samples as such either using sequence-specific or conformation-dependent antibodies or a combination of both. (B) *Seed amplification-based assay.* In these assays, oligomers (seeds) in the samples are amplified at the expense of monomers as a substrate following a seeding-nucleation mechanism allowing the detection of amplified aggregates.

oligomers have been used. During the past years, multiple conformation-specific antibodies have been developed against Aβ and α-Syn oligomers. However, only some of them have been employed for the detection of Aβ and α-Syn oligomers in body fluids, and convincingly or consistently demonstrated increased levels of Aβ and α-Syn oligomers in AD and PD patients as compared to healthy controls.[117–119] Due to the heterogeneous nature of Aβ and α-Syn oligomers in shape, size, and structures, the most relevant oligomers associated with disease pathogenesis are still unclear and are being investigated. The main limitation of using conformation-specific antibody is that it detects only a subset of oligomers that may not be relevant for disease pathology. Thus, it is likely that ELISAs based on conformation-specific antibodies may not detect disease-relevant Aβ and α-Syn oligomers.

ELISAs using the same antibody for capture and detection: This approach has been frequently used to detect Aβ and α-Syn oligomers in body fluids. The use of the same antibody for capture and detection excludes identification of monomers which possess only one epitope. Only those oligomers that contain more than one copy of the same epitope are detected. Therefore, dimers that contain two copies of the same epitope are the smallest species to be detected by this ELISA. Several monoclonal antibodies against the sequence of Aβ and α-Syn have been used to measure the oligomers of these proteins.[91,120–123] A plethora of studies have shown higher levels of Aβ oligomers in AD and MCI patients as compared to nondemented healthy controls. Interestingly, studies have found a strong negative correlation of Aβ oligomer level with MMSE (MINI-Mental State Examination) score, suggesting a higher level of Aβ oligomers in patients with lower cognitive disabilities.[91,123]

Surface-based fluorescence intensity distribution analysis (sFIDA): This method resembles conventional sandwich ELISA. Here, Aβ-specific capture antibody is immobilized on a functionalized glass surface. Aβ oligomers in CSF are allowed to bind to capture antibody. Subsequently, Aβ oligomers are labeled with at least two fluorescence-labeled Aβ antibodies. Since the capture and detection antibodies share the same epitope or an overlapping epitope, Aβ monomers bound to capture antibodies are not recognized by the detection antibody. To obtain the results, the surface is then imaged by dual-color microscopy. The signal intensities above background noise in both channels are counted and correlated with the concentration of Aβ oligomers in the sample. A study using sFIDA showed higher levels of Aβ oligomers in CSF of AD patients as compared to age-matched controls, and the levels of Aβ oligomers correlated negatively with the MMSE score.[124–126]

Luminex assay: A Luminex platform assay is adapted from a sandwich ELISA using the same monoclonal antibody for capturing and detecting Aβ oligomers. In this assay, Magplex microspheres labeled with a monoclonal antibody against the N-terminal sequence of Aβ, BAN50 as a capture antibody, and biotinylated BAN50 as a detection antibody, are incubated with samples containing Aβ oligomers on a 96-well plate. Following incubation, the plate is read on a FLEXMAP 3D platform. A Luminex-based assay detects elevated levels of Aβ oligomers in CSF of AD patients as compared to controls.[127]

Nanoparticle-based assay: This method has been used to detect small oligomers of Aβ and shown to have an increased sensitivity over commercial ELISA. In this assay, gold nanoparticles coated with hundreds of DNA barcodes and magnetic beads are labeled with Aβ conformation-specific antibody. Both the labeled nanoparticles and the beads are incubated with CSF samples containing Aβ oligomers, resulting in the formation of a gold nanoparticles-oligomers-magnetic beads complex. Subsequently, these complexes are separated using a magnet, and the total number of DNA barcodes are determined. Since gold nanoparticles are coated with hundreds of DNA barcodes, this augments the detection of oligomers in the samples. This assay also demonstrated increased levels of Aβ oligomers in CSF samples from probable AD patients.[128]

Immunoprecipitation-based assay: This method is specifically designed to precipitate Aβ oligomers from blood plasma samples. To accomplish this, magnetic beads are labeled with Aβ-specific antibody and subsequently incubated with plasma samples. The Aβ species immunocaptured on magnetic beads are then separated using a magnet, and then incubated with fluorescence-labeled Aβ antibody. Finally, these immunocomplexes, upon labeling with fluorescence-labeled Aβ antibody, are detected by flow cytometry. The detection of Aβ oligomers using this method discriminated plasma samples of AD patients from nondemented controls.[129]

Fluorescence resonance energy transfer (FRET) combined with flow cytometry: Detection of Aβ oligomers in this method is based on FRET, followed by the detection of FRET events utilizing flow cytometry. Aβ oligomers present in CSF are labeled with two different fluorescence-labeled Aβ antibodies (N-terminal Aβ-specific 6E10-Alexa-Fluor 594 as an acceptor; central region of Aβ-specific 4G8-Alexa-Fluor-488 as a donor). A FRET event only occurs when the two antibodies, bound to an oligomer, are in proximity (<10 nm). Upon the excitation of Alexa-488, the resulting emission fluorescence elicits FRET events that are captured by flow cytometry. The use of two antibodies greatly increases the specificity of this assay for the detection of Aβ oligomers over Aβ monomers. Although this method is very specific toward the detection of Aβ oligomers in CSF samples of nondemented individuals, no correlation has been found between the concentration of Aβ oligomers and age.[130]

Quartz crystal microbalance with dissipation (QCM-D) assay: QCM-D is a mechanical technique that measures quartz resonant frequency changes due to the addition of mass to the surface-modified crystal as well as surface film rigidity. The QCM-D assay utilizes the same antibody for capture and signal amplification to improve specificity toward the Aβ oligomers similar to sandwich ELISA. In this assay, partially reduced fragments of bapineuzumab Aβ antibodies are covalently immobilized on a silica crystal. The addition of mass on the surface of silica leads to a decrease in frequency which is monitored using the QCM-D instrument. In a final step, the labeled bapineuzumab antibody is incubated with samples containing the Aβ oligomers. This results in the formation of an immunocomplex in a sandwich manner. On readout, the amount of Aβ oligomer present is directly proportional to the fluorescent signal.[131]

Single molecule array (Simoa) assay: Simoa, a digital bead-based sandwich ELISA, is designed to quantify the concentration of low abundant analytes in samples. It uses bapineuzumab-coated beads for capture and biotin-conjugated bapineuzumab antibody for detection. The use of the same antibody for both capture and detection ensures the detection of higher-order Aβ species. To assay, capture beads are incubated with samples containing Aβ oligomers, followed by incubation with detection antibody conjugated with biotin. The immunocomplexes bound to beads are loaded into an array of femtoliter-sized wells on a Simoa disc. This ensures the occupation of one bead in each well. Florescence images are taken to determine and count the number of active wells in the array which correlates with the concentration of analytes in the samples. The active well displays increased fluorescence over time.[131]

Seed amplification-based assay

The formation of oligomers is the first molecular event in Alzheimer's and Parkinson's diseases and thought to occur decades before the accumulation of insoluble fibrils in the brain. Recently, significant efforts have been made to detect these early misfolded oligomers in biological fluids. However, due to low concentration, their detection by direct assays has never attained significant sensitivity and specificity over current clinical diagnosis. Most of the direct seed detection assays rely on specific antibodies for the detection of oligomers. However, due to the heterogeneous nature of these oligomers, the development of antibodies against most physiologically relevant ones is a challenge. As a consequence, these assays cannot always differentiate between on-pathway and off-pathway oligomers.

To overcome these hurdles, two ultrasensitive seed amplification assays, protein misfolding cyclic amplification (PMCA) assay, and real-time quacking-induced conversion (RT-QuIC), have been recently developed for the detection of misfolded oligomers in biological fluids through amplification of the misfolding and aggregation process in vitro.[132–135] More recently, a similar assay has been developed to detect oligomers in the CSF of patients, i.e., the HANdai amyloid burst inducer (HANABI) assay. Although the underlying principle of HANABI and RT-QuIC are similar, their methodologies differ. Considering that both techniques have been extensively reviewed and compared elsewhere, the basic principle and applications of PMCA will be emphasized.[136, 137]

Development of PMCA assays for the detection of Aβ and α-Syn oligomers

The basic principle of the PMCA relies on a seeding-nucleation mechanism (Fig. 2). PMCA is conceptually similar to DNA amplification by polymerase chain reaction (PCR) that combines two phases. In the first phase, the samples containing minute amounts of misfolded oligomers and an excess of monomeric protein are incubated to induce growth of the polymers by converting soluble monomeric protein to polymeric misfolded proteins (fibrils). In the second phase, the sample is subjected to mechanical fragmentation (sonication or shaking) to break down the polymers, multiplying the number of seeding-competent seeds. After each cycle, the number of seeds increases exponentially, resulting in large quantities of misfolded aggregates[138–140] (Fig. 3). PMCA has been widely used to detect small amounts of misfolded prion protein (PrP^{sc}) in biological fluids (e.g., CSF, Blood, and Urine).[141–145] It has been shown to detect the equivalent of a single particle of misfolded PrP oligomers.[146] Since the molecular kinetics involved in the misfolding of prion protein is similar to the misfolding of Aβ and α-Syn, we sought to adapt the PMCA technology for the detection of low levels of misfolded Aβ and α-Syn oligomers present in biological fluids.[133–135]

In a traditional PMCA for the detection of PrP^{Sc}, normal brain homogenates are most commonly used as a source of PrP^{c} substrate, and sonication is used for the fragmentation process. At the conclusion of this step, the presence of amplified PrP^{sc} at the end of the PMCA is detected by western blot (WB) after proteinase K (PK) digestion.[138] Although traditional PMCA has been proven to be highly sensitive toward the detection of PrP^{Sc}, the assay is time consuming and cannot be automated for real-time monitoring of protein

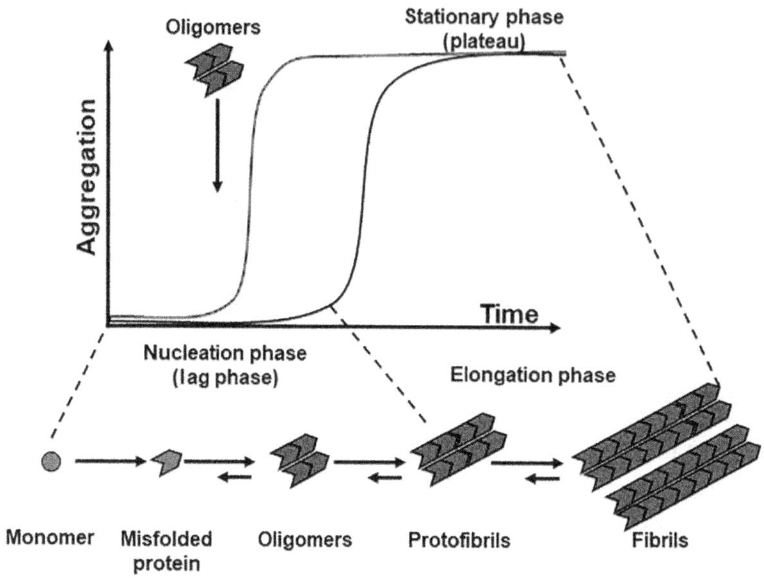

Fig. 2 A seeding-nucleation model of protein aggregation. The protein misfolding and aggregation in neurodegenerative diseases, including AD and PD, follow a seeding-nucleation mechanism. In this model, monomers misfold into misfolded oligomers that act as "seeds" to catalyze the polymerization of fibrils by recruiting the new monomers into the seeds. The formation of oligomers is a slow process in which the first stable misfolded structures are evolved, referred to as the nucleation phase (lag phase). Once the oligomers are formed, new monomers are incorporated into the oligomers, and fibrils grow exponentially in the elongation phase leading to a stationary phase (a plateau). The addition of preformed "seeds or oligomers" dramatically accelerates the polymerization process and thus shortens the lag phase.

aggregation. To overcome this limitation, we modified the traditional PMCA format, resulting in two novel assays termed Aβ-PMCA and α-Syn-PMCA. Both these assays exhibit high sensitivity and specificity. In these assays, synthetic Aβ peptide in Aβ-PMCA and recombinant α-Syn protein in α-Syn-PMCA are used as substrates, and sonication is replaced by intermittent shaking. For readout, an amyloid binding dye (Thioflavin-T) is added to the sample to quantitate the amount of aggregation using fluorescence. Since each peptide or protein requires a defined condition for misfolding and aggregation, it is necessary to optimize the experimental conditions for each targeted peptide or protein (e.g., temperature, pH, ionic strength, peptide, and protein concentration).[133–135]

Methodologies of Aβ-PMCA and α-Syn-PMCA and data interpretation

In Aβ-PMCA, the samples are incubated with seed-free Aβ monomers (2 μM) in an aggregation buffer (0.1 M Tris-Cl, pH 7.4 and 5 μM ThT) at 22°C, whereas in α-Syn-PMCA, samples are incubated with recombinant α-Syn monomers (1 mg/mL) in a buffer (100 mM PIPES-NaOH pH 6.5, 500 mM NaCl, and 5 μM ThT) at 37°C. Both assays are

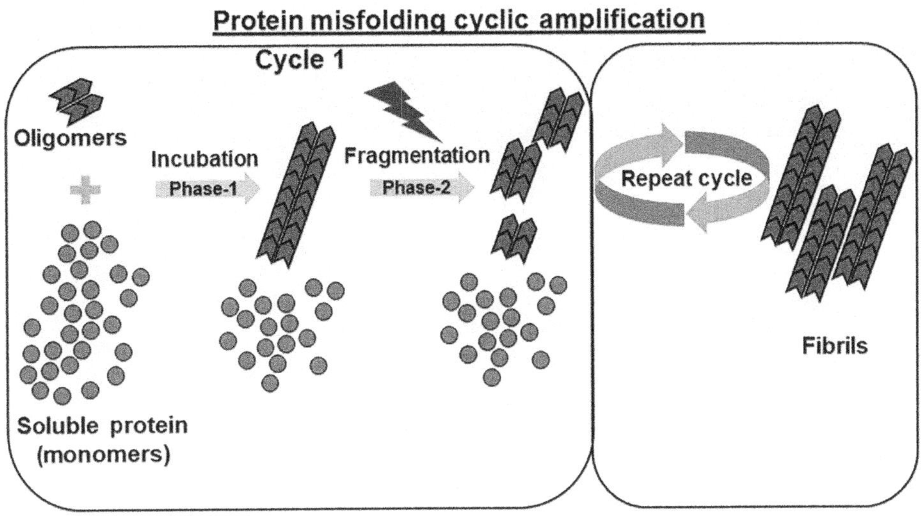

Fig. 3 The principle of protein misfolding cyclic amplification (PMCA) assay. PMCA enables ultrasensitive detection of misfolded aggregates through the amplification of the misfolding and aggregation process in vitro. PMCA is conceptually similar to DNA amplification by polymerase chain reaction (PCR) that combines two phases. In the first phase, the samples containing minute amounts of misfolded oligomers (seeds) and an excess of monomeric proteins are incubated to induce growth of the fibrils by converting soluble monomeric protein to polymeric misfolded proteins (fibrils). In the second phase, the sample is subjected to mechanical fragmentation (sonication or shaking) to break down the fibrils, multiplying the number of seeding-competent seeds. After each cycle, the number of seeds increases exponentially, resulting in large quantities of misfolded aggregates.

conducted in a 96-well plate and are subjected to a cyclic shaking (1 min at 500 rpm followed by 29 min no shanking). The fluorescence of ThT that is enhanced after binding to fibrils is recorded periodically upon excitation at 435 nm and emission at 485 nm (Fig. 4A).[133–135] The ThT fluorescence signal over time gives a sigmoidal curve with distinct phases—a lag phase (slow nucleation phase), an elongation phase (fast exponential phase), and a plateau (stationary phase) (Figs. 2 and 4B). The lag phase is highly influenced by experimental conditions (e.g., concentration of protein, agitation speed, pH, temperature, concentration of seed, etc.). Ideally, the fluorescence of ThT is proportional to the mass of the fibrillar aggregates formed overtime. However, ThT fluorescence can also be affected by distinct conformations of fibrillar aggregates. The differences in the kinetics of aggregation between different samples are evaluated by estimating various kinetic parameters, including the lag phase, the T50, and maximum fluorescence. The lag phase is defined as the time required to reach a ThT fluorescence which is higher than 5 times the background value of the buffer alone. The T50 corresponds to the time which is required to reach 50% of the maximum aggregation which is primarily governed by the amount of seed (Fig. 4B). T50 is used to measure the amount of seed quantity in

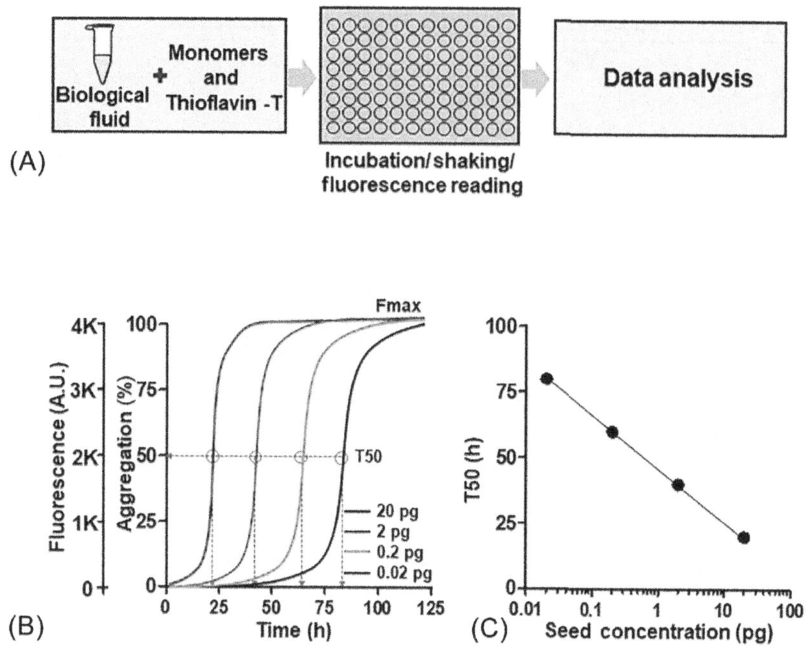

Fig. 4 Methodology of PMCA assay and data analysis. (A) A schematic representation of the PMCA assay, from sample preparation to data analysis. (B) The representative aggregation kinetics obtained after performing α-Syn-PMCA with soluble α-Syn monomers (1 mg/mL) in the presence of different concentrations of seeds (20, 2, 0.2, and 0.02 pg). (C) Time to reach 50% of aggregation (T50) in each condition (in the presence of 20, 2, 0.2, and 0.02 pg seeds), as shown in (B), is plotted as a function of the logarithm of the amount of α-Syn seeds. AU, arbitrary units; pg, picogram; Fmax, maximum fluorescence at the plateau of the aggregation curve; T50, time to reach 50% of maximum aggregation; and h, time in hours.

a sample using a known concentration of seed as a standard (Fig. 4C). The maximum fluorescence (Fmax) is the value at which plateau of ThT fluorescence is reached.

Aβ-PMCA and α-Syn-PMCA assays: Utility for diagnosis, monitoring of disease progression, and therapeutic intervention

One of the greatest hurdles for developing disease-modifying therapy for Alzheimer's and Parkinson's diseases is the lack of early diagnosis to support clinical evaluation, monitor disease progression, and identify affected individuals before they display clinical symptoms. The early diagnosis of these diseases is of utmost importance for the development of disease-modifying or preventive therapies before the irreversible brain damage has already occurred. Since the first pathological event is the formation of oligomers, before the accumulation of protein aggregates, much of our effort to increase sensitivity of the current assays was focused on the detection of the "oligomer phase" in biological fluids. With development, both Aβ-PMCA and α-Syn-PMCA assays have exhibited remarkable performance.[133–135] Aβ-PMCA has differentiated AD from other neurological and nonneurological diseases with high sensitivity and specificity (Fig. 5A).[133] Similarly, α-Syn-PMCA has shown outstanding performance toward differentiating PD and related synucleinopathies (multiple system atrophy, MSA; and dementia with Lewy bodies, DLB) from other neurological disorders and healthy individuals (Fig. 5B).[134] Interestingly, α-Syn-PMCA not only is able to detect α-Syn oligomers at the symptomatic stage but it can also detect them at the presymptomatic stage of the disease, showing its potential application in early diagnosis of patients.[134] More recently, we have shown that α-Syn-PMCA offers the ability to distinguish conformationally distinct α-Syn oligomers implicated in PD and MSA, allowing the differential diagnosis of two closely related diseases that pose challenges in clinical diagnosis particularly at the early stage of the disease (Fig. 5C).[135] This property of PMCA assays will illuminate the roles of distinct oligomers in disease heterogeneity as well as their roles at different stages of the disease. Further, these assays will have application in the discovery of new therapeutics and their assessment in clinical trials.

PMCA assays depend on the amplification of functional oligomers, the concentrations of which influence the time to reach 50% of aggregation (T50) in a dose-dependent manner. Based on the T50, the use of PMCA assays permits quantification of the amount of oligomers in biological fluids (Fig. 4C). Further, a defined relationship between T50 and clinical progression can be determined as measured by clinical scores at different stages of the disease (e.g., Hoehn and Yahr scale that describes different stages of PD and AD; Unified Parkinson's Disease Rating Scale that is used to monitor the course

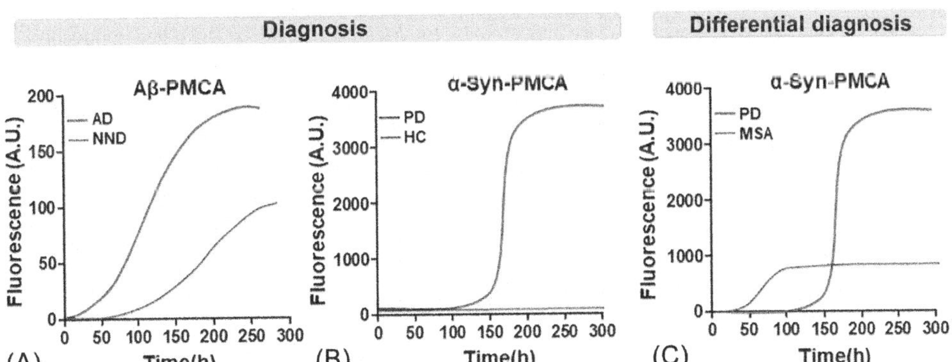

Fig. 5 Performance of PMCA assays in diagnosis and differential diagnosis. (A) The representative aggregation curves obtained after performing Aβ-PMCA with CSF samples from AD patients *(red line)* and patients with the nonneurodegenerative disease *(blue line)*. (B) The representative aggregation curves obtained after performing α-Syn-PMCA with CSF samples from PD patients *(red line)* and healthy individual *(blue line)*. (C) The representation of distinct aggregation kinetics obtained after performing α-Syn-PMCA with CSF samples from PD *(red line)* and multiple system atrophy *(blue line)* patients. AU, arbitrary units; and h, time in hours.

of PD).[134] Of utmost importance, these kinetic studies may prove useful for determining the effectiveness of drugs in disease-modifying clinical trials that target the formation of oligomers.

Conclusion

The diagnosis of neurologic disorders such as AD and PD is based on clinical symptoms that are thought to be preceded by a significant loss of nerve cells. Thus, reliable biomarkers are needed that can detect disease at an earlier stage, aid in clinical diagnosis, monitor disease progression, and assess therapeutic intervention. While current diagnostic biomarkers for both AD and PD have been in clinical use for some time, the detection of misfolded oligomers in CSF and plasma have emerged as potentially more useful. In this regard, results from Aβ-PMCA and α-Syn-PMCA assays are promising advances. Both assays can detect oligomers in CSF with a high sensitivity and specificity. Further, they offer the ability to detect oligomers at the preclinical stage of the disease and the possibility of discriminating between similar neurodegenerative diseases. Besides their role in diagnosis, these assays should facilitate the monitoring of disease progression as well as the efficacy of ongoing therapy with approved drugs and those that are under development. While both assays have been validated utilizing CSF samples, their utility will be exponentially enhanced when it is shown that they can successfully interrogate more accessible biological fluids such as blood, saliva, and tears.

Acknowledgments

We thank the American Parkinson Disease Association for funding to MS, the Michael J. Fox Foundation for Parkinson's Research and NIH to CS, and Hanna Hovren at the Center for Neurologic Study for logistical support.

Conflict of interest disclosure

Dr. Soto and Dr. Shahnawaz are inventors on patents covering the use of Aβ-PMCA and α-Syn-PMCA for the diagnosis of Alzheimer's and Parkinson's diseases. Dr. Soto is an inventor on several patents related to the PMCA technology and is currently the founder, chief scientific officer, and member of the Board of Directors of Amprion Inc., a biotech company focusing on the commercial utilization of PMCA for high-sensitive detection of misfolded protein aggregates implicated in a variety of neurodegenerative diseases. The University of Texas Health Science Center at Houston owns some patent applications related to the PMCA technology that have been licensed to Amprion Inc.

References

1. Gelb DJ, Oliver E, Gilman S. Diagnostic criteria for Parkinson disease. *Arch Neurol*. 1999;56(1):33–39.
2. Tolosa E, Wenning G, Poewe W. The diagnosis of Parkinson's disease. *Lancet Neurol*. 2006;5(1):75–86.

3. Dubois B, Feldman HH, Jacova C, et al. Advancing research diagnostic criteria for Alzheimer's disease: the IWG-2 criteria. *Lancet Neurol.* 2014;13(6):614–629.
4. Santacruz KS, Swagerty D. Early diagnosis of dementia. *Am Fam Physician.* 2001;63(4):703–708.
5. Fearnley JM, Lees AJ. Ageing and Parkinson's disease: substantia nigra regional selectivity. *Brain.* 1991;114(Pt 5):2283–2301.
6. Franklin W, Lowell FC. Sounding board. Unapproved drugs in the practice of medicine. *N Engl J Med.* 1975;292(20):1075–1077.
7. Mayeux R. Biomarkers: potential uses and limitations. *NeuroRx.* 2004;1(2):182–188.
8. Naylor S. Biomarkers: current perspectives and future prospects. *Expert Rev Mol Diagn.* 2003; 3(5):525–529.
9. Hall S, Surova Y, Ohrfelt A, Zetterberg H, Lindqvist D, Hansson O. CSF biomarkers and clinical progression of Parkinson disease. *Neurology.* 2015;84(1):57–63.
10. Jimenez-Jimenez FJ, Alonso-Navarro H, Garcia-Martin E, Agundez JA. Cerebrospinal fluid biochemical studies in patients with Parkinson's disease: toward a potential search for biomarkers for this disease. *Front Cell Neurosci.* 2014;8:369.
11. Kozauer N, Katz R. Regulatory innovation and drug development for early-stage Alzheimer's disease. *N Engl J Med.* 2013;368(13):1169–1171.
12. Parnetti L, Castrioto A, Chiasserini D, et al. Cerebrospinal fluid biomarkers in Parkinson disease. *Nat Rev Neurol.* 2013;9(3):131–140.
13. Ballard C, Gauthier S, Corbett A, Brayne C, Aarsland D, Jones E. Alzheimer's disease. *Lancet.* 2011;377 (9770):1019–1031.
14. Haass C, Kaether C, Thinakaran G, Sisodia S. Trafficking and proteolytic processing of APP. *Cold Spring Harb Perspect Med.* 2012;2(5):a006270.
15. Selkoe DJ. The molecular pathology of Alzheimer's disease. *Neuron.* 1991;6(4):487–498.
16. Lesne S, Kotilinek L, Ashe KH. Plaque-bearing mice with reduced levels of oligomeric amyloid-beta assemblies have intact memory function. *Neuroscience.* 2008;151(3):745–749.
17. Lue LF, Kuo YM, Roher AE, et al. Soluble amyloid beta peptide concentration as a predictor of synaptic change in Alzheimer's disease. *Am J Pathol.* 1999;155(3):853–862.
18. McLean CA, Cherny RA, Fraser FW, et al. Soluble pool of Abeta amyloid as a determinant of severity of neurodegeneration in Alzheimer's disease. *Ann Neurol.* 1999;46(6):860–866.
19. Resende R, Ferreiro E, Pereira C, de Resende OC. Neurotoxic effect of oligomeric and fibrillar species of amyloid-beta peptide 1-42: involvement of endoplasmic reticulum calcium release in oligomer-induced cell death. *Neuroscience.* 2008;155(3):725–737.
20. Klein WL, Stine Jr. WB, Teplow DB. Small assemblies of unmodified amyloid beta-protein are the proximate neurotoxin in Alzheimer's disease. *Neurobiol Aging.* 2004;25(5):569–580.
21. Shankar GM, Li S, Mehta TH, et al. Amyloid-beta protein dimers isolated directly from Alzheimer's brains impair synaptic plasticity and memory. *Nat Med.* 2008;14(8):837–842.
22. Bertram L, Tanzi RE. The genetic epidemiology of neurodegenerative disease. *J Clin Invest.* 2005; 115(6):1449–1457.
23. Frisardi V, Santamato A, Cheeran B. Parkinson's disease: new insights into pathophysiology and rehabilitative approaches. *Parkinsons Dis.* 2016;2016:3121727.
24. Mazzoni P, Shabbott B, Cortes JC. Motor control abnormalities in Parkinson's disease. *Cold Spring Harb Perspect Med.* 2012;2(6):a009282.
25. Postuma RB, Berg D, Adler CH, et al. The new definition and diagnostic criteria of Parkinson's disease. *Lancet Neurol.* 2016;15(6):546–548.
26. Teive HA, Bertucci DF, Munhoz RP. Unusual motor and non-motor symptoms and signs in the early stage of Parkinson's disease. *Arq Neuropsiquiatr.* 2016;74(10):781–784.
27. Mahlknecht P, Seppi K, Poewe W. The concept of prodromal Parkinson's disease. *J Parkinsons Dis.* 2015;5(4):681–697.
28. Wu Y, Le W, Jankovic J. Preclinical biomarkers of Parkinson disease. *Arch Neurol.* 2011;68(1):22–30.
29. Goedert M, Spillantini MG, Del TK, Braak H. 100 years of Lewy pathology. *Nat Rev Neurol.* 2013;9 (1):13–24.
30. Kalia LV, Lang AE. Parkinson's disease. *Lancet.* 2015;386(9996):896–912.

31. Postuma RB, Aarsland D, Barone P, et al. Identifying prodromal Parkinson's disease: pre-motor disorders in Parkinson's disease. *Mov Disord*. 2012;27(5):617–626.
32. Baba M, Nakajo S, Tu PH, et al. Aggregation of alpha-synuclein in Lewy bodies of sporadic Parkinson's disease and dementia with Lewy bodies. *Am J Pathol*. 1998;152(4):879–884.
33. Goedert M. Alpha-synuclein and neurodegenerative diseases. *Nat Rev Neurosci*. 2001;2(7):492–501.
34. Spillantini MG, Schmidt ML, Lee VM, Trojanowski JQ, Jakes R, Goedert M. Alpha-synuclein in Lewy bodies. *Nature*. 1997;388(6645):839–840.
35. Spillantini MG, Crowther RA, Jakes R, Hasegawa M, Goedert M. Alpha-synuclein in filamentous inclusions of Lewy bodies from Parkinson's disease and dementia with lewy bodies. *Proc Natl Acad Sci U S A*. 1998;95(11):6469–6473.
36. Tan JM, Wong ES, Lim KL. Protein misfolding and aggregation in Parkinson's disease. *Antioxid Redox Signal*. 2009;11(9):2119–2134.
37. Dawson TM, Dawson VL. Rare genetic mutations shed light on the pathogenesis of Parkinson disease. *J Clin Invest*. 2003;111(2):145–151.
38. Polymeropoulos MH, Lavedan C, Leroy E, et al. Mutation in the alpha-synuclein gene identified in families with Parkinson's disease. *Science*. 1997;276(5321):2045–2047.
39. Jakes R, Spillantini MG, Goedert M. Identification of two distinct synucleins from human brain. *FEBS Lett*. 1994;345(1):27–32.
40. Spinelli KJ, Taylor JK, Osterberg VR, et al. Presynaptic alpha-synuclein aggregation in a mouse model of Parkinson's disease. *J Neurosci*. 2014;34(6):2037–2050.
41. Clayton DF, George JM. The synucleins: a family of proteins involved in synaptic function, plasticity, neurodegeneration and disease. *Trends Neurosci*. 1998;21(6):249–254.
42. Marques O, Outeiro TF. Alpha-synuclein: from secretion to dysfunction and death. *Cell Death Dis*. 2012;3:e350.
43. Ottolini D, Cali T, Szabo I, Brini M. Alpha-synuclein at the intracellular and the extracellular side: functional and dysfunctional implications. *Biol Chem*. 2017;398(1):77–100.
44. Varkey J, Isas JM, Mizuno N, et al. Membrane curvature induction and tubulation are common features of synucleins and apolipoproteins. *J Biol Chem*. 2010;285(42):32486–32493.
45. Han H, Weinreb PH, Lansbury Jr. PT. The core Alzheimer's peptide NAC forms amyloid fibrils which seed and are seeded by beta-amyloid: is NAC a common trigger or target in neurodegenerative disease? *Chem Biol*. 1995;2(3):163–169.
46. Theillet FX, Binolfi A, Bekei B, et al. Structural disorder of monomeric alpha-synuclein persists in mammalian cells. *Nature*. 2016;530(7588):45–50.
47. Weinreb PH, Zhen W, Poon AW, Conway KA, Lansbury Jr. PT. NACP, a protein implicated in Alzheimer's disease and learning, is natively unfolded. *Biochemistry*. 1996;35(43):13709–13715.
48. Bae EJ, Ho DH, Park E, et al. Lipid peroxidation product 4-hydroxy-2-nonenal promotes seeding-capable oligomer formation and cell-to-cell transfer of alpha-synuclein. *Antioxid Redox Signal*. 2013;18(7):770–783.
49. Hashimoto M, Hsu LJ, Xia Y, et al. Oxidative stress induces amyloid-like aggregate formation of NACP/alpha-synuclein in vitro. *Neuroreport*. 1999;10(4):717–721.
50. Li W, West N, Colla E, et al. Aggregation promoting C-terminal truncation of alpha-synuclein is a normal cellular process and is enhanced by the familial Parkinson's disease-linked mutations. *Proc Natl Acad Sci U S A*. 2005;102(6):2162–2167.
51. Muchowski PJ. Protein misfolding, amyloid formation, and neurodegeneration: a critical role for molecular chaperones? *Neuron*. 2002;35(1):9 12.
52. Ruzafa D, Hernandez-Gomez YS, Bisello G, Broersen K, Morel B, Conejero-Lara F. The influence of N-terminal acetylation on micelle-induced conformational changes and aggregation of alpha-Synuclein. *PLoS One*. 2017;12(5):e0178576.
53. Sharon R, Bar-Joseph I, Frosch MP, Walsh DM, Hamilton JA, Selkoe DJ. The formation of highly soluble oligomers of alpha-synuclein is regulated by fatty acids and enhanced in Parkinson's disease. *Neuron*. 2003;37(4):583–595.
54. Uversky VN, Li J, Fink AL. Evidence for a partially folded intermediate in alpha-synuclein fibril formation. *J Biol Chem*. 2001;276(14):10737–10744.
55. Volles MJ, Lansbury Jr. PT. Zeroing in on the pathogenic form of alpha-synuclein and its mechanism of neurotoxicity in Parkinson's disease. *Biochemistry*. 2003;42(26):7871–7878.

56. Ahn TB, Kim SY, Kim JY, et al. Alpha-synuclein gene duplication is present in sporadic Parkinson disease. *Neurology*. 2008;70(1):43–49.
57. Chartier-Harlin MC, Kachergus J, Roumier C, et al. Alpha-synuclein locus duplication as a cause of familial Parkinson's disease. *Lancet*. 2004;364(9440):1167–1169.
58. Ibanez P, Bonnet AM, Debarges B, et al. Causal relation between alpha-synuclein gene duplication and familial Parkinson's disease. *Lancet*. 2004;364(9440):1169–1171.
59. Kruger R, Kuhn W, Muller T, et al. Ala30Pro mutation in the gene encoding alpha-synuclein in Parkinson's disease. *Nat Genet*. 1998;18(2):106–108.
60. Pasanen P, Myllykangas L, Siitonen M, et al. Novel alpha-synuclein mutation A53E associated with atypical multiple system atrophy and Parkinson's disease-type pathology. *Neurobiol Aging*. 2014; 35(9):2180–2185.
61. Proukakis C, Dudzik CG, Brier T, et al. A novel alpha-synuclein missense mutation in Parkinson disease. *Neurology*. 2013;80(11):1062–1064.
62. Singleton AB, Farrer M, Johnson J, et al. Alpha-synuclein locus triplication causes Parkinson's disease. *Science*. 2003;302(5646):841.
63. Valente EM, Abou-Sleiman PM, Caputo V, et al. Hereditary early-onset Parkinson's disease caused by mutations in PINK1. *Science*. 2004;304(5674):1158–1160.
64. Zarranz JJ, Alegre J, Gomez-Esteban JC, et al. The new mutation, E46K, of alpha-synuclein causes Parkinson and Lewy body dementia. *Ann Neurol*. 2004;55(2):164–173.
65. Caughey B, Lansbury PT. Protofibrils, pores, fibrils, and neurodegeneration: separating the responsible protein aggregates from the innocent bystanders. *Annu Rev Neurosci*. 2003;26:267–298.
66. Helwig M, Klinkenberg M, Rusconi R, et al. Brain propagation of transduced alpha-synuclein involves non-fibrillar protein species and is enhanced in alpha-synuclein null mice. *Brain*. 2016;139 (Pt 3):856–870.
67. Winner B, Jappelli R, Maji SK, et al. In vivo demonstration that alpha-synuclein oligomers are toxic. *Proc Natl Acad Sci U S A*. 2011;108(10):4194–4199.
68. Haass C, Selkoe DJ. Soluble protein oligomers in neurodegeneration: lessons from the Alzheimer's amyloid beta-peptide. *Nat Rev Mol Cell Biol*. 2007;8(2):101–112.
69. Lesne SE, Sherman MA, Grant M, et al. Brain amyloid-beta oligomers in ageing and Alzheimer's disease. *Brain*. 2013;136(Pt 5):1383–1398.
70. Sperling RA, Aisen PS, Beckett LA, et al. Toward defining the preclinical stages of Alzheimer's disease: recommendations from the National Institute on Aging-Alzheimer's Association workgroups on diagnostic guidelines for Alzheimer's disease. *Alzheimers Dement*. 2011;7(3):280–292.
71. Bucciantini M, Giannoni E, Chiti F, et al. Inherent toxicity of aggregates implies a common mechanism for protein misfolding diseases. *Nature*. 2002;416(6880):507–511.
72. Gaig C, Marti MJ, Ezquerra M, Rey MJ, Cardozo A, Tolosa E. G2019S LRRK2 mutation causing Parkinson's disease without Lewy bodies. *J Neurol Neurosurg Psychiatry*. 2007;78(6):626–628.
73. Outeiro TF, Putcha P, Tetzlaff JE, et al. Formation of toxic oligomeric alpha-synuclein species in living cells. *PLoS One*. 2008;3(4):e1867.
74. Lesne S, Koh MT, Kotilinek L, et al. A specific amyloid-beta protein assembly in the brain impairs memory. *Nature*. 2006;440(7082):352–357.
75. Mc Donald JM, Savva GM, Brayne C, et al. The presence of sodium dodecyl sulphate-stable Abeta dimers is strongly associated with Alzheimer-type dementia. *Brain*. 2010;133(Pt 5):1328–1341.
76. Colosimo C, Hughes AJ, Kilford L, Lees AJ. Lewy body cortical involvement may not always predict dementia in Parkinson's disease. *J Neurol Neurosurg Psychiatry*. 2003;74(7):852–856.
77. Parkkinen L, Pirttila T, Alafuzoff I. Applicability of current staging/categorization of alpha-synuclein pathology and their clinical relevance. *Acta Neuropathol*. 2008;115(4):399–407.
78. Chen L, Jin J, Davis J, et al. Oligomeric alpha-synuclein inhibits tubulin polymerization. *Biochem Biophys Res Commun*. 2007;356(3):548–553.
79. Danzer KM, Haasen D, Karow AR, et al. Different species of alpha-synuclein oligomers induce calcium influx and seeding. *J Neurosci*. 2007;27(34):9220–9232.
80. Lambert MP, Barlow AK, Chromy BA, et al. Diffusible, nonfibrillar ligands derived from Abeta1-42 are potent central nervous system neurotoxins. *Proc Natl Acad Sci U S A*. 1998;95(11):6448–6453.

81. Nasstrom T, Fagerqvist T, Barbu M, et al. The lipid peroxidation products 4-oxo-2-nonenal and 4-hydroxy-2-nonenal promote the formation of alpha-synuclein oligomers with distinct biochemical, morphological, and functional properties. *Free Radic Biol Med*. 2011;50(3):428–437.
82. Shankar GM, Bloodgood BL, Townsend M, Walsh DM, Selkoe DJ, Sabatini BL. Natural oligomers of the Alzheimer amyloid-beta protein induce reversible synapse loss by modulating an NMDA-type glutamate receptor-dependent signaling pathway. *J Neurosci*. 2007;27(11):2866–2875.
83. Tetzlaff JE, Putcha P, Outeiro TF, et al. CHIP targets toxic alpha-synuclein oligomers for degradation. *J Biol Chem*. 2008;283(26):17962–17968.
84. Walsh DM, Klyubin I, Fadeeva JV, et al. Naturally secreted oligomers of amyloid beta protein potently inhibit hippocampal long-term potentiation in vivo. *Nature*. 2002;416(6880):535–539.
85. Dahlgren KN, Manelli AM, Stine Jr. WB, Baker LK, Krafft GA, LaDu MJ. Oligomeric and fibrillar species of amyloid-beta peptides differentially affect neuronal viability. *J Biol Chem*. 2002;277 (35):32046–32053.
86. Kuperstein I, Broersen K, Benilova I, et al. Neurotoxicity of Alzheimer's disease Abeta peptides is induced by small changes in the Abeta42 to Abeta40 ratio. *EMBO J*. 2010;29(19):3408–3420.
87. Ono K, Condron MM, Teplow DB. Structure-neurotoxicity relationships of amyloid beta-protein oligomers. *Proc Natl Acad Sci U S A*. 2009;106(35):14745–14750.
88. Cheng IH, Scearce-Levie K, Legleiter J, et al. Accelerating amyloid-beta fibrillization reduces oligomer levels and functional deficits in Alzheimer disease mouse models. *J Biol Chem*. 2007;282 (33):23818–23828.
89. Hsia AY, Masliah E, McConlogue L, et al. Plaque-independent disruption of neural circuits in Alzheimer's disease mouse models. *Proc Natl Acad Sci U S A*. 1999;96(6):3228–3233.
90. Meilandt WJ, Cisse M, Ho K, et al. Neprilysin overexpression inhibits plaque formation but fails to reduce pathogenic Abeta oligomers and associated cognitive deficits in human amyloid precursor protein transgenic mice. *J Neurosci*. 2009;29(7):1977–1986.
91. Fukumoto H, Tokuda T, Kasai T, et al. High-molecular-weight beta-amyloid oligomers are elevated in cerebrospinal fluid of Alzheimer patients. *FASEB J*. 2010;24(8):2716–2726.
92. Gessel MM, Bernstein S, Kemper M, Teplow DB, Bowers MT. Familial Alzheimer's disease mutations differentially alter amyloid beta-protein oligomerization. *ACS Chem Nerosci*. 2012;3(11):909–918.
93. Ono K, Condron MM, Teplow DB. Effects of the English (H6R) and Tottori (D7N) familial Alzheimer disease mutations on amyloid beta-protein assembly and toxicity. *J Biol Chem*. 2010;285 (30):23186–23197.
94. Hardy J, Selkoe DJ. The amyloid hypothesis of Alzheimer's disease: progress and problems on the road to therapeutics. *Science*. 2002;297(5580):353–356.
95. Moore BD, Rangachari V, Tay WM, Milkovic NM, Rosenberry TL. Biophysical analyses of synthetic amyloid-beta(1-42) aggregates before and after covalent cross-linking. Implications for deducing the structure of endogenous amyloid-beta oligomers. *Biochemistry*. 2009;48(49):11796–11806.
96. Stefani M. Biochemical and biophysical features of both oligomer/fibril and cell membrane in amyloid cytotoxicity. *FEBS J*. 2010;277(22):4602–4613.
97. Stefani M. Structural polymorphism of amyloid oligomers and fibrils underlies different fibrillization pathways: immunogenicity and cytotoxicity. *Curr Protein Pept Sci*. 2010;11(5):343–354.
98. Cleary JP, Walsh DM, Hofmeister JJ, et al. Natural oligomers of the amyloid-beta protein specifically disrupt cognitive function. *Nat Neurosci*. 2005;8(1):79–84.
99. Reed MN, Hofmeister JJ, Jungbauer L, et al. Cognitive effects of cell-derived and synthetically derived Abeta oligomers. *Neurobiol Aging*. 2011;32(10):1784–1794.
100. Finder VH, Vodopivec I, Nitsch RM, Glockshuber R. The recombinant amyloid-beta peptide Abeta1-42 aggregates faster and is more neurotoxic than synthetic Abeta1-42. *J Mol Biol*. 2010;396 (1):9–18.
101. Lublin AL, Gandy S. Amyloid-beta oligomers: possible roles as key neurotoxins in Alzheimer's disease. *Mt Sinai J Med*. 2010;77(1):43–49.
102. Takahashi RH, Almeida CG, Kearney PF, et al. Oligomerization of Alzheimer's beta-amyloid within processes and synapses of cultured neurons and brain. *J Neurosci*. 2004;24(14):3592–3599.
103. Giasson BI, Duda JE, Quinn SM, Zhang B, Trojanowski JQ, Lee VM. Neuronal alpha-synucleinopathy with severe movement disorder in mice expressing A53T human alpha-synuclein. *Neuron*. 2002;34(4):521–533.

104. Janezic S, Threlfell S, Dodson PD, et al. Deficits in dopaminergic transmission precede neuron loss and dysfunction in a new Parkinson model. *Proc Natl Acad Sci U S A*. 2013;110(42):E4016–E4025.
105. Kahle PJ, Neumann M, Ozmen L, et al. Selective insolubility of alpha-synuclein in human Lewy body diseases is recapitulated in a transgenic mouse model. *Am J Pathol*. 2001;159(6):2215–2225.
106. Periquet M, Fulga T, Myllykangas L, Schlossmacher MG, Feany MB. Aggregated alpha-synuclein mediates dopaminergic neurotoxicity in vivo. *J Neurosci*. 2007;27(12):3338–3346.
107. Choi BK, Choi MG, Kim JY, et al. Large alpha-synuclein oligomers inhibit neuronal SNARE-mediated vesicle docking. *Proc Natl Acad Sci U S A*. 2013;110(10):4087–4092.
108. Colla E, Coune P, Liu Y, et al. Endoplasmic reticulum stress is important for the manifestations of alpha-synucleinopathy in vivo. *J Neurosci*. 2012;32(10):3306–3320.
109. Emmanouilidou E, Stefanis L, Vekrellis K. Cell-produced alpha-synuclein oligomers are targeted to, and impair, the 26S proteasome. *Neurobiol Aging*. 2010;31(6):953–968.
110. Kaufmann TJ, Harrison PM, Richardson MJ, Pinheiro TJ, Wall MJ. Intracellular soluble alpha-synuclein oligomers reduce pyramidal cell excitability. *J Physiol*. 2016;594(10):2751–2772.
111. Lindersson E, Beedholm R, Hojrup P, et al. Proteasomal inhibition by alpha-synuclein filaments and oligomers. *J Biol Chem*. 2004;279(13):12924–12934.
112. Plotegher N, Gratton E, Bubacco L. Number and brightness analysis of alpha-synuclein oligomerization and the associated mitochondrial morphology alterations in live cells. *Biochim Biophys Acta*. 2014;1840(6):2014–2024.
113. Ryan BJ, Hoek S, Fon EA, Wade-Martins R. Mitochondrial dysfunction and mitophagy in Parkinson's: from familial to sporadic disease. *Trends Biochem Sci*. 2015;40(4):200–210.
114. Schulz-Schaeffer WJ. The synaptic pathology of alpha-synuclein aggregation in dementia with Lewy bodies, Parkinson's disease and Parkinson's disease dementia. *Acta Neuropathol*. 2010;120(2):131–143.
115. Conway KA, Rochet JC, Bieganski RM, Lansbury Jr. PT. Kinetic stabilization of the alpha-synuclein protofibril by a dopamine-alpha-synuclein adduct. *Science*. 2001;294(5545):1346–1349.
116. Paleologou KE, Kragh CL, Mann DM, et al. Detection of elevated levels of soluble alpha-synuclein oligomers in post-mortem brain extracts from patients with dementia with Lewy bodies. *Brain*. 2009;132(Pt 4):1093–1101.
117. Majbour NK, Vaikath NN, van Dijk KD, et al. Oligomeric and phosphorylated alpha-synuclein as potential CSF biomarkers for Parkinson's disease. *Mol Neurodegener*. 2016;11:7.
118. Vaikath NN, Majbour NK, Paleologou KE, et al. Generation and characterization of novel conformation-specific monoclonal antibodies for alpha-synuclein pathology. *Neurobiol Dis*. 2015;79:81–99.
119. Williams SM, Schulz P, Sierks MR. Oligomeric alpha-synuclein and beta-amyloid variants as potential biomarkers for Parkinson's and Alzheimer's diseases. *Eur J Neurosci*. 2016;43(1):3–16.
120. El-Agnaf OM, Salem SA, Paleologou KE, et al. Detection of oligomeric forms of alpha-synuclein protein in human plasma as a potential biomarker for Parkinson's disease. *FASEB J*. 2006;20(3):419–425.
121. Park MJ, Cheon SM, Bae HR, Kim SH, Kim JW. Elevated levels of alpha-synuclein oligomer in the cerebrospinal fluid of drug-naive patients with Parkinson's disease. *J Clin Neurol*. 2011;7(4):215–222.
122. Tokuda T, Qureshi MM, Ardah MT, et al. Detection of elevated levels of alpha-synuclein oligomers in CSF from patients with Parkinson disease. *Neurology*. 2010;75(20):1766–1772.
123. Holtta M, Hansson O, Andreasson U, et al. Evaluating amyloid-beta oligomers in cerebrospinal fluid as a biomarker for Alzheimer's disease. *PLoS One*. 2013;8(6):e66381.
124. Funke SA, Birkmann E, Henke F, et al. Single particle detection of Abeta aggregates associated with Alzheimer's disease. *Biochem Biophys Res Commun*. 2007;364(4):902–907.
125. Funke SA, Birkmann E, Henke F, et al. An ultrasensitive assay for diagnosis of Alzheimer's disease. *Rejuvenation Res*. 2008;11(2):315–318.
126. Wang-Dietrich L, Funke SA, Kuhbach K, et al. The amyloid-beta oligomer count in cerebrospinal fluid is a biomarker for Alzheimer's disease. *J Alzheimers Dis*. 2013;34(4):985–994.
127. Herskovits AZ, Locascio JJ, Peskind ER, Li G, Hyman BT. A Luminex assay detects amyloid beta oligomers in Alzheimer's disease cerebrospinal fluid. *PLoS One*. 2013;8(7):e67898.
128. Georganopoulou DG, Chang L, Nam JM, et al. Nanoparticle-based detection in cerebral spinal fluid of a soluble pathogenic biomarker for Alzheimer's disease. *Proc Natl Acad Sci U S A*. 2005;102(7):2273–2276.

129. Santos AN, Simm A, Holthoff V, Boehm G. A method for the detection of amyloid-beta1-40, amyloid-beta1-42 and amyloid-beta oligomers in blood using magnetic beads in combination with flow cytometry and its application in the diagnostics of Alzheimer's disease. *J Alzheimers Dis.* 2008;14(2):127–131.
130. Santos AN, Torkler S, Nowak D, et al. Detection of amyloid-beta oligomers in human cerebrospinal fluid by flow cytometry and fluorescence resonance energy transfer. *J Alzheimers Dis.* 2007;11(1): 117–125.
131. Hwang SS, Chan H, Sorci M, et al. Detection of amyloid beta oligomers toward early diagnosis of Alzheimer's disease. *Anal Biochem.* 2019;566:40–45.
132. Fairfoul G, McGuire LI, Pal S, et al. Alpha-synuclein RT-QuIC in the CSF of patients with alpha-synucleinopathies. *Ann Clin Transl Neurol.* 2016;3(10):812–818.
133. Salvadores N, Shahnawaz M, Scarpini E, Tagliavini F, Soto C. Detection of misfolded Abeta oligomers for sensitive biochemical diagnosis of Alzheimer's disease. *Cell Rep.* 2014;7(1):261–268.
134. Shahnawaz M, Tokuda T, Waragai M, et al. Development of a biochemical diagnosis of Parkinson disease by detection of alpha-Synuclein misfolded aggregates in cerebrospinal fluid. *JAMA Neurol.* 2017;74(2):163–172.
135. Shahnawaz M, Mukherjee A, Pritzkow S, et al. Discriminating alpha-synuclein strains in Parkinson's disease and multiple system atrophy. *Nature.* 2020;578(7794):273–277.
136. Kakuda K, Ikenaka K, Araki K, et al. Ultrasonication-based rapid amplification of alpha-synuclein aggregates in cerebrospinal fluid. *Sci Rep.* 2019;9(1):6001.
137. Paciotti S, Bellomo G, Gatticchi L, Parnetti L. Are we ready for detecting alpha-synuclein prone to aggregation in patients? The case of "protein-misfolding cyclic amplification" and "real-time quaking-induced conversion" as diagnostic tools. *Front Neurol.* 2018;9:415.
138. Castilla J, Saa P, Morales R, Abid K, Maundrell K, Soto C. Protein misfolding cyclic amplification for diagnosis and prion propagation studies. *Methods Enzymol.* 2006;412:3–21.
139. Soto C, Saborio GP, Anderes L. Cyclic amplification of protein misfolding: application to prion-related disorders and beyond. *Trends Neurosci.* 2002;25(8):390–394.
140. Soto C, Anderes L, Suardi S, et al. Pre-symptomatic detection of prions by cyclic amplification of protein misfolding. *FEBS Lett.* 2005;579(3):638–642.
141. Castilla J, Saa P, Soto C. Detection of prions in blood. *Nat Med.* 2005;11(9):982–985.
142. Concha-Marambio L, Pritzkow S, Moda F, et al. Detection of prions in blood from patients with variant Creutzfeldt-Jakob disease. *Sci Transl Med.* 2016;8(370):370ra183.
143. Moda F, Gambetti P, Notari S, et al. Prions in the urine of patients with variant Creutzfeldt-Jakob disease. *N Engl J Med.* 2014;371(6):530–539.
144. Saa P, Castilla J, Soto C. Ultra-efficient replication of infectious prions by automated protein misfolding cyclic amplification. *J Biol Chem.* 2006;281(46):35245–35252.
145. Saborio GP, Permanne B, Soto C. Sensitive detection of pathological prion protein by cyclic amplification of protein misfolding. *Nature.* 2001;411(6839):810–813.
146. Chen B, Morales R, Barria MA, Soto C. Estimating prion concentration in fluids and tissues by quantitative PMCA. *Nat Methods.* 2010;7(7):519–520.

CHAPTER 13

3D Alzheimer's disease in a dish: Implications for drug discovery

Luisa Quinti, Joseph Park, Emma Brand, Rudolph E. Tanzi, and Doo Yeon Kim
Genetics and Aging Research Unit, MassGeneral Institute for Neurodegenerative Disease, McCance Center for Brain Health, Massachusetts General Hospital, Harvard Medical School, Charlestown, MA, United States

Introduction

Alzheimer's disease (AD) is the most common form of dementia. AD currently affects more than 5.8 million people in the United States and the number of patients is projected to greatly increase over time.[1] However, no therapeutic option is currently available for AD patients, except for a few relief remedies. Recent failures in clinical trials clearly demonstrate a gap in our understanding of AD pathogenesis and call for more unbiased approaches for developing AD therapeutics.

AD starts with short-term memory impairments, gets worse over time, and leads to complete loss of cognition. Early-onset familial AD (FAD) strikes in the fourth or fifth decade of life, while sporadic late-onset AD (SAD) typically develops after the age of 70. FAD forms are rare, autosomal-dominant, and are caused by fully penetrant mutations either in the amyloid β (Aβ) precursor protein (*APP*), presenilin 1 (*PSEN1*), or presenilin 2 (*PSEN2*) genes.[2] SAD is the commonly occurring form of the disease, usually due to multifactorial genetic and environmental risk factors.[3] The pathological hallmarks of AD, shared by FAD and SAD, are senile plaques and neurofibrillary tangles (NFTs), accompanied by progressive neurodegeneration. Amyloid plaques are extracellular amyloid aggregates, consisting of ~4 kDa Aβ peptides, which are generated from APP through a sequential proteolytic cleavage by β- and presenilin/γ-secretase. NFTs consist of aggregated and highly phosphorylated forms of the microtubule-associated protein tau (p-tau). In AD, p-tau accumulates in atypical cellular compartments, such as the cell body and dendrites. Neuroinflammatory changes, including activation of glial cells (gliosis) and increases in pro-inflammatory cytokines, are another important pathological marker of AD pathogenesis. Blood-brain barrier (BBB) disruption is commonly observed in the brains of AD patients and may significantly contribute to AD pathogenesis.[4] The relative importance and interplay between these pathological events remain a subject of intense research.

Neurotherapeutics in the Era of Translational Medicine
https://doi.org/10.1016/B978-0-12-816475-4.00014-8

In the last decade, transgenic mice overexpressing *APP* or *APP/PSEN1* with FAD mutations have been used as the standard AD model for basic mechanistic studies and drug discovery.[5–8] However, these mouse models do not convincingly show NFTs nor the robust neurodegeneration found in human AD patients, despite strong Aβ deposition, pronounced gliosis, and synaptic deficits.[5–8] According to the "amyloid cascade hypothesis," abnormal accumulation of pathogenic Aβ species starts a pathogenic cascade that leads to hyperphosphorylation and aggregation of tau, and finally, neuronal death.[2,3] The recent failures of anti-Aβ therapies in human clinical studies, which were conversely highly effective in AD mouse models, underscore the limitation of AD mouse models in comprehensively recapitulating human AD pathologies.[9,10] Testing drugs in AD mouse models is also expensive and time-consuming and therefore not suitable for unbiased drug screening of compound libraries.

Advances in stem cell technology have made it possible to generate human AD neurons harboring FAD mutation(s).[8,11–14] Induced-pluripotent stem cell (iPSC) technology can nowadays provide human neurons expressing the identical genetic data of AD patients. These novel human neural cell culture models allow one to comprehensively recapitulate the AD pathogenic cascade in a human brain-like environment. We have shown that transgenic human neural stem cells that express APP or APP/PSEN1 with multiple FAD, combined with our Matrigel-based three-dimensional (3D) cell culture system, recapitulate robust AD pathology including Aβ aggregation and deposition (Aβ plaques), and NFTs.[15,16] Importantly, these models demonstrate Aβ-driven NFT pathology for the first time.[15,17] Multiple groups have also shown that human iPSC-derived neurons with FAD mutation(s) display Aβ and tau pathology in two-dimensional (2D) and 3D culture conditions.[11,18–27] However, most of 2D FAD iPSC models display a rather mild AD pathology without neurodegenerative phenotypes and a limited tau pathology only with selected FAD mutations.[22,23] Interestingly, AD iPSCs that are differentiated in 3D brain organoid models have more robust Aβ and tau pathology,[20,21,28,29] possibly due to the accelerating local accumulation of pathogenic Aβ, in agreement with our 3D AD Matrigel-based cellular models.

This chapter aims to provide an overview of the recent advances in 3D modeling of AD using human neural cells and their implications for drug discovery. We will describe the basic concepts of 3D AD human cellular models including 3D brain organoid models and our engineered 3D culture models and their impacts on mechanistic studies and drug discovery. We will introduce the advantages of our engineered 3D culture models that are designed to comprehensively recapitulate the different aspects of AD pathogenesis occurring in patient brains. We will also show our ongoing efforts to develop 3D AD culture platforms validated for unbiased and high-throughput drug screening. Finally, we will discuss perspectives and challenges of applying 3D AD cellular models in AD drug discovery.

Innovation and evolution of 3D AD culture models

Engineered 3D cellular models versus self-assembling 3D brain organoid models

Recent advances in cell culture technology and human stem cell biology enable scientists to miniaturize human brain structures in 3D. Conventional 3D culture technology uses natural or synthesized hydrogels, or other biocompatible materials such as extracellular matrix (ECM)-like 3D scaffolds, to maintain 3D neural networks that are composed of either neurons or neurons with supporting glial cells.[8,15,30–33] Cell numbers, cell composition, and the environment in these conventional 3D models can be precisely controlled. Therefore, these systems are highly reproducible and can be easily arrayed to fit into high-throughput drug-screening platforms (Fig. 1). However, it has been challenging to recapitulate complex brain structures where multiple brain cell types are present.

In contrast, brain organoid models are taking advantage of the self-assembling and -organizing properties of human neural stem cells to generate more complicated structures that mimic specific regions of human brains.[34,35] Most of the current brain organoid models still use hydrogel to support the development of brain structures in 3D, where neural precursor cells actively migrate and differentiate regionally to form a complex neural-glial network, mimicking brain development in vivo. Therefore, these models have advantages in recapitulating complex brain structures when compared to conventional 3D cell culture models, where cells remain in the originally implanted position (Fig. 1). However, due to these complex structures, it is challenging to generate homogeneous and reproducible systems, two essential requirements for developing them into mid- and high-throughput drug screening platforms (Fig. 1).

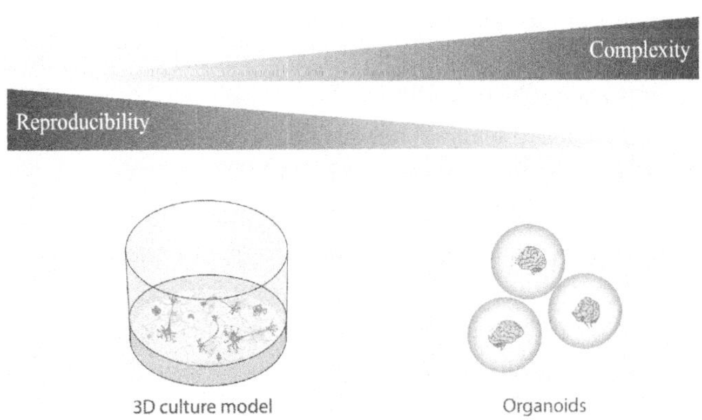

Fig. 1 Engineered 3D cell culture model vs self-assembling 3D brain organoid model.

To overcome these limitations in both engineered 3D cell culture and self-assembling brain organoid models, we have developed modified 3D AD cellular systems that can recapitulate different stages of the AD pathogenic cascade, by combining various microfluidic designs (Fig. 2).[16,36–39] These modified models have been proven helpful in reproducing different stages of the AD pathogenic cascade including neuroinflammation/neurodegeneration and blood-brain barrier damages, which have not been feasible with conventional 3D culture systems. Furthermore, these new 3D models are useful for both basic mechanistic studies and drug discovery (Fig. 2).

Recapitulating Aβ-triggered tau pathology using 3D human neuron-astrocyte cell culture models

Over the last three decades, the Aβ hypothesis of AD has reigned, providing the foundation for numerous basic studies and clinical trials.[2,40–42] According to this hypothesis, excessive accumulation of pathogenic Aβ species triggers a pathogenic cascade that includes synaptic deficits, altered neuronal activity, inflammation, oxidative stress, neuronal injury, hyperphosphorylation of tau triggering NFTs, and lastly, neuronal death and dementia.[2,40,42–44] However, the Aβ hypothesis has not been fully validated and several major unresolved issues remain.[45–47] One of the major unresolved issues of the Aβ hypothesis is to demonstrate a direct causal link between Aβ and NFTs.[45–47] Transgenic AD mouse models carrying single or multiple human familial AD mutations in APP and/or PSEN1, used as standard AD models for basic mechanistic studies and drug screening, do not develop NFTs or the robust neurodegeneration observed in human patients, despite strong Aβ deposition.[46,48,49] Only triple transgenic mouse models show both Aβ plaques and NFTs. Tau pathology in these mouse models mainly depends on the overexpressed frontotemporal dementia (FTD)-associated P301L *MAPT* (tau) mutant forms, not on β-amyloid accumulation.[50] The recent failures of anti-β-amyloid therapies in humans, which were successful in mouse models, may be explained by the limitation of AD mouse models in comprehensively modeling human AD pathologies.[51,52]

In 2014, we developed a unique 3D AD cellular model based on human neural progenitor cells overexpressing APP and PSEN1 with human FAD mutations (Fig. 2). To achieve high levels of toxic Aβ species (comparable to the levels in brains of AD patients), we overexpressed human APP or APP together with PSEN1, harboring multiple familial AD mutations, in immortalized human fetal neural progenitor cells.[15] Similar strategies were previously applied in mouse AD models to accelerate the severity and the onset of Aβ deposition in rodent brains.[45–47] However, in standard 2D cell cultures, secreted Aβ species readily disperse into the relatively large volume of culture media, and therefore do not develop robust Aβ accumulation nor Aβ-triggered tau pathology.[8] Conversely, we observed that the 3D culture environment can robustly accelerate Aβ accumulation and, more importantly, Aβ-induced tau pathology (NFTs) by limiting the diffusion of pathogenic Aβ species. For most of our 3D culture models, we chose BD Matrigel (BD

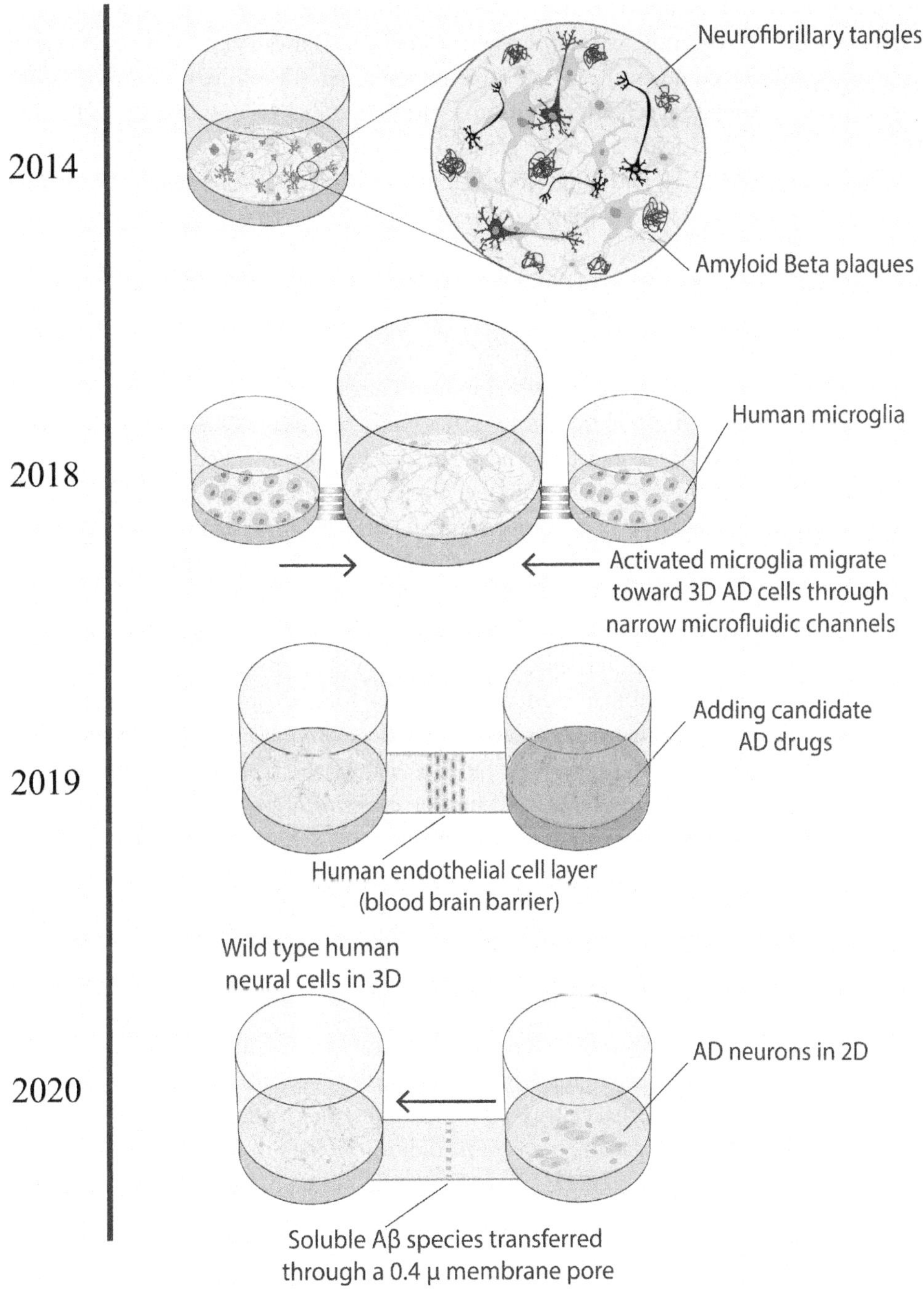

Fig. 2 Evolution of 3D cell culture models of AD: toward more comprehensive cellular models of AD pathogenesis.

Biosciences) as a 3D support matrix since it contains high levels of brain extracellular matrix proteins.[50] Recently, we further demonstrated a clear causal relationship between pathogenic Aβ42 accumulation and NFT pathology using 3D culture models with single clonal human neural stem cell lines expressing various familial AD mutations.[17] These 3D AD cellular models provide a unique platform to screen and test novel AD drugs that block Aβ-triggered NFT pathology in a human brain-like environment. This cannot be achieved with standard AD transgenic mouse models.

3D Neuron-astrocyte-microglia triculture model of Alzheimer's disease: Modeling neuroinflammation and neurodegeneration in AD

Activation of glial cells (the so-called gliosis) is one of the major pathological hallmarks of AD.[2,40–42] Increasing evidence strongly implicates a role for innate immunity and neuroinflammatory mechanisms in the etiology and pathogenesis of AD etiology and pathogenesis.[53–58] Multiple AD-associated genetic variants have been linked to innate immune pathways in the microglia.[59–62] The sustained activation of microglia results in a chronic neuroinflammatory response and increased production of proinflammatory cytokines, such as tumor necrosis factor (TNF)-α and interleukin 1 beta (IL-1β), which play an important role in the progression of AD pathology. While our 3D human neuron-astrocyte AD models can provide a valid system to study the Aβ-triggered tau pathology, it does not contain human microglia, an essential component of innate immunity and neuroinflammation in the human brain (Fig. 2). Therefore, we introduced a new microfluidic 3D culture system that recapitulates interactions among neurons, astrocytes, and microglia in an AD-like environment,[16] which enables the study of human microglia recruitment, inflammatory responses, and neuronal loss as observed in the brain of AD patients.[63] In this 3D triculture system, human stem-cell-derived neurons and astrocytes are grown in a 3D culture matrix in the center of a microfluidic device as shown in our original 3D AD cellular models. Subsequently, human microglial cells were added at specific time-points of interest into a peripheral reservoir and only activated microglia could migrate through narrow microfluidic channels, attracted by Aβ, cytokine/chemokines, and other soluble factors that are secreted by 3D AD cells, in contrast to control cells (Fig. 2).[16]

This triculture system is a timely approach to study glial effects, based on ongoing or recently published research, ranging from genetic risk factors to biological mechanistic studies, which suggests that glial cells and their inflammatory pathways are the key mechanisms of Alzheimer's disease pathogenesis. In addition to their independent roles in neuroinflammation, there is increasing evidence for interaction between microglia and astrocytes in destructive cell-cell crosstalk: i.e., the secretion of cytokines by microglia can activate astrocytes, causing them to lose some of their physiological functions and become toxic to neurons.[48–53,59] Increased secretion of a broad range of inflammatory molecules such as monocyte chemoattractant protein-1 (MCP-1/CCL2) attracts more microglial infiltration in 3D cultures. Interestingly, interferon (IFN)-γ can only be detected in 3D AD (not 2D) cultures suggesting that 3D cultures promote a more complex and presumably more disease-relevant inflammatory microenvironment.[16] Furthermore, cellular death of neurons and astrocytes occurs mostly in close proximity to microglial cells, providing convincing evidence that microglia may be critical mediators of neural death. We also have found that the neurotoxic effects of microglia are mediated by toll-like receptor 4 (TLR4) and IFN-γ, which are developed only in the 3D AD environment: inhibition of these co-stimulated pathways, either using a TLR4 antagonist/shRNA knockdown or by treatment of IFN-γ neutralizing antibody, reduces cellular damage. This results in lower levels of TNF-α and nitric oxide (NO) in the culture media, but does not affect microglial infiltration, suggesting that TNF-α and NO are neurotoxic factors released by active microglia.[16]

In summary, this 3D AD triculture models can provide a unique platform to test candidate drugs that block neuroinflammation and inflammation-related neuronal death in an AD brain-like environment (Fig. 2). Although the current model may not be able to fully recapitulate chronic inflammatory conditions as present in AD and other neurodegenerative diseases, it is helpful in identifying drugs that can tame the overactivation and inflammatory responses of microglia cells in AD brain-like conditions. We are currently working on adapting this triculture system for high-throughput cell culture formats incorporating 384- or 96-well plates.

3D Blood-brain barrier model of Alzheimer's disease

The blood-brain barrier (BBB) plays a vital role in the maintenance of brain homeostasis. It strictly restricts the passage of molecules from the brain vasculature into the brain via its high transendothelial electrical resistance and low paracellular and transcellular permeability.[64] Specialized brain endothelial cells, astrocytes, pericytes, neurons, and microglia contribute synergistically to the functional properties of the BBB. Emerging evidence has shown that the dysfunctional BBB is a key factor in the cause and consequences of AD.[65–67] Cellular models of BBB have been used to study the role of pathogenic Aβ accumulation and its clearance.[68] We developed 3D BBB models of AD, by combining 3D microfluidic BBB models and our 3D AD neuron-astrocyte model[39] (Fig. 2). This model recapitulates several key events of BBB dysfunction observed in AD patients including decreases in claudins and VE-cadherins, increased expression of matrix-metalloproteinase-2 and reactive oxygen species, deposition of β-amyloid (Aβ) peptides at the vascular endothelium (like in cerebral amyloid angiopathy, CAA), and increased BBB permeability. This 3D AD BBB cellular model can provide a platform for the screening of new drugs that can ameliorate Aβ-mediated BBB disruption in a AD brain-like environment. It can also be used to rapidly screen BBB-permeable candidate AD drugs, which can preserve the integrity of neurons and glial cells (Fig. 2).

3D Non-cell-autonomous model of AD: Modeling sporadic AD-like environment

The majority of cellular and animal models of AD harbor FAD mutations to recapitulate AD brain-like conditions.[5,8] Our 3D culture models also depend on overexpression of AD genes (APP and/or PS1) with FAD mutations to induce robust accumulation of pathogenic Aβ42 species (Fig. 2).[7,8,14,17,33] However, these models may not fully recapitulate the sporadic form of the disease since more than 95% of AD patients (mostly with late-onset SAD) do not express familial AD mutations. According to the Aβ hypothesis, SAD is caused by gradual accumulation of pathogenic Aβ species, possibly due to decreased Aβ clearance, whereas in early-onset familial AD forms, Aβ accumulation is more abrupt, driven by mutations that increase the relative levels of pathogenic Aβ species.[2,3]

Recently, we have published a 3D non-cell-autonomous model of AD, which manifests robust Aβ and tau pathology in human neural cells that do not contain FAD mutations[17] (Fig. 2). In this non-cell-autonomous model, 3D-differentiated non-AD gene-expressing (naïve) human neural cells are cocultured with 2D-differentiated neurons carrying FAD mutations in separate, but communicating, compartments for 6–12 weeks. Over time, pathogenic, soluble Aβ species from the familial 2D AD cells gradually accumulate in the 3D naïve cells (Fig. 2). After 12 weeks of co-differentiation, we detected a dramatic accumulation of detergent-insoluble pathogenic phosphorylated tau species only in the 3D naïve cells cocultured with 2D cells secreting pathogenic Aβ42. The increase of insoluble p-tau proteins did not occur when 2D AD cells secreting nonpathogenic Aβ40 were used, demonstrating for the first time the pathogenic role of Aβ42 isoform in triggering robust NFT pathology in human neurons.[17] In this model, the tau/NFT pathology developed in wild-type human neurons without familial AD or FTD mutations. Thus, our 3D non-cell-autonomous models recapitulate Aβ-triggered tau pathology similar to that in SAD patient brains. These will be useful for testing AD drug candidates in physiologically relevant conditions.

The impact of 3D AD cellular models on drug discovery

Selecting suitable 3D culture models for drug screening: Mechanism of action, reproducibility, cost and time considerations

Current 3D human AD cellular models have multiple advantages over conventional AD transgenic mouse models for mid- and large-scale drug screening efforts. However, it is still challenging to choose an appropriate 3D human cellular model for drug screening. One of the major criteria is the mechanism of action of target drugs. As explained in the previous sections, different 3D AD cellular models recapitulate different aspects of disease pathology. For example, 3D AD triculture models are suitable for screening and testing drugs that modulate the microglia or neuroinflammatory cascades that lead to reduced synaptic and neuronal damages (Fig. 2). For drugs targeting Aβ-triggered neuronal deficits that lead to tau and NFT pathology, our original and non-cell-autonomous 3D AD neuron-astrocyte models can be a better choice to identify candidate drugs. For drugs modulating Aβ-mediated BBB deficits, our 3D AD BBB models can be used (Fig. 2).

Cost and assay time are also important factors when choosing 3D cellular models for drug screening. Fig. 3 compares the approximate cost and time for screening ~1700 FDA-approved compounds in transgenic AD mice, 3D brain organoids, and our engineered 3D human neural culture models (the estimates only calculate the cost of maintaining the culture models for drug testing; without analysis or human cost included). If all the FDA-approved compounds are screened in a transgenic mouse model of AD, such

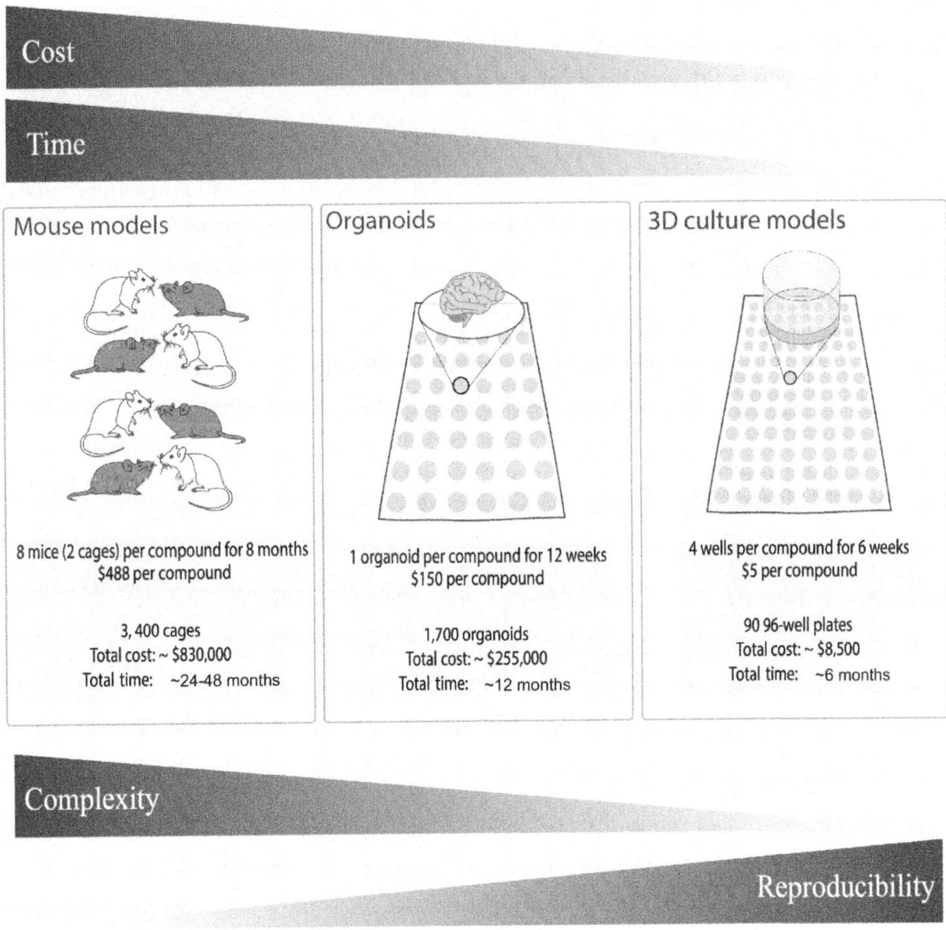

Fig. 3 Time and cost comparison for the screening of 1700 FDA-approved drugs in transgenic AD mouse models, AD organoids, and 3D AD cell cultures.

as the 5xFAD, with eight mice per compound bred for 8 months, the cost for the maintenance of the mouse colony amounts to $488 per compound for primary screening. About 3400 cages of mice are required to screen the entire library, a number that cannot be handled by most facilities over a short period of time. The screening of the entire library is therefore expected to last over 24–48 months. In the instance of 3D human brain organoids, the cost is estimated to be >$150 per compound (assuming one organoid for each compound), given the complexity and length of the protocol required for their growth[34,69,70] (Fig. 3). The price would be expected to increase for primary screening since more than 1700 organoids would initially have to be generated to reduce the organoid to organoid heterogeneity.[71] The time required to complete the screening of FDA-approved compounds is expected to be at least 12 months. Lastly, the screening of the 1700 FDA-approved compounds at one single concentration in our 3D AD cellular models costs about $5 per compound and ~6 months to complete. Due to the scalability of high-throughput plate formats and relatively simple culture conditions (Fig. 3), it would require ninety 96-well plates of 3D AD cell cultures with four replicates per compound. Twelve plates can be handled at one time and therefore the whole screen can be completed in approximately 6 months, based on our own experience with drug screening. This clearly suggests a huge advantage of our 3D culture models in cost-effectiveness.

Another criterion for choosing the appropriate model is reproducibility. The majority of complex brain organoids models are not ready for high-throughput drug library screening due to their heterogeneity which results from their dependence on self-assembling and self-organization (Fig. 1).[71] In contrast, engineered 3D culture models show much higher reproducibility and therefore are a better fit for large-scale drug library screening. Our original 3D neuron-astrocyte AD models were generated with mixed heterogeneous human neural progenitor cells.[15] To improve their reproducibility, we recently reported on the development of a 3D AD cellular model based on clonal AD neural stem cell lines. This would be useful for large-scale drug screening.[17]

Optimizing 3D AD neuron-astrocyte culture model in a large-scale drug screening platform

Since 2014, we have been focusing on adapting our 3D AD cellular models into 96- and 384-well based drug screening systems. With these, libraries of approximately 1000 compounds can be tested in a reasonable amount of time and at an affordable cost (Fig. 3). Fig. 4 shows a summary diagram of the protocol that we developed for screening a mid- and large-scale drug library utilizing our 3D AD neuron-astrocyte model. Since these 3D neuron-astrocyte AD cultures are effective in modeling Aβ-triggered tau pathology, this protocol focuses on analyzing the impact of candidate drugs on the accumulation of pathogenic Aβ and phosphorylated tau in soluble and insoluble fractions using multiplexed Meso Scale Discovery ELISAs (Fig. 4). An automated liquid dispenser is used to streamline daily operations (Fig. 5A).

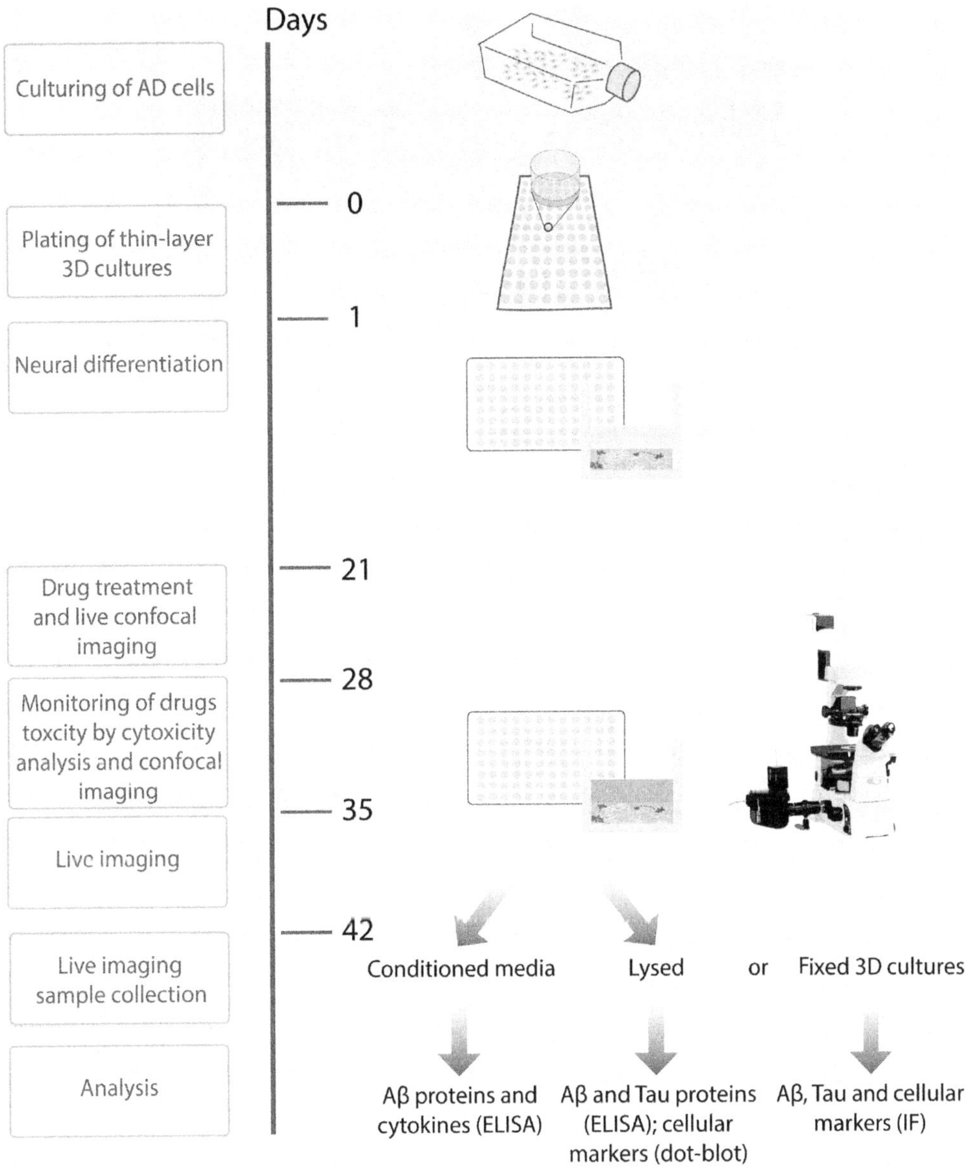

Fig. 4 A protocol for unbiased drug library screening with 3D AD neuron-astrocyte cell culture models.

The optimized drug-screening protocol with the 3D cultures involves plating human neural cells in a thin 3D Matrigel matrix in growth media on day 0. The cell cultures are switched to differentiation media on day 1 (Fig. 4). This improves long-term gel adherence to the plates and gel intactness. Approximately 1/3 of the differentiation media is subsequently changed twice per week, while the AD phenotype develops and reaches a

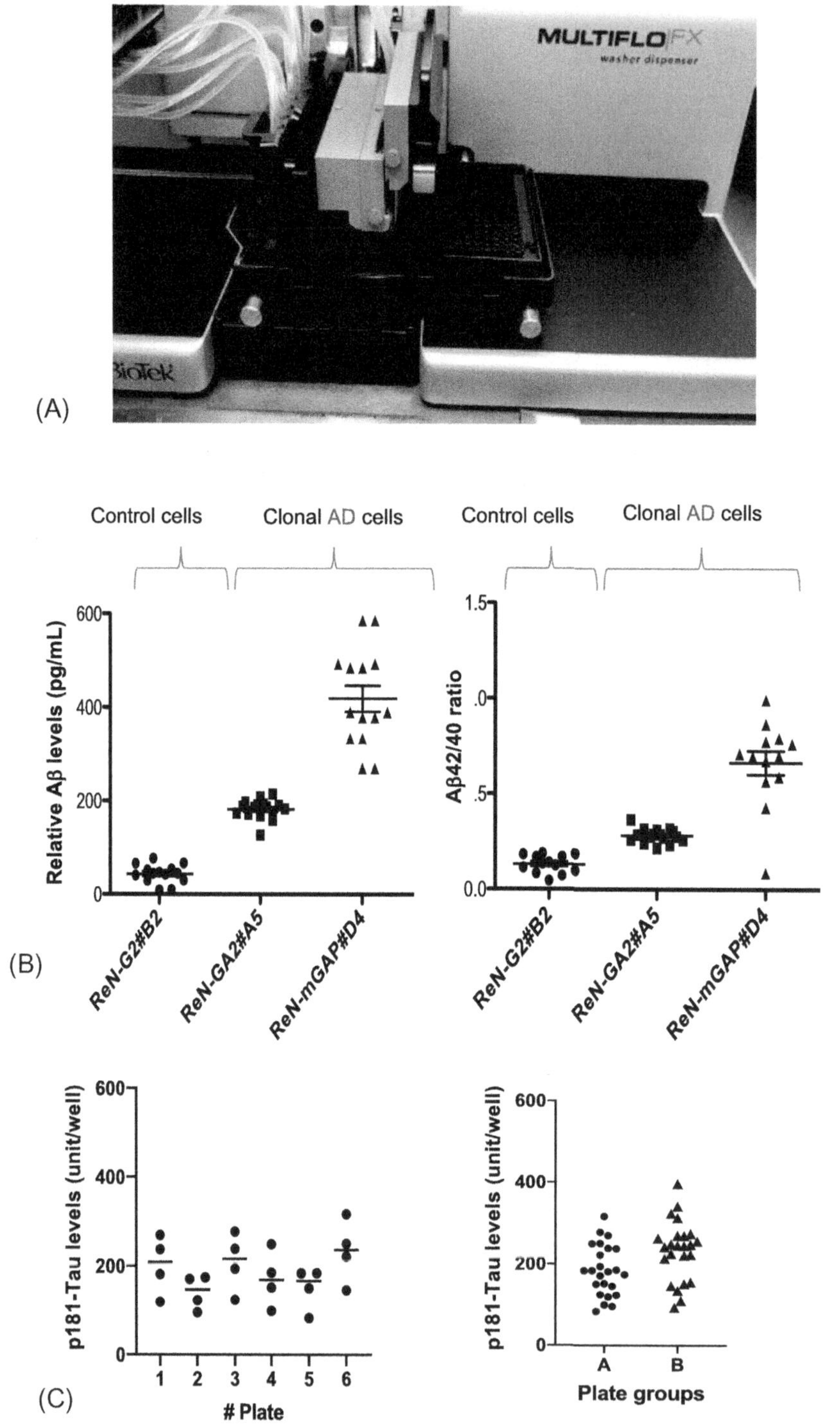

Fig. 5 Developing a reproducible drug screen platform with 3D cell cultures. (A) Automated liquid dispenser used to streamline daily cell culture operations. (B) Aβ42 *(left panel)* and Aβ42/40 *(right panel)* in several wells of 3D AD and non-AD neural human cell cultures. (C) *Left panel*, concentration of phosphorylated Tau in four wells (A6, B6, C6, D6) from six different 96-well plates, plated at the same time; *Right panel*, Phosphorylated Tau levels in two groups of six plates from two independent experiments.

stage that can be impacted by potentially therapeutic drugs. As described earlier, single clonal human AD stem cell lines were used for drug screening to reduce well-to-well fluctuation and batch number variation.[17] We found that clonal cell lines maintain, over time, stable expression of pathogenic Aβ species[17] and display high reproducibility in Aβ and tau pathology after >6 weeks of 3D differentiation. The high number of plates required to complete the whole screen therefore does not represent an obstacle considering the low heterogeneity among 3D cultures in different plates. Fig. 5B shows the concentration of pathogenic Aβ in the lysed 3D gels from different wells in the same 96-well plates (pooled samples). As expected, Aβ42 concentrations (left panel) and the ratio between pathogenic Aβ42 versus nonpathogenic Aβ40 (right panel) display comparable levels, depending on cell types. Fig. 5C, right panel compares the accumulation of pathogenic phosphorylated Tau (p181Thr-tau)[69,70] concentrations from two sets of six 96-well plates prepared in two separate experiments. The analysis of the lysed gels also occurred separately. There is statistical difference among the concentration of phosphorylated Tau measured in the two groups of plates; however; the range of concentrations is the same. Low inter-plate variation and the presence of a control, positioned in the same wells within each plate, allows a reliable comparison among drugs screened in the same plate as well as in different plates.

Another advantage of using clonal AD cell lines is that we can test drugs in 3D AD cell culture models with differential Aβ42/40 ratios (Fig. 5B). As we demonstrated in our recent publication,[17] 3D models with differential Aβ42/40 ratios show different AD pathologies including NFT. The difference in Aβ expression in the 3D cultures is reflected in the different kinetics at which the AD pathology develops, typically between 2 and 6 weeks from the beginning of differentiation. Drugs that alter the course of the disease before, during, or after its development, can therefore be identified depending on the cell line chosen for drug screening, the level of AD pathology when drug treatment begins and, finally, the length of the treatment interval.

In a typical screen, plated 3D cultures are differentiated for 3 weeks and then treated with compounds, twice per week in differentiation media for the following 3 weeks. In a primary drug-screening run in 96-well plates, one control and 19 novel compounds are tested in each plate, with four replicates for each compound. Using 384-well plates, one control and up to 21 novel compounds can be reliably tested in each plate, with six replicates for each compound. The use of 384-well plates does not dramatically increase the screening throughput when compared to 96-well plates: the extensive manual handling of the plates during the 6, or more, weeks that the cells are kept in cultures becomes, in fact, a limiting factor. The costs for drug screenings run in 384-well plates are however lower than those in 96-well plates due to the miniaturized volumes. Three-hundred and eighty-four-well plates should, therefore, be preferred when screening large libraries.

Monitoring of cell viability and drug toxicity

To assess the effect of the tested compounds while the screening is taking place, plates are live-imaged the day before treatment starts and weekly thereafter, using a confocal microscope equipped with image analysis software. Taking advantage of constitutively expressed eGFP fluorescence in clonal cells, a 20X image is recorded at three fixed positions in each well and compared over time. An example of this analysis is shown in Fig. 6. Images of a whole 96-well plate prior to the start of drug screening, and after 7 and 21 days of drug treatment, are shown in Fig. 6A–C, respectively. Compound 13 is found to progressively reduce over time the endogenous fluorescence of cell cultures, signaling its toxicity at the tested concentration. Cell viability and drug toxicity can be further quantified over time with neurite counting. Other live functional assays can also be

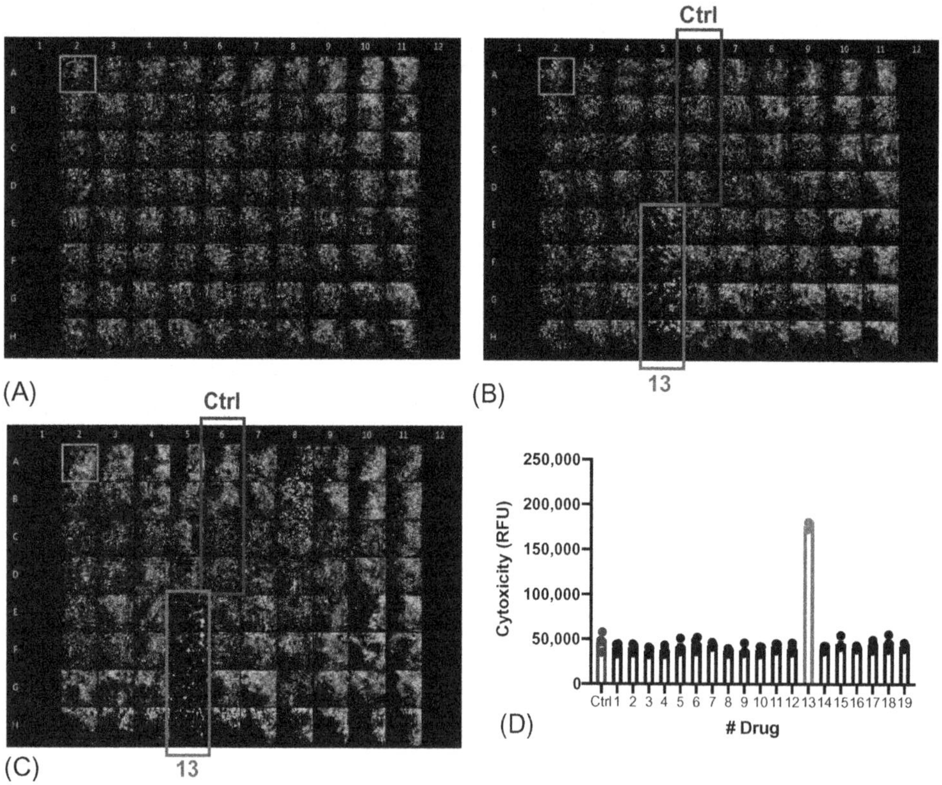

Fig. 6 Monitoring of compounds toxicity during drug screening. (A) Live imaging of a 96-well plate with differentiated neural 3D cultures before drug screening, obtained by automated confocal microscopy. (B, C) Live imaging of the same plate respectively 7 and 21 days after drug treatment has started. The wells treated with the control compound (Ctrl) are shown in *blue*. The wells treated with a toxic compound, 13, are shown in *red*. (D) Measurement of cytotoxicity in the conditioned media 7 days after drug screening has started.

undertaken with 3D cultures to gain more information about drug effects. To further assess the toxicity of screened drugs and more readily exclude false hits, cytotoxicity assays are run on the conditioned media collected 7 days after beginning of drug treatment. Media from 3D neural cultures treated with compound 13 show increased cytotoxicity when compared to that from the other 18 drugs and the control compound, tested on the same plate (Fig. 6D). Screened compounds that show toxicity by both cytotoxicity analysis and live imaging are removed from further analysis.

Endpoint analysis

As shown in Fig. 4, at the end of drug treatment, conditioned media is collected from each well and analyzed for soluble proteins of interest, e.g., Aβ42, Aβ40, cytokines, etc., with ELISA assays. The 3D gels remaining after media removal from 96-well plates can be analyzed using two methods: biochemical analysis based on immunoassays and immunofluorescence. When the drug-screening analysis is based on immunoassays, 3D gels are lysed and the levels of proteins of interest, such as Aβ and Tau, are determined by ELISAs. Fig. 7A shows the protein concentrations of p181-Tau, total Tau, and Aβ42 in the lysed control wells from a representative 96-well plate. The levels of pathogenic

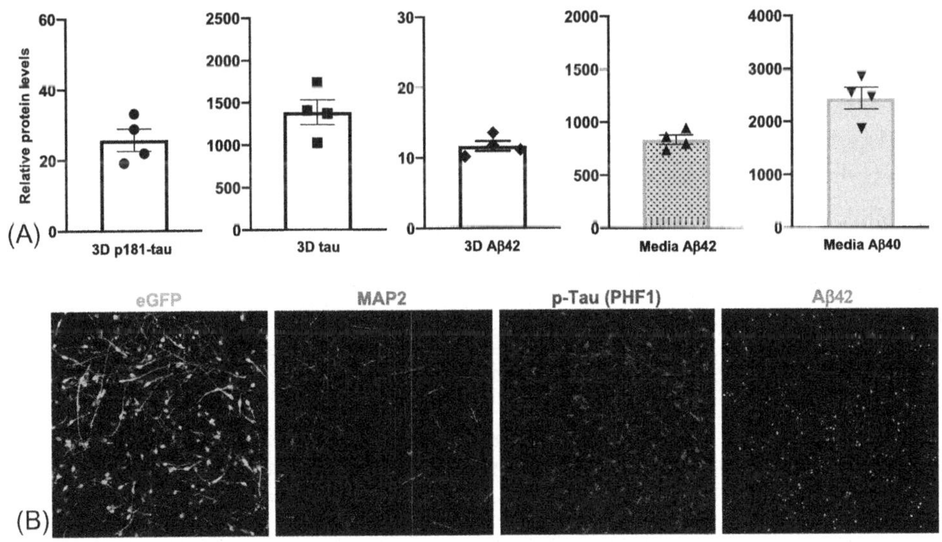

Fig. 7 End of treatment analysis with 3D AD cell cultures. (A) p181-Tau, Tau, and Aβ42 concentrations in representative 3D AD cell cultures lysates. The grey bars show the representative levels of Aβ42 and Aβ40 respectively in the conditioned media. (B) Immunofluorescence analysis performed on fixed 3D cultures. Cells are visualized using their endogenous fluorescence (in *green*). Cellular markers like MAP2 (in *red*) are analyzed together with proteins of interest such as phosphorylated Tau (in *blue*) and Aβ42 (in *yellow*).

Aβ42 and nonpathogenic Aβ40 in the conditioned media (gray bars) from the same wells is also reported. Cellular markers such as microtubule-associated protein 2 (MAP2), actin, glial fibrillary acidic protein (GFAP), etc. are evaluated by dot-blot analysis.

If immunofluorescence is the preferred readout of screening, the 3D gels remaining after media removal are fixed and stained to determine the levels of Aβ and Tau species, as well as control proteins such as MAP2, actin, and GFAP. Fig. 7B shows representative examples of immunofluorescence analysis. 3D AD cultures are shown in green while the staining for the differentiation marker MAP2 is in red. Phosphorylated Tau is shown in blue and Aβ42 in yellow. Assisted image-processing tools are applied to identify hits with this method. For drug screenings run in 384-well plates, immunofluorescence is the preferred method of analysis.

Accelerating AD drug discovery using human 3D culture models of AD

3D human cellular models carry great potential to expand the range of features that can be established in culture, enabling higher-order investigation of brain development and functionality. Recently, multiple 3D human brain models of AD have been developed which recapitulate key pathogenic features of AD in a human brain-like environment. These new human cellular models provide exciting tools for mechanistic studies and, importantly, facilitate high-throughput drug screening in comparison with standard transgenic AD mouse models. Although these organoid models generate broad cellular diversity and to a degree architecturally mimic AD, additional progress is needed to achieve enough heterogeneity and reproducibility for large-scale library screening (Fig. 1). In spite of their limitations for precisely reconstructing relevant AD brain regions, these models do simulate pathogenic cascades of the disease and importantly are highly reproducible and scalable for high-throughput drug screening applications (Fig. 2). As we described in this book chapter, some of these 3D models are ready for unbiased high-throughput drug screening. Considering the current failures of clinical trials,[34,71] usually based on limited targets, this unbiased approach to drug screening is timely and likely to result in the identification of new therapeutic targets (Fig. 8).

Previously, most AD candidate drugs were preselected based on mechanistic hypothesis or chemical library screening using biochemical assays or simple cellular models (i.e., APP cleavage, Aβ aggregation, phosphorylation or degradation of tau protein, etc.) (Fig. 8, left panel).[7] The validated targets were then tested in AD transgenic mice to evaluate their efficacy on AD pathogenic markers such as synaptic deficits, Aβ and/or tau accumulation, cognitive deficits, and potential toxicity in vivo. This process generally took multiple years to complete and was very costly (Fig. 3). Nevertheless, most of the promising AD drug candidates that worked in biochemical, cell culture, and AD transgenic models lacked efficacy when tested in human clinical trials.[9] In comparison, 3D human cellular models are relatively cheaper and faster (Figs. 3 and 8, right panel).

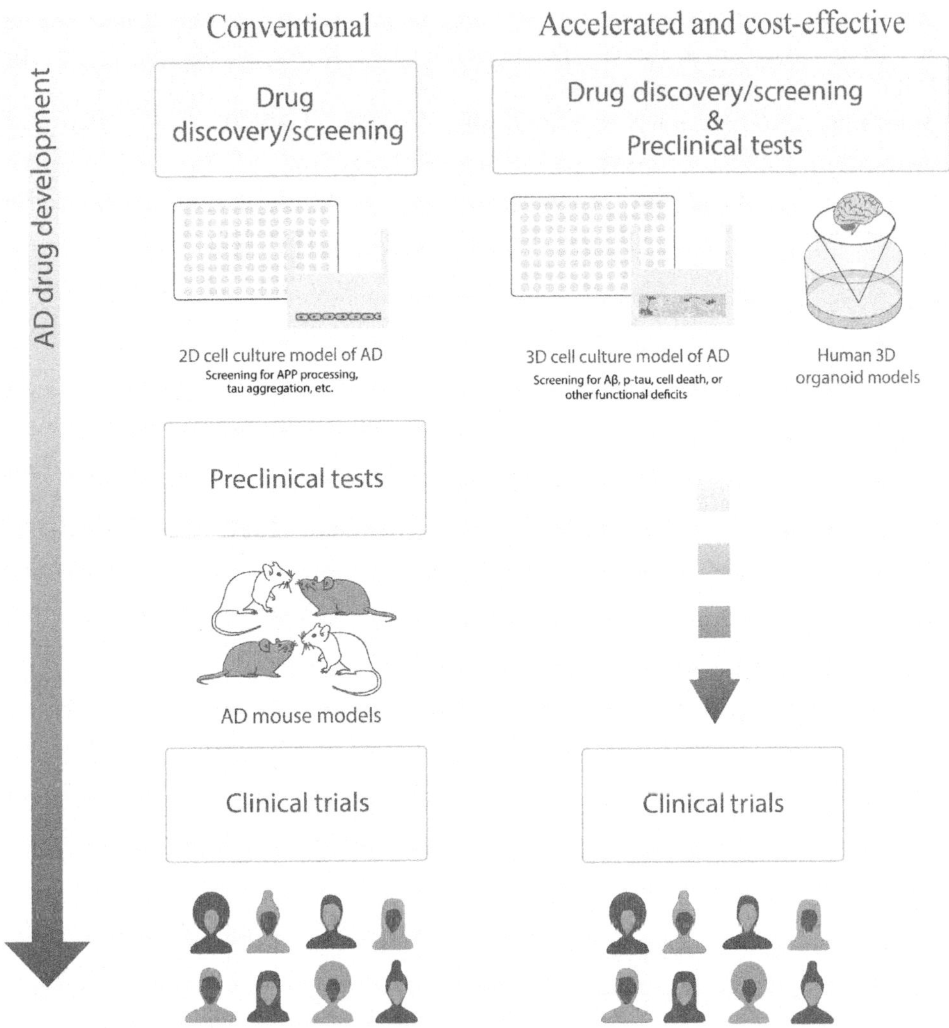

Fig. 8 Accelerating AD drug discovery using 3D Alzheimer's in a dish model. *Left panel*, Conventional AD drug screening pipeline. AD candidate drugs are preselected based on biochemical assays or cellular model tests. Validated hits are then tested in AD transgenic mice before patient trials. *Right panel*, 3D human cellular models drug screening pipeline. Direct screening of safety-validated drug libraries may lead to patient trials without revalidation in animals, a relatively cheaper and faster model.

Direct screening of candidate drugs in human AD brain models conceivably could advance to patient trials without revalidation in AD animal models. Of course, current 3D human brain models cannot provide some crucial information regarding *in vivo* toxicity, pharmacokinetics, and cognitive benefits. Further, by using drug libraries that are validated for human safety (i.e., FDA-approved drugs), we may be able to reduce the drug discovery and development process by skipping animal tests (Fig. 8, right panel). In addition to unbiased screening, human 3D cellular models can be also used to validate candidate AD drugs that are not suitable for testing in current AD mouse models. For example, the Aβ-driven tau/NFT phenotype cannot be directly modeled with the current AD mouse models, which do not express endogenous human tau isoforms.[5,7,8] Therefore, drug candidates that are targeting Aβ-driven tau pathology can be tested in our 3D AD neuron-astrocyte models or 3D AD non-cell-autonomous models as described in Fig. 3.

Conclusion

It is not surprising that much excitement is currently being generated around the development of 3D human cellular models of the human brain with the promise of exploring both normal development and the complex neurobiology of neuropsychiatric disorders in the dish. Although much work remains to be done, it is highly likely that over the next few years, we will witness a dramatic increase in the use of 3D AD cellular models for screening and testing new candidate AD drugs. Due to the technical limitations of current 3D AD cellular models, their application may be limited to early-stage drug development, replacing early screening and testing of candidate AD drugs. As we explained in this chapter, multiple factors need to be considered when choosing the appropriate 3D cellular models for drug testing and chemical library screening. Recent advances in 3D culture technology make it possible to consistently recapitulate cell-cell interactions and circuit connectivity among diverse brain cell types and functional alterations in diseased conditions. Accordingly, in the near future, these advanced human brain models may allow industries to proceed to human clinical trials without undertaking extensive animal studies. Finally, similar cell scaffold models may have application for the development of pharmaceuticals in the instance of other neurodegenerative diseases. Overall, this could be a leap forward in understanding and treating diseases of the nervous system, which currently have not enjoyed the therapeutic advances that have characterized modern medicine.

References

1. Alzheimer's Association. 2020 Alzheimer's disease facts and figures. *Alzheimers Dement.* 2020;16:391–460. https://doi.org/10.1002/alz.12068.
2. Tanzi RE, Bertram L. Twenty years of the Alzheimer's disease amyloid hypothesis: a genetic perspective. *Cell.* 2005;120:545–555. https://doi.org/10.1016/j.cell.2005.02.008.
3. Selkoe DJ, Hardy J. The amyloid hypothesis of Alzheimer's disease at 25 years. *EMBO Mol Med.* 2016;8:595–608. https://doi.org/10.15252/emmm.201606210.
4. Sweeney MD, Sagare AP, Zlokovic BV. Blood-brain barrier breakdown in Alzheimer disease and other neurodegenerative disorders. *Nat Rev Neurol.* 2018;14(3):133–150. https://doi.org/10.1038/nrneurol.2017.188.
5. Götz J, Bodea L-G, Goedert M. Rodent models for Alzheimer disease. *Nat Rev Neurosci.* 2018;19:583–598. https://doi.org/10.1038/s41583-018-0054-8.

6. Sasaguri H, Nilsson P, Hashimoto S, et al. APP mouse models for Alzheimer's disease preclinical studies. *EMBO J.* 2017;36:e201797397–2487. https://doi.org/10.15252/embj.201797397.
7. Choi SH, Kim YH, Quinti L, Tanzi RE, Kim DY. 3D culture models of Alzheimer's disease: a road map to a "cure-in-a-dish". *Mol Neurodegener.* 2016;11:75. https://doi.org/10.1186/s13024-016-0139-7.
8. D'Avanzo C, Aronson J, Kim YH, Choi SH, Tanzi RE, Kim DY. Alzheimer's in 3D culture: challenges and perspectives. *Bioessays.* 2015;37:1139–1148. https://doi.org/10.1002/bies.201500063.
9. Oxford AE, Stewart ES, Rohn TT. Clinical trials in Alzheimer's disease: a hurdle in the path of remedy. *Int J Alzheimers Dis.* 2020;2020:5380346–13. https://doi.org/10.1155/2020/5380346.
10. Cummings J, Lee G, Ritter A, Sabbagh M, Zhong K. Alzheimer's disease drug development pipeline: 2019. *Alzheimer's Dement.* 2019;5:272–293. https://doi.org/10.1016/j.trci.2019.05.008.
11. Kwart D, Gregg A, Scheckel C, et al. A large panel of isogenic APP and PSEN1 mutant human iPSC neurons reveals shared endosomal abnormalities mediated by APP β-CTFs, Not Aβ. *Neuron.* 2019;104:256–270.e5. https://doi.org/10.1016/j.neuron.2019.07.010.
12. de Leeuw S, Tackenberg C. Alzheimer's in a dish-induced pluripotent stem cell-based disease modeling. *Transl Neurodegener.* 2019;8:21. https://doi.org/10.1186/s40035-019-0161-0.
13. Goldstein LSB, Reyna S, Woodruff G. Probing the secrets of Alzheimer's disease using human-induced pluripotent stem cell technology. *Neurotherapeutics.* 2015;12:121–125. https://doi.org/10.1007/s13311-014-0326-6.
14. Choi SH, Kim YH, D'Avanzo C, Aronson J, Tanzi RE, Kim DY. Recapitulating amyloid β and tau pathology in human neural cell culture models: clinical implications. *US Neurol.* 2015;11:102–105. https://doi.org/10.17925/USN.2015.11.02.102.
15. Choi SH, Kim YH, Hebisch M, et al. A three-dimensional human neural cell culture model of Alzheimer's disease. *Nature.* 2014;515:274–278. https://doi.org/10.1038/nature13800.
16. Park J, Wetzel I, Marriott I, et al. A 3D human triculture system modeling neurodegeneration and neuroinflammation in Alzheimer's disease. *Nat Neurosci.* 2018;21:941–951. https://doi.org/10.1038/s41593-018-0175-4.
17. Kwak SS, Washicosky KJ, Brand E, et al. Amyloid-β42/40 ratio drives tau pathology in 3D human neural cell culture models of Alzheimer's disease. *Nat Commun.* 2020;11:1377. https://doi.org/10.1038/s41467-020-15120-3.
18. Arber C, Toombs J, Lovejoy C, et al. Familial Alzheimer's disease patient-derived neurons reveal distinct mutation-specific effects on amyloid beta. *Mol Psychiatry.* 2019;8:595. https://doi.org/10.1038/s41380-019-0410-8.
19. Meyer K, Feldman HM, Lu T, et al. REST and neural gene network dysregulation in iPSC models of Alzheimer's disease. *Cell Rep.* 2019;26:1112–1127.e9. https://doi.org/10.1016/j.celrep.2019.01.023.
20. Raja WK, Mungenast AE, Lin Y-T, et al. Self-organizing 3D human neural tissue derived from induced pluripotent stem cells recapitulate Alzheimer's disease phenotypes. *PLoS One.* 2016;11:e0161969. https://doi.org/10.1371/journal.pone.0161969.
21. Lee H-K, Velazquez Sanchez C, Chen M, et al. Three dimensional human neuro-spheroid model of Alzheimer's disease based on differentiated induced pluripotent stem cells. *PLoS One.* 2016;11:e0163072. https://doi.org/10.1371/journal.pone.0163072.
22. Moore S, Evans LDB, Andersson T, et al. APP metabolism regulates tau proteostasis in human cerebral cortex neurons. *Cell Rep.* 2015;11:689–696. https://doi.org/10.1016/j.celrep.2015.03.068.
23. Muratore CR, Rice HC, Srikanth P, et al. The familial Alzheimer's disease APPV717I mutation alters APP processing and Tau expression in iPSC-derived neurons. *Hum Mol Genet.* 2014;23:3523–3536. https://doi.org/10.1093/hmg/ddu064.
24. Sproul AA, Sproul AA, Jacob S, et al. Characterization and molecular profiling of PSEN1 familial Alzheimer's disease iPSC-derived neural progenitors. *PLoS One.* 2014;9:e84547. https://doi.org/10.1371/journal.pone.0084547.
25. Woodruff G, Young JE, Martinez FJ, et al. The Presenilin-1 ΔE9 mutation results in reduced γ-secretase activity, but not total loss of PS1 function, in isogenic human stem cells. *Cell Rep.* 2013;5:974–985. https://doi.org/10.1016/j.celrep.2013.10.018.
26. Kondo T, Asai M, Tsukita K, et al. Modeling Alzheimer's disease with iPSCs reveals stress phenotypes associated with intracellular Aβ and differential drug responsiveness. *Cell Stem Cell.* 2013;12:487–496. https://doi.org/10.1016/j.stem.2013.01.009.

27. Israel MA, Yuan SH, Bardy C, et al. Probing sporadic and familial Alzheimer's disease using induced pluripotent stem cells. *Nature*. 2012;482:216–220. https://doi.org/10.1038/nature10821.
28. Gonzalez C, Armijo E, Bravo-Alegria J, Becerra-Calixto A, Mays CE, Soto C. Modeling amyloid beta and tau pathology in human cerebral organoids. *Mol Psychiatry*. 2018;23:2363–2374. https://doi.org/10.1038/s41380-018-0229-8.
29. Lin Y-T, Seo J, Gao F, et al. APOE4 causes widespread molecular and cellular alterations associated with Alzheimer's disease phenotypes in human iPSC-derived brain cell types. *Neuron*. 2018;98:1141–1154.e7. https://doi.org/10.1016/j.neuron.2018.05.008.
30. Kim YH, Choi SH, D'Avanzo C, et al. A 3D human neural cell culture system for modeling Alzheimer's disease. *Nat Protoc*. 2015;10:985–1006. https://doi.org/10.1038/nprot.2015.065.
31. Yu YJ, Kim YH, Na K, et al. Hydrogel-incorporating unit in a well: 3D cell culture for high-throughput analysis. *Lab Chip*. 2018;18:2604–2613. https://doi.org/10.1039/C8LC00525G.
32. Madhusudanan P, Raju G, Shankarappa S. Hydrogel systems and their role in neural tissue engineering. *J R Soc Interface*. 2020;17:20190505. https://doi.org/10.1098/rsif.2019.0505.
33. Jorfi M, D'Avanzo C, Kim DY, Irimia D. Three-dimensional models of the human brain development and diseases. *Adv Healthc Mater*. 2017;7:1700723. https://doi.org/10.1002/adhm.201700723.
34. Lancaster MA, Renner M, Martin C-A, et al. Cerebral organoids model human brain development and microcephaly. *Nature*. 2013;501:373–379. https://doi.org/10.1038/nature12517.
35. Qian X, Song H, Ming G-L. Brain organoids: advances, applications and challenges. *Development*. 2019;146:dev166074. https://doi.org/10.1242/dev.166074.
36. Jorfi M, D'Avanzo C, Tanzi RE, Kim DY, Irimia D. Human neurospheroid arrays for in vitro studies of Alzheimer's disease. *Sci Rep*. 2018;8:153. https://doi.org/10.1038/s41598-018-20436-8.
37. Park J, Wetzel I, Dréau D, Cho H. 3D Miniaturization of human organs for drug discovery. *Adv Healthc Mater*. 2018;7:1700551. https://doi.org/10.1002/adhm.201700551.
38. Park J, Baik SH, Mook-Jung I, Irimia D, Cho H. Mimicry of central-peripheral immunity in Alzheimer's disease and discovery of neurodegenerative roles in neutrophil. *Front Immunol*. 2019;10:2231. https://doi.org/10.3389/fimmu.2019.02231.
39. Shin Y, Choi SH, Kim E, et al. Blood-brain barrier dysfunction in a 3D in vitro model of Alzheimer's disease. *Adv Sci (Weinh)*. 2019;6:1900962. https://doi.org/10.1002/advs.201900962.
40. Hardy J. The amyloid hypothesis of Alzheimer's disease: progress and problems on the road to therapeutics. *Science*. 2002;297:353–356. https://doi.org/10.1126/science.1072994.
41. Tanzi RE. A brief history of Alzheimer's disease gene discovery. *J Alzheimers Dis*. 2013;33(Suppl 1):S5–13. https://doi.org/10.3233/JAD-2012-129044.
42. Karran E, Mercken M, Strooper BD, De Strooper B. The amyloid cascade hypothesis for Alzheimer's disease: an appraisal for the development of therapeutics. *Nat Rev Drug Discov*. 2011;10:698–712. https://doi.org/10.1038/nrd3505.
43. Selkoe D. Alzheimer's disease is a synaptic failure. *Science*. 2002;298:789–791.
44. Benilova I, Karran E, De Strooper B. The toxic Aβ oligomer and Alzheimer's disease: an emperor in need of clothes. *Nat Neurosci*. 2012;15:349–357. https://doi.org/10.1038/nn.3028.
45. Morris GP, Clark IA, Vissel B. Inconsistencies and controversies surrounding the amyloid hypothesis of Alzheimer's disease. *Acta Neuropathol Commun*. 2014;2:135. https://doi.org/10.1186/s40478-014-0135-5.
46. Armstrong RA. A critical analysis of the "amyloid cascade hypothesis" *F N*. 2014;3:211–225. https://doi.org/10.5114/fn.2014.45562.
47. Herrup K. The case for rejecting the amyloid cascade hypothesis. *Nat Neurosci*. 2015;18:794–799. https://doi.org/10.1038/nn.4017.
48. Götz J, Ittner LM. Animal models of Alzheimer's disease and frontotemporal dementia. *Nat Rev Neurosci*. 2008;9:532–544. https://doi.org/10.1038/nrn2420.
49. Chin J. Selecting a mouse model of Alzheimer's disease. *Methods Mol Biol*. 2011;670:169–189. https://doi.org/10.1007/978-1-60761-744-0_13.
50. Oddo S, Caccamo A, Shepherd JD, et al. Triple-transgenic model of Alzheimer's disease with plaques and tangles: intracellular Abeta and synaptic dysfunction. *Neuron*. 2003;39:409–421.

51. De Strooper B. Lessons from a failed γ-secretase Alzheimer trial. *Cell*. 2014;159:721–726. https://doi.org/10.1016/j.cell.2014.10.016.
52. Henley DB, Sundell KL, Sethuraman G, Dowsett SA, May PC. Safety profile of semagacestat, a gamma-secretase inhibitor: IDENTITY trial findings. *Curr Med Res Opin*. 2014;30:2021–2032. https://doi.org/10.1185/03007995.2014.939167.
53. McQuade A, Blurton-Jones M. Microglia in Alzheimer's disease: exploring how genetics and phenotype influence risk. *J Mol Biol*. 2019;431:1805–1817. https://doi.org/10.1016/j.jmb.2019.01.045.
54. Moir RD, Lathe R, Tanzi RE. The antimicrobial protection hypothesis of Alzheimer's disease. *Alzheimers Dement*. 2018;14:1602–1614. https://doi.org/10.1016/j.jalz.2018.06.3040.
55. Marioni RE, Harris SE, Zhang Q, et al. GWAS on family history of Alzheimer's disease. *Transl Psychiatry*. 2018;8:1204. https://doi.org/10.1038/s41398-018-0150-6.
56. Hansen DV, Hanson JE, Sheng M. Microglia in Alzheimer's disease. *J Cell Biol*. 2018;217:459–472. https://doi.org/10.1083/jcb.201709069.
57. Heneka MT, Golenbock DT, Latz E. Innate immunity in Alzheimer's disease. *Nat Immunol*. 2015;16:229–236. https://doi.org/10.1038/ni.3102.
58. Zhang Y, Chen K, Sloan SA, et al. An RNA-sequencing transcriptome and splicing database of glia, neurons, and vascular cells of the cerebral cortex. *J Neurosci*. 2014;34:11929–11947. https://doi.org/10.1523/JNEUROSCI.1860-14.2014.
59. Kleinberger G, Yamanishi Y, Suarez-Calvet M, et al. TREM2 mutations implicated in neurodegeneration impair cell surface transport and phagocytosis. *Sci Transl Med*. 2014;6:243ra86. https://doi.org/10.1126/scitranslmed.3009093.
60. Griciuc A, Serrano-Pozo A, Parrado AR, et al. Alzheimer's disease risk gene CD33 inhibits microglial uptake of amyloid beta. *Neuron*. 2013;78:631–643. https://doi.org/10.1016/j.neuron.2013.04.014.
61. Griciuc A, Patel S, Federico AN, et al. TREM2 acts downstream of CD33 in modulating microglial pathology in Alzheimer's disease. *Neuron*. 2019;103:820–835.e7. https://doi.org/10.1016/j.neuron.2019.06.010.
62. Efthymiou AG, Goate AM. Late onset Alzheimer's disease genetics implicates microglial pathways in disease risk. *Mol Neurodegener*. 2017;12:43. https://doi.org/10.1186/s13024-017-0184-x.
63. Marx V. Neuroscience models: choose your dimension. *Nat Methods*. 2018;15:863–866. https://doi.org/10.1038/s41592-018-0191-z.
64. Ballabh P, Braun A, Nedergaard M. The blood-brain barrier: an overview: structure, regulation, and clinical implications. *Neurobiol Dis*. 2004;16:1–13. https://doi.org/10.1016/j.nbd.2003.12.016.
65. Sweeney MD, Sagare AP, Zlokovic BV. Blood-brain barrier breakdown in Alzheimer disease and other neurodegenerative disorders. *Nat Rev Neurol*. 2018;14:133–150. https://doi.org/10.1038/nrneurol.2017.188.
66. Montagne A, Nation DA, Sagare AP, et al. APOE4 leads to blood-brain barrier dysfunction predicting cognitive decline. *Nature*. 2020;581:71–76. https://doi.org/10.1038/s41586-020-2247-3.
67. Nation DA, Sweeney MD, Montagne A, et al. Blood-brain barrier breakdown is an early biomarker of human cognitive dysfunction. *Nat Med*. 2019;25:270–276. https://doi.org/10.1038/s41591-018-0297-y.
68. Sivandzade F, Cucullo L. In-vitro blood-brain barrier modeling: a review of modern and fast-advancing technologies. *J Cereb Blood Flow Metab*. 2018;38:1667–1681. https://doi.org/10.1177/0271678X18788769.
69. Chi KR. *Orchestrating Organoids*. https://www.the-Scientist.com https://doi.org/10.3791/52483.
70. Lancaster MA, Knoblich JA. Generation of cerebral organoids from human pluripotent stem cells. *Nat Protoc*. 2014;9:2329–2340. https://doi.org/10.1038/nprot.2014.158.
71. Quadrato G, Brown J, Arlotta P. The promises and challenges of human brain organoids as models of neuropsychiatric disease. *Nat Med*. 2016;22:1220–1228. https://doi.org/10.1038/nm.4214.

CHAPTER 14

Improving clinical trial efficiency with machine learning models of disease progression

Danielle Beaulieu, Albert A. Taylor, Dustin Pierce, Jonavelle Cuerdo, Mark Schactman, Mike Keymer, and David L. Ennist
Origent Data Sciences, Inc., Vienna, VA, United States

Introduction

Ten years from now, the establishment of the PRO-ACT (*P*ooled *R*esource *O*pen-Access *A*myotrophic Lateral Sclerosis *C*linical *T*rials) database will be viewed as a watershed moment in the development of therapeutics for not just amyotrophic lateral sclerosis (ALS) but also for all neurodegenerative diseases. Indeed, the approach taken in establishing PRO-ACT may eventually serve as a template for improving our understanding of all diseases that plague humankind. The PRO-ACT database was developed by Prize4Life in collaboration with the Neurological Clinical Research Institute (NCRI) at Massachusetts General Hospital and the Northeast ALS Consortium with funding from the ALS Therapy Alliance.[1] It was launched in December 2012 with the longitudinal records of 8635 ALS clinical trial participants.[2] In a massive 2-year undertaking, datasets from 17 phase II and III ALS clinical trials conducted between 1990 and 2010 were collected, de-identified, aggregated, cleaned, and harmonized using a common data structure and made openly available to the global research and development community without cost. Importantly, the PRO-ACT database continues to be actively curated and a major update was completed in December 2015 with the addition of six additional ALS clinical trials. It now includes the clinical trial records of over 10,700 patients who participated in 23 phase II and III trials. Prize4Life and the NCRI shared the 2013 Bio-IT World Best Practices Award in the clinical and health IT category for jointly developing PRO-ACT.[3,4]

Prize4Life was founded as a nonprofit organization in 2007 by Avichai Kremer, who was diagnosed with ALS at age 29 during his first year at Harvard Business School.[5] The stated objective of Prize4Life is to accelerate the discovery of treatments and a cure for ALS through an inducement prize model designed to drive innovation and expand human resources devoted to studying ALS by attracting "new minds" to the field.[6] Rather than awarding grants, the inducement prize model followed by Prize4Life was

Neurotherapeutics in the Era of Translational Medicine
https://doi.org/10.1016/B978-0-12-816475-4.00005-7

intended to reward participants who completed a challenge, like the Longitude Rewards of 1714 offered by the British government to determine longitude while at sea,[7] the Orteig Prize, won by Charles Lindbergh in 1927, that was offered to aviators to fly across the Atlantic Ocean from New York to Paris,[8] the 2004 Ansari XPrize, won by the Tier One project, that was awarded to the first nongovernmental organization to launch a reusable manned spacecraft twice within 2 weeks,[9] and the 2009 Netflix Prize, won by the collaborative BellKor's Pragmatic Chaos team, for the best algorithm to predict user movie ratings.[10] Modern-day challenges, run on the Internet, capture the imagination, competitive urges, and collaborative instincts of data scientists and subject matter experts in remote locations across academic fields in a structure that provides for a transparent, rigorous evaluation of computational methods.

To promote and publicize the launch of the PRO-ACT ALS database, Prize4Life teamed up with the DREAM Project (*D*ialogue for *R*everse *E*ngineering *A*ssessments and *M*ethods) to organize the $50,000 DREAM Phil Bowen ALS Prediction Prize4Life Challenge, which was conducted on the InnoCentive crowdsourcing platform[11] and sponsored by *Nature, The Economist* and *Popular Science*. InnoCentive has deep roots in drug development, as it was founded in 2001 by two Eli Lilly employees as a way to farm out some of its most difficult product development challenges.[12]

The goal of interventional clinical trials is to determine whether a treatment significantly improved an outcome for a patient as compared to what would have occurred had the patient not received the treatment. Since we cannot simultaneously know how a disease would have progressed in an individual with and without a given treatment, clinical science has developed the randomized, placebo-controlled trial (RCT,[13]) first conducted in 1946.14 This framework has been successful and numerous outstanding therapies have been developed over the last 70 years. However, diseases with heterogeneous rates of progression represent populations for which the current state of the art for RCTs is suboptimal. Examples of such diseases include ALS, Alzheimer's disease, Huntington's disease, and Parkinson's disease. ALS provides a classic illustrative case: two of the most famous ALS patients were Lou Gehrig and Stephen Hawking. Gehrig died within 2 years of playing his final game for the Yankees, while Hawking lived with ALS for over five decades. In over 20 years of clinical experimentation in ALS, there have been over 60 failures in clinical development and only two approved medications.15–17 One of the major causes for the history of failed clinical trials is believed to be the marked heterogeneity in disease progression.

ALS is a relentlessly progressive neurodegenerative disease affecting both upper and lower motor neurons leading to muscle weakness, paralysis, and death, most often 3–5 years following disease onset. Hidden in this description is wide variability in disease progression; some patients succumb within a year while others live with the disease for more than 10 years (Fig. 1, Ref. [18]). This extreme heterogeneity in disease progression results in an inherently high variability that makes it difficult to demonstrate statistically

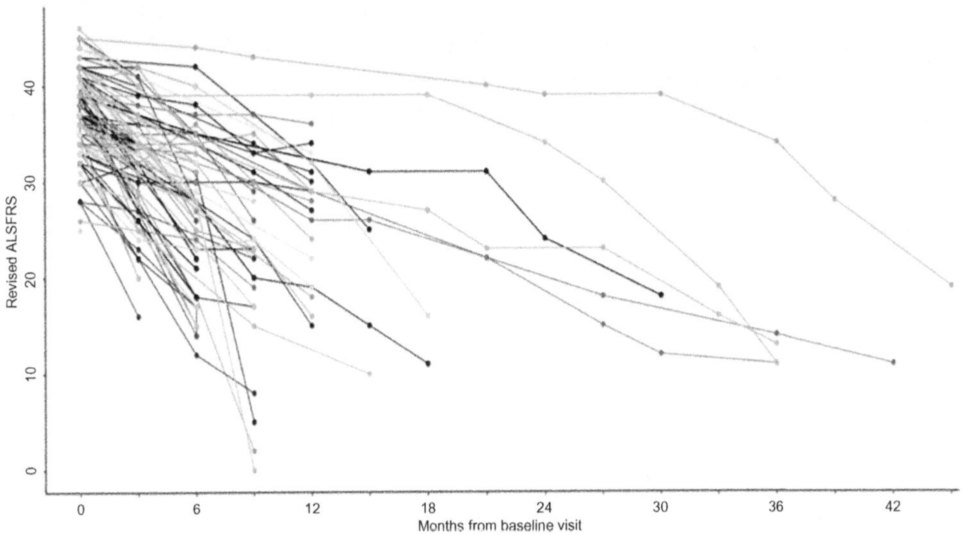

Fig. 1 Heterogeneity of disease progression in ALS as measured by the revised ALS Functional Rating Scale (ALSFRS-R) score over time from baseline of 100 randomly selected PRO-ACT patients with death records. The ALSFRS-R is a 48-point scale composed of four domains (bulbar, respiratory, fine motor, and gross motor function) that measures a patient's ability to perform a number of daily functions. Lower scores are indicative of diminished functional status. Similar spaghetti plots have been presented by Swinnen and Robberecht[20] and Fournier and Glass.[18]

significant therapeutic effects in ALS clinical trials. A prime motivation for developing PRO-ACT and conducting the ALS Prediction Challenge was the need to better understand the heterogeneity of the disease with the hope that tools to improve ALS clinical trials could eventually be developed. The ability to prospectively predict the rate of disease progression was seen as having the possibility of increasing the likelihood of bringing an ALS treatment into clinical practice.[19]

Using a PRO ACT subset consisting of the records of 1822 patients (broken down into 918 given to contestants for model training, 279 used as an interim test set, and 625 as the final validation set), contestants were challenged to develop an ALS disease progression model using 3 months of run-in data to predict the slope of change in patient functional capacity 9 months later as measured by the original ALS Functional Rating Scale (ALSFRS[21]). As the slope of the Functional Rating Scale is a commonly used primary endpoint in ALS clinical trials, its prediction would be directly relevant to the design and simulation of ALS clinical trials. Solutions offered by contestants needed to outperform fellow contestants as well as a model created by the organizers. Contestants could test their algorithms up to 100 times using the interim test set, with the results appearing online on a leaderboard during the 10 weeks allotted to the Challenge. The performance of models submitted by contestants was ranked using root-mean-square deviation

(RMSD) and the Pearson correlation coefficient. When the dust settled, the Challenge had drawn 1073 contestants from 64 countries and the $50,000 prize was split between three teams; teams from Stanford (using the Bayesian additive regression tree algorithm,[22]) and Sentrana, Inc. (using the random forest algorithm,[23]), a Washington, DC based marketing analytics company were declared cowinners and a third team from the University of Zurich was awarded a second prize.[24] Winning contestants were invited to present at the 2012 RECOMB Conference on Regulatory and Systems Genomics, with Dream Challenges sponsored by the International Society for Computational Biology[25,26] and the results were published in Nature Biotechnology.[19] Key findings reported in the paper included the observation that the two best algorithms outperformed a panel of a dozen ALS clinicians from seven countries. Furthermore, the Challenge identified several unusual predictors, including uric acid, creatinine, and blood pressure, and, perhaps most tantalizingly, the authors estimated that using the predictive algorithms in future ALS trials could lower sample sizes on the order of 20%.

Sentrana had entered the Challenge at the urging of a Vice President of one of their food service clients, who has an adult son with ALS. With no prior experience in disease modeling, Sentrana Principal Scientist Liuxia Wang, PhD, and Quantitative Modeler Guang "Eric" Li had set out to create a solution for this Challenge using random forest algorithms and a mixed model approach to better identify predictors of ALS progression. Wang and Li approached the Challenge by assuming that a nonlinear causal relationship exists between the disease progression rate and the initial condition. They then used the different dimensions of available data to develop new variables to better predict the progression of ALS by looking for patterns behind the data, such as relationships between different "body parts." One example of how they rearranged the variables was to subdivide the 10 component sub-scores of the ALSFRS into five body part-related categories (face, hand, trunk, legs, and chest). They found that the rapidity or stability of the face-related scores (which include tests of speech and swallowing) best predicted the overall disease trajectory and that a decline in chest-related scores signaled the end stage of the disease. The new variables created using this common-sense approach, rather than a purely medical or statistical approach, when used along with other available variables allowed for better prediction of disease progression.

The Sentrana team selected a random forest method due to the relative accuracy of the learning algorithm and its ability to utilize interactions among a variety of predictor variables.[23] The random forest algorithm uses a series of decision trees made from random samplings of the available data. After multiple iterations, the results are aggregated for greater predictive accuracy. Additionally, rather than using a single random forest model, Wang and Li opted to use a mixed model to incorporate multiple random forest models in order to mitigate the potential impact of inconsistent or missing measurements across certain important predictor variables. The multiple random forest models developed for each subset of data were then combined for the final model. Much of the model development rationale as well as the code developed for the Challenge are available in the supplementary information that accompanied the Nature Biotechnology publication.[19]

When interviewed[26] shortly after the announcement of the winning entries, Dr. Wang stated: "There are two key areas our research can inform: the possibility of stages in the disease and the relationship between body parts. We hope the questions we raised around these two areas will help guide the direction of new research," and Mr. Li added: "It's very exciting to use something that we do every day and apply those techniques in other really meaningful areas, especially considering the treatment limitations for this disease."

Nearly as soon as the results were announced[27] the CEO of Sentrana began receiving calls from pharma and biotech companies interested in the potential use of the ALSFRS disease progression model to improve drug development clinical trials for ALS. The use of machine learning for drug development was not in the Sentrana business plan, but realizing that there appeared to be an innovative new drug development application for machine learning, the Sentrana CEO decided to spin off a new company to explore this innovative application. Together with Topspin Labs, a northern Virginia spinoff incubator specializing in developing technologies that are beyond the parent company's core technologies, Sentrana created Origent Data Sciences, Inc. in 2013 to develop this concept.

Once angel investor funding had been secured and the beginnings of a team assembled, our nascent start-up set about transferring the relevant technology from Sentrana. Technology transfer was efficiently accomplished by embedding the Origent team with the Sentrana team on the same floor of a downtown Washington, DC office building located two blocks from the White House and a block from the (in)famous K Street center of the Washington DC lobbying industry. Most people imagine that DC and the surrounding areas (Maryland and Northern Virginia suburbs) contain very few industries beyond government, lobbying, and the legal profession, but with the NIH, the FDA, strong biotechnology and information technology sectors, and numerous top-ranked universities and medical schools, the region is well positioned to support a vibrant biostatistics and data science community. In fact, Washington DC boasts more than 10 active regular data science meetings[28] and information technology has been identified as a critical component of its economy.[29]

Since the Prize4Life Challenge, the Origent team has published a number of manuscripts[19,30–36] and presented our work at numerous scientific meetings.[34,35,37–51] With the help of several ALS Association and NIH grants, we now have a dozen ALS models and are branching out into other neurodegenerative diseases, including Alzheimer's disease, Parkinson's disease, and Huntington's disease. We have used predictive models to improve clinical trial randomization methods[30] and collaborated on diverse projects including improving study power,[37] predicting disease progression in clinic patients,[34,35] predicting vital capacity,[31,43] ALS resistance,[41] patient stratification,[30, 48] and analyzing the effects of diaphragm pacing.[45, 47] These real-world experiences with disparate datasets have made us thoroughly aware of the challenges in applying machine learning (ML) models to patient disease progression and in using the models in drug and device development. While much work by others has gone into developing ALS disease progression models,[24,52–67] most have focused on using the models in the context of clinical decision aids. Our focus is on developing models and applications to increase the efficiency of drug, biologic, and device clinical trials.[30–33]

ALS models

On the surface, it initially seemed that predicting the slope of ALSFRS change may not result in optimal results. Knowing that disease progression does not necessarily follow a linear course, we reasoned that it may be more accurate over different time periods to predict the functional score itself rather than the slope of change. Three individuals may have the same baseline score, reach a common future time point, and thus have the same overall slope from baseline to that point in time, but they likely reached that point by three dramatically different paths. Individual A may indeed have followed a linear path, but individual B may have started from a plateau that at some point followed an accelerated path and individual C may have declined dramatically shortly after the baseline time point and then reached a plateau. We reasoned that overall it would be preferable to make predictions for the ALSFRS score itself for every trial participant. If the slope were needed for the outcome in a given study, we would then simply divide by the number of months. We proceeded to modify the winning model to predict ALSFRS rather than slope, then used RMSD to compare the accuracy of the original model predicting slope to the new model predicting ALSFRS. Using several hundred randomly selected records from PRO-ACT, we calculated the slope for the score model by dividing by the number of elapsed months at the endpoint and calculated the score for the slope model by multiplying by the number of elapsed months (rounded to two decimal places). Subsequenty, the accuracy of scores and slopes generated both ways were compared. In both scenarios, it turned out that using predicted ALSFRS score was 5% to 10% more accurate than predicting slope—a significant improvement over a model that had just triumphed in an international model development contest. The Sentrana Team may have been declared the sole Challenge winners if they had thought of this little trick!

The Sentrana model was actually a combination, or in data science lingo, an "ensemble," of four separate models (Ref. [19] supplementary material), the use of which depends on the availability of baseline features. Model 1, the most accurate model included baseline forced vital capacity (FVC), model 2 included uric acid, model 3 included the diagnosis delta (days between symptom onset and diagnosis), and finally, in the absence of FVC, uric acid, and diagnosis delta, a base model was used. The logical way to combine the models is to first use the most accurate model, followed by the second most accurate model, etc., until the base model is invoked. This type of model construction is useful in competitive challenges such as the ALS Prediction Challenge when it becomes important to have the best performance metric possible. An alternative is to impute these values, but the Challenge dataset was relatively small and significant imputation of key predictors would have had a deleterious effect on performance. On the other hand, having a complicated ensemble for a production model is difficult because it can be computationally expensive. In addition, it is difficult to assure quality when multiple models for the same outcome need to be maintained, assessed, and validated.

As a result of these considerations, we at Origent have decided to develop models that include a set of predictors that are widely and commonly collected at baseline in ALS clinical trials, based on mining the PRO-ACT database. These predictors include demographics (age, sex), patient characteristics (height, weight, riluzole use), ALS history (site of symptom onset, days since symptom onset, days since diagnosis), ALSFRS-R scores (including answers to individual questions), vital capacity, and vital signs (systolic and diastolic blood pressure, pulse, respiratory rate, temperature). Since we have found uric acid, creatinine, and phosphorus to be useful predictors, we also encourage trialists and clinicians to collect these labs, the idea being that sometime in the future these values will be common enough in datasets to be useful for modeling purposes.

The original random forest model built for the ALS Prediction Challenge utilized a 90-day run-in period to make predictions 9 months later. While occasionally used, particularly in autologous stem cell trials where the cells need to be manufactured, discussions with a number of key opinion leaders shortly after the company was launched revealed that the use of a run-in period in ALS trials is not ideal. The consensus was that it would be far preferable to be able to make predictions from a baseline point without the use of a 90-day run-in period, the reason being that a patient should start the trial as soon as possible after being deemed eligible. Furthermore, given the limited time an ALS patient often has left to live, any delay in enrolling a participant in a trial is grossly unfair to the patient. We first tried to simply eliminate the variables that depended on the day 90 time point. Interestingly, the accuracy of the models degraded only slightly, but we were suddenly confronted with serious bias issues. Most of the features that depended on day 90 were slopes from day 90 to the baseline day of the trial, so we deduced that if we could replace these slopes, we may be able improve the bias situation. Fortunately, "day of symptom onset" is recorded for nearly all ALS patients and, furthermore, this key date, while it is an estimate based on the patients' and their families' recollections, is included for most of the patient records in PRO-ACT. A detailed comparison of run-in versus baseline models for predicting vital capacity is included in Jahandideh et al.[31] We found that the two models had similar accuracy as measured by RMSD and R^2, but the baseline model that used slopes generated from symptom onset demonstrated less bias. The date of symptom onset has turned out to be a key time point for ALS model development.

Table 1 ALS models, specifications, and validation. Models most likely to be used in ALS clinical trials include the ALSFRS-R, percent expected vital capacity regression models, and the time to 50% vital capacity and survival time-to-event models. Shown are the number of unique PRO-ACT patients included in each model, number of variables in the final gradient boosting machine model, and either the root-mean-square deviation (RMSD) for regression models or the area under the curve (AUC) for time-to-event models. Number of records for regression models and number of events for time-to-event models are also shown.

Regression models	Model specifications			PRO-ACT validation
	Unique patients	Records	Variables	RMSD
ALSFRS-R	3059	26944	19	4.74
Bulbar sub-score	6053	52892	14	1.584
Fine motor sub-score	6032	52670	17	1.88
Gross motor sub-score	6037	52712	18	1.738
Respiratory sub-score	2850	24502	32	1.692
Vital capacity	5729	37707	21	0.596
Percent expected vital capacity	5229	33110	16	14.606
Time-to-event models	**Unique patients**	**Events**	**Variables**	**AUC**
50% vital capacity	4604	1927	27	0.867
Survival	5831	1751	30	0.815
Loss of speech	5506	1361	31	0.902
Gastrostomy	5833	1141	32	0.859
Wheelchair use	5536	1920	23	0.847

The models described in Table 1, our current models, were all generated using a set of features available at baseline without the use of a run-in period. Slopes used in these models were generated assuming 100% function at the day of symptom onset.

Since the contest we have changed the functional score model to predict the revised ALSFRS (ALSFRS-R[68]), the current metric for tracking function in ALS patients. This switch was greatly facilitated by the December 2015 PRO-ACT update which added over 2000 patient records, most of which included the ALSFRS-R. Generally speaking, machine learning models crave data, so we investigated whether model accuracy would improve if we imputed the revised, 12-question ALSFRS-R from the several thousand records in PRO-ACT that included the original 10-question ALSFRS. It turned out that the model was more accurate and less biased if we limited the training to patient records that used the ALSFRS-R. As a result, our current ALSFRS-R functional score model is a 19-variable regression model that includes the 26,944 longitudinal records of 3059 ALS patients. In addition, we have developed a number of models that predict various facets of ALS disease progression, including survival as well as ALSFRS-R, which together have been the most common endpoints for clinical trials in ALS. It also became clear to us that we should develop a model that predicts respiratory decline because most ALS patients die of respiratory complications[69] and a number of recent ALS clinical trials, including the *tirasemtiv* and *reldesemtiv* trials conducted by Cytokinetics, have used a respiratory endpoint as the primary outcome measure.[70,71] These decisions were affirmed by the recent FDA guidance on developing drugs for ALS, where the Agency listed ALSFRS-R, respiratory function, and mortality as endpoints that could be used to establish the effectiveness of a new therapeutic.[72] Finally, on a recommendation from Drs. Jonathan Glass and Christina Fournier of the Emory ALS Clinic, we developed several time-to-event models which, in their experience, are questions frequently asked by ALS patients and their caregivers and are of potential clinical importance for treating and counseling patients. In addition to survival, these models include time to 50% expected vital capacity, which is used by Medicare in the United States as one of the specific requirements to determine coverage of noninvasive ventilation (NIV) and is used by many US ALS physicians as a time to initiate NIV use,[73] time to loss of speech as reflected by a score of one or lower on question 1 of the ALSFRS-R, time to gastrostomy (a switch from question 5A to 5B in the ALSFRS-R), and time to wheelchair use (a score of one or lower on ALSFRS-R question 8). Thus, to date we have developed a dozen ALS models that make predictions for multiple ALS outcomes (Table 1). Early iterations of the ALSFRS-R,[19,34–36] vital capacity,[31] and survival models[30] have been published. The primary consideration for the selection of models was the use of the dependent variables as clinically meaningful endpoints in phase IIb and 3 clinical trials.[74] These considerations led us to the development of applications for ALSFRS-R (used in *edaravone* studies[75]), survival (used in *riluzole* studies[76]), and respiratory function (used in *carnitine* trial[77]; and *tirasemtiv* trials[78,79]) as these are most likely to have drug development implications.

The current generation of models has been developed using similar methodology. An initial random forest model,[23] trained using about 200 available features, was used as a variable reduction step. Features with a Gini index greater than 0.5% (a reflection of how often the feature was used in the model, and hence its variable importance) were used to train a gradient boosting machine model[80,81] that was then validated for internal consistency using 10-fold cross validation (Hastie et al.[82] p. 241) and generalizability using external, contemporary ALS datasets (Steyerberg[83] p. 304 & ff). With support from the ALS Association and in collaboration with Cytokinetics, Inc., we have been extremely fortunate to be able to work with the placebo arms of the *tirasemtiv* BENEFIT-ALS[79] and VITALITY-ALS[78] datasets. These two trials spanned times typical of ALS clinical trials; 6 months in the case of BENEFIT-ALS and 1 year in the case of VITALITY-ALS. For time-to-event models, a Cox proportional hazards model was used as the loss function to train the GBM model. The outputs of the regression models are the values of the outcome variables at future time points and the outcomes of the time-to-event models are log-hazard coefficients.

Performance of the regression models (Table 1) was evaluated by RMSD and model bias was assessed by bootstrap analysis of mean prediction error. We sampled prediction error from the population of patient predictions with replacement and calculated a mean prediction error. This process was repeated 5000 times thus establishing mean error distributions and 95% confidence intervals for all models. Models were considered biased if a mean error of zero was not included within the 95% confidence interval (Hastie et al.[82] p. 249). RMSD is somewhat akin to a standard deviation and can be used in a similar way to judge a model. We have found that models with an RMSD that is 15% or lower than the scale under investigation are useful in the applications we have been developing. Given these bias and accuracy standards, the ALSFRS-R, bulbar, gross motor, and percent expected vital capacity regression models all showed acceptable accuracy and passed the bias test using both the internal 10-fold cross validations and the external BENEFIT-ALS and VITALITY-ALS placebo datasets.[84]

Performance of the time-to-event models (Table 1) was evaluated using area under the receiver operating characteristic curve (AUC), discrimination, and calibration (Steyerberg[83] p. 259 & ff). Discrimination, evaluated using the C-statistic, is a measure of how well a predictive model can distinguish individuals with an outcome from those without the outcome. We used discrimination to ask whether the models could stratify patients into slowly, moderately, and rapidly progressing patients. Calibration is a measure of the agreement between observed and predicted outcomes. To analyze calibration, we converted the log-hazards generated by the models into probability curves over time then compared the predicted probability curves to the observed Kaplan-Meier plots. The time-to-event models demonstrated acceptable accuracy (AUC $\geq$0.75) and passed the discrimination test, indicating that they can be used to stratify patients. The ability to stratify patients has allowed us to address the heterogeneity issue inherent to the disease progression of a group of ALS patients. This set of models represents the most complete collection of externally validated predictive models and therefore offers a credible approach to developing applications for ALS drug development, including stratification and controlling for the multiple confounding factors common to ALS trials. We set out to understand whether methods incorporating machine learning-based predicted outcomes are better than traditional methods at randomization, covariate adjustment, enrichment, and historical controls, and whether a machine learning approach can provide a novel method of subgroup analysis.

Applications for drug development

The models are merely a start to understanding how to use predictive analytics to improve the efficiency of drug development. By "improving the efficiency of drug development," we specifically mean being able to run quicker clinical trials that enroll fewer patients and have a better chance of being successful. From this perspective, how we use the models, that is, the application of the models, is of paramount importance. We focused on a central question that needed to be solved. Circling back to the spaghetti plot (Fig. 1): Can the predictive models be used to effectively stratify ALS patients into more homogeneous groups (Fig. 2)?

Randomization and covariate adjustment

Both theoretically and practically speaking, the best way to deal with confounding factors is to make sure they are evenly distributed between the study arms through randomization.85 As pointed out in the recently published FDA draft guidance on adjusting for covariates,86 treatment group baseline imbalances in prognostic factors increase the variability of treatment effect estimates and lower statistical power of a clinical study. Such arm balancing has been a major problem in neurodegenerative diseases. For example, a typical ALS clinical trial can be expected to have 25% of the trial population die within 1 year while simultaneously having a similar percentage of patients who progress very slowly.18,87 It is not difficult to understand that these simple observations have profound implications for the success of ALS clinical trials. They increase sample sizes, reduce study power, and increase the chance of arm balance failures, both in the balance of confounding features at the beginning of the trial, and also in the balance of clinically relevant outcomes and the primary endpoint at the end of the trial, which is what we should really care about! Often in recent years, clinical trialists have been left wondering whether the arms of their RCTs were balanced with respect to disease progression and mortality.

It is therefore incumbent on the trialist to identify features that successfully create strata that balance the dual confounds of deaths and patients who functionally progress slowly *not* at the beginning of the experiment, which is really an irrelevant time point, but at the end of the experiment when comparisons between study arms are due to be made. Ideally, the balance of deaths and slow progressors to be achieved should be between the placebo arm and the counterfactual of the treatment arm at the *end* of the trial.

To test whether trial randomization could be improved using strata generated by predicted outcome compared to traditional methods, we simulated a series of virtual parallel arm trials of sample sizes of 44 to 448 patients using randomly selected PRO-ACT records.[30] Depending on the size of the trial, we performed between 1000 (for $n=448$) and 10,000 (for $n=44$) simulations (Table 2). We compared randomization failure rates using simple randomizations generated by a random number generator to strata based on riluzole use and bulbar onset (traditional randomization) to stratification based on predicted survival. In our literature review, we discovered that a random number generator,[88–96] followed by the use of onset site/riluzole,[97,98] were the most common methods used to generate strata in recent ALS clinical trials. Randomization failures were defined as statistically significant differences ($P \leq 0.05$) between trial arms on any given characteristic and we assessed the performance of the traditional versus predicted

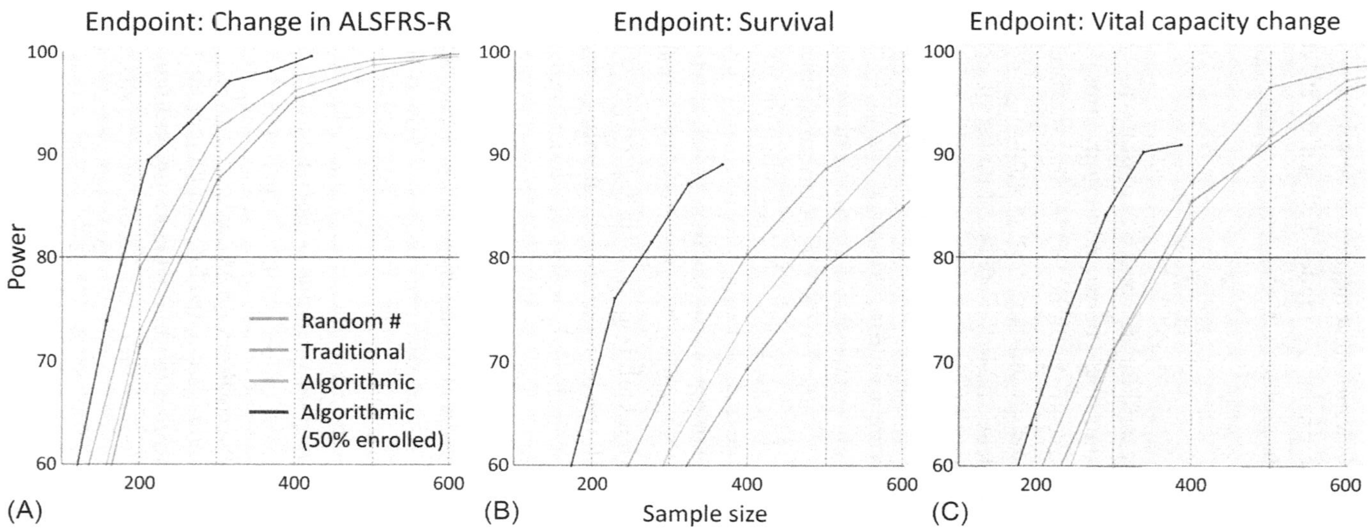

Fig. 2 ALS clinical trial power simulations of randomization and covariate adjustment. 1000 clinical trials were simulated using three stratification methods (simple, *gray*; onset site/riluzole, *blue*; and predicted, *green*) followed by the use of predictions to additionally enrich *(red)* for homogeneous clinical trial participants. Data are summarized in Table 3. Simulations for (A) 30% slowing of ALSFRS-R decline; (B) 2.5-month life extension; and (C) 30% slowing of decline of vital capacity.

Table 2 Summary of trial arm balance failure reductions comparing the use of strata generated by use of riluzole and site of onset ("traditional stratification") to strata for the indicated quantiles created using predicted survival. The percent reductions in arm balance failures by predicted survival relative to traditional stratification averaged 20% to 27% depending on the predicted strata (far right column).

	% Reduction of failures by algorithm								
Quantile	FVC	FVC % predicted	ALSFRS-R	FVC change	FVC % Pred slope	Survival	ALSFRS-R slope	ALSFRS-R change	Average
≤20th, >20th to <90th, ≥90th	39%	34%	38%	8%	10%	25%	4%	4%	20%
≤25th, >25th to <75th, ≥75th	50%	36%	46%	7%	13%	40%	6%	11%	26%
≤33rd, >33rd to <66th, ≥66th	47%	42%	36%	16%	16%	38%	10%	13%	27%
≤50th and >50th	38%	28%	37%	10%	10%	27%	10%	9%	21%
	← Balance of baseline features →			← Balance of outcome features →					

randomizations relative to a simple randomization using a random number generator that assigned patients to the two arms of the simulations. Finally, and most significantly, because we had the full patient records with outcomes available, we could compare the randomizations at both the baseline and outcome time points.

The results showed a marked improvement (Table 2). On average, with a 27% reduction, tertiles showed the greatest improvement in randomization failures. In terms of randomizing baseline features (forced vital capacity (FVC), percent predicted forced vital capacity, and ALSFRS-R), randomizing by predicted survival strata outperformed traditional (randomization by site of onset and riluzole use) methodology by 28%–50%. We examined both baseline parameters and, as this is a retrospective analysis with known outcomes, we took advantage of the ability to also assess randomization of outcome parameters (i.e., change in FVC, the slope of percent expected FVC, survival, ALSFRS-R slope, and the change in ALSFRS-R). These were also improved over traditional methods, ranging from 4% to 40% fewer randomization errors. As expected, because the strata were defined using predicted survival, the improvement was most marked for randomization of survival. The other outcomes (FVC and ALSFRS-R) also showed an improvement because these are among the predictors included in the survival model.

In the recently issued guidance on ALS drug development,[72] the FDA recognizes that an increasing number of ALS predictors are being identified and, in addition to their use in randomization, is encouraging their use as prespecified covariates in the statistical analysis plan. Toward this end, we wanted to examine the use of predicted outcomes as covariates in the analysis, so we conducted power simulations for three endpoints of wide interest as ALS trial outcomes—ALSFRS-R, survival, and vital capacity (Fig. 2). We conducted the power analysis in three ways, using a random number generator (gray in Fig. 2; this was the most common method in our literature review), using the traditional onset site/riluzole covariates compared previously in the randomization analysis (blue in Fig. 2), and using the predicted tertiles (green in Fig. 2) except that the predicted tertiles used matched the outcome endpoint under analysis (i.e., for the change in ALSFRS-R endpoint, predicted change in ALSFRS-R tertiles were used to create the strata used for randomization and covariate adjustment and similarly for survival and vital capacity). Importantly, the simulations performed for ALS can be extended to other diseases, including Alzheimer's disease and Parkinson's disease (Fig. 3). The data clearly show that the algorithmic randomizations with covariate adjustment, across different endpoints and diseases, result in smaller simulated sample sizes for reaching 80% power at the 0.05 significance level (Figs. 2 and 3 and Table 3).

For the change in ALSFRS-R outcome, the sample size reduction over the best traditional method was 15.2%, for survival the reduction was 14%, and for vital capacity the reduction was 8.6% (Table 3). If we consider the survival outcome, where the algorithmic method outperformed the onset site/riluzole use covariates by 65 patients, the smaller number of patients would result in a trial that costs $2.8 million less using the estimated phase III per patient trial costs of $43.5K/patient[99] and lasts about 5 fewer months (using 12 patients/month as an average ALS trial enrollment rate) to execute, thereby bringing successful therapeutics onto the market and improving the lives of patients 5 months sooner.

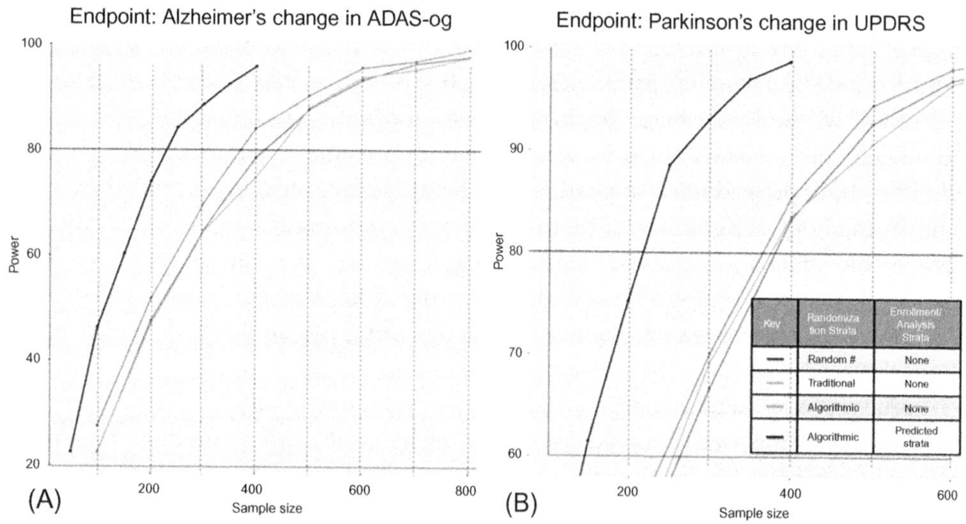

Fig. 3 One thousand clinical trials were simulated using three stratification methods (simple, *gray*; traditional, *blue*; and predicted, *green*) followed by the use of algorithmic methods to additionally enrich *(red)* for homogeneous clinical trial participants predicted to progress at a more similar rate. Data are summarized in Table 3. (A) Alzheimer's disease, endpoint used was 40% slowing in ADAS-cog. (B) Parkinson's disease, endpoint used was 40% slowing in UPDRS.

Table 3 Summary of traditional and predicted covariates used to simulate ALS, Parkinson's and Alzheimer's disease trials. In each case, the indicated sample sizes are needed to detect the simulated treatment effect noted in the second row of the table at 80% power with $P \leq 0.05$.

	ALS			Parkinson's	Alzheimer's
Endpoint	ALSFRS-R change	Survival	Vital capacity (L)	UPDRS	ADAS-cog
Simulated treatment effect	30% Slowing of decline	2.5-month extended life	30% Slowing of decline	40% Slowing of decline	40% Slowing of decline
SIMPLE randomization, without covariate adjustment					
Randomization starata:	Random # Generator	Random # Generator	Random # Generator	Random # Generator	Random # Generator
Required sample size:	254 patients	517 patients	362 patients	380 patients	415 patients
TRADITIONAL CONFOUNDERS used for randomization and covariate adjustment					
Randomization and covariate adjustment strata:	Riluzole/onset site	Riluzole/onset site	Riluzole/onset site	Levadopa/H&Y score	MMSE/AD severity
Required sample size:	244 patients	463 patients	372 patients	375 patients	439 patients
PREDICTED OUTCOMES used for randomization and covariate adjustment					
Randomization and covariate adjustment strata:	Predicted Δ ALSFRS-R 360 days (tertiles)	Predicted survival (tertiles)	Predicted Δ vital capacity 360 days (tertiles)	Predicted Δ UPDRS 360 days (tertiles)	Predicted Δ ADAS-cog 360 days (tertiles)
Required sample size:	207 patients	398 patients	331 patients	365 patients	371 patients
Sample size reduction from best non-algorithmic simulation:	*−37 patients*	*−65 patients*	*−31 patients*	*−10 patients*	*−44 patients*
Sample size % reduction from best	*−15.2%*	*−14.0%*	*−8.6%*	*−2.7%*	*−10.6%*

Power is defined as the probability that a false null hypothesis, the hypothesis that there is no significant difference between treatment arms, will be rejected. So, when a study is powered to 80%, it means that there is still a one in five chance that the outcome of the study will accept the null hypothesis (i.e., no difference between study arms) even if there actually is a treatment effect. When we showed the simulations in Fig. 2 and Table 3 to the FDA at a Type C meeting with Cytokinetics conducted under the *tirasemtiv* IND, the FDA suggested that it would be preferable to "improve the chance of success" by boosting power rather than lowering sample size, at least for the first several applications of these methods. Lowering the sample size too much runs the danger of underpowering a study, a serious ethical breach. On the other hand, if a study is powered and sample size determined using a conventional approach, but the analysis of the study is adjusted using predicted outcomes as covariates, it is likely that study power will increase as a direct result of the use of the predicted outcomes as covariates. In Fig. 4, rather than lowering sample size, an ALS study with an endpoint of increasing survival by 2.5 months simulated to 80% power using simple randomization can be powered to close to 90% using predicted survival as the covariate in the analysis. Power may not increase as much as indicated by the

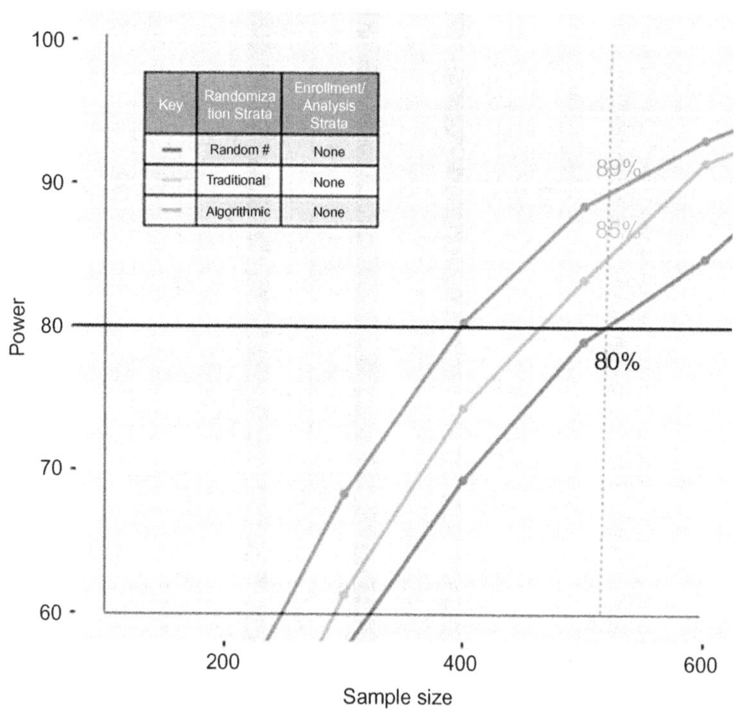

Fig. 4 Power boost afforded by the use of predicted survival as a covariate to adjust the survival analysis of a simulated ALS trial. Study conducted as in Fig. 2B. Study power can be boosted by the use of predicted outcomes in the analysis to about 89% over a study simulated to 80% power with simple randomization.

simulations, but it would be highly unlikely that power would be adversely affected. So, for now, we agree with this conservative approach. The use of predicted covariates can improve clinical trial efficiency by boosting study power and decreasing clinical trial sample sizes, lowering the overall costs of a trial, and decreasing trial duration.

Predicted outcomes can also be used effectively to adjust the analysis of the Combined Assessment of Function and Survival (CAFS[100]). The FDA points out in the ALS clinical trials Guidance[72] that patient deaths can confound the analysis of functional endpoints in ALS clinical trials. To counteract this issue, the FDA recommends an integrated assessment of function and survival such as CAFS. CAFS compares each trial participant's outcome to others in the trial in a series of pairwise comparisons based on both function as measured by the ALSFRS-R scale, and survival. For each pairwise comparison, a trial participant is assigned a score and then the summed scores are ranked for all participants. The score is assigned based on the change in ALSFRS-R from baseline and the survival endpoint where a shorter survival corresponds to a lower rank. In the pairwise comparisons a score of 1 is assigned when the subject has a better function or reached the survival endpoint later than the comparison. A score of 0 is assigned when the subject and the comparison have the same function or reached the survival endpoint at the same time. A score of −1 is assigned when the subject has a worse function or reached the survival endpoint earlier than the comparison. The mean rank score for each treatment group is calculated. A higher mean CAFS score indicates a better group outcome.

CAFS can be analyzed using nonparametric rank analysis of covariance, adjusting for predicted ALSFRS-R and predicted survival determined for each study subject. The ALSFRS-R covariate used in the adjustment can be the baseline ALSFRS-R score or predicted change in ALSFRS-R score at 1 year while the survival covariate can be the baseline percent expected vital capacity or the predicted log-hazard of survival. In Fig. 5 we compared the degree to which the mean squared error (MSE) is reduced in

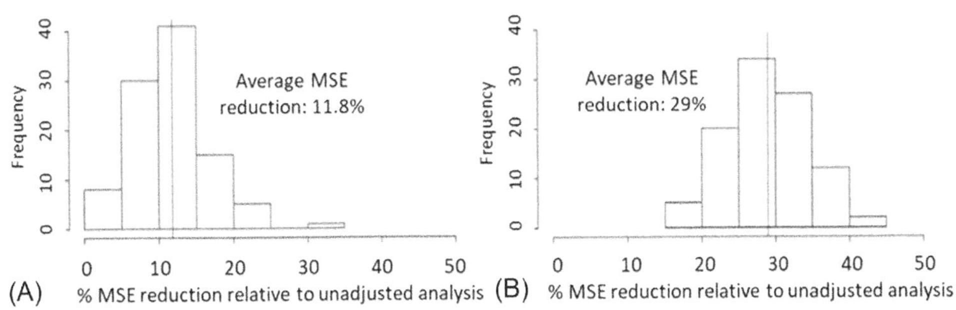

Fig. 5 Mean square error reductions relative to unadjusted analysis of CAFS simulations using (A) traditional covariates (baseline ALSFRS-R and baseline percent expected vital capacity) compared to the use of (B) predicted covariates (predicted 1-year change in ALSFRS-R and survival). One hundred simulations with a sample size of 200 randomly drawn PRO-ACT patients were each analyzed without adjustment, using traditional covariates and predicted covariates.

simulations using traditional covariates (baseline ALSFRS-R and baseline percent expected vital capacity) or predicted covariates (predicted 1-year ALSFRS-R and log-hazard of survival). MSE is a measure of the variance, or noise in an analysis. Covariate adjustment should lower the MSE if the covariates effectively explain some of the noise. As shown in Fig. 5, compared to an unadjusted analysis, the use of predicted outcomes as covariates in the analysis can lower the MSE by 29% compared to the 11.8% lowering afforded by the use of traditional covariates.

So, why is covariate adjustment using predicted outcomes more effective than covariate adjustment using a single (or even two) baseline prognostic factors? Fig. 6 shows the relative importance of the feature set used in the ALSFRS-R model (Fig. 6A) and the survival model (Fig. 6B). In developing these models, we originally considered over 200 features and culled these down to 19 features for the ALSFRS-R model and 30 for the survival model. In essence, the outputs of the models represent the relative contributions to predicting the outcome of interest of the 19 most important features in the case of the ALSFRS-R model and the 30 most important features in the case of the survival model. In the CAFS example above, the traditional covariates baseline ALSFRS-R and baseline percent expected vital capacity are merely two of the features used in the predictive models. Using the predicted outcomes as covariates to adjust the analysis explains some of the variation in the outcomes, thus adding greater precision to the evaluation of the therapeutic effect.

Enrichment

A prognostic enrichment strategy designed to reduce the heterogeneity of a clinical trial or its analysis can be achieved a number of ways. As suggested by the red lines in the simulations shown in Figs. 2 and 3, a study or its prespecified analysis could, for example, include only those patients predicted to be most likely to progress at an average rate during the period of the study. Cutoffs for study inclusion could be generated using PRO-ACT (in the case of an ALS trial) as a reference set and then applied to patients at study enrollment. As presented in Table 4, a study that enrolled only the 50% predicted to be the most likely to succumb to the disease would result in a study design with 199 fewer patients compared to a study with broad inclusion criteria. Alternately, in an approach discussed in the FDA guidance on ALS clinical trials,[72] a study could include a broader patient population, but the primary endpoint of the study could be designed *a priori* to include only the patients most likely to die during the course of the study. This latter design would be more expensive, but it would also be more inclusive, thus affording more patients the opportunity to be involved in a potentially beneficial study. It would also give very useful information on the safety profile of the test agent in a broader population and, while the broader group may not demonstrate statistically significant effect sizes, it may inform whether there are possible treatment effects outside of the group used in the primary analysis.

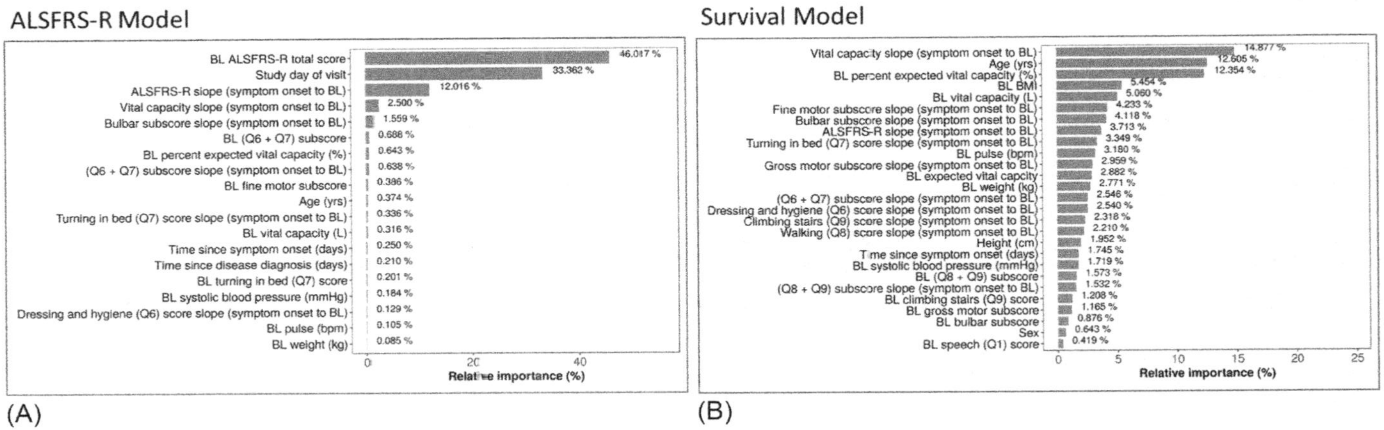

Fig. 6 Variable importance of features used in (A) ALSFRS-R model and (B) survival model.

Table 4 Enrichment. Summary of the addition of an enrichment strategy to simulations of randomization and covariate adjustment using predicted outcomes for ALS, Parkinson's, and Alzheimer's disease trials.

	ALS			**Parkinson's**	**Alzheimer's**
Endpoint	ALSFRS-R change	Survival	Vital capacity (L)	UPDRS	ADAS-cog
Simulated treatment effect	30% Slowing of decline	2.5-month extended life	30% Slowing of decline	40% Slowing of decline	40% Slowing of decline
PREDICTED OUTCOMES **used for randomization, covariate adjustment and enrichment**					
Randomization and covariate adjustment strata:	Predicted survival (tertiles)	Predicted survival (tertiles)	Predicted survival (tertiles)	Predicted Δ UPDRS 360 days (tertiles)	Predicted Δ ADAS-cog 360 days (tertiles)
Enrollment strata:	50% fastest ALSFRS-R	50% highest death risk	40% Average	50% fastest UPDRS	50% fastest ADAS-cog
Required sample size:	178 patients	264 patients	268 patients	216 patients	229 patients
Sample size reduction from best non-algorithmic simulation:	*−66 patients*	*−199 patients*	*−94 patients*	*−159 patients*	*−186 patients*
Sample size % reduction from best	*−27%*	*−43%*	*−26%*	*−42%*	*−45%*
Sample size % reduction from adding Enrollment strategy:	*−12%*	*−29%*	*−17%*	*−40%*	*−34%*

By way of illustration, another possible stratification protocol is depicted in Fig. 7. It is based on the sequential application of the survival and ALSFRS-R disease progression models to identify rapidly, slowly, and average progressing patients. Using a randomly selected group of 425 records from PRO-ACT that were not used in model training, we first predicted the 10% of patients most likely to die in one year using the survival model (Fig. 7, blue, fast progressors), followed by identifying the 25% remaining slowest predicted ALSFRS-R progressing patients at 1 year (Fig. 7, green, slow progressors). The remaining group (Fig. 7, red) we defined as the average progressors. Plotted in Fig. 7 are the actual Kaplan-Meier curves for the three groups.

The initial group of 425 patients that was used to conduct this in silico experiment had a median survival of 16.9 months and an average decline in the ALSFRS-R scale of 1.06 pts/mo. With a median survival of 15.9 months and an average ALSFRS-R decline of 1.18 pts/mo, the group we defined as average progressors ($n=286$) was very similar to the starting group. In contrast, the slowly progressing group did not reach a median survival and declined at 0.44 pts/mo while the rapidly progressing group had a median survival of 5.8 months and declined at 1.80 pts/mo.

We simulated two clinical trials (Table 5) that compared the initial group of 425 patients to the average selected group of patients in 1:1 RCT trials with a 30±10%

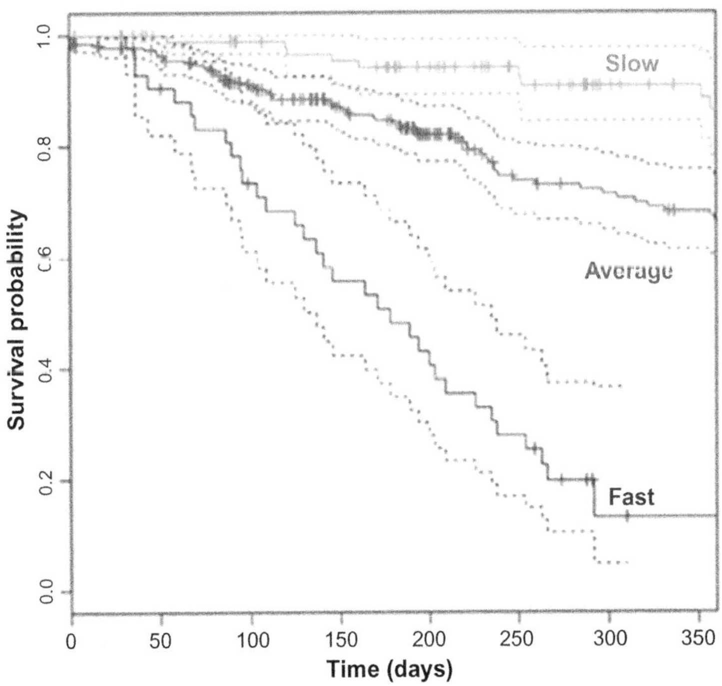

Fig. 7 Stratification of ALS patients using predicted survival and disease progression. 425 randomly selected PRO-ACT records were selected and sequentially stratified at day 0 by predicted survival (10% most likely to die, *blue*), predicted ALSFRS-R progression (25% slowest, *green*) and average progressors (remainder, *red*). The observed Kaplan-Meier survival curves of each group are plotted.

Table 5 Clinical trial simulation without (second column) and with (third column) stratification. The "Full Set," randomly drawn from PRO-ACT, was stratified into predicted slowly, average, and rapidly progressing subgroups and two simulated 1:1 RCTs each with a 30%±10% slower decline in ALSFRS-R for the "treatment arm" vs the "placebo arm" were modeled using the full set and the average progressing subgroup. In this hypothetical scenario, the trial that stratified for average progressors needed 139 fewer patients, saved \$6.1 million, was nearly a year shorter and came to a positive result.

Enrollment method:	Full set	Average progressors
Enrollable candidates:	425	286 (139 fewer)
Random 1:1 assignment to "treatment" or "placebo":	✓	✓
Percent ALSFRS-R improvement of "treatment" or "placebo":	30±10%	30±10%
T test	P = 0.361, ns	P = 0.024
Patient trial costs at \$43.5K/patient :	\$18.5M	\$12.4M
Patient trial cost savings:	–	\$6.1M
Trial enrollment period (12 treatments/month):	35.4 mo	23.8 mo
Time to marketing decision:	–	11.6 mo sooner

improvement in the rate of ALSFRS-R decline in the "treatment" vs the "placebo" arm. In this simulation, the trial built around the selected average progressing group of patients achieved significance while the original, heterogeneous starting group of 425 patients failed to reach significance. The trial using the selected average progressing patients included 139 fewer participants, was estimated to cost $6.1 million less, came to a decision nearly a year sooner, and, most importantly, came to a positive result.

Thus, this procedure effectively isolates two widely acknowledged confounders of ALS clinical trials (deaths and slowly progressing patients) while preserving the average characteristics of ALS patients as a group, the very definition of reducing heterogeneity in a clinical trial population. While this protocol has been developed for ALS, the concept of predicting key confounding features is readily extensible as a platform to other diseases, including diseases with similar drug development issues such as Alzheimer's, Parkinson's, and Huntington's diseases.

Virtual controls

An application that is intuitive is something we call "virtual controls" that we intend to use to compare arms in early trials that otherwise lack a control arm. We use the term to distinguish our approach from historical controls, the reason being that virtual controls represent the predicted counterfactuals for the participants in a trial. In contrast, historical controls are the real, observed outcomes of a group of patients who actually existed at a previous point in time.

We can assess the accuracy of the predictions using the residuals, that is, by subtracting the predicted values from the observed values (Fig. 8). With a perfect model, the predictions would be the same as the observations and the residuals would be zero. As shown in Fig. 8, the residuals are very close to zero at monthly time points through the first year following the baseline date. Root-mean-square deviation (RMSD) is a measure of the difference between the observed and predicted values. The lower the RMSD, the more accurate the model. Empirically, we have found that RMSDs less than 15% of the scale being measured are useful in drug development applications. In this case, this translates to RMSD values less than or equal to 7.2 for the 48-point ALSFRS-R scale. As can be seen, while the RMSDs creep up a bit with time, they remain below 7.2 and the average RMSD of the ALSFRS-R model is 4.732, less than 10% of the ALSFRS-R scale.

As evidenced by the closeness of the residuals to zero, the virtual control application can be used to estimate a treatment effect in early phase I or phase II clinical trials that lack control arms. The treatment effect is simply the residual at the end of the trial (Fig. 9). This is very useful, valuable information because it gives an estimate of the possible treatment effect in an actual clinical trial situation, not an extrapolation from in vitro or animal experiments. If clinical experts agree that the estimated treatment effect is great enough to confer clinical benefit, then the sponsor can proceed with the clinical development

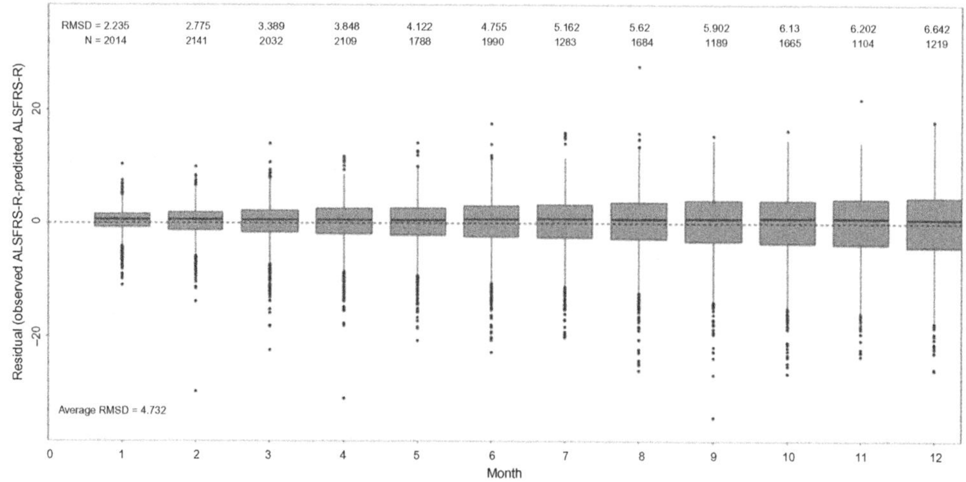

Fig. 8 Residuals for ALSFRS-R model through 12 months. For each boxplot, the top and bottom of the box correspond to the 75th and 25th percentiles, respectively, and the center line represents the 50th percentile. The lower and upper whiskers extend from the box to the smallest or largest value no further than 1.5 times the interquartile range (IQR). Data three times the IQR or more above the third quartile or three times the IQR or more below the first percentile are considered outliers. The boxplot displays the distribution of residuals between observations and predictions made by the model through 1 year. The root-mean-square deviation (RMSD) was calculated for each month and as an average over all observations.

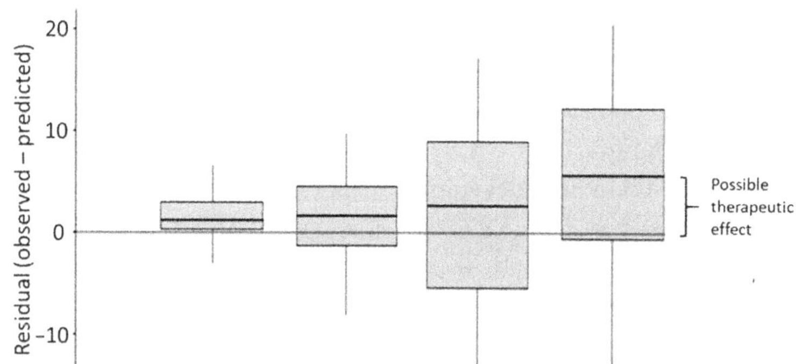

Fig. 9 Mock-up showing how residual might be used to determine a possible therapeutic effect. The residuals (observed minus predicted outcome) for a hypothetical 1-year study are plotted as boxplots. A possible therapeutic effect is simply the residual at the 1-year time point. This figure can be used to model a subsequent, definitive trial.

program by first simulating, then conducting a randomized controlled trial (RCT). In addition to early trials that lack a control arm, the virtual control may also be useful to compare the relative efficacy of doses, schedules, and formulations, or in device trials where it is difficult to conduct an RCT, in open label extensions, or in trials where it may be ethically challenging to have a true placebo. In all these situations, the virtual control can be used to inform decision-making for subsequent clinical trials.

The FDA has indicated to us that such uses in nonregistration trials can be helpful in planning registration trials. The FDA has also clearly indicated that they discourage trials where the control arm is based entirely on predictions, which they consider a type of historical control. A significant reason is the tendency for predicted outcomes to be subject to type 1 errors, also known as false positive findings. On an individual basis, a prediction can be thought of as the most likely outcome for that individual. However, some individuals may in fact be outliers. As a result, whereas an average prediction can be close to the observed average outcome for a group of patients, it is more difficult to generate an estimate of the variance, or noise. Whereas the tendency to generate false positives makes it difficult to use virtual controls to identify efficacious therapies with certainty, a negative result should cast doubt on the therapeutic and could be used as rationale for ending a project early.

In the recent guidance on ALS drug development,[72] the FDA stated that historical controls may be of value in future trial designs and they have indicated to us that they could see virtual controls being used to supplement small concurrent placebo arms. Generating useful variance estimates is an active area of research for us. The most likely initial uses of predicted outcomes as virtual controls in registration trials will be to boost the power of an RCT that has a small concurrent placebo arm. For the time being, virtual controls will be useful in providing an estimate of the treatment effect and in generally informing subsequent clinical development.

Detectable effect cluster (DEC) analysis

An ALS trial composed of participants with a wide range of disease progression profiles may be expected to have a high variance such that a positive signal among the treated patients cannot be detected over the noise. Thus, the conclusion may be reached that the drug does not work, when, in fact, the opposite is true. This false negative may cause the drug development program to be scrapped, resulting in the loss of hundreds of millions of dollars already invested, and, more significantly, a drug with potential to alleviate human suffering would be abandoned. Current subgroup analysis methods rely on parsing trial participants[101–106] based on discrete clinical characteristics, biomarkers, or covariates available at baseline with the aim of possibly identifying subgroups that display greater or lesser benefit relative to the full analysis set. The primary constraint with this approach is that the variance of highly heterogeneous and multifactorial diseases cannot be adequately captured using a small subset of baseline variables, hence these stratification methods fail to group patients into more homogeneous subgroups. Statisticians may, for example, define groups on the basis of different combinations of sex, age, and/or baseline clinical values. This approach is limited and inherently biased in that it depends on the opinion and ability of the statistician and clinical consultants to tease out subgroups with baseline differences. It is also laborious and time consuming in that for an analysis, a given parameter is varied while keeping other parameters constant (e.g., vital capacity can be varied in 5% increments beginning at 65% and going to 100%). The analysis is repeated by varying another parameter, and so on. Importantly, investigator bias in picking the parameters and ranges to test is an issue. The result is frequently a haphazard, incomplete, and biased analysis. Many possible subgroups are never analyzed due to time (including computational time) and cost constraints as well as investigator bias. With luck, a subgroup with a combination of treatment effect and variance that yields a statistically significant effect size may be discovered.

Our approach is radically different—using the outputs of our validated models, subgrouping is based on machine-learning generated predicted outcomes for the end of the study. We rank order all patients by predicted outcome, then, starting with small subgroups, systematically expand the patients included in each subgroup using nearest neighbors, i.e., the most similar patients, until all patients are included in the analysis. The result is a series of overlapping subgroups of increasing size that are defined by thresholds based on predicted outcomes. The method is computationally efficient and rapidly derived without investigator bias and ultimately includes all the nearest neighbor analyses possible for a given trial. Since the method groups patients that are most similar, it stands a good chance of generating subgroups with significantly less noise than the full analysis set, giving the therapy a better chance to demonstrate an effect size that will yield a significant p-value. This innovative approach to subgroup analyses can be used to methodically identify patient subgroups, "hot spots," within a failed clinical trial that could have formed the basis for a successful trial. We call this tool Detectable Effect Cluster (DEC) analysis.

DEC analysis has broad implications for the design and analysis of clinical trials. We have thus far focused on one of its applications, namely, subgroup analysis for the purpose of identifying patient subgroups within the context of a failed clinical trial that could have formed the basis for a successful trial had the trial been limited to the identified subgroup. This is essentially a form of "drug rescue" as the drug candidate would likely have been abandoned following the unsuccessful trial. But DEC analysis is not limited to a drug rescue scenario. It could, for example, be used with broad inclusion criteria in a placebo-controlled phase II trial primarily designed to test the safety of a new therapeutic and secondarily to identify a subgroup with a promising therapeutic profile. This application is well suited for use in a platform trial, a seamless trial, or an adaptive design trial. As in the rescue scenario, the identified subgroup would be used to define the inclusion criteria for a registration trial. This second scenario also satisfies the growing desire of both patients and the clinical trial community to design trials that are more inclusive. Finally, DEC analysis could be combined with "virtual controls" in a phase I or II safety study conducted without a control arm. In this case, the comparator for DEC analysis would be a virtual control composed of the predicted outcomes for each patient. This final scenario is quite speculative and prone to type 1 error, but it does point to the potential novel uses of DEC analysis. We believe that the drug rescue scenario is the most likely initial application of DEC analysis and we find significant interest in discussions with academic clinical trialists and commercial drug developers.

For DEC analysis, a regression or time-to-event model is used to rank order trial participants by predicted outcome (e.g., predicted change in ALSFRS-R at 1 year or predicted risk of time to 50% vital capacity, respectively). Starting with a small percentile of the study population, risk groups using upper and lower prediction thresholds are defined by systematically expanding each risk group in small increments until all participants are included in the cohort. For the 50 subgroups possible starting with 2% of the study population and expanding using the nearest additional 2%, a series of overlapping subgroups generate 1250 possible risk groups that can be analyzed. A matrix of upper versus lower thresholds is then plotted in which each block is derived using distinct upper and lower risk group percentile limits (Fig. 10). Each subgroup square includes patients who were assigned to the treatment or placebo arms, thus, for each subgroup created in this way, we perform a statistical analysis that can be plotted as a "heat map" to reveal combinations of

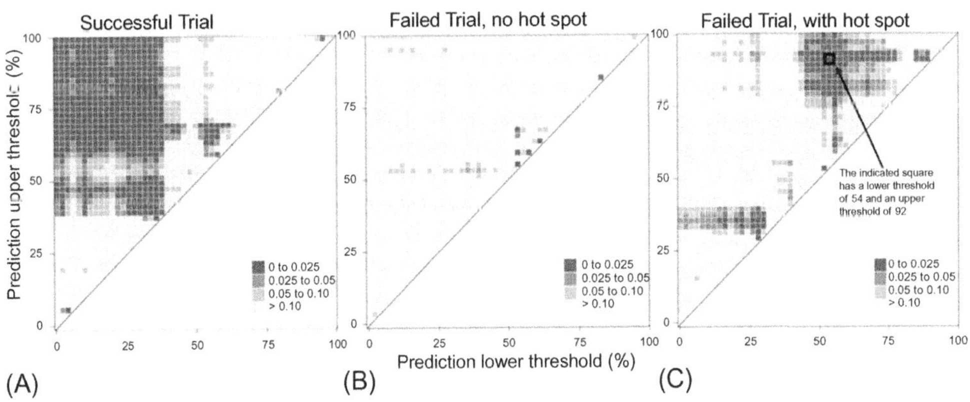

Fig. 10 Detectable effect cluster (DEC) analysis. Three trial scenarios were modeled to understand the range of DEC analysis use cases. This encompassed a range of scenarios including the application of a subthreshold treatment effect as well as a number of positive and negative controls. This figure shows three sample cases in which a significant, detectable treatment effect was applied (A), a subthreshold treatment was applied (B), and a scenario in which a universal treatment effect was applied but was only detectable within a narrow subpopulation with a decreased variance compared to the broader population (C).

treatment effect and root-mean-square error (RMSE) that result in statistically significant effect sizes ($P<0.05$).

Fig. 10 illustrates a number of concepts important to understanding DEC analysis. Fig. 10A shows a successful trial in that the upper left-hand square, representing 100% of the trial participants (i.e., the full analysis set, FAS), was statistically significant. It also illustrates that the trial was quite inefficient in that it could have succeeded with far fewer participants. Any of the red squares below and/or to the right of the upper left corner would have resulted in a positive trial (e.g., a subgroup with a lower percentile threshold of 25% and an upper of 75% would have been positive while including only half of the FAS). Fig. 10B, on the other hand, was a failure and only had small, scattered red squares, most likely due to random effects. In contrast, what we are looking for in a failed trial with a subgroup that demonstrated a significant treatment effect is depicted in Fig. 10C. This study was also a failure when the FAS is examined, but it demonstrated a robust hot spot with threshold percentile limits of 54% and 92% (38% of the FAS). This method presents the possibility to rescue a drug that otherwise would have been discarded. In this scenario, a confirmatory trial would be conducted using the thresholds identified in the earlier, failed trial as inclusion criteria. We have recently been awarded an NIH grant to test this methodology in a number of recent failed ALS clinical trials.

Conclusions and future directions

The applications outlined above have an impact on all phases of clinical drug development (Fig. 11). Virtual controls can begin to give clinical trialists an idea of whether a therapeutic is efficacious as early as during uncontrolled phase I and II studies. In addition, virtual controls will be useful in situations where a blinded clinical trial is difficult such as in some device studies, or ethically challenging as in some interventions that require difficult surgical procedures. Virtual controls will also be useful in the analysis of open-label extension studies and in phase IV post-marketing studies. The single most important use of the virtual control may very well be in estimating drug treatment effects from early,

	Phase I: Safety	Phase II: Safety dose finding efficacy	Phase III: Confirm efficacy & safety	Phase IV: Open Label extension
Virtual controls	✓	✓		✓
Enrichment	✓	✓	✓	
Randomization		✓(IIb)	✓	
Covariate adjustment		✓(IIb)	✓	
Subgroup analysis		✓(IIb)	✓	

Fig. 11 Impact of predictions on clinical development phases I through IV.

uncontrolled trials. Prediction-based improvements to trial enrichment, randomization, covariate adjustment, and subgroup analyses all depend on the ability of the models to stratify patients and will find their most important uses in phase II and III randomized controlled clinical trials.

The view into the future that these predictive algorithms provide is revolutionizing all aspects of drug development, including randomization, covariate adjustment, clinical trial simulation, virtual controls, and subgroup analysis. Trials designed using these tools are faster, cheaper, use fewer patients, and have greater power. Using these tools, clinical trials will reach better decisions faster while saving time, money, and patient resources. Progress in developing new therapies for ALS, neurodegenerative diseases, and multiple other indications will be dramatically improved as these methods are implemented in the conduct of clinical trials.

Acknowledgments

The Pooled Resource Open-Access ALS Clinical Trials Consortium Data used in the preparation of this article were obtained from the Pooled Resource Open-Access ALS Clinical Trials (PRO-ACT) Database. As such, the following organizations within the PRO-ACT Consortium contributed to the design and implementation of the PRO-ACT database and/or provided data, but did not participate in the analysis of the data or the writing of this report: (i) Neurological Clinical Research Institute, MGH; (ii) Northeast ALS Consortium; (iii) Novartis; (iv) Prize4Life; (v) Regeneron Pharmaceuticals, Inc.; (vi) Sanofi; and (vii) Teva Pharmaceutical Industries, Ltd.

References

1. Prize4Life Israel, Neurological Clinical Research Institute. Pooled Resource Open-Access ALS Clinical Trials Database. Accessed September 1, 2017 from Massachusetts General Hospital/NCRI website:https://nctu.partners.org/ProACT; 2017
2. Atassi N, Berry J, Shui A, et al. The PRO-ACT database. *Neurology*. 2014;83(19):1719–1725. https://doi.org/10.1212/WNL.0000000000000951.
3. Bio-IT World. Best Practices Awards Winners Named. Downloaded from Bio-IT World website on July 3, 2019: http://www.bio-itworld.com/2013/4/10/best-practices-winners.html; 2013.
4. Borfitz D. PRO-ACT: Bigger and Better ALS Database Open for Mining. *Clin Res News*. 2013; Retrieved July 2, 2019 from: http://www.clinicalinformaticsnews.com/clin/2013/5/10/pro-act-bigger-better-als-database-open-mining.html.
5. Smith S. ALS: offering prizes for a cure, Avi Kremer makes a business of survival. *New York Times*. May 30, 2007; Retrieved July 2, 2019 from: https://www.nytimes.com/2007/05/30/health/30iht-snals.5924646.html?pagewanted=all.
6. Prize4Life. About Prize4Life. Retrieved July 2, 2019 from: https://prize4life.org.il/en/about-prize4life/; 2019.
7. Howse D. Britain's Board of Longitude: the Finances, 1714–1828. *Mariner's Mirror*. 1998;84(4):400–417. https://doi.org/10.1080/00253359.1998.10656713.
8. Williams M. History of Challenges: The Orteig Prize (1919-1927). Retrieved July 2, 2019 from: https://www.herox.com/crowdsourcing-news/428-history-of-challenges-the-orteig-prize-1919-1927; 2019.
9. Howell E. History of Challenges: Ansari XPRIZE. Retrieved July 2, 2019 from: https://www.herox.com/crowdsourcing-news/324-history-of-challenges-ansari-xprize; 2019.

10. Tösche A, Jahrer M, Bell RM. The BigChaos Solution to the Netflix Grand Prize. Retrieved July 2, 2019 from: https://www.netflixprize.com/assets/GrandPrize2009_BPC_BigChaos.pdf; 2009.
11. InnoCentive. InnoCentive and Prize4Life Launch Prize Competition to Unlock Mysteries of Disease Progression in Amyotrophic Lateral Sclerosis (ALS) Patients. Downloaded from Innocentice Website on July 3, 2019. https://www.innocentive.com/innocentive-and-prize4life-launch-prize-competition-to-unlock-mysteries-of-disease-progression-in-amyotrophic-lateral-sclerosis-als-patients/; 2012.
12. Roush W. *InnoCentive Raises $6.5 Million for Innovation Network: "Ready for Prime Time," says CEO in Our Q&A*. xconomy, May 7.
13. Spieth PM, Kubasch AS, Penzlin AI, Illigens BM-W, Barlinn K, Siepmann T. Randomized controlled trials—a matter of design. *Neuropsych Dis Treat*. 2016;12:1341–1349. https://doi.org/10.2147/NDT.S101938.
14. Bhatt A. Evolution of clinical research: a history before and beyond James Lind. *Perspect Clin Res*. 2010;1:6–10. 21829774.
15. Habib AA, Mitsumoto H. Emerging drugs for amyotrophic lateral sclerosis. *Expert Opin Emerg Drugs*. 2011;16:537–558. https://doi.org/10.1517/14728214.2011.604312.
16. Mitsumoto H, Brooks BR, Silani V. Clinical trials in amyotrophic lateral sclerosis: why so many negative trials and how can trials be improved? *Lancet Neurol*. 2014;13:1127–1138. https://doi.org/10.1016/S1474-4422(14)70129-2.
17. Petrov D, Mansfield C, Moussy A, Hermine O. ALS clinical trials review: 20 years of failure. Are we any closer to registering a new treatment? *Front Aging Neurosci*. 2017;9:68–78. https://doi.org/10.3389/fnagi.2017.00068.
18. Fournier C, Glass JD. Modeling the course of amyotrophic lateral sclerosis. *Nat Biotechnol*. 2015;33(1):45–47. https://doi.org/10.1038/nbt.3118.
19. Küffner R, Zach N, Norel R, et al. Crowdsourced analysis of clinical trial data to predict amyotrophic lateral sclerosis progression. *Nat Biotechnol*. 2015;33(1):51–57. https://doi.org/10.1038/nbt.3051.
20. Swinnen B, Robberecht W. The phenotypic variability of amyotrophic lateral sclerosis. *Nat Rev Neurol*. 2014;10(11):661–670. https://doi.org/10.1038/nrneurol.2014.184.
21. Cedarbaum JM, Stambler N. Performance of the ALS Functional Rating Scale (ALSFRS) in multicenter clinical trials. *J Neurol Sci*. 1997;152(Suppl l):1–9. https://doi.org/10.1016/S0022-510X(97)00237-2.
22. Chipman HA, Georg EI, McCulloch RE. BART: Bayesian additive regression trees. *Ann Appl Stat*. 2010;4(1):266–298. https://doi.org/10.1214/09-aoas285.
23. Breiman L. Random forests. *Mach Learn J*. 2001;45:5–32. https://doi.org/10.1023/A:1010933404324.
24. Hothorn T, Jung HH. RandomForest4Life: a random forest for predicting ALS disease progression. *Amyotroph Lateral Scler Frontotemporal Degener*. 2014;15(5-6):444–452. https://doi.org/10.3109/21678421.2014.893361.
25. Li G. Discovering new variables and improving the prediction of ALS progression. In: *Presented at the 2012 RECOMB Conference on Regulatory and Systems Genomics, November 13, 2012*; 2012 Available at the ISCB website (search Guang Li): https://www.iscb.org/recomb-mm/media-recombrsg2012.
26. Origent. Sentrana Researchers Presented Their Winning Research From the Prize4Life ALS Prediction Prize Challenge. Downloaded July 3, 2019 from Origent Data Sciences website: https://www.origent.com/sentrana-presents-at-recomb-conference/; 2013.
27. InnoCentive. Prize4Life Announces $50,000 ALS Prediction Prize Winners. Downloaded from Innocentice website on July 3, 2019. https://www.innocentive.com/2819-2/; 2012.
28. DC Economic Partnership. Data Science and Analytics. https://wdcep.com/dc-industries/data-science-analytics/; 2019. Accessed 4 July 2019.
29. Cushman & Wakefield. Tech Cities 2.0. Downloaded July 4, 2019 from Cushman & Wakefield website:http://info.cushmanwakefield.com/l/263412/2018-09-26/b39wj?_ga=2.71769814.433233501.1562253911-1238816938.1562253911; 2018.
30. Berry JD, Taylor AA, Beaulieu D, et al. Improved stratification of ALS clinical trials using predicted survival. *Ann Clin Transl Neurol*. 2018;5(4):474–485. https://doi.org/10.1002/acn3.550.
31. Jahandideh S, Taylor AA, Beaulieu D, et al. Longitudinal modeling to predict vital capacity in amyotrophic lateral sclerosis. *Amyotroph Lateral Scler Frontotemporal Degener*. 2018;19:294–302. https://doi.org/10.1080/21678421.2017.1418003.

32. Nicholson K, Chan J, Macklin EA, et al. Pilot trial of inosine to elevate urate levels in amyotrophic lateral sclerosis. *Ann Clin Transl Neurol.* 2018; https://doi.org/10.1002/acn3.671 Published online Nov 12.
33. Schoenfeld DA, Finkelstein DM, Macklin E, et al. Design and analysis of a clinical trial using previous trials as historical control. *Clin Trials.* 2019; https://doi.org/10.1177/1740774519858914 First Published online 1 Jul 2019.
34. Taylor AA, Fournier C, Polak M, et al. Predicting disease progression in amyotrophic lateral sclerosis. *Ann Clin Transl Neurol.* 2016;3:866–875. https://doi.org/10.1002/acn3.348.
35. Taylor AA, Fournier C, Polak M, et al. Predicting disease progression for ALS clinic patients. In: *27th International Symposium on ALS/MND, 2016 December 9, Dublin, Ireland*, vol. 6. F1000Research; 2016:4. poster. https://doi.org/10.7490/f1000research.1113586.1).
36. Zach N, Ennist DL, Taylor AA, et al. Being PRO-ACTive: what can a clinical trial database reveal about ALS? *Neurotherapeutics.* 2015;12(2):417–423. https://doi.org/10.1007/s13311-015-0336-z.
37. Beaulieu D, Taylor AA, Conklin A, Cuerdo J, Keymer M, Ennist DL. Increasing study power using a machine learning approach. In: *Joint Statistical Meeting (JSM) of the American Statistical Association (ASA) in Vancouver, British Columbia, Canada, July 30, 2018*, vol. 7. F1000Research; 2018:1785. poster. https://doi.org/10.7490/f1000research.1116287.1.
38. Beaulieu D, Taylor AA, Cuerdo J, Conklin A, Keymer M, Ennist DL. Machine learning applications for increasing the efficiency of ALS clinical trials. In: *17th Annual NEALS Meeting, Tampa, FL, USA, October 3, 2018*, vol. 7. F1000Research; 2018:1782. poster. https://doi.org/10.7490/f1000research.1116286.1.
39. Beaulieu D, Cuerdo J, Taylor AA, et al. Validation of a suite of machine learning models using the longitudinal VITALITY-ALS data set. In: *29th International Symposium on ALS/MND; Glasgow, Scotland; Dec 5, 2018.* vol. 7; 2018:1944. F1000Research. poster. https://doi.org/10.7490/f1000research.1116358.1.
40. Beaulieu D, Taylor AA, Conklin A, et al. Detectable effect cluster analysis: a novel machine-learning based clinical trial subgroup analysis tool. In: *18th Annual NEALS Meeting, Tampa, FL, USA, October 2019*, F1000Research; 2019 submitted (poster). https://f1000research.com/posters/8-1883.
41. Bedlack RS, Ennist DL, Taylor A, Keymer M. ALS resistance is regional and not explained by demographics, medications or labs. In: *27th International Symposium on ALS/MND, 2016 December 9, Dublin, Ireland*, vol. 5: F1000Research; 2016:2917. poster. https://doi.org/10.7490/f1000research.1113569.1.
42. Fournier CN, Taylor AA, Ennist DL. Enriched clinical trial cohorts improve study power. In: *16th Annual NEALS Meeting, Tampa, FL, USA, October 4, 2017*, vol. 6. F1000Research; 2017:1868. poster. https://doi.org/10.7490/f1000research.1114995.1.
43. Jahandideh S, Taylor AA, Bian A, et al. Machine learning model for the prediction of slow vital capacity. In: *27th International Symposium on ALS/MND, 2016 December 9, Dublin, Ireland*, vol. 6. F1000Research; 2016:8. poster https://doi.org/10.7490/f1000research.1113590.1.
44. Jahandideh S, Ennist DL. Machine learning models for the assessment of potential ALS biomarkers. In: *28th International Symposium on ALS/MND, 2017 December 9, Boston, MA*, vol. 7. F1000Research; 2018:322. poster. https://doi.org/10.7490/f1000research.1115313.1.
45. Miller RG, Ennist DL, Jenkins L, Thompson J, Fritz M, Moore DH. Novel trial design in a clinical study of Diaphragm Pacing (DPS) for ALS. In: *27th International Symposium on ALS/MND, 2016 December 9, Dublin, Ireland*, vol. 6. F1000Research; 2016:9. poster. https://doi.org/10.7490/f1000research.1113591.1.
46. Schoenfeld D, Küffner R, Macklin E, et al. The proper use of historical controls in ALS trials. In: *27th International Symposium on ALS/MND, 2016 December 9, Dublin, Ireland*, vol. 5. F1000Research; 2016:2904. poster. https://doi.org/10.7490/f1000research.1113562.1.
47. Taylor AA, Miller R, Onders R, Ennist DL. Analysis of function and survival in ALS patients with diaphragm pacing using virtual controls. In: *26th International Symposium on ALS/MND, 2015 December 11, Orlando, FL, USA*, vol. 5. F1000 Research; 2015:120. poster. https://doi.org/10.7490/f1000research.1111282.1).
48. Taylor AA, Jahandideh S, Beaulieu D, Keymer M, Ennist DL. In silico stratification of ALS patients using machine learning algorithms. In: *27th International Symposium on ALS/MND, 2016 December 9, Dublin, Ireland*, vol. 6. F1000Research; 2017:3. poster. https://doi.org/10.7490/f1000research.1113585.1.

49. Taylor AA, Jahandideh S, Beaulieu D, Keymer M, Ennist DL. Machine learning models for the clinical development of gene and cell therapies. In: *20th Annual Meeting of the American Society for Cell and Gene Therapy, May 10, 2017*, vol. 6. F1000Research; 2017:1866. poster. https://doi.org/10.7490/f1000research.1114991.1).
50. Taylor AA, Beaulieu D, Jahandideh S, et al. Validation of predictive ALS machine learning models with a contemporary, external dataset and application to trial simulations. In: *16th Annual NEALS Meeting, Tampa, FL, USA, October 4, 2017*, vol. 6. F1000Research; 2017:1867. poster (also presented at ALS/MND 2017 in Boston and AAN 2018 in LA). https://doi.org/10.7490/f1000research.1114992.1.
51. Taylor AA, Beaulieu D, Jahandideh S, et al. Validation of predictive ALS machine learning models with a contemporary, external dataset and application to trial simulations. In: *28th International Symposium on ALS/MND, 2017 December 9, Boston, MA*, vol. 7. F1000Research; 2018:322. poster (also presented at NEALS 2017 in Boston and AAN 2018 in LA. https://doi.org/10.7490/f1000research.1115313.1.
52. Beaulieu-Jones BK, Moore JH. Missing data imputation in the electronic health record using deeply learned autoencoders. In: Altman RB, et al. ed. *Pacific Symposium on Biocomputing 2017*: 2017:207–218.
53. Elamin M, BEDE P, Montuschi A, Pender N, Chio A, Hardiman O. Predicting prognosis in amyotrophic lateral sclerosis: a simple algorithm. *J Neurol*. 2015;262(6):1447–1454. https://doi.org/10.1007/s00415-015-7731-6.
54. Gomeni R, Fava M, Pooled Resource Open-Access ALS Clinical Trials Consortium. Amyotrophic lateral sclerosis disease progression model. *Amyotroph Lateral Scler Frontotemporal Degener*. 2014;15(1–2):119–129.
55. Haverkamp LJ, Appel V, Appel SH. Natural history of amyotrophic lateral sclerosis in a database population. Validation of a scoring system and a model for survival prediction. *Brain*. 1995;118:707–719. https://doi.org/10.1093/brain/118.3.707.
56. Huang Z, Zhang H, Boss J, et al. Complete hazard ranking to analyze right-censored data: an ALS survival study. *PLoS Comput Biol*. 2017;13(12):e1005887. https://doi.org/10.1371/journal.pcbi.1005887.
57. Kaufmann P, Levy G, Thompson JL, et al. The ALSFRS-R predicts survival time in an ALS clinic population. *Neurology*. 2005;64(1):38–43. https://doi.org/10.1212/01.WNL.0000148648.38313.64.
58. Kimura F, Fujimura C, Ishida S, et al. Progression rate of ALSFRS-R at time of diagnosis predicts survival time in ALS. *Neurology*. 2006;66(2):265–267. https://doi.org/10.1212/01.wnl.0000194316.91908.8a.
59. Kollewe K, Mauss U, Krampfl K, Petri S, Dengler R, Mohammadi B. ALSFRS-R score and its ratio: a useful predictor for ALS-progression. *J Neurol Sci*. 2008;275(1-2):69–73. https://doi.org/10.1016/j.jns.2008.07.016.
60. Labra J, Menon P, Byth K, Morrison S, Vucic S. Rate of disease progression: a prognostic biomarker in ALS. *J Neurol Neurosurg Psychiatry*. 2016;87(6):628–632. https://doi.org/10.1136/jnnp-2015-310998.
61. Lo Coco D, Marchese S, La Bella V, Piccoli T, Lo Coco A. The amyotrophic lateral sclerosis functional rating scale predicts survival time in amyotrophic lateral sclerosis patients on invasive mechanical ventilation. *Chest*. 2007;132(1):64–69. https://doi.org/10.1378/chest.06-2712.
62. Lunetta C, Lizio A, Melazzini MG, Maestri E, Sansone VA. Amyotrophic Lateral Sclerosis Survival Score (ALS-SS): a simple scoring system for early prediction of patient survival. *Amyotroph Lateral Scler Frontotemporal Degener*. 2015;17(1-2):93–100. https://doi.org/10.3109/21678421.2015.1083585.
63. Magnus T, Beck M, Giess R, Puls I, Naumann M, Toyka KV. Disease progression in amyotrophic lateral sclerosis: predictors of survival. *Muscle Nerve*. 2002;25(5):709–714. https://doi.org/10.1002/mus.10090.
64. Ong M-L, Tan PF, Holbrook JD. Predicting functional decline and survival in amyotrophic lateral sclerosis. *PLoS One*. 2017;12(4):e0174925.
65. Simon NG, Turner MR, Vucic S, et al. Quantifying disease progression in amyotrophic lateral sclerosis. *Ann Neurol*. 2014;76(5):643–657. https://doi.org/10.1002/ana.24273.
66. Watanabe H, Atsuta N, Nakamura R, et al. Factors affecting longitudinal functional decline and survival in amyotrophic lateral sclerosis patients. *Amyotroph Lateral Scler Frontotemporal Degener*. 2015;16(3-4):230–236. https://doi.org/10.3109/21678421.2014.990036.
67. Westeneng HJ, Debray TPA, Visser AE, et al. Prognosis for patients with amyotrophic lateral sclerosis: development and validation of a personalised prediction model. *Lancet Neurol*. 2018; https://doi.org/10.1016/S1474-4422(18)30089-9. pii: S1474-4422(18)30089-9.

68. Cedarbaum JM, Stambler N, Malta E, et al. The ALSFRS-R: a revised ALS functional rating scale that incorporates assessments of respiratory function. *J Neurol Sci.* 1999;169(1-2):13–21. https://doi.org/10.1016/s0022-510x(99)00210-5.
69. Niedermeyer S, Murn M, Choi PJ. Respiratory failure in amyotrophic lateral sclerosis. *Chest.* 2019;155(2):401–408. https://doi.org/10.1016/j.chest.2018.06.035.
70. ClinicalTrials.gov. A Study to Evaluate Efficacy, Safety and Tolerability of CK-2127107 in Patients With Amyotrophic Lateral Sclerosis (ALS). Retrieved July 5, 2019 from clinicaltrials.gov website: https://clinicaltrials.gov/ct2/show/NCT03160898?term=reldesemtiv&cond=als&rank=1; 2019.
71. Shefner JM, Cudkowicz ME, Hardiman O, et al. A phase III trial of tirasemtiv as a potential treatment for amyotrophic lateral sclerosis. *Amyotroph Lateral Scler Frontotemporal Degener.* 2019;1–9. https://doi.org/10.1080/21678421.2019.1612922.
72. FDA. Amyotrophic Lateral Sclerosis: Developing Drugs for Treatment Guidance for Industry. Guidance for Industry. Retrieved October 15, 2019 from FDA website: https://www.fda.gov/media/130964/download; 2019.
73. Heiman-Patterson TD, Cudkowicz ME, De Carvalho M, et al. Understanding the use of NIV in ALS: results of an international ALS specialist survey. *Amyotroph Lateral Scler Frontotemporal Degener.* 2018;19(5-6):331–341. https://doi.org/10.1080/21678421.2018.1457058.
74. Paganoni S, Cudkowicz M, Berry JD. Outcome measures in amyotrophic lateral sclerosis clinical trials. *Clin Investig (Lond).* 2014;4(7):605–618. https://doi.org/10.4155/CLI.14.52.
75. Abe K, Itoyama Y, Sobue G, et al. Confirmatory double-blind, parallel-group, placebo-controlled study of efficacy and safety of edaravone (MCI-186) in amyotrophic lateral sclerosis patients. *Amyotroph Lateral Scler Frontotemporal Degener.* 2014;15(7-8):610–617. https://doi.org/10.3109/21678421.2014.959024.
76. Miller RG, Mitchell JD, Moore DH. Riluzole for amyotrophic lateral sclerosis (ALS)/motor neuron disease (MND). *Cochrane Database Syst Rev.* 2012;(3):CD001447. https://doi.org/10.1002/14651858.CD001447.pub3.
77. Beghi E, Pupillo E, Bonito V, et al. Randomized double-blind placebo-controlled trial of acetyl-L-carnitine for ALS. *Amyotroph Lateral Scler Frontotemporal Degener.* 2013;14(5-6):397–405. https://doi.org/10.3109/21678421.2013.764568.
78. Andrews JA, Cudkowicz ME, Hardiman O, et al. VITALITY-ALS, a phase III trial of tirasemtiv, a selective fast skeletal muscle troponin activator, as a potential treatment for patients with amyotrophic lateral sclerosis: study design and baseline characteristics. *Amyotroph Lateral Scler Frontotemporal Degener.* 2018;19(3-4):259–266. https://doi.org/10.1080/21678421.2018.1426770 Erratum in: Amyotroph Lateral Scler Frontotemporal Degener. 2018 May;19(3-4):321. doi: 10.1080/21678421.2018.1446685.
79. Shefner JM, Wolff AA, Meng L, et al. A randomized, placebo-controlled, double-blind phase IIb trial evaluating the safety and efficacy of tirasemtiv in patients with amyotrophic lateral sclerosis. *Amyotroph Lateral Scler Frontotemporal Degener.* 2016;17:426–435. https://doi.org/10.3109/21678421.2016.1148169.
80. Friedman JH. Greedy function approximation: a gradient boosting machine. *Ann Stat.* 2001;29:1189–1232. https://doi.org/10.1214/aos/1013203451.
81. Natekin A, Knoll A. Gradient boosting machines, a tutorial. *Front Neurorobot.* 2013;7. https://doi.org/10.3389/fnbot.2013.00021.
82. Hastie T, Tibshirani R, Friedman J. *The Elements of Statistical Learning: Data Mining, Inference, and Prediction.* 2nd ed. New York: Springer; 2009. Available at: https://web.stanford.edu/~hastie/Papers/ESLII.pdf.
83. Steyerberg EW. *Clinical Prediction Models: A Practical Approach to Development, Validation and Updating.* New York: Springer; 2009.
84. Taylor AA, Beaulieu D, Jahandideh S, et al. Validation of predictive ALS machine learning models with a contemporary, external dataset and application to trial simulations. *F1000 Res.* 2018;7:322. https://doi.org/10.7490/f1000research.1115313.1. https://f1000research.com/posters/7-322.
85. Armitage P. Fisher, Bradford Hill, and Randomization. *Int J Epidemiol.* 2003;32:925–928. Discussion 945-948. https://doi.org/10.1093/ije/dyg286.
86. FDA. Adjusting for Covariates in Randomized Clinical Trials for Drugs and Biologics With Continuous Outcomes. Guidance for Industry. DRAFT GUIDANCE. Retrieved October 20, 2019 from FDA website: https://www.fda.gov/media/123801/download; 2019.

87. Traxinger K, Kelly C, Johnson BA, Lyles RH, Glass J. Prognosis and epidemiology of amyotrophic lateral sclerosis: analysis of a clinic population, 1997–2011. *Neurol Clin Pract.* 2013;3:313–320. https://doi.org/10.1212/CPJ.0b013e3182a1b8ab.
88. Aggarwal SP, Zinman L, Simpson E, et al. Safety and efficacy of lithium in combination with Riluzole for treatment of amyotrophic lateral sclerosis: a randomised, double-blind, placebo-controlled trial. *Lancet Neurol.* 2010;9:481–488. https://doi.org/10.1016/S1474-4422(10)70068-5.
89. Cudkowicz ME, Shefner JM, Schoenfeld DA, et al. Trial of celecoxib in amyotrophic lateral sclerosis. *Ann Neurol.* 2006;60:22–31. https://doi.org/10.1002/ana.20903.
90. Cudkowicz ME, Shefner JM, Schoenfeld DA, et al. A randomized, placebo-controlled trial of topiramate in amyotrophic lateral sclerosis. *Neurology.* 2003;61:456–464. https://doi.org/10.1212/WNL.61.4.456.
91. Jacobs TL, Brown DL, Baek J, Migda EM, Funckes T, Gruis KL. Trial of early noninvasive ventilation for ALS: a pilot placebo-controlled study. *Neurology.* 2016; https://doi.org/10.1212/WNL.0000000000003158.
92. Miller RG, Moore 2nd DH, Gelinas DF, et al. Phase III randomized trial of gabapentin in patients with amyotrophic lateral sclerosis. *Neurology.* 2001;56:843–848. https://doi.org/10.1212/WNL.56.7.843-a.
93. Ryberg H, Askmark H, Persson LI. A double-blind randomized clinical trial in amyotrophic lateral sclerosis using lamotrigine: effects on CSF glutamate, aspartate, branched-chain amino acid levels and clinical parameters. *Acta Neurol Scand.* 2003;108:1–8. https://doi.org/10.1034/j.1600-0404.2003.00111.x.
94. Scelsa SN, MacGowan DJ, Mitsumoto H, et al. A pilot, double-blind, placebo-controlled trial of indinavir in patients with ALS. *Neurology.* 2005;64:1298–1300. https://doi.org/10.1212/01.WNL.0000156913.24701.72.
95. Shefner JM, Cudkowicz ME, Schoenfeld D, et al. A clinical trial of creatine in ALS. *Neurology.* 2004;63:1656–1661. https://doi.org/10.1212/01.WNL.0000142992.81995.F0.
96. Nagata E, Ogino M, Iwamoto K, et al. ALS Consortium Investigators. Bromocriptine mesylate attenuates amyotrophic lateral sclerosis: a phase 2a, randomized, double-blind, placebo-controlled research in Japanese patients. *PLoS One.* 2016;11(2):e0149509. https://doi.org/10.1371/journal.pone.0149509.
97. Cudkowicz ME, van den Berg LH, Shefner JM, et al. Dexpramipexole versus placebo for patients with amyotrophic lateral sclerosis (EMPOWER): a randomised, double-blind, Phase III trial. *Lancet Neurol.* 2013;12:1059–1067. https://doi.org/10.1016/s1474-4422(13)70221-7.
98. Miller R, Bradley W, Cudkowicz M, et al. Phase II/III randomized trial of TCH346 in patients with ALS. *Neurology.* 2007;69:776–784. https://doi.org/10.1212/01.wnl.0000269676.07319.09.
99. PhRMA. Biopharmaceutical Industry-Sponsored Clnical Trials: Growing State Economies. Retrieved October 29, 2019 from PhRMA website: https://www.phrma.org/-/media/TEConomy_PhRMA-Clinical-Trials-Impacts.pdf; 2019.
100. Berry JD, Miller R, Moore DH, et al. The combined assessment of function and survival (CAFS): a new endpoint for ALS clinical trials. *Amyotroph Lateral Scler Frontotemporal Degener.* 2013;14:162–168. https://doi.org/10.3109/21678421.2012.762930.
101. Brookes ST, Whitely E, Egger M, Smith GD, Mulheran PA, Peters TJ. Subgroup analyses in randomized trials: risks of subgroup-specific analyses; power and sample size for the interaction test. *J Clin Epidemiol.* 2004;57(3):229–236. https://doi.org/10.1016/j.jclinepi.2003.08.009.
102. EMA. Guideline on the Investigation of Subgroups in Confirmatory Clinical Trials. Retrieved from EMA website Dec 30, 2018. Available at: https://www.ema.europa.eu/documents/scientific-guideline/draft-guideline-investigation-subgroups-confirmatory-clinical-trials en.pdf; 2014.
103. FDA. Guidance for Industry: E9 Statistical Principles for Clinical Trials. Retrieved from FDA website Jan 2, 2019. Available at: https://www.fda.gov/downloads/drugs/guidancecomplianceregulatoryinformation/guidances/ucm073137.pdf; 1998.
104. FDA. Guidance for Industry: Enrichment Strategies for Clinical Trials to Support Approval of Human Drugs and Biological Products. Retrieved December 15, 2018. Available at: https://www.fda.gov/downloads/drugs/guidancecomplianceregulatoryinformation/guidances/ucm332181.pdf; 2012.
105. Ruberg SJ, Shen L. Personalized medicine: four perspectives of tailored medicine. *Statist Biopharm Res.* 2015;7:214–229. https://doi.org/10.1080/19466315.2015.1059354.
106. Tanniou J, van der Tweel I, Teerenstra S, Roes KCB. Subgroup analyses in confirmatory trials: time to be specific about their purposes. *BMC Med Res Methodol.* 2016;16:20–34. https://doi.org/10.1186/s12874-016-0122-6.

CHAPTER 15

Positron emission tomography imaging agents for evaluating the pathologic features of Alzheimer's disease and drug development

Yiyun Henry Huang[a], Hongmei Jia[b], and Weiqi Bao[c]
[a]PET Center, Department of Radiology and Biomedical Imaging, Yale University School of Medicine, New Haven, CT, United States
[b]Ministry of Education Key Laboratory of Radiopharmaceuticals, College of Chemistry, Beijing Normal University, Beijing, China
[c]PET Center, Huashan Hospital, Fudan University, Shanghai, China

Introduction

Alzheimer's disease (AD) is a debilitating progressive neurodegenerative disorder prevalent in older populations. According to the World Alzheimer's Report 2016, 47 million people live with dementia worldwide.[1] This number is likely to increase to about 152 million by 2050. Alzheimer's disease (AD) is the most prevalent cause of dementia, accounting for 60%–80% of all dementia cases.[2] Dementia exerts a tremendous economic burden on societies. In 2018, the global societal cost of dementia was estimated to reach 1 trillion US dollars a year. This is expected to double by 2030.[3]

AD is characterized histopathologically by the presence and abundance of extracellular amyloid-beta (Aβ) plaques and intracellular neurofibrillary tangles (NFTs) composed of hyperphosphorylated tau proteins in brain tissue. However, the molecular pathology of AD is complex and many hypotheses regarding its etiology and pathogenesis have been proposed. In past decades, the amyloid cascade hypothesis,[4] and later the amyloid-β oligomer (AβO) hypothesis, dominated thinking about the disorder (for reviews, see Refs. 5,6). Subsequently, the tau protein hypothesis has captured significant attention.[7] Moreover, the cholinergic[8–11] and inflammation hypotheses,[12] and most recently, the mitochondrial cascade[13] and mitochondria-associated endoplasmic reticulum (ER) membrane (MAM) hypotheses[14] have been put forward. Accumulating data have indicated that the cognitive decline in AD occurs continuously over a long period of time and that progression of AD pathologic biomarkers is also a continuous process that begins long before clinical symptoms emerge. Therefore, the current prevailing theory considers AD as a continuum rather than three distinct clinically defined entities,

Neurotherapeutics in the Era of Translational Medicine
https://doi.org/10.1016/B978-0-12-816475-4.00013-6

i.e., preclinical AD, mild cognitive impairment (MCI), and dementia as proposed in 2011.[15–17]

Radioligand-based molecular imaging techniques such as positron emission tomography (PET) imaging offer a means for noninvasive in vivo assessment of diverse protein target expression and function in living subjects. With the aid of specific radioligands, PET imaging can provide a quantitative measurement of different biomarkers in AD. Thus, applications of PET imaging biomarkers make it possible to visualize and quantify AD pathophysiological processes, to monitor AD progression, and to accelerate AD drug development.[18–21]

Recently, the United States National Institute of Aging (NIA) and the Alzheimer's Association (AA) published a research framework for AD, i.e., the NIA-AA Research Framework, which is based on the current understanding of the AD continuum and grounded on a biomarker-based definition of AD.[22] Biomarkers are grouped into three types: Aβ deposition, pathologic tau, and neurodegeneration [AT(N)] based on the nature of the pathologic process that each measures. Biomarkers of Aβ plaques (labeled "A") are represented by cortical amyloid PET ligand binding or decreased cerebrospinal fluid (CSF) $A\beta_{42}$. Biomarkers of fibrillar tau (labeled "T") are associated with elevated CSF phosphorylated tau (p-tau) and cortical tau PET ligand binding. Biomarkers of neurodegeneration or neuronal injury [labeled "(N)"] are correlated with CSF total tau (t-tau), hypometabolism as measured by PET using [^{18}F]fluorodeoxyglucose ([^{18}F]FDG, or FDG), and brain atrophy detected on magnetic resonance imaging (MRI). The AT(N) system is flexible in that new biomarkers can be added to the three existing AT(N) groups, and new biomarker groups beyond AT(N) can also be added when they become available and validated. The AT(N) criteria allow the staging of AD based solely on biomarkers instead of the clinical consequences of the disease (i.e., symptoms/signs). Owing to the fact that biomarkers are the best window to look into the biology of AD pathologic changes, the NIA-AA Research Framework facilitates the consideration of target engagement and trial outcomes and provides guidance for uniform reporting of biological outcomes in clinical trials. This is critical to building an understanding of the relationship between drug mechanisms and their biological consequences.[23]

In this chapter, after a brief discussion on the criteria for effective radiotracers for brain imaging, we review the development of unique imaging agents for evaluating the pathologic features of AD, tracking AD progression, and monitoring therapeutic interventions and facilitating drug development based on the AT(N) research framework.

Characteristics of an effective brain PET imaging agent

Effective PET radioligands should be able to provide reliably measurable binding signals in vivo within a reasonable scan time (e.g., 120 min for ^{11}C; 240 min for ^{18}F). Due to the usually low capacity of the target protein in the brain and presence of the blood-brain barrier (BBB), the development of effective PET radioligands for imaging target proteins in the brain is especially challenging, because of the demanding, and sometimes conflicting requirements that have to be fulfilled by the candidate radioligand. Nonetheless, the important requisite properties for an effective brain imaging agent have been well recognized and established in the field of PET radioligand development (for reviews, see Refs. 24–26). These requirements include (1) appropriate binding affinity (K_i 0.01–10 nM based on the density of the target protein) and high binding selectivity (e.g., >100 times) for the target; (2) feasible radiochemistry to enable preparation of the radioligand in high yield using limited synthetic steps; (3) ability to cross the BBB into the brain (moderate lipophilicity without efflux transporter substrate behavior) and achieve adequate levels of brain uptake (e.g., standardized uptake value (SUV) > 1); (4) appropriate metabolic profile with no production of radioactive metabolites able to cross into the brain and confound the binding signals; (5) appropriate pharmacology (e.g., saturable and displaceable binding and appropriate distribution pattern in the brain that reflects the regional expression levels of the target); (6) acceptable toxicology and dosimetry profile without production of adverse effects by the radioligand at the administered dose; (7) appropriate pharmacokinetic profile in the brain (i.e., reversible binding characteristics observed within the usable time window of the radioligand to enable reliable measurement of physiological parameters from imaging data through quantitative kinetic modeling); (8) low nonspecific binding and good specific binding signals in the brain (e.g., displaying sufficiently high (e.g., >0.5) regional nondisplaceable binding potential (BP_{ND}) to allow for reliable quantification); (9) sufficiently high free fraction of the radioligand in plasma (f_P) to allow for reliable measurement.

PET imaging agents for AD biomarkers

Imaging agents for β-amyloid

Description of the agents

Amyloid-beta (β-amyloid, or Aβ) plaques in the brain are one of the hallmark features of AD as confirmed by postmortem evaluation and at one time were considered to be the principal actor in the pathogenesis of AD. In the last two decades, many PET imaging agents for Aβ plaques have been developed and proved effective in detecting Aβ deposits in vivo (for reviews see Refs. 27–30). The first radiotracer used for imaging Aβ in AD patients was ^{11}C-labeled Pittsburgh Compound-B ([^{11}C]PIB, or PIB) developed at the University of Pittsburgh (Fig. 1). This radiotracer is the most studied Aβ PET imaging agent up to date and still remains the 'gold standard' for imaging Aβ pathology.

[^{11}C]PIB is a neutral and lipophilic derivative of thioflavin-T, an in vitro staining agent for the AD brain. This benzothiazole-based radiotracer can cross the blood-brain barrier and bind to Aβ plaques with high affinity and specificity in vivo. The clearance of

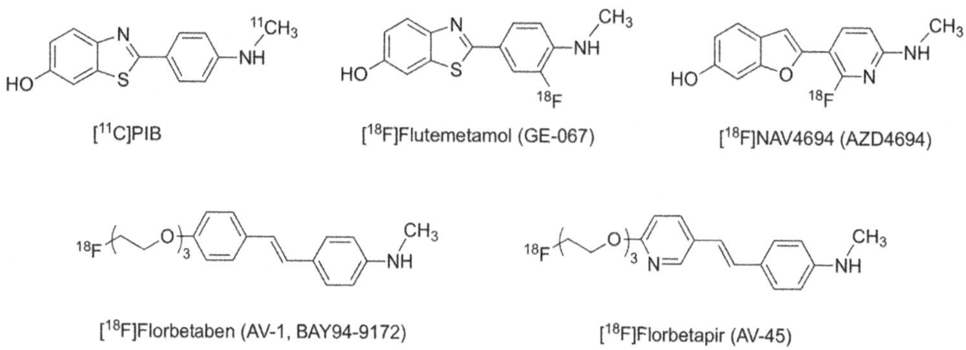

Fig. 1 Chemical structures of Aβ-specific PET radioligands.

nonspecific binding is fast, allowing imaging at high contrast. Its first study of in vivo Aβ imaging was initiated in 2002 and published in 2004.[31] Since then, a number of PET imaging studies with this radiotracer have been conducted in subjects with normal aging and MCI, as well as patients with autosomal dominant AD, early-onset familial AD, and non-AD dementia (for reviews see Refs. 32–34).

The regional brain binding of [^{11}C]PIB is highest in the frontal cortex, cingulate gyrus, precuneus, striatum, parietal cortex, and lateral temporal cortex. The regional retention of [^{11}C]PIB reflects the regional density of Aβ plaques in the AD brain as reported at autopsy and measured by the quantification of immunohistochemical staining of brain slices. The highly fluorescent PIB analogue, 6-CN-PIB, almost exclusively labeled Aβ-containing structures including compact/cored, diffuse/amorphous, neuritic, and non-neuritic Aβ plaques, with intense labeling of all compact/cored plaques, weaker labeling of diffuse cortical, striatal, and hippocampal plaques, and the complete absence of diffuse amyloid labeling in the cerebellum. The retention of [^{11}C]PIB in healthy individuals is associated with a greater risk of cognitive decline and a faster rate of brain atrophy.[34] Longitudinal studies have indicated that Aβ accumulation is a slow process that lasts over 20–30 years. Healthy elderly subjects and subjects with MCI with increased [^{11}C]PIB retention present faster rates of Aβ deposition, which was significantly associated with faster memory decline and disease progression. MCI subjects with high [^{11}C]PIB binding exhibit much more frequent conversion to AD dementia than those with low [^{11}C]PIB binding. [^{11}C]PIB PET imaging can stratify cognitively normal healthy subjects into amyloid-positive and amyloid-negative subtypes for inclusion in the clinical trials of anti-Aβ therapeutics and can be utilized to monitor therapeutic outcomes. Of concern, however, is the finding that the degree of [^{11}C]PIB retention appears to display high intragroup variability among healthy elderly subjects with MCI, and patients with AD, i.e., elevated cortical retention of [^{11}C]PIB does not always correlate with the severity of dementia in AD patients.

Due to the short radioactive half-life (20.4 min) of the carbon-11 in [^{11}C]PIB, central production and off-site distribution for imaging applications are not possible. Therefore, radioligands labeled with the longer half-life radionuclide fluorine-18 (radioactive half-life of 109.8 min) were developed for commercialization and wide availability in the clinics. Currently, three ^{18}F-labeled PET radiotracers (Fig. 1) have been approved by the United States Food and Drug Administration (FDA) and commercially available for clinical use, including [^{18}F]florbetapir ([^{18}F]AV-45, Amyvid), [^{18}F]florbetaben ([^{18}F]AV-1, [^{18}F]BAY94–9172, NeuraCeq), and [^{18}F]flutemetamol ([^{18}F]GE-067, Vizamyl) (Fig. 1). Another ^{18}F-labeled radiotracer [^{18}F]NAV4694 ([^{18}F]AZD4694) is in phase 3 clinical trial (NCT01886820). From the structure point of view, [^{18}F]florbetapir and [^{18}F]florbetaben are stilbene-based radiotracers, derived from the prototypical Aβ-staining agents Congo Red and Chrysamine G, while [^{18}F]flutemetamol (3′-[^{18}F]PIB, also known as [^{18}F]GE067) is an analog of [^{11}C]PIB, derived from the Aβ-staining agent thioflavin-T. Interestingly, [^{18}F]florbetaben, the first ^{18}F-labeled PET agent developed for imaging Aβ, was approved by the FDA in March 2014, much later than [^{18}F]florbetapir (approved by the FDA in April 2012) and [^{18}F]flutemetamol (approved in October 2013). Marketing authorizations were granted by the European Medicines Agency for [^{18}F]florbetapir in 2013,[35] and for [^{18}F]florbetaben[36] and [^{18}F]flutemetamol[37] in 2014.

Among the three approved ^{18}F-labeled amyloid radiotracers, [^{18}F]florbetapir is the most widely investigated and displays faster brain kinetics than the other two radiotracers.[38] In a group of individuals with a life expectancy of no more than 6 months, [^{18}F]florbetapir showed a sensitivity and specificity of 92% and 100%, respectively, in detecting a significant degree of Aβ plaques. Further, Aβ accumulation detected by in vivo PET imaging correlated with amyloid pathology measured from brain tissue postmortem.[39] In AD, frontal, temporal, occipital, parietal, cingulate, and precuneus cortical areas all show increased Aβ retention as measured by this ligand compared to healthy controls.

A meta-analysis indicated that the three approved ^{18}F-labeled amyloid radiotracers showed no marked differences in diagnostic accuracy.[28] They performed better when used to discriminate between patients with AD and healthy controls. Amyloid PET with ^{18}F-labeled radiotracers, in addition to routine assessment in patients with cognitive impairment, has a significant effect on diagnosis, diagnostic confidence, and drug treatment plan.

Clinical utility

During the last two decades, great advances have been achieved in PET imaging of Aβ plaques. The successful PET amyloid radiotracers show detectable cortical uptake with high sensitivity and specificity when a moderate to severe burden of Aβ plaques is present and provide a quantitative measure of the insoluble cortical Aβ load in vivo. Amyloid imaging offers the preclinical detection of Aβ and distinguishes AD from non-AD dementia in patients with mild or atypical symptoms. The applications for amyloid imaging include confirmation or exclusion of AD, differential diagnosis of dementia, particularly for the distinction of AD from frontotemporal dementia, and earlier diagnosis of the disease. Moreover, amyloid PET imaging has allowed the investigation of the correlation between Aβ burden and cognitive function from normal aging to AD. Regarding clinical utility, appropriate use criteria (AUCs) have been well established. Amyloid imaging changes diagnosis in about 30% of cases, increases diagnostic confidence in about 60% of cases, and changes patient management in about 60% of cases.[29] Most importantly, amyloid PET imaging can monitor the biological effects of anti-Aβ drugs and provide the direct evidence for the effectiveness of anti-Aβ therapeutics in clearing Aβ from the brain and thus help test the amyloid cascade hypothesis of AD.[40]

Differential diagnosis

In general, cortical Aβ retention detected by PET imaging correlates well with immunohistochemical staining of amyloid plaques in brain biopsy or autopsy tissue.[41–44] This qualifies the use of Aβ PET imaging as a tool for in vivo detection of cerebral Aβ deposition. In the routine of daily practice, Aβ PET imaging provides valuable information for distinguishing between AD dementia and dementing disorders associated with non-Aβ pathologies such as frontotemporal lobe dementia (FTLD). This is far more informative than the clinical observations provided by neuropsychiatric assessments and conventional structural imaging modalities.[45]

However, it is noteworthy that cognitively normal subjects, including younger ones, are occasionally found to be amyloid positive as judged from PET imaging, suggesting that cognitive decline is more likely due to factors other than amyloid pathology. In dementia with Lewy bodies (DLB) or other comorbid situations, it has been emphasized that an amyloid-positive PET merely reveals the existence of amyloid pathology in vivo and does not necessitate a diagnosis of AD.[46–48] Conversely, a negative amyloid scan does not completely exclude the possibility of AD,[49] since evidence of neurodegeneration in the absence of amyloid pathology challenges the proposed AD progression scheme.[17]

Conversion from MCI to AD and correlation with cognition

The ability of amyloid PET imaging to predict conversion from MCI due to AD, the prodromal stage of the disease defined by different diagnostic guidelines,[16,50] to AD dementia, has been extensively investigated. Roughly 70% of amyloid-positive MCI patients convert to AD dementia within 3 years.[51] As compared to [^{18}F]FDG PET, amyloid PET imaging has higher sensitivity but relatively lower specificity.[52] This is consistent with previous experience since amyloid accumulation commences at least a decade before the worsening of synaptic activity and brain function to a clinically significant level where amyloid deposition reaches a plateau.[53,54] Correlation of Aβ retention with cognitive performance has been shown to be greater in MCI and cognitively intact subjects than in AD dementia patients,[55–57] which could also be explained by the hypothesized plateau model. Therefore, it is not surprising that regional hypometabolism and Aβ deposition in the temporoparietal regions are closely associated with each other whereas those in the frontal lobe, a region affected only in advanced AD, are not.[58,59]

The centiloid scaling project for standardization of amyloid PET imaging data

As the utilization of amyloid PET imaging in clinical trials and disease research multiplies, the urgent need for inter-tracer standardization and for multicenter collaboration and longitudinal comparison drove the launch of the centiloid scaling project.[60] According to the concept, one institute can follow a multistep regime to create a scaling from 0 (young healthy controls) to 100 (typical AD patients) using its own amyloid PET data.[61] In this way, a universal cutoff value could then be directly or indirectly applied in multicenter and/or longitudinal studies to allow for comparison of imaging results across study sites using different radiotracers. The study group of the centiloid project has now made progress in the derivation and verification of a converting formula, enabling the translation of non-PIB Aβ PET semi-quantitative values to standardized PIB counterpart.[62–65]

Summary

In the past decades, the amyloid cascade hypothesis has been the focus of AD pathogenesis, AD drug development, and PET imaging agent development. The availability of Aβ PET imaging agents and its applications in AD research have indicated that amyloid pathology may be a high-risk factor for future cognitive decline. However, increasing evidence indicates that cortical amyloid is not specific for the presence of cognitive symptoms, thus affecting the positive predictive value of Aβ PET imaging. Among populations without dementia, the prevalence of cerebral amyloid pathology as determined by Aβ PET imaging or CSF findings is associated with age,[66] e.g., 33% of healthy elderly individuals have significant levels of Aβ deposition without apparent clinical symptoms.[67] Therefore, Aβ deposition alone cannot explain AD pathogenesis and progression. Repeated failures of clinical trials for many anti-Aβ drug candidates have dampened the hope for their efficacy as disease-modifying therapeutics and shifted the interest to tau pathology for the development of therapeutics and PET imaging agents. Further, to gain a more comprehensive view on the impact of AD treatment, more attention should be paid not only to the AT(N) biomarker data but also to emerging biomarkers, in particular biomarkers specifically linked to the mechanisms of the intervention.[68] Nonetheless, it should be kept in mind that Aβ PET imaging will still remain the gold standard to investigate disease mechanisms as it provides information regarding the topography of Aβ lesions. Although there have been no successful anti-Aβ drugs up to date, Aβ PET imaging has provided useful outcome measures for anti-Aβ therapeutics in clinical trials (see below).

Imaging agents for the tau protein

The tau protein is a microtubule-associated protein and plays a vital role in neuronal stability and transport of axonal nutrients. It is normally highly soluble in its native monomeric form and primarily expressed by neurons in the brain. It is alternatively spliced at the *N*-terminal domain (N) and microtubule-binding repeat domain (R) of the microtubule-associated protein tau (MAPT) gene on chromosome 17 to form six distinct isoforms (0N3R, 0N4R, 1N3R, 1N4R, 2N3R, and 2N4R). The neuropathology of different tauopathies exhibits varying isoform composition of the different intracellular tau inclusions, distinct neuroanatomical distribution, and relative amount of tau inclusions. AD is the most common neurodegenerative tauopathy featured with the accumulation of pathological misfolded tau protein. Neurofibrillary tangles (NFTs) are aggregates of hyperphosphorylated tau protein and contain both the 3R and 4R isoforms accumulating in distinct spatial and temporal patterns. Unlike Aβ, the stages of tau pathology appear to correlate well with the progression of cognitive impairment.[69]

In vivo tau PET imaging can provide a powerful tool to investigate the tauopathy and shed further light on the time course of tau accumulation in relation to other biomarkers and clinical symptomatology. [^{18}F]FDDNP was the first PET tracer reported to be able to visualize tau pathology in AD.[70] However, it displays a poor binding affinity for NFTs[71] and a lack of specificity and selectivity for tau over other β-sheet structures.[72] Over the past few years, a number of radiotracers have been developed for tau-specific PET imaging (for reviews, see Refs. 73–77). The first generation tau radiotracers have been evaluated extensively, and second-generation radiotracers have been developed and in various phases of clinical trials. The chemical structures of first- and second-generation tau PET tracers are depicted in Figs. 2 and 3, respectively, and discussed below.

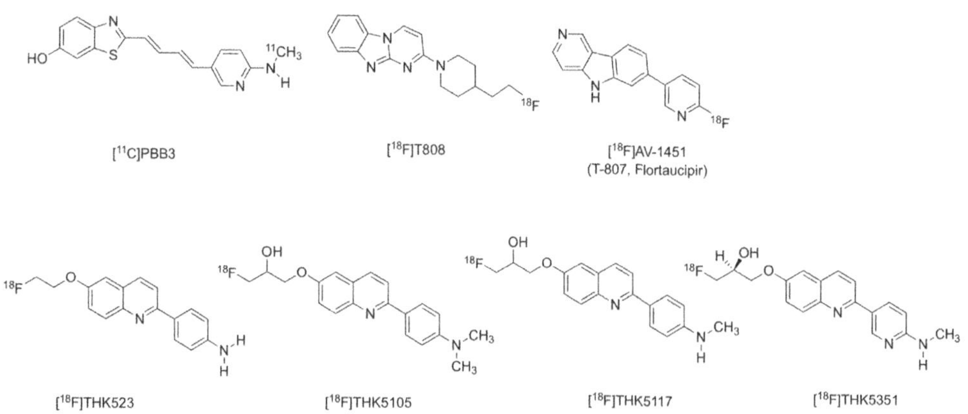

Fig. 2 Chemical structures of the first generation tau PET radioligands.

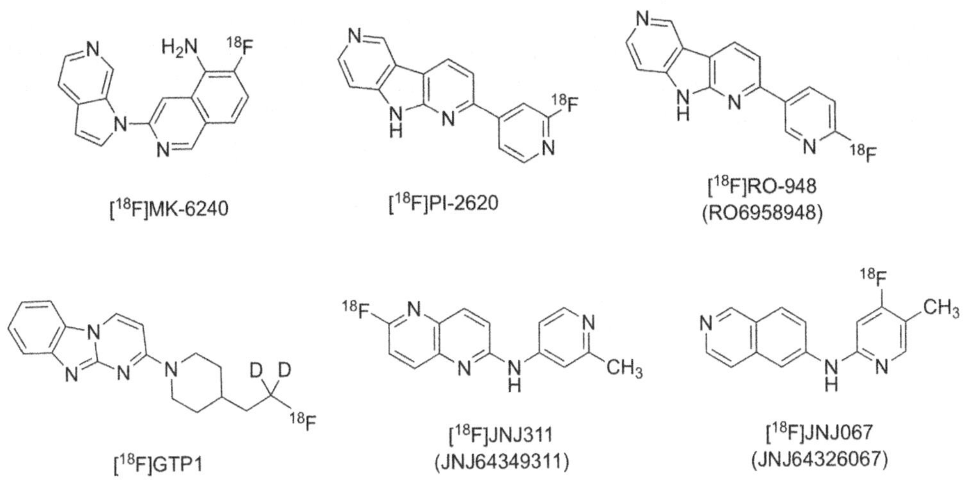

Fig. 3 Chemical structures of the second generation tau PET radioligands.

First-generation tau imaging agents

Among the first generation tau radiotracers, [^{11}C]PBB3, a pyridinated benzothiazole, was shown to be able to detect in vivo tau inclusions both in AD and non-AD tauopathies such as corticobasal degeneration (CBD). In AD patients, specific [^{11}C]PBB3 binding was observed in the Cornu Ammonis 1 (CA1) area and subiculum regions in the hippocampus with a high density of fibrillar tau aggregates. However, the occurrence of a brain-penetrant radioactive metabolite, low specific binding, and off-target binding preclude its further validation for clinical use.[78]

Several tau radiotracers were developed from arylquinoline derivatives. [^{18}F]THK523 (formerly [^{18}F]BF242) was the first ^{18}F-labeled radiotracer from this family. However, this tracer showed high white matter binding and unfavorable pharmacokinetics.[79] Similar to [^{18}F]THK523, [^{18}F]THK5105, and [^{18}F]THK5117 also displayed high nonspecific binding in subcortical white matter.[80] [^{18}F]THK5351 is the single (*S*)-enantiomer and pyridine derivative of [^{18}F]THK5117 with lower lipophilicity than [^{18}F]THK5117. In preclinical studies, [^{18}F]THK5351 demonstrated a high and selective binding affinity for tau aggregates, low binding to white matter, and rapid kinetics, which resulted in a higher signal-to-background ratio than [^{18}F]THK5117. In addition, autoradiography indicated no binding to Aβ plaques, α-synuclein, or TAR DNA-binding protein 43 (TDP43) deposits, suggesting that [^{18}F]THK5351 binds to NFTs with high selectivity.[81] In first-in-human PET imaging studies in AD patients, [^{18}F]THK5351 demonstrated faster kinetics and higher contrast, with prominent retention in the mesial temporal lobe and lateral temporal cortex and lower retention in the subcortical white matter. A comparative study of [^{18}F]THK5317 ([^{18}F]-(*S*)-THK5117) and [^{18}F]THK5351 showed more favorable pharmacokinetic and imaging characteristics for [^{18}F]THK5351.[82] However, both [^{18}F]THK5317 and [^{18}F]THK5351 displayed off-target binding to monoamine oxidase B (MAO-B) in the brain.[83]

[^{18}F]AV-1451 ([^{18}F]Flortaucipir, or formerly [^{18}F]T807) is one of the most widely investigated radiotracers for PET imaging of tau aggregates. In a recent report, patients with Braak IV or greater stage were found to have elevated [^{18}F]AV-1451 uptake. Further, antemortem PET findings of elevated [^{18}F]AV-1451 uptake in AD patients were strongly correlated with postmortem quantitative neuropathology measurements of tau burdens in the hippocampal and temporal lobe.[84] With this critical piece of evidence, [^{18}F]AV-1451 has now become the first tau tracer approved by the FDA, under the brand name Tauvid, for PET imaging detection of NFTs (https://www.fda.gov/drugs/drug-approvals-and-databases/drug-trial-snapshot-tauvid).

[^{18}F]AV-1451 displayed a high affinity for tau aggregates in the form of classic paired helical filament (PHF)-tau tangles containing all six isoforms of tau present in AD brains. However, it displayed a relatively low binding affinity to tau aggregates predominantly made of the 4R isoforms that adopt a structure of straight filaments (SF).[85,86] [^{18}F]AV-1451 was shown to have properties suitable for tau quantification with PET in human subjects, and its uptake in AD brain tracked well with expected tau pathology distribution throughout the stages of AD.[87] [^{18}F]AV-1451 appeared to bind preferentially to neurofibrillary tangles with late tau deposition, suggesting that its binding may not completely reflect the early stage of tau progression.[88] An autoradiographic evaluation indicated high specific binding of this radiotracer to "3R+4R" tau present in AD brains with minimal binding to non-AD type tau, including 3R tau in Pick's disease or 4R tau in progressive supranuclear palsy (PSP) and CBD.[88] PET imaging also indicated a high binding affinity of [^{18}F]AV-1451 for PHF-tau pathology in AD brain and low affinity for SF-tau in PSP.[85,88,89] Another report suggested that [^{18}F]AV-1451 is likely to delineate patients with AD and other tauopathies from healthy subjects.[90] In a large scale clinical trial of 719 participants that included cognitively normal controls and patients with MCI, AD dementia, and non-AD neurodegenerative diseases, [^{18}F]AV-1451 uptake in the medial-basal and lateral temporal cortex was shown to have 89.9% sensitivity and 90.6% specificity in differentiating AD dementia from non-AD diseases.[91] [^{18}F]AV-1451, however, may have limited utility for selective and reliable detection of tau aggregates in vivo in non-AD tauopathies such as PSP.[85] The other limitation of [^{18}F]AV-1451 is its substantial off-target binding to monoamine oxidase A (MAO-A) and MAO-B,[92] vessels, iron-associated regions, substantia nigra, calcifications in the choroid plexus, and neuromelanin- and melanin-containing cells.[86,88,93] Most recently, [^{18}F]AV-1451 was confirmed to be a weak inhibitor of MAO-A and MAO-B, as MAO inhibitors can alter its binding in in vivo PET imaging.[94]

Second-generation tau imaging agents

Off-target binding to molecular structures other than tau protein deposits limited the clinical validity of first-generation radiotracers, especially in diseases where tau deposition is expected in areas with a high load of MAO-B (e.g., CBD and PSP). The second-generation tracers showed a lower signal, or no signal, in MAO-B-rich area.

One of the promising second-generation tau tracers is [^{18}F]MK-6240, a pyridine isoquinoline amine derivative (Fig. 3). It showed a high binding affinity for tau aggregates specifically in AD brain tissue,[95] but not for neuronal and glial tau aggregates in non-AD tauopathies such as Pick's disease, PSP, CBD, or chronic traumatic encephalopathy (CTE), or for TDP-43, Aβ, and α-synuclein.[96] Off-target binding to MAO-A and MAO-B was weak, or absent. However, it appeared to exhibit strong off-target binding to neuromelanin- and melanin-containing cells.[96] PET imaging with [^{18}F]MK-6240 demonstrated favorable brain kinetics and high binding levels in brain regions with a plausible pattern for NFT deposition in AD subjects.[97,98] The total distribution volume of this tracer was two- to three-fold higher in neocortical and medial temporal brain regions of AD subjects than healthy controls.[97] No apparent off-target binding was found in the basal ganglia and choroid plexus, although binding in the meninges was seen in some AD patients.[98]

[^{18}F]PI-2620 (Fig. 3) is another recently developed radioligand with good selectivity and no off-target binding to Aβ or MAO-A and MAO-B.[99] The first study in healthy controls and AD patients showed fast kinetics, excellent signal-to-noise contrast, and test-retest reproducibility of standardized uptake ratio (SUVR).[100] Further, uptake of [^{18}F]PI-2620 in neocortical regions was found to significantly correlate with the degree of cognitive impairment.[101]

Three other potential tau tracers [^{11}C]RO-963, [^{11}C]RO-643, and [^{18}F]RO-948 (previously referred to as [^{11}C]RO6924963, [^{11}C]RO6931643, and [^{18}F]RO6958948, respectively) were recently evaluated in amyloid-positive AD patients and controls.[102] [^{18}F]RO-948 was found to demonstrate more favorable properties for the characterization of tau pathology in AD. The favorable kinetic properties of [^{18}F]RO-948 was confirmed in a large cohort of young and old healthy subjects and AD patients, with less off-target binding typically found for [^{18}F]AV-1451 in the basal ganglia, thalamus and choroid plexus, although binding to neuromelanin was suspected.[103] In a head-to-head comparison with [^{18}F]AV-1451, [^{18}F]RO-948 demonstrated faster kinetics and greater binding in the medial temporal lobe, but lower off-target binding.[104] Interestingly, uptake in the off-target region of basal ganglia increased with age for both tracers.

[^{18}F]GTP1 is derived from [^{18}F]T808, a first-generation radiotracer with intense uptake in the skull due to defluorination in vivo.[105] [^{18}F]GTP1 exhibited high affinity and selectivity for tau pathology with no measurable binding to Aβ plaques or MAO-B in AD tissues, or binding to other tested proteins at an affinity predicted to impede imaging interpretation. Its pharmacokinetic and imaging profiles were demonstrated to be favorable: good brain uptake, rapid tissue washout, and excellent test-retest reproducibility. SUV ratio (SUVR) from 60 to 90 min following administration appeared to be a robust surrogate measure for [^{18}F]GTP1 specific binding to tau. In a large group of cognitively normal subjects and patients in the prodromal, mild, and moderate AD stages, uptake of [^{18}F]GTP1 in the temporal lobe of AD patients, along with cortical volume, was found to strongly correlate with cognitive performance.[106] Nonetheless, [^{18}F]GTP1 may still possess off-target binding to MAO-B in vivo similar to that of [^{18}F]T808 since it is a di-deuterated derivative of the first-generation tau tracer [^{18}F]T808.[107]

[^{18}F]JNJ-311 ([^{18}F]JNJ64349311) and its analog [^{18}F]JNJ-067 ([^{18}F)JNJ-64326067) are the newest additions to the tau tracer family and structurally different from other second-generation tau tracers.[108–110] [^{18}F]JNJ-067 has a higher affinity for aggregated tau than [^{18}F]JNJ-311 and improved pharmacokinetic properties with no apparent off-target binding. In humans, [^{18}F]JNJ-067 displayed fast kinetics and retention in the temporal lobe of AD patients.[111] No apparent off-target binding was observed.

Clinical research findings from tau imaging and correlation with clinical information and other biomarkers

As opposed to the globally elevated pattern seen in brain amyloid PET of AD dementia patients, the spatiotemporal distribution of tau tracer uptake appears to follow a typical neuropathological sequence of spreading (Braak's stage) that is in strong correlation with the clinical phenotype of the affected brain regions, but not with regional amyloid retention.[112,113] In parallel to postmortem immunohistochemical findings, in vivo tau deposition starts from the medial temporal regions, i.e., hippocampus and entorhinal cortex (Braak stage I–II), and then spreads to the adjacent neocortices (Braak stage III–IV), and finally to the entire brain (Braak stage V–VI).[75] According to current experience with available tau tracers, AD is not likely to be the diagnosis of a tau-negative individual, while non-tauopathies can be basically excluded from the cause of cognitive decline for a tau-positive individual.[114] Non-AD tauopathies such as FTLD, PSP, and CBD, however, manifest different distribution patterns from AD, involving different subregions in the brain stem and basal ganglia as well as cortical regions, which could be of additional differential diagnostic value in the clinical context.[114]

According to the amyloid cascade hypothesis, neurodegeneration characterized by misfolded tau tangle aggregation is the downstream event secondary to major amyloid deposition.[115] Yet without the plateau effect observed in amyloid PET, tau retention (in terms of both quantity and scope) is strongly correlated with both cross-sectional and longitudinal memory performance in a wide range of subjects from advanced AD dementia patients to cognitively intact elderly controls.[87,116] This quantitative and sequential connection demonstrated that this tangle spreading trajectory reflects neuronal function as well as disease progression. The same spatiotemporal relationship can also be observed in non-AD tauopathies.[75]

As for the clinical variants of sporadic AD, tau binding patterns are not necessarily the same. For instance, while temporal lobe deposition correlates well with memory performance, frontal lobe retention strongly correlates with execution and global cognition.[117–119]

Recently, tau deposition in cognitively intact healthy elderly subjects was revealed by PET and confirmed by autopsy results. This phenomenon was termed as primary age-related tauopathy (PART) and believed to produce only minimal, if any, clinical symptoms.[120] In the absence of amyloid, PART alone is believed to be insufficient to develop memory decline. Its relationship to suspected non-Alzheimer's pathophysiology (SNAP) has also been proposed, presumably underlying preclinical abnormalities during the development of AD.[121]

Hypometabolism in brain regions that are commonly affected in advanced AD, including the posterior cingulate cortex, precuneus, and temporoparietal association cortex, represents synaptic dysfunction of the neurons in the course of AD. Cortical tau tracer deposition can be seen in these same areas, establishing a typical topographic "yin-yang" offset between the two biomarkers. Correlation between these two grows stronger as the tau burden increases.[122] On the other hand, multi-modality studies also reveal that the correspondence between frontal hypometabolism and medial temporal tau retention is present independent of amyloid deposition, which can be interpreted as a result of aging.

The relationship between tau aggregation and cortical atrophy as demonstrated by structural MRI is similar to that between tau aggregation and hypometabolism. However the former was found to be weaker than the latter, especially in conditions other than mild AD.[123,124] In addition to local correlation, tau deposition in the parietal lobe and precuneus can also be correlated to medial temporal lobe atrophy.[119] These observations seem plausible due to the fact that substantial structural alterations are often related to the later-stage disease.

In general, levels of CSF phosphorylated tau (p-tau, representing misfolded tangle formation) and total tau (t-tau, representing neurodegeneration) levels are found to be concordant with in vivo tau tracer binding.[75,125] However, discordant results have also been found in small subgroups possibly due to varying detectability around the CSF threshold and varying acceleration of tau tangle formation in different disease stages.[126,127] Although it is generally believed that elevated CSF tau levels precede detectable levels of tau tangle formation by imaging, tau PET has the advantage of visualization and topographic quantitation capacity, as well as relative non-invasiveness.

In a recently published longitudinal and cross-sectional study of tau- and amyloid-PET imaging combined with structural MRI, only the presence or accumulation of tau was found to be a predictor of cognitive impairment, while the presence or accumulation of amyloid appeared to be a predictor of more severe tau-associated cognitive deficits. These findings suggest that cognitive decline and neurodegeneration in AD may be primarily driven by the onset and spread of tau pathology.[128]

Summary

Over the last few years, notable progress has been made in the development of tau PET tracers for imaging applications, resulting in the recent approval of [^{18}F]AV-1451 by the US FDA for PET imaging detection of NFTs. Nonetheless, the current batch of available tau PET tracers is not optimal and has their limitations, including the inability to differentiate the isoforms of tau aggregates, and off-target binding that may complicate the interpretation of PET imaging data.

Coming from different compound families, current tau tracers exhibit affinities for various MAPT isoforms and tangle structures, leading to distinctive topographic binding patterns in the same tauopathy. The development of PET radiotracers highly selective for isoforms and tangle structures (e.g., 3R or 4R-selective) will aid the validation of tau imaging as an effective biomarker which ideally will correlate better with synaptic function and cognitive performance and allow the differentiation of AD from other tauopathies. Recent cryo-electron microscopic structure elucidation of tau tangles from the AD brain may provide new insights into the binding interactions between the various tracers and neurofibrillary tangles. This will further our understanding of ligand structures on binding behavior and the consequent relationship between subtype (3R/4R/3R + 4R, PHF/SF, etc.) affinity and inter-tracer cerebral distribution discrepancies, thereby hopefully leading to new perspectives and advances on the design of novel, subtype-selective radiotracers for tau imaging in AD.[129]

The other prominent issue with the current batch of tau radiotracers is their off-target binding to other proteins and structures in the brain.[130] MAO-A and MAO-B are the most frequent off-target binding sites whose secreting neurons highly overlap with Parkinsonism-related brain regions.[88,114,131] Choroid plexus is a frequent tissue of apparent "off-target" binding, although the exact nature and mechanism of this binding have yet to be fully understood,[132] and there is even debate whether the suspected "off-target" binding in the choroid plexus may actually be reflecting true tau binding.[132,133] It is also postulated that melanin, neuromelanin, mineralized structures, and hemorrhagic lesions can cause off-target binding of the current tau tracers in various brain regions. Off target binding of the first-generation radiotracers, primarily to MAO-B, limits their clinical utility, while off-targets of the second-generation radiotracers are yet to be fully delineated. More recently, in silico modeling has identified a binding site for all the current tau radiotracers on MAO-B, which is the same binding site for the reversible MAO-B inhibitor safinamide.[107] The first-generation tau PET tracers except ^{18}F-THK5317 exhibit binding affinity to MAO-B, while the second-generation tracers ([^{18}F]MK-6240, [^{18}F]PI-2620, [^{18}F]RO-948, and [^{18}F]JNJ-067) appear to interact less with MAO-B.

In the field of PET imaging for neurodegeneration, the selectivity of tau tracers over other pathological proteins such as Aβ, α-synuclein, and TDP-43 would always require careful evaluation and validation. The similarity in the secondary/tertiary structures of the binding sites in these proteins makes it difficult to develop a truly selective and specific probe, not to mention the complexity introduced by potential comorbidity of misfolded proteins in various tauopathies. These complexities pose substantial challenges to clinical differentiation of tauopathies, and potential post-treatment evaluation.[132]

In summary, in the fast-advancing area of PET imaging agents for tau, several radioligands have been developed and evaluated. The first-generation tau tracers have been shown to have substantial off-target binding, especially to MAO-B and some of them have faded away (in particular the THK series of radioligands). The second-generation radiotracers display improved pharmacological, pharmacokinetic, and imaging properties and are being validated in clinical trials. Although some issues remain, especially isoform selectivity and other potential off-target binding, the applications of these radiotracers in AD imaging are expected to provide new information on the time course and topography of tau tangles in AD, as well as their correlation with cognitive performance. Longitudinal and cross-sectional multi-target and multi-modality studies will help to further elucidate the role of tau in the course of AD, as well as its interaction and relationship with Aβ, synaptic dysfunction, brain atrophy, and other pathological biomarkers. Another important application of tau PET imaging is for patient selection and endpoint measurements in phase 2 and phase 3 clinical trials of disease-modifying anti-tau treatments (see below).[134]

Imaging agent for neurodegeneration-[^{18}F]FDG

As a radionuclide labeled analog of glucose, the major metabolic substrate of neurons, [^{18}F]FDG has long been used in CNS disorders to reflect neuronal degeneration and injury and is designated as one of the biomarkers for neurodegeneration (N) in the NIA-AA research framework of AT(N) for Alzheimer's disease.[22] In fact, the first report of this most widely used tracer was a brain imaging study,[135] although now the majority of its clinical applications is for oncological purposes. The characteristic AD pathology[136] and neurodegenerative changes[137,138] are associated with cortical hypometabolism demonstrated in premortem FDG PET imaging. Clinical application of FDG PET in AD lies mainly in differential diagnosis from other causes of dementia, as well as treatment effect evaluation of disease-modifying or progression-slowing therapies. The clinical value of FDG PET has also been investigated for its predictive ability for conversion to AD dementia of high-risk subjects assumed to be in the prodromal or asymptomatic stage.

Clinical utility

AD dementia patients

The characteristic manifestation of FDG PET in AD dementia is hypometabolism in the posterior part of the brain including the posterior cingulate cortex (PCC), precuneus

(PrC), and parietotemporal association cortices such as the angular gyrus. These regions are the most discriminating components of AD-specific cerebral hypometabolism as well as strong indicators of disease severity and progression. It is of significant differential value that PCC and PrC are substantially spared in other types of senile dementia including FTLD, DLB, Parkinson's disease dementia (PDD), and vascular dementia (VaD). The hippocampus and entorhinal cortex are involved in the earliest stages of AD according to neuropathological findings based on NFTs formation.[139] However, glucose metabolism in these areas cannot be readily distinguished between AD patients and normal controls. This is largely due to the frequent presence of medial temporal hypometabolism in normal aging, and the partial volume effect (PVE) resulted from regional cortical atrophy. Frontal lobe hypometabolism is often associated with advanced disease, executive dysfunction, or atypical behavior in AD cases, although it can also be observed in normal aging. Occipital lobe hypometabolism is related to the posterior cortical atrophy (PCA) subtype of AD, but care should be taken when differentiating dementia cases because this is also a prominent sign of DLB.

MCI patients

Early differentiation of MCI due to AD is crucial since early intervention may be beneficial in slowing disease progression.[140,141] It should be noted, though, that the etiology of MCI is heterogeneous, meaning that approximately half of the patients convert to dementia other than AD, or do not convert at all.[67,142,143] Therefore, whether FDG PET has discriminative and predictive ability for MCI patients is of great clinical importance.

Hypometabolism revealed by FDG PET in typical AD-affected brain regions including the inferior parietal lobe, precuneus, and posterior cingulate cortex is present early in prodromal AD, namely the MCI stage. In a meta-analysis comparing the accuracy of three different imaging modalities, FDG PET (sensitivity = 88.8%, specificity = 84.9%) was found to exhibit higher sensitivity and higher specificity than cerebral blood flow single-photon emission computed tomography (SPECT) (sensitivity = 83.8%, specificity = 70.4%) and structural MRI (sensitivity = 72.8%, specificity = 81%) in terms of predicting short-term conversion to AD dementia.[144] Interestingly, FDG PET performs better in excluding nonconverters than [^{11}C]PIB Aβ PET (specificity: 74.0% vs 56.2%), while its sensitivity is lower (sensitivity: 78.7% vs 93.5%).[145] This could be explained by the compensatory mechanism of preserved cerebral synaptic function against amyloid burden. Aside from the advantage of accurate short-term predictive ability, FDG PET can also exclude other etiologies underlying MCI such as FTLD and DLB, in contrast to other biomarkers including CSF $A\beta_{1-42}$ and t-tau/p-tau assays.[146]

The value of FDG PET in MCI has been acknowledged in existing diagnostic criteria.[15,50] Meanwhile, future research in this field is encouraged to overcome current limitations in terms of methodology normalization, gold standard verification, and effectiveness/economy evaluation.[146]

High-risk asymptomatic subjects

People with subjective cognitive decline (SCD) rather than objective evidence of cognitive impairment, subjects with amyloidosis, and family members of AD patients expressing *PSEN1/2* or *APP* genetic mutation are reported to have higher tendencies to advance to AD dementia.[147–149] It is hypothesized that neurodegenerative alterations could have been silently progressing in these cases.

Various confounding factors lead to heterogeneity in the underlying etiology of SCD, resulting in controversy regarding the regional metabolic manifestation in this subgroup.[150–152] Although Aβ positive asymptomatic subjects, as defined by CSF $A\beta_{1-42}$ or Aβ PET imaging, are perceived to have a higher risk of converting to AD in a life-long period, the utility of FDG PET reflecting neurodegeneration in this population is still not as evident as that in MCI patients, due to its poor performance in the prediction of short-term conversion.[148] Further, glucose metabolic abnormalities prior to the onset of clinical symptoms have been observed in asymptomatic carriers of mutated alleles of *APP* and *PSEN1/2* genes,[153,154] individuals who may be at risk of ultimately developing dementia. Nonetheless, the relationship between cerebral hypometabolism and time-to-conversion cannot be easily concluded in the absence of well-designed longitudinal studies.

In the presence of the above mentioned facts, it is not recommended to clinically utilize FDG PET to evaluate asymptomatic subjects with only one risk factor for diagnostic or prognostic purposes. Whether FDG PET is useful for individuals with multiple risk factors requires further investigation.[153,155]

Treatment monitoring

Based on the fact that glucose utilization is an indicator of overall brain function, FDG PET has been used for outcome assessment of AD therapeutic agents in multiple clinical trials (see below).[156–158] Global as well as sub-global or regional FDG uptake alterations following physiological, pharmacological, or surgical intervention have been observed in multiple clinical studies.[159–167]

Summary

On the whole, FDG PET provides a biomarker for neurodegeneration—the (N) in the AT(N) framework of AD diagnosis. It also offers the potential for treatment optimization should a disease-modifying therapy become available.[168,169] This would have the effect of lessening the overall healthcare expenditures associated with dementing illnesses.[170,171]

Although some critical issues in the use of FDG PET relating to AD have been resolved, there are still many unresolved issues that require further clarification using larger and better-designed clinical trials.[172]

Imaging agents for neuroinflammation biomarkers

Multiple lines of evidence have highlighted the involvement of neuroinflammation in the etiology of AD.[173,174] While inflammation is likely to be protective in the instance of CNS injury or infection, chronic inflammation is likely to be detrimental to both neurons and non-neuronal tissue, leading to neurodegeneration and loss of cognitive functions.[175,176] If, as suggested, neuroinflammation underlying AD starts from the earliest stages without any obvious clinical symptoms and lasts to the end stage of the disease, it is thought that AD dementia could be postponed by therapeutic interventions that modulate this process.[177–180] In this regard, multiple pathophysiological factors that might be the targets of therapeutic intervention include complement pathways, cytokine activation, oxidant stress, and microglial activation.[181,182] PET imaging agents designed to visualize the neuroinflammatory processes are thus inspired by these targets.[183,184]

The 18-kDa translocator protein (TSPO) is a mitochondrial protein, formerly referred to as the peripheral benzodiazepine receptor (PBR). Under normal circumstances, its expression is limited to microglial cells in specific brain regions. These cells comprise approximately 15% of non-neuronal cells in the CNS. In response to infection and traumatic injury, TSPO is upregulated in activated microglia.[185] Soluble and fibrillary Aβ can induce a similar effect, and since activated microglia co-localizes in areas of Aβ deposition and brain atrophy, TSPO is a prominent target for imaging of in vivo inflammatory processes in AD.[186–188]

Imaging agents to measure activation of microglia and astrocytes

[^{11}C]PK11195 was the first TSPO tracer developed for human PET imaging,[189] but it remains controversial whether its binding is correlated with the amyloid burden.[190–192] It is uncertain if this can be explained by its insensitivity due to high nonspecific binding and thus low specific binding signals in vivo. To overcome the disadvantages of [^{11}C]PK11195, a series of second-generation TSPO tracers were developed and evaluated, including [^{11}C]PBR28, [^{11}C]DAA1106, [^{18}F]DPA714, [^{18}F]FEPPA, [^{18}F]FEMPA, and [^{18}F]FEDAA1106 (Fig. 4).[184,193,194] Although the majority of these second-generation radioligands are more sensitive than [^{11}C]PK11195, their uptake was found to be regulated by *TSPO* gene rs6971 polymorphism, and therefore phenotyping is needed to assess individual binding affinity for accurate between-group comparisons.[195]

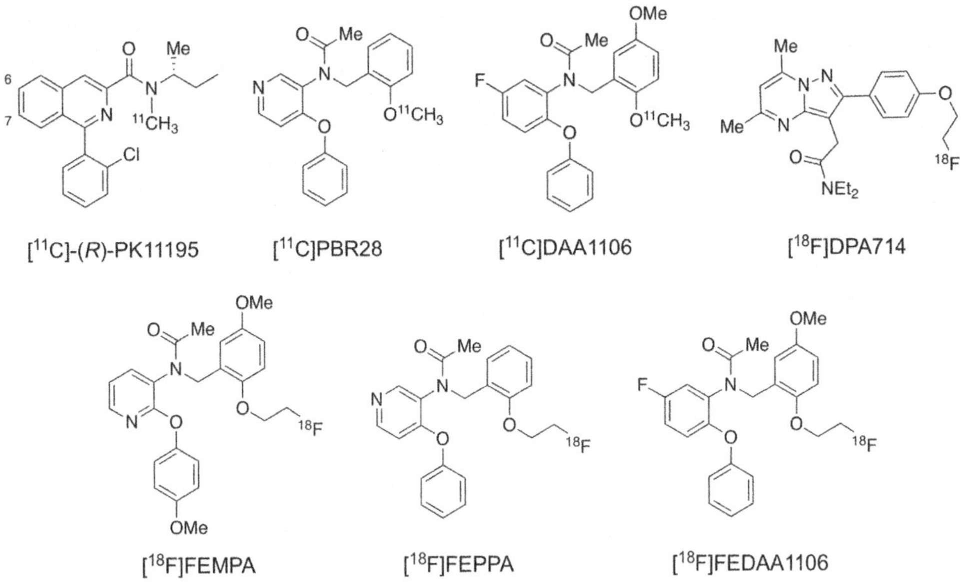

Fig. 4 Chemical structures of TSPO PET radioligands.

Similar to the paradigm of TSPO imaging to assess microglial activation, monoamine oxidase-B (MAO-B) is found to be elevated in reactive astrocytes and thus chosen as the target for in vivo imaging. The selective MAO-B tracer [^{11}C]deuterium-*L*-deprenyl ([^{11}C]DED) has been applied to studies of neurodegenerative disorders including AD in humans.[196]

Clinical research findings

In vivo TSPO imaging, in general, has been found to discriminate between AD dementia patients and normal controls.[189,197–201] The majority of the multi-modality studies report that regional TSPO binding correlates positively with [^{11}C]PIB retention, and negatively with FDG uptake and cortical volume.[190,192,202,203] In most cases, the extent of TSPO binding is associated with baseline cognitive performance and deterioration over time.[190,192,201,202,204] Elevated TSPO binding in advanced AD dementia patients is seen in various cortical regions following an anticipated distribution pattern.[189,200] Apart from global discrimination, regional TSPO binding is found to correlate with attenuated brain function. In one study, researchers uncover an inverse correlation between visuospatial function and [^{18}F]FEPPA binding in the parietal and posterior internal capsule, as well as the correlation between language ability and binding in the latter region.[198] Interestingly, early-onset AD (EOAD) is reported to be associated with higher TSPO binding than late-onset AD (LOAD), especially in the frontal and parietal cortices, suggesting greater microglial activation under the former condition.[202] In addition, a longitudinal study indicates that the TSPO binding of MCI converters to AD dementia is drastically different from that of nonconverters.[201] These results are concordant with previous observations that chronic microglial activation impacts AD progression.

However, the results have not been consistent for TSPO imaging in AD. For example, TSPO binding and cortical amyloid retention or neuronal metabolism do not always correlate as expected.[191,204,205] Further, it has been reported that TSPO PET imaging fails to discriminate healthy subjects from patients with MCI or even advanced dementia.[191,204–206] Likewise, a significant association between TSPO binding and memory performance or disease severity is not always detected.[197,205] Single-nuclide polymorphism of rs6971 has been discovered to regulate TSPO binding in vivo and may have contributed to the discrepant findings thus far, as binding adjusted to individual genotyping has been shown to increase discriminating accuracy.[198] In addition to doubts about the sensitivity of current TSPO tracers to detect subtle alterations in the prodromal stages of AD, recent data appear to relate higher binding of [^{18}F]DPA714 to slower cognitive decline. This may suggest a neuroprotective role of microglial activation in the early phase of the disease.[200]

According to PET studies with both [^{11}C]DED and [^{11}C]PIB, astrocytosis is most pronounced in amyloid-positive MCI subjects, followed by advanced AD patients, amyloid-negative MCI subjects, and healthy controls.[196,207] This intriguing PET finding suggests that astrocytosis may be diminished after prodromal AD converts to dementia. This is supported by a postmortem autoradiographic study showing the highest binding of MAO-B radioligand in the earliest Braak stages.[208]

Summary

It is generally acknowledged that neuroinflammation plays an active part in the course of AD. Nonetheless, in light of the complexity of both the process and components, it remains unsettled whether neuroinflammation is beneficial at the beginning and harmful thereafter. Further, it is unsettled whether TSPO imaging and MAO-B imaging reflect true activation of microglia and astrocytes. A breakthrough is required to answer these questions. For example, selective radioligands discriminating between M1 type microglia, which release neurotoxic substances, or M2, which release protective cytokines, will be valuable, although their development is expected to be challenging. Further, longitudinal studies, in contrast to cross-sectional ones, should illuminate the topography of microglial activation. Finally, multi-target studies that elucidate the chronology and association of neuroinflammation with synaptic dysfunction, energetic imbalance, metabolic alterations, and the presence of misfolded proteins will be insightful.

Imaging agents for the cholinergic system

The basal/rostral forebrain cholinergic pathways play important roles in a variety of neuropsychological functions including attention, consciousness, and memory processing.[209] Currently, cholinergic enhancement therapies constitute the basis of AD-treatment strategies.[9,210] Cholinergic depletion with resulting defects in neuronal plasticity has been confirmed in autopsy studies of AD patients.[211,212] Accordingly, PET imaging that targets the cholinergic system could interrogate the roles of cholinergic axonopathy and its interaction/relation with other pathophysiological components in AD.

Acetylcholinesterase (AChE) is the major target of AD medication, most of which are acetylcholinesterase inhibitors (AChEIs). These include galantamine, rivastigmine, donepezil, tacrine, and memantine, which block AChE and inhibit the hydrolysis of ACh. This in turn increases ACh levels in the synaptic cleft. *N*-[^{11}C]Methyl-4-piperidinyl propionate ([^{11}C)PMP) and *N*-[^{11}C]methyl-4-piperidyl acetate ([^{11}C]MP4A) are two selective substrates for AChE that have been developed for in vivo imaging (Fig. 5). Another presynaptic cholinergic PET tracer [^{18}F]FEOBV, selective for the vesicular ACh transporter (VAChT), has also been tested in people.[213] Postsynaptic acetylcholine receptors can be classified into nicotinic acetylcholine receptors (nAChRs) and muscarinic acetylcholine receptors (mAChRs). Radiotracers available for $\alpha_4\beta_2$ nAChR include 2-[^{18}F]F-A-85380 (2-[^{18}F]FA), [^{18}F]AZAN, and [^{18}F]flubatine (also known as [^{18}F]NCFHEB) (Fig. 5).[214–216]

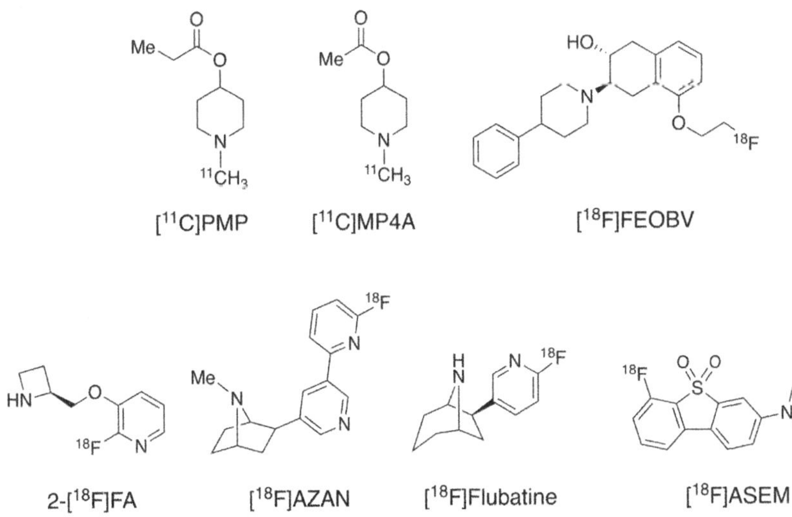

Fig. 5 Chemical structures of PET radioligands for the cholinergic system.

Clinical research findings

PET imaging of AChE using [^{11}C]PMP and [^{11}C]MP4A reveals reduced cortical binding in AD patients compared to healthy controls, especially in regions innervated with cholinergic projections.[217,218] AChE binding is lower in AD patients than in healthy controls, in parallel with VAChT reduction, and is further inhibited by treatment with AChE inhibitors including donepezil, rivastigmine, and galantamine.[218–222] However, this induced inhibition does not affect nAChR binding, suggesting that allosteric modulation maintains ACh signaling by upregulating nAChR but not AChE.[223] The decline in AChE binding induced by treatment is more evident in the frontal cortex than in the temporoparietal cortices and is correlated to improvement in frontal lobe functions. There is no correlation between AChE hydrolysis and hypometabolism in the posterior cingulate cortex. There is, however, a link between a higher hydrolysis rate and APOE4 positivity.[218,224]

The binding of the VAChT tracer [^{18}F]FEOBV is found to be lower in AD than in healthy controls in a recent multi-tracer study involving [^{18}F]FDG and [^{18}F]NAV4694,[213] and among these ligands, [^{18}F]FEOBV appears to be the most sensitive. Further, [^{18}F]FEOBV binding correlates positively to the minimal mental state examination (MMSE) and Montreal cognitive assessment (MoCA) scores. A more recent study attributes basal forebrain degeneration to loss of cortico-amygdalar cholinergic input as demonstrated by reductions in [^{18}F]FEOBV binding.[225]

Nicotinic $\alpha_4\beta_2$ AChR binding is not only lower in AD and MCI subjects than in cognitively normal volunteers, but also lower in MCI converters to AD dementia than in nonconverters.[226–228] Binding measures in the AD-affected brain regions are demonstrated to be significantly correlated with cognitive functions.[228–231] Frontal [^{11}C]PIB retention has been shown to be inversely correlated with 2-[^{18}F]FA binding in the medial frontal cortex and nucleus basalis magnocellularis.[232] Using [^{18}F]flubatine, a new-generation $\alpha_4\beta_2$ tracer with more favorable brain kinetics, receptor deficiency was found within the basal forebrain-cortical and septo-hippocampal cholinergic projections in AD.[233] Meanwhile, episodic memory, working memory, and executive functions were found to be well-correlated with levels of $\alpha_4\beta_2$ AChR in the corresponding cortices.

Summary

Cholinergic dysfunction plays a critical role in AD. PET imaging with various cholinergic biomarkers has revealed the connection between cholinergic dysfunction and AD. In view of the fact that current clinically approved treatments are focusing mainly on the AChE system, cholinergic PET might be useful for screening drug-sensitive/insensitive candidates and could help to monitor treatment response. Further research, especially longitudinal, multi-target imaging studies will help to shed further light on the correlation of cholinergic biomarkers with AD pathogenesis and progression, therapeutic response to AD medications, as well as their association with other pathophysiological biomarkers of the disease.

While most of the previous human PET studies on postsynaptic nAChRs used $\alpha_4\beta_2$ nAChR-targeted radiotracers, growing interest has focused on new targets including the α_7 nAChR and muscarinic acetylcholine receptors (mAChRs). The α_7 subtype of nAChR is also believed to play an important role in AD pathophysiology, as it binds to $A\beta_{1-42}$ with high affinity, mediates Aβ-induced tau phosphorylation, and contributes to neuronal cell death.[234,235] Agonists of α_7 nAChR inhibit this binding and prevent $A\beta_{1-42}$-induced toxicity.[236] In addition, the α_7 nAChR is also expressed on microglia and astrocytes and involved in AD-associated neuroinflammation and neurodegeneration.[237,238] Hence, the α_7 nAChR appears to be a prime target in AD investigation. However, there have been no suitable radiotracers for PET imaging of the α_7 nAChR in humans until recently, when [^{18}F]ASEM and [^{18}F]DBT-10 (Fig. 5), two regioisomers of the same chemical formula, were discovered to provide adequate α_7-specific binding signals in nonhuman primates and humans.[239–243] Initial PET imaging with the α_7 nAChR selective radioligand [^{18}F]ASEM in healthy controls and MCI subjects indicated an elevation of this receptor subtype with healthy aging, and in MCI, in a direction opposite of that for $\alpha_4\beta_2$ nAChR in AD.[244,245] Considering the spectrum of disease from MCI to AD, PET imaging studies of the α_7 nAChR should provide additional insight into the role of this receptor subtype in AD pathogenesis and progression, as well as the effectiveness of α_7 nAChR agonists in AD treatment.

Imaging agents for synaptic proteins

Synapses are critical for neurotransmission in neuron-to-neuron interaction, the deficiency of which would result in neuronal dysfunction and consequent occurrence of neuropsychiatric symptoms including amnesia, apathy, and executive dysfunction. Loss of synapses is a pathologic hallmark of Alzheimer's disease[246–248] and correlates strongly with cognitive impairment.[247,249–252] Reductions in synapses are seen in the neocortex and limbic system in MCI, a prodromal form of AD, and other dementias.[253–255] Accumulation of toxic Aβ oligomers is believed to lead to the loss of synapses and presynaptic proteins in MCI patients.[252,255–257] More recent research also suggests an emerging role for tau mediated toxicity at the synapses.[258,259]

The synaptic vesicle glycoprotein 2 (SV2) is one of the synaptic vesicle proteins highly expressed on the presynaptic membranes. Among the three forms of SV2, only SV2A is extensively expressed in glutamatergic and γ-aminobutyric acid (GABA) neurons throughout the CNS.[260] The pharmacological effect of SV2A is linked to the inhibition of presynaptic calcium channels and neuronal excitability, thus explaining the pharmacological mechanism of SV2A ligand levetiracetam as an antiepileptic medication.[261]

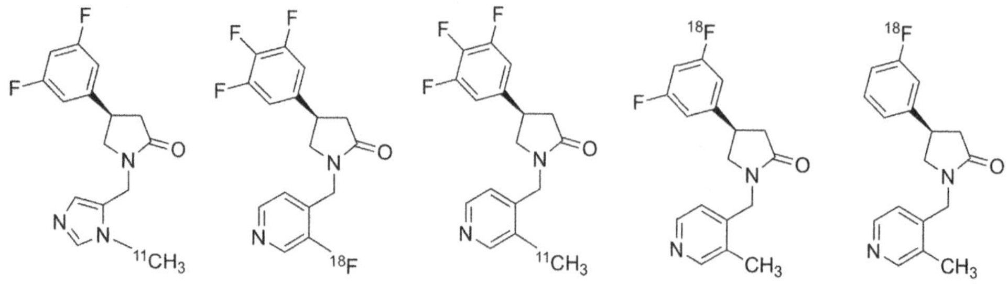

[^{11}C]UCB-A [^{18}F]UCB-H [^{11}C]UCB-J [^{18}F]SynVesT-1 (SDM-8, MNI-1126) [^{18}F]SynVesT-2 (SDM-2)

Fig. 6 Chemical structures of SV2A PET radioligands.

Traditionally, synapses can merely be observed by electron microscopy and immunohistochemistry in biopsy and autopsy tissue samples. Recently, a series of novel PET radiotracers selectively targeting SV2A has been developed (Fig. 6),[262–265] enabling non-invasive observation and quantification of synaptic density in vivo. Studies in nonhuman primates and humans demonstrate that [^{11}C]UCB-J displays favorable in vivo pharmacokinetic properties, binding selectivity, and excellent imaging characteristics.[264,266] A second SV2A radiotracer, [^{18}F]UCB-H, has also been shown to possess good pharmacokinetic properties in humans, albeit the binding signal is smaller in magnitude.[267] The availability of SV2A radiotracers for PET imaging has excited scientists in the biomedical imaging field, as it affords a promising tool for early and precise detection of synaptic changes in neurologic disorders, including AD.[268,269]

Clinical research findings

A recent SV2A PET study using [^{11}C]UCB-J in cohorts of MCI/AD patients and age-matched cognitively intact elderly subjects revealed a significant synaptic loss in the hippocampus and entorhinal cortex (44% and 27%, respectively).[270] The 44% reduction in hippocampal synaptic density (much greater than the 22% hippocampal volume loss) was more profound than the 27% decrease in entorhinal synaptic density (in parallel to the 25% entorhinal volume loss).[270] Moreover, statistically significant correlations were found between hippocampal synaptic density and episodic memory (Logical Memory II and Rey Auditory Verbal Learning Test, $r=0.56$, $P=0.01$) and global function (Clinical Dementia Rating Sum of Boxes, CDR-SB, $r=-0.61$, $P=0.003$).[270] These findings suggest that [^{11}C]UCB-J can capture the early pathological alterations in the hippocampus and entorhinal cortices. These brain structures are not readily analyzed by FDG PET because their small sizes make it difficult to detect metabolic changes due to partial volume loss, despite their early involvement in AD, i.e., Braak Stage I–II.[139] An additional study with the other SV2A radiotracer [^{18}F]UCB-H corroborated the findings with [^{11}C]UCB-J, demonstrating a significant reduction of SV2A in the hippocampus of AD patients (31%). Further, an 11%–18% reduction in the basal forebrain and anterior/dorsomedial thalamus was correlated with cognitive decline measured by MMSE.[271] A recent [^{11}C]UCB-J PET study with a larger cohort of patients with MCI and AD (34 patients vs 10 in the earlier report) further confirmed loss of synaptic density, most significant in the hippocampus and entorhinal cortex, but also present in the parahippocampal, amygdala, lateral temporal, prefrontal, posterior cingulate/precuneus, lateral parietal, and pericentral cortices.[272] These results are consistent with postmortem findings of more synaptic loss in the regions of the hippocampus, frontal cortex, cingulate gyrus, entorhinal, and temporal cortices.[273] Further, regions with reduced SV2A binding found by PET generally correspond with regional volume loss detected by MRI analysis and better correlate with CDR-SB ($r = -0.54$, $P = 0.00003$) and episodic memory ($r = 0.56$, $P = 0.00001$) than in the smaller sample reported earlier.[270,272] Finally, a recent PET imaging study with both the second generation tau tracer [^{18}F]MK-6240 and [^{11}C]UCB-J in amnestic mild cognitive impairment (aMCI) patients revealed that tau accumulation and synaptic loss in the medial temporal lobe (hippocampus, parahippocampal gyrus) are positively correlated, and both in turn are correlated with cognitive impairment.[274]

Summary

SV2A PET has great promise as a biomarker for the non-invasive detection of synaptic density/function, which can serve as a useful tool for studying neuropsychiatric disorders,[268,275] and in particular, as a direct and more stable biomarker for synaptic function and neurodegeneration [the (N) in the AT(N) research framework] in AD.[22,269] This is based on the fact that SV2A PET is presumed to be less sensitive to cofounding factors such as blood glucose level, stimulation, and medication which additionally affect FDG PET, a surrogate measure of synaptic function.[276–278] Due to the ubiquitous distribution of synaptic vesicles throughout the CNS, a highly selective molecular probe such as SV2A PET tracer can spot the subtle alteration of synaptic density both quantitatively and topographically.

With the help of multi-target imaging, the correlation between synaptic loss and other pathophysiological features related to AD can be further understood. For example, decreased energetics of the brain as measured by glucose utilization can now be correlated with synaptic loss, and the regulation of receptors (e.g., mGluR$_5$, σ_2R, and α_7 nAChR) by soluble Aβ oligomers can be more clearly elucidated.[279–281] Since pathological synaptic alterations emerge in the early stage of the disease and throughout its course, many unanswered questions and controversial issues would be hopefully settled with the applications PET imaging biomarkers for SV2A, Aβ, and tau in cross-sectional and longitudinal studies, for example, the time sequence and topographical pattern of synaptic loss, Aβ deposition, and tau pathology along the AD progression stages, and which biomarker may serve as the most stable and useful tool for diagnosis of AD at the earliest, asymptomatic stage. SV2A PET imaging will likely play a role in the early detection of AD, the differential diagnosis of dementing illnesses, and the ability to predict the conversion of prodromal stage subjects to more advanced disease states. Many studies using SV2A PET imaging in AD and other neurodegenerative diseases are ongoing, and results are emerging rapidly in publications. The introduction of additional SV2A radiotracers such as [^{18}F]SynVesT-1 (also known as [^{18}F]SDM-8 and [^{18}F]MNI-1126) and [^{18}F]SynVesT-2 (also known as [^{18}F]SDM-2) should facilitate these studies since they are suitable for central production and distribution for use in off-site imaging facilities (Fig. 6).[275,282–284] Further, PET imaging with these SV2A radiotracers will be useful as an endpoint measure in AD drug clinical trials.

Use of PET imaging biomarkers in AD drug development

Drug development is a lengthy and costly endeavor. It is estimated that it takes 10–12 years and costs close to 1 billion dollars to advance a single drug from discovery to the clinic.[285] This is a significant burden on the pharmaceutical industry, and in turn, on society as a whole, and it may have contributed to the fact that despite the rapid advance in molecular biology in the last decades and the identification of an increasing number of druggable targets, the number of approved drugs has been decreasing, not increasing. One way to contain the rising cost of drug development is to incorporate biomarkers and surrogate endpoints in drug discovery and development. PET has been increasingly recognized as a powerful imaging modality that facilitates the identification of biomarkers useful for the development of CNS therapeutics.[286,287]

Over the last two decades, PET imaging biomarkers have become an especially powerful and indispensable tool in the development of drugs for AD treatment and are widely used in the selection of trial participant, target engagement assessment, and proof of mechanism studies, as well as in the monitoring of treatment outcomes.[288] Treatments for AD include disease-modifying therapies (DMTs), cognitive enhancement, and treatments for neuropsychiatric symptoms. Among these therapeutic drugs, about two-thirds of the current AD drug development pipelines are DMTs.[289–291] Since DMTs must intervene in the basic biology of AD and modify disease progression, objective biomarkers have become an essential endpoint in clinical trials. Increasingly, PET-based biomarkers have served as an inclusion criterion for participant selection, and/or outcome measures in large-scale phase 2 and phase 3 clinical trials. Examples of anti-amyloid agents in phase 2 and 3 clinical trials with the use of PET imaging biomarkers are shown in Table 1, while Table 2 lists clinical trials of anti-tau agents that incorporate PET imaging biomarkers. These extensive lists clearly illustrate the importance of PET imaging biomarkers in AD drug development. PET imaging studies with Aβ biomarkers have repeatedly demonstrated the clearance of Aβ plaques from AD patients enrolled in clinical trials of anti-Aβ antibodies.[292–294] Unfortunately, cognitive improvement has not been

Table 1 Examples of anti-Aβ agents in phase 2 and 3 clinical trials with the use of PET imaging biomarker

No.	Drug	Mechanism	Clinical trial ID	Current status	Sponsor	Study start date	Completion date	Imaging agents/ purpose
1	Aducanumab (Monoclonal antibody, mAb)	Anti-Aβ	NCT02484547	3: T	Biogen	September 15, 2015	August 5, 2019	Aβ PET for inclusion
			NCT02477800	3: T		August 13, 2015	August 8, 2019	
			NCT03639987	2: T		December 20, 2018	July 30, 2019	
2	Bapineuzumab (mAb)	Anti-Aβ	NCT01254773	2: C	Janssen	December 2010	March 2013	Aβ PET for inclusion
			NCT00575055	3: C		December 2007	April 2012	MRI for inclusion, PET imaging and biochemical biomarkers for outcomes
			NCT00676143	3: T	Pfizer	January 2008	November 2012	[^{11}C]PIB Aβ PET for outcome measures
3	CAD106 (aβ vaccine) & CNP520 (BACE inhibitor)	Anti-Aβ	NCT02565511	2/3: T	Novartis	November 30, 2015	April 30, 2020	Aβ and tau PET for outcome measures
4	Candesartan (Angiotensin receptor blocker)	Anti-Aβ, Neuroprotection	NCT02646982	2: R	Emory University	June 2016	September 2021	[^{11}C]PIB Aβ PET for inclusion
5	CNP520 (BACE inhibitor)	Anti-Aβ	NCT03131453	2/3: C	Novartis	August 3, 2017	March 26, 2020	Aβ and tau PET for outcome measures

Continued

Table 1 Examples of anti-Aβ agents in phase 2 and 3 clinical trials with the use of PET imaging biomarker—cont'd

No.	Drug	Mechanism	Clinical trial ID	Current status	Sponsor	Study start date	Completion date	Imaging agents/purpose
6	Crenezumab (mAb)	Anti-Aβ	NCT03114657	3: T	Hoffmann-La Roche	March 29, 2017	June 11, 2019	Aβ PET for inclusion
			NCT02670083	3: T		March 22, 2016	May 31, 2019	
			NCT03977584	2: R		June 10, 2019	February 25, 2022	[^{18}F]GTP1 tau PET for outcome measures
			NCT01998841	2: A	Genentech, Inc.	December 20, 2013	February 25, 2022	Aβ PET and FDG PET for outcome measures
7	Elenbecestat (BACE inhibitor)	Anti-Aβ	NCT02956486	3: A	Eisai Co., Ltd.	October 20, 2016	November 21, 2023	Aβ PET for outcome measures
			NCT03036280	3: A	Eisai Co., Ltd.	December 29, 2016	November 21, 2023	Aβ PET for outcome measures
			NCT02322021	2: C	Eisai Co., Ltd.	November 26, 2014	December 10, 2019	Aβ PET for *inclusion*
8	Gantenerumab (mAb)	Anti-Aβ	NCT03443973	3: R	Hoffmann-La Roche	August 22, 2018	March 3, 2023	Aβ PET and tau PET for outcome measures.
			NCT03444870	3: R		June 6, 2018	November 2 , 2023	
9	Gantenerumab/ Solanezumab (mAb)	Anti-Aβ	NCT01760005	2/3: C	Washington University	December 2012	March 6, 2021	[^{11}C]PIB and [^{18}F] florbetapir Aβ PET, [^{18}F] flortaucipir tau PET and FDG PET for outcome measures

10	Lanabecestat (BACE inhibitor)	Anti-Aβ	NCT02783573	3: T	AstraZeneca	July 1, 2016	September 28, 2018	[^{18}F] Florbetapir Aβ PET for outcome measures
11	Solanezumab (mAb)	Anti-Aβ	NCT02760602	3: T	Eli Lilly & Co.	June 2016	May 2017	[^{18}F] Florbetapir Aβ PET for inclusion, [^{18}F] Florbetapir Aβ PET and [^{18}F]AV-1451 tau PET for outcome measures
			NCT01900665	3: T		July 2013	February 2017	[^{18}F] Florbetapir Aβ PET for both inclusion and outcome measures.
			NCT02008357	3: A		February 28, 2014	January 16, 2023	
12	CT1812 (Sigma-2 receptor antagonist)	Anti-Aβ	NCT03493282	1/2: A	Cognition Therapeutics	April 20, 2018	March 1, 2021	[^{11}C]PIB Aβ PET for inclusion, [^{11}C]UCB-J SV2A PET for outcome measures

Continued

Table 1 Examples of anti-Aβ agents in phase 2 and 3 clinical trials with the use of PET imaging biomarker—cont'd

No.	Drug	Mechanism	Clinical trial ID	Current status	Sponsor	Study start date	Completion date	Imaging agents/ purpose
13	ID1201 (α-secretase enhancer)	Anti-Aβ	NCT03363269	2: C	IlDong Pharmaceutical Co. Ltd.	April 12, 2016	April 17, 2019	Aβ PET for inclusion
14	LY3002813 (mAb)	Anti-Aβ	NCT03367403	2: A	Eli Lilly & Co.	December 18, 2017	November 12, 2021	[^{18}F] Florbetapir Aβ PET and [^{18}F] flortaucipir tau PET for both inclusion and outcome measures
15	Octagam 10% (human immunoglobulin)	Anti-Aβ	NCT03319810	2: C	Sutter Health	January 4, 2018	July 19, 2018	[^{18}F] Florbetapir Aβ PET for both inclusion and outcome measures
16	Rasagiline (MAO-B inhibitor)	Anti-Aβ, neuroprotection	NCT02359552	2: C	The Cleveland Clinic	May 2015	January 4, 2019	FDG PET for both inclusion and outcome measures
17	Semagacestat (γ-secretase inhibitor)	Anti-Aβ	NCT00594568 NCT00762411	3: C	Eli Lilly & Co.	March 2008 September 2008	May 2011 April 2011	FDG and [^{18}F]AV-45 Aβ PET for outcome measures

18	UB-311 (Active immunotherapy)	Anti-Aβ	NCT03531710	2: T	United Neuroscience Ltd.	August 10, 2018	October 31, 2019	[18F]AV-45 Aβ PET for outcome measures
19	LY3202626 (BACE Inhibitor)	Anti-Aβ	NCT02791191	2: T	Eli Lilly & Co.	June 16, 2016	July 2, 2018	[18F]AV-1451 tau PET for outcome measures
20	UB-311 (Aβ vaccine)	Anti-Aβ	NCT03531710	2: T	United Neuroscience Ltd.	August 10, 2018	October 31, 2019	[18F]AV-45 Aβ PET for outcome measures
21	AZD0530 (Saracatinib, Fyn inhibitor)	Anti-Aβ	NCT02167256	2: C	Yale University	December, 2014	February 27, 2018	[18F] Florbetapir Aβ PET for inclusion and FDG PET for outcome measures

A: active, not recruiting; C: completed; E: enrolling by invitation; R: recruiting; T: terminated.

Table 2 Examples of anti-tau agents in phase 2 and 3 clinical trials with PET imaging biomarkers

No.	Drug	Mechanism	Trial ID	Current status	Sponsor	Study start date	Completion date	Imaging agents/ purpose
1	AADvac1 (Active immunotherapy)	Anti-tau	NCT02579252	2: C	Axon Neuroscience	March 2016	June 2019	FDG PET for outcome measures
2	Tilavonemab (ABBV-8E12, anti-tau mAb)	Anti-tau	NCT02880956	2: A	AbbVie	October 17, 2016	July 28, 2021	Aβ PET for inclusion
3	BIIB092 (mAb)	Anti-tau	NCT03352557	2: A	Biogen	May 3, 2018	March 26, 2024	Aβ PET for inclusion
4	LY3303560 (mAb)	Anti-tau	NCT03518073	2: A	Eli Lilly & Co.	April 30, 2018	October 22, 2021	tau PET for outcome measures
5	Semorinemab (mAb)	Anti-tau	NCT03289143	2: A	Genentech	October 4, 2017	June 21, 2022	$[^{18}F]$GTP1 tau PET for outcome measures
			NCT03828747	2: R	Genentech	January 30, 2019	August 30, 2023	
6	TRx0237 (Tau aggregation inhibitor)	Anti-tau	NCT03446001	3: R	TauRx Therapeutics	January 10, 2018	December 2022	Aβ PET for inclusion, FDG PET for outcome measures
			NCT01689246	3: C	TauRx Therapeutics	January 2013	November 2015	FDG PET for outcome measures

A: active, not recruiting; C: completed; E: enrolling by invitation; R: recruiting; T: terminated.

observed in these studies.[295] Ultimately, this has cast doubt on the validity of the amyloid hypothesis in AD.[296] Nonetheless, Biogen will seek FDA approval of its anti-Aβ antibody aducanumab for AD treatment based on the enhancement of cognitive function in a subgroup of patients treated with the highest dose of aducanumab.[297]

Given the generally disappointing clinical trial results with anti-Aβ antibodies and the resulting consensus of targeting AD treatment for the early stages of the disease, the focus has now shifted to the development of drugs targeting tau protein aggregates, as well as soluble Aβ oligomers, through removal or prevention of their binding to neurons.[298] As these agents enter clinical trials and suitable tau PET imaging agents become available, AD imaging biomarkers such as Aβ, tau, and FDG PET undoubtedly will continue to play an important role in trial participant selection and outcome assessment.

In summary, the utilization of PET imaging biomarkers in CNS drug development, and in particular AD drug development, has become a mainstay in clinical trials to validate target engagement, mechanism of action, and treatment efficacy. This trend is expected to continue in the drug development arena.

Conclusion and perspectives

Alzheimer's disease, characterized by impairments of cognitive functions and memory loss, is an extremely complex neurodegenerative disorder involving multiple pathophysiological processes, from loss of synapses and neurons, to neuroinflammation, to deposition of Aβ plaques and neurofibrillary tau tangles. Early diagnosis is the key to effectively treat the disease before the manifestation of clinical symptoms.

In the last two decades, PET imaging agents targeting various pathologic and physiological biomarkers in AD have greatly contributed to the investigation and diagnosis of this disease. In particular, the development of Aβ imaging agents has allowed the differentiation of AD and non-AD dementia, the tracking of Aβ burden over the temporal course of the disease, the stratification of patients for clinical trials, and the monitoring of biological effects of therapeutic agents. Results from PET imaging of Aβ burden in moderate to severe AD dementia have been found to largely parallel glucose hypometabolism as imaged by FDG PET. However, findings in healthy elderly subjects and patients with MCI are more variable, and correlation with FDG hypometabolism and cognitive function is relatively weak.

Radiotracers for tau protein aggregates, the other pathologic hallmark of AD, have multiplied over the last few years, and the second-generation tracers with improved pharmacokinetic and binding profiles are in the early stage of clinical validation. Initial results have indicated the promise of tau PET imaging in AD by showing that aggregation of tau, but not Aβ, tracks well with brain atrophy and memory performance.[128] Although the first-generation tau tracer [^{18}F]AV-1451 ([^{18}F]flortaucipir, Tauvid) has gained FDA approval for the detection of neurofibrillary tangles, the value of tau PET imaging in the early diagnosis and progression monitoring of AD remains to be fully established.

As cholinergic dysfunction, synaptic loss, and microglial activation have all been thought to precede the formation of neurofibrillary tangles in AD, applications of PET radiotracers for these targets, together with those for Aβ and tau, should help unravel the chronological and topographic progression of these biomarkers and identify the most appropriate one(s) for truly early diagnosis of the disease and monitoring of its progression.

Measurement of synaptic density through PET imaging of the synaptic biomarker SV2A opens a new avenue for the investigation of neurodegenerative diseases.[268,269,299] Initial results in MCI and AD dementia patients have uncovered synaptic losses in the hippocampus and cortex and their correlation with cognitive dysfunction. Preliminary results from our laboratory (unpublished results) and others[274] have also indicated an excellent correlation between synaptic loss measured by SV2A PET and tau protein aggregates from healthy aging to AD dementia that is consistent with postmortem findings, indicating the potential of SV2A PET imaging for early detection of synaptic loss and AD progression. We presume that PET imaging with SV2A agents will be able to detect synaptic changes and cognitive abnormalities in the prodromal and preclinical stages of AD since the loss of synaptic density has been shown to be closely correlated with cognitive impairment. This will need to be proven in large-scale clinical trials with cognitively normal subjects, subjects with MCI, and AD dementia patients.

PET imaging biomarkers have played an important role in AD drug development. This trend will continue, as newer imaging biomarkers such as tau (Table 2) and SV2A PET (NCT03493282, Table 1) are increasingly incorporated into clinical trials of next-generation AD drugs as outcome measures. It is expected that applications of PET imaging biomarkers in drug clinical trials will continue to greatly facilitate the development of neurotherapeutics for the treatment of neurologic and psychiatric disorders including AD.

References

1. Prince M, Comas-Herrera A, Knapp M, Guerchet M, Karagiannidou M. *World Alzheimer Report 2016: Improving Healthcare for People Living With Dementia Coverage, Quality and Costs Now and in the Future.* London: Alzheimer's Disease International; 2016.
2. Podcasy JL, Epperson CN. Considering sex and gender in Alzheimer disease and other dementias. *Dialogues Clin Neurosci.* 2016;18:437–446.
3. Patterson C. *Alzheimer's Disease International. World Alzheimer Report 2018: The State of the Art of Dementia Research: New Frontiers.* London: Alzheimer's Disease International; 2018 p. 48.
4. Hardy JA, Higgins GA. Alzheimer's disease: the amyloid cascade hypothesis. *Science.* 1992;256:184–185.
5. Cline EN, Bicca MA, Viola KL, Klein WL. The amyloid-β oligomer hypothesis: beginning of the third decade. *J Alzheimers Dis.* 2018;64:S567–S610.
6. Viola KL, Klein WL. Amyloid β oligomers in Alzheimer's disease pathogenesis, treatment, and diagnosis. *Acta Neuropathol.* 2015;129:183–206.

7. Grundke-Iqbal I, et al. Abnormal phosphorylation of the microtubule-associated protein tau (tau) in Alzheimer cytoskeletal pathology. *Proc Natl Acad Sci U S A*. 1986;83:4913–4917.
8. Davies P, Maloney AJF. Selective loss of central cholinergic neurons in Alzheimer's disease. *Lancet*. 1976;2:1403.
9. Bartus RT, Dean III RL, Beer B, Lippa AS. The cholinergic hypothesis of geriatric memory dysfunction. *Science*. 1982;217:408–414.
10. Coyle JT, Price DL, DeLong MR. Alzheimer's disease: a disorder of cortical cholinergic innervation. *Science*. 1983;219:1184–1190.
11. Aigner TG. Pharmacology of memory: cholinergic-glutamatergic interactions. *Curr Opin Neurobiol*. 1995;5:155–160.
12. McGeer PL, McGeer EG. The amyloid cascade-inflammatory hypothesis of Alzheimer disease: implications for therapy. *Acta Neuropathol*. 2013;126:479–497.
13. Swerdlow RH. Mitochondria and mitochondrial cascades in Alzheimer's disease. *J Alzheimers Dis*. 2018;62:1403–1416.
14. Area-Gomez E, Schon EA. On the pathogenesis of Alzheimer's disease: the MAM hypothesis. *FASEB J*. 2017;31:864–867.
15. Albert MS, et al. The diagnosis of mild cognitive impairment due to Alzheimer's disease: recommendations from the National Institute on Aging-Alzheimer's Association workgroups on diagnostic guidelines for Alzheimer's disease. *Alzheimers Dement*. 2011;7:270–279.
16. McKhann GM, et al. The diagnosis of dementia due to Alzheimer's disease: recommendations from the National Institute on Aging-Alzheimer's Association workgroups on diagnostic guidelines for Alzheimer's disease. *Alzheimers Dement*. 2011;7:263–269.
17. Sperling RA, et al. Toward defining the preclinical stages of Alzheimer's disease: recommendations from the National Institute on Aging-Alzheimer's Association workgroups on diagnostic guidelines for Alzheimer's disease. *Alzheimers Dement*. 2011;7:280–292.
18. Barthel H, Seibyl J, Sabri O. The role of positron emission tomography imaging in understanding Alzheimer's disease. *Expert Rev Neurother*. 2015;15:395–406.
19. Declercq LD, Vandenberghe R, Van Laere K, Verbruggen A, Bormans G. Drug development in Alzheimer's disease: the contribution of PET and SPECT. *Front Pharmacol*. 2016;7:88.
20. Arora A, Bhagat N. Insight into the molecular imaging of Alzheimer's disease. *Int J Biomed Imaging*. 2016;2016:7462014.
21. Lloret A, et al. When does Alzheimer's disease really start? The role of biomarkers. *Int J Mol Sci*. 2019;20:5536.
22. Jack Jr. CR, et al. NIA-AA research framework: toward a biological definition of Alzheimer's disease. *Alzheimers Dement*. 2018;14:535–562.
23. Cummings J. The National Institute on Aging-Alzheimer's Association framework on Alzheimer's disease: application to clinical trials. *Alzheimers Dement*. 2019;15:172–178.
24. Laruelle M, Slifstein M, Huang Y. Relationships between radiotracer properties and image quality in molecular imaging of the brain with positron emission tomography. *Mol Imaging Biol*. 2003;5:363–375.
25. Pike VW. Considerations in the development of reversibly binding PET radioligands for brain imaging. *Curr Med Chem*. 2016;23:1818–1869.
26. McCluskey SP, Plisson C, Rabiner EA, Howes O. Advances in CNS PET: the state-of-the-art for new imaging targets for pathophysiology and drug development. *Eur J Nucl Med Mol Imaging*. 2020;47:451–489.
27. Mathis CA, Lopresti BJ, Ikonomovic MD, Klunk WE. Small-molecule PET tracers for imaging proteinopathies. *Semin Nucl Med*. 2017;47:553–575.
28. Morris E, et al. Diagnostic accuracy of ^{18}F amyloid PET tracers for the diagnosis of Alzheimer's disease: a systematic review and meta-analysis. *Eur J Nucl Med Mol Imaging*. 2016;43:374–385.
29. Barthel H, Sabri O. Clinical use and utility of amyloid imaging. *J Nucl Med*. 2017;58:1711–1717.
30. Cistaro A, Alongi P, Caobelli F, Cassalia L. Radiotracers for amyloid imaging in neurodegenerative disease: state-of-the-art and novel concepts. *Curr Med Chem*. 2018;25:3131–3140.
31. Klunk WE, et al. Imaging brain amyloid in Alzheimer's disease with Pittsburgh compound-B. *Ann Neurol*. 2004;55:306–319.

32. Ikonomovic MD, et al. Post-mortem correlates of *in vivo* PiB-PET amyloid imaging in a typical case of Alzheimer's disease. *Brain*. 2008;131:1630–1645.
33. Mathis CA, Mason NS, Lopresti BJ, Klunk WE. Development of positron emission tomography β-amyloid plaque imaging agents. *Semin Nucl Med*. 2012;42:423–432.
34. Rowe CC, Villemagne VL. Brain amyloid imaging. *J Nucl Med*. 2011;52:1733–1740.
35. European Medicines Agency. Amyvid: florbetapir (^{18}F). Authorisation details. https://www.ema.europa.eu/en/medicines/human/EPAR/amyvid; 2013. Accessed 22 December 2019.
36. Anonymous, European Medicines Agency. Neuraceq: florbetaben (18F). Authorisation details. https://www.ema.europa.eu/en/medicines/human/EPAR/neuraceq; 2014. Accessed 22 December 2019.
37. European Medicines Agency. Vizamyl: flutemetamol (^{18}F). Authorisation details. https://www.ema.europa.eu/en/medicines/human/EPAR/vizamyl; 2014. Accessed 22 December 2019.
38. Wong DF, et al. In vivo imaging of amyloid deposition in Alzheimer disease using the radioligand ^{18}F-AV-45 (florbepapir F 18). *J Nucl Med*. 2010;51:913–920.
39. Clark CM, et al. Cerebral PET with florbetapir compared with neuropathology at autopsy for detection of neuritic amyloid-β plaques: a prospective cohort study. *Lancet Neurol*. 2012;11:669–678.
40. Mathis CA, Lopresti BJ, Klunk WE. Impact of amyloid imaging on drug development in Alzheimer's disease. *Nucl Med Biol*. 2007;34:809–822.
41. Clark CM, et al. Cerebral PET with florbetapir compared with neuropathology at autopsy for detection of neuritic amyloid-beta plaques: a prospective cohort study. *Lancet Neurol*. 2012;11:669–678.
42. Rinne JO, et al. [^{18}F]Flutemetamol PET imaging and cortical biopsy histopathology for fibrillar amyloid β detection in living subjects with normal pressure hydrocephalus: pooled analysis of four studies. *Acta Neuropathol*. 2012;124:833–845.
43. Curtis C, et al. Phase 3 trial of flutemetamol labeled with radioactive fluorine 18 imaging and neuritic plaque density. *JAMA Neurol*. 2015;72:287–294.
44. Sabri O, et al. Florbetaben PET imaging to detect amyloid beta plaques in Alzheimer's disease: phase 3 study. *Alzheimers Dement*. 2015;11:964–974.
45. Rowe CC, et al. Imaging of amyloid β in Alzheimer's disease with ^{18}F-BAY94-9172, a novel PET tracer: proof of mechanism. *Lancet Neurol*. 2008;7:129–135.
46. Johnson KA, et al. Imaging of amyloid burden and distribution in cerebral amyloid angiopathy. *Ann Neurol*. 2007;62:229–234.
47. Ossenkoppele R, et al. Prevalence of amyloid PET positivity in dementia syndromes: a meta-analysis. *JAMA*. 2015;313:1939–1949.
48. Petrou M, et al. Amyloid deposition in Parkinson's disease and cognitive impairment: a systematic review. *Mov Disord*. 2015;30:928–935.
49. Jack Jr. CR, et al. Suspected non-Alzheimer disease pathophysiology—concept and controversy. *Nat Rev Neurol*. 2016;12:117–124.
50. Dubois B, et al. Advancing research diagnostic criteria for Alzheimer's disease: the IWG-2 criteria. *Lancet Neurol*. 2014;13:614–629.
51. Okello A, et al. Conversion of amyloid positive and negative MCI to AD over 3 years: an ^{11}C-PIB PET study. *Neurology*. 2009;73:754–760.
52. Teipel S, et al. Multimodal imaging in Alzheimer's disease: validity and usefulness for early detection. *Lancet Neurol*. 2015;14:1037–1053.
53. Jack Jr. CR, Holtzman DM. Biomarker modeling of Alzheimer's disease. *Neuron*. 2013;80:1347–1358.
54. Jack Jr. CR, et al. Brain β-amyloid load approaches a plateau. *Neurology*. 2013;80:890–896.
55. Pike KE, et al. β-Amyloid imaging and memory in non-demented individuals: evidence for preclinical Alzheimer's disease. *Brain*. 2007;130:2837–2844.
56. Villemagne VL, et al. Aβ deposits in older non-demented individuals with cognitive decline are indicative of preclinical Alzheimer's disease. *Neuropsychologia*. 2008;46:1688–1697.
57. Pike KE, et al. Cognition and beta-amyloid in preclinical Alzheimer's disease: data from the AIBL study. *Neuropsychologia*. 2011;49:2384–2390.
58. Edison P, et al. Amyloid, hypometabolism, and cognition in Alzheimer disease: an [^{11}C]PIB and [^{18}F]FDG PET study. *Neurology*. 2007;68:501–508.

59. Cohen AD, et al. Basal cerebral metabolism may modulate the cognitive effects of Aβ in mild cognitive impairment: an example of brain reserve. *J Neurosci.* 2009;29:14770–14778.
60. Klunk WE, et al. The Centiloid Project: standardizing quantitative amyloid plaque estimation by PET. *Alzheimers Dement.* 2015;11:1–15.
61. Rowe CC, et al. Standardized expression of ^{18}F-NAV4694 and ^{11}C-PiB beta-amyloid PET results with the centiloid scale. *J Nucl Med.* 2016;57:1233–1237.
62. Rowe CC, et al. ^{18}F-Florbetaben PET β-amyloid binding expressed in Centiloids. *Eur J Nucl Med Mol Imaging.* 2017;44:2053–2059.
63. Battle MR, et al. Centiloid scaling for quantification of brain amyloid with [^{18}F]flutemetamol using multiple processing methods. *EJNMMI Res.* 2018;8:107.
64. Bourgeat P, et al. Implementing the centiloid transformation for ^{11}C-PiB and β-amyloid ^{18}F-PET tracers using CapAIBL. *NeuroImage.* 2018;183:387–393.
65. Navitsky M, et al. Standardization of amyloid quantitation with florbetapir standardized uptake value ratios to the Centiloid scale. *Alzheimers Dement.* 2018;14:1565–1571.
66. Jansen WJ, et al. Prevalence of cerebral amyloid pathology in persons without dementia: a meta-analysis. *JAMA.* 2015;313:1924–1938.
67. Rowe CC, et al. Amyloid imaging results from the Australian imaging, biomarkers and lifestyle (AIBL) study of aging. *Neurobiol Aging.* 2010;31:1275–1283.
68. Cummings J, Ritter A, Zhong K. Clinical trials for disease-modifying therapies in Alzheimer's disease: a primer, lessons learned, and a blueprint for the future. *J Alzheimers Dis.* 2018;64:S3–S22.
69. Giannakopoulos P, et al. Tangle and neuron numbers, but not amyloid load, predict cognitive status in Alzheimer's disease. *Neurology.* 2003;60:1495–1500.
70. Shin J, Kepe V, Barrio JR, Small GW. The merits of FDDNP-PET imaging in Alzheimer's disease. *J Alzheimers Dis.* 2011;26(Suppl. 3):135–145.
71. Thompson PW, et al. Interaction of the amyloid imaging tracer FDDNP with hallmark Alzheimer's disease pathologies. *J Neurochem.* 2009;109:623–630.
72. Shoghi-Jadid K, et al. Localization of neurofibrillary tangles and beta-amyloid plaques in the brains of living patients with Alzheimer disease. *Am J Geriatr Psychiatry.* 2002;10:24–35.
73. Dani M, Brooks DJ, Edison P. Tau imaging in neurodegenerative diseases. *Eur J Nucl Med Mol Imaging.* 2016;43:1139–1150.
74. Okamura N, et al. Advances in the development of tau PET radiotracers and their clinical applications. *Ageing Res Rev.* 2016;107–113.
75. Leuzy A, et al. Tau PET imaging in neurodegenerative tauopathies-still a challenge. *Mol Psychiatry.* 2019;24:1112–1134.
76. Lois C, Gonzalez I, Johnson KA, Price JC. PET imaging of tau protein targets: a methodology perspective. *Brain Imaging Behav.* 2019;13:333–344.
77. Wang YT, Edison P. Tau imaging in neurodegenerative diseases using positron emission tomography. *Curr Neurol Neurosci Rep.* 2019;19:45.
78. Kimura Y, et al. PET quantification of tau pathology in human brain with ^{11}C-PBB3. *J Nucl Med.* 2015;56:1359–1365.
79. Villemagne VL, et al. In vivo evaluation of a novel tau imaging tracer for Alzheimer's disease. *Eur J Nucl Med Mol Imaging.* 2014;41:816–826.
80. Okamura N, et al. Non-invasive assessment of Alzheimer's disease neurofibrillary pathology using ^{18}F-THK5105 PET. *Brain.* 2014;137:1762–1771.
81. Harada R, et al. ^{18}F-THK5351: a novel PET radiotracer for imaging neurofibrillary pathology in Alzheimer disease. *J Nucl Med.* 2016;57:208–214.
82. Betthauser T, et al. In vivo comparison of tau radioligands ^{18}F-THK-5351 and ^{18}F-THK-5317. *J Nucl Med.* 2017;58:996–1002.
83. Ng KP, et al. Monoamine oxidase B inhibitor, selegiline, reduces ^{18}F-THK5351 uptake in the human brain. *Alzheimers Res Ther.* 2017;9:25.
84. Lowe VJ, et al. Tau-positron emission tomography correlates with neuropathology findings. *Alzheimers Dement.* 2020;16:561–571.

85. Marquie M, et al. Pathological correlations of [F-18]-AV-1451 imaging in non-alzheimer tauopathies. *Ann Neurol.* 2017;81:117–128.
86. Marquie M, et al. Validating novel tau positron emission tomography tracer [F-18]-AV-1451 (T807) on postmortem brain tissue. *Ann Neurol.* 2015;78:787–800.
87. Wooten DW, et al. Pharmacokinetic evaluation of the tau PET radiotracer ^{18}F-T807 (^{18}F-AV-1451) in human subjects. *J Nucl Med.* 2017;58:484–491.
88. Lowe VJ, et al. An autoradiographic evaluation of AV-1451 tau PET in dementia. *Acta Neuropathol Commun.* 2016;4:58.
89. Smith R, et al. Tau neuropathology correlates with FDG-PET, but not AV-1451-PET, in progressive supranuclear palsy. *Acta Neuropathol.* 2017;133:149–151.
90. Sander K, et al. Characterization of tau positron emission tomography tracer [^{18}F]AV-1451 binding to postmortem tissue in Alzheimer's disease, primary tauopathies, and other dementias. *Alzheimers Dement.* 2016;12:1116–1124.
91. Ossenkoppele R, et al. Discriminative accuracy of [^{18}F]flortaucipir positron emission tomography for Alzheimer disease vs other neurodegenerative disorders. *JAMA.* 2018;320:1151–1162.
92. Vermeiren C, et al. The tau positron-emission tomography tracer AV-1451 binds with similar affinities to tau fibrils and monoamine oxidases. *Mov Disord.* 2018;33:273–281.
93. Choi JY, et al. Off-target ^{18}F-AV-1451 binding in the basal ganglia correlates with age-related iron accumulation. *J Nucl Med.* 2018;59:117–120.
94. Drake LR, et al. Identification of AV-1451 as a weak, nonselective inhibitor of monoamine oxidase. *ACS Chem Neurosci.* 2019;10:3839–3846.
95. Hostetler ED, et al. Preclinical characterization of ^{18}F-MK-6240, a promising PET tracer for in vivo quantification of human neurofibrillary tangles. *J Nucl Med.* 2016;57:1599–1606.
96. Aguero C, et al. Autoradiography validation of novel tau PET tracer [F-18]-MK-6240 on human postmortem brain tissue. *Acta Neuropathol Commun.* 2019;7:37.
97. Lohith TG, et al. Brain imaging of Alzheimer dementia patients and elderly controls with ^{18}F-MK-6240, a PET tracer targeting neurofibrillary tangles. *J Nucl Med.* 2019;60:107–114.
98. Betthauser TJ, et al. In vivo characterization and quantification of neurofibrillary tau PET radioligand ^{18}F-MK-6240 in humans from Alzheimer disease dementia to young controls. *J Nucl Med.* 2019;60:93–99.
99. Kroth H, et al. Discovery and preclinical characterization of [^{18}F]PI-2620, a next-generation tau PET tracer for the assessment of tau pathology in Alzheimer's disease and other tauopathies. *Eur J Nucl Med Mol Imaging.* 2019;46:2178–2189.
100. Bullich S, et al. Evaluation of dosimetry, quantitative methods, and test-retest variability of ^{18}F-PI-2620 PET for the assessment of tau deposits in the human brain. *J Nucl Med.* 2020;61:920–927.
101. Mueller A, et al. Tau PET imaging with ^{18}F-PI-2620 in patients with Alzheimer disease and healthy controls: a first-in-humans study. *J Nucl Med.* 2020;61:911–919.
102. Wong DF, et al. Characterization of 3 novel tau radiopharmaceuticals, ^{11}C-RO-963, ^{11}C-RO-643, and ^{18}F-RO-948, in healthy controls and in Alzheimer subjects. *J Nucl Med.* 2018;59: 1869–1876.
103. Kuwabara H, et al. Evaluation of ^{18}F-RO-948 PET for quantitative assessment of tau accumulation in the human brain. *J Nucl Med.* 2018;59:1877–1884.
104. Smith R, et al. Head-to-head comparison of tau positron emission tomography tracers [^{18}F]flortaucipir and [^{18}F]RO948. *Eur J Nucl Med Mol Imaging.* 2020;47:342–354.
105. Sanabria Bohorquez S, et al. [^{18}F]GTP1 (Genentech Tau Probe 1), a radioligand for detecting neurofibrillary tangle tau pathology in Alzheimer's disease. *Eur J Nucl Med Mol Imaging.* 2019;46: 2077–2089.
106. Teng E, et al. Cross-sectional associations between [^{18}F]GTP1 tau PET and cognition in Alzheimer's disease. *Neurobiol Aging.* 2019;81:138–145.
107. Murugan NA, et al. Cross-interaction of tau PET tracers with monoamine oxidase B: evidence from in silico modelling and in vivo imaging. *Eur J Nucl Med Mol Imaging.* 2019;46:1369–1382.
108. Declercq L, et al. Preclinical evaluation of ^{18}F-JNJ64349311, a novel PET tracer for tau imaging. *J Nucl Med.* 2017;58:975–981.

109. Rombouts FJR, et al. Discovery of *N*-(4-[^{18}F]fluoro-5-methylpyridin-2-yl)isoquinolin-6-amine (JNJ-64326067), a new promising tau positron emission tomography imaging tracer. *J Med Chem.* 2019;62:2974–2987.
110. Rombouts FJ, et al. Discovery of *N*-(pyridin-4-yl)-1,5-naphthyridin-2-amines as potential tau pathology PET tracers for Alzheimer's disease. *J Med Chem.* 2017;60:1272–1291.
111. Schmidt ME, et al. Clinical evaluation of [^{18}F]JNJ-64326067, a novel candidate PET tracer for the detection of tau pathology in Alzheimer's disease. *Eur J Nucl Med Mol Imaging.* 2020. https://doi.org/10.1007/s00259-020-04880-1.
112. Rabinovici GD, Jagust WJ. Amyloid imaging in aging and dementia: testing the amyloid hypothesis in vivo. *Behav Neurol.* 2009;21:117–128.
113. Villemagne VL, Dore V, Burnham SC, Masters CL, Rowe CC. Imaging tau and amyloid-β proteinopathies in Alzheimer disease and other conditions. *Nat Rev Neurol.* 2018;14:225–236.
114. Bischof GN, Endepols H, van Eimeren T, Drzezga A. Tau-imaging in neurodegeneration. *Methods.* 2017;130:114–123.
115. Jack Jr. CR, et al. Tracking pathophysiological processes in Alzheimer's disease: an updated hypothetical model of dynamic biomarkers. *Lancet Neurol.* 2013;12:207–216.
116. Johnson KA, et al. Tau positron emission tomographic imaging in aging and early Alzheimer disease. *Ann Neurol.* 2016;79:110–119.
117. Ossenkoppele R, et al. Tau PET patterns mirror clinical and neuroanatomical variability in Alzheimer's disease. *Brain.* 2016;139:1551–1567.
118. Bejanin A, et al. Tau pathology and neurodegeneration contribute to cognitive impairment in Alzheimer's disease. *Brain.* 2017;140:3286–3300.
119. Shimada H, et al. Association between Aβ and tau accumulations and their influence on clinical features in aging and Alzheimer's disease spectrum brains: a [^{11}C]PBB3-PET study. *Alzheimers Dement.* 2017;6:11–20.
120. Crary JF, et al. Primary age-related tauopathy (PART): a common pathology associated with human aging. *Acta Neuropathol.* 2014;128:755–766.
121. Jack Jr. CR. PART and SNAP. *Acta Neuropathol.* 2014;128:773–776.
122. Whitwell JL, et al. Imaging correlations of tau, amyloid, metabolism, and atrophy in typical and atypical Alzheimer's disease. *Alzheimers Dement.* 2018;14:1005–1014.
123. Sepulcre J, et al. *In vivo tau*, amyloid, and gray matter profiles in the aging brain. *J Neurosci.* 2016;36:7364–7374.
124. Iaccarino L, et al. Local and distant relationships between amyloid, tau and neurodegeneration in Alzheimer's disease. *Neuroimage Clin.* 2018;17:452–464.
125. Mattsson N, et al. Comparing ^{18}F-AV-1451 with CSF t-tau and p-tau for diagnosis of Alzheimer disease. *Neurology.* 2018;90:e388–e395.
126. Murray ME, et al. Clinicopathologic and ^{11}C-Pittsburgh compound B implications of Thal amyloid phase across the Alzheimer's disease spectrum. *Brain.* 2015;138:1370–1381.
127. Thal DR, et al. [^{18}F]flutemetamol amyloid positron emission tomography in preclinical and symptomatic Alzheimer's disease: specific detection of advanced phases of amyloid-β pathology. *Alzheimers Dement.* 2015;11:975–985.
128. La Joie R, et al. Prospective longitudinal atrophy in Alzheimer's disease correlates with the intensity and topography of baseline tau-PET. *Sci Transl Med.* 2020;12:eaau5732.
129. Fitzpatrick AWP, et al. Cryo-EM structures of tau filaments from Alzheimer's disease. *Nature.* 2017;547:185–190.
130. Lemoine L, Leuzy A, Chiotis K, Rodriguez-Vieitez E, Nordberg A. Tau positron emission tomography imaging in tauopathies: the added hurdle of off-target binding. *Alzheimers Dement.* 2018;10:232–236.
131. Okamura N, et al. The development and validation of tau PET tracers: current status and future directions. *Clin Transl Imaging.* 2018;6:305–316.
132. Passamonti L, et al. ^{18}F-AV-1451 positron emission tomography in Alzheimer's disease and progressive supranuclear palsy. *Brain.* 2017;140:781–791.

133. Ikonomovic MD, Abrahamson EE, Price JC, Mathis CA, Klunk WE. [F-18]AV-1451 positron emission tomography retention in choroid plexus: more than "off-target" binding. *Ann Neurol.* 2016;80: 307–308.
134. Giacobini E, Gold G. Alzheimer disease therapy—moving from amyloid-β to tau. *Nat Rev Neurol.* 2013;9:677–686.
135. Phelps ME, et al. Tomographic measurement of local cerebral glucose metabolic rate in humans with (F-18)2-fluoro-2-deoxy-D-glucose: validation of method. *Ann Neurol.* 1979;6:371–388.
136. McGeer PL, et al. Comparison of PET, MRI, and CT with pathology in a proven case of Alzheimer's disease. *Neurology.* 1986;36:1569–1574.
137. Mielke R, et al. Regional cerebral glucose metabolism and postmortem pathology in Alzheimer's disease. *Acta Neuropathol.* 1996;91:174–179.
138. Scholl M, et al. Time course of glucose metabolism in relation to cognitive performance and postmortem neuropathology in Met146Val PSEN1 mutation carriers. *J Alzheimers Dis.* 2011;24:495–506.
139. Braak H, Braak E. Frequency of stages of Alzheimer-related lesions in different age categories. *Neurobiol Aging.* 1997;18:351–357.
140. Vellas B, Andrieu S, Sampaio C, Wilcock G, European Task Force. Disease-modifying trials in Alzheimer's disease: a European task force consensus. *Lancet Neurol.* 2007;6:56–62.
141. Molinuevo JL, Berthier ML, Rami L. Donepezil provides greater benefits in mild compared to moderate Alzheimer's disease: implications for early diagnosis and treatment. *Arch Gerontol Geriatr.* 2011;52:18–22.
142. Bennett DA, Wilson RS, Boyle PA, Buchman AS, Schneider JA. Relation of neuropathology to cognition in persons without cognitive impairment. *Ann Neurol.* 2012;72:599–609.
143. Frisoni GB, et al. Strategic roadmap for an early diagnosis of Alzheimer's disease based on biomarkers. *Lancet Neurol.* 2017;16:661–676.
144. Yuan Y, Gu ZX, Wei WS. Fluorodeoxyglucose-positron-emission tomography, single-photon emission tomography, and structural MR imaging for prediction of rapid conversion to Alzheimer disease in patients with mild cognitive impairment: a meta-analysis. *AJNR Am J Neuroradiol.* 2009;30:404–410.
145. Zhang S, et al. Diagnostic accuracy of ^{18}F-FDG and ^{11}C-PIB-PET for prediction of short-term conversion to Alzheimer's disease in subjects with mild cognitive impairment. *Int J Clin Pract.* 2012;66: 185–198.
146. Arbizu J, et al. Clinical utility of FDG-PET for the clinical diagnosis in MCI. *Eur J Nucl Med Mol Imaging.* 2018;45:1497–1508.
147. Bateman RJ, et al. Autosomal-dominant Alzheimer's disease: a review and proposal for the prevention of Alzheimer's disease. *Alzheimers Res Ther.* 2011;3:1.
148. Villemagne VL, et al. Longitudinal assessment of Aβ and cognition in aging and Alzheimer disease. *Ann Neurol.* 2011;69:181–192.
149. Wolfsgruber S, et al. Differential risk of incident Alzheimer's disease dementia in stable versus unstable patterns of subjective cognitive decline. *J Alzheimers Dis.* 2016;54:1135–1146.
150. Scheef L, et al. Glucose metabolism, gray matter structure, and memory decline in subjective memory impairment. *Neurology.* 2012;79:1332–1339.
151. Brugnolo A, et al. Metabolic correlates of rey auditory verbal learning test in elderly subjects with memory complaints. *J Alzheimers Dis.* 2014;39:103–113.
152. Van Der Gucht A, et al. Complementarity of visual and voxel-based FDG-PET analysis to detect MCI-like hypometabolic pattern in elderly patients with hypertension and isolated memory complaints. *Acta Radiol.* 2015;56:980–989.
153. Mosconi L, et al. Hypometabolism and altered cerebrospinal fluid markers in normal apolipoprotein E E4 carriers with subjective memory complaints. *Biol Psychiatry.* 2008;63:609–618.
154. Benzinger TL, et al. Regional variability of imaging biomarkers in autosomal dominant Alzheimer's disease. *Proc Natl Acad Sci U S A.* 2013;110:E4502–E4509.
155. Vannini P, et al. Hippocampal hypometabolism in older adults with memory complaints and increased amyloid burden. *Neurology.* 2017;88:1759–1767.
156. Hoyer S. The brain insulin signal transduction system and sporadic (type II) Alzheimer disease: an update. *J Neural Transm (Vienna).* 2002;109:341–360.

157. Landau SM, et al. Associations between cognitive, functional, and FDG-PET measures of decline in AD and MCI. *Neurobiol Aging*. 2011;32:1207–1218.
158. Herholz K. Use of FDG PET as an imaging biomarker in clinical trials of Alzheimer's disease. *Biomark Med*. 2012;6:431–439.
159. Nordberg A, et al. Tacrine restores cholinergic nicotinic receptors and glucose metabolism in Alzheimer patients as visualized by positron emission tomography. *Neurobiol Aging*. 1992;13:747–758.
160. Heiss WD, Kessler J, Mielke R, Szelies B, Herholz K. Long-term effects of phosphatidylserine, pyritinol, and cognitive training in Alzheimer's disease. A neuropsychological, EEG, and PET investigation. *Dementia*. 1994;5:88–98.
161. Mega MS, et al. Cognitive and metabolic responses to metrifonate therapy in Alzheimer disease. *Neuropsychiatry Neuropsychol Behav Neurol*. 2001;14:63–68.
162. Potkin SG, et al. Brain metabolic and clinical effects of rivastigmine in Alzheimer's disease. *Int J Neuropsychopharmacol*. 2001;4:223–230.
163. Tune L, et al. Donepezil HCl (E2020) maintains functional brain activity in patients with Alzheimer disease: results of a 24-week, double-blind, placebo-controlled study. *Am J Geriatr Psychiatry*. 2003;11:169–177.
164. Schmidt R, et al. Longitudinal multimodal imaging in mild to moderate Alzheimer disease: a pilot study with memantine. *J Neurol Neurosurg Psychiatry*. 2008;79:1312–1317.
165. Tzimopoulou S, et al. A multi-center randomized proof-of-concept clinical trial applying [^{18}F]FDG-PET for evaluation of metabolic therapy with rosiglitazone XR in mild to moderate Alzheimer's disease. *J Alzheimers Dis*. 2010;22:1241–1256.
166. Craft S, et al. Intranasal insulin therapy for Alzheimer disease and amnestic mild cognitive impairment: a pilot clinical trial. *Arch Neurol*. 2012;69:29–38.
167. Smith GS, et al. Increased cerebral metabolism after 1 year of deep brain stimulation in Alzheimer disease. *Arch Neurol*. 2012;69:1141–1148.
168. Laforce Jr. R, et al. The value of PET in mild cognitive impairment, typical and atypical/unclear dementias: a retrospective memory clinic study. *Am J Alzheimers Dis Other Dement*. 2010;25: 324–332.
169. Elias A, Woodward M, Rowe CC. Management impact of FDG-PET in dementia: results from a tertiary center memory clinic. *J Alzheimers Dis*. 2014;42:885–892.
170. Banerjee S, Wittenberg R. Clinical and cost effectiveness of services for early diagnosis and intervention in dementia. *Int J Geriatr Psychiatry*. 2009;24:748–754.
171. Getsios D, Blume S, Ishak KJ, Maclaine G, Hernandez L. An economic evaluation of early assessment for Alzheimer's disease in the United Kingdom. *Alzheimers Dement*. 2012;8:22–30.
172. Garibotto V, et al. Clinical validity of brain fluorodeoxyglucose positron emission tomography as a biomarker for Alzheimer's disease in the context of a structured 5-phase development framework. *Neurobiol Aging*. 2017;52:183–195.
173. McGeer EG, McGeer PL. Neuroinflammation in Alzheimer's disease and mild cognitive impairment: a field in its infancy. *J Alzheimers Dis*. 2010;19:355–361.
174. Morales I, Guzman-Martinez L, Cerda-Troncoso C, Farias GA, Maccioni RB. Neuroinflammation in the pathogenesis of Alzheimer's disease. A rational framework for the search of novel therapeutic approaches. *Front Cell Neurosci*. 2014;8:112.
175. Wyss-Coray T, Mucke L. Inflammation in neurodegenerative disease—a double-edged sword. *Neuron*. 2002;35:419–432.
176. Mrak RE, Griffin WS. Glia and their cytokines in progression of neurodegeneration. *Neurobiol Aging*. 2005;26:349–354.
177. Vehmas AK, Kawas CH, Stewart WF, Troncoso JC. Immune reactive cells in senile plaques and cognitive decline in Alzheimer's disease. *Neurobiol Aging*. 2003;24:321–331.
178. Hoozemans JJ, et al. Maximal COX-2 and ppRb expression in neurons occurs during early Braak stages prior to the maximal activation of astrocytes and microglia in Alzheimer's disease. *J Neuroinflammation*. 2005;2:27.
179. McGeer PL, Schulzer M, McGeer EG. Arthritis and anti-inflammatory agents as possible protective factors for Alzheimer's disease: a review of 17 epidemiologic studies. *Neurology*. 1996;47:425–432.

180. Hoozemans JJ, Veerhuis R, Rozemuller JM, Eikelenboom P. Soothing the inflamed brain: effect of non-steroidal anti-inflammatory drugs on Alzheimer's disease pathology. *CNS Neurol Disord Drug Targets*. 2011;10:57–67.
181. Barger SW, Harmon AD. Microglial activation by Alzheimer amyloid precursor protein and modulation by apolipoprotein E. *Nature*. 1997;388:878–881.
182. Akiyama H, et al. Inflammation and Alzheimer's disease. *Neurobiol Aging*. 2000;21:383–421.
183. Zimmer ER, et al. Tracking neuroinflammation in Alzheimer's disease: the role of positron emission tomography imaging. *J Neuroinflammation*. 2014;11:120.
184. Varley J, Brooks DJ, Edison P. Imaging neuroinflammation in Alzheimer's disease and other dementias: recent advances and future directions. *Alzheimers Dement*. 2015;11:1110–1120.
185. Streit WJ, Mrak RE, Griffin WS. Microglia and neuroinflammation: a pathological perspective. *J Neuroinflammation*. 2004;1:14.
186. McGeer PL, Itagaki S, Boyes BE, McGeer EG. Reactive microglia are positive for HLA-DR in the substantia nigra of Parkinson's and Alzheimer's disease brains. *Neurology*. 1988;38:1285–1291.
187. Venneti S, et al. PK11195 labels activated microglia in Alzheimer's disease and *in vivo* in a mouse model using PET. *Neurobiol Aging*. 2009;30:1217–1226.
188. Schilling T, Eder C. Amyloid-β-induced reactive oxygen species production and priming are differentially regulated by ion channels in microglia. *J Cell Physiol*. 2011;226:3295–3302.
189. Cagnin A, et al. In-vivo measurement of activated microglia in dementia. *Lancet*. 2001;358:461–467.
190. Edison P, et al. Microglia, amyloid, and cognition in Alzheimer's disease: an [11C](R)PK11195-PET and [11C]PIB-PET study. *Neurobiol Dis*. 2008;32:412–419.
191. Wiley CA, et al. Carbon 11-labeled Pittsburgh compound B and carbon 11-labeled (R)-PK11195 positron emission tomographic imaging in Alzheimer disease. *Arch Neurol*. 2009;66:60–67.
192. Yokokura M, et al. In vivo changes in microglial activation and amyloid deposits in brain regions with hypometabolism in Alzheimer's disease. *Eur J Nucl Med Mol Imaging*. 2011;38:343–351.
193. Calsolaro V, Edison P. Neuroinflammation in Alzheimer's disease: current evidence and future directions. *Alzheimers Dement*. 2016;12:719–732.
194. Edison P, Donat CK, Sastre M. *In vivo* imaging of glial activation in Alzheimer's disease. *Front Neurol*. 2018;9:625.
195. Owen DR, et al. An 18-kDa translocator protein (TSPO) polymorphism explains differences in binding affinity of the PET radioligand PBR28. *J Cereb Blood Flow Metab*. 2012;32:1–5.
196. Carter SF, et al. Evidence for astrocytosis in prodromal Alzheimer disease provided by ^{11}C-deuterium-L-deprenyl: a multitracer PET paradigm combining ^{11}C-Pittsburgh compound B and ^{18}F-FDG. *J Nucl Med*. 2012;53:37–46.
197. Yasuno F, et al. Increased binding of peripheral benzodiazepine receptor in Alzheimer's disease measured by positron emission tomography with [^{11}C]DAA1106. *Biol Psychiatry*. 2008;64:835–841.
198. Suridjan I, et al. *In-vivo* imaging of grey and white matter neuroinflammation in Alzheimer's disease: a positron emission tomography study with a novel radioligand, [^{18}F]-FEPPA. *Mol Psychiatry*. 2015;20:1579–1587.
199. Varrone A, et al. Positron emission tomography imaging of the 18-kDa translocator protein (TSPO) with [^{18}F]FEMPA in Alzheimer's disease patients and control subjects. *Eur J Nucl Med Mol Imaging*. 2015;42:438–446.
200. Hamelin L, et al. Early and protective microglial activation in Alzheimer's disease: a prospective study using ^{18}F-DPA-714 PET imaging. *Brain*. 2016;139:1252–1264.
201. Kreisl WC, et al. ^{11}C PBR28 binding to translocator protein increases with progression of Alzheimer's disease. *Neurobiol Aging*. 2016;44:53–61.
202. Kreisl WC, et al. *In vivo* radioligand binding to translocator protein correlates with severity of Alzheimer's disease. *Brain*. 2013;136:2228–2238.
203. Kreisl WC, et al. Distinct patterns of increased translocator protein in posterior cortical atrophy and amnestic Alzheimer's disease. *Neurobiol Aging*. 2016;51:132–140.
204. Okello A, et al. Microglial activation and amyloid deposition in mild cognitive impairment: a PET study. *Neurology*. 2009;72:56–62.

205. Schuitemaker A, et al. Microglial activation in Alzheimer's disease: an (R)-[^{11}C]PK11195 positron emission tomography study. *Neurobiol Aging*. 2013;34:128–136.
206. Golla SS, et al. Quantification of [^{18}F]DPA-714 binding in the human brain: initial studies in healthy controls and Alzheimer's disease patients. *J Cereb Blood Flow Metab*. 2015;35:766–772.
207. Rodriguez-Vieitez E, et al. Comparison of early-phase ^{11}C-deuterium-l-deprenyl and ^{11}C-Pittsburgh compound B PET for assessing brain perfusion in Alzheimer disease. *J Nucl Med*. 2016;57: 1071–1077.
208. Gulyas B, et al. Activated MAO-B in the brain of Alzheimer patients, demonstrated by [^{11}C]-L-deprenyl using whole hemisphere autoradiography. *Neurochem Int*. 2011;58:60–68.
209. Perry E, Walker M, Grace J, Perry R. Acetylcholine in mind: a neurotransmitter correlate of consciousness? *Trends Neurosci*. 1999;22:273–280.
210. Schliebs R, Arendt T. The cholinergic system in aging and neuronal degeneration. *Behav Brain Res*. 2011;221:555–563.
211. Bierer LM, et al. Neurochemical correlates of dementia severity in Alzheimer's disease: relative importance of the cholinergic deficits. *J Neurochem*. 1995;64:749–760.
212. Craig LA, Hong NS, McDonald RJ. Revisiting the cholinergic hypothesis in the development of Alzheimer's disease. *Neurosci Biobehav Rev*. 2011;35:1397–1409.
213. Aghourian M, et al. Quantification of brain cholinergic denervation in Alzheimer's disease using PET imaging with [^{18}F]-FEOBV. *Mol Psychiatry*. 2017;22:1531–1538.
214. Horti AG, Wong DF. Clinical perspective and recent development of PET radioligands for imaging cerebral nicotinic acetylcholine receptors. *PET Clin*. 2009;4:89–100.
215. Wong DF, et al. PET imaging of high-affinity $\alpha 4\beta 2$ nicotinic acetylcholine receptors in humans with ^{18}F-AZAN, a radioligand with optimal brain kinetics. *J Nucl Med*. 2013;54:1308–1314.
216. Sabri O, et al. First-in-human PET quantification study of cerebral $\alpha 4\beta 2^*$ nicotinic acetylcholine receptors using the novel specific radioligand (−)-[^{18}F]Flubatine. *NeuroImage*. 2015;118: 199–208.
217. Iyo M, et al. Measurement of acetylcholinesterase by positron emission tomography in the brains of healthy controls and patients with Alzheimer's disease. *Lancet*. 1997;349:1805–1809.
218. Kuhl DE, et al. In vivo mapping of cerebral acetylcholinesterase activity in aging and Alzheimer's disease. *Neurology*. 1999;52:691–699.
219. Kuhl DE, et al. Limited donepezil inhibition of acetylcholinesterase measured with positron emission tomography in living Alzheimer cerebral cortex. *Ann Neurol*. 2000;48:391–395.
220. Shinotoh H, et al. Effect of donepezil on brain acetylcholinesterase activity in patients with AD measured by PET. *Neurology*. 2001;56:408–410.
221. Kaasinen V, et al. Regional effects of donepezil and rivastigmine on cortical acetylcholinesterase activity in Alzheimer's disease. *J Clin Psychopharmacol*. 2002;22:615–620.
222. Kadir A, et al. PET imaging of the in vivo brain acetylcholinesterase activity and nicotine binding in galantamine-treated patients with AD. *Neurobiol Aging*. 2008;29:1204–1217.
223. Maelicke A, et al. Allosteric sensitization of nicotinic receptors by galantamine, a new treatment strategy for Alzheimer's disease. *Biol Psychiatry*. 2001;49:279–288.
224. Eggers C, Herholz K, Kalbe E, Heiss WD. Cortical acetylcholine esterase activity and ApoE4-genotype in Alzheimer disease. *Neurosci Lett*. 2006;408:46–50.
225. Schmitz TW, et al. Longitudinal Alzheimer's degeneration reflects the spatial topography of cholinergic basal forebrain projections. *Cell Rep*. 2018;24:38–46.
226. Nordberg A, et al. Decreased uptake and binding of ^{11}C-nicotine in brain of Alzheimer pa tients as visualized by positron emission tomography. *J Neural Transm Park Dis Dement Sect*. 1990;2:
215–224.
227. Nordberg A, Lundqvist H, Hartvig P, Lilja A, Langstrom B. Kinetic analysis of regional (S)(−)^{11}C-nicotine binding in normal and Alzheimer brains—in vivo assessment using positron emission tomography. *Alzheimer Dis Assoc Disord*. 1995;9:21–27.

228. Kendziorra K, et al. Decreased cerebral $\alpha4\beta2^*$ nicotinic acetylcholine receptor availability in patients with mild cognitive impairment and Alzheimer's disease assessed with positron emission tomography. *Eur J Nucl Med Mol Imaging*. 2011;38:515–525.
229. Ellis JR, et al. Relationship between nicotinic receptors and cognitive function in early Alzheimer's disease: a 2-[^{18}F]fluoro-A-85380 PET study. *Neurobiol Learn Mem*. 2008;90:404–412.
230. Sabri O, Kendziorra K, Wolf H, Gertz HJ, Brust P. Acetylcholine receptors in dementia and mild cognitive impairment. *Eur J Nucl Med Mol Imaging*. 2008;35(Suppl 1):S30–S45.
231. Okada H, et al. Alterations in $\alpha4\beta2$ nicotinic receptors in cognitive decline in Alzheimer's aetiopathology. *Brain*. 2013;136:3004–3017.
232. Maas T, Eidenmuller J, Brandt R. Interaction of tau with the neural membrane cortex is regulated by phosphorylation at sites that are modified in paired helical filaments. *J Biol Chem*. 2000;275: 15733–15740.
233. Sabri O, et al. Cognitive correlates of $\alpha4\beta2$ nicotinic acetylcholine receptors in mild Alzheimer's dementia. *Brain*. 2018;141:1840–1854.
234. Wang HY, et al. β-Amyloid$_{1-42}$ binds to $\alpha7$ nicotinic acetylcholine receptor with high affinity. Implications for Alzheimer's disease pathology. *J Biol Chem*. 2000;275:5626–5632.
235. Wang HY, Li W, Benedetti NJ, Lee DH. α_7 nicotinic acetylcholine receptors mediate β-amyloid peptide-induced tau protein phosphorylation. *J Biol Chem*. 2003;278:31547–31553.
236. Buckingham SD, Jones AK, Brown LA, Sattelle DB. Nicotinic acetylcholine receptor signalling: roles in Alzheimer's disease and amyloid neuroprotection. *Pharmacol Rev*. 2009;61:39–61.
237. Conejero-Goldberg C, Davies P, Ulloa L. Alpha7 nicotinic acetylcholine receptor: a link between inflammation and neurodegeneration. *Neurosci Biobehav Rev*. 2008;32:693–706.
238. Wang H, et al. Nicotinic acetylcholine receptor $\alpha7$ subunit is an essential regulator of inflammation. *Nature*. 2003;421:384–388.
239. Hillmer AT, et al. PET imaging evaluation of [^{18}F]DBT-10, a novel radioligand specific to α_7 nicotinic acetylcholine receptors, in nonhuman primates. *Eur J Nucl Med Mol Imaging*. 2016;43: 537–547.
240. Horti AG. Development of [^{18}F]ASEM, a specific radiotracer for quantification of the $\alpha7$-nAChR-nAChR with positron-emission tomography. *Biochem Pharmacol*. 2015;97:566–575.
241. Horti AG, et al. ^{18}F-ASEM, a radiolabeled antagonist for imaging the $\alpha7$-nicotinic acetylcholine receptor with PET. *J Nucl Med*. 2014;55:672–677.
242. Wong DF, et al. Human brain imaging of $\alpha7$ nAChR with [^{18}F]ASEM: a new PET radiotracer for neuropsychiatry and determination of drug occupancy. *Mol Imaging Biol*. 2014;16:730–738.
243. Hillmer AT, et al. PET imaging of α_7 nicotinic acetylcholine receptors: a comparative study of [^{18}F] ASEM and [^{18}F]DBT-10 in nonhuman primates, and further evaluation of [^{18}F]ASEM in humans. *Eur J Nucl Med Mol Imaging*. 2017;44:1042–1050.
244. Coughlin JM, et al. The distribution of the $\alpha7$ nicotinic acetylcholine receptor in healthy aging: an *in vivo* positron emission tomography study with [^{18}F]ASEM. *NeuroImage*. 2018;165:118–124.
245. Coughlin JM, et al. High availability of the $\alpha7$-nicotinic acetylcholine receptor in brains of individuals with mild cognitive impairment: a pilot study using ^{18}F-ASEM PET. *J Nucl Med*. 2020;61: 423–426.
246. Scheff SW, DeKosky ST, Price DA. Quantitative assessment of cortical synaptic density in Alzheimer's disease. *Neurobiol Aging*. 1990;11:29–37.
247. Terry RD, et al. Physical basis of cognitive alterations in Alzheimer's disease: synapse loss is the major correlate of cognitive impairment. *Ann Neurol*. 1991;30:572–580.
248. Scheff SW, Price DA. Alzheimer's disease-related alterations in synaptic density: neocortex and hippocampus. *J Alzheimers Dis*. 2006;9:101–115.
249. DeKosky ST, Scheff SW. Synapse loss in frontal cortex biopsies in Alzheimer's disease: correlation with cognitive severity. *Ann Neurol*. 1990;27:457–464.
250. DeKosky ST, Scheff SW, Styren SD. Structural correlates of cognition in dementia: quantification and assessment of synapse change. *Neurodegeneration*. 1996;5:417–421.
251. Hamos JE, DeGennaro LJ, Drachman DA. Synaptic loss in Alzheimer's disease and other dementias. *Neurology*. 1989;39:355–361.

252. Robinson JL, et al. Perforant path synaptic loss correlates with cognitive impairment and Alzheimer's disease in the oldest-old. *Brain*. 2014;137:2578–2587.
253. Masliah E, et al. Altered expression of synaptic proteins occurs early during progression of Alzheimer's disease. *Neurology*. 2001;56:127–129.
254. Masliah E, et al. Synaptic and neuritic alterations during the progression of Alzheimer's disease. *Neurosci Lett*. 1994;174:67–72.
255. Pham E, et al. Progressive accumulation of amyloid-beta oligomers in Alzheimer's disease and in amyloid precursor protein transgenic mice is accompanied by selective alterations in synaptic scaffold proteins. *FEBS J*. 2010;277:3051–3067.
256. Beeri MS, et al. Synaptic protein deficits are associated with dementia irrespective of extreme old age. *Neurobiol Aging*. 2012;33:1125.e1–1125.e8.
257. Wei W, et al. Amyloid beta from axons and dendrites reduces local spine number and plasticity. *Nat Neurosci*. 2010;13:190–196.
258. Pooler AM, Noble W, Hanger DP. A role for tau at the synapse in Alzheimer's disease pathogenesis. *Neuropharmacology*. 2014;76(Pt A):1–8.
259. Wang Y, Mandelkow E. Tau in physiology and pathology. *Nat Rev Neurosci*. 2016;17:5–21.
260. Dong M, et al. SV2 is the protein receptor for botulinum neurotoxin A. *Science*. 2006;312:592–596.
261. Vogl C, Mochida S, Wolff C, Whalley BJ, Stephens GJ. The synaptic vesicle glycoprotein 2A ligand levetiracetam inhibits presynaptic Ca^{2+} channels through an intracellular pathway. *Mol Pharmacol*. 2012;82:199–208.
262. Mercier J, et al. Discovery of heterocyclic nonacetamide synaptic vesicle protein 2A (SV2A) ligands with single-digit nanomolar potency: opening avenues towards the first SV2A positron emission tomography (PET) ligands. *ChemMedChem*. 2014;9:693–698.
263. Estrada S, et al. [^{11}C]UCB-A, a novel PET tracer for synaptic vesicle protein 2A. *Nucl Med Biol*. 2016;43:325–332.
264. Nabulsi NB, et al. Synthesis and preclinical evaluation of ^{11}C-UCB-J as a PET tracer for imaging the synaptic vesicle glycoprotein 2A in the brain. *J Nucl Med*. 2016;57:777–784.
265. Becker G, et al. Pharmacokinetic characterization of [^{18}F]UCB-H PET radiopharmaceutical in the rat brain. *Mol Pharm*. 2017;14:2719–2725.
266. Finnema SJ, et al. Imaging synaptic density in the living human brain. *Sci Transl Med*. 2016;8:348ra396.
267. Bahri MA, et al. Measuring brain synaptic vesicle protein 2A with positron emission tomography and [^{18}F]UCB-H. *Alzheimers Dement*. 2017;3:481–486.
268. Rabiner EA. Imaging synaptic density: a different look at neurologic diseases. *J Nucl Med*. 2018;59: 380–381.
269. Colom-Cadena M, et al. The clinical promise of biomarkers of synapse damage or loss in Alzheimer's disease. *Alzheimers Res Ther*. 2020;12:21.
270. Chen MK, et al. Assessing synaptic density in Alzheimer disease with synaptic vesicle glycoprotein 2A positron emission tomographic imaging. *JAMA Neurol*. 2018;75:1215–1224.
271. Bastin C, et al. In vivo imaging of synaptic loss in Alzheimer's disease with [^{18}F]UCB-H positron emission tomography. *Eur J Nucl Med Mol Imaging*. 2020;47:390–402.
272. Mecca AP, et al. In vivo measurement of widespread synaptic loss in Alzheimer's disease with SV2A PET. *Alzheimers Dement*. 2020;16:974–982.
273. de Wilde MC, Overk CR, Sijben JW, Masliah E. Meta-analysis of synaptic pathology in Alzheimer's disease reveals selective molecular vesicular machinery vulnerability. *Alzheimers Dement*. 2016;12: 633–644.
274. Vanhaute H, et al. In vivo synaptic density loss is related to tau deposition in amnestic mild cognitive impairment. *Neurology*. 2020. 95:e545–e553.
275. Cai Z, Li S, Matuskey D, Nabulsi N, Huang Y. PET imaging of synaptic density: a new tool for investigation of neuropsychiatric diseases. *Neurosci Lett*. 2019;691:44–50.
276. Herholz K. Imaging of dementia. In: Hodler J, Von Schulthess GK, Zollikofer CL, eds. *Diseases of the Brain, Head & Neck, Spine*. Milan: Springer; 2008:215–218. Chapter 33.
277. Burns CM, et al. Higher serum glucose levels are associated with cerebral hypometabolism in Alzheimer regions. *Neurology*. 2013;80:1557–1564.

278. Ishibashi K, Onishi A, Fujiwara Y, Ishiwata K, Ishii K. Relationship between Alzheimer disease-like pattern of ^{18}F-FDG and fasting plasma glucose levels in cognitively normal volunteers. *J Nucl Med.* 2015;56:229–233.
279. Izzo NJ, et al. Alzheimer's therapeutics targeting amyloid beta 1-42 oligomers II: sigma-2/PGRMC1 receptors mediate Abeta 42 oligomer binding and synaptotoxicity. *PLoS One.* 2014;9:e111899.
280. Lombardo S, Maskos U. Role of the nicotinic acetylcholine receptor in Alzheimer's disease pathology and treatment. *Neuropharmacology.* 2015;96:255–262.
281. Haas LT, Strittmatter SM. Oligomers of amyloid β prevent physiological activation of the cellular prion protein-metabotropic glutamate receptor 5 complex by glutamate in Alzheimer disease. *J Biol Chem.* 2016;291:17112–17121.
282. Li S, et al. Synthesis and in vivo evaluation of a novel PET radiotracer for imaging of synaptic vesicle glycoprotein 2A (SV2A) in nonhuman primates. *ACS Chem Neurosci.* 2019;10:1544–1554.
283. Constantinescu CC, et al. Development and in vivo preclinical imaging of fluorine-18-labeled synaptic vesicle protein 2A (SV2A) PET tracers. *Mol Imaging Biol.* 2019;21:509–518.
284. Cai Z, et al. Synthesis and preclinical evaluation of an ^{18}F-labeled synaptic vesicle glycoprotein 2A PET imaging probe: [^{18}F]SynVesT-2. *ACS Chem Neurosci.* 2020;11:592–603.
285. DiMasi JA, Hansen RW, Grabowski HG. The price of innovation: new estimates of drug development costs. *J Health Econ.* 2003;22:151–185.
286. Hargreaves RJ, et al. Optimizing central nervous system drug development using molecular imaging. *Clin Pharmacol Ther.* 2015;98:47–60.
287. Gunn RN, Rabiner EA. Imaging in central nervous system drug discovery. *Semin Nucl Med.* 2017;47:89–98.
288. Cummings J. The role of biomarkers in Alzheimer's disease drug development. *Adv Exp Med Biol.* 2019;1118:29–61.
289. Cummings J, Lee G, Mortsdorf T, Ritter A, Zhong K. Alzheimer's disease drug development pipeline: 2017. *Alzheimers Dement.* 2017;3:367–384.
290. Cummings J, Lee G, Ritter A, Zhong K. Alzheimer's disease drug development pipeline: 2018. *Alzheimers Dement.* 2018;4:195–214.
291. Cummings J, Lee G, Ritter A, Sabbagh M, Zhong K. Alzheimer's disease drug development pipeline: 2019. *Alzheimers Dement.* 2019;5:272–293.
292. Rinne JO, et al. ^{11}C-PiB PET assessment of change in fibrillar amyloid-β load in patients with Alzheimer's disease treated with bapineuzumab: a phase 2, double-blind, placebo-controlled, ascending-dose study. *Lancet Neurol.* 2010;9:363–372.
293. Ostrowitzki S, et al. Mechanism of amyloid removal in patients with Alzheimer disease treated with gantenerumab. *Arch Neurol.* 2012;69:198–207.
294. Sevigny J, et al. The antibody aducanumab reduces Aβ plaques in Alzheimer's disease. *Nature.* 2016;537:50–56.
295. van Dyck CH. Anti-amyloid-β monoclonal antibodies for Alzheimer's disease: pitfalls and promise. *Biol Psychiatry.* 2018;83:311–319.
296. Selkoe DJ, Hardy J. The amyloid hypothesis of Alzheimer's disease at 25 years. *EMBO Mol Med.* 2016;8:595–608.
297. Schneider L. A resurrection of aducanumab for Alzheimer's disease. *Lancet Neurol.* 2020;19:111–112.
298. Walsh DM, Selkoe DJ. Aβ oligomers—a decade of discovery. *J Neurochem.* 2007;101:1172–1184.
299. Heurling K, et al. Synaptic vesicle protein 2A as a potential biomarker in synaptopathies. *Mol Cell Neurosci.* 2019;97:34–42.

CHAPTER 16

Repurposing FDA-approved pharmaceuticals: The development of Nuedexta

Richard A. Smith, Randall Kaye, and Gerald Yakatan
Center for Neurologic Study, La Jolla, CA, United States

Introduction

The drug libraries that reside in-house in large pharmaceutical companies have long been mined as a source of new drug candidates. Compound libraries for high-throughput screening are also available from multiple commercial sources, and thousands of new compounds are added to these libraries annually. If a promising compound is found, this eliminates one step (chemical synthesis) along the development pathway, but the obstacles remaining are still substantial. Another avenue toward shortening the developmental timeline and its inherent costs involves the repurposing of existing pharmaceuticals. Drugs approved for other uses have already been tested in humans, so detailed information is available regarding their pharmacokinetics, toxicity, pharmacology, and formulation. Recognizing the inherent value of this, the National Center for Advancing Translational Sciences (NCATS), established by the NIH in 2011, has begun to evaluate a library of 2500 approved drugs and 1000 investigational compounds that have been approved for human clinical testing with the intent of discovering novel therapies for rare diseases.

To our knowledge, the success of repurposing existing pharmaceuticals has not yet been established. But the development of Nuedexta, a combination product incorporating dextromethorphan and quinidine, must rank as one of its earliest triumphs.[1] The story goes back to the 1950s when dextromethorphan (DM) was first synthesized in Switzerland and subsequently patented in the United States. The drug is an enantiomer of an opioid (levorphanol). It is a guess, but its subsequent use as an antitussive was likely tested because of the known suppressant effect of codeine on cough. Accordingly, its principle use has been as a constituent of cough syrup, primarily in the United States, where it is sold over-the-counter.

No one would have predicted that DM would have resurfaced as a potential neuroprotective agent. Due to the work of John Olney, who formulated the concept of excitotoxicity,[2] DM was screened in a high-throughput assay for its ability to block

Neurotherapeutics in the Era of Translational Medicine
https://doi.org/10.1016/B978-0-12-816475-4.00001-X

glutamate by a research group at Stanford. Following up on this discovery, which was patented, DM was demonstrated to be neuroprotective in a murine stroke model, presumably because of its ability to block the excitatory effect of glutamate released as a result of an anoxic insult.[3,4] At the time, it was considered that the neurologic consequences of a stroke were irreversible except in the surrounding areas (penumbra) where collateral circulation could mitigate the damage. If the effect in the animal model could be replicated in humans, this would have represented a major advance in the treatment of cerebrovascular disease. It didn't take long for this opportunity to be tested. The first effort to do so was undertaken by Merck which pioneered the development of MK-801, a noncompetitive NMDA receptor antagonist.[5] While the results in preclinical studies generated considerable attention as a potential treatment for stroke,[6] the clinical development was terminated at the Phase 2 stage due to side effects.[7] The side effects included disorientation and hallucinations. Not to be deterred, Hoffmann LaRoche licensed Stanford University's relevant drug patents, after which they initiated a stroke treatment trial using dextrorphan (DX), a molecule similar to DM which had lost its patent status by then. Once again, these clinical studies never progressed beyond clinical trials due to the same set of side effects that occurred in the MK-801 studies. Any reasonable person would have concluded that a pharmaceutical strategy that targeted the NMDA receptor was fraught with uncertainty and likely to be a waste of time and resources. Herein lies the tale which has led to the issuance of multiple pharmaceutical patents and ultimately to the development of Nuedexta which was approved by the FDA in 2010 for the treatment of emotional lability (PBA).

Metabolism and pharmacokinetics of dextromethorphan

The Nuedexta story starts with a unique understanding of the pharmacokinetic profile of DM. Surprisingly, the pharmacokinetic behavior of drugs was rather late to be appreciated. For a number of reasons, DM proved to be a useful tool for studying drug metabolism. Its principal advantages as a metabolic probe were its safety and high therapeutic index.[8] It was found upon ingestion that the drug was extensively degraded in the liver as a result of first-pass metabolism, which is mediated by Cytochrome P450 2D6 (CYP2D6). This enzyme, along with CYP3A4/5 isozymes, accounts for a significant amount of the metabolic processing of drugs that occurs in the liver[9] (see Fig. 1).

Because of the importance of the liver in detoxifying potential neurotoxins, we wondered at the outset whether mutations in the cytochrome oxidative system were a risk factor for the development of ALS and other neurodegenerative diseases. To explore this, we administered a test dose of DM 8–12 h before obtaining a urine collection. By determining the ratio of DM to DX in these samples, subjects were classified as being fast, intermediate, or slow metabolizers based on the arbitrary convention that a ratio ≤ 0.3 identified a rapid metabolizer.[10] Currently, it is more efficient to utilize a DNA probe

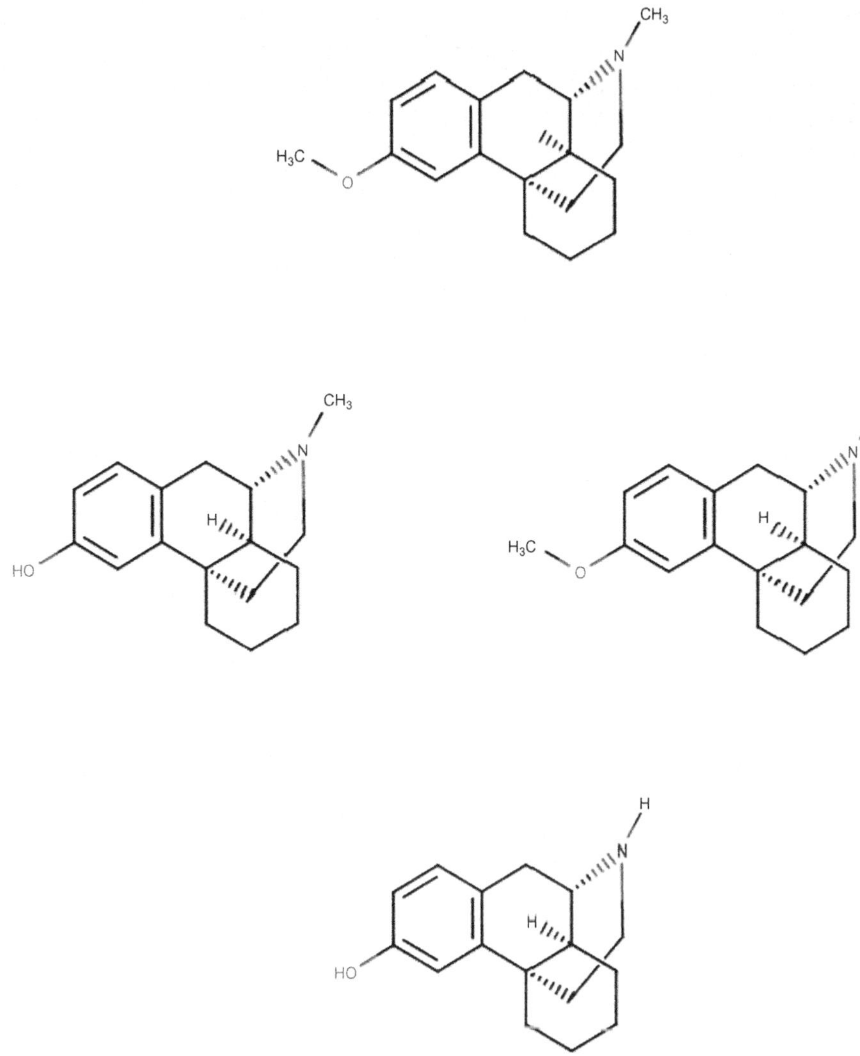

Fig. 1 Metabolism of dextromethorphan. Dextromethorphan (DM) is extensively metabolized in over 90% of the population by the hepatic CYP2D6 enzyme, which catalyzes the *O*-demethylation of DM to its main metabolite dextrorphan (DX). An alternate pathway is mediated primarily by CYP3A4 and *N*-demethylation to form 3-methoxymorphinan.

to determine the metabolic genotype in clinical studies. Whatever the method, it has been concluded that most Caucasians, approximately 93%, rapidly metabolize DM using the CYP2D6 pathway. The proportion of rapid metabolizers is even higher in some subpopulations, primarily African American and Chinese.

As regards ALS, we found the distribution of slow vs medium vs fast metabolizers to be the same as in the general population. But this did not deter our initial thinking that dextromethorphan alone might be a useful treatment in a subgroup of ALS patients who could not effectively metabolize the drug. Although identifying these patients and implementing a treatment trial would be a laborious task, we developed a protocol for undertaking such a clinical trial in collaboration with an investigator at Henry Ford Hospital in Detroit. However, this notion was soon put aside when we realized that a number of drugs, including quinidine, can inhibit the action of cytochrome oxidases. And, at the time, we were similarly surprised that grapefruit can have a similar effect on CYP3A. But it was not surprising that these effects on the liver were generally considered to be potential liabilities by the medical community, leading either to adverse side effects or to unwelcome drug interactions. As this review will reveal, this metabolic inhibitory property of a drug may actually be advantageous in certain circumstances.

At this point, the reader understandably would be curious to understand the logic of blocking the CYP2D6 enzyme, considering the prior results with Merck's MK-801 and Hoffmann LaRoche's experience with dextrorphan. Since, as noted previously, the use of DM as a cough suppressant dates to the 1950s, one of us (R.S.) surmised that DM alone must be relatively well tolerated. As a practicing neurologist, he had never observed any patient suffering from apparent DM toxicity, and neither had colleagues with whom he had discussed this matter. Further, he was unaware of any reports in the literature which detailed this, although he was aware of a subculture of drug abusers who were taking large doses of cough syrup.

Confident that DM could be administered safely, the next step was to figure out how to develop a combination product that could convert rapid metabolizers into slow ones so as to mimic nature. To accomplish this, representative members of a variety of drug classes were tested for their ability to inhibit the CYP2D6 enzyme. A publication by Inaba et al.[11] suggested that quinidine would be the best pharmacologic candidate to do this. He reported it was 10-fold more potent than any of the drugs that were tested. But considering quinidine's principal use as an anti-arrhythmic drug, it did not seem like an ideal drug partner for DM. This objection had to be overcome if we were to develop a viable product. Since our goal was to ultimately conduct an ALS treatment trial using DM, our pharmacokinetic studies were conducted in this population of patients. Subjects were administered standard doses of representative classes of drugs such as benzodiazepine, a tricyclic antidepressant, etc. for 1 week. But in the instance of quinidine, administered at a dose ranging from 600 to 1600 mg/day for the treatment of atrial fibrillation, the daily dose was reduced to a presumably safer dose of 150 mg twice daily at the outset. The principal concern about the use of quinidine is that one can induce a fatal cardiac arrhythmia (torsades de pointes) due to QT interval prolongation as determined by an electrocardiogram. To minimize this possibility, subjects with this abnormality were excluded from our trial. Contrary to what has been published, the only drug that significantly inhibited the CYP2D6 enzyme in our study was quinidine which converted all 13 subjects to a slow metabolizer status.

Having established that it was possible to convert rapid metabolizers into slow metabolizers, we next set out to gain further insight into the pharmacokinetic properties of quinidine-induced inhibition of CYP2D6. In the beginning, it was assumed that the blood levels of quinidine would be stabilized after dosing the drug for 1 week. On this background, the dose of DM was gradually increased at 1 week intervals, commencing with the lowest dose of 15 mg/day to a maximum dose of 60 mg twice a day. Each week, approximately 3 h after the last dose of DM, a blood sample was drawn. For our purposes, one could not have hoped for a better result: a linear relationship was definitively established between the dose of DM and the resulting plasma levels (see Fig. 2).

These results were published in the Journal of Clinical Pharmacology and Therapeutics,[12] and led to the filing of our first patent in 1991: A Method for Enhancing the Systemic Delivery of Dextromethorphan for the Treatment of Neurological Disorders (5,166,207).

As a side note, a group in Boston who were aware of our strategy to protect DM using quinidine embarked on a study to demonstrate that this was unnecessary on the assumption that one could accomplish the same thing by administering ultra-high doses of DM. They appeared to be oblivious to the concerns, as noted above, that this would likely not be well-tolerated. Noting that both DM and DX were demonstrated to be neuroprotective in in vitro studies at brain concentrations as low as 20 μM, they reasoned that they could achieve this level by the systemic administration of DM as high as 10 mg/kg/day.

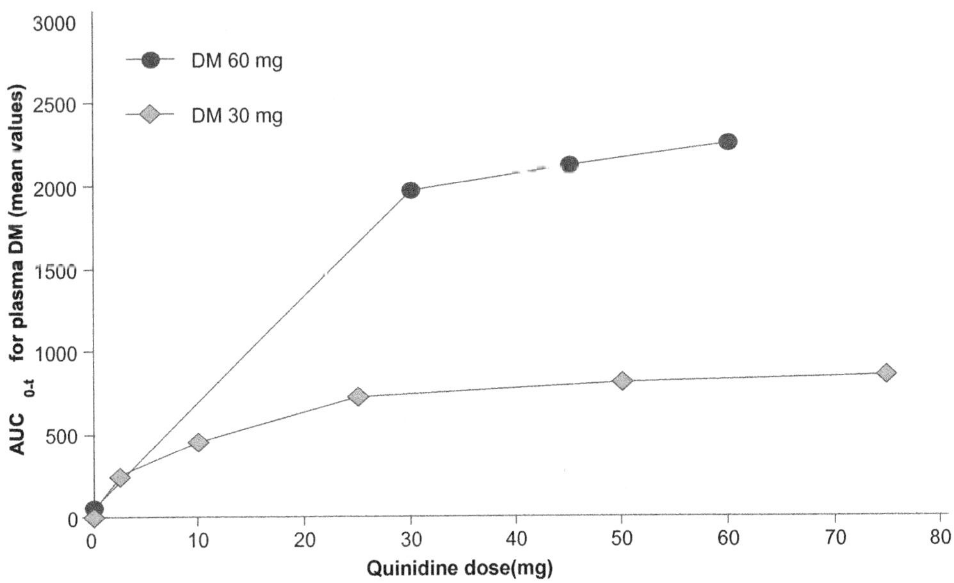

Fig. 2 Low doses of quinidine boost dextromethorphan plasma levels. In this experiment, low doses of quinidine (under 25 mg b.i.d.) resulted in a linear increase in plasma levels of DM. Steady-state plasma levels of DM were achieved after 1 week of twice-daily dosing with the DM and quinidine combination.

Before reaching the maximum dose, three of their five patients experienced hallucinations, and interestingly, plasma DM levels were only detectable in one patient.[13] Seemingly more encouraging was the finding that the mean concentration of DX in plasma was 9.8 μM at steady state, suggesting that it might be possible to achieve brain levels of DX approximating the 20 μM concentration that was demonstrated to be neuroprotective in the rodent model. However, it was apparent that this goal was unattainable when CSF levels of DX were shown to be only 5% of the plasma levels as would be expected since DX is conjugated in the liver to glucuronide, and unable to cross the blood-brain barrier. As so often occurs in the literature, the investigators interpreted their failed experiment as a successful result, to wit, that it would be preferable to treat neurologic patients with mega-doses of DM than to employ a strategy that protected DM from metabolic degradation. Undeterred, we were encouraged to pursue our idea that a combination product containing DM and an inhibitor of DM metabolism would have a better chance of being safe and effective.

The development of Nuedexta

Preclinical studies

Given our limited resources, it was apparent that we would need to find a corporate partner to finance the clinical development of the combination product. This proved to be a daunting exercise. One of our first presentations was to a local venture capital firm. Following this, we anticipated that a term sheet would arrive shortly. But that failed to materialize, and after numerous corporate presentations, mostly in the United States but also in Europe, it appeared that we were at a standstill. Discussing this with a former colleague who had taken a position at Parke Davis, he opined that the corporate resistance we encountered was due to a bias against developing products "not invented here." As the reader might surmise, this chapter would not have been written had not one of us (G. Y.) recognized the potential of our discovery which ultimately was in-licensed to Avanir Pharmaceuticals where he served as its CEO. In retrospect, their licensing decision was likely unique in that Avanir, a local company, arranged to meet subjects who had been involved in the Center for Neurologic Study's clinical trials. Even during Phase 1 studies, patients with ALS reported clinical benefits, most notably palliation of inappropriate emotionality which is associated with a number of neurologic disorders.[14] In short, what patients had to say was far more effective than our formal corporate presentations.

Once underway, Avanir's elegant studies, conducted by Pope et al.,[15] established the minimum dose of quinidine required to maximally block CYP2D6. To do this, they placed rapid metabolizers on a fixed dose of DM. In the first cohort, they found that a single dose of quinidine at a dose of 75 mg/day converted approximately 60% of subjects to slow metabolizers. Five doses of quinidine at a dose of 25 mg b.i.d. for 3 days converted all subjects. Even a miniscule dose of quinidine (2.5 mg b.i.d. for 7 days) resulted in a

12-fold increase in the mean peak plasma levels of DM. Based on these results, the company's initial combination product was formulated to contain equal amounts of DM and quinidine sulfate (30 mg).

A potential obstacle to the development of Nuedexta was the report in the journal Science by John Olney et al. that chronicled the occurrence of "spongiform" lesions in the brains of rodents treated with an NMDA antagonist.[16] Understandably, the FDA considered this a red flag leading to the requirement that dextromethorphan be studied to determine if it too was associated with similar pathology. At a meeting of the neuropharmacology division of the FDA which several of us attended, we anticipated that this would be one of the subjects that would be discussed. Our considered response was that this outcome, in the instance of dextromethorphan, was unlikely since the drug had long been in the marketplace and one would have reasonably presumed that pathologists would have noticed these pathologic abnormalities long ago. This argument was particularly compelling because persons who were "slow metabolizers," about 3% of the population, would have been exposed to high blood levels of DM as a result of taking cough syrup; in other words, nature had already conducted the experiment. Although this argument may have allayed some of the agency's concerns, they still obligated Avanir to undertake toxicology studies to investigate this issue. In addition to the "Olney" toxicology studies, Avanir was required to perform toxicology studies on DM and quinidine separately and in combination. In today's 505(b)(2) regulatory environment, such studies may not have been required because the constituent drugs in the combination were previously approved.

Corporate development path

By the time Avanir acquired the right to develop DM/Q, investigators at the Center for Neurologic Study had been issued use patents for a number of clinical indications, including the treatment of cough, pain, dermatitis, addiction, and labile emotionality (pseudobulbar affect, PBA). Of all these indications, the best human clinical data available was the effect of DM/Q on PBA. However, from a corporate point of view, it was not clear which of these patented indications would make the most sense to study first. Firstly, most physicians were not familiar with the diagnosis of PBA. In fact, they often misdiagnosed the condition, thinking instead that patients were depressed. And secondly, even when they were aware of the diagnosis, they often considered that it was a trivial matter relative to an underlying life-threatening disease (e.g., ALS). And further, the markets for conditions such as cough, pain, and dermatitis were likely to be far more lucrative. But after deliberation, it was ultimately decided to initiate the first treatment trial in patients with pseudobulbar affect associated with neurologic disorders. The fact that no drugs had been approved for the treatment of PBA proved to be the deciding factor in the choice to proceed with this indication.

Pseudobulbar affect: PBA

Although emotional lability has been recognized by medical practitioners for ages, the earliest references to the disorder that we are aware of are attributable to Charles Darwin and George Huntington. In a little-known book, The Emotions in Man and Animals, which was published in 1872, Charles Darwin wrote, "*Certain conditions that cause wasting such as senility have a special tendency to evoke weeping.*" However, the nomenclature associated with the condition may have been an impediment to its recognition by the public and physicians. Historically, the medical literature has referred to the condition as pseudobulbar affect, i.e., PBA, which implied that the problem was localized to the cerebral hemispheres rather than the lower brain. To be more descriptive, some authors have referred to the condition as pathological laughing and crying, the unflattering term emotional incontinence, and more commonly emotional lability. In regards to the latter term, emotional lability has long been used by psychiatrists to describe rapid mood changes associated with bipolar illness. Adding to the confusion was an effort to rebrand the condition with the moniker involuntary emotional expression disorder (IEED). The problem with the latter nomenclature was that such a diagnosis might have been applied to a variety of conditions, e.g., road rage. As such, a diagnosis of IEED would have strayed far from the common understanding that PBA is associated with structural changes in the brain associated with a variety of neurologic disorders. Specifically, the condition is associated with ALS, multiple sclerosis, stroke, Parkinson's disease, traumatic brain injury, Alzheimer's, etc.[17–22] The incidence can vary. In the instance of ALS, it has been reported to be as high as 50%, whereas in multiple sclerosis, the incidence is much lower (10%).[17,18] And even the presentation can vary. In the former disease, inappropriate tearfulness is more evident than inappropriate laughter, whereas, in MS, the reverse is true. And the condition is not only manifested by tearfulness and laughter, but by the occurrence of labile anger and frustration. There is general agreement that episodes occur involuntarily, are of sudden onset, and unpredictable in that the precipitating circumstances are not often apparent. In the literature, there is disagreement as to whether the manifestations of PBA, i.e., tearfulness or laughter, are congruent with the subject's mood.

Although PBA is common, it understandably goes unrecognized in some instances since its manifestations can easily be interpreted a result of a depressive disorder, as noted above. Adding to the confusion, drugs such as tricyclic antidepressants have been reported to palliate PBA, leading physicians to incorrectly assume that their diagnosis of depression was correct.[23] Until the approval of Nuedexta in 2010, there were no drugs that had been approved for the treatment of PBA. Prior treatment trials involved a small number of patients, were not placebo-controlled, and generated mixed results, i.e., they were more akin to case reports than to formal clinical trials. Further, the effectiveness of these agents, e.g., SSRIs, tricyclic antidepressants, and even L-dopa, were not assessed using validated measurements, such as the Center for Neurologic Study's lability scale (the CNS-LS).[24,25]

The anatomical substrate underlying PBA

The holy grail of neuroscience would be to understand the basis for complex behaviors, including their anatomical and physiologic substrates. Thus far, this has eluded our understanding in the instance of human happiness, spirituality, morality, etc. In contrast, we understand much more about what underlies PBA. The initial effort to make sense of this was undertaken in the early 20th century by Kinnier Wilson who proposed a release of cortical inhibition stemming from a lesion in the brainstem. More recently, Parvizi and Damassio concluded that PBA is a disconnection syndrome that disrupts contextual information processed by cortical association areas that are regulated in part by the cerebellum. In their single case report published in Brain in 2001,[26] they observed that lesions, assessed by MRI, in a patient with PBA were restricted to the cerebellum and brain stem. This report greatly influenced one of us (R.S.), who concluded that sigma-1 receptors, known to localize to these anatomical areas, were likely to be implicated in the occurrence of PBA. Prior to this report, Musacchio et al. had found that dextromethorphan preferentially bound to the brainstem and cerebellum,[27] at the time leading to the idea that there was a DM receptor in the brain. Following up on this, several of us collaborated with Linda Werling at George Washington University, to determine the receptor binding characteristics of DM. Along with binding to the NMDA receptor, she found that DM was indeed a sigma-1 agonist.[28] Somewhat later, sigma-1 receptors were reported to preferentially decorate neurons in the brainstem and cerebellum[29] (see Fig. 3).

Taken together, these studies strongly supported the notion that the sigma receptor and likely its agonist (tryptamine) play a critical role in the control of emotional expression. This is further supported by the robust treatment effect of Nuedexta in patients with PBA, as reviewed herein. Further studies will be required to determine if this effect is predominantly due to sigma-1 receptor activation or to other known receptors, namely the NMDA and serotonin pathways.

The measurement of emotional lability

In 1993 Robinson et al. developed the Pathologic Laughter and Crying Scale (PLACS), the first scale to quantitively assess the occurrence of symptoms associated with PBA.[30] Their rater-administered scale was validated in a group of stroke patients who manifested PBA. In the instance of clinical trials, a self-assessment scale is advantageous because it eliminates the need for establishing inter-rater reliability. This motivated us to develop the Center for Neurologic Study Lability Scale (CNS-LS), a self-report instrument. In its original form, included in Herndon's standard reference The Handbook of Neurologic Rating Scales,[31] the CNS-LS is a 15-item scale that interrogates the occurrence of inappropriate tearfulness, laughter, anger, and frustration. The clinical and research utility of a shortened version of this scale has been demonstrated in numerous clinical trials (see Fig. 4).

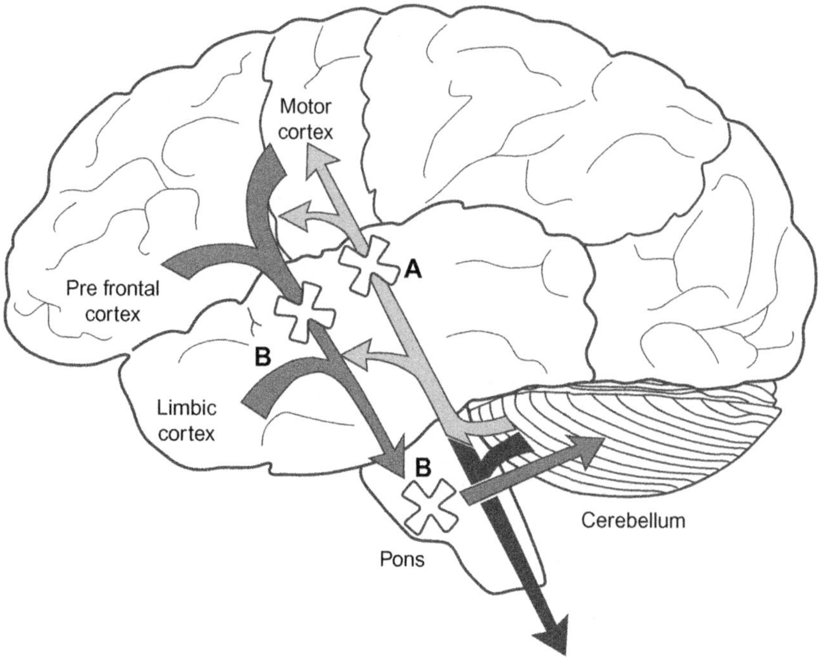

Fig. 3 The anatomy of emotional lability. Emotional lability is caused by structural damage to brain networks underlying the motor expression of emotion. The disorder is proposed to be a disinhibition syndrome, resulting from lesions to two main brain pathways: (A) corticobulbar degeneration that effectively releases cortical inhibition of brainstem centers that organize laughing/crying responses and (B) cerebro-ponto-cerebellar lesions that interrupt cerebellar communication with higher association cortex (e.g., limbic and prefrontal cortex), thus disrupting cerebellar adjustment of crying/laughing to the appropriate situation. Large X symbols indicate potential lesions along these pathways.

The principal aspects of emotional lability—its frequency, intensity, degree of voluntary control, and its inappropriateness—are reliably quantitated by the CNS-LS. In separate studies involving patients with ALS and MS, an increase of one point in the scale represented approximately a 12% increase in the episode rate of PBA.[24,25] In these populations, responses to the scale exhibited a Gaussian distribution. In a separate study involving 77 ALS patients, 46 of whom had been diagnosed by a clinician as exhibiting emotional lability, the CNS-LS score was significantly higher than in the patients who did not exhibit emotionality.[1] The results of this study allowed for the following conclusion: a CNS-LS score of 13 represents the cut-off for determining PBA.

PBA vs depression

While PBA and depression can exist concomitantly, it isn't difficult to distinguish between them. Unfortunately, this is often not the case. Most likely this is due to lack

Never	Rarely	Occasionally	Frequently	Most of the time
1	2	3	4	5

Questions

I find that even when I try to control my laughter I am often unable to do so

I find that I am easily overcome by laughter

There are times when I won't be thinking of anything happy or funny at all, but then I'll suddenly be overcome by funny or happy thoughts

Others have told me that I seem to become amused very easily or that I seem to become amused about things that aren't really funny

I find myself crying very easily

There are times when I feel fine one minute, and then I'll become tearful the next over something small or for no reason at all

I find that even when I try to control my crying I am often unable to do so

A CNS-LS score of 13 or higher may suggest PBA.

Fig. 4 The Center for Neurologic Study Emotional Lability Scale (CNS-LS). The CNS-LS is a short, self-administered questionnaire completed by patients and designed to provide a quantitative measure of the perceived frequency of pseudobulbar affect (PBA) episodes. A score of 13 or higher may indicate the presence of PBA.

of familiarity with the former condition. Although the DSM-5 (Diagnostic and Statistical Manual of Mental Disorders, the bible of psychiatric diagnosis) advises that the occurrence of tearfulness should not automatically lead to a diagnosis of a depressive disorder, no mention is made of the occurrence of PBA as an alternative diagnostic entity. The distinction between depression and PBA from a practical point of view centers on whether the occurrence of emotionality is congruent or not with the patient's underlying emotional state. In most circumstances, a patient exhibiting PBA will report that he or she is crying or laughing without apparent reason. Persons with depression will usually identify a loss of self-esteem or a life-changing event as the source of their affect. But in the instance of labile emotionality (PBA), neurologists have long appreciated that the condition is associated with underlying neurological disorders, e.g., Alzheimer's disease.[22] Further, our own studies have confirmed that the occurrence of labile emotionality in patients with neurologic disease is usually due to PBA. To avoid the diagnostic confusion between depression and PBA, all patients in the recently completed Nuedexta PBA treatment trials were administered either the Hamilton or Beck depression scales. It turned out that only a few subjects in these groups scored above the cutoff for having depression.

The amyotrophic lateral sclerosis and multiple sclerosis PBA treatment trials

Although two medications have been approved for the treatment of ALS,[32,33] for the most part, symptomatic treatment is the mainstay of therapy. In this regard, providing nutritional and ventilatory support have the most impact. Arguably, palliation of PBA, which occurs in approximately 50% of cases, is another example in which treatment has a profound impact on the quality of life for patients and indirectly, their family and acquaintances. The fact that there was no approved medication for the treatment of PBA influenced Avanir's decision to undertake a treatment trial to remedy this circumstance (as a side note, the company conducted a survey that led them to conclude that PBA was one of the most bothersome symptoms in the instance of ALS because it interfered with patients' relationships with loved ones).

The design of the PBA treatment trial was complicated by the fact that Nuedexta is a combination product involving two ingredients: dextromethorphan and quinidine. In this instance, the FDA mandated that each individual component as well as the combination product be tested for efficacy. The study was conducted at academic centers in the United States commencing in 2001.[34] All subjects met the criteria for the diagnosis of ALS according to established guidelines. In addition, they all exhibited PBA as determined by the referring physician and this diagnosis was confirmed by their scores on the CNS-LS (i.e., scores $\geq$13). Further, none of them were depressed as determined by their scores on the Hamilton rating scale. Concomitant medications such as antidepressants or drugs that were known to interfere with the CYP2D6 or 3A4 enzymes were prohibited. The medication was administered in capsules that contained either 30 mg of DM, an equivalent amount of quinidine, or in the instance of the combination product, 30 mg each of both ingredients. These were taken twice daily at approximately 12-h intervals for 28–30 days. A total of 140 patients were enrolled, half of whom took DM/Q, whereas half of the remaining subjects took either quinidine or DM. The outcome measures included the CNS-LS, a quality of life measure, and the number of episodes of PBA as chronicled in a daily diary.

As expected, the concentration of DM in patients taking the combination product was markedly elevated (18-fold higher) than in subjects taking DM alone, and the concentration of DX was 3.3-fold lower in these subjects than in the DM group, confirming that quinidine converted subjects to slow metabolizers. Following adjustment for prospectively defined covariates, the adjusted mean reduction of the CNS-LS in the DM/Q group was 3.29 points greater than in the DM or quinidine limbs of the trial ($P=0.001$ or better). Comparable effects favoring the combination treatment were noted for the episode count and in the quality of life measure (see Fig. 5).

A regulatory requirement to obtain approval for treating conditions such as PBA is that the putative therapy be tested in at least two diseases. Since, as noted previously,

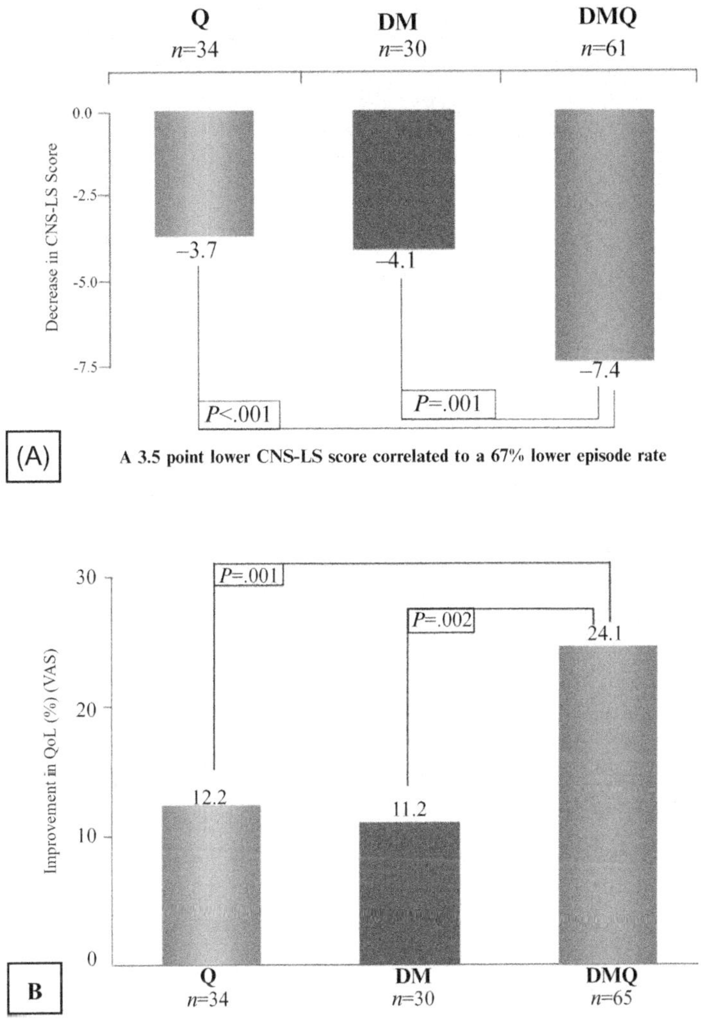

Fig. 5 ALS trial results. Improvements in Center for Neurologic Study Lability Scale (CNS-LS) and quality of life scores (QoL) in ALS patients (n = a number of subjects) treated with either dextromethorphan (D), quinidine (Q), or the combination of dextromethorphan and quinidine (DM/Q). A calculated P-value of ≤ 0.05 is statistically significant in the instance of a treatment trial.

the reported incidence of PBA in multiple sclerosis is about 10%, and the disease is far more common than ALS, the treatment of PBA in MS represents an unmet medical need. Further, for the most part, MS is a chronic disease that pursues a lengthy course. Although it was expected that MS patients would respond favorably to treatment with DM/Q, this was not certain since the occurrence of inappropriate laughter in MS patients may be disproportionate to the occurrence of tearfulness in patients with other neurologic disorders. On this background, Avanir Pharmaceuticals initiated a treatment trial at research

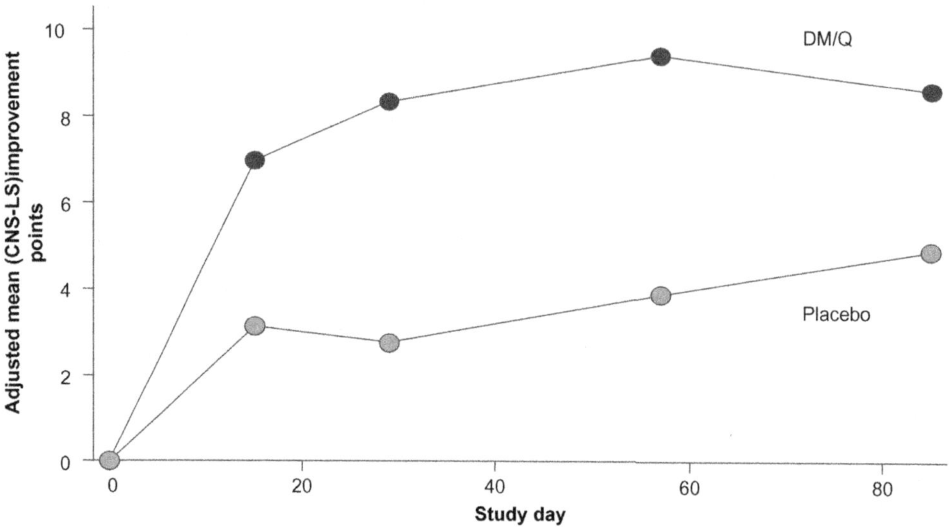

Fig. 6 Improvement in CNS-LS score by study day observed in the emotional lability multiple sclerosis efficacy trial. Values shown are least-square means, adjusted for baseline levels and center effects. At each time point, DM/Q patients exhibited a greater decrease in CNS-LS score compared with the placebo group ($P < 0.0001$).

sites both in the United States and Israel. The study was similar to the ALS study but different in that the combination product was tested against a placebo, and the study was conducted over a 3-month period, versus the 1-month ALS study.[35] And since patients with multiple sclerosis experience a variety of pain syndromes, it was of interest to determine whether treatment reduced the discomfort associated with these.

Enrollment in the year and a half study commenced at the end of 2002. Of the 150-treated patients, half received DM/Q. As in the earlier study, there was a significant reduction in CNS-LS scores compared to placebo ($p < 0.001$) (see Fig. 6).

At each of the four-time points measured over the course of the study, the CNS-LS scores demonstrated improvement compared to placebo, indicating the durability of the treatment effect. Further, there was a significant difference in the number of subjects who experienced complete remission in their PBA symptoms compared to placebo. And of further interest was the finding that there was a direct relationship between the concentration of DM in the plasma and the treatment effect. Additionally, the intensity of pain experienced by subjects in the treatment group was decreased, although no effort was made to document the variety of pain syndromes in the study.

Safety and tolerability

Whereas it was assumed that the FDA would approve Nuedexta for the treatment of emotional lability, the agency concluded that the drug was "approvable" (a term used by FDA at the time that indicated a possible path forward for approval versus "not approvable") but not approved because it was concerned about the potential adverse effects of quinidine on cardiac function. In both of the aforementioned clinical trials, the mean QTc interval in the treated groups was 7.5 ms. The worry was that this could lead to the generation of a torsades de pointes arrhythmia, with possibly fatal consequences.

Reformulation of Nuedexta

In the Approvable Letter of 2006, the FDA raised safety concerns related to the original 30/30 mg dosage formulation of Nuedexta. Subsequently, Avanir officials met with the agency to address their concerns. In preparation for the meeting, the company developed a briefing package that included a sophisticated PK/PD model that was generated from their initial treatment trials employing the original dose formulation (DM/Q 30/30 mg), as well as a statistical analysis addressing the utility of the secondary efficacy endpoint in these studies. Although the prediction from the PK/PD model led to the conclusion that it would be possible to reduce the QTc interval threshold below 5 ms, as well as sustain clinical efficacy with the revised formulation, the company was obligated to conduct a confirmatory Phase III study—the STAR trial. Regrettably, completion of the nonclinical and clinical work added an additional 2 years to the development process. The STAR trial, a randomized, double-blind, placebo-controlled, multicenter international study, enrolled 270 ALS and MS patients in 2007. Two active dose regimens (DM/Q 30/10 mg and 20/10 mg) were compared with a placebo. The 3 month study was followed by a 3 month open-label limb. The primary efficacy analysis, a comparison of the treatment arms with the placebo, was based on changes in the crying/laughing episode rate, as recorded in patient diaries. Secondary endpoints included (1) Center for Neurologic Study-Lability Scale (CNS-LS) score, (2) Neuropsychiatric Inventory Questionnaire (NPI-Q), (3) SF-36 Health Survey, (4) Beck Depression Inventory (BDI-II), and (5) Pain Rating Scale score (MS patients only). Standard safety and tolerability measures were employed in the study (adverse events, lab chemistries, etc.). Fortunately, the study was successful from both a safety and efficacy standpoint.

At the end of 2009, an updated regulatory filing was submitted to the FDA. It contained twice the number of patient exposures that had been included in the original NDA: over 1600 subjects exposed and over 500 patient-years of exposure to the components of Nuedexta over the many years of its clinical development. Avanir believed that their response would lead to approval, which was ultimately attained in October 2010, nearly 4 years to the day of the original 2007 Approvable Letter.

Nuedexta Alzheimer's agitation study

Agitation and aggression are prevalent in patients with dementia, ranging from approximately 10% in patients with mild cognitive impairment to more than 30% in those with severe dementia.[36] Agitation can be defined as a state of purposeless and poorly organized psychomotor activity characterized by aggressive verbal and physical manifestations as well as behaviors such as pacing, restlessness, and repetitive vocalizations. This is distressing for patients, family members, and caregivers, and it additionally places the patient at higher risk for an accelerated disease course, as well as early death. At some point, the management of patients in their homes becomes too burdensome, leading to the decision to relinquish care to others (nursing home or dedicated dementia facility). Even in this environment, management is challenging. At one time, agitated patients were restrained or heavily medicated with drugs such as Haldol. However, some years ago it was realized that this and similar medications resulted in excessive sedation, falls, and their use has been associated with tardive dyskinesia, parkinsonism, and metabolic syndrome. Worse is the occurrence of increased mortality which has resulted in warnings by the FDA regarding the treatment of agitated patients. This is not the case with citalopram, that effectively palliates Alzheimer's disease-associated agitation,[37] but comes with a cost: QTc interval prolongation and further cognitive decline.

On this background, Avanir Pharmaceuticals investigated the ability of Nuedexta to ameliorate agitation associated with Alzheimer's disease.[38] Their initial Phase 2 study randomized 220 patients in a multicenter, double-blind trial, 194 of whom completed the study. Using a sequential parallel comparison design, subjects were randomized in a 3:4 ratio to receive either DM/Q (93 patients) or placebo ($n=127$). In the second stage, the DM/Q patients continued while those receiving placebo were stratified by their response and rerandomized in a 1:1 ratio to DM/Q or placebo. The primary end point was the change from baseline on the Neuropsychiatric Inventory (NPI) agitation/aggression domain, which ranges from a score of 0 (no symptoms) up to 12 (daily symptoms with marked severity).

In the first limb of the trial, the agitation/aggression scores were reduced from 7.1 to 3.8 with DM/Q and from 7.0 to 5.3 with placebo, a statistically significant reduction ($P<0.001$). In the second stage, DM/Q treatment resulted in a reduction of agitation/aggression scores from 5.8 to 3.8, and from 6.7 to 5.8 with placebo, which is also a statistically significant intergroup difference ($P=0.02$). Importantly, DM/Q treatment was not associated with cognitive impairment or significant QTc prolongation. The most common adverse events included falls, diarrhea, and urinary tract infections, which were slightly higher in the DM/Q group.

Second generation dextromethorphan-quinidine (AVP-786) formulation for Alzheimer's agitation

In late 2012, Avanir began the clinical development of a second-generation drug for Alzheimer's agitation. Like its predecessor Nuedexta, AVP-786 is a combination product containing dextromethorphan and quinidine, but in this case, the dextromethorphan is deuterated. In a deuterated drug, one or more of the compound's hydrogen atoms are substituted with deuterium. It is a naturally occurring, stable, and nonradioactive isotope of hydrogen, which forms strong bonds with carbon. This increased bond strength can positively impact the ADME properties of a drug.[39] In comparison to their nondeuterated brethren, deuterated drugs may exhibit lower rates of metabolism and longer half-lives, thereby resulting in the need for less frequent dosing. Further, deuteration may lower drug toxicity as a result of decreased metabolite formation, and it may minimize drug-drug interactions. Because the size and shape of deuterium are basically identical to that of hydrogen, deuteration is generally not expected to affect the biochemical potency and selectivity of a drug. The first patent in the United States for deuterated molecules was granted in the 1970s. Currently, several pharmaceutical companies base their drug development strategy on the deuterium modification of existing drugs or compounds with established pharmacologic activity. One of these, Concert Pharmaceuticals, collaborated with Avanir to develop AVP-786.

Subsequently, Avanir initiated a variety of studies to compare the pharmacokinetic behavior of deuterated dextromethorphan alone and in combination with low dose quinidine to identify a combination product that would achieve the same pharmacologic effects as Nuedexta (AVP-923). Follow-up studies were conducted in healthy volunteers to evaluate the drug-drug interactions with AVP-786. Toward the end of 2015, the FDA fast-tracked AVP-786 for the treatment of agitation associated with Alzheimer's disease. Soon thereafter, Otsuka Pharmaceuticals, which had previously acquired Avanir, began enrolling patients in TRIAD-1 and TRIAD-2 (Phase 3 trials) to evaluate the efficacy of the deuterated drug combination in Alzheimer's patients. The difference between the two trials was that one employed a single dose of the drug combination and the other compared two doses to placebo. The primary outcome measure in both studies was a change in the Cohen Mansfield Agitation Inventory (CMAI) scale. Secondary outcomes included the neuropsychiatric inventory (NPI) scale, the agitation component of the ADCS-CGIC scale, and other global, quality of life, resource utilization, and cognitive measures. While both doses demonstrated improvement, only one was statistically significant. Similar to results with the nondeuterated DM/Q combination, the most common adverse events in patients receiving AVP-786 versus placebo were falls, urinary tract infection, headache, and diarrhea. Overall mortality during the study was low and none of the deaths were considered related to treatment. A follow-up study, concluded in 2019, did not meet its primary and key secondary endpoints. The results of a similar worldwide phase 3 trial are expected to be available in 2024.

The ALS speech and swallowing trial

Even before Nuedexta was approved for PBA, patients and their families who were involved in the various clinical trials noticed that the drug enhanced the ability to speak and swallow and handle oral secretions. One of the most compelling, unsolicited testimonials came from the wife of a patient who had recently died of ALS. Throughout her husband's long illness, during which one would have expected that he would have lost most of his motor functions, his wife reported that "his speech and swallowing remained normal until 1 week before his death." This, as well as many other anecdotal reports, were too numerous to ignore. Reflecting on this, it appeared that science offered an explanation for these observations. Since motor neurons in the brainstem are known to be decorated with sigma-1 receptors,[40] it seemed reasonable that a sigma-1 agonist, in this case, DM, might enhance bulbar function in patients with motor neuron disease by facilitating the function of oral, pharyngeal, and lingual musculature, i.e., the motor substrate for speech and swallowing.

Having been convinced that we were on to something, we set about trying to prove this. When considering a clinical trial design, the first challenge was to figure out how to come up with endpoints to assess bulbar function. To do this, we extensively reviewed the literature and talked to many of the authors of these publications. In addition to recommending the use of traditional measures such as timed speech and swallowing, some recommended tests appeared to be too cumbersome for inclusion in a multicenter treatment trial. For example, as regards speech, it was suggested that speech pathologists analyze recordings taken at intervals for a number of criteria such as intelligibility, nasality, etc. Even more daunting was the recommendation that subjects be instrumented with an endoscope for the purpose of studying various phases of glutition.

Although we considered the advice of experts, we ultimately decided to develop and validate a self-report scale which was similar in design to the Center for Neurologic Study Lability Scale (CNS-LS), a standard assessment tool for measuring emotional lability.[34,35] The Center for Neurologic Study Bulbar Function Scale (CNS-BFS) consists of three domains of bulbar function which interrogate speech, swallowing, and the ability to manage oral secretions.[41] Each domain includes seven statements which the subject is asked to rate on a scale of 1–5, the highest score indicating the maximum loss of function. The scale was validated in 120 ALS patients using conventional statistical tools. The CNS-BFS was highly correlated with a clinician's diagnosis of impaired bulbar function. Similarly, the individual subscales were highly predictive of a corresponding clinical diagnosis of impaired speech, swallowing, and salivary impairment. To assess construct validity, the CNS-BFS was compared to the bulbar subscale of the rater-administered ALSFRS-R. These two scales were also highly correlated ($r = -0.90$). As regards factor structure, items within each domain demonstrated strong internal consistency. Finally, test/retest scores for the CNS-BFS were reliable over a 2-week interval.

Getting the proposed treatment trial off, the ground was an additional challenge. Fortuitously, the ALS Association established a "Treat ALS" award which in 2012 was awarded to one of us (R.S.) for the purpose of confirming patients' anecdotal reports of enhanced bulbar function coincident with the use of Nuedexta. The trial was unique in several respects.[42] Firstly, we chose to conduct the trial over a finite period (70 days) in contrast to most ALS trials which have typically extended over 12–18 months. We did this primarily because we knew Nuedexta acts quickly in the instance of PBA. While it may have been a leap, we assumed Nuedexta would have a similar effect on speech and swallowing. Secondly, we employed a crossover design which is ideal for short-term studies and statistically powerful since subjects serve as their own controls. And finally, we elected to limit the study to 60 patients. A priori, we knew that the trial, if successful, would not answer all the questions that might need to be addressed. The design would not allow us to comment on the duration of the treatment effect or whether treatment altered the course of ALS. The trial was conducted at seven sites that belong to a consortium of ALS centers (NEALS). In advance, the CNS-BFS was selected as the principal endpoint. Secondary endpoints included the use of the rater-administered ALSFRS-R, visual analog scales, and timed assessments of speech and swallowing. Following a baseline visit to establish the patients' eligibility for inclusion in the study, subjects were randomly assigned to commence treatment with either Nuedexta or placebo. At the end of the first 30-day limb of treatment, they were switched to the opposite treatment regimen (following a 10–14 day washout period). At the beginning and end of each treatment period, patients filled out the self-report scales and a clinical evaluator administered the other tests. Following each visit, all data were entered into an electronic database. At the completion of the study, a research administrator confirmed that the database faithfully reflected the research record. Following this, the database was "locked" and analyzed by a statistician. The results were striking: there was a highly significant difference between the placebo and treated groups as measured by the CNS-BFS and the bulbar component of the ALSFRS-R. Interestingly, Nuedexta proved to be equally effective in patients with or without associated labile emotionality (PBA) (see Fig. 7).

While both speech and swallowing times improved, neither of these were statistically significant. From our point of view, this convincingly demonstrated that a patient-derived measure was the most effective means of assessing a treatment effect in this study. Had we relied solely on timed measures, the trial would have failed, leading us or others to suggest the necessity of conducting a larger and longer study. But since the initial study convincingly demonstrated the effect of Nuedexta on speech and swallowing, a larger and longer study would be required to convince the FDA to approve Nuedexta for the treatment of bulbar impairments in persons with ALS. Thus far, support for a definitive Phase 3 study has not been forthcoming.

The approval of Nuedexta

Before Nuedexta was approved in October of 2010, market manipulators were constantly spreading rumors about approvability of the drug. In fact, early on the day of approval, Avanir's stock plummeted based on a rumored "leak" from the FDA that the drug had been turned down (popularly called a bear raid, setting false rumors to drive the price down and reward short sellers). But since public companies are not allowed by regulators to withhold information that is germane to investors, Avanir issued a statement

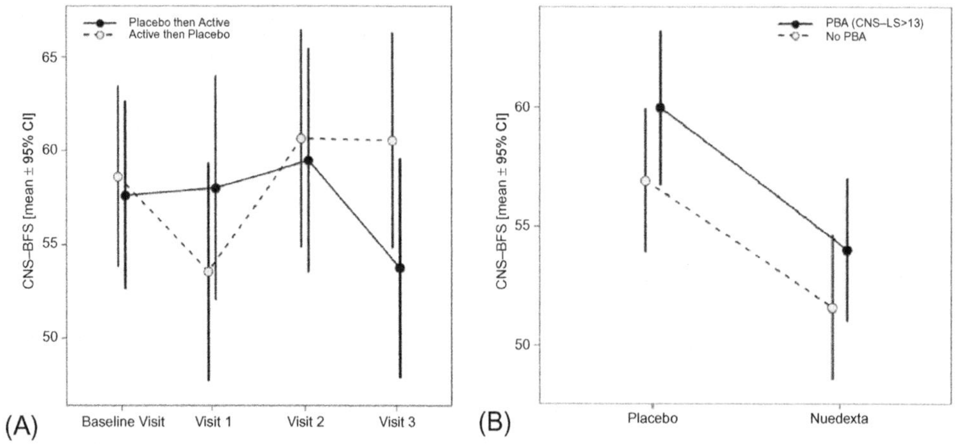

Fig. 7 Effect of Nuedexta vs placebo on changes in the Center for Neurologic Study Bulbar Function Scale (CNS-BFS), the study's primary outcome measure. (A) Mean CNS-BFS scores for each treatment arm (i.e., patients treated with placebo first and then switched to Nuedexta vs patients placed on Nuedexta initially and later switched to placebo) were calculated at the end of the clinical trial. Measurements were obtained at baseline and during three subsequent clinical visits. The crossover effect is apparent: CNS-BFS scores declined following the period of DMQ treatment. (B) CNS-BFS scores vs pseudobulbar affect (PBA) status. Mean baseline-adjusted CNS-BFS total scores ±95% confidence intervals (CI) are displayed on the *y*-axis. DMQ treatment improved bulbar functioning irrespective of whether patients had PBA.

to the effect that they had not received any communication from the FDA. Pursuant to this, the stock price rose, but at the time of the market's closing, it was down approximately 15% from its opening price. Promptly, presumably at the end of the work day, the FDA issued a statement that the drug had been approved. At the time, one of us (R.S.) was on a train headed to New York, in somewhat gloomy spirits given the aforementioned events of the day, when his phone began to ring. The first caller was his son, who was following the news, and immediately thereafter the second caller (R.K.) reported that he would have called earlier, but understandably he first "had to speak to his mom and dad."

Once Nuedexta was approved, as might be expected, Avanir Pharmaceutical initiated a vigorous marketing campaign to educate both patients and physicians about the treatment options for patients living with PBA. Over time, this initiated an outpouring of public commentary that focused on the cost of the medication and the means employed to promote it in the marketplace. In this regard, a New York Times article noted that Nuedexta is one of the most promoted pharmaceutical products on television and that some physicians who routinely prescribed the drug were receiving large consulting fees. Further, this and a follow-up article in JAMA Internal Medicine[43] raised concerns about the off-label use of Nuedexta. The implication was that Avanir and its parent company Otsuka were guilty of unethical practices. While these practices might have accounted for a small percentage of sales, it appears that none of the reporters spent any time speaking with physicians to explore alternative reasons for prescribing the drug. For example, patients with ALS who suffer from impaired speech and swallowing might be expected to be highly motivated to take a drug that enhances these functions. And considering the lack of treatment for this condition, one would expect that physicians would be highly motivated to offer the best available treatment despite the fact that they would be prescribing for an off-label indication. And for sure, nurses and physicians taking care of demented patients who are institutionalized would opt to treat patients who exhibit aggressive behavior with a drug that had been reported in a respected medical journal to be effective and relatively safe versus alternative treatment such as the use of Haldol. Physicians have long been warned that Haldol, which carries a black box warning, is associated with increased mortality when prescribed to patients with dementia.

As might be expected, Avanir countered these public criticisms, stating: "We vehemently oppose any mischaracterization of how Avanir and other members of the pharmaceutical industry interact with the medical community, including physicians that educate other health care professionals about medicines. Patients and doctors want options for treatment, and pharmaceutical companies engage physicians in ways that are both ethical and appropriate to convey the state of medical conditions and FDA-approved treatments that could otherwise remain unknown to patients in need. Such relationships allow research companies, physicians, and other health care professionals to share first-hand knowledge and raise awareness of medicines that could improve patient conditions." Depending on one's point of view, the company's statement could be seen as either self-serving or factual.

This tale would not be complete without commenting on the effort of five generic pharmaceutical companies to invalidate the patents that were foundational to the development of Nuedexta. This assault was spearheaded under the cover of the ANDA (Abbreviated New Drug Application) regulations established by Congress. This legislation permits companies to apply to the FDA for permission to market any drug that has been approved and whose patent has expired. However, a company may challenge the validity of an existing patent, and if successful, seek approval for a generic version of the marketed product. The only means to interfere with this process is for the patent holder to file a suit in federal court to protect its intellectual property. This is costly, especially in the case of a small pharmaceutical company like Avanir. For certain, the generic companies consider the ANDA option part of their business model since, if successful, they can market a drug at a fraction of the current cost and without having to have undergone years of development expense. The company that has done all of the work is particularly vulnerable because their records—for example, laboratory notebooks, grants, emails, publications, and internal communications—are subject to deposition once the matter is in litigation. Having provided their opponent with a massive volume of documents, one can't exclude that lurking in one of these files is a note or an email that could suggest that their patents are based on prior work and thus invalid. Knowing these risks, and the huge expense of defending their investment, it's common practice for the generics to offer a *deal* to their beleaguered foe. For example, their settlement terms might include their willingness to allow the proprietary drug to enjoy exclusivity for several years before it attains generic status. In spite of a unified assault by five companies, Avanir ultimately prevailed in federal court.

Having defended its patents, a number of potential pharma partners considered acquiring the company. However, it was not until the results of the agitation study, described above, that Avanir announced that it had agreed to be acquired by Otsuka Pharmaceuticals. The transaction, unanimously approved by the boards of directors of both companies, was valued at approximately $3.5 billion.

Conclusion

To a considerable degree, the development of Nuedexta followed a circuitous path that involved the interplay of basic science, clinical research, serendipity, intuition, and dogged persistence. It's certain that Nuedexta and its successors will find future clinical applications, as in the instance of aspirin, which is useful for a medley of conditions (headache, arthritis, stroke, pain, etc.). Similarly, it isn't much of a stretch to assume that Nuedexta or its look-alikes might be useful in the instance of substance abuse, autism, learning disorders, and depression, to name a few such medical conditions. At the minimum, anecdotal reports lead to this conclusion, as does basic science which also supports such an opinion. Not certain is whether industry is prepared to invest time and resources to explore this. In part, this is due to regulatory constraints that have to be overcome to obtain approval for new indications. In the current environment, market forces rather than unmet medical needs often hold sway, so it is uncertain that the full therapeutic potential of Nuedexta will ever be realized. While the story of Nuedexta, as told herein, is not a parable of Biblical proportion, it does affirm that there is still a place for human ingenuity in the drug discovery and development process.

Acknowledgments

Considering the years over which the work leading to the development and ultimate approval of Nuedexta occurred, it would be necessary to acknowledge so many persons that the singular contributions of some might be lost in the process. First and foremost, we acknowledge the support of families and of participants in the clinical trials that were conducted worldwide. Further, as will be evident to the reader, the studies summarized herein relied upon the indispensable participation of neurologists, clinical research coordinators, statisticians, and their support staffs. Needed institutional support was provided by colleagues and the Board of Directors at the Center for Neurologic Study and Avanir Pharmaceutical Corporation. As noted in the manuscript, R.S. was a recipient of a "Treat ALS" award in 2012 (the ALS Association). If you were involved in any way in the work described herein and happen to read this, we extend our heartfelt thanks.

References

1. Smith RA. Dextromethorphan/quinidine: a novel dextromethorphan product for the treatment of emotional lability. *Expert Opin Pharmacother.* 2006;7(18):2581–2598.
2. Olney JW. Glutamate, a neurotoxic transmitter. *J Child Neurol.* 1989;4(3):218–226.
3. Choi DW. Dextrorphan and dextromethorphan attenuate glutamate neurotoxicity. *Brain Res.* 1987;403 (2):333–336.
4. Tortella FC, Pellicano M, Bowery NG. Dextromethorphan and neuromodulation: old drug coughs up new activities. *Trends Pharmacol Sci.* 1989;10(12):501–507.
5. Woodruff GN, Foster AC, Gill R, Kemp JA, Wong EH, Iversen LL. The interaction between MK-801 and receptors for N-methyl-D-aspartate: functional consequences. *Neuropharmacology.* 1987;26(7b): 903–909.
6. Buchan AM, Slivka A, Xue D. The effect of the NMDA receptor antagonist MK-801 on cerebral blood flow and infarct volume in experimental focal stroke. *Brain Res.* 1992;574(1–2):171–177.
7. Lee JM, Zipfel GJ, Choi DW. The changing landscape of ischaemic brain injury mechanisms. *Nature.* 1999;399(6738 Suppl):A7–14.
8. Kupfer A, Schmid B, Preisig R, Pfaff G. Dextromethorphan as a safe probe for debrisoquine hydroxylation polymorphism. *Lancet.* 1984;2(8401):517–518.
9. Gorski JC, Jones DR, Wrighton SA, Hall SD. Characterization of dextromethorphan N-demethylation by human liver microsomes. Contribution of the cytochrome P450 3A (CYP3A) subfamily. *Biochem Pharmacol.* 1994;48(1):173–182.
10. Schmid B, Bircher J, Preisig R, Kupfer A. Polymorphic dextromethorphan metabolism: co-segregation of oxidative O-demethylation with debrisoquin hydroxylation. *Clin Pharmacol Ther.* 1985;38(6): 618–624.
11. Inaba T, Tyndale RE, Mahon WA. Quinidine: potent inhibition of sparteine and debrisoquine oxidation in vivo. *Br J Clin Pharmacol.* 1986;22(2):199–200.
12. Zhang Y, Britto MR, Valderhaug KL, Wedlund PJ, Smith RA. Dextromethorphan: enhancing its systemic availability by way of low-dose quinidine-mediated inhibition of cytochrome P4502D6. *Clin Pharmacol Ther.* 1992;51(6):647–655.
13. Hollander D, Pradas J, Kaplan R, McLeod HL, Evans WE, Munsat TL. High-dose dextromethorphan in amyotrophic lateral sclerosis: phase I safety and pharmacokinetic studies. *Ann Neurol.* 1994;36(6): 920–924.
14. Smith R, Moore S, Gresham L, Manley P, Licht J. The treatment of affective lability in ALS patients with dextromethorphan. *Neurology.* 1995;45:A330.
15. Pope LE, Khalil MH, Berg JE, Stiles M, Yakatan GJ, Sellers EM. Pharmacokinetics of dextromethorphan after single or multiple dosing in combination with quinidine in extensive and poor metabolizers. *J Clin Pharmacol.* 2004;44(10):1132–1142.
16. Collins RC, Olney JW. Focal cortical seizures cause distant thalamic lesions. *Science.* 1982;218 (4568):177–179.
17. Gallagher JP. Pathologic laughter and crying in ALS: a search for their origin. *Acta Neurol Scand.* 1989; 80(2):114–117.
18. Feinstein A, Feinstein K, Gray T, O'Connor P. Prevalence and neurobehavioral correlates of pathological laughing and crying in multiple sclerosis. *Arch Neurol.* 1997;54(9):1116–1121.
19. House A, Dennis M, Molyneux A, Warlow C, Hawton K. Emotionalism after stroke. *BMJ.* 1989;298 (6679):991–994.
20. Kaschka WP, Meyer A, Schier KR, Froscher W. Treatment of pathological crying with citalopram. *Pharmacopsychiatry.* 2001;34(6):254–258.
21. Zeilig G, Drubach DA, Katz-Zeilig M, Karatinos J. Pathological laughter and crying in patients with closed traumatic brain injury. *Brain Inj.* 1996;10(8):591–597.
22. Starkstein SE, Migliorelli R, Teson A, et al. Prevalence and clinical correlates of pathological affective display in Alzheimer's disease. *J Neurol Neurosurg Psychiatry.* 1995;59(1):55–60.
23. Schiffer RB, Herndon RM, Rudick RA. Treatment of pathologic laughing and weeping with amitriptyline. *N Engl J Med.* 1985;312(23):1480–1482.

24. Smith RA, Berg JE, Pope LE, Thisted RA. Measuring pseudobulbar affect in ALS. *Amyotroph Lateral Scler Other Motor Neuron Disord*. 2004;5(Suppl 1):99–102.
25. Smith RA, Berg JE, Pope LE, Callahan JD, Wynn D, Thisted RA. Validation of the CNS emotional lability scale for pseudobulbar affect (pathological laughing and crying) in multiple sclerosis patients. *Mult Scler*. 2004;10(6):679–685.
26. Parvizi J, Anderson SW, Martin CO, Damasio H, Damasio AR. Pathological laughter and crying: a link to the cerebellum. *Brain*. 2001;124(Pt 9):1708–1719.
27. Musacchio JM, Klein M, Canoll PD. Dextromethorphan and sigma ligands: common sites but diverse effects. *Life Sci*. 1989;45(19):1721–1732.
28. Werling LL, Keller A, Frank JG, Nuwayhid SJ. A comparison of the binding profiles of dextromethorphan, memantine, fluoxetine and amitriptyline: treatment of involuntary emotional expression disorder. *Exp Neurol*. 2007;207(2):248–257.
29. Maurice T, Urani A, Phan VL, Romieu P. The interaction between neuroactive steroids and the sigma1 receptor function: behavioral consequences and therapeutic opportunities. *Brain Res Brain Res Rev*. 2001;37(1–3):116–132.
30. Robinson RG, Parikh RM, Lipsey JR, Starkstein SE, Price TR. Pathological laughing and crying following stroke: validation of a measurement scale and a double-blind treatment study. *Am J Psychiatry*. 1993;150(2):286–293.
31. Herndon RM. Generic and general use scales. In: Herndon RM, ed. *Handbook of Neurologic Rating Scales*. 2nd ed. New York, New York: Demos Medical Publishing; 2006:11–13.
32. Miller RG, Mitchell JD, Moore DH. Riluzole for amyotrophic lateral sclerosis (ALS)/motor neuron disease (MND). *Cochrane Database Syst Rev*. 2012;3:CD001447.
33. Writing Group; Edaravone (MCI-186) ALS 19 Study Group. Safety and efficacy of edaravone in well defined patients with amyotrophic lateral sclerosis: a randomised, double-blind, placebo-controlled trial. *Lancet Neurol*. 2017;16(7):505–512.
34. Brooks BR, Thisted RA, Appel SH, et al. Treatment of pseudobulbar affect in ALS with dextromethorphan/quinidine: a randomized trial. *Neurology*. 2004;63(8):1364–1370.
35. Panitch HS, Thisted RA, Smith RA, et al. Randomized, controlled trial of dextromethorphan/quinidine for pseudobulbar affect in multiple sclerosis. *Ann Neurol*. 2006;59(5):780–787.
36. Livingston G, Barber J, Marston L, et al. Prevalence of and associations with agitation in residents with dementia living in care homes: MARQUE cross-sectional study. *BJPsych Open*. 2017;3(4):171–178.
37. Porsteinsson AP, Drye LT, Pollock BG, et al. Effect of citalopram on agitation in Alzheimer disease: the CitAD randomized clinical trial. *JAMA*. 2014;311(7):682–691.
38. Cummings JL, Lyketsos CG, Peskind ER, et al. Effect of dextromethorphan-quinidine on agitation in patients with Alzheimer disease dementia: a randomized clinical trial. *JAMA*. 2015;314(12):1242–1254.
39. Shao L, Hewitt MC. The kinetic isotope effect in the search for deuterated drugs. *Drug News Perspect*. 2010;23(6):398–404.
40. Mavlyutov TA, Epstein ML, Andersen KA, Ziskind-Conhaim L, Ruoho AE. The sigma-1 receptor is enriched in postsynaptic sites of C-terminals in mouse motoneurons. An anatomical and behavioral study. *Neuroscience*. 2010;167(2):247–255.
41. Smith RA, Macklin EA, Myers KJ, et al. Assessment of bulbar function in amyotrophic lateral sclerosis: validation of a self-report scale (Center for Neurologic Study Bulbar Function Scale). *Eur J Neurol*. 2018;25(7):907–e66.
42. Smith R, Pioro E, Myers K, et al. Enhanced bulbar function in amyotrophic lateral sclerosis: the Nuedexta treatment trial. *Neurotherapeutics*. 2017;.
43. Fralick M, Sacks CA, Kesselheim AS. Assessment of use of combined dextromethorphan and quinidine in patients with dementia or Parkinson disease after US Food and Drug Administration approval for pseudobulbar affect. *JAMA Intern Med*. 2019;179(2):224–230.

CHAPTER 17

Bioethics in the era of translational medicine

Rajesh Sampath[a], Meryem Guler[b], and Richard A. Smith[c]
[a]Heller School for Social Policy and Management, Brandeis University, Waltham, MA, United States
[b]Cogent Family Healthcare, Mountain View, CA, United States
[c]Center for Neurologic Study, La Jolla, CA, United States

Introduction

Humans have long articulated ethical principles to guide their conduct. This may have originated with the Ten Commandments or independently in Asia based on Confucian principles of benevolence and kindheartedness which are the mainstay of traditional Chinese medicine. In the fifth century BCE, two seminal ethical forces emerged: the articulation of the Hippocratic Oath and the Declaration of Human Rights by the Persian emperor, Cyrus the Great, and his son and successor, Darius. Although not well documented it is fair to assume that over the centuries these affirmations would have greatly influenced the professional conduct of practitioners of the medical arts. Centuries later, Maimonides augmented these principles with his emphasis on the humanistic practice of medicine. Less familiar than the Hippocratic Oath, it is cited herein:

> *The Eternal Providence has appointed me to watch over the life and health of Your creatures. May the love for my art actuate me at all times; may neither avarice nor miserliness, nor thirst for glory or for a great reputation engage my mind; for the enemies of truth and philanthropy could easily deceive me and make me forgetful of my lofty aim of doing good to Your children.*
>
> *May I never see in the patient anything but a fellow creature in pain.*
>
> *Grant me the strength, time and opportunity always to correct what I have acquired, always to extend its domain; for knowledge is immense and the spirit of man can extend indefinitely to enrich itself daily with new requirements. Today he can discover his errors of yesterday and tomorrow he can obtain a new light on what he thinks himself sure of today.*
>
> *Oh, God, You have appointed me to watch over the life and death of Your creatures; here am I ready for my vocation and now I turn unto my calling.*

A further milestone, three centuries later, was the publication of *The Code of Medical Ethics* authored by Thomas Percival.[1] This was the first known use of the term "medical ethics" which, more or less, became the byword for the Hippocratic Oath. The first use of the term "bioethics" was articulated in 1926 by Fritz Jahr.[2] While the term's dictionary definition is "a discipline dealing with the ethical implications of biological research and

Neurotherapeutics in the Era of Translational Medicine
https://doi.org/10.1016/B978-0-12-816475-4.00011-2

applications especially in medicine," in the vernacular its use has often been synonymous with the term "medical ethics."

One of the earliest known challenges to medical ethics was deliberated by the French Academy of Science in 1887 when Louis Pasteur demonstrated that vaccination of a rabies patient was effective.[3] This was considered a major medical advance but aroused ethical concerns primarily due to the fact that Pasteur had grossly overstated the results of his preclinical studies. However, this proved to be a minor objection considering the horrors of the twentieth century which exposed the frailty of medical ethics and the willingness of nation states to undertake experimentation on human beings that violated all accepted standards of medical conduct.

While one would hope that the long-standing tradition of medical ethics would have reached a point of full articulation with the advent of biotechnology such has not proved to be the case. The ensuing discussion provides historical context and identifies contemporary issues that deserve further reflection, possibly leading to a further codification of generally agreed upon bioethical principles. This chapter does not propose to offer fixed ethical principles on what to do in such cases. Rather, our intent is to provide both a historical and contemporary overview of ethical questions that one anticipates will inform public policy in the biotechnological age.

Human experimentation

During World War II horrific experiments were undertaken at concentration camps, ostensibly for the purpose of advancing the German war effort as well as Nazi racial ideology. Nazi physicians and concentration camp staff forced prisoners to participate in radical and hazardous experiments, leading to death or disability.

The experiments exposed captives to mustard gas, freezing temperatures, anoxia, and electroshock. Surgical experimentation involved the transplantation of bone, muscle, and nerves. The most infamous practitioner of these procedures was Josef Mengele who became known as the "Angel of Death." Of particular note was his interest in twin studies. One of each pair was variously subjected to blood transfusions, amputation, and exposure to a variety of pathogens.

At the end of World War II, physicians were held responsible for crimes against humanity in what became known as the Doctors' Trial. Seven of the accused received death sentences, others received prison sentences from 10 years to life, and seven were acquitted. The accused faced four primary charges[4]:

1. Conspiracy to commit war crimes and crimes against humanity as described in counts 2 and 3.
2. War crimes: performing medical experiments, without the subjects' consent, on prisoners of war and civilians of occupied countries, in the course of which experiments the defendants committed murders, brutalities, cruelties, tortures, atrocities, and other

inhuman acts. Also, planning and performing the mass murder of prisoners of war and civilians of occupied countries, stigmatized as aged, insane, incurably ill, deformed, and so on, by gas, lethal injections, and diverse other means in nursing homes, hospitals, and asylums during the Euthanasia Program and participating in the mass murder of concentration camp inmates.

3. Crimes against humanity: committing crimes described under count 2 also on German nationals.
4. Membership in a criminal organization, the SS.

Ultimately, these trials led to the development of the Nuremberg Code of medical ethics, a set of 10 principles of research ethics for human experimentation, as follows[5]:

1. The voluntary consent of the human subject is absolutely essential.
2. The experiment should be such as to yield fruitful results for the good of society, unprocurable by other methods or means of study, and not random and unnecessary in nature.
3. The experiment should be so designed and based on the results of animal experimentation and a knowledge of the natural history of the disease or other problem under study that the anticipated results will justify the performance of the experiment.
4. The experiment should be so conducted as to avoid all unnecessary physical and mental suffering and injury.
5. No experiment should be conducted where there is an a priori reason to believe that death or disabling injury will occur; except, perhaps, in those experiments where the experimental physicians also serve as subjects.
6. The degree of risk to be taken should never exceed that determined by the humanitarian importance of the problem to be solved by the experiment.
7. Proper preparations should be made and adequate facilities provided to protect the experimental subject against even remote possibilities of injury, disability, or death.
8. The experiment should be conducted only by scientifically qualified persons. The highest degree of skill and care should be required through all stages of the experiment of those who conduct or engage in the experiment.
9. During the course of the experiment the human subject should be at liberty to bring the experiment to an end if he has reached the physical or mental state where continuation of the experiment seems to him to be impossible.
10. During the course of the experiment the scientist in charge must be prepared to terminate the experiment at any stage, if he has probable cause to believe, in the exercise of the good faith, superior skill, and careful judgment required of him that a continuation of the experiment is likely to result in injury, disability, or death to the experimental subject.

Although the precepts imbued within the Hippocratic Oath and the more recent Nuremberg Code have seemingly guided the medical professions in current times, no nation or organization has acknowledged either of these as legally enforceable.

Among the Axis Powers, Japan also conducted experiments involving surgeries on living humans that included dismemberment and bacterial inoculations. Their "Units" subjected civilians and captured prisoners of war to biological and chemical weapon tests. Inexplicably, following the surrender of Japan, General Douglas MacArthur offered immunity from prosecution to Shiro Ishii, the medical director of Unit 731, which was the most notorious of these facilities.[6] This was offered in the interest of national security, ostensibly to block Soviet access to the results of the human experiments conducted by the Japanese. To their credit, the Soviets prosecuted some Unit 731 members during its Khabarovsk War Crime Trials.

One would be remiss to conclude that the egregious perversion of medical ethics during World War II was an isolated phenomenon. However, the record does not support this. With the participation of the US Public Health Service (PHS) hundreds of mental patients in Guatemala were subjected to experiments to determine the transmissibility and effects of syphilis and gonorrhea. These studies were undertaken at the country's National Mental Health Hospital. In 2010 Hillary Clinton, US Secretary of State, and President Barack Obama acknowledged these experiments as crimes against humanity.[7]

Tuskegee syphilis experiments

The "Tuskegee Study of Untreated Syphilis in the Negro Male," conducted for 40 years (1932–1972) by the PHS in Alabama, was among the most notorious violations of medical ethics in the twentieth century. There were three notable features of this study. First, all the participants were African American men. Second, participants were essentially uninformed of the nature of the research. Third, the study continued even after it was known that syphilis could be successfully treated through the use of penicillin. It has been noted that the only information that was provided for the subjects was that they were being treated for "bad blood," a term that was attributed to a variety of disorders including syphilis.[8]

Even after the Tuskegee experiments were widely reported, John Heller Jr. from the PHS defended the ethics of the undertaking, stating "the longer the study, the better the ultimate information we would derive." Ultimately, the study was terminated following a review by several federal agencies. Subsequently, the National Association for the Advancement of Colored People (NAACP) filed a class action lawsuit on behalf of the participants. This resulted in a financial settlement for the surviving participants and their family members and included free medical treatment for the living.

In response to the revelations about Tuskegee, the US Congress passed the National Research Act (1974) and established the Office for Human Research Protection to provide oversight for the conduct of clinical trials. This led to the requirement that all human trials be subject to the supervision of Institutional Review Boards (IRBs), a critical advance in the protection of human subjects. These locally empowered groups, by statute, are required to include laypeople and their deliberations are anchored in principles laid down by the Declaration of Helsinki. In 1997, President Bill Clinton conducted a ceremony at the White House that included surviving Tuskegee participants. His commendable and overdue apology follows:

> *What was done cannot be undone. But we can end the silence. We can stop turning our heads away. We can look at you in the eye and finally say on behalf of the American people, what the United States government did was shameful, and I am sorry... To our African American citizens, I am sorry that your federal government orchestrated a study so clearly racist.*

Ultimately the federal government, likely as an act of contrition, established a National Center for Bioethics in Research and Health Care at Tuskegee University. Opened in 1999, the Center explores issues that underlie research and medical care of African Americans and other underserved people.

Declarations of Geneva and Helsinki

Following the Nuremberg Trials, the World Medical Association (WMA) thought it necessary to revisit the Hippocratic Oath with the intent of generating a contemporary version. This resulted in what is known as the Declaration of Geneva (1948)[9]:

> *As a member of the medical profession:*
>
> *I solemnly pledge to dedicate my life to the service of humanity;*
>
> *The health and well-being of my patient will be my first consideration;*
>
> *I will respect the autonomy and dignity of my patient;*
>
> *I will maintain the utmost respect for human life;*
>
> *I will not permit considerations of age, disease or disability, creed, ethnic origin, gender, nationality, political affiliation, race, sexual orientation, social standing, or any other factor to intervene between my duty and my patient;*
>
> *I will respect the secrets that are confided in me, even after the patient has died;*
>
> *I will practice my profession with conscience and dignity and in accordance with good medical practice;*
>
> *I will foster the honor and noble traditions of the medical profession;*
>
> *I will give to my teachers, colleagues, and students the respect and gratitude that is their due;*
>
> *I will share my medical knowledge for the benefit of the patient and the advancement of healthcare;*
>
> *I will attend to my own health, well-being, and abilities in order to provide care of the highest standard;*
>
> *I will not use my medical knowledge to violate human rights and civil liberties, even under threat;*
>
> *I make these promises solemnly, freely, and upon my honor.*

Approximately 16 years later, the WMA once again felt the need to make the first of seven revisions to the Declaration. The initial revisions primarily focused on the conduct of clinical research, as follows[10]:

> *Medical research involving human subjects must conform to generally accepted scientific principles, be based on a thorough knowledge of the scientific literature, other relevant sources of information, and adequate laboratory and, as appropriate, animal experimentation. The welfare of animals used for research must be respected.*
>
> *The design and performance of each research study involving human subjects must be clearly described and justified in a research protocol. The protocol should contain a statement of the ethical considerations involved and should indicate how the principles in this Declaration have been addressed. The protocol should include information regarding funding, sponsors, institutional affiliations, potential conflicts of interest, incentives for subjects and information regarding provisions for treating and/or compensating subjects who are harmed as a consequence of participation in the research study. In clinical trials, the protocol must also describe appropriate arrangements for post-trial provisions.*

It would not be unreasonable for persons familiar with the violations of bioethics in the twentieth century to think that such violations were from a bygone era. However, this is clearly not the case, as can be seen through the following incidents.

Prisoner experiments: 1940s to 2010s

Irrespective of the growing consensus that human experimentation is governed by ethical principles, human experimentation on prisoners in the United States is filled with egregious examples, albeit this brief summary is incomplete.[11]

Under the auspices of highly regarded academic institutions a variety of infectious agents (malaria, hepatitis, etc.) were administered to prisoners, ostensibly to gain insight into the natural history of these infections. In the pursuit of cancer research, malignant cells were administered to convicts at Ohio State Prison in a study undertaken by investigators at the Sloan-Kettering Institute. Similarly, the US Atomic Energy Commission conducted high-dose radiation experiments on inmates. Incentives to participate included payments and other privileges. Studies of this sort continued as late as 1973 in Oregon. Consent for these experiments was grossly inadequate as to the long-term risks to the subjects. Upon investigation, it was found that one investigator "didn't warn them of all the dangers because I didn't want to frighten them." And many years later another investigator claimed that nothing wrong had been done although the US Department of Energy called this research "ethically flawed, even by the standards at the time."

As a variant on the use of inmates as human guinea pigs, prisoner populations were considered to be an ideal group to conduct Phase I pharmaceutical trials. At one point 90% of investigational drugs were initially tested on prisoners. When these practices were publicly commented upon, the protection of human subjects was deemed less noteworthy than the practice of revising prisoners' sentences as an incentive for participation. Not surprisingly, these privileges were more likely to be accorded to whites than African Americans, thus revealing the profound racial disparity at the heart of these decisions.

University of Pennsylvania: 1999

When gene therapy trials began in the early 1990s, they were under strict federal regulations. However, by the end of the 1990s, the NIH had cut back its oversight in response to lobbying by the biotechnology industry which complained about undue restrictions and patient communities which demanded "faster research." Ultimately, a failure of oversight dramatically tarnished the luster of gene therapy. When Jesse Gelsinger, an 18-year-old with an ornithine transcarbamylase (OTC) gene mutation, was enrolled in a gene therapy treatment trial, what at the time seemed like a heroic effort to replace a gene shortly resulted in a fatal outcome. Subsequently, the FDA terminated all gene therapy trials at the University of Pennsylvania which was shown to have violated numerous FDA regulations.[12] Among these was a failure to notify the FDA or research subjects that preclinical studies had often resulted in deaths of subhuman primates that had been injected with the viral vector containing the human OTC gene. Of equal importance, the FDA was concerned that the biotech company involved in the study had unduly influenced the research and one of the investigators owned its stock.

Johns Hopkins: 2001

The death of Ellen Roche, a healthy volunteer in an asthma study at Johns Hopkins, illustrates another ethical failure. As noted in the Declaration of Helsinki: "Medical research involving human subjects must conform to generally accepted scientific principles, be based on a thorough knowledge of the scientific literature, other relevant sources of information, and adequate laboratory and, as appropriate, animal experimentation." The researchers in this instance were using normal volunteers to study the effects of airway constrictions so as to better understand asthma. To induce these airway changes subjects inhaled hexamethonium along with another drug to simulate an asthmatic attack. Hexamethonium had not been approved by the FDA for this purpose and the researchers had not adequately surveyed the literature germane to its effects on human subjects. Further, the IRB failed to appreciate that this research would have required approval by the FDA which would undoubtedly have required the investigators to conduct preclinical studies prior to conducting their research in humans.[13]

Cloning

Cloning genes

The cloning of genes in the 1970s ushered in the era of biotechnology. Almost immediately it was recognized that the new technology posed a Faustian bargain in that the risks to humankind might outweigh the benefits. The principal concern was that a gene introduced into a bacterium could lead to fatal consequences if it resulted in a novel organism that threatened human or animal health. To minimize this possibility a group of scientists met in 1975 at the Asilomar Conference Center in California. The result was the publication of guidelines recommending that all such experiments be conducted in containment facilities to assure that the engineered bacterium could not escape into the environment. These regulations have been relaxed, but it is not certain that the Asilomar principles are going to prevent careless or unscrupulous scientists from putting society at risk.

Cloning animals

Dolly the sheep, the first mammal to be cloned, ushered in a new era of reproductive medicine. Inevitably, the implications of this feat have raised ethical concerns. The process used to create Dolly is referred to as somatic cell nuclear transfer (SCNT). This involves the extraction of the host egg's cell nucleus which is replaced with the nucleus from another cell. As the "reprogrammed" cloned cell replicates, an early-stage embryo is produced which may either be used to generate stem cells or be implanted into a host mother.

With the cloning of macaque monkeys, a species within the human family tree, it is now certain that it is possible to clone human beings. The implications of this for the future is already the basis for science fiction.

While some nations have outlawed human cloning, an argument has been made for the use of this technology to generate tissues and whole organs for therapeutic purposes. For example, normal skin grown from a burn patient could be generated using SCNT. A graft utilizing this tissue would not be rejected by the patient's immune system.

Gene editing in humans

If one pondered the emerging concerns of bioethicists, the future is now, as can be appreciated in the following instance.

In 2017 He Jiankui, a biophysics researcher, edited a gene that codes for the human immunodeficiency virus (HIV) receptor in the ova of a woman whose husband was HIV positive. The intent was to shield the unborn infants of these parents from being infected inadvertently in utero with HIV. The twins, Lulu and Nana, were the first-ever genetically edited humans.[14]

After presentation of their work at an international conference, an uproar resulted within the scientific and medical ethics communities, ultimately leading to the suspension of He's research by the Chinese authorities. In December 2019, He was found guilty of unethical conduct, fined, and sentenced to 3 years in prison. Concurrently, a number of bodies including the National Institutes of Health of the United States condemned He and his team for violating international ethical norms. In addition to a concern about ethical standards, the research itself was criticized since it was felt that the need for inactivating the HIV receptor (CCR5) was unconvincing and such procedures might lead to off-target effects. However, at this juncture it is certain that a number of stakeholders will make a legitimate claim to their right to alter the genetic heritage of their offspring. Putting aside for the moment the scientific objections for doing so, the principle moral concern will be the instance when parents insist on altering their offspring for the purpose of enhancing their cosmetic, cognitive, or physical abilities.

Access to care

A corporate strategy for acquiring commonly used pharmaceuticals and subsequently increasing their price has been a topic of recent public concern and condemnation. Some of these concerns have been allayed by regulatory action which has facilitated the introduction of generic substitutes. However, for the purpose of this discussion, the introduction of a new generation of pharmaceuticals such as monoclonal antibodies, antisense therapeutics, and gene therapies whose costs are astronomical raise new concerns. On one hand, the development of these products often represent risky ventures. For example, in the instance of drugs developed for the treatment of neurodegenerative diseases the time to market averages about 15 years, the cost approaches a billion dollars, and the clinical trial failure rate has led some companies to cease development in this area.

On the other hand, drugs such as Zolgensma and Spinraza, both recently approved for the treatment of spinal muscular atrophy—a fatal motor disease in infants, cost patients/families or third-party carriers millions of dollars.[15] For health-care systems this and a myriad of other 21st century treatments are budgetary nightmares. And depending on the health-care system, these treatments are not universally available to all citizens.

In addition to the cost of medications the organization of health care, depending on the country, is another area of concern. In third-world countries access to 21st-century therapies are often limited to a privileged few. Even in the United States there are conflicts between the welfare of patients and the welfare of corporate interests that limit the access to care. In spite of vigorous public discussion, a "Hippocratic Oath" for business interests has yet to be formulated.

Precision medicine

Publication of the first draft of the human genome seemingly launched a new health-care paradigm: the advent of personalized medicine. Almost immediately a rhetorical skirmish followed. Some expressed concern that reducing a disease to its molecular signature was ironic in that it is depersonalizing. Commercial interests suggested that "stratified medicine" would be a more appropriate substitute since it is customary in industry to stratify populations in clinical trials, but this suggestion was discarded due to concerns about its discriminatory implications.[16] Ultimately, after considering "genomic medicine," "personalized medicine" was rebranded as "precision medicine."

Among the recent notable examples of precision medicine, the case of a child with Batten's disease strikingly illustrates the power of this new treatment paradigm.[17] Batten's disease is a rare and fatal neurodegenerative lysosomal storage disorder which inexorably results in blindness, mental deterioration, seizures, etc. At 6 years of age the aforementioned child was hospitalized after exhibiting loss of balance and impairment of speech and swallowing. Following the identification of a pathogenic transposon that interfered with the normal expression of the *CLN7* gene, an antisense therapeutic was designed and subsequently administered intrathecally following the example of Nusinersen which had earlier been approved for the treatment of spinal muscular atrophy. At the time of publication, objective treatment results were reported.

While acknowledging the medical and scientific prowess that was brought to bear in an effort to treat the above case, such efforts understandably lead to a number of ethical considerations. These are almost too numerous to enumerate. In the instance of a rapidly progressing disease involving unique genetic alterations, it is often necessary to suspend a number of normal drug-development processes. Accordingly, one cannot guarantee the safety of the drug or be assured of its pharmacokinetic properties, having not subjected them to rigorous preclinical evaluation. Further, to date there have been examples when wealthy families have been able to support N-of-1 treatments that may or may not benefit others but have involved extensive resource allocation.

Philosophical discourses informing bioethics

At the time this chapter was being written the world was in the throes of a pandemic of seismic proportions. One result is that health-care systems throughout the world have faced medical choices that have primarily been restricted to battlefield theaters, namely triaging who should or should not be treated due to limited resources. To avoid leaving these decisions specifically to nurses and doctors, ethicists have proposed policies that could be evenly applied. Although not privy to these deliberations, it is certain that they have relied upon long-established ethical principles and their current interpretations.

While at times life and death decisions fit nicely into a "utilitarian" conception of ethics, some philosophers, for example Ivan Illich, consider contemporary values to be a rupture from the historical past when shamans and medicine men seemingly offered a more holistic view of illness and its care.

But if our understanding of life and death are about to change it has been noted that we risk creating states that belong to the realm of neither life nor death. Such an example would be the artificial prolongation of life, perhaps to the point of interminability. Theological conceptions of "eternal life" would have no bearing in this instance because one cannot speak of life after a real physical death per se or an eternal being with no birth or death; rather the possibility of giving birth and erasing death speaks to another conception entirely. However, given that most societies assume principles of noninterference when it comes to their most cherished cultural, religious, and moral norms as they apply to legal and medical decisions, the biotechnology age challenges moral norms that may not readily give way to a global consensus.

One way to respond to moral and ethical conundrums can be found in John Rawls's classic, *A Theory of Justice* (1971). He pondered on the ability of individuals to self-reflexively and deliberately examine the assumptions underpinning the rightness of their decision-making. His answer to the search for consensus led him to suggest the need for "reflective equilibrium." From this perspective, there is a point when equally valid yet contrasting judgments align to the effect that consensus is achieved by avoiding uncritical and hasty decisions while decisively making them. Considering the contemporary divisiveness on so many issues this may be the fork in the road not yet taken.

Conclusion

For certain, no codification of bioethics will be sufficient to keep pace with the moral challenges that humanity will face as a consequence of the inexorable advances in medical innovation and experimentation. It is likely that this review of bioethics, admittedly limited in scope, will be seen as being a quaint exposition of the subject in the near future. Each of the areas discussed herein, for example, the cost of care and access to it or the reengineering of animals and humans, will be provocative subjects for future moral deliberation. One can only hope that societies will make room for such discussions. If the past is any example it is not at all certain this will be the case. However, if a global dialogue does not lead to agreed-upon norms, choices (or lack thereof) could be made now that cause irreversible consequences for future generations.

Acknowledgments

The authors greatly appreciate the indispensable editorial assistance of Hanna Hovren and Kathleen Myers, PhD, and the critical commentary provided by Andrew Feenberg, PhD and Tania Schweig. This chapter is dedicated to the memory of Yonim Schweig whose short life was anchored in ethical concerns which tragically motivated his last noble act.

References

1. Patuzzo S, Goracci G, Ciliberti R. Thomas percival. Discussing the foundation of medical ethics. *Acta Biomed*. 2018;89(3):343–348.
2. Rincic I, Muzur A. Fritz Jahr: the invention of bioethics and beyond. *Perspect Biol Med*. 2011;54(4): 550–556.
3. Gelfand T. 11 January 1887, the day medicine changed: Joseph Grancher's defense of Pasteur's treatment for rabies. *Bull Hist Med*. 2002;76(4):698–718.
4. U.S. Holocaust Memorial Museum. *Trials of War Criminals before the Nuremberg Military Tribunals under Control Council Law No. 10, Nuremberg October 1946–April 1949*. Washington, DC: US GPO; 1949–1953.
5. *Trials of War Criminals before the Nuremberg Military Tribunals under Control Council Law No. 10*. Washington, DC: U.S. Government Printing Office; 1949181–182.
6. Brody H, Leonard SE, Nie J-B, Weindling P. U.S. responses to Japanese wartime inhuman experimentation after world war II. *Camb Q Healthc Ethics*. 2014;23(2):220–230.
7. Rodriguez MA, García R. First, do no harm: the US sexually transmitted disease experiments in Guatemala. *Am J Public Health*. 2013;103(12):2122–2126.
8. Feenberg A. Ethical aspects of experimental treatment. In: Smith RA, ed. *Handbook of Amyotrophic Lateral Sclerosis*. New York, NY: Marcel Dekker, Inc.; 1992:347–362
9. World medical association declaration of Geneva. *Afr Health Sci*. 2017;17(4):1203.
10. Parsa-Parsi RW. The revised declaration of Geneva: a modern-day physician's pledge. *JAMA*. 2017;318 (20):1971–1972.
11. Hornblum AM. They were cheap and available: prisoners as research subjects in twentieth century America. *BMJ*. 1997;315(7120):1437–1441.
12. Bishop T. Gene therapy trials shut down at University of Pennsylvania following patient death. Available from: https://www.wsws.org/en/articles/2000/03/gene-m13.html; 2000.
13. Savulescu J, Spriggs M. The hexamethonium asthma study and the death of a normal volunteer in research. *J Med Ethics*. 2002;28:3–4.
14. Schaefer GO. Toward realism about genetic enhancement. *Am J Bioeth*. 2019;19(7):28–30.
15. Starner CI, Gleason PP. Spinal muscular atrophy therapies: ICER grounds the price to value conversation in facts. *J Manag Care Spec Pharm*. 2019;25(12):1306–1308.
16. Juengst E, McGowan ML, Fishman JR, Settersten Jr. RA. From "personalized" to "precision" medicine: the ethical and social implications of rhetorical reform in genomic medicine. *Hastings Cent Rep*. 2016;46(5):21–33.
17. Kim J, Hu C, Moufawad El Achkar C, et al. Patient-customized oligonucleotide therapy for a rare genetic disease. *N Engl J Med*. 2019;381(17):1644–1652.

CHAPTER 18

Government efforts to promote the development of neurotherapeutics for the treatment of neurological diseases

Petra Kaufmann[a] **and Clinton B. Wright**[b]
[a]VC R&D Translational Medicine R&D, AveXis, A Novartis Company, San Diego, CA, United States
[b]Division of Clinical Research, National Institute of Neurological Disorders and Stroke, National Institutes of Health, Bethesda, MD, United States

Registries and patient-focused resources

Registries are databases that collect information on individuals with the goal to promote research. Registries can provide researchers with epidemiological information; they can connect patients with potential research opportunities; and they can support the collection of data about disease characteristics and progression. The National Center for Advancing Translational Sciences Rare Disease Registry provides guidance on how to implement registries, including instructions, examples, and shared resources (RaDaR, https://registries.ncats.nih.gov/).

The United States (US) Center for Disease Control and Prevention (CDC) can support translational research in neurology by establishing disease registries. An example is a registry aimed at better capturing the incidence and prevalence of ALS (www.cdc.gov/als). The registry collects data on demographics, and occupational and environmental exposures. Based on this registry, the prevalence of ALS in the US was estimated at 5 in 100,000 Americans.[1] Analysis of registry data allows studying the potential risk factors for ALS. The registry can also provide information on opportunities for research participation to interested participants.

The Office of Rare Diseases Research at the National Institute of Health has developed a Toolkit for Patient-Focused Therapy Development. It provides online access to resources that can help patient groups advance medical research. The toolkit is organized along a "Therapy Development Roadmap" with five stages, entitled: (1) Getting Started, (2) Discovery, (3) Preparing for Clinical Trials, (4) Clinical Trials and FDA Review, and (5) After FDA Approval (https://ncats.nih.gov/toolkit).

Neurotherapeutics in the Era of Translational Medicine
https://doi.org/10.1016/B978-0-12-816475-4.00012-4

National Institute of Health (NIH) NeuroBioBank

Researchers need access to tissue specimens from patients with neurological diseases who have died and chosen to donate their remains to science. The National Institute of Neurological Disorders and Stroke (NINDS) therefore supports a tissue repository, the NIH NeuroBioBank. This initiative connects several institutions "virtually" across the United States that collaborate to provide researchers with tissues of different types. It is important to note that there is not only a sufficient quantity of tissue available to researchers, but also that the specimens are of high quality based on the speed and methods by which they were harvested, transported, and stored (https://neurobiobank.nih.gov).

Clinical research networks

Rare diseases clinical research network (RDCRN)

The NINDS, in collaboration with the NIH Office of Rare Diseases Research (ORDR), has funded several networks that allow investigators to share protocols, data, and gather natural history information collaboratively, to study phenotype/genotype correlations and to develop and optimize outcome measures and biomarkers. Examples of neurological diseases that have benefited from this network support include frontotemporal dementia, autonomic disorders, brain vascular malformations, developmental synaptopathies, dystonias, inherited neuropathies, lysosomal diseases, mitochondrial diseases, Rett syndrome and related disorders, and amyotrophic lateral sclerosis (ALS). Each consortium consists of clinical sites in the United States and other regions (when needed) that collaborate on a group of diseases and agree to share data. Research teams are required to partner with patient organizations to receive funding. In addition to the clinical sites, there is a central Data Management and Coordinating Center with the goal to streamline electronic data capture, to enhance data quality, and to facilitate downstream data sharing (www.rarediseasenetwork.org).

NeuroNEXT

The "Network for Excellence in Neuroscience Clinical Trials," or NeuroNEXT, is funded by the NINDS to conduct studies of treatments for neurological diseases such as ALS through partnerships with academia, private foundations, and industry. NeuroNEXT was created with the goal of conducting high-quality, impactful clinical studies using an efficient platform to accelerate progress (Lancet Editorial).

NeuroNEXT consists of Cooperative Agreements (U01 funding mechanism) that fund a Clinical Coordinating Center (CCC), a Data Coordinating Center (DCC), and 25 clinical sites with broad geographical representation. The network is designed to conduct biomarker studies and phase 2 clinical trials efficiently in pediatric and adult populations. Operational efficiency is achieved through the use of master clinical trial agreements, a single centralized Institutional Review Board (IRB), and harmonized standard operating procedures. Investment in this infrastructure has resulted in shortened clinical trial start-up times. The first NeuroNEXT protocol reviewed by the central IRB achieved full approval to allow participant enrollment within 56 days and went from grant award to the first patient visit in less than 4 months (Kaufmann et al.). Similarly, electronic data capture (EDC) has streamlined data collection through the use of standardized data elements and case report forms, shortening the average time from funding to database release from 8.6 months in nonnetwork studies to 3.6 months in NeuroNEXT studies. Participant recruitment in NeuroNEXT studies also is more successful than in nonnetwork studies for similar indications (www.NeuroNEXT.org). As of Fall 2019 the NeuroNEXT network has completed five trials, with four that are ongoing or in the start-up phase. These trials address a wide range of diseases and conditions, including spinal muscular atrophy, progressive multiple sclerosis, myasthenia gravis, acute ischemic stroke, Huntington disease, glioblastoma multiforme, Fragile X syndrome, GNE myopathy, and cryptogenic peripheral neuropathy.

StrokeNet

The NIH StrokeNet was established to conduct small and large clinical trials and biomarker studies to advance acute stroke treatment, prevention, recovery, and rehabilitation following a cerebrovascular event (CVE), including those in pediatric populations. This network of 29 regional centers across the United States involves more than 500 hospitals and rehabilitation centers and is designed to serve as the infrastructure pipeline for the evaluation of new potential treatments (phase 2) for patients who have suffered a cerebrovascular event or are at risk for one as well as test the efficacy of therapies (phase 3) and examine comparative effectiveness of treatments already in the clinic. The geographical reach of the network is broad, with roughly half the US population within a 40-mile radius of an NINDS-funded StrokeNet site. To date, the network has completed four clinical trials, two of them entirely within the network, across the areas of intracerebral hemorrhage, endovascular therapy, and stroke recovery. An additional six phase 2 and 3 trials are ongoing and three are in the initial start-up phase; in addition, a separately funded biomarker validation study is embedded in an ongoing phase 3 trial. Efficiencies brought about by such innovations as master trial agreements and central IRB oversight are two important goals of the network (www.nihstrokenet.org).

Early-phase pain investigation clinical network (EPPIC-Net)

The opioid crisis motivated Congress to allocate $500 million to the NIH Helping to End Addiction Long-term initiative (HEAL) https://heal.nih.gov/ to address substance misuse, addiction, and pain. The NINDS dedicates $250 million per year to understanding acute and chronic pain as well as developing optimal treatments. The Foundation for NIH has supported extensive discussions between various stakeholders, including FDA, industry, and academic investigators to understand and promote the development of novel nonaddictive therapeutics for acute and chronic pain. One result, at the request of industry, was to streamline the review of clinical trial applications as well as the application process itself. This led to the development of the innovative EPPIC-Net program (https://www.ninds.nih.gov/Current-Research/Trans-Agency-Activities/NINDS-Role-HEAL-Initiative/NINDS-Role-HEAL-Initiative-EPPIC) that combines the following: awards that fund the clinical trial network and the funding mechanism that supports the clinical trials (Other Transactional Authority, abbreviated OTA). The OTA mechanism facilitates the importation of potential therapeutics (small molecules, devices, and biologics) into the network, utilizing a rolling submission that permits review of a scaled down preliminary application within 2 months. Preliminary applications receiving a "green light" move into the second stage of review. To encourage applicants without NIH grant-writing experience, asset holders are then paired with an NIH contractor that creates a detailed "dossier" describing the preclinical, toxicology, and dose finding data for the proposed asset. Dossiers are completed within 2 months and reviewed, with approved assets moving to the third and final stage. In the final stage, NIH approves the asset holder to work with the network clinical and data coordinating centers to develop a clinical trial protocol, in cooperation with NIH program staff. Approved protocols in the third stage of review lead to funding using Other Transaction Authority (OTA) in which case funds are provided to the clinical coordinating center for subcontracting with the sites that are to perform the clinical trial. The intellectual property of the asset holder is preserved throughout this process using Cooperative Research and Development Agreements (CRADA), Memorandums of Understanding, and other instruments. The goal is for the time from preliminary application submission to trial funding to be 6–8 months, representing a much shorter timeline than afforded by traditional NIH funding mechanisms (usually 18 months or longer).

Clinical trial readiness for rare neurological and neuromuscular diseases

NINDS PAR-19-220 is a cooperative agreement funding mechanism that aims to support clinical studies that address critical needs for clinical trial readiness in rare neurological and neuromuscular diseases. These studies should result in clinically validated biomarkers and clinical outcome assessment measures appropriate for use in upcoming clinical trials. This is a collaborative project between the NINDS and the Office of Rare Diseases Research to enhance the quality and increase the likelihood of success of clinical trials in these rare diseases (https://grants.nih.gov/grants/guide/pa-files/PAR-19-220.html).

Clinical trial support

In addition to research funded through its clinical trial networks, the NINDS supports a large portfolio of clinical trials investigating potential interventions for neurological diseases. Private sector investment has often contributed to advancing NINDS-supported clinical trials testing new products or new indications for approved products. However, many trials evaluate non-pharmacological interventions or evaluations of "repurposed" products so that private sector investment could not be generated. Public funding therefore remains an important source of support for such trials which can impact the standard of care.

The NIH also provides a valuable resource for information on clinical trials. Patients can learn about participation in a clinical trial at www.nih.gov/health/clinicaltrials. A large database of clinical trials can be found at www.clinicaltrials.gov. This database fulfills several purposes, primarily providing a public record of clinical trials that have been completed, are planned or are ongoing. Researchers, patients, and the public can find information on interventions, trial designs, sponsors, sites, status of clinical trials, and, when applicable, results, thus improving transparency and accountability. Registration of US phase 2 and 3 trials is mandatory, as is the posting of results in a timely manner with fines imposed in cases of noncompliance (https://www.fda.gov/regulatory-information/search-fda-guidance-documents/civil-money-penalties-relating-clinicaltrialsgov-data-bank). The database has also been enhanced to allow searching for trial participation opportunities: patients or clinicians can use search terms such as the name of a particular disease, and also refine the search for geographic relevance by adding a search term indicating their location (www.clinicaltrials.gov).

NINDS conference support

Scientific meetings are often a key initial step in advancing treatments for neurological diseases. Developing new treatments typically requires the collaboration of experts from several disciplines. A conference can be an important way to launch a project, or to accelerate the momentum. Such conferences can be supported with an R13 grant to investigators, and also take the form of a Cooperative Conference Grant (U13), in which case NINDS program staff is an active partner in the development and implementation of the conference (http://grants1.nih.gov/grants/funding/r13/index.htm).

NINDS clinical trials methodology course

The NINDS-funded Clinical Trials Methodology Course has been designed to provide junior investigators with the knowledge and practical skills needed for clinical research. The course provides in-person and online learning opportunities focused on early-phase clinical trials. It covers the process of developing and implementing a clinical research project from formulating an appropriate scientific question, identifying a suitable trial design, building on existing preclinical and clinical data, obtaining funding, and working toward project implementation (https://nett.umich.edu/training/ctmc/ctmc-course-overview).

FDA-NINDS coordination

The NINDS and three FDA Centers have developed memorandums of understanding (MoUs) covering multiple modalities (small molecules, biologics, devices, and natural products). The following MoUs have been established between NINDS and the Center for Biologics Evaluation and Research (CBER) (https://www.fda.gov/about-fda/domestic-mous/mou-225-18-017), the Center For Drug Evaluation and Research (CDER) (https://www.fda.gov/about-fda/domestic-mous/mou-225-19-027), and the Center For Devices and Radiological Health (CDRH) (https://www.fda.gov/about-fda/domestic-mous/mou-225-13-023) to coordinate and collaborate with each other to protect and improve public health. To achieve this, the NINDS and the FDA share expertise, resources, and relationships in order to increase the capability and readiness of each organization to respond to situations that may arise. These MOUs provide a cross-agency conduit to work collaboratively to improve efficiency of the submission and review process for applications requesting funding from NIH/NINDS to support preclinical or clinical studies relevant to subsequent submissions to the FDA. These agreements provide opportunities for the NIH and the FDA to discuss topics of common interest, such as FDA's managed regulatory review process and NIH/NINDS's extramural, federally funded translational research initiatives and programs. The trans-agency interactions facilitate NINDS's ability to respond to NIH-investigators' general questions about regulatory issues and to reinforce the message of when and how to seek FDA guidance on therapy development and filings for human testing.

FDA guidance

To facilitate clinical trials in neurological diseases, the FDA may issue guidance for investigators in the field. Examples of this include the guidance for ALS that the FDA issued in 2019. Their recommendations recognize the etiological heterogeneity of ALS and the need to study subgroups for some interventions. Mortality is considered an appropriate endpoint, as are functional scales such as the ALSFRSr. The FDA guidance encourages innovation in endpoint development and trial design (Food and Drug Administration Guidance, ALS). Other recently released draft and final directives address the following topics: Alzheimer's Disease: Developing Drugs for Treatment, Drugs for Treatment of Partial Onset Seizures: Full Extrapolation of Efficacy from Adults to Pediatric Patients 4 Years of Age and Older, Duchenne Muscular Dystrophy and Related Dystrophinopathies: Developing Drugs for Treatment, and Migraine: Developing Drugs for Acute Treatment.

Summary

The burden of neurological diseases is increasing, in part because they are more prevalent in the elderly, a growing segment of the US population.[2] Many of these disorders are lacking effective treatment options.

Several publicly funded initiatives are therefore aimed at promoting the clinical development of neurotherapeutics. These initiatives include registries, biobanks, resources and networks for observational and interventional studies, methodology course, conferences, and other resources. Many of these initiatives promote the collaboration of researchers, clinicians, and patient groups.

While over half of the NIH's research funding goes toward basic research, clinical research is an important step toward making sure that the advances in basic neuroscience ultimately benefit patients with neurological diseases.

References

1. Mehta P, Kaye W, Raymond J, et al. Prevalence of amyotrophic lateral sclerosis – United States, 2015. *MMWR Morb Mortal Wkly Rep*. 2018;67(46):1285–1289.
2. Gooch CL, Pracht E, Borenstein AR. The burden of neurological disease in the United States: a summary report and call to action. *Annal Neurol*. 2017;81(4):479–484.

Further reading

Kaufmann P, O'Rourke PP. Central institutional review board review for an academic trial network. *Acad Med*. 2015;90(3):321–323.

Heemskerk J, Farkas R, Kaufmann P. Neuroscience networking: linking discovery to drugs. *Neuropsychopharmacology*. 2012;37:287–288.

The Lancet Neurology. NeuroNEXT: accelerating drug development in Neurology. *Lancet Neurol*. 2012;11(2):119.

https://www.fda.gov/regulatory-information/search-fda-guidance-documents/amyotrophic-lateral-sclerosis-developing-drugs-treatment.

https://www.fda.gov/media/92233/download.

https://www.fda.gov/media/89829/download.

https://www.fda.gov/media/110903/download.

CHAPTER 19

Innovative funding models for neurotherapeutics

Neta Zach[a] **and Melanie Leitner**[b]
[a]Prize4Life, Haifa, Israel
[b]Accelerating NeuroVentures, LLC, Boston, MA, United States

Introduction

Need for innovation in funding

Since the days of Vannevar Bush,[a] director of the US Office of Scientific Research and Development in the late 1940s, and the change in budgetary priorities in the 1960s, the National Institutes of Health has largely funded early-stage biological and medical discovery, with a significant contribution from the philanthropic sector. Later stage therapeutic development costs have largely been funded by industry. Over the past 20 years, the costs of drug development have ballooned from the "800 Million Dollar Pill"[b] of the early 2000s to the mind-blowing cost today, estimated to be approximately $2.5B.[1] Regardless of the exact cost of any particular program, the development of a pharmaceutical targeting a neurologic disorder is more costly than one which targets a non-neurologic indication, and the time for approval of a CNS product far exceeds that for other indications. There are multiple reasons for these increased costs, ranging from the high number of failures due to limited understanding of the CNS pathophysiology associated with a given disease to the lack of translatable in vitro and in vivo models, and the lack of sensitive clinical outcome measures. Addressing many of these problems (patient heterogeneity, insensitive outcome measures, inadequate drug development tools) requires substantial upfront investment, which traditional funding sources do not provide.

As an increasing number of large pharma companies have exited the CNS space due to these myriad challenges (and the sense that non-CNS indications offer better opportunities), NIH and voluntary health organizations have struggled to fill the void using traditional grant-based models. In response, newer funding models have been sought and applied.

[a] https://www.nsf.gov/od/lpa/nsf50/vbush1945.htm.
[b] https://www.amazon.com/800-Million-Pill-Merrill-Goozner/dp/0520246705.

Neurotherapeutics in the Era of Translational Medicine
https://doi.org/10.1016/B978-0-12-816475-4.00018-5

New funding models

Over the past 20 years foundations, patients, and other interested parties have launched a number of new funding models to accelerate therapeutic development in the CNS space. Three of the most effective and widely embraced are the Incentive Prize model, venture philanthropy model, and the newest funding model, disease-focused venture-capital-based investment funds.

Incentive prizes are a form of crowdsourcing intended to focus attention and resources on very specific outcomes.[c] Unlike recognition prizes, which are awarded retrospectively based on past accomplishments, incentive or inducement prizes are launched prospectively in the hope of being awarded once a well-defined and measurable set of objectives has been achieved. Following the success of the Ansari X-Prize for commercial space flight in the early 2000s, prize-based efforts, including those focusing on the CNS, have grown in popularity over the past 15 years.

Venture philanthropy can be defined as a funding model in which nonprofit, frequently disease-focused, organizations fund *internally managed* ventures aligned with organizational objectives with the potential to achieve financial returns, which can then be reinvested per their mission. This is in contrast to funding grants, which typically have no potential for financial return. While early efforts like the Alzheimer's Drug Discovery Foundation pioneered this model over 20 years ago, venture philanthropy efforts in the CNS therapeutic space have accelerated over the past 10 years largely driven by the well-publicized success of the Cystic Fibrosis Foundation's efforts, which resulted in the 2012 approval and launch of ivacaftor (trade name Kalydeco), developed by Vertex Pharmaceuticals, Inc.[d]

Disease-focused venture funds are a specialized type of stand-alone investment fund run by experienced investment professionals *outside of* and frequently unaffiliated with a nonprofit entity, designed to attract impact/mission-oriented investors with an interest in both return and furtherance of a social goal targeting a disease of interest. There has been a growing increase in the launch of such funds over the past 5 years, including the launch of the Dementia Discovery Fund in 2015 and the ALS Investment Fund in 2016.

Given the timelines for developing new therapies, it is still too early to tell if the advent of these novel funding models will lead to a surge in the number of effective CNS-oriented therapeutics. However, most of the chapters in this book describe recent results driven, in part, by these models.

[c] https://www.mckinsey.com/business-functions/strategy-and-corporate-finance/our-insights/using-prizes-to-spur-innovation.

[d] https://alo.mit.edu/wp-content/uploads/2020/05/SSRN-id3376673.pdf.

Incentive prizes

History

In business, there is something known as Bill Joy's Law, which states: "No matter who you are, most of the smartest people work for someone else." Prizes are part of a larger movement to tap into this "wisdom of the crowd" or crowd-sourcing, to solve a wide variety of challenges from engineering feats to data-mining discoveries. Incentive prizes have a long history, going back at least as far as the Longitude Prizes (to determine longitude at sea) starting in the 16th century (with Britain's Longitude Prize awarded to John Harrison in 1773).[2] There was also the food preservation prize issued by Napoleon in 1795,[e] which brought about the advent of modern food preservation techniques, and the 1919 Orteig Prize[3] that stimulated Lindberg to cross the Atlantic, jump-starting the commercial aviation industry. Over the past 15+ years, incentive prizes have seen a resurgence in popularity driven largely by the awarding of the XPrize for suborbital space flight in 2004 and the subsequent founding of the XPrize Foundation. In 2010, the US Congress passed legislation that gave every federal agency the authority to support incentive prizes of up to $50 million.[f] As of July 2018, Executive Branch agencies had sponsored more than 840 incentive prizes, and 97 of these challenges were completed in 2019 alone.[g]

Incentive prizes and neurotherapeutics

The first large-scale effort to apply the Incentive Prize model to a CNS-related challenge was the brainchild of a young Harvard Business School (HBS) student, Avi Kremer, who was diagnosed with ALS. As reviewed earlier in this book, ALS is a devastating and complicated disease, which has been thwarting science and medicine's best efforts for years. In 2004, when Avi first received his diagnosis, little was known about the underlying cause of ALS other than that approximately 2% of cases were caused by a previously identified gene.

Avi Kremer and some HBS classmates launched the 501(c)3 nonprofit organization, Prize4Life, after he was diagnosed with ALS in his first semester at HBS at the age of 29. Inspired by the recent well-publicized success of the Ansari XPrize, Avi and his cofounders decided to embrace the incentive prize model to speed-up ALS drug development. Their reasons were based on several principles:

- That effective drug development for ALS was not an intractable problem; breakthroughs were possible and could come from anywhere

[e] https://www.npr.org/sections/money/2012/03/01/147751097/why-napoleon-offered-a-prize-for-inventing-canned-food.

[f] https://ssir.org/articles/entry/the_promise_of_incentive_prizes.

[g] https://www.challenge.gov/archive/2019/.

- That particular identifiable leverage points/obstacles once solved would have a major impact on the field
- That new minds and new ideas had the important potential to accelerate ALS drug development, and
- That the incentive prize model might accelerate the entire process

Importantly, Prize4Life was not trying to supplant the existing grant-based efforts of other foundations and government agencies. Rather, the organization hoped to use the increased attention, new minds, and new resources attracted by the incentive prize model to supplement and complement the existing range of resources focused on ALS. Prize4Life hoped to mimic the success of the Ansari XPrize and use large monetary prizes to complement basic research by both drawing attention to particular problems/research needs and by capturing the imagination of many different people, including researchers, patients, donors, and the media.

Prize4Life found that the newly rediscovered incentive prize model appealed to new donors, particularly those who were interested in supporting outcome-based philanthropy and who were less likely to support traditional research grant mechanisms. At the time, the majority of Prize4Life's major donors were not involved in funding other grant-making ALS organizations, demonstrating the ability of the prize model to attract new sources of funding. The organization's experience further showed that launching incentive prizes in ALS did not have any adverse effects on the resources available for grants (a concern held at the time by some key opinion leaders in the field).

Launching prizes

With novel funding in hand, Prize4Life then aimed to identify the impediments to more rapid and effective ALS drug development and prioritize those that could potentially be accelerated using the incentive prize model. The team convened a roundtable of leaders in the ALS academic, clinical, and industry communities who worked with the founding members of Prize4Life to identify several key research challenges. Given Prize4Life's intention to launch incentive prizes, these needed to be challenges where solutions would (a) meet the previously defined criteria necessary for an effective incentive prize, i.e., broadly understandable, readily measurable, and right-sized given the size of Prize4Life's purse, (b) have a direct impact on ALS drug development, and (c) open the door to increased interest and investment from drug companies and thereby have a major impact for ALS patients. Following this exercise, Prize4Life launched two \$1M prizes.

The \$1M ALS Biomarker Prize[h] was launched in November of 2006 with the goal of accelerating the development of a biomarker—an inexpensive and easy-to-use tool that could accurately measure the progression of ALS in patients. Such a tool would make clinical trials cheaper, faster, and more appealing to industry. In 2007, Prize4Life awarded

[h] https://www.fiercebiotech.com/biotech/prize4life-awards-1m-als-biomarker-prize.

five $15,000 "thought" prizes to encourage promising biomarker concepts, including submissions from a chemist, a small biotech company, a plant biologist, and a dermatologist. In 2009, Prize4Life awarded two $50,000 "progress" prizes to researchers who had made significant progress toward meeting the explicitly defined prize criteria. In 2010, Prize4Life's Scientific Advisory Board unanimously voted to award the $1M prize to Dr. Seward Rutkove of Beth Israel Deaconess Medical Center (one of the two "progress" prize winners) for his development of electrical impedance myography (EIM). EIM has since been shown to have the potential to enhance clinical trials in Duchenne muscular dystrophy, spinal muscular atrophy, facioscapulohumeral muscular dystrophy, and a variety of other neuromuscular conditions. Subsequently, Dr. Rutkove has formed a company to design, build, and market a dedicated medical EIM device, and an advanced version of the original pilot device is currently being reviewed by the FDA for regulatory approval. Multiple companies are currently considering EIM for inclusion in their forthcoming clinical trials, fulfilling the promise of the ALS Biomarker Prize.

Beyond the impact of incentivizing the development of a novel outcome measure to facilitate ALS drug development, the prize also achieved several of Prize4Life's other important goals:

- New minds: Over 1000 "solvers" from more than 20 countries signed up to compete in the challenge, bring new ideas, disciplines, and methods to the field of ALS, an orphan disease
- New funding: The prize helped facilitate Dr. Rutkove's development of EIM as a biomarker for ALS. This work was further supported by NIH research and SBIR grants, a NeuroNEXT grant, and funding from several drug companies to test EIM in their early phase trials totaling more than $6M in funding. The prize also helped catalyze and attract funding to the formation of a start-up company, Myolex, dedicated to commercializing medical-grade EIM devices

Following on the heels of launching the ALS Biomarker Prize, the $1M ALS Treatment Prize was[i] launched in 2008. The goal of this prize was to spur preclinical development efforts for ALS by incentivizing the development of a treatment that could reliably extend the lifespan of ALS mouse models by 25%, thereby filling the ALS drug development pipeline with promising therapeutic candidates. The $1M Prize was awarded in 2017 to a team based in Paris, France for a gene-therapy based approach.[4] The prize-winning work had previously been independently validated by the ALS Therapy Development Institute.

Similar to the ALS Biomarker Prize, the ALS Treatment prize also attracted new minds, with over 40 research teams from academia and industry participating in the challenge. While the ALS Treatment Prize was only awarded relatively recently, it has already resulted in the filing of seven patents[j] and a collaboration agreement with

[i] https://www.prweb.com/releases/2017/02/prweb14028039.htm.

[j] https://patents.justia.com/inventor/martine-barkats.

the pharmaceutical company AveXis,[k] a subsidiary of Novartis. In spite of not winning the prize, many of the participating teams have also continued to advance their therapeutic efforts, leveraging the expert feedback they received from the prize judges to improve the design and interpretation of their preclinical studies.

Inspired by the success of the ALS Biomarker Prize, and the interest in the ALS Treatment Prize, starting in 2012 Prize4Life set out to extend its incentive prize portfolio by leveraging another activity it was leading, the development of the Pooled Resource Open-Access ALS Clinical Trial (PRO-ACT) database of completed ALS clinical trials.[5] Prior to the launch of the database in 2012, and the launch of a major data update in 2015, Prize4Life launched two data challenges:

The ALS Prediction Prize[l] was a $50,000 award, launched in 2012 in collaboration with the IBM-affiliated crowdsourcing data initiative DREAM. This data prize invited participants to utilize 3months' worth of anonymized ALS patient data from the PRO-ACT database to predict disease progression at a year, a goal with important implications for reducing patient heterogeneity enabling more effective ALS clinical trial design, as well as the potential to improve understanding of disease pathology through the identification of novel predictors. The prize attracted over 1000 participants from 63 countries, with over 90% of participants indicating they had little to no awareness of ALS prior to the challenges.

In November of 2012, two teams split the prize awarded for the challenge, Team Sentrana, from the DC-based Scientific Marketing Company of the same name, and a team of two Stanford students, a math major and a lawyer/mechanical engineer.[m] The winning algorithms outperformed the predictions of 12 expert ALS clinicians and incentivized the Sentrana team to launch a start-up company, Origent Data Sciences, dedicated to the application of advanced analytics to ALS and other neurodegenerative diseases. As described elsewhere in this book, Origent has since collaborated with more than 10 companies utilizing its algorithms to improve ALS clinical trial design and data analysis, and received multiple grants from NIH and ALS nonprofit organizations to further enhance its analytic methods.

The ALS Stratification Prize,[n] a $28,000 prize that launched in 2015 in collaboration with DREAM and Sage Bionetworks, asked participants to use the PRO-ACT dataset to identify distinct subgroups of ALS patients, another key need to disentangle the heterogeneity of the disease and enable drug development. The challenge attracted over 70 entrants from 30 teams across 15 countries and was sponsored by leading research-based biopharmaceutical companies Biogen and Eli Lilly and Company. IBM also supported

[k] https://www.institut-myologie.org/en/2018/09/13/martine-barkats-and-aav9-smn-from-discovery-to-treatment/.

[l] https://www.innocentive.com/2819-2/.

[m] https://www.innocentive.com/2819-2/.

[n] https://sagebionetworks.org/in-the-news/prize4life-sage-bionetworks-and-dream-launch-an-open-science-challenge-to-find-clues-to-understanding-heterogeneity-in-patients-with-amyotrophic-lateral-sclerosis-als-2/.

the challenge with computational resources. The winners were three teams from the National Cheng Kung University in Taiwan, The University of Michigan, and The University of Illinois.[6]

Following Prize4Life's successes, the concept of incentive prizes has continued to have an impact throughout the world of CNS therapeutic discovery and development, including the following notable examples:

2016: **The Melbourne University AES-MathWorks-NIH Seizure Prediction Challenge**, which invited the development of algorithms to improve seizure prediction using long-term electrical brain activity recordings from patients with the implantable NeuroVista Seizure Advisory System. This challenge attracted 646 participants from 478 teams around the world[7] and led to the submission of more than 10,000 algorithms. The challenge was so successful, it was followed by an additional series of challenges around predicting Sudden Unexpected Death in Epilepsy (**SUDEP**).[o]

2017: **The Parkinson's disease Digital Biomarker DREAM Challenge**,[p] which was a challenge designed to identify digital Parkinson's biomarkers. This challenge attracted over 440 data experts from six continents.

2018: **The iCare-AD/ADRD Eureka Challenge**[q] by the National Institute of Aging at the NIH invited teams to develop technology-based applications, fostering connections between relevant stakeholders to either use existing technology or to develop new technology applications, to improve dementia care coordination and/or care navigation. This challenge led to the development of 33 solutions, with the leading approach utilizing visual mapping techniques to preserve independent function and improve quality-of-life for individuals with dementia.

The 2020 BEAT-PD (Biomarker and Endpoint Assessment to Track Parkinson's disease) DREAM Challenge[r] (just awarded earlier this year) was a collaboration between the Michael J Fox Foundation, DREAM, and Sage Bionetworks, which built on the 2017 digital biomarker challenge and invited data-savvy individuals to use wearable sensor data from Parkinson's disease patients to address the question whether measures of parkinsonian symptoms and disease severity, as assessed using digital wearable sensors, could be improved.

Recently, as a demonstration of the immediate relevancy the prize model can have, MIT invited scientists to participate in a COVID-19 challenge series, including developing prediction models for which patient populations are most likely to have neurological manifestations of COVID-19.[s]

[o] https://innocentive.wazoku.com/#/challenge/a4b65d625b5b416cba6eea2cbc642927?searchIndex=4.

[p] https://www.businesswire.com/news/home/20180117006187/en.

[q] https://www.nia.nih.gov/icare-ad-adrd-eureka-challenge-winners.

[r] https://www.michaeljfox.org/news/mjff-and-sage-bionetworks-announce-winners-beat-pd-dream-challenge.

[s] https://covid19challenge.mit.edu/.

Benefits of the incentive prize model (see Fig. 1)

As the example of Prize4Life proved, the incentive prize funding model has significant potential to drive the development of neurotherapeutics. Tom Kalil, the former Deputy Head of the Office of Science and Technology Policy under the Obama administration, whose office was responsible for raising the visibility of prizes within the government, eloquently captured the potential of this type of funding model in a recent interview, stating I "believe that a well-designed incentive prize[t] can enable the sponsor to:

- Set a goal without having to decide in advance which team or approach is most likely to be successful
- Pay only for results
- Leverage investment that can exceed the value of the prize purse
- Shine a spotlight on a problem
- Encourage fresh approaches by reaching beyond the "usual suspects"
- Change people's views about what is possible"

Challenges of the incentive prize model

While prizes can be a powerful stimulator of creative problem solving, they present a number of challenges and are not as simple as just putting up a big cash prize for the answer to a question (as learned by the Hedge Fund founders of the Gotham Cancer Prize, which in 2008 awarded $1M[u] for the best idea to "cure cancer" and then never awarded the prize again). Research from Harvard Business School's Karim Lakhani suggests that in order to be successful, some particular aspects of incentive prize design must be met, and the XPrize Foundation has distilled some of these "design" elements down to the following (see Fig. 2):

- First, the need behind the prize/the description of the solution (what is desired and necessary in order to win the prize) must be clear, compelling, and easily understood by anyone, regardless of their technical expertise, so that the broadest number of teams, including both experts and nondomain experts, will want and be able to compete
- Second, the solution must be objectively measurable by the right (expert) judges, so that the winning solution is transparent to all
- Last, the prize must be attractive enough to the potential solvers to incentivize participation.

As Karim Lakhani is paraphrased in The Atlantic[v] "Vague questions are ignored, good questions go unanswered without sufficient rewards, and if you don't have proper oversight to evaluate the answers, crowd-sourcing is just one big useless guessing game."

[t] https://ssir.org/articles/entry/the_promise_of_incentive_prizes.

[u] https://www.reuters.com/article/us-cancer-prize/dna-idea-wins-1-million-cancer-prize-idUSN3141663120080401.

[v] https://www.theatlantic.com/magazine/archive/2014/01/finding-the-next-edison/355747/.

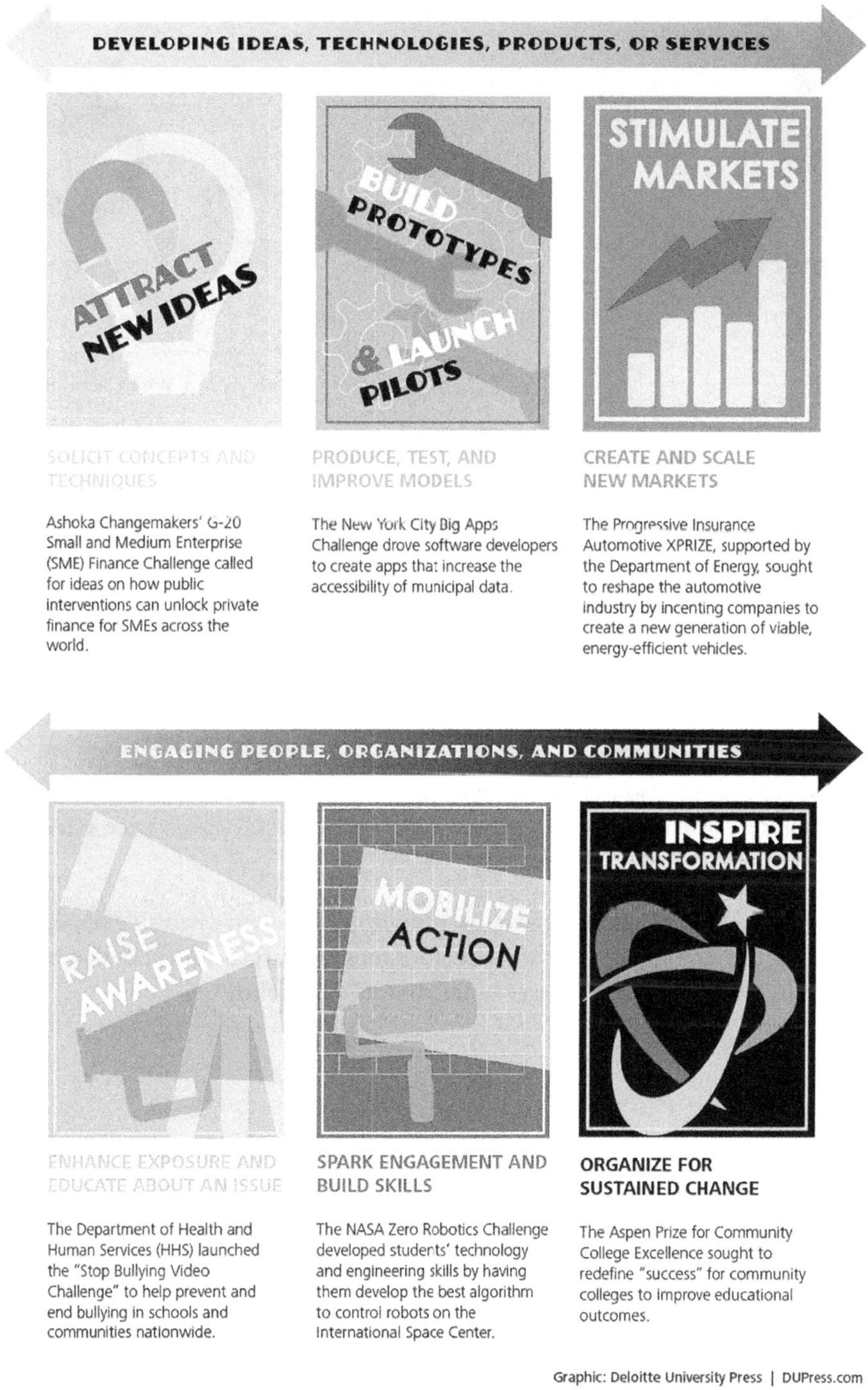

Fig. 1 Major outcomes for incentive prize design. *(Parker, et al. The craft of incentive prize design. Deloitte Insights. Deloitte University Press; 2014, www2.deloitte.com/us/en/insights/topics/social-impact/the-craft-of-incentive-prize-design.html.)*

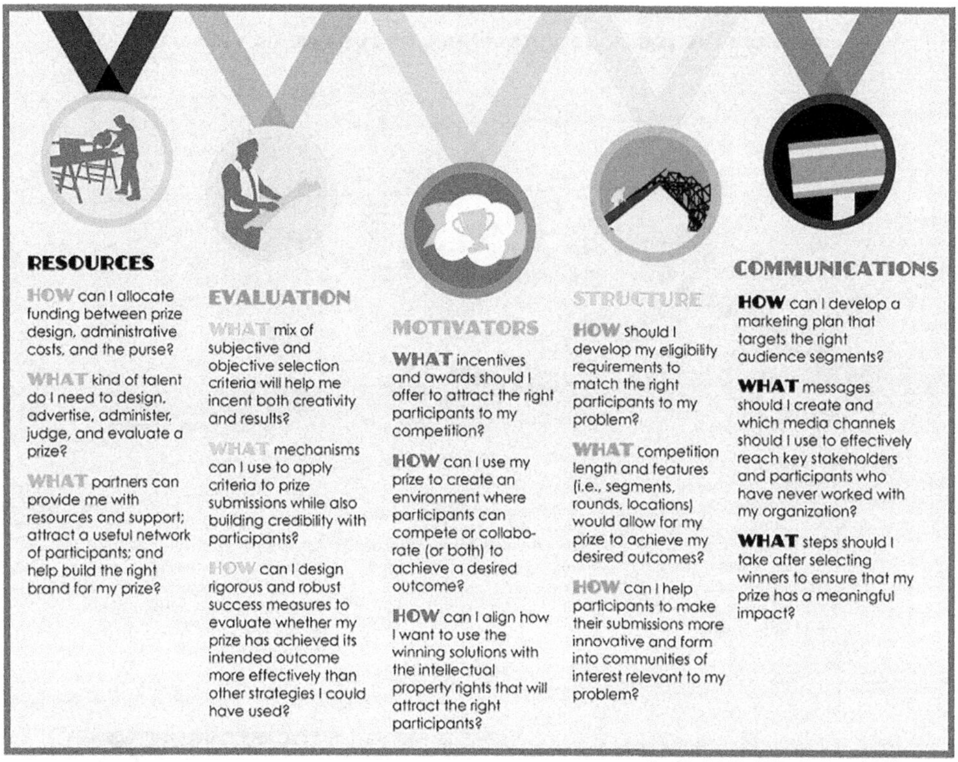

Graphic: Deloitte University Press | DUPress.com

Fig. 2 Getting started with the elements of prize design. *(Parker, et al. The craft of incentive prize design. Deloitte Insights. Deloitte University Press; 2014, www2.deloitte.com/us/en/insights/topics/social-impact/the-craft-of-incentive-prize-design.html.)*

While prizes have been one innovative funding approach to catalyzing breakthroughs in CNS drug development, another model for funding neurotherapeutics that has grown in importance over the past 10 years is the venture philanthropy model.

Venture philanthropy

Venture philanthropy is a more recent funding model on the scene than crowdsourcing-based prizes, with the earliest adopters in the development of CNS therapeutics getting their start in the mid-1990s. The concept of "venture philanthropy" arose from that of venture capitalism, which invests money from various third-party sources in typically high-risk investment areas[8] like biotechnology.

As mentioned earlier, as drug development costs have continued to rise, an increasing number of companies are reducing or halting investments in risky, early-stage programs in favor of later-stage opportunities with a greater likelihood of success. This is particularly true for CNS programs where promising drug targets often languish in a translational gap, i.e., the period after they have shown early signs of efficacy in test tubes and/or mice, but before they have been through any testing in humans. This phase of drug development has been called the "Valley of Death" (a term attributed to former NIH Director Elias Zerhouni) because it is seen as too early/high risk to attract funding from venture investors or pharmaceutical companies. These typically prefer to fund drugs further along on the development path. As articulated more than 10 years ago by workshop cochair Timothy Coetzee of the National Multiple Sclerosis Society, as part of a workshop on venture philanthropy organized by the National Academies of Science and the Institute of Medicine, "The risks are simply too high for the private sector to tackle alone. Recognition of this new reality is what is driving voluntary health organizations into venture philanthropy."[8]

Venture philanthropy efforts targeting health-related outcomes grew out of two parallel aligned trends: an increase in (and comfort around) nonprofits working more directly with the biopharma industry (and vice versa), and a rise in a more activist approach to philanthropy among wealthy, frequently entrepreneurial, individuals. In a 2014 FasterCures survey of 250 disease-specific nonprofit organizations, about 60% of respondents had governance policies that permitted investing in for-profit biotechnology companies.[w] Venture philanthropy has continued to grow as a funding model of interest in the CNS space, sparked largely by the experience of a highly successful non-CNS related initiative in Cystic Fibrosis.

The Cystic Fibrosis Foundation

According to their website,[x] the Cystic Fibrosis Foundation is the world's leading mission-driven organization involved in the search for a cure for cystic fibrosis (CF), an orphan, genetic, multisystem disease that affects more than 30,000 Americans. Although not the first organization to utilize the model, given their enormous success, the Cystic Fibrosis Foundation is largely credited with pioneering the venture philanthropy model to transform the drug development landscape for a rare disease. While the foundation was first founded in the 1950s, in 1998 it launched the Therapeutics Development Program[9] and then a nonprofit drug development affiliate, Cystic Fibrosis Foundation Therapeutics, Inc. (CFFT)[y] in order to pursue its interest in directly funding and working with companies to pursue R&D for cystic fibrosis-targeted therapies.

Over the course of a decade, starting in 2001, CFFT directed $150 million[z] to fund CF programs in development at Vertex Pharmaceuticals, a Boston-based biotechnology

[w] https://alo.mit.edu/wp-content/uploads/2020/05/SSRN-id3376673.pdf.

[x] https://www.cff.org/.

[y] https://alo.mit.edu/wp-content/uploads/2020/05/SSRN-id3376673.pdf.

[z] https://alo.mit.edu/wp-content/uploads/2020/05/SSRN-id3376673.pdf.

firm. The agreement with Vertex included a provision for the CF Foundation to receive a percentage of royalties on future sales of successful CF drugs resulting from the work it was co-funding. In 2014, CFFT sold the rights to its future Vertex royalties to New York City-based Royalty Pharma for $3.3 billion,[10] an impressive return on the Foundation's initial investment that it was able to reinvest in pursuit of its mission. More importantly for the Cystic Fibrosis Foundation's mission, Vertex has now developed a suite of highly effective drugs for cystic fibrosis, transforming the disease landscape and serving as a model for other venture philanthropy efforts.

Alzheimer's Drug Discovery Foundation[aa]

While the Cystic Fibrosis Foundation is perhaps the most widely known example of effective venture philanthropy, an example of a well-known CNS-targeted venture philanthropy is that of the Alzheimer's Drug Discovery Foundation (ADDF). Founded in 1995, the ADDF funds both academic and industry-driven preclinical studies (including toxicology studies) all the way up through early-stage proof-of-concept clinical trials relevant to Alzheimer's disease. The ADDF applies a venture philanthropy model and typically takes an ownership position in the early-stage companies it funds and then reinvests any revenues generated back into its programs. To date, the ADDF has awarded more than $150 million to fund over 626 Alzheimer's drug discovery programs and clinical trials in 19 countries.[ab]

Challenges of the venture philanthropy model

Of the three models discussed in this chapter, at present the venture philanthropy is perhaps the most widely embraced, as it is seen by the philanthropic community as an attractive funding vehicle, with the opportunity for a return on invested funds (partially, fully, or even with interest). The thinking goes that if a commercial venture fails, the charity/philanthropist would be in no worse a financial position than if they had initially made a grant to that venture. On the other hand, if the commercial venture produces a large financial return, the investment proceeds could go back to the charitable entity to further its goals.[ac] Thus, such mission-related investing offers the potential for a regenerative funding stream, providing organizations with additional funding for their missions. However, there are a number of challenges facing the venture philanthropy model.

Conflict of interest

Perhaps the largest challenge facing nonprofit organizations interested in utilizing the venture philanthropy model is that of conflict of interest. This challenge can take two

[aa] alzdiscovery.org.

[ab] alzdiscovery.org.

[ac] Trusts and Trustees Pozen article.

forms: the appearance of a conflict of interest and an actual conflict of interest. Given their nonprofit status, organizations must be very careful when investing in for-profit entities, and in fact the Internal Revenue Code of the United States specifically imposes special restrictions on investments by private foundations because they are perceived as having a relatively high potential for conflicts of interest.[ad] Organizations interested in pursuing venture philanthropy must be vigilant to ensure their investment activities don't conflict with their other goals/commitments, given that many foundations include multiple goals within their missions including supporting patients, lobbying government, funding research through grantmaking, and providing education.

As an example of the appearance of a conflict of interest, if a commercial entity in which an organization has directly invested charges high prices to the patients represented by the nonprofit, the organization might appear conflicted with respect to drug profits. As an example of an actual conflict of interest, if a nonprofit organization invests in a company it may also be tempted to provide grant funding to that same company in order to increase the likelihood of success of its investment. Typically, to the extent possible, patient-related organizations prefer to avoid even the appearance of financial interest in a commercial product they have funded and that their patient community will use. This is one of the main reasons why the Cystic Fibrosis Foundation sold off its royalty stake in Vertex, even though from an investment perspective the company was doing very well.

Lack of investment expertise

A second challenge is that typically foundations are not run by/nor do they include investment professionals among their staff, so the expertise necessary to both select a strong portfolio of investments and to effectively structure investments may be lacking. Investment decisions are driven by a wide variety of factors beyond simply the quality of the science and the interest in the indication (e.g., the composition of the team, the competitive landscape, the structure of the financing, etc.), but these other considerations may not be considered or weighted as highly by typical nonprofit professionals without investment expertise (and certainly they may not be appreciated by a foundation's constituents). In fact, even at the Cystic Fibrosis Foundation, out of nearly a hundred agreements, only a very small number actually resulted in a significant financial return to the Foundation.[ae] In addition, unlike the Cystic Fibrosis Foundation, most nonprofit organizations simply do not have the financial resources required to sustain a portfolio, i.e., "many shots on goal" strategy for investing.

[ad] Trusts and Trustees Pozen article.

[ae] https://alo.mit.edu/wp-content/uploads/2020/05/SSRN-id3376673.pdf.

Limited control

Third, because of their nonprofit status and the regulations surrounding investing in for profit entities, unlike disease-specific investment funds, foundations are unable to take Board seats. They have limited control over drug development strategy and are prohibited from engaging in the commercial plans of their investments. This has the potential to reduce the likelihood of return.

Like prizes, venture philanthropy continues to mature as a nonprofit funding model, increasingly embraced by more foundations and philanthropists. Given both the success and the limitations of this funding model, over the past 5 years multiple players within the CNS space have begun to consider the benefits of a third-related funding model for drug development that of disease-specific venture funds.

Disease-specific venture funds

Given the perceived challenges to CNS therapeutic development mentioned earlier in this chapter, traditional venture capital (VC) investment in drug development efforts targeting diseases of the CNS declined by 40% in the 2009–13 period compared with the 2004–08 period.[11] Partially in response to this pull-back, starting in the 2010s with the creation of the Cure Duchenne Ventures fund and the Dementia Discovery Fund, a new type of targeted VC fund began to appear on the R&D landscape. These disease-focused funds were part of a larger movement known as impact investing, a movement which has also led to the rise of social impact funds. Impact investments are broadly defined by the Global Impact Investing Network (GIIN) as investments "[made] into companies, organizations, and funds with the intention to generate social and environmental impact alongside a financial return."[10]

These CNS disease-focused impact funds were designed to identify, fund, and support primarily private companies developing therapies for defined disease areas where there was the perception that traditional investment was lacking or less than needed. The vision was that given deep expertise in the disease area of focus, these funds would be well positioned to advance the disease-relevant therapeutic opportunities with the greatest chance of success and to use a diversification approach to de-risk the investments in a way that nontargeted funds could not.

The ALS Investment Fund

One recent example of a disease-focused social impact VC launched to accelerate the development of neurotherapeutics is the ALS Investment Fund (ALS-IF). The ALS Investment Fund was launched in 2016 to support and finance companies with at least one internal drug or diagnostic program in ALS. ALS-IF was conceptualized by three ALS patients[af] and initially supported by two foundations: the Qurit Alliance[ag] and the ALS Foundation of the Netherlands,[ah] who recognized that ALS therapeutic efforts had historically struggled to raise venture capital and that the pipeline for ALS therapeutic programs had finally grown deep enough to allow for portfolio diversification. The founders shared a belief that entrepreneurship would drive progress in developing efficacious ALS therapies and that the most efficient and proven model for incentivizing and advancing the efforts of those entrepreneurs was the venture capital model.

Unfortunately, in spite of almost a decade of enormous progress, broadly focused venture capital funds have traditionally avoided ALS as an indication because of a continuing lack of understanding of the disease and some high-profile failures of clinical studies over the past decade. Drug development for ALS remains a challenge, given that the origin of the disease for most people is still largely unknown. This has made it difficult to develop targeted therapies. Therapies also have failed because of problematic animal studies, faulty patient selection, and/or difficulties in accurately measuring disease progression during small clinical studies. A lack of knowledge and tools has been a major impediment. But through studying these failures, and through exciting new research discoveries, recently the field has begun to understand much more about the disease. The output of ALS-related scientific research has been booming; between 1925 and 2000, there were 25,000 publications referencing ALS;[ai] between 2000 and 2020, this number doubled to more than 50,000, with more than 10,000 of these new publications since 2015. New genes have been discovered accounting for the majority of inherited cases of ALS. There has been an increase in the number of researchers dedicated to ALS and a resulting increase in the number of biomarker efforts and new animal models. All of this bodes well for the future of ALS drug discovery and development.

Rationale for ALS-IF

While foundations and the US government have been providing resources for ALS research and discovery for years, given the price tag to approve a new CNS therapy, moving potential therapeutic discoveries through clinical development requires more concentrated resources than any single entity within the ALS ecosystem could provide. Unlike the Cystic Fibrosis Foundation situation, where funding resources, patient support, and clinical infrastructure were all consolidated within a single

[af] Garmt van Soest, Bernard Muller, and Robbert-Jan Stuit.
[ag] http://www.qurit.org/.
[ah] www.als.nl.
[ai] https://pubmed.ncbi.nlm.nih.gov/?term=ALS&sort=date.

organization, the ALS landscape is far more fragmented, with no single entity capable of pursuing multiple well-funded shots on goal and the necessary portfolio derisking approach comparable to a VC. Given the structure of the ALS ecosystem, and the relatively advanced state of the pipeline, the ALS Investment Fund's founders believed treatments for the disease could best be accelerated using a disease-specific yet diversified investment fund approach.

Foundations as a springboard

In order to launch the ALS Investment Fund, it was first necessary for the founders to raise enough capital to make multiple sizeable investments, particularly as they initially sought to target companies with clinical-stage ALS programs. Although the ALS Investment Fund is a for-profit venture fund, the support of the nonprofit community was key to its establishment. The ALS Foundation of the Netherlands[aj] served as a cornerstone investor and was pivotal in providing some of the initial capital and contacts to the newly created fund, as well as providing a key endorsement as ALS-IF sought to build up a set of Limited Partner (LP) investors. With the imprimatur provided by the ALS Foundation of the Netherlands, the ALS Investment Fund was able to raise the rest of its initial $25M fund from a mix of family foundation investors, corporate investors, and other high net-worth investors. To date, the fund has invested in five companies across a range of targets, modalities, geographic regions, and stages of development and is currently raising new funding as it prepares to make its next set of investments.

Dementia Discovery Fund

Another example of a CNS-focused VC fund is the Dementia Discovery Fund (DDF), which was formed in 2015 when SV Health Investors won the bid to become the new fund's Manager through a competitive selection process. The DDF is a venture capital fund that invests in projects and companies to discover and develop[ak] novel, effective, disease-modifying therapeutics for dementia. Since completing its fundraising in 2018, with £250 million ($350 million at the time of completion), the DDF has invested around £150m with 17 companies and projects currently active in the portfolio. Seven pharmaceutical companies (Biogen, Eli Lilly and Company, GSK, Johnson & Johnson, Otsuka (Astex), Pfizer, and Takeda), multiple nonprofit organizations and philanthropists (including AARP, Bill Gates, and Alzheimer's Research UK),

[aj] www.als.nl.

[ak] https://www.biospace.com/article/dementia-discovery-fund-annual-forum-reviews-progress-in-discovery-of-novel-dementia-therapeutics.

the UK's Department of Health and Social Care, and a variety of other entities have invested in the DDF to date.

Benefits of disease-focused funds

While it is still early in the life-cycle of these CNS-focused VC funds, some of the benefits the disease-focused social impact investment fund model has demonstrated thus far include:

- Involvement of educated investment professionals with years of experience and strong networks in the disease/area of focus. This is beneficial for the portfolio company, other members of funding syndicates (typically necessary because of high capital needs in biotech), as well as the fund's own investors
- Ability to identify promising investments early on and stay committed during various development stages
- Diversification across a portfolio of approaches (potentially enabling future polypharmacy)
- Social impact goals appeal to a range of investors who are looking for more than just a return

Challenges of disease-focused funds

Like incentive prizes and venture philanthropy, disease-focused funds also face a number of challenges, including time to return, breadth of the pipeline, and confusion with venture philanthropy.

Time to return

Similar to prizes, most venture capital funds have a defined duration. Unlike prizes, however, investments are upfront and at risk. The typical lifespan of a venture fund, once the funding has been raised, is to spend the first 3–5 years making investments and then planning for exits from these within 10 years of launching the fund. This time horizon can limit the types of investments a fund can make. Some disease-focused funds have run into problems where the most promising science is still at a very early stage, and it is a challenge to identify assets that will reach an exit within the necessary timeline. This time-related limitation also relates to a second challenge for disease-focused VC funds.

Breadth of the pipeline

Given the pressure to exit within 10 years, the need to manage risk through diversification, and the availability of other funding models, it does not make sense to launch a VC fund in the absence of a robust pipeline of drug development programs at different levels of maturity. This means that diseases of high unmet need but limited scientific and/or medical understanding of disease pathology and/or limited attention from industry are likely not good diseases around which to launch a fund. If there is a lack of diversification within the drug development pipeline, then a traditional grant-based or potentially a venture philanthropy-based funding model is more appropriate to accelerate therapeutic development.

Confusion with venture philanthropy

Given that disease-specific funds have a social impact mission and that they are a relatively new form of funding, many companies and investors confuse such funds with the more widely known venture philanthropy approach. This can be problematic when companies mistakenly reach out to funds looking for nondilutive funding with few strings attached. It may also be problematic for disease-specific foundations who may erroneously suspect that disease-specific funds are competing with them for dollars. Further, philanthropists are used to thinking about their investment dollars and their philanthropic dollars completely separately. Accordingly, they may confuse disease-specific funds with philanthropic efforts rather than perceiving them as investment opportunities aligned with their social impact goals.

As disease-specific venture funds become more common and familiar across the therapeutic innovation ecosystem, the limitations and opportunities of this funding model will become more clearly defined. Over time, those interested in funding neurotherapeutic development will better understand how and when it is best to apply this new tool in their funding arsenal.

Conclusions

Innovation in funding models has been very important in accelerating the development of neurotherapeutics. Over the past 25 years, as large pharma and traditional private equity investment in CNS drug development has ebbed and flowed, multiple creative new funding models have emerged to offset some of the dollars draining away. Incentive prizes, venture philanthropy, and disease-focused investment funds have been the most innovative and successful models pursued to date. For those individuals and organizations interested in broadly furthering the development of neurotherapeutics, it is important to highlight that each of these models is most effective under a particular set of circumstances; none of them is one size fits all.

As shown in Fig. 3, the model comparison chart below:

- Prizes require less upfront capital, typically have the shortest timeline to key milestones and are the most effective at getting new innovation (new minds) to enter into a field to address a targeted challenge. However, prizes are unable to return capital to an organization, are only applicable for certain well-defined questions/problems, and

	Incentive Prizes	Venture Philanthropy	Disease Specific Funds
Amount of capital required	0–20M	at least 100K	at least 10M
Attract new minds to a problem			
Generate financial returns for the organization			
Help cross the "Valley of Death"			
Enable precompetitive efforts			
Directly enable commercial efforts			
Potential for conflicts of interest with organizational mission			
Require deep domain knowledge for success			
Degree of control over outcome			
Entity able to implement funding model	Non-profits and/or individual philanthropists	Non-profits and/or individual philanthropists	For-profit venture capital funds (foundations and philanthropists can be investors in funds but don't control/run the funds)
Time horizon until success	1–2 years	10+ years	0–10 years (lifetime of fund)
Success metrics	media, minds, new money	Advancing mission>return	Advancing mission=return

Fig. 3 Neurotherapeutics Funding Model Comparison Figure.

are most relevant for accelerating the earlier stage aspects of drug development (e.g., biomarker/outcome measure development, preclinical efforts, data-driven trial design, etc.) and much less effective at accelerating commercial-stage aspects of drug development.

- Venture philanthropy efforts typically require more capital to launch than prizes but less capital than funds, are the most effective for getting early-stage individual companies and/or projects across the valley of death (and therefore often have the greatest appeal to biopharma), and unlike prizes have potential to return capital to philanthropic individuals and organizations. However, venture philanthropy has the highest risk of engendering organizational conflicts of interest, the longest timelines to return, and gives funding organizations the least amount of control over outcomes.
- Disease-focused funds have the greatest likelihood of helping promising drugs cross the finish line to commercialization and of returning significant capital to organizations and philanthropists. Their advantage is that they are typically run by seasoned investment professionals, include later-stage companies in their portfolio, and,

based on their size, are able to diversify their commitments and have limited but significant opportunities to control their investments, i.e., by serving as advisors or corporate board members. However, funds require the most capital to establish, are often more relevant for later-stage development efforts (once companies are preparing to enter the clinic), and unlike prizes and venture philanthropy, must be independent of a philanthropic organization.

Each of these three innovative funding models has promise for furthering the development of future neurotherapeutics. Given the enormous and increasing burden to society posed by brain disease, and the recent decision by numerous large pharma companies to exit the CNS space, the need for these and other innovative funding models will only grow over the coming decades. As this book highlights, breakthroughs in neurotherapeutics are closer now than ever before, so hopefully, this will inspire continued commitment on the part of philanthropists, entrepreneurs, and nonprofit professionals to apply and develop these and other innovative funding strategies.

Acknowledgments

NZ currently works for Takeda Pharmaceuticals; MLL previously worked for Prize4Life and currently works with the ALS Investment Fund.

References

1. TUFTS. *CSDD Study*.
2. Quill H. *John Harrison: The Man Who Found Longitude*. London: John Baker Publishers; 1966.
3. Bak R. *The Big Jump—Lindbergh and the Great Atlantic Air Race*. Hoboken: John Wiley & Sons; 2011. ISBN 978-0-471-47752-5.
4. Grazia Biferi M, Cohen-Tannoudji M, Cappelletto A, Giroux B, Roda M, Astord S, et al. A new AAV10-U7-mediated gene therapy prolongs survival and restores function in an ALS mouse model. *Mol Ther*. 2017;25(9):2038–2052.
5. Atassi N, Berry J, Shui A, Zach N, Sherman A, Sinani E, et al. The PRO-ACT database: design, initial analyses, and predictive features. *Neurology*. 2014;83(19):1719–1725. https://doi.org/10.1212/WNL.0000000000000951 Epub 2014 Oct 8. PMID: 25298304.
6. Kueffner R, Zach N, Bronfeld M, Norel R, Atassi N, Balagurusamy V, et al. Stratification of amyotrophic lateral sclerosis patients: a crowdsourcing approach. *Sci Rep*. 2019;9(1):690. https://doi.org/10.1038/s41598-018-36873-4.
7. Baldassano SN, Brinkmann BH, Ung H, et al. Crowdsourcing seizure detection: algorithm development and validation on human implanted device recordings. *Brain*. 2017;140(6):1680–1691.
8. 2008 National Academies of Science Institute of Medicine: Venture Philanthropy Strategies to Support Translational Research: Workshop Summary.
9. Higgins RF, LaMontagne S, Kazan B. *Vertex Pharmaceuticals and the Cystic Fibrosis Foundation: Venture Philanthropy Funding for Biotech*. HBS No. 9-808-005, revised July 3, 2013 Boston, MA: Harvard Business School Publishing; 2014, p.5.
10. Hamermesh. *HBS Case N9-818-068*.
11. Thomas D, Wessel C. *Venture Funding of Therapeutic Innovation: A Comprehensive Look at a Decade of Venture Funding of Drug R&D*. Washington, DC: Biotechnology Industry Organization; 2015.

Index

Note: Page numbers followed by *f* indicate figures, *t* indicate tables, and *b* indicate boxes.

B

C

D

E

I

P

U

V

W

Z